**Proceedings of the
Ninth GAMM-Conference
on Numerical Methods
in Fluid Mechanics**

Ed. by Jan B. Vos,
Arthur Rizzi
and Inge L. Ryhming

Notes on Numerical Fluid Mechanics (NNFM)   Volume 35

Series Editors: Ernst Heinrich Hirschel, München
Kozo Fujii, Tokyo
Bram van Leer, Ann Arbor
Keith William Morton, Oxford
Maurizio Pandolfi, Torino
Arthur Rizzi, Stockholm
Bernard Roux, Marseille
(Adresses of the Editors: see last page)

Volume 8   Vectorization of Computer Programs with Applications to Computational Fluid Dynamics (W. Gentzsch)

Volume 11   Advances in Multi-Grid Methods (D. Braess / W. Hackbusch / U. Trottenberg, Eds.)

Volume 12   The Efficient Use of Vector Computers with Emphasis on Computational Fluid Dynamics (W. Schönauer / W. Gentzsch, Eds.)

Volume 13   Proceedings of the Sixth GAMM-Conference on Numerical Methods in Fluid Mechanics (D. Rues / W. Kordulla, Eds.)

Volume 14   Finite Approximations in Fluid Mechanics (E. H. Hirschel, Ed.)

Volume 15   Direct and Large Eddy Simulation of Turbulence (U. Schumann / R. Friedrich, Eds.)

Volume 17   Research in Numerical Fluid Dynamics (P. Wesseling, Ed.)

Volume 18   Numerical Simulation of Compressible Navier-Stokes Flows (M. O. Bristeau / R. Glowinski / J. Periaux / H. Viviand, Eds.)

Volume 19   Three-Dimensional Turbulent Boundary Layers – Calculations and Experiments (B. van den Berg / D. A. Humphreys / E. Krause / J. P. F. Lindhout)

Volume 20   Proceedings of the Seventh GAMM-Conference on Numerical Methods in Fluid Mechanics (M. Deville, Ed.)

Volume 21   Panel Methods in Fluid Mechanics with Emphasis on Aerodynamics (J. Ballmann / R. Eppler / W. Hackbusch, Eds.)

Volume 22   Numerical Simulation of the Transonic DFVLR-F5 Wing Experiment (W. Kordulla, Ed.)

Volume 23   Robust Multi-Grid Methods (W. Hackbusch, Ed.)

Volume 26   Numerical Solution of Compressible Euler Flows (A. Dervieux / B. van Leer / J. Periaux / A. Rizzi, Eds.)

Volume 27   Numerical Simulation of Oscillatory Convection in Low-Pr Fluids (B. Roux, Ed.)

Volume 28   Vortical Solutions of the Conical Euler Equations (K. G. Powell)

Volume 29   Proceedings of the Eighth GAMM-Conference on Numerical Methods in Fluid Mechanics (P. Wesseling, Ed.)

Volume 30   Numerical Treatment of the Navier-Stokes Equations (W. Hackbusch / R. Rannacher, Eds.)

Volume 31   Parallel Algorithms for Partial Differential Equations (W. Hackbusch, Ed.)

Volume 32   Adaptive Finite Element Solution Algorithm for the Euler Equations (R. A. Shapiro)

Volume 33   Numerical Techniques for Boundary Element Methods (W. Hackbusch, Ed.)

Volume 34   Numerical Solutions of the Euler Equations for Steady Flow Problems (A. Eberle / A. Rizzi / E. H. Hirschel)

Volume 35   Proceedings of the Ninth GAMM-Conference on Numerical Methods in Fluid Mechanics (J. B. Vos / A. Rizzi / I. L. Ryhming, Ed.)

Volumes 1 to 7, 9 to 11, 16, 24 and 25 are out of print.

Proceedings of the Ninth GAMM-Conference on Numerical Methods in Fluid Mechanics

Lausanne, September 25–27, 1991

Edited by
Jan B. Vos, Arthur Rizzi,
and Inge L. Ryhming

Die Deutsche Bibliothek - CIP-Einheitsaufnahme

Conference on Numerical Methods in Fluid Mechanics
< 09, 1991, Lausanne> :
Proceedings of the Ninth GAMM-Conference on
Numerical Methods in Fluid Mechanics: Lausanne,
September 25–27, 1991 / ed. by Jan B. Vos... – Braunschweig;
Wiesbaden: Vieweg, 1992
  (Notes on numerical fluid mechanics; Vol. 35)
  ISBN 3-528-07635-6
NE: Vos, Jan B. [Hrsg.]; Gesellschaft für Angewandte
  Mathematik und Mechanik; GT

All rights reserved
© Friedr. Vieweg & Sohn Verlagsgesellschaft mbH, Braunschweig / Wiesbaden, 1992

Vieweg is a subsidiary company of the Bertelsmann Publishing Group International.

No part of this publication may be reproduced, stored in a retrieval system or transmitted, mechanical, photocopying or otherwise, without prior permission of the copyright holder.

Produced by W. Langelüddecke, Braunschweig
Printed on acid-free paper
Printed in Germany

ISSN 0179-9614
ISBN 3-528-07635-6

Preface

This volume contains the proceedings of the Ninth GAMM-Conference on Numerical Methods in Fluid Mechanics, held at the Ecole Polytechnique Fédérale de Lausanne, Switzerland, on September 25-27, 1991. This conference, as well as the preceding eight ones, was organized by the GAMM Committee on Numerical Methods in Fluid Dynamics. It was probably also the last one in this successful series of conferences, since in the future, a bi-annual European meeting on Computational Fluid Dynamics will be organized by ECCOMAS, a new organization, representing all existing European professional societies working in this field.

The conference was attended by about 100 registered participants coming from all corners of the world. These proceedings contain the written version of the 56 papers presented during the meeting. In order to eliminate all kinds of errors, omissions, spelling mistakes, etc. these papers have been reviewed by an ad hoc scientific committee. As a result of this work a good deal of the papers were sent back to the authors for correction. This procedure has somewhat delayed the publication of this volume. We feel, however, that this has been a worth while effort.

The subjects treated during the meeting represent well current interests in CFD. For instance, multigrid and multiblock techniques for viscous as well as inviscid 3D flows were presented. Similarly, finite rate chemistry hypersonic flows still attracts many scientists due to the continuation of the Hermes Project.

Even before the preparation of this meeting got under way, I myself became seriously ill. My duties on the technical level were assumed by my close collaborator J.B. Vos, as well as by my co-chairman A. Rizzi. The practical details concerning the meeting itself were taken care of by Mrs. Christina Berg and Mrs. Monique Dubois. I would especially like to express my sincere thanks and appreciation to these four persons for their devoted effort in the running of this meeting.

Financially, the conference was supported by the following sponsors: Cray Research, EPFL, Gesellschaft für Angewandte Mathematik und Mechanik, Gb.Sulzer AG, IBM, NEC France. This support is gratefully acknowledged.

Lausanne, March 3, 1992

I.L.Ryhming, conference co-chairman and co-editor

Contents

Page

GENERAL TOPICS — 1

Arnal M. and Friedrich R.
On the effects of spatial resolution and the subgrid-scale modeling in the large eddy simulation of a recirculating flow — 3

Rizk Y. M. and Gee K.
Unsteady high-angle-of-attack flowfield simulation around the F-18 aircraft — 14

Van Dyke M.
Is computer extension of series a part of CFD ? — 24

FLOW OVER DELTAWINGS — 33

Ceresola N.
Validation of a computer code for the numerical simulation of turbulent flows past delta wings — 35

Srinivasan S., Eliasson P. and Rizzi A.
Hypersonic laminar flow computations over a blunt leading edged delta wing at three different chord Reynolds numbers — 46

Van den Berg J.I., Hoeijmakers H.W.M. and Sytsma H.A.
Study into the limits of an Euler equation method applied to leading-edge vortex flow — 55

MULTIGRID TECHNIQUES — 67

Catalano L.A., de Palma P., Deconinck H. and Napolitano M.
Explicit multigrid smoothing for multidimensional upwinding of the Euler equations — 69

Gaspar C. and Jozsa J.
Two-dimensional Lagrangian flow simulation using fast, quadtree-based adaptive multigrid solver — 79

Koren B. and Hemker P.W.
Multi-D upwinding and multigridding for steady Euler flow computations — 89

Oosterlee C.W. and Wesseling P.
A multigrid method for discretization of the incompressible Navier-Stokes equations in general coordinates — 99

VISCOUS HYPERSONIC FLOWS — 107

Bellucci V. and Grasso F.
Numerical solution of viscous hypersonic flows in chemical non equilibrium — 109

Fey M. and Jeltsch R.
Influence of numerical diffusion in high temperature flow   119

Schroeder W. and Hartmann G.
Robust computation of 3D viscous hypersonic flow problems   128

Wüthrich S., Perrel F., Sawley M.L. and Lafon A.
Comparison of thin layer Navier-Stokes and coupled Euler/boundary layer calculations of non-equilibrium hypersonic flows   138

MULTI BLOCK TECHNIQUES FOR VISCOUS COMPRESSIBLE FLOWS   149

Bergman C.M. and Vos J.B.
A multiblock flow solver for viscous compressible flows   151

Elmiligui A., Cannizzaro F., Melson N.D. and von Lavante E.
A three dimensional multigrid multiblock multistage time stepping scheme for the Navier-Stokes equations   160

Fatica M. and Grasso F.
Numerical solution of compressible viscous flows using multidomain techniques   170

INCOMPRESSIBLE NAVIER STOKES METHODS   181

Bottaro A.
Three-dimensional Navier-Stokes computations of spatially developing Goertler vortices   183

Deng G.B., Ferry M., Piquet J., Queutey P. and Visonneau M.
New fully coupled solution of the Navier-Stokes equations   191

Marx Y.P.
Evaluation of the artificial compressibility method for the solution of the incompressible Navier-Stokes equations   201

VISCOUS SUPERSONIC FLOWS   211

Borrel M., d'Espiney P. and Jouet C.
Three-dimensional thin-layer and space-marching Navier-Stokes computations using an implicit muscl approach: comparison with experiments and Euler computations   213

Leyland P., Richter R. and Neve T.
High speed flows over compression ramps   223

Secretan Y. and Thomann H.H.
A simple triangular element for the compressible Navier-Stokes equations   237

Von Lavante E. and Groenner J.
Semiimplicit schemes for solving the Navier-Stokes equations — 247

INCOMPRESSIBLE FLOWS IN COMPLEX GEOMETRIES — 257

Arakawa C., Qian Y., Samejina M., Matsuo Y. and Kubota T.
Turbulent flow simulation of Francis water runner with pseudo-compressibility — 259

Kost A., Bai L., Mitra N.K. and Fiebig M.
Calculation procedure for unsteady incompressible 3-D flows in arbitrarily shaped domains — 269

Lecheler S. and Fruehauf H.H.
A fully implicit 3-D Euler-solver for accurate and fast turbomachinery flow calculation — 279

Lube G. and Auge A.
Galerkin/least-squares approximations of incompressible flow problems — 289

METHODS USING UNSTRUCTURED GRIDS — 299

Riemslagh K. and Dick E.
A Runge-Kutta TVD finite volume method for steady Euler equations on adaptive unstructured grids — 301

Stolcis L. and Johnston L.J.
Computation of the viscous flow around multi-element aerofoils using unstructured grids — 311

Vilsmeier R. and Hänel D.
Adaptive solutions of the conservation equations on unstructured grids — 321

CONVECTIVE-ADVECTIVE FLOWS — 331

Akiba Y., Doi S. and Kuwahara K.
Analysis of natural convection with large temperature difference using a fast matrix solver — 333

Pourquié M. and Nieuwstadt F.T.M.
Comparison of some numerical schemes for the advection of a passive positive scalar on a coarse grid — 343

Rousse D.R. and Baliga B.R.
Numerical method for conduction-convection-radiation heat transfer in two-dimensional irregular geometries — 353

MESH GENERATION AND ADAPTATION — 363

Fischer J.
Sensors for self-adapting grid generation in viscous flow computations — 365

LeVeque R.J. and Walder R.
Grid alignment effects and rotated methods for computing complex flows in astrophysics — 376

Mokry M.
Adaptive mesh coupling of Euler equation and boundary layer solutions for transonic airfoils — 386

Suzuki M.
An attempt to the surface grid generation based on unstructured grid — 396

TURBULENT COMPRESSIBLE FLOWS — 405

Drikakis D. and Tsangaris S.
Laminar and turbulent viscous compressible flows using improved flux vector splittings — 407

Hanine F., Kourta A. and Haminh H.
Supersonic turbulent boundary layer with pressure gradients — 417

Johnston L.J.
Transonic aerofoil performance by solution of the Reynolds-averaged Navier-Stokes equations with a one-equation turbulence model — 427

COMPLEX FLOWS — 437

Bai X.S. and Fuchs L.
Numerical simulation of wind tunnel flows past a flame-holder model — 439

Chargy D., Larrouturou B. and Loriot M.
Numerical simulation of H2-O2 diffusion flames in a rocket engine — 449

Li X.K.
A modified semi-implicit method for two-phase flow problems — 459

Tu J.Y and Fuchs L.
Numerical calculations of flows in curved-duct intake ports and combustion chambers — 469

HYPERSONIC FLOWS — 479

Müller B., Niederdrenk P. and Sobieczky H.
Hypersonic flow simulation by marching on an adapting grid — 481

Mundt C.
Calculation of viscous hypersonic flows in chemical non-equilibrium — 491

Valorani M. and di Giacinto M.
Adaptive mesh refinement for unsteady, nonequilibriumn, high speed flows — 501

Yee H.C. and Sweby P.K.
On reliability of the time-dependent approach to obtaining steady-state numerical solutions — 512

DIRECT SIMULATION OF TURBULENT FLOWS — 521

Adams N.A., Sandham N.D. and Kleiser L.
A method for direct numerical simulation of compressible boundary-layer transition — 523

Lê T.H., Ryan J. and Dang Tran K.
Direct simulation of incompressible, viscous flow through a rotating square channel — 533

Wörner M. and Grötzbach G.
Analysis of semi-implicit time integration schemes for direct numerical simulation of turbulent convection in liquid metals — 542

NUMERICAL TECHNIQUES FOR COMPRESSIBLE FLOWS — 553

Di Mascio A. and Favini B.
A two-step Godunov-type scheme for multidimensional compressible flows — 555

Rossow C.C.
Flux balance splitting with rotated differences: a second order accurate cell vertex upwind scheme — 567

Sidilkover D., Giannakouros J. and Karniadakis G.E.
Hybrid spectral element methods for shock wave calculations — 577

Toro E.F.
Viscous flux limiters — 592

Contents in alphabetical order

Page

Adams N.A., Sandham N.D. and Kleiser L.
A method for direct numerical simulation of compressible boundary-layer transition — 523

Akiba Y., Doi S. and Kuwahara K.
Analysis of natural convection with large temperature difference using a fast matrix solver — 333

Arakawa C., Qian Y., Samejina M., Matsuo Y. and Kubota T.
Turbulent flow simulation of Francis water runner with pseudo-compressibility — 259

Arnal M. and Friedrich R.
On the effects of spatial resolution and the subgrid-scale modeling in the large eddy simulation of a recirculating flow — 3

Bai X.S. and Fuchs L.
Numerical simulation of wind tunnel flows past a flame-holder model — 439

Bellucci V. and Grasso F.
Numerical solution of viscous hypersonic flows in chemical non equilibrium — 109

Bergman C.M. and Vos J.B.
A multiblock flow solver for viscous compressible flows — 151

Borrel M., d'Espiney P. and Jouet C.
Three-dimensional thin-layer and space-marching Navier-stokes computations using an implicit muscl approach: comparison with experiments and Euler computations — 213

Bottaro A.
Three-dimensional Navier-Stokes computations of spatially developing Goertler vortices — 183

Catalano L.A., de Palma P., Deconinck H. and Napolitano M.
Explicit multigrid smoothing for multidimensional upwinding of the Euler equations — 69

Ceresola N.
Validation of a computer code for the numerical simulation of turbulent flows past delta wings — 35

Chargy D., Larrouturou B. and Loriot M.
Numerical simulation of H_2-O_2 diffusion flames in a rocket engine — 449

Deng G.B., Ferry M., Piquet J., Queutey P. and Visonneau M.
New fully coupled solution of the Navier-Stokes equations   191

Di Mascio A. and Favini B.
A two-step Godunov-type scheme for multidimensional compressible flows   555

Drikakis D. and Tsangaris S.
Laminar and turbulent viscous compressible flows using improved flux vector splittings   407

Elmiligui A., Cannizzaro F., Melson N.D. and von Lavante E.
A three dimensional multigrid multiblock multistage time stepping scheme for the Navier-Stokes equations   160

Fatica M. and Grasso F.
Numerical solution of compressible viscous flows using multidomain techniques   170

Fey M. and Jeltsch R.
Influence of numerical diffusion in high temperature flow   119

Fischer J.
Sensors for self-adapting grid generation in viscous flow computations   365

Gaspar C. and Jozsa J.
Two-dimensional Lagrangian flow simulation using fast, quadtree-based adaptive multigrid solver   79

Hanine F., Kourta A. and Haminh H.
Supersonic turbulent boundary layer with pressure gradients   417

Johnston L.J.
Transonic aerofoil performance by solution of the Reynolds-averaged Navier-Stokes equations with a one-equation turbulence model   427

Koren B. and Hemker P.W.
Multi-D upwinding and multigridding for steady Euler flow computations   89

Kost A., Bai L., Mitra N.K. and Fiebig M.
Calculation procedure for unsteady incompressible 3-D flows in arbitrarily shaped domains   269

Lê T.H., Ryan J. and Dang Tran K.
Direct simulation of incompressible, viscous flow through a rotating square channel   533

Lecheler S. and Fruehauf H.H.
A fully implicit 3-D Euler-solver for accurate and fast turbomachinery flow calculation   279

LeVeque R.J. and Walder R.
Grid alignment effects and rotated methods for computing complex flows in astrophysics   376

Leyland P., Richter R. and Neve T.
High speed flows over compression ramps   223

Li X.K.
A modified semi-implicit method for two-phase flow problems   459

Lube G. and Auge A.
Galerkin/least-squares approximations of incompressible flow problems   289

Marx Y.P.
Evaluation of the artificial compressibility method for the solution of the incompressible Navier-Stokes equations   201

Mokry M.
Adaptive mesh coupling of Euler equation and boundary layer solutions for transonic airfoils   386

Müller B., Niederdrenk P. and Sobieczky H.
Hypersonic flow simulation by marching on an adapting grid   481

Mundt C.
Calculation of viscous hypersonic flows in chemical non-equilibrium   491

Oosterlee C.W. and Wesseling P.
A multigrid method for discretization of the incompressible Navier-Stokes equations in general coordinates   99

Pourquié M. and Nieuwstadt F.T.M.
Comparison of some numerical schemes for the advection of a passive positive scalar on a coarse grid   343

Riemslagh K. and Dick E.
A Runge-Kutta TVD finite volume method for steady Euler equations on adaptive unstructured grids   301

Rizk Y. M. and Gee K.
Unsteady high-angle-of-attack flowfield simulation around the F-18 aircraft   14

Rossow C.C.
Flux balance splitting with rotated differences: a second order accurate cell vertex upwind scheme   567

Rousse D.R. and Baliga B.R.
Numerical method for conduction-convection-radiation heat transfer in two-dimensional irregular geometries   353

Schroeder W. and Hartmann G.
Robust computation of 3D viscous hypersonic flow problems   128

Secretan Y. and Thomann H.H.
A simple triangular element for the compressible Navier-Stokes equations   237

Sidilkover D., Giannakouros J. and Karniadakis G.E.
Hybrid spectral element methods for shock wave calculations   577

Srinivasan S., Eliasson P. and Rizzi A.
Hypersonic laminar flow computations over a blunt leading edged delta wing at three different chord Reynolds numbers   46

Stolcis L. and Johnston L.J.
Computation of the viscous flow around multi-element aerofoils using unstructured grids   311

Suzuki M.
An attempt to the surface grid generation based on unstructured grid   396

Toro E.F.
Viscous flux limiters   592

Tu J.Y and Fuchs L.
Numerical calculations of flows in curved-duct intake ports and combustion chambers   469

Valorani M. and di Giacinto M.
Adaptive mesh refinement for unsteady, nonequilibriumn, high speed flows   501

Van den Berg J.I. , Hoeijmakers H.W.M. and Sytsma H.A.
Study into the limits of an Euler equation method applied to leading-edge vortex flow   55

Van Dyke M.
Is computer extension of series a part of CFD ?   24

Vilsmeier R. and Hänel D.
Adaptive solutions of the conservation equations on unstructured grids 321

Von Lavante E. and Groenner J.
Semiimplicit schemes for solving the Navier-Stokes equations 247

Wörner M. and Grötzbach G.
Analysis of semi-implicit time integration schemes for direct numerical simulation of turbulent convection in liquid metals 542

Wüthrich S., Perrel F., Sawley M.L. and Lafon A.
Comparison of thin layer Navier-Stokes and coupled Euler/boundary layer calculations of non-equilibrium hypersonic flows 138

Yee H.C. and Sweby P.K.
On reliability of the time-dependent approach to obtaining steady-state numerical solutions 512

GENERAL TOPICS

ON THE EFFECTS OF SPATIAL RESOLUTION AND SUBGRID-SCALE MODELING IN THE LARGE EDDY SIMULATION OF A RECIRCULATING FLOW

M. Arnal and R. Friedrich

Lehrstuhl für Fluidmechanik

Technical University of Munich

Federal Republic of Germany

SUMMARY

The influence of spatial resolution and subgrid-scale modeling on the results of large eddy simulations (LES) of a turbulent, recirculating flow is discussed. A series of simulations of the high Reynolds number turbulent flow over a backward-facing step have been performed and compared with experimental data. In these calculations the grid spacing and spanwise (homogeneous) dimension of the computational domain have been systematically varied. Two different eddy-viscosity-type, subgrid-scale (SGS) models have been employed to model the fine scale effects of turbulence in the flow. It is shown that the time-averaged flow variables are profoundly influenced by the dimensions of the computational domain in the spanwise direction. Results obtained with the two SGS models are qualitatively similar although distinct differences exist.

INTRODUCTION

Turbulent flows which undergo separation and reattachment occur in a wide range of engineering and industrial applications. In such practical problems it is important to understand the fluid behavior in the recirculation and reattachment zones. New insight into the physics of these flows can lead to improved methods of heat transfer augmentation and turbulent mixing.

In the present study we consider the turbulent flow over a rearward-facing step as a representative example of such recirculating flows. This flow is inherently unsteady and three-dimensional and is further complicated by the extra strain rates in the mixing layer and reattachment regions. It was one of the test cases for the 1980-81 Stanford Conference on Complex Turbulent Flows [1] and is again a test case for the current Collaborative Testing of Turbulence Models [2].

Although its geometry is simple and the separation line is fixed, the accurate numerical prediction of the flow is still very difficult. Calculations made with statistical turbulence models, even those applying second-order closure, consistently fail to predict such basic quantities as the mean reattachment length, X_r. As shown by Peric et al. [3], not all the deficiencies are due to modelling and much improvement can be achieved by using sophisticated numerical techniques and high spatial resolution.

The large-eddy simulation technique provides an insightful alternative to the statistical modelling approach to turbulent flows. As will be shown in the present paper it is capable of accurately reproducing the mean flow and statistical quantities of experimental studies. It can also predict distributions of statistical quantities which are still not amenable to measurement (e.g. pressure-velocity and pressure-strain correlations). However, the accurate prediction of average flow quantities is only one advantage of the LES technique. In addition, it makes the investigation of the instantaneous turbulence structure, as in Friedrich and Arnal [4] and the unsteady characteristics of the flow possible.

The present study is part of a program to develop the LES technique into a tool for investigating turbulent flows of engineering interest. As the LES technique is applied to flows of increasing complexity, the dependency of the predictions on the spatial grid resolution and SGS model becomes more of a concern. The present study was undertaken to address these issues in a systematic manner. In Part 1, the influence of the spatial grid resolution and varying the dimensions of the computational domain is discussed. For the results presented, a single SGS model due to Schumann [5] was used so that the grid and geometry influences could be focussed on. Once an optimal grid and computational domain is determined, a comparison of simulations of the step flow using two different SGS models can be made in Part 2. For these calculations the domain geometry, mesh-cell size and the Reynolds number are all identical so that the influence of the models can be isolated.

THE NUMERICAL METHOD

In applying the LES technique, the time-dependent, three-dimensional grid scale (GS) quantities are computed directly, while the unresolved, subgrid scale (SGS) structures and their influence on the flow field are modelled. The transport equations for the resolvable (GS) flow quantities are derived by lowpass filtering the conservation equations for an incompressible fluid. Filtering by integration over a control volume of the equidistant, Cartesian grid leads directly to a spatially second-order accurate difference scheme which conserves mass exactly, Schumann [5]. The resulting difference equations have the following non-dimensional form:

$$\delta_j{}^j\overline{v_j} = 0 , \qquad (1)$$

$$\frac{\partial}{\partial t}({}^i\overline{v_i}) + \delta_j({}^j\overline{v_j}{}^j\overline{v_i} + {}^j\overline{v'_j v'_i} + {}^v\overline{p}\delta_{ji} - {}^j\overline{\tau_{ji}}) = 0 , \qquad (2)$$

where ${}^j\overline{\tau_{ji}}$ is the surface-averaged viscous stress tensor. The decomposition of a velocity into a resolvable, GS component, ${}^j\overline{v_i}$ and an unresolved, SGS component, v'_i produces the SGS stresses, $-{}^j\overline{v'_j v'_i}$. These stresses arise from the non-linear convection terms in the momentum equations and must be modelled in terms of known GS quantities.

In the present study, two eddy-viscosity-type models which have been used in previous studies are compared. The first model is based on the eddy-viscosity concept proposed by Schumann [5]. The SGS stresses are assumed to consist of two components, a fluctuating (locally isotropic) part which vanishes in the mean, and a statistical (inhomogeneous) part which is important in regions of strong mean shear:

$$-{}^j\overline{v'_j v'_i} + \frac{1}{3}\delta_{ji}{}^k\overline{v'_k v'_k} = 2\mu_{iso}({}^j\overline{S_{ji}} - \langle {}^j\overline{S_{ji}}\rangle) + 2\mu_{inh}\langle {}^j\overline{S_{ji}}\rangle. \qquad (3)$$

In equation (3) the filtered deformation tensor is defined as ${}^j\overline{S_{ji}} = 1/2(\delta_j{}^i\overline{v_i} + \delta_i{}^j\overline{v_j})$ and the angular brackets ($\langle\rangle$) denote statistical averages. The eddy viscosity coefficients μ_{iso} and μ_{inh} consist of length scales related to the mesh cell size and velocity scales related to the filtered, fluctuating and mean deformation tensors, respectively.

The second model due to Smagorinsky [6] and Lilly [7] has found wide use in meteorological applications. Here the SGS stresses are assumed to be proportional to the local instantaneous deformation tensor:

$$-{}^j\overline{v'_j v'_i} + \frac{1}{3}\delta_{ji}{}^k\overline{v'_k v'_k} = 2\mu_t({}^j\overline{S_{ji}}). \qquad (4)$$

In the Smagorinsky-Lilly model the eddy viscosity coefficient μ_t also contains a length scale related to the mesh cell size and a velocity scale related to the deformation tensor, ${}^j\overline{S_{ji}}$.

The equations are integrated in time on a staggered grid using a second-order accurate leap-frog scheme. An averaging time step is taken every 50 steps to counter the tendency for $2\Delta t$-oscillations to develop. The use of Chorin's [8] projection method leads to a discretized Poisson equation for the pressure field. A Fourier transformation in the spanwise direction reduces the three-dimensional problem to a series of independent two-dimensional Helmholtz problems in an irregular domain. The 2-D problems are solved exactly using the Cyclic Reduction Algorithm for computational domains which are box-like. To accommodate the simply-connected domain of the present study the Capacitance Matrix Technique [9] is employed. This essentially leads to the need to solve the Poisson equation twice for a box-like domain.

BOUNDARY AND INITIAL CONDITIONS

For spatially developing flows which are not periodic in the main flow direction, the specification of boundary conditions is of critical importance. At the inlet plane the instantaneous velocity vector and SGS energy must be specified at each time step in the simulation. In the current investigations the inlet conditions are taken from separate large-eddy simulations of fully-developed flow in a plane channel. For these precursor simulations the Reynolds number, SGS model and spatial and temporal resolution are identical with the corresponding values for the step flow calculations.

At the outlet plane the concept of Richter et al. [10] that the turbulence is "frozen" is adopted. This leads to a simplified convection equation for the instantaneous velocity vector at the outlet plane:

$$\frac{\partial}{\partial t}(\overline{v_i}) + \langle \overline{v_x} \rangle \delta_x \overline{v_i} = \langle \overline{v_x} \rangle \delta_x \langle \overline{v_i} \rangle. \qquad (5)$$

In equation (5) the convection velocity is the local mean streamwise velocity, $\langle \overline{v_x} \rangle$. A linear extrapolation is used to determine the mean streamwise velocity gradient and constant extrapolation is used for the remaining two mean velocity components.

Since the flow is statistically two-dimensional and stationary in the mean periodic conditions are applied for all flow variables. The spanwise dimension of the computational domain has turned out to be very influential on the development of the mean flow field.

At solid boundaries, the limitation of spatial resolution makes special formulations for the boundary conditions necessary. For the staggered variable arrangement employed, conditions for the normal velocity and wall shear-stress must be specified. At the wall the normal velocity is set to zero. Following Schumann [5], the wall shear-stress components are assumed to be proportional to and in phase with the wall-nearest tangential velocity components. In specifying the spanwise wall shear-stress, the proposal of Piomelli et al. [11] has been implemented. In this case, the proportionality factor relating the spanwise velocity and wall shear-stress is taken to be that computed for the local mean flow direction.

Initial conditions for the filtered velocity vector must be specified in the entire computational domain. These are taken from a separate two-dimensional steady state calculation of the flow field using a standard k-ε model of turbulence. The instantaneous fluctuating velocity field is obtained by adding random numbers weighted with $(2k/3)^{1/2}$ to the mean velocity component. The entire velocity field is then made divergence free by a single application of the Poisson solver.

The basic geometry of the backward-facing step flow field is illustrated in Figure 1. It consists of a 1:2 sudden expansion with the inlet plane located 4 step heights upstream of the step. The Reynolds number of the simulated flow, based on the step height, H, and the inlet bulk velocity, is $Re_b = 1.55 \cdot 10^5$. In every case the unsteady flow was simulated for more than 900 characteristic time scales, where the time scale is defined as $T_c = H/U_b$. This was ample time for the flow to reach a statistically stationary state independent of the initial conditions.

THE INFLUENCE OF SPATIAL RESOLUTION

In the first series of calculations we have systematically investigated the influence of the grid resolution and mesh-cell shape (ratio of Δx: Δy: Δz) on the predicted mean flow results. Calculations were performed on grids of $80 \times 16 \times 16$, $160 \times 32 \times 32$ and $320 \times 64 \times 48$ for a computational domain of $20H \times 4H \times 2H$ in the x-, y- and z-directions, respectively. Predictions of the mean flow field made on the three grids are compared with the experimental measurements of Tropea [12] and Durst and Schmitt

[13] in Figures 2 and 3. The velocities and Reynolds stresses have been normalized with the mean centerline velocity U_0, of the inlet channel. In the figures profiles of the mean streamwise velocity and the Reynolds shear stress are shown for 3 different streamwise positions up- and downstream of reattachment. The predicted reattachment lengths were $X_r = 7.2$ for the finest grid, $X_r = 7.3$ for the middle grid and $X_r = 7.7$ for the coarse grid, compared with the experimental value of $X_r = 8.5$. The difference in the predicted mean reattachment length for the two finer grids is less than 2%, indicating for this quantity at least the resolution of the middle grid is quite adequate. The disagreement with the experimental value is therefore not likely to be a result of an incomplete grid resolution. In general, for Figures 2 and 3 the variations in the predicted profiles are larger between the coarse and medium grids than between the medium and fine grids.

As a further indication of the essentially grid-independent nature of the LES results we compare the dimensionless, total-shear force acting on the channel walls in the streamwise direction:

$$F_{\tau x} = \int_T \tau_{xz}\, dxdy + \int_B \tau_{xz}\, dxdy,$$

where the subscripts T and B refer to the top and bottom walls respectively. In calculating the shear force we consider only the top and bottom surfaces downstream of the step. The values of $F_{\tau x}$ for the three grids are given in Table 1. With grid refinement the precentage change in the force decreases from 4.3% to 1.7%, indicating the value of $F_{\tau x}$ is converging. Although one must be careful in referring to LES results as being grid independent except in the limit of a direct numerical simulation (DNS), we have found results which appear to converge with grid refinement well before the DNS limit is reached.

Table 1 Grid dependence of the integrated shear force, $F_{\tau x}$.

	Grid	$F_{\tau x}$	% Change
1.	$80 \times 16 \times 16$	45.44	–
2.	$160 \times 32 \times 32$	47.39	+ 4.3
3.	$320 \times 64 \times 48$	46.22	+ 1.7

In investigating the effects of the grid resolution on the predictions, we have also evaluated the influence of the mesh-cell shape on the results. In this case we have found large variations in the mean flow predictions in changing from cells stretched in the streamwise and spanwise directions to cells which are cubic. The predicted mean streamwise velocity profiles for two cases are compared with experimental measurements in Figure 4. For the predictions made with a cubic mesh ($\Delta x = \Delta y = \Delta z$) the reattachment length agrees very well with the experimental value ($X_r = 8.6$ vs. 8.5). In contrast, the predictions made with a stretched grid ($\Delta x = \Delta y = \Delta z/2$) significantly underpredict the reattachment length ($X_r = 7.3$ vs. 8.5). Similarly, computed results for a grid midway between the two ($\Delta x = \Delta y = 3\Delta z/2$) gives a reattachment length of $X_r = 7.9$, which lies between the previous two values. Additional computations with a uniform distribution of cubic cells confirm this trend. Calculations made on grids with cubic mesh-cells yield mean flow field results in closer agreement with experimental results than those made with stretched cells.

The strong dependence of the results on the relative dimensions of the mesh cell is connected to the process of filtering the governing equations over a cell volume. The velocity components which are normal to a larger cell-surface will not be as well resolved as those normal to a smaller surface. This, in turn, leads to higher fluctuations about the mean for the better resolved velocity components and turbulent transport which is skewed with respect to its true value. In the present calculations this leads to higher streamwise velocity fluctuations (u'') compared with the cross-stream component (w''). This in turn, leads to an overprediction of the Reynolds stress in the region of high shear downstream of the step and a mean reattachment length which is shorter than experimentally observed. As a result of these calculations we note that a refined grid with cubic cells gives the best overall agreement with experimental measurements of mean-flow quantities.

Finally, we discuss the influence on the flow predictions of the periodic boundary conditions in the spanwise direction. For spanwise widths of $2H$, $4H$, $8H$ and $16H$, calculations were performed keeping the spatial and temporal resolution identical in each case. Variations of three to four step-heights in the mean reattachment length X_r were observed. The variation in the predicted reattachment length with the spanwise (homogeneous) dimension is illustrated in Figure 5. The reason for the strong influence of the boundaries in the spanwise direction can be seen in Figure 6. This figure shows instantaneous vector plots of the velocity field in the plane nearest to the wall where reattachment occurs for spanwise widths of $4H$ and $8H$. In the two simulations the mean reattachment line varied between $X_r = 7.6$ and $X_r = 7.3$ for channel widths of $4H$ and $8H$, respectively. The presence of the large fluid structures can be seen on the velocity fields in the reattachment region. The size of these structures is on the order of $2H$ in width in the narrower channel and $3H$ in the wider channel. The width of the calculation domain acts to limit the size of such structures as well as their movement in the lateral direction. These limitations will, in turn, influence global flow quantities such as the reattachment length. An examination of the results shown in Figure 5 indicates that calculations in a domain with a spanwise width of $6H$ or more will be little affected by the periodic boundary conditions.

A COMPARISON OF SGS MODELS

In the previous section the influence of the mesh resolution and dimensions of the computational domain were examined. Using these results, obtained with a single SGS model, an optimal computational domain size and grid distribution can be chosen for further calculations. In this section the results of calculations with two different SGS models are discussed. Fine-grid solutions consisting of $192 \times 64 \times 16$ cells in a computational domain of $24H \times 8H \times 2H$ in the x-, y- and z-directions, respectively have been performed.

Qualitatively, the statistical quantities predicted with the two eddy-viscosity-type SGS models are remarkably similar. Quantitatively however, there are differences to be noted. For example, the mean reattachment length predicted using the Smagorinsky-Lilly (SL) model was $X_r = 9.2H$, compared with the Schumann (S) model, $X_r = 8.6H$. The difference between the two is approximately 7%. Figure 7 illustrates the mean velocity profiles predicted in the two simulations compared with local instantaneous velocity profiles. Note that the largest departures from the mean occur in the high shear regions of the mixing layer, upstream of reattachment. In both plots the mean reattachment length is marked by an arrow.

Figure 8 compares contours of the Reynolds shear stress, $\langle -u"w" \rangle$ predicted in the two cases. The fields are qualitatively very similar with the maximum values differing by less than 10%. Note that the streamwise location of the maximum compares well for the two models when normalized by the mean reattachment length: $X_{uw,max} = 0.88 X_r$ (SL model) and $X_{uw,max} = 0.85 X_r$ (S model). This position in both simulations corresponds closely with the location of maximum mean shear.

The figures portraying the distribution of the Reynolds shear stress for the two sets of calculations are illustrative of other statistical quantities. In general, the differences between the predictions using the two models are small (under 10%). It does not appear that the greater computational effort involved in employing the Schumann SGS model is justified by the results.

CONCLUSIONS

The effects of grid resolution and variations in the computational domain have been investigated in the large-eddy simulation of the flow over a backward-facing step. Essentially grid independent results were obtained for a Reynolds number of $Re_b = 150,000$, based on the inlet channel bulk velocity. The influence of the spanwise (homogeneous) dimension of the computational domain was found to play a crucial role in determining the global features of the predicted flow field. The periodic boundary conditions at these boundaries act to limit the spanwise dimension and movement of the large scale structures present in the flow. This in turn, can lead to large variations in the reattachment length.

A comparison of two widely-used, eddy-viscosity-type SGS models has been made. Although differing in details, the predicted statistical quantities are in good agreement with each other. For this particular flow at least, it does not appear that the increased computational effort required for the Schumann model is justified by the results.

ACKNOWLEDGEMENTS

The authors are grateful for financial assistance provided by the German Research Society (DFG).

REFERENCES

1. Kline, S.J., Cantwell, B.J. and Lilley, G.M. (eds.) "Proc. 1980-1 AFOSR-HTTM-Stanford Conf. on Complex Turbulent Flows", Stanford University, Stanford, CA, Vols. I-III (1982).
2. Bradshaw, P., Launder, B.E. and Lumley, J.L.: "Collaborative Testing of Turbulence Models", AIAA Paper No. AIAA-91-0215 (1991).
3. Peric, M., Rüger, M. and Scheuerer, G.: "A finite volume multigrid method for calculating turbulent flows", Proc. 7th Symp. on Turbulent Shear Flows, Vol. 1, Stanford University, Stanford, CA (1989), pp. 7.3.1-7.3.6.
4. Friedrich, R. and Arnal, M.: "Analysing turbulent backward-facing step flow with the lowpass-filtered Navier-Stokes equations", J. Wind Eng. and Ind. Aero., **35** (1990) pp. 101-128.
5. Schumann, U.: "Subgrid scale model for finite difference simulations of turbulent flows in plane channels and annuli", J. Comp. Phys., **18** (1975) pp. 376-404.
6. Smagorinsky, J.S.: "General circulation experiments with the primitive equations", Mon. Weather Rev., **91** (1963) pp. 99-164.
7. Lilly, D.K.: "The representation of small-scale turbulence in numerical simulation experiments", Proceedings of the IBM Scientific Computing Symposium on Environmental Sciences, IBM Form No. 320-1951 (1967) pp. 195-210.
8. Chorin, A.J.: "Numerical solution of the Navier-Stokes equations", Math. of Comp., **22** (1968) pp. 745-762.
9. Schmitt, L. and Friedrich, R.: "Large-eddy simulation of turbulent backward facing step flow", Notes on Numerical Fluid Mechanics **20**, Vieweg Verlag Braunschweig 1988, pp. 355-362.
10. Richter, K., Friedrich, R. and Schmitt, L.: "Large-eddy simulation of turbulent wall boundary layers with pressure gradient", Proc. 6th Symp. on Turbulent Shear Flows, Paul Sabatier University, Toulouse, France (1987), pp. 22.3.1-22.3.7.
11. Piomelli, U., Ferziger, J., Moin, P. and Kim, J.: "New approximate boundary conditions for large-eddy simulations of wall-bounded flows", Phys. Fluids A, **1** (1989) pp. 1061-1068.
12. Tropea, C.: "Die turbulente Strömung in Flachkanälen und offenen Gerinnen", Dissertation, University of Karlsruhe, Karlsruhe, Germany (1982).
13. Durst, F. and Schmitt, F.S.: "Experimental study of high Reynolds number backward-facing step flow", Proc. 5th Symp. on Turbulent Shear Flows, Cornell University, Ithaca, NY (1985).

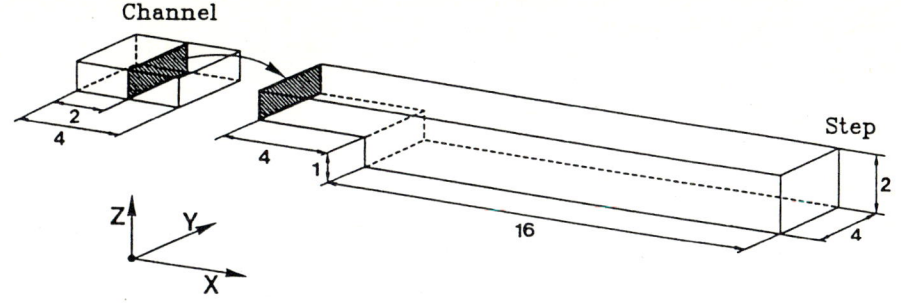

Fig. 1. Geometry of the backward-facing step flow.

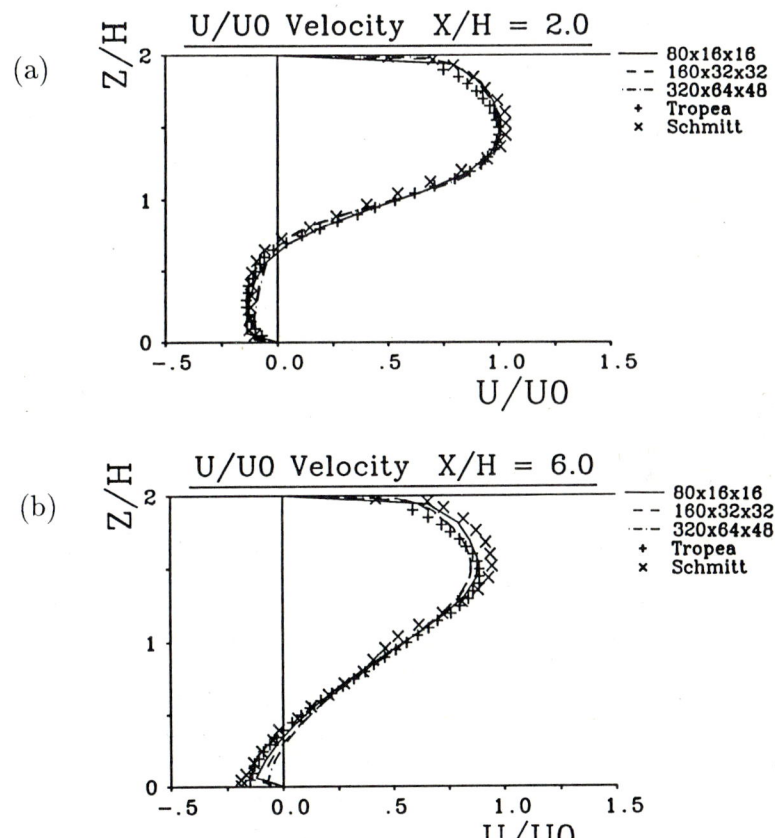

Fig. 2. Mean streamwise U-velocity profiles normalized with the inlet channel centerline velocity, U_0. Effect of grid refinement on predictions. Streamwise locations measured from the step position.

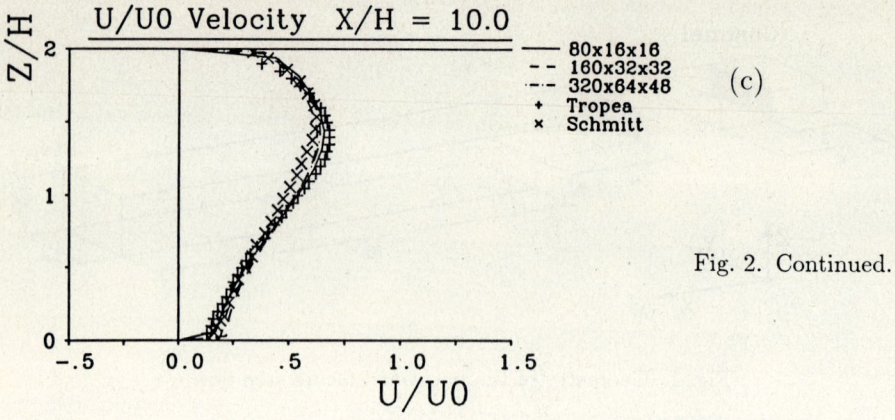

Fig. 2. Continued.

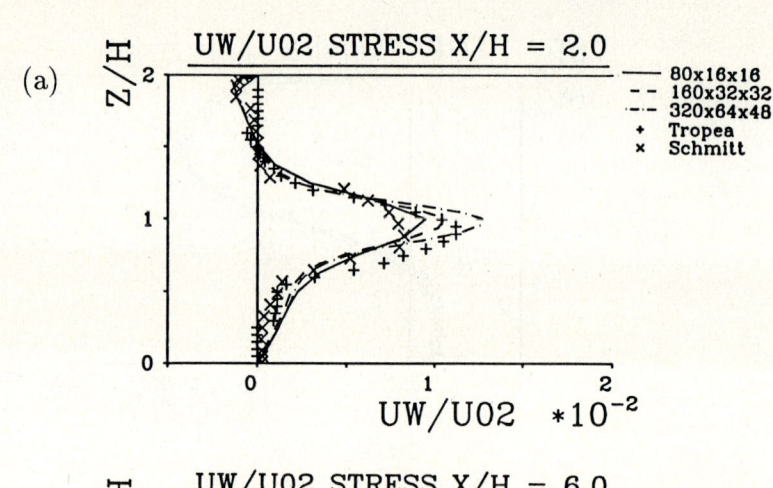

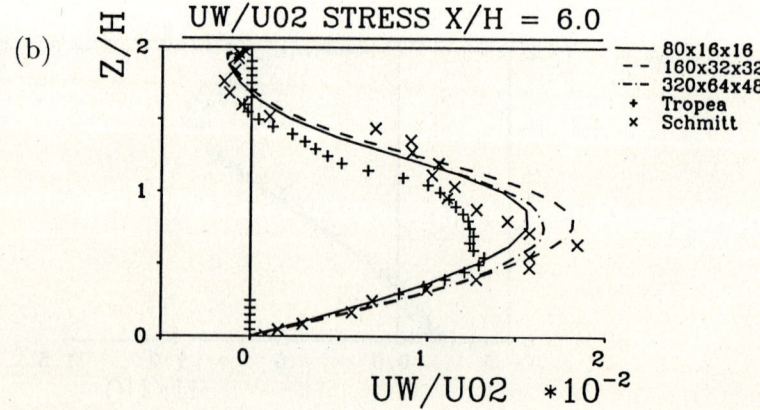

Fig. 3. Reynolds shear-stress ($\langle -u''w'' \rangle$) normalized with the inlet channel centerline velocity, U_0. Effect of grid refinement on predictions. Streamwise locations measured from the step position.

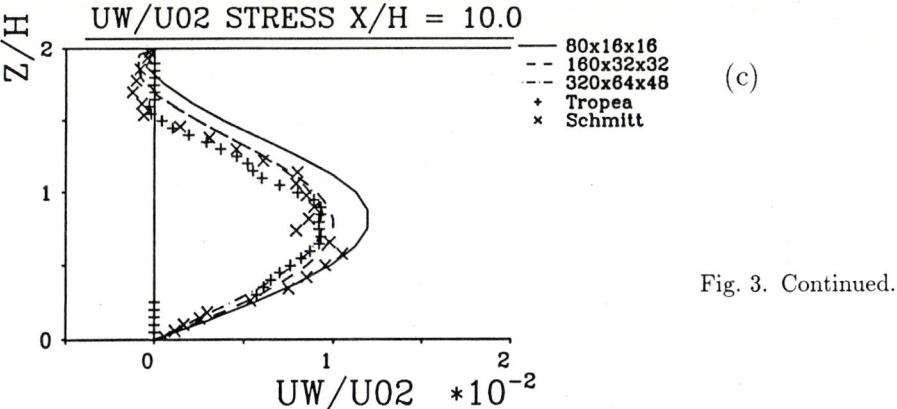

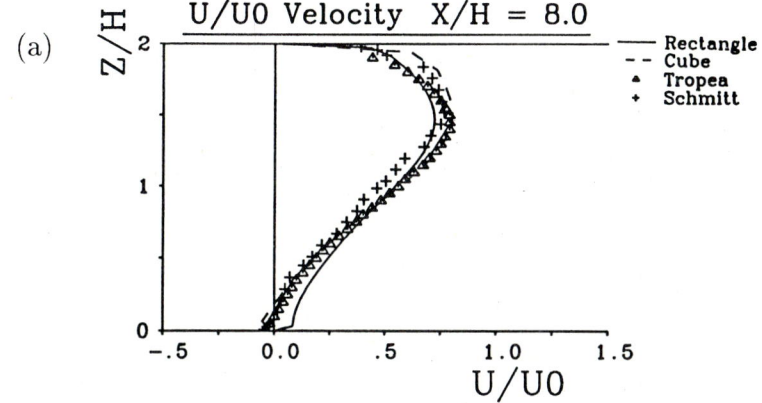

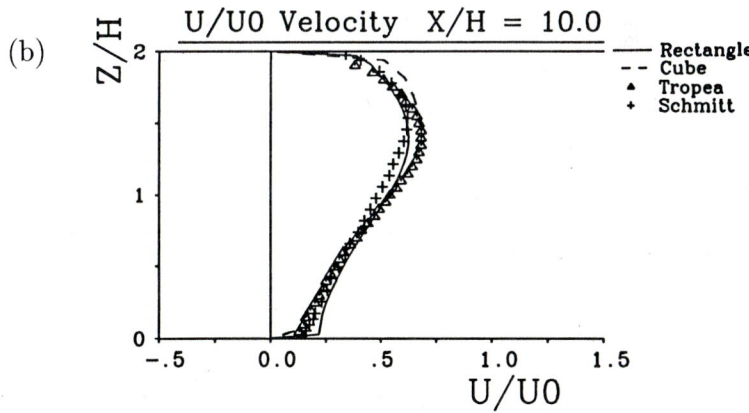

Fig. 3. Continued.

Fig. 4. Mean streamwise U-velocity profiles normalized with the inlet channel centerline velocity, U_0. Effect of mesh-cell shape on predictions. Streamwise locations just upstream and downstream of the measured reattachment length, $X_r = 8.6$.

11

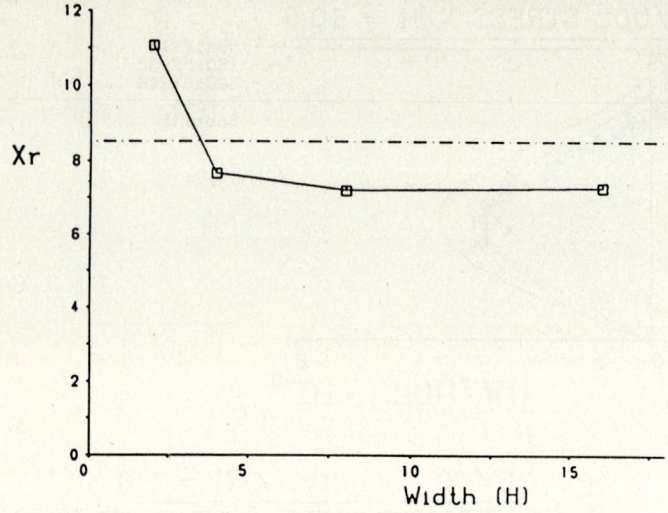

Fig. 5. Variations in the mean reattachment length, X_r as a function of the spanwise dimension of the computational domain.

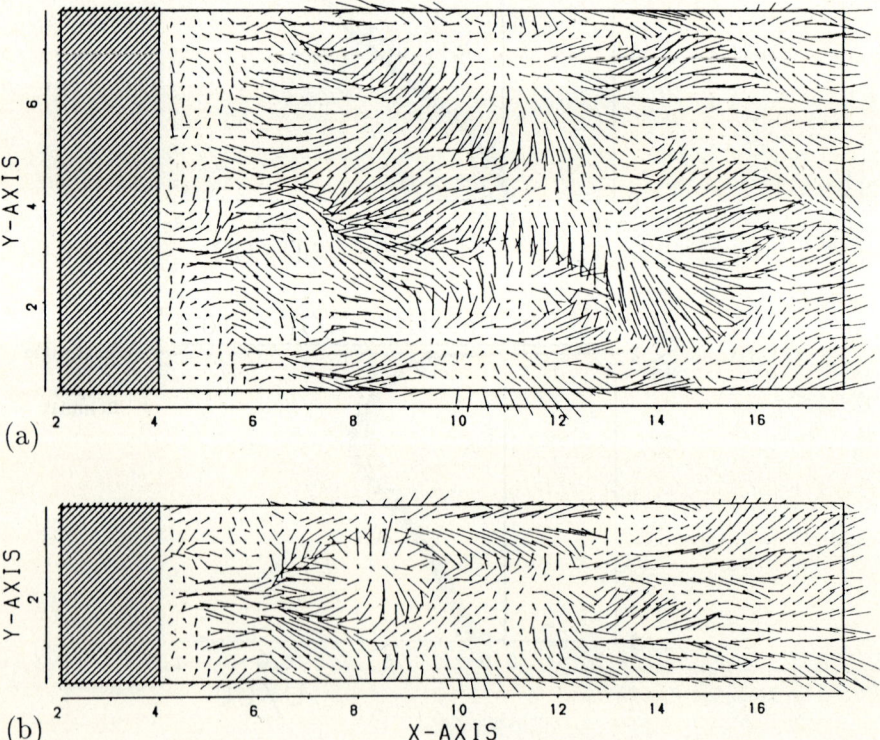

Fig. 6. Instantaneous velocity field in the $x - y$ plane nearest the reattachment surface.   a) Spanwise width $= 8H$. b) Spanwise width $= 4H$.

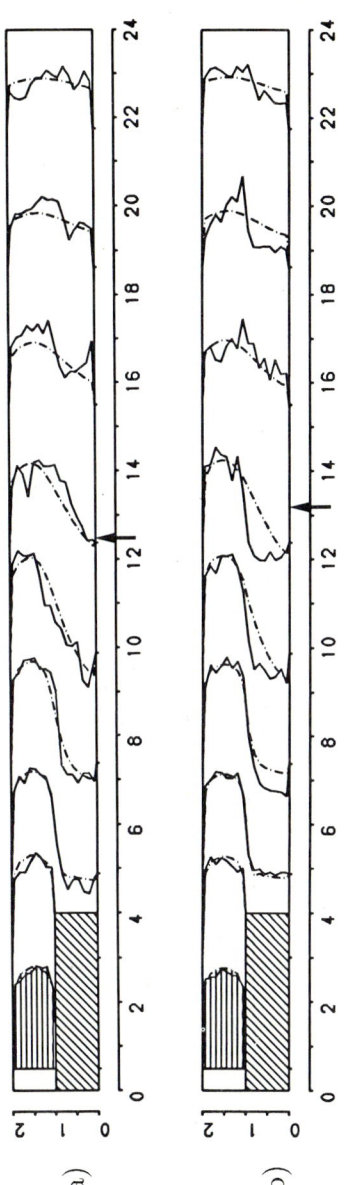

Fig. 7. Mean streamwise velocity profiles compared with local instantaneous velocity profiles. a) Schumann model. b) Smagorinsky-Lilly model.

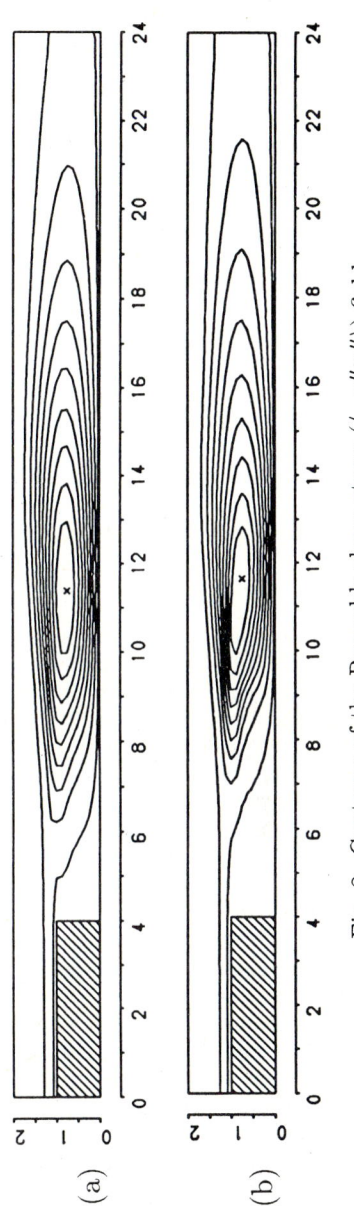

Fig. 8. Contours of the Reynolds shear-stress ($\langle -u''w'' \rangle$) field. a) Schumann model. b) Smagorinsky-Lilly model.

13

Unsteady High-Angle-Of-Attack Flowfield Simulation Around the F-18 Aircraft

Yehia M. Rizk
NASA Ames Research Center, Moffett Field, CA 94035

Ken Gee
MCAT Institute, Moffett Field, CA 94035

SUMMARY

This paper describes a numerical procedure for simulating the unsteady viscous flow around the F-18 aircraft at high angles of attack. A generalized overset zonal grid scheme is used to decompose the computational space around the complete aircraft, including deflected control surfaces. The grids around various components of the aircraft are created numerically using a three-dimensional hyperbolic grid generation procedure. The Reynolds-averaged Navier-Stokes equations are integrated using a time-accurate, implicit procedure. Turbulent flowfield results at 30 degrees angle of attack show the details of the flowfield structure, including the unsteadiness created by the vortex breakdown and the resulting fluctuating airloads exerted on the vertical tail. The computed results agree fairly well with flight data for surface pressure, surface flow pattern, vortex breakdown location, and tail load fluctuations frequency.

I. INTRODUCTION

A more thorough understanding of the high-angle-of-attack regime is needed to improve the high alpha performance of current and future aircraft. The NASA High Alpha Technology Program is devoted to developing an extensive data base for the flowfield around the F-18 aircraft through wind tunnel testing of sub- and full-scale models, flight testing on the High Alpha Research Vehicle (HARV) shown in Fig. 1, and CFD simulation. The objective of this paper is to describe a computational method for predicting the steady and unsteady external flowfield around the complete F-18 aircraft at large incidence. Emphasis is directed towards studying the inherent unsteadiness downstream of the vortex breakdown and the resulting unsteady loads on the twin vertical tails.

The task of simulating the flowfield around the F-18 represents a challenge to current CFD methods and computer resources because it involves both complicated physics and complex geometry. The complicated physics associated with high-angle-of-attack vortical flows involves massive separation, vortex interaction, vortex breakdown and the resulting unsteadiness in the flowfield. The complex geometry of the F-18 represents a challenge to grid generation because of the many closely coupled components (e.g., wing and empennage) and the use of the control surfaces (leading- and trailing-edge flaps and horizontal stabilator) to enhance the aircraft maneuverability. In addition, the wing leading edge extension (LEX) generates a strong vortex which enhances the wing lift at high angles of attack. The twin vertical tails are canted to intercept the high energy flow in the vortex to increase their effectiveness. However, at moderate to high angles of attack, the LEX vortex bursts and induces buffeting to the tails.

The current numerical technique is based on the overset [1], or Chimera [2,3] scheme. This technique acts as the framework for including the effects of different components of the aircraft. A generalized overset/patch zonal grid interfacing method [4,5] is added to the basic Chimera scheme to increase its range of application. The grids around the main components of the aircraft are created numerically using a three-dimensional hyperbolic [6] grid generation code. The equations governing fluid motion, represented by the Reynolds-averaged Navier-Stokes equations, are integrated using a partially upwind, two-factored procedure [7].

The numerical method is briefly described in Section II, while the results and a comparison with flight data are described in Section III.

II. NUMERICAL METHOD

The numerical simulation procedure is based on the use of the overset zonal scheme (Chimera) because of its considerable flexibility. The grid generation task is made easier because the scheme does not require neighboring grids to match along any common surface. Instead, it is merely required that two adjacent grids overlap each other. Moreover, overset grids become more attractive when simulating movable control surfaces scheduled with the angle of attack, or for flows where one grid is in relative motion with respect to the other grids. Therefore, the overset scheme would be a suitable tool to extend the present work to couple the aerodynamic and the structural response of a flexible tail due to buffeting. For complex junctions, such as fuselage-strake-wing, it would be advantageous to remove the requirement of overlap along body surfaces. Unlike overset grids, patch grids do not neccessarily require grid overlapping, but they are generally more restrictive since they require zonal boundaries to match along common surfaces. For the current simulation, the generalized Chimera grid interfacing scheme [4,5] is used to maintain the flexibility of overset grids while relaxing the requirement of grid overlapping along body surfaces.

A three-dimensional body-fitted mesh can be generated numerically using the hyperbolic [6] procedure. The hyperbolic procedure has the desirable feature of allowing the user to specify an arbitrary point distribution along the body surface from which a body-normal grid is automatically created. Compared to other numerical grid generation procedures, the hyperbolic scheme is less costly in terms of computer time and memory, and the resulting grid can be made orthogonal or close to orthogonal everywhere. Since the grid is obtained by marching outward from the body surface, the hyperbolic scheme has a drawback in its inability to prescribe an exact location for the grid outer boundary. However, the use of the overset scheme does not require the exact matching of outer grid surfaces, thus, there is no need to specify an exact location for the outer boundary.

The nature of the separated flow in the high-angle-of-attack regime mandates the solution of the Navier-Stokes equations. A simpler set of equations, like the Euler equations, would not be suitable for simulating massively separated regions. Any Navier-Stokes code written for a single grid can be readily adapted to work with the overset scheme. The two-factored F3D code [7] was selected because it is implicit and can be run in either a time-accurate or steady state mode.

Geometry Definition and Approximations

It would be extremely difficult to model all the details of the F-18 geometry. Therefore, some simplifications are made to facilitate the simulation process. Small protuberances (e.g., antenna fairings) and small components (such as missile rails, etc.) are neglected because they are not expected to have a significant effect on the overall flowfield. The flow through the inlet is excluded in the current work. The use of overset gridding will allow the eventual inclusion of the internal flow through the engines by adding grids to simulate the inlet region and the exhaust plume.

The surface grid for the current computational model consists of the forebody, LEX, faired-over inlet, wing, deflected leading-edge flaps, vertical and horizontal tails, and an idealized boundary layer divertor vent. The fuselage surface definition was given in terms of cross sectional cuts. A fairing was added as shown in Fig. 2 to cover the inlet, and part of the boundary layer divertor vent and a sting was added to the back end of the computational model. The surface of the wing and the empennage are defined from given NACA cross-sections with prescribed taper, twist, thickness, etc. The leading-edge flap surface grid was obtained by deflecting the wing surface grid around the hinge line, which is located at the 20% chord position. A small spanwise gap (with thickness of about .005 chord) was assumed between the wing root section and the flap root section. The gap between the inboard and outboard flaps as well as any gaps along the hinge line are not simulated. The surface grid for the horizontal tail trim position was obtained by rotating the tail around its root mid-chord.

Domain Decomposition

The main advantage of using the overset method is to reduce the grid generation procedure from the difficult task of creating a single grid around the entire configuration to the simpler task of creating grids around relatively simpler components separately. Moreover, like other zonal methods, the overset method provides the capability to use different grid densities, different sets of governing equations, or different turbulence models in different regions, depending on physical considerations. In addition, zoning could be used to reduce the run-time computer core memory requirement through the use of auxiliary memory.

The grid system for the F-18 consists of ten grids shown in Fig. 2. The fuselage (which consists of the forebody, LEX, faired-over inlet, aft end, and sting) is gridded separately from the other components. Furthermore, to reduce the memory requirement, two grids, instead of one, were created around the fuselage. Separate grids were created for the outer mesh, wing, flaps, gap between the fuselage and flaps, B. L. vent, horizontal and vertical tails. A detailed discussion of the grid system is given in Ref. [5]. The hyperbolic grid generation code was used to generate the two fuselage grids, outer mesh, wing, flap and tails grids. Special treatment for the sharp LEX leading edge, which uses one-sided differencing at discontinuities is discussed in Ref. [4]. The horizontal tail grid was distorted near the tail root to make it coincide with the surface of the fuselage. The gap grid were created algebraically through transfinite interpolation from the wing and flap grids. Similarly, the vent grid was created from the fuselage grid. Finally, communication among all interconnecting grids is established using the Pegasus code [2].

Governing Equations

For high-Reynolds-number flow, the use of body-fitted coordinates allows employing the thin-layer approximation in the outward direction to simplify the full Reynolds-averaged Navier-Stokes equations [8]. However, in order to treat non-body-conforming grids and to maintain flexibility in zoning, the thin-layer approximation is extended to three directions. The governing equations take the following form:

$$\partial_\tau \widehat{Q} + \partial_\xi \widehat{F} + \partial_\eta \widehat{G} + \partial_\zeta \widehat{H} = (\partial_\xi \widehat{R} + \partial_\eta \widehat{S} + \partial_\zeta \widehat{T})/Re$$

where $\widehat{Q}$ represents the dependent variable vector while $\widehat{F}$, $\widehat{G}$, and $\widehat{H}$ are the inviscid flux vectors associated with the ξ, η, and ζ directions, respectively. The viscous terms in ξ, η and ζ have been collected into the vectors $\widehat{R}$, $\widehat{S}$ and $\widehat{T}$, respectively. In accordance with the thin-layer approximation, $\widehat{R}$, $\widehat{S}$ and $\widehat{T}$ do not contain any cross derivative terms. The above equations are numerically integrated using the code F3D which is an implicit two-factored scheme that uses central differencing in the η and ζ directions and upwind differencing in the ξ direction. Details of the numerical procedure are given in Ref. [3,7].

For time-accurate calculations, the time step Δt is the same for all grid points, while for steady state calculations, local time stepping is used to speed up the convergence by allowing Δt to vary from point to point according to a function of the Jacobian of the coordinate transformation, J. Such a variation results in using values of Δt on the order of 1.0 away from the body surface, and Δt on the order of 0.01 near the body surface where the mesh spacing is finest.

Boundary Conditions

There are two types of boundary conditions in the F3D/Chimera scheme. The first type deals with physical boundaries, while the second type deals with grid communication as a result of domain decomposition. Both types of boundary conditions are imposed explicitly.

The physical boundary conditions are enforced with calls to modular routines. Various grid topologies are treated through input control parameters to the code. The boundary conditions include viscous wall conditions which could be applied on one or more sides of the computational domain. Also, outflow, axis-averaging, wing-cut-averaging, or symmetry boundary conditions can be enforced on any boundary surface or a subset of it.

For the grid interface boundary conditions, trilinear interpolation across zones is used to obtain the flow variables at boundary points resulting from domain decomposition. Boundary points could lie within the field points of another grid (overset mode) or on the boundary of another grid (patch mode). In both modes, three-dimensional interpolation is used to maintain generality. In the patch mode, the interpolated values are averaged with a zeroth-order extrapolation from the interior domain. Communications is done in a patch mode when the overlap is zero, while communication in overset mode is used for overlap values greater than half a cell size. A weighted average is used for overlap ratios between zero and half a cell size.

Turbulence Modeling

There is no universal model which can be used to simulate the turbulence on all parts of the aircraft, due to the difference in physics in each region. As mentioned before, zonal schemes allow the use of different turbulence models in different zones. The algebraic turbulence model of Baldwin and Lomax [8] has been proven to be simple and fairly accurate for attached boundary layer flow. Different modifications to this widely used model are available. An example of this is the Degani-Schiff [9] modification to account for crossflow separation.

The flow on the LEX is dominated by the effects of the sharp leading edge, which fixes the location of the primary crossflow separation line. This means that, the flow in the LEX region will not be too sensitive to the turbulence model. However it is important to use the crossflow separation modification of Ref. [9] on the forebody because the flow there is more sensitive to the turbulence model. The LEX vortex inhibits flow separation on the majority of the upper side of the wing. Therefore, the use of the Baldwin-Lomax model would be justified. However, the outer layer search cut-off distance was set to a small value in a manner similar to Panaras and Steger [10] to ensure that the high vorticity in the LEX vortex does not result in a high turbulent eddy viscosity on the upper side of the wing.

A crude turbulence model was devised to be used in the gap region between the LEX and wing leading-edge flap and inside the boundary layer divertor vent. This model uses the Baldwin-Lomax model on two opposite walls. A blending function was used to smooth out the variations of the eddy viscosity in the middle of the two walls. No modification for corner regions was used. In the future, other models will be tried. However, for the time being, only the rough effects of the flow inside the gap and vent on the overall flowfield are being simulated.

III. RESULTS

The present numerical procedure was used to compute the turbulent flow around the F-18 at a Mach number of 0.243, an angle of attack 30.3°, and a Reynolds number (based on the wing mean aerodynamic chord) of 11×10^6. These conditions were selected to correspond to flight-test conditions of the HARV. At this angle of attack, the inboard and outboard leading-edge flaps are deflected 33° nose down and the horizontal tail is deflected 7° nose down. Bilateral symmetry was assumed and ten grids consisting of a total of about 0.9 million points around half the configuration were used. The code was run in the non-time-accurate mode for 1000 steps and in a time accurate mode for an additional 8600 time steps. The solution required less than 8 MW of core memory and about 40 seconds per time step on the CRAY-YMP.

The computed surface pressure distributions are compared to the HARV flight-test data [11] in Fig. 3. The first five stations shown are located on the fuselage forebody, while the last three stations are on the LEX (Fig. 1). It is seen that there is a fairly good agreement between the computations and flight data. The overall pressure distribution along the body surface is shown in Fig. 4. Figure 5a shows a flight photo of the LEX vortex as visualized with smoke and the surface flow pattern visualized with the tufts. The numerical solution in Fig. 5b shows particle traces and limiting-surface streamlines on the LEX, wing and deflected leading-edge flaps. There is a good agreement between the computations and the flight-test for the LEX vortex burst location and the surface flow pattern.

Both flight data and computations indicate that the flow is steady ahead of the burst point. However, the burst point location and the spiral flow structure downstream is highly unsteady. This unsteady flowfield of the burst vortex generates an unsteady loading on the vertical tail as seen from the history of the tail total bending moment in Fig. 6. This time history was obtained by using a constant time step of 0.005. This value of time step was determined by stability and is nondimensionalized based on wing root chord and free stream speed of sound. Figure 6 indicates that after an initial transient, the time history shows a pattern of near periodic fluctuations.

In order to study the frequency content of the unsteadiness, a Fast Fourier Transform was performed after each 2^n steps, where n is an integer. The Fourier Transforms for two consecutive intervals corresponding to 4096 steps ($n = 12$) and 8192 steps ($n = 13$) are shown in Fig. 7. This corresponds to intervals of about .29 and .58 seconds of actual time, respectively. Both intervals yield a dominant frequency in the range of 15-20 Hz. It is also seen that increasing the interval reveals an additional low frequency and a wider frequency band centered around the dominant frequency. It is obvious that carrying out the computations further will give a better definition of the low frequency. It must be noted that the first natural frequency of the vertical tail [12] is 15 Hz (first bending mode). Therefore, tail buffeting is induced because the airloads dominant frequency is very close to the first natural frequency of the tail. The additional low frequency is not expected to play a significant role in buffeting because it is considerably less than the lowest natural frequency.

Figure 8 shows a comparison between the predicted airloads frequency and the frequency measured by sub-scale wind tunnel [13] and HARV flight test [14] at angles of attack of about 30 degrees. The wind tunnel frequency was obtained using pressure transducers and represents the induced unsteady airloads frequency. The flight frequency was obtained from an accelerometer and represents the structural response of the vertical tail. The wind tunnel data shows that the frequency is linearly proportional to the tunnel speed. The tunnel speed was scaled to the full-scale dimensions to allow for comparison with flight data and CFD. The comparison shows that there is a good agreement between the computations, wind tunnel, and flight data.

Finally, in order to check the time accuracy of the code, the time step was reduced from 0.005 to 0.0025 and the computations were repeated for about five nondimensional time units. Figure 9 shows that reducing the time step does not alter the time history, indicating that the time variations are being well resolved and no higher frequencies are present.

IV. CONCLUSIONS

A numerical procedure for simulating the unsteady flow around the F-18 aircraft at high angles of attack was outlined. The method employs a generalized overset zonal scheme which has the flexibility needed for zoning a complex configuration like the F-18. Time-accurate computations were performed to examine the unsteady nature of the flowfield downstream of the vortex breakdown point and the resulting unsteady loads on the vertical tail. A Fourier analysis revealed that the dominant frequency is very close to the first natural frequency of the tail. The predicted frequency agrees fairly well with measured frequency. This work represents the first step in studying the tail buffet phenomenon computationally, and will be extended to include the structural response of a flexible vertical tail.

Acknowledgements

The authors would like to thank Dr. Lewis Schiff of NASA Ames Research Center for helpful suggestions.

REFERENCES

[1] Atta, E. H., "Component Adaptive Grid Interfacing for Three Dimensional Transonic Flows About Aircraft Configuration," AIAA Paper 81-0382 (1981).

[2] Benek, J. A., Steger, J. L., Dougherty, F. C., and Buning P. G., "Chimera: A Grid Embedding Technique," AEDC-TR-85-64, (1986).

[3] Buning, P. G., Chiu, I. T., Obayashi, S., Rizk, Y. M., and Steger, J. L., "Numerical Simulation of the Integrated Space Shuttle in Ascent," AIAA Paper 88-4359 (1988).

[4] Rizk, Y. M., Schiff, L. B., and Gee, K., "Numerical Simulation of the Viscous Flow Around a Simplified F/A-18 At High Angles of Attack," AIAA Paper 90-2999 (1990).

[5] Rizk, Y. M., and Gee, K., "Numerical Prediction of the Unsteady Flowfield Around the F-18 Aircraft at Large Incidence ," AIAA Paper 91-0020 (1991).

[6] Steger, J. L and Rizk, Y. M., "Generation of Three Dimensional Body Fitted Coordinates Using Hyperbolic Partial Differential Equations," NASA TM 86753, (1985).

[7] Ying, S. X., Steger, J. L., Schiff, L. B., and Baganoff, D., "Numerical Simulation of Unsteady, Viscous, High Angle of Attack Flows Using a Partially Flux-Split Algorithm," AIAA Paper 86-2179 (1986).

[8] Baldwin, B. S. and Lomax, H., "Thin Layer Approximation and Algebraic Model for Separated Turbulent Flow," AIAA Paper 78-257 (1978).

[9] Degani, D. and Schiff, L. B., "Computation of Turbulent Supersonic Flows Around Pointed Bodies Having Crossflow Separation," *J. Comp. Phys.* **66** (1986), 173–196.

[10] Panaras, A.G. and Steger, J.L. "A Thin-Layer Solution of the Flow About a Prolate Spheriod," *Z. Flugwiss, Weltraumforsch.*, **12** (1988), pp. 173-180.

[11] Fisher, D. F., Banks, D. W., and Richwine, D. M., "F-18 High Alpha Research Vehicle Surface Pressure :Initial In-Flight Results and Correlation With Flow Visualization and Wind Tunnel Data," NASA TM 101724 (1990).

[12] Graham, A. D., and Watters, K. C., "Full Scale Fatigue Testing of the F/A-18 Empennage," The Australian Aeronautical Conference, (1989).

[13] Martin, C. A., and Thompson, D. H., " Scale Model Measurements of Fin Buffet Due to Vortex Bursting on F/A-18 ," AGARD Fluid Dynamics Specialists Meeting, Toulouse, France, Paper 12, (1991).

[14] Del Frate, J. H., Freudinger, L. C., and Kehoe, M. W., "F-18 Tail Buffet," High Alpha Technology Program Workshop (1989).

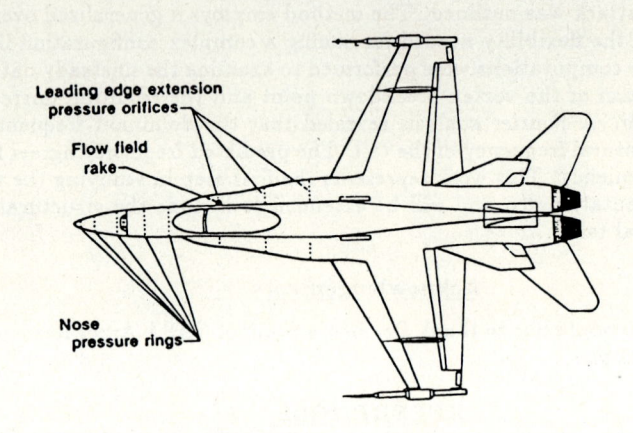

Fig. 1. Top view of the F-18 HARV.

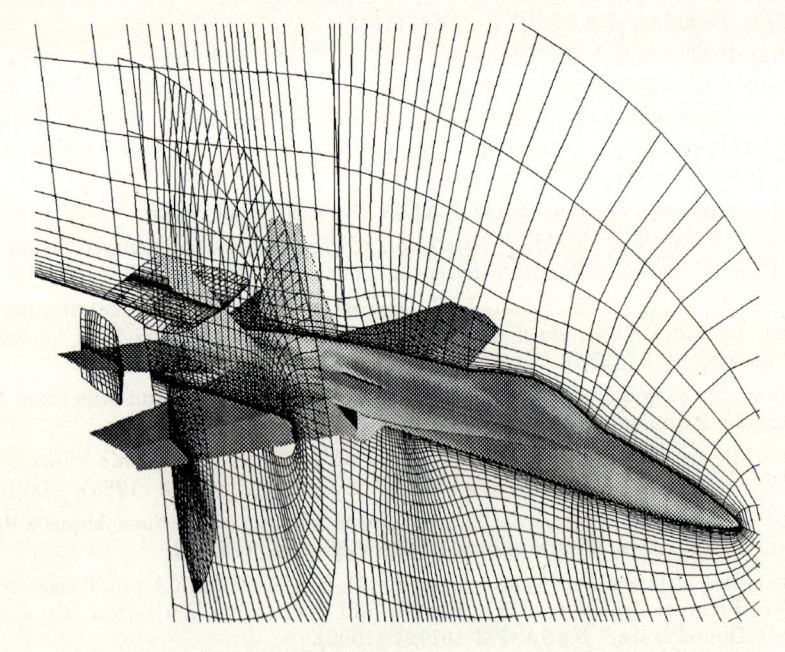

Fig. 2. Overall view of the grid system.

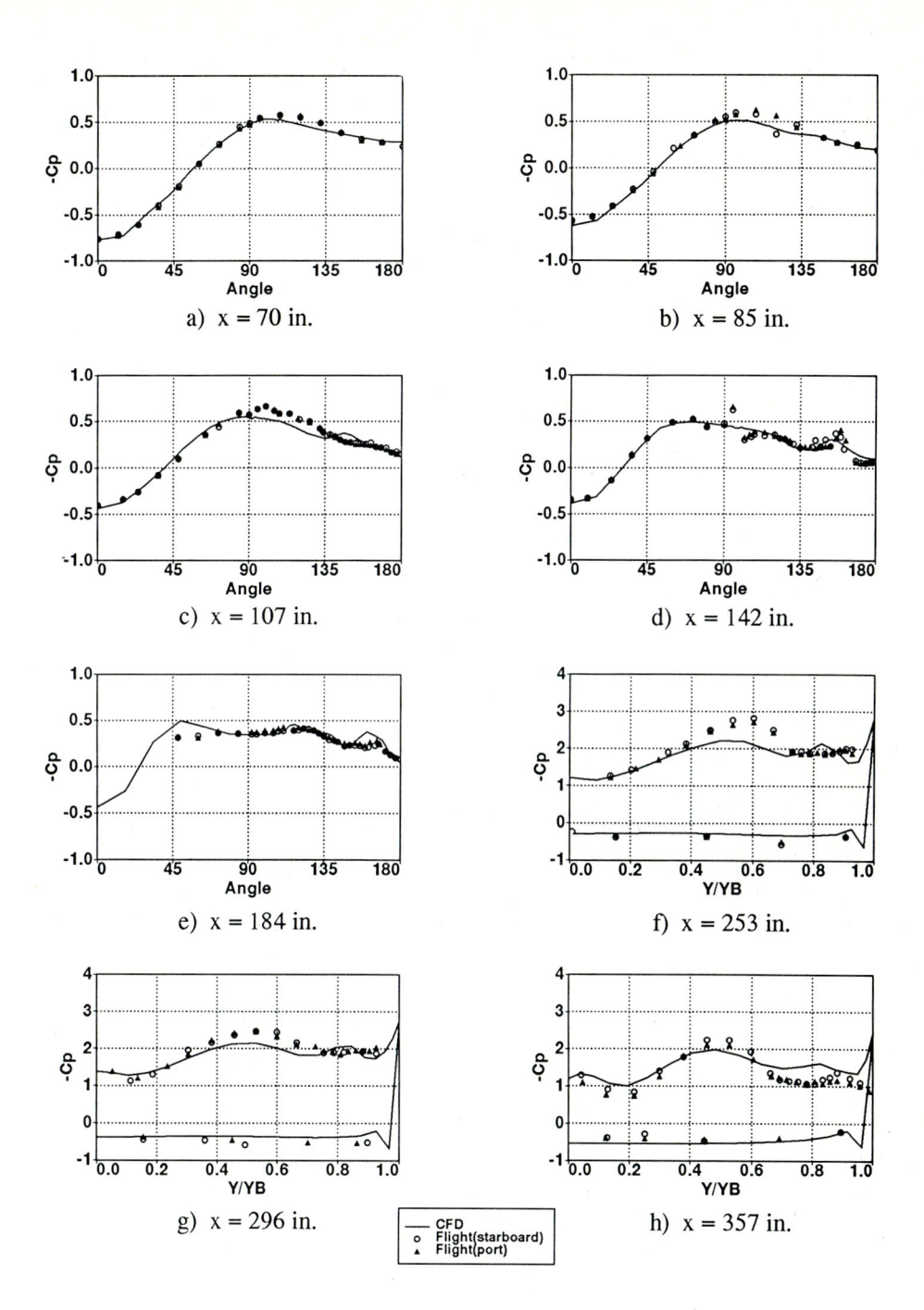

Fig. 3. Cp comparison between computations and flight data.

Fig. 4. Computed surface pressure.

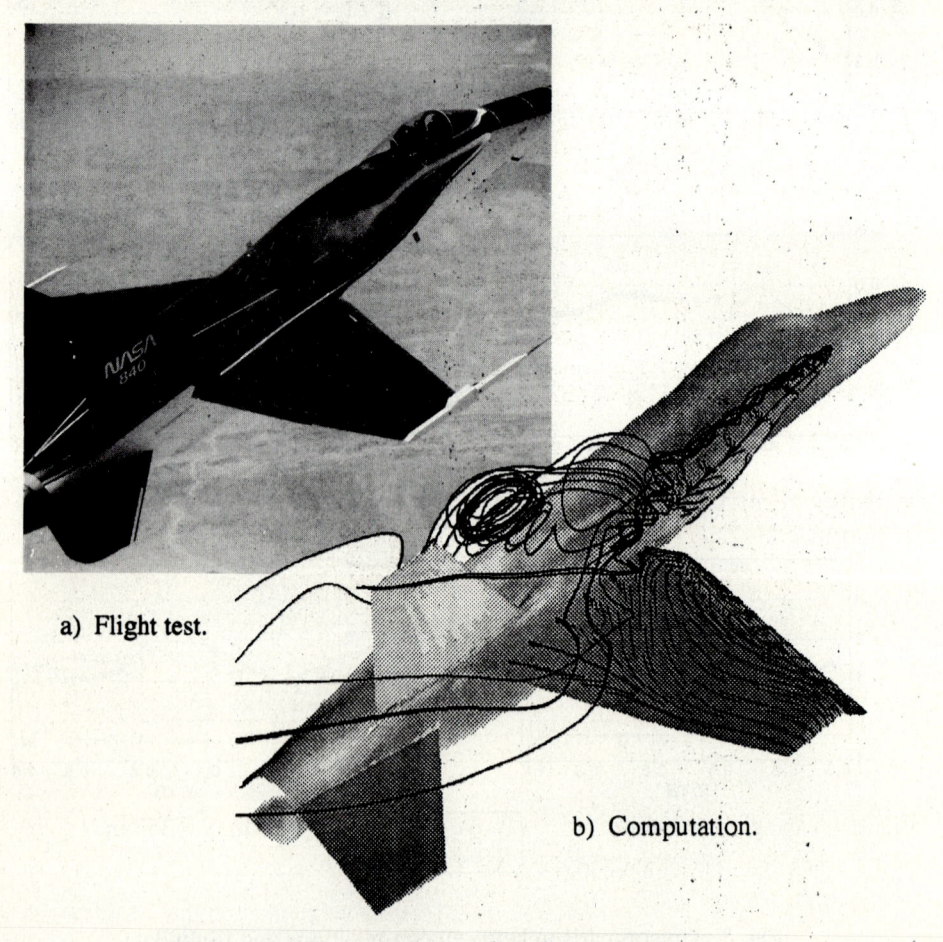

a) Flight test.

b) Computation.

Fig. 5. Particle traces and surface flow pattern.

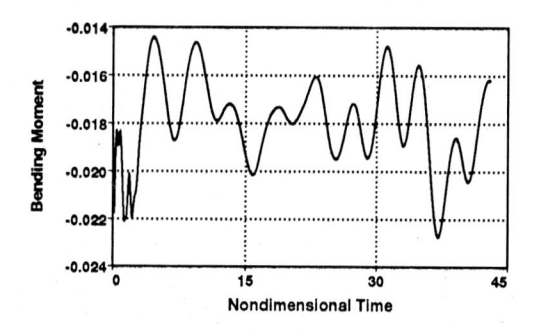

Fig. 6. Time history of vertical tail total bending moment.

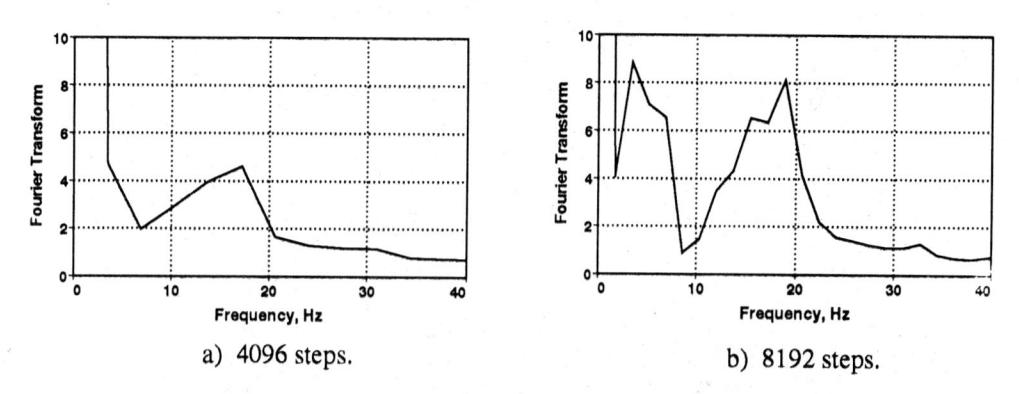

a) 4096 steps.

b) 8192 steps.

Fig. 7. Fourier Transform of the vertical tail total bending moment.

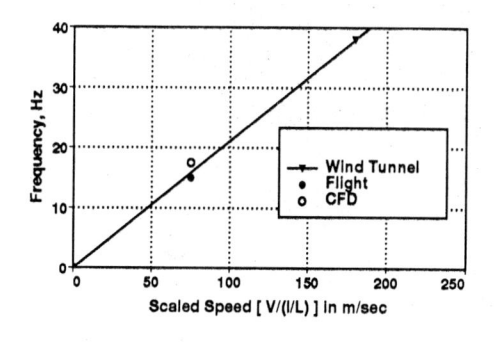

Fig. 8. Comparison between predicted and measured frequency.

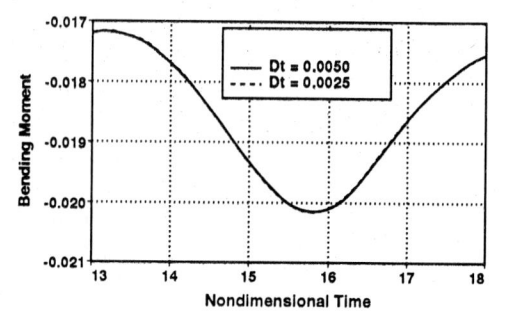

Fig. 9. Effect of time step on vertical tail load history.

Is Computer Extension of Series a Part of CFD?

Milton VAN DYKE
Division of Applied Mechanics, Stanford University
Stanford, CA 94305, USA

Summary

A path midway between pure analysis and sheer number-crunching is provided by the technique of extending a regular perturbation series to high order by computer, and then analyzing and improving the results. This three-step process has, during the past two decades, been applied to perhaps a hundred problems in many branches of fluid mechanics. However, there has been no great rush to follow this route. Some researchers assert that it is not a part of computational fluid mechanics. Here we survey the technique using the flat-plate boundary layer as a simple example, consider the reasons for its relative neglect, and discuss prospects for the future.

1. Introduction

I have for more than twenty years been obsessed with the idea of extending a regular perturbation series in fluid mechanics to high order by computer [1], [2]. From workers in the statistical mechanics of critical phenomena [3], [4] we can borrow a battery of effective techniques for analyzing and improving a power series. In favorable cases this approach yields results that are scarcely obtainable by any other method.

Thus the three-step process of computer extension, analysis, and improvement of a power series would seem to represent a fruitful blend of analysis and computation. In the study of critical phenomena it has long been accepted as a technique that is "competitive with other numerical approaches," and one that "has an enormous future potential in the modern supercomputing environment" [5]. My title raises the question of whether it has gained similar acceptance in fluid mechanics. I have been perplexed over the years to see that it has not. And so I of course ask, "Why not?"

2. The development of computer extension

Because the technique of computer extension is not yet well known, I shall briefly outline its history in fluid mechanics, and comment on a simple example.

With a successful perturbation scheme it is usually sufficient to calculate just the first approximation. However, long series exert a fascination to which several natural philosophers of the nineteenth century succumbed. Thus Stokes, at the age of thirty, calculated the third approximation for plane periodic progressive waves in deep water; and then thirty years later, in preparing his collected works for publication, he extended the series to fifth order. Such laborious hand calculations often contain errors. Although Stokes's five-term expansion is flawless, Wilton tried in 1914 to extend it to ten terms, but committed an error at eighth order. Fascination became an obsession for the French astronomer Delaunay, who in the middle of the last century spent some twenty years computing the motion of the moon to ninth order in each of five small parameters.

This situation changed with the advent of the computer, which made it possible to extend a perturbation series to high order, and without error. In fluid mechanics the first step was taken in 1952, when Munk and Rawling [6] wired the IBM 604 Calculating Punch to compute the successive coefficients of the Janzen-Rayleigh expansion in powers of M^2 for subsonic flow past a circle. The first four terms had been found previously by hand, and Munk and Rawling added the fifth term in eight hours of handling and machine time. In 1978 the same calculation [7] required one-third of a second on the then current IBM 3032 computer; and last year, though the computing time increases as the sixth power of the number of terms, Guttmann and Thompson [8] calculated 30 terms on a Cyber 990.

If the first approximation is not adequate, adding higher-order terms usually gives little direct improvement. However, a great deal of information is concealed in the higher coefficients; and much of it can often be extracted by analyzing the coefficients to unveil the structure of the solution, and on that basis applying a variety of devices for improving the utility of the series. Those two additional steps were introduced into statistical mechanics in the late 1950s [9], but made their way into fluid mechanics only in the 1970s.

3. Illustrative example: the Blasius problem

We can survey many features of computer extension by outlining its application to a simple problem. Though most fluid-mechanical problems involve partial differential equations, it suffices to consider a single nonlinear ordinary equation. In Prandtl's boundary-layer theory, the stream function for uniform flow past a semi-infinite plate has, in the normalization of Schlichting [10], the self-similar form $\psi = \sqrt{(\nu x U)} f(\eta)$, where $\eta = y\sqrt{(U/\nu x)}$. This yields the two-point boundary-value problem

$$2f''' + ff'' = 0, \qquad f(0) = f'(0) = 0, \quad f'(\infty) = 1. \tag{1}$$

This is simplified to an initial-value problem by taking advantage of a second similitude—that $k f(k\eta)$ is also a solution of the differential equation and the two initial conditions. We therefore solve the surrogate problem

$$2F''' + FF'' = 0, \qquad F(0) = F'(0) = 0, \quad F''(0) = 1, \tag{2}$$

and then find the desired value of the skin friction from

$$f''(0) = [F'(\infty)]^{-3/2}. \tag{3}$$

Hand computation. The first step in extending a series by computer is to calculate as many terms as may reasonably be found by hand. This serves to suggest the form of the general term, and can also be used later to check the computer program. For equation (1) an expansion in powers of η was calculated by Blasius [11]. For the surrogate problem (2) the boundary conditions suggest the first approximation $F(\eta) \approx \eta^2/2$, and substituting into the problem and iterating yields

$$F = \frac{1}{2}\eta^2 - \frac{1}{240}\eta^5 + \frac{11}{161\,280}\eta^8 - \frac{5}{4\,257\,792}\eta^{11} + \frac{9299}{464\,950\,886\,400}\eta^{14} - \ldots \tag{4}$$

At this stage it is clear that the form of the expansion is simply

$$F(\eta) = \sum_{n=0}^{\infty} c_n \eta^{3n-1}. \tag{5}$$

Multiple summations appear in more complicated problems. For example, in the Janzen-Rayleigh expansion treated by Munk and Rawling the velocity potential is a triple sum of powers of M^2, inverse powers of the radius r, and cosines of odd multiples of the polar angle θ.

Recursion relations. Having deduced the form of the expansion, our second step is to substitute it into the problem and equate like quantities. In the Blasius problem we simply equate like powers of η; whereas in the Janzen-Rayleigh problem one must equate not only powers of M^2 but also like powers of r and like Fourier components. This requires such manipulations as replacing the product of two sums by a double sum and then summing along diagonals.

In some problems this algebraic labor is so great that one must delegate it to the computer, using a symbol-manipulation language such as Reduce, Macsyma, or Mathematica. However, experience shows that in such problems even a large computer can extend the series by only a few terms before it is engulfed by "intermediate expression swell." We disregard problems of that sort, and consider only those in which it is possible to carry out by hand the derivation of the recursion relations for the successive coefficients, complicated and convoluted though they may be. For the Blasius problem we easily find that each new coefficient is given by a "convolution sum" of its predecessors:

$$c_n = \frac{-1}{2(3n-1)(3n-2)(3n-3)} \sum_{k=1}^{n-1} (3k-1)(3k-2) c_k c_{n-k}, \qquad n = 2, 3, \ldots \quad (6)$$

Computer program. The next step is to write a computer program to carry out the arithmetic in the recursion relation. For that purpose Fortran or a similar language is appropriate. The heart of any such program is a set of DO loops, nested only two deep in our simple example, but six or more in more complicated problems.

Equation (4) shows that all our coefficients are rational fractions (though they may be irrational in other problems). Nevertheless, we have always chosen to carry out the computations in floating-point arithmetic. Working with exact rational fractions would be much slower; and to comprehend the results one must eventually convert to decimals. Of course the penalty paid is that the accuracy of successive coefficients is gradually eroded by accumulation of round-off error. Consequently, we work in double- or even quadruple-precision arithmetic.

It is convenient to introduce a scale factor S into the expansion, replacing equation (5) by

$$F(\eta) = \sum_{n=0} d_n \left(\frac{\eta}{S}\right)^{3n-1}. \qquad (7)$$

[which adds a factor S^{-1} to the right-hand side of the equation for d_n corresponding to equation (6)]. This rescaling serves two purposes. First, S can be chosen to delay overflow or underflow. Second (as suggested privately by H. Takagi) it can be used to assess the loss of accuracy resulting from accumulated round-off error. Two computations carried out with different values of S (whose ratio is not an integral power of 2) will involve completely different roundings, so their results may be regarded as correct to as many figures as they have in common. This test has shown that the coefficients listed below (computed on a Macintosh Plus) are correct to more figures than are given:

n	c_n	n	c_n	n	c_n
1	5.0000 00000 E-01	23	1.4636 36153 E-40	138	-5.1305 59318 E-246
2	-4.1666 66667 E-03	24	-2.3921 78230 E-42	139	8.3869 59318 E-246
3	6.8204 36508 E-05	25	3.9098 52471 E-44	140	-1.3710 32759 E-249
4	-1.1743 17581 E-06				
5	1.9999 96187 E-08	49	5.1762 65383 E-87	1349	1.5407 E-2409
6	-3.3536 61721 E-10	50	-8.4615 14367 E-89	1350	-2.5187 E-2411

Nearest singularity. In analyzing the coefficients, even though the perturbation quantity is purely real in a physical problem—and usually positive, as in our Blasius problem—it is essential to consider its complex plane. The reason is that the structure of an analytic function defined by a power series is evident only in the complex plane. In particular, the singularities that are the backbone of the function often lie off the real axis.

The first step in analyzing the coefficients is to examine the ultimate pattern of signs, because it indicates the direction of the nearest singularity. In our Blasius problem the signs alternate from the very outset (through at least 1350 terms!) This pattern means that the nearest singularity lies on the negative axis of η^3, by analogy with the simple model of a pole at -1 in the complex plane of ε:

$$\frac{1}{1+\varepsilon} = 1 - \varepsilon + \varepsilon^2 - \varepsilon^3 + \ldots + (-\varepsilon)^n + \ldots, \qquad |\varepsilon| < 1. \qquad (8)$$

In other problems the signs are often fixed, corresponding to a physically meaningful nearest singularity on the positive axis. Occasionally they are of longer period, quasi-periodic, or "random," corresponding to a complex-conjugate nearest pair of singularities.

The second step is to estimate the distance to the nearest singularity—the radius of convergence R. It is given by d'Alembert's ratio test as the limit of the ratio of coefficients $|c_{n-1}/c_n|$. Domb and Sykes have observed [9] that it is better to work with the reciprocal ratio $|c_n/c_{n-1}|$, and consider its variation with $1/n$. The reason is that most singularities encountered in physical problems are of *algebraico-logarithmic* form, the function behaving locally like

$$g(\varepsilon) \sim \text{const.} \begin{cases} (R \pm \varepsilon)^\alpha, & \alpha \neq 0, 1, 2, \ldots \\ (R \pm \varepsilon)^\alpha \ln(R \pm \varepsilon), & \alpha = 0, 1, 2, \ldots \end{cases} \qquad (9)$$

and the binomial theorem shows that for these functions the reciprocal ratios have a linear dependence on $1/n$:

$$\frac{c_n}{c_{n-1}} = -\frac{1}{R}\left(1 - \frac{1+\alpha}{n}\right). \qquad (10)$$

A graph of $|c_n/c_{n-1}|$ versus $1/n$—the *Domb-Sykes plot*—gives for the Blasius problem an estimate of the limit $1/R = 0.01635$. Hence the series converges for $|\eta^3| < R = 61.16$.

The Domb-Sykes plot provides as a bonus an estimate of the *nature* of the nearest singularity from the limiting slope. Here the asymptote is clearly horizontal, corresponding to an exponent $\alpha = -1$; the nearest singularity is a simple pole.

Another way of exploring the structure of the solution is to form Padé approximants from the power series. That is particularly useful when the nearest singularities lie off the real axis (in which case the Domb-Sykes plot is useless) or when more distant singularities are of interest.

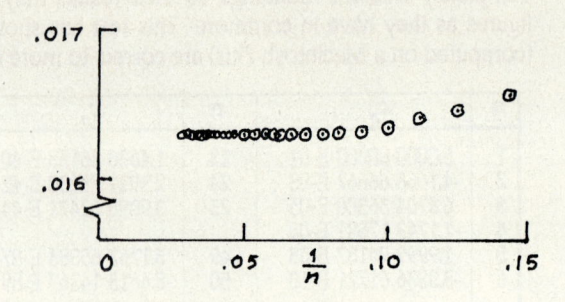

Fig. 1. Domb-Sykes plot for Blasius series

Improving a series. Physicists have given us a battery of techniques, special, and general for improving a series, especially after the nearby singularities have been explored. I mention just two techniques applicable to our Blasius problem.

A special technique that happens to apply here is reversion of series—interchanging the independent and dependent variables. Richardson observed [12] that if the series (4) is first differentiated, then the reverted series

$$\eta = F' + \frac{1}{48} F'^4 + \frac{1}{840} F'^7 + \frac{1}{11\,520} F'^{10} + \frac{2099}{296\,524\,800} F'^{13} + \ldots \quad (11)$$

converges for the entire range of physical interest, $0 < \eta < \infty$. The fixed signs indicate a nearest singularity on the positive axis of F'^3, and a new Domb-Sykes plot shows a weak logarithmic singularity indicating that $1/F'^3$ approaches 0.1103 for large η. This yields for the original problem the value $f''(0) = 0.332$ of the textbooks.

A general technique when the nearest singularity lies on the negative axis is to map it away to infinity using an Euler transformation. Here, knowing that F' must be bounded at $\eta = \infty$, we re-expand equation (4) in powers of $\eta^3/(R + \eta^3)$ to find

$$F'(\eta) = \left(\frac{\eta^3}{61.16 + \eta^3}\right)^{1/3} \left[3.9399 - 3.7068\left(\frac{\eta^3}{61.16 + \eta^3}\right) + 2.2233\left(\frac{\eta^3}{61.16 + \eta^3}\right)^2 + \ldots\right]. \quad (12)$$

This appears to be a purely asymptotic series, with coefficients that change sign about every seven terms and decrease in magnitude until the thirty-fifth term, beyond which they grow rapidly. Estimating its sum for $\eta = \infty$ by the device of averaging the closest successive maxima and minima yields $F'(\infty) \approx 2.082$, and then equation (3) gives $f''(0) = 0.333$.

4. Some successes of computer extension

Computer extension has produced at least a few unqualified successes in fluid mechanics, of which we mention just two examples. Schwartz [13] extended Stokes's five-term hand computation of periodic deep-water waves to 117 terms. Thus he uncovered errors in Wilton's longer hand computation, and showed that Stokes was mistaken in thinking that his series converges up to the sharp-crested highest wave, but that it could be recast to do so. That work has spawned many papers treating a variety of nonlinear wave problems [14].

Another application in a quite different branch of fluid motion was to the convergence of a spherical or cylindrical shock wave. Guderley showed by local analysis [15] that the final collapse is self-similar, with an exponent that can be found only in the course of solution—a sort of nonlinear eigenvalue problem. Guttmann and I [16] treated the global problem of a spherical or cylindrical cavity containing quiescent gas that begins to contract with high radial speed, driving an axisymmetric shock wave inward to collapse at the center and rebound. We were able to refine Guderley's local solution, and discovered that the exponents in higher terms are equally spaced.

5. Some controversial results

That the technique of computer extension has not been more enthusiastically embraced by the fluid-mechanics fraternity may be attributed in part to its having led to conclusions that are surprising, paradoxical, and controversial—and in at least one case evidently wrong. Chief among these is a series of attempts to extract from an expansion in powers of Reynolds number the limiting behavior for steady laminar flow at infinite Reynolds number. I found this attempt to be altogether successful for the model problem of linearized Oseen flow past a sphere [1]. However, when I applied it to the fully developed flow through a loosely coiled pipe [18], 25 terms of the series led to the conclusion that the friction grows asymptotically as the one-quarter power of the Dean number (the product of the Reynolds number and the square root of the coiling ratio), whereas four different crude boundary-layer analyses and several careful finite-different solutions predict a square-root growth. That behavior is also confirmed by several experiments, though the recent careful measurements of Ramshankar & Sreenivasan [18] lie closer to my one-quarter power.

The worrying possibility that this disagreement arises from numerical errors in the series has been eliminated by Herbert's recent independent recalculation [19]. However, he has extended the computation from 25 to 37 terms, and finds the result tending away from the one-quarter power toward the square root. Thus it may well turn out that the resolution of the puzzle is simply that my extrapolation from zero to infinite Reynolds number was too rash, and my conclusion erroneous.

Another apparently less rash extrapolation—because it extended over only a finite range—has apparently already proved erroneous. Guttmann and I [7] extended the Janzen-Rayleigh expansion for the circle in the hope of resolving the forty-year-old *transonic controversy* : can an airfoil exhibit, above its critical Mach number, a continuous range of shock-free supersonic flow. From 29 terms of the series in power of M^2 we concluded that shock-free flow over the circle does exist until the free-stream Mach exceeds its critical value by about 1.1 percent. However, Guttmann & Thompson [8] have added one more term, improved the accuracy, and refined the analysis, and demonstrate that there is no convincing evidence for the existence of a range of shock-free flow.

6. Prospects for the future

Despite the possible missteps just described, computer extension of series seems to me a useful technique with a promising future in computational fluid dynamics. Compared with its use in statistical mechanics, it is in its infancy, with many more applications needed to establish a solid basis of experience. Several specific aspects of the method need to be studied, and new techniques of analysis devised. I give here a shopping list of five of the most urgent and important matters:

1. Series with long-period or "random" signs. Most series in fluid mechanics display either an ultimately fixed pattern of signs, corresponding to a nearest singularity on the positive axis of the perturbation quantity, or alternating signs, corresponding to a singularity on the negative axis. A few display a period of four, indicating a complex-conjugate pair on the imaginary axis. But in an interesting minority, patterns of period 3, 4, 5, and so on up to at least 14 have been discerned, either strictly or with periodic skips; and in other cases the pattern appears to be random. Techniques more effective than mapping or Padé approximants must be developed for performing numerical analytic continuation through and beyond such pairs of singularities. A promising new technique is the method of *differential approximants* [4].

2. Analysis and improvement of the entire field. It is a shortcoming of existing analyses in fluid mechanics that we examine at one time only a single quantity—usually some global property of the solution. We need to learn when different properties of the solution have the same or different singularities in the complex plane, and to develop methods for treating the entire solution at once.

3. Purely asymptotic series. It is remarkable that most series considered in applied mechanics have finite or occasionally infinite radius of convergence. However, the useful approximation of *slow variations*—exmplified by flow through a tube of slowly varying radius—appears almost always to yield purely asymptotic series (having zero radius of convergence). . Methods of summing such series, such as the *terminants* of Dingle [20] should be refined and generalized—in particular, to series that also have longer or random sign patterns, like the Euler-transformed Blasius series of equation (12).

4. Multiple power series. In the field of statistical mechanics, which serves as our guide, double and multiple expansions have only recently been considered. According to Guttmann [4], the only techniques so far applied are generalizations of Padé approximants. It is unfortunate that ratio methods, relying as they do on the theory of complex variables, become much less transparent for even a double power series. This is virgin territory in fluid mechanics, which it will be interesting to explore.

5. Series in other than integral powers. Singular perturbations usually yield expansions that are more complicated than power series. They often involve fractional as well as integral powers, logarithms, or exponentials of the perturbation quantity. Soon symbol-manipulation languages will be capable of generating extended series of this sort. They no doubt have zero radius of convergence; so that if effective techniques can be developed for summing asymptotic series in integral powers, we must then try to generalize them to these more complicated forms.

This work was supported by the National Science Foundation under Grant CTS-88-21460, administered by Dr. Stephen C. Traugott.

References

[1] Van Dyke, M.: Extension of Goldstein's series for the Oseen drag of a sphere. *J. Fluid Mech.* **44** (1970) 365-372

[2] Van Dyke, M.: Computer-extended series, *Ann. Rev. Fluid Mech.* **16** (1984) 237-309, Annual Reviews Inc, Palo Alto

[3] Gaunt, D. S. & Guttmann, A. J.: Asymptotic analysis of coefficients, *Phase Transitions and Critical Phenomena* **3** (1974) 287-309, Academic Press

[4] Guttmann, A. J.: Asymptotic analysis of power series, *Phase Transitions and Critical Phenomena* **13** (1989) 1-234, Academic Press

[5] Adler, J., Meir, Y., Amnon, A., Harris, A. B. & Klein, L.: Low-concentration series in general dimensions, *J. Statistical Phys.* **58** (1990) 511-538

[6] Munk, M. & Rawling, G.: Calculation of compressible subsonic flow past a circular cylinder, *US Nav. Ordnance Lab. NAVORD Rep. 2477* (1952)

[7] Van Dyke, M. & Guttmann, A. J.: Subsonic potential flow past a circle and the transonic controversy, *J. Austral. Math. Soc., Ser. B* **24** (1983) 243-261

[8] Guttmann, A. J. & Thompson, C. J.: Subsonic potential flow and the transonic controversy, *To be published.*

[9] Domb, C. & Sykes, M. F.: On the susceptibility of a ferromagnetic above the Curie point, *Proc. Roy. Soc. London A* **240** (1957) 214-228

[10] Schlichting, H.: *Boundary-layer Theory*, 7th ed. (1979), McGraw-Hill

[11] Blasius, H.: Grenzschichten in Flüssigkeiten mit kleiner Reibung, *Z. Math. u. Phys.* **56** (1908) 1-37; English transl. *NACA Tech. Memo. 1256*

[12] Richardson, S.: On Blasius's equation governing flow in the boundary layer on a flat plate, *Proc. Camb. Philos. Soc.* **74** (1978) 179-184

[13] Schwartz, L. W.: Computer extension and analytic continuation of Stokes' expansion for gravity waves, *J. Fluid Mech.* **62** (1974) 553-578

[14] Schwartz, L. W. & Fenton, J. D.: Strongly nonlinear waves, *Ann. Rev. Fluid Mech.* **44** (1982) 39-60, Annual Reviews Inc, Palo Alto

[15] Guderley, G.: Starke kugelige und zylindrische Verdichtungsstösse in der Nähe des Kugelmittelpunktes bzw. der Zylinderachse, *Luftfahrtforschung* **19** (1942) 302-312

[16] Van Dyke, M. & Guttmann, A. J.: The converging shock wave from a spherical or cylindrical piston, *J. Fluid Mech.* **120** (1982) 451-462

[17] Van Dyke, M.: Extended Stokes series: laminar flow through a loosely coiled pipe, *J. Fluid Mech.* **86** (1978) 129-145

[18] Ramshankar, R. & Sreenivasaan, K. R.: A paradox concerning the extended Stokes series solution for the pressure drop in coiled pipes, *Phys. Fluids* **31** (1988) 1339-1347

[19] Herbert, T. & Feng, H. Y.: Flows in a loosely coiled pipe and their stability, *Bull. Amer. Phys. Soc.* **34** (1989) 2319

[20] Dingle, R. B.: *Asymptotic Expansions: Their Derivation and Interpretation* (1973) Academic Press

FLOW OVER DELTAWINGS

Validation of a computer code for the simulation of turbulent flows past wings

Nicola Ceresola
Alenia Aeronautica
Corso Marche,41 10146 Torino Italy

SUMMARY

Numerical simulations of turbulent flows past a delta wing are performed using a finite differences, implicit, factored thin-layer Navier-Stokes code. Detailed comparisons are made with experimental data. The prediction capability of the algebraic Baldwin-Lomax turbulence model in case of vortical flows is investigated. The main problem resulted into finding the appropriate lengthscales of the shear layer in the regions of vortical flow.

This investigation has been carried out under contract with the Italian Ministry of Defence.

INTRODUCTION

The vortical flows past delta wings, at moderate or high incidence, have became an important subject of experimental and theoretical research. In particular, CFD has became an important tool for wing design in the aerospace industry, due to the advancements in numerical simulation techniques. To gain insight on the advantages and limitations of such methods, and to validate them on a range of flow conditions, a programme of experimental and numerical investigation on vortical flows past delta wings was conducted. Within this program, the numerical data generated by Euler and Navier-Stokes computations past a fixed wing geometry were compared with the experiments.

In the present paper the results of the computations of viscous turbulent flows past a 65° swept cropped delta wing are reported, and compared with wind tunnel data obtained at NLR[1,2]. The wing chosen for numerical testing has a sharp leading edge, so the uncertainty about the location of the primary separation is eliminated and it is easier to understand the effect of the turbulence model. The code NAVIER3D, developed in ALENIA, is employed to solve the Reynolds-averaged, thin-layer Navier-Stokes equations. The numerical method is based on the finite difference, implicit, factored scheme of Pulliam and Chaussee[3]. Diagonalization of the Jacobian matrices on the implicit side is performed. Space-centered differences are employed, along with nonlinear second and fourth order numerical damping. To improve the efficiency of the memory management on a vector computer, an explicit procedure is applied in one space direction (see[4]), retaining the factored scheme on the mesh surfaces defined by the other two coordinates. An algebraic eddy viscosity turbulence model is employed, namely the Baldwin-Lomax model, which is the most widely used for industrial applications in external aerodynamics. It is known, however, that the Baldwin-Lomax model, in its original form, is unable to give correct values of the eddy viscosity in case of vortical flows[5,6]. Modifications had therefore to be made in order to obtain predictions that are suitable for engineering purposes.

MATHEMATICAL MODEL

Thin-layer Navier-Stokes equations in generalized coordinates

It is a common practice, having to deal with high Reynolds number flows, to write the Navier-Stokes equations in the 'thin layer' approximation, which consists in neglecting the viscous stress components parallel to the body surface, keeping the perpendicular ones. This allows to employ high aspect ratio meshes near the solid surfaces, to correctly resolve the shear layers. In the present implementation, the solid wall coincides with an $\eta = constant$ surface, so that the diffusion in the ξ, ζ mesh directions is neglected. The equations in body-fitted curvilinear coordinates are (the hat symbol indicates the contravariant quantities):

$$\partial_t \hat{Q} + \partial_\xi \hat{E} + \partial_\eta \hat{F} + \partial_\zeta \hat{G} = Re^{-1} \partial_\eta \hat{S} \qquad (1)$$

where

$$\hat{Q} = J^{-1} \begin{pmatrix} \rho \\ \rho u \\ \rho v \\ \rho w \\ e \end{pmatrix} \qquad \hat{E} = J^{-1} \begin{pmatrix} \rho U \\ \rho u U + \xi_x p \\ \rho v U + \xi_y p \\ \rho w U + \xi_z p \\ U(e+p) \end{pmatrix}$$

$$\hat{F} = J^{-1} \begin{pmatrix} \rho V \\ \rho u V + \eta_x p \\ \rho v V + \eta_y p \\ \rho w V + \eta_z p \\ V(e+p) \end{pmatrix} \qquad \hat{G} = J^{-1} \begin{pmatrix} \rho W \\ \rho u W + \zeta_x p \\ \rho v W + \zeta_y p \\ \rho w W + \zeta_z p \\ W(e+p) \end{pmatrix}.$$

J is the Jacobian determinant of the mapping and

$$U = \xi_x u + \xi_y v + \xi_\zeta w$$

$$V = \eta_x u + \eta_y v + \eta_\zeta w$$

$$W = \zeta_x u + \zeta_y v + \zeta_\zeta w$$

are the contravariant components of the velocity.
The viscous stress terms are defined as

$$\hat{S} = J^{-1} \begin{pmatrix} 0 \\ \mu m_1 u_\eta + (\mu/3) m_2 \eta_x \\ \mu m_1 v_\eta + (\mu/3) m_2 \eta_y \\ \mu m_1 w_\eta + (\mu/3) m_2 \eta_z \\ \mu m_1 m_3 + (\mu/3) m_2 (\eta_x u + \eta_y v + \eta_\zeta w) \end{pmatrix}$$

where

$$m_1 = \eta_x^2 + \eta_y^2 + \eta_\zeta^2$$

$$m_2 = \eta_x u_\eta + \eta_y v_\eta + \eta_\zeta w_\eta$$

$$m_3 = (u^2 + v^2 + w^2)_\eta / 2 + Pr^{-1}(\gamma - 1)^{-1}(a^2)_\eta.$$

The turbulence model

An algebraic eddy viscosity model is included to approximate the effects of the turbulence. It is the well-proven Baldwin-Lomax model, which is mostly appropriate for attached and moderately separated flows. It is based on the assumption that the flow is locally in equilibrium., i.e. the velocity profiles are locally self-preserving. The turbulent region is divided into two zones, a wall layer where the strength of the turbulent stresses depends on the distance from the solid surface, and an outer zone where it is roughly constant. The eddy viscosity in the inner layer is given by the Prandtl mixing length theory and it is expressed by:

$$\mu_{inner} = \rho \, l^2 \, |\omega|$$

where the length scale l is proportional to the distance from the wall and the velocity scale is proportional to the length scale times the vorticity.
In the outer region the eddy viscosity is proportional to the wake function

$$F_{wake} = y_{max} F_{max} \, .$$

In the standard B-L formulation, F_{max} is taken to be the absolute maximum of a function $F(y)$ along the local normal to the wall, while y_{max} is the corresponding distance from the solid surface. In attached or mildly separated flows, y_{max} is the characteristic lengthscale of the shear layer. It is known, however, that in vortical-type flows the procedure gives the lengthscale of the external main vortex instead of the boundary layer one, causing totally erroneous predictions. A modification suggested by Degani and Schiff [4] is to take the first relative maximum of $F(y)$. This results to be effective in the vortex-dominated part of the flow. In the case of delta wing flows, it underpredicts the eddy viscosity levels where viscous effects are dominant, i.e. in the region of secondary separation. A blending of the two formulations was then needed, taking the standard form near the leading edge, and switching to the modified one near the line of detachment of the secondary vortex.
In the wake region the standard wake formulation was employed.

NUMERICAL ALGORITHM

Basic computational scheme

A backward Euler discretization in time is applied to Eq.(1). Local linearization of the fluxes is performed, obtaining an equation in 'delta form'. The left hand side is then factored and written in the approximate form at time $t = n$:

$$L_\xi L_\eta L_\zeta \Delta \hat{Q} = -h(\partial_\xi \hat{E}^n + \partial_\eta \hat{F}^n + \partial_\zeta \hat{G}^n - Re^{-1} \partial_\eta \hat{S}^n) \qquad (2)$$

where $h = \Delta t$, $\Delta \hat{Q}^n = \hat{Q}^{n+1} - \hat{Q}^n$ and the factors at the right hand side are given by

$$L_\xi = I + h \partial_\xi \hat{A}^n$$

$$L_\eta = I + h \partial_\eta \hat{B}^n - h Re^{-1} \partial_\eta \hat{M}^n$$

$$L_\zeta = I + h \partial_\zeta \hat{C}^n$$

where the flux Jacobian matrices $\hat{A} = \partial \hat{E}/\partial \hat{Q}$, $\hat{B} = \partial \hat{F}/\partial \hat{Q}$, $\hat{C} = \partial \hat{G}/\partial \hat{Q}$ and $\hat{M} = \partial \hat{S}/\partial \hat{Q}$ are defined in [3].
Nonlinear second and fourth-order numerical damping is applied. The damping term for each mesh direction is scaled by the corresponding eigenvalue, to add the minimum possible numerical viscosity. The dissipation terms are calculated implicitly, including them also in the left-hand side.
The factorized left-hand side of (2) is the product of 5X5 block pentadiagonal operators, the inversion of each requiring $o(5NM^3/3)$ operations, where M is the order of the blocks. For steady state calculations, Pulliam and Chaussee[3] suggested to decouple

the equations by diagonalization of the flux Jacobians. The problem is so reduced to the inversion of scalar pentadiagonal matrices, requiring only $o(5NM)$ operations. The factors in (2) take the form

$$L_\xi = T_\xi [I + h\partial_\xi \Lambda_\xi] T_\xi^{-1}$$
$$L_\eta = T_\eta [I + h\partial_\eta \Lambda_\eta] T_\eta^{-1} \qquad (3)$$
$$L_\zeta = T_\zeta [I + h\partial_\zeta \Lambda_\zeta] T_\zeta^{-1}$$

where $\Lambda_{\xi,\eta,\zeta}, T_{\xi,\eta,\zeta}$ are the diagonal eigenvalues and the eigenvector matrices, respectively.

A modification to the factored scheme.

A modification to the 3D factored scheme was carried out by neglecting, in the left-hand side of (2), the factor containing the differences in the ζ direction and expressing in an alternate way the ζ derivatives at the right-hand side. A similar approach is followed in [4] for the solution of the Euler equations.
Following this technique, a full iteration consists of two sweeps, respectively in the positive and negative ζ direction. At each $\zeta = constant$ station, eq.(2) is inverted, with the transversal differences expressed as

$$Q_{k+1}^n - Q_{k-1}^{n+1}$$

and

$$Q_{k+1}^{n+1} - Q_{k-1}^n$$

in the forward and backward sweep respectively. Eq.(2) so becomes

$$T_\xi [I + h\partial_\xi \Lambda_\xi] T_\xi^{-1} T_\eta [I + h\partial_\eta \Lambda_\eta] T_\eta^{-1} \Delta \hat{Q}^n = RHS^\pm$$

where the plus or minus sign indicates forward or backward ζ-differences in the backward or forward sweep respectively.
The present technique allows a much more efficient memory management, requiring only three or five planes to be kept stored in memory at the same time, while the original factorization required the whole geometry and flowfield to be stored.

Data flow organization.

For a 2D Navier-Stokes solver, requiring frequently over a million meshpoints to get a meaningful solution, it is of extreme importance to efficiently handle the I/O transfer between the memory and the disk storage; in fact, the performance is often determined not by the speed of the vector pipes, but by the I/O bottleneck of the computer.
In the present case, the choice was made to split the data into a number of "slices", each corresponding to a $\zeta = constant$ computational plane; the solution algorithm requires the presence in memory only of three or five planes at a time. Once the solution has been computed at a $\zeta = constant$ station, the corresponding data are saved on disk, while the next plane is read in the marching direction. Concurrent I/O is made, in order to reduce the overhead.

RESULTS

The computations were made on the cropped delta wing geometry, which is sketched in Fig.1. The wing has a sweep of $65°$ and sharp leading edge. The choice of a wing with sharp leading edge eliminates the uncertainties relative to the location of the primary vortex separation, allowing a meaningful comparison between experimental data and numerical results. It was found that a critical parameter to be considered for such a comparison is the location and strength of the secondary vortex, from which the location of the primary one appears to depend. A C-O 161X63X40 grid was employed, with 91 points along the wing profile, 2X35 in the wake, 40 in spanwise direction and 63 in the normal direction. The minimum spacing near the wall was fixed to 3E-05 times the local chordlength.

Mach = .15, Alpha = 10°, Re = 2 millions

A study of the influence of the turbulence model was conducted for this case. The Reynolds number was 2 millions, and the transition was fixed at 20% of the root chord. In fig.2 the spanwise Cp distributions are shown, that were obtained applying the standard Baldwin-Lomax turbulence model, at 80% of the root chord. The large amount of eddy viscosity that was erroneously predicted resulted in a dramatic flattening of the pressure levels. The reason is in the computed location of the absolute maximum of the wake function, which results well outside the boundary layer, in the main vortex itself. The first correction that was applied followed the suggestion of Degani and Schiff [3], to take in account only the first maximum of the wake function. The results are shown in Fig.3. A pressure peak due to an overprediction of the strength of the secondary vortex appears near the leading edge. We argued that in the region of secondary separation, where viscous effects are still dominating the physics of the flow, the standard Baldwin-Lomax model had to be applied. Applying this criterion, we obtained the results shown in Fig.4. In Fig.5 the experimental and computed wall Cp distributions on nine spanwise sections are plotted, the thicker line representing the experimental data. The disagreement in the location and strength of the vortex core up to 30% of the root chord seems to be due to the interference with the fuselage in the experimental model (see[7]). The Cp contours on the leeward side may be seen in Fig.6, the difference between two levels being 0.1. From the skin friction lines in Fig.7 it is easy to detect both the line of separation of the secondary vortex and the line of reattachment of the primary one.

Mach = .5, Alpha = 20°, Re = 9 millions

This case was computed as a test to assess the validity of the modified turbulence model at higher angle of attack. The transition line was fixed at 10% of the root chord. From the skin friction lines in Fig.8, the main characteristics of the flow appears to be the presence of a stronger secondary vortex, displacing inboard the primary one. The primary vortex spans over most of the surface of the wing, the reattachment region being located quite close to the plane of symmetry. The Cp values at respectively 45%, 60% and 80% of the root chord are reported in Figs.9 to 11. In he region of secondary separation, the pressure appears to be reasonably well predicted; however, especially at 80%, the computed primary vortex is stronger and more outboard than the measured one. A comparative view of the measured and computed Cp distributions is shown in Fig.12. A general sketch of the flowfield is depicted in Fig.13, where the Cp isolevels are shown at four mesh cuts, respectively at 45%, 60%, 80% and 100% of the root chord. Also an idea of the 3D evolution of the primary and secondary vortices is given by the computed particle trajectories in Fig.14.

Mach = .85, Alpha = 10°, Re = 9 millions

Figs.15, 16 and 17 show the wall Cp distributions at 45%, 60% and 80% respectively of the root chord. The solid line refers to the computational data. A synthetic view of the experimental vs. computed results at all the nine experimental stations may be seen in Fig.18. In general, the computed suction peak is stronger and more outboard than the experimental one. The Cp contours on the leeward surface are depicted in Fig. 19. A comparison between the calculated skin friction lines and the oil flow visualizations is made in Fig.20. A substantial qualitative agreement seems to result.

CONCLUSIONS

A Navier-Stokes solver has been applied to the simulation of viscous turbulent flows past a 65° swept cropped delta wing. The solution algorithm has been designed to ensure efficient memory management. A detailed comparison between numerical and experimental results has been made. An appropriate modification to the Baldwin-Lomax turbulence model was found to be necessary; in fact, difficulties were encountered into find the appropriate lengthscale of the shear layer in a vortical flow. The computed data compare quite favourably with the experiment in the leading edge region; the location of the main vortex, however, is placed in most cases more outboard with respect to the measurements. Improvements to the prediction capabilities may be expected from the application of a turbulence model that calculates directly the lengthscales, such as a two-equations model. Another important improvement may be the development of a mesh adaption procedure, to better resolve the region of the vortex core.

BIBLIOGRAPHY

[1] Proceedings of the Symposium on International Vortex Flow Experiment and Euler Code Validation, Stockolm, Oct.1-3 1986, ed. by A. Elsenaar and G. Eriksson

[2] Elsenaar, A., Hoejmakers, H.W.M.: *An experimental study of the flow over a sharp-edged delta wing at subsonic, transonic and supersonic speeds.*
AGARD-FDP Symposium on Vortex Flow Aerodynamics, Scheveningen, Oct.1-4, 1990.

[3] Pulliam, T.H., And Chaussee, D.S.: *A diagonal form of an implicit factorization algorithm.*
J.Comp.Phys., vol. 39 n. 2, 1981, pp.347-363.

[4] Degani, D., And Schiff, L.B.: *Computation of supersonic viscous flows around pointed bodies at large incidence.*
AIAA 83-0034

[5] Sankar, L.N., And Al.: *Euler solutions for transonic flow past a fighter wing.*
J.Aircraft vol.24, n.1

[6] Hilgenstock, A., And Vollmers, H.: *On the simulation of compressible turbulent flows past delta wing, wing-body and delta wing-canard.*
AGARD-FDP Symposium on Vortex Flow Aerodynamics, Scheveningen, Oct. 1-4, 1990.

[7] Williams, B.R., Kordulla, W., Borsi, M., Hoejmakers, H.W.M.: *Comparison of solution of various Euler solvers and one Navier-Stokes solver for the flow about a sharp-edged delta wing.*
AGARD-FDP Symposium on Vortex Flow Aerodynamics, Scheveningen, Oct. 1-4, 1990.

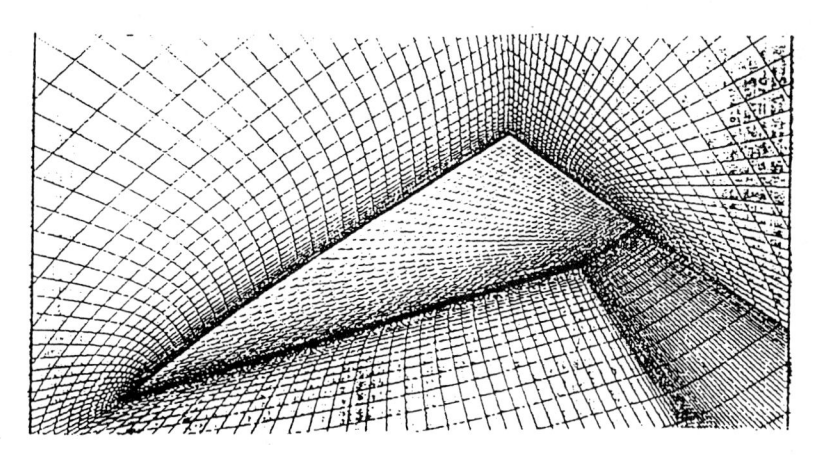

Fig.1 - Model geometry and C-O mesh surfaces.

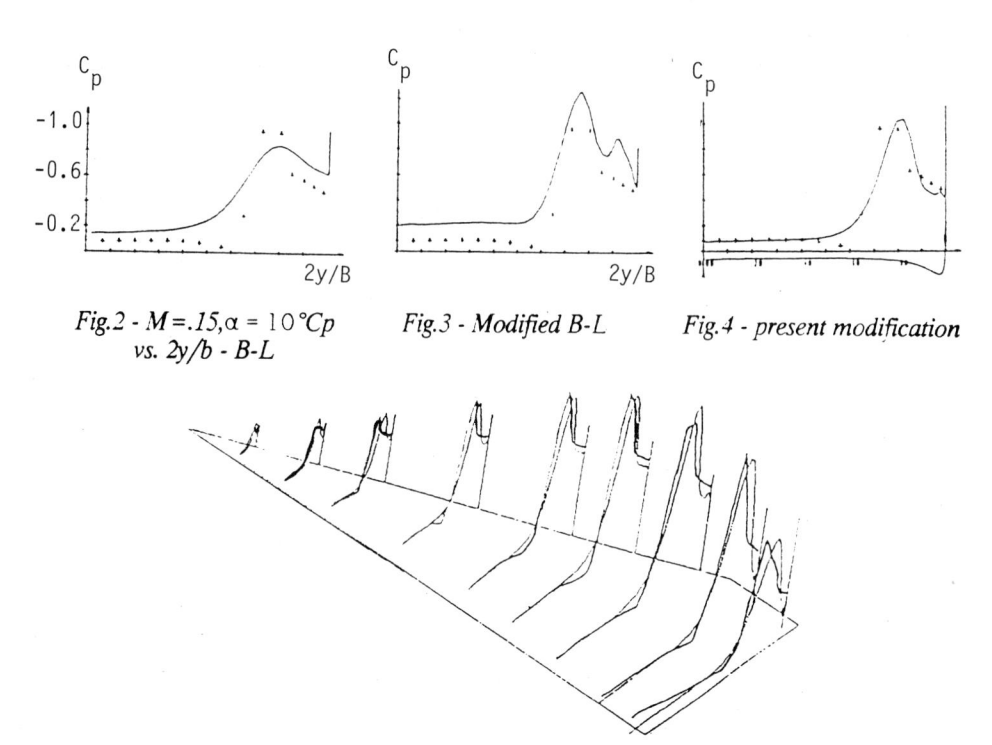

Fig.2 - $M=.15, \alpha = 10°$ C_p vs. $2y/b$ - B-L

Fig.3 - Modified B-L

Fig.4 - present modification

Fig.5 - Cp vs. 2y/b on nine spanwise stations.

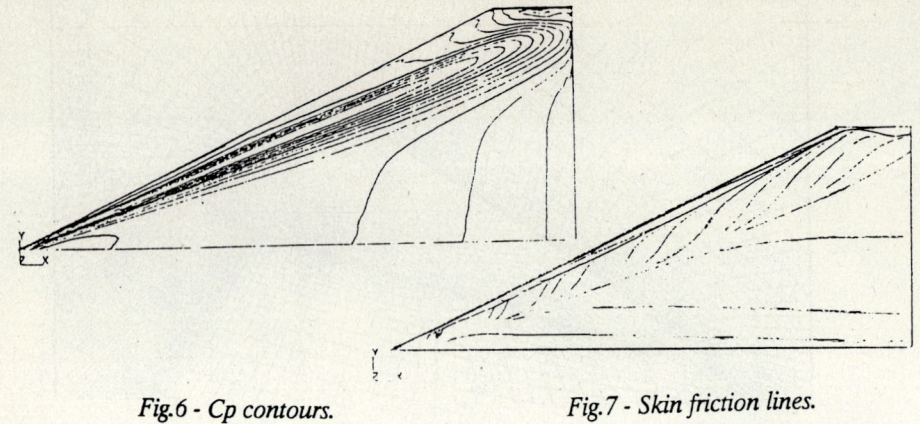

Fig.6 - Cp contours. *Fig.7 - Skin friction lines.*

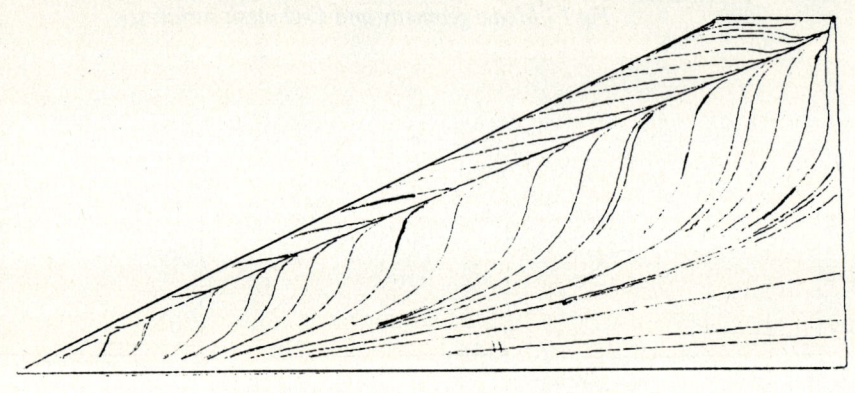

Fig.8 - M =.5, α = 20°: Skin friction lines.

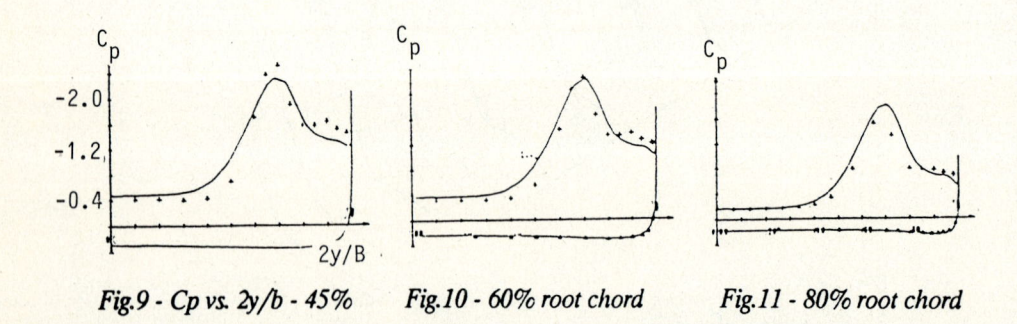

Fig.9 - Cp vs. 2y/b - 45% *Fig.10 - 60% root chord* *Fig.11 - 80% root chord*

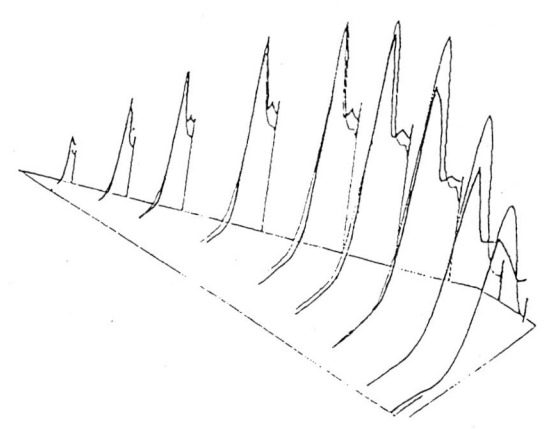

Fig.12 - Cp vs. 2y/b on nine spanwise stations.

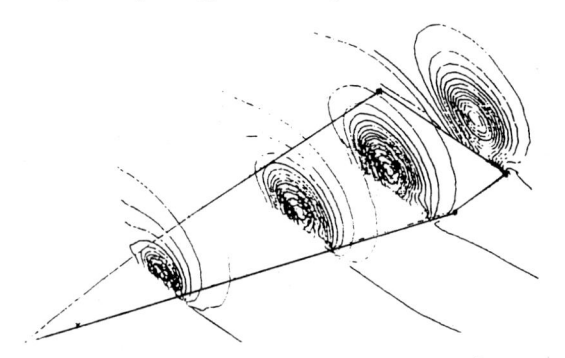

Fig.13 - Cp contours on 30%, 60%, 100% spanwise cuts.

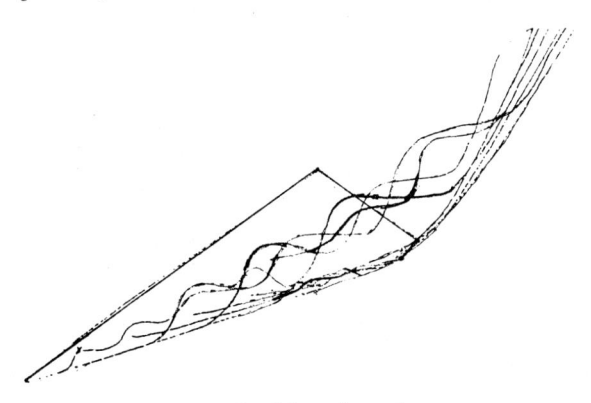

Fig.14 - Particle trajectories.

Fig.15 - M = .85, α = 10°45%

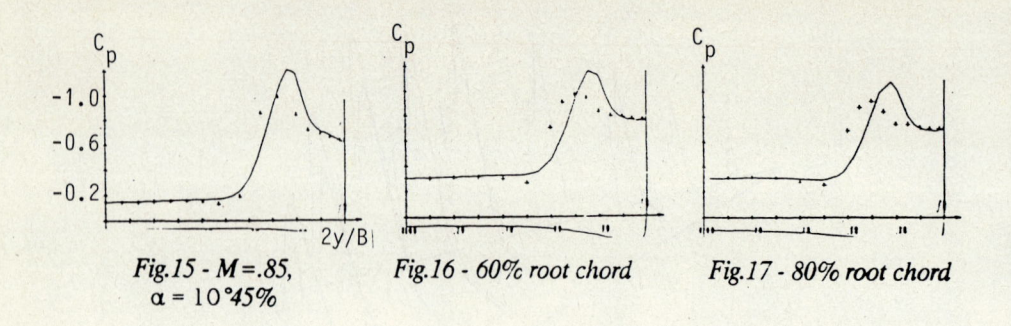

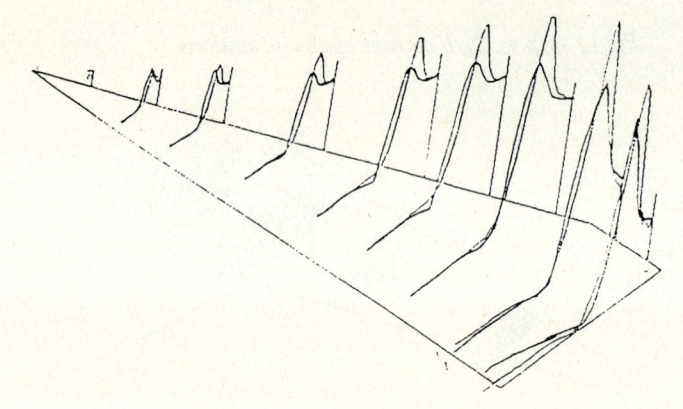

Fig.16 - 60% root chord

Fig.17 - 80% root chord

Fig.18 - Cp vs. 2y/b on nine spanwise stations.

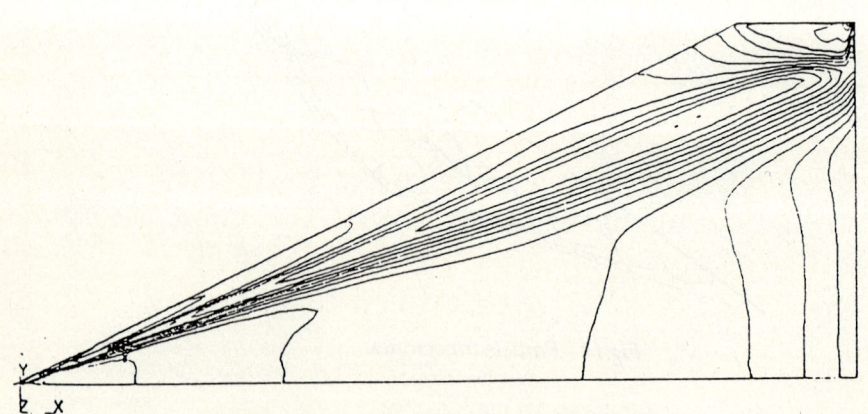

Fig.19 - Cp contours on the leeward surface.

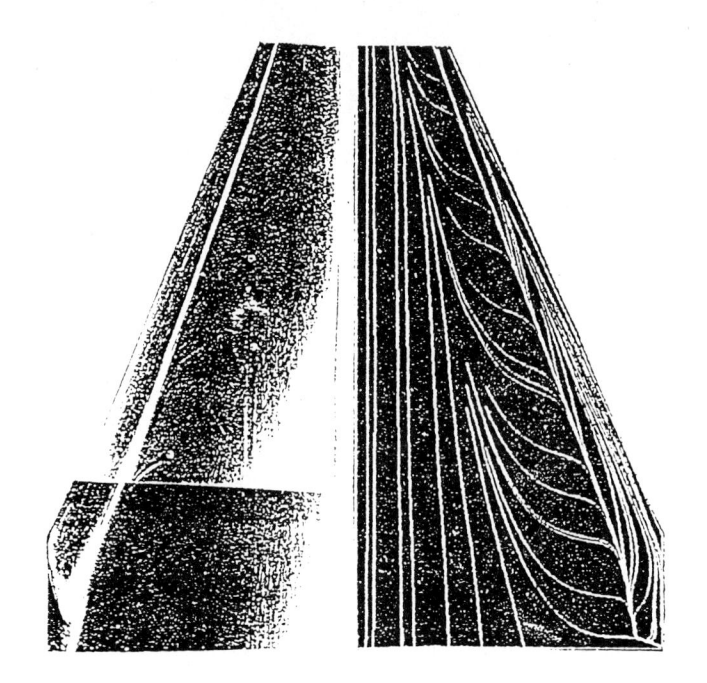

Fig.20 - Skin friction lines and oil flow visualizations.

Hypersonic Laminar Flow Computations Over a Blunt Leading Edged Delta Wing At Three Different Chord Reynolds Numbers

Shivakumar Srinivasan, Peter Eliasson, and Arthur Rizzi
FFA, The Aeronautical Research Institute of Sweden
S-16111 Bromma, Sweden

Summary

The hypersonic flow past a blunt leading edged and a blunt nosed delta wing has been simulated numerically by solving the three dimensional Navier-Stokes equation. The laminar flow simulations have been performed for three different chord Reynolds numbers. The Navier-Stokes equations have been solved using the explicit finite volume four stage Runge-Kutta scheme. The intent is to reach a reasonable understanding of the separation characteristics of the flow on the leeside. It has been observed that in the case of blunt edged delta wings at high angles of attack, the hypersonic flow is dominated by a shear layer that separates just past the blunt leading edge forming a more distributed vortical region over the wing, rather than a concentrated vortex structure as observed at lower speeds. As the Reynolds number increases, the separation characteristic on the lee side of the wing changes from a primary separation near the wing mid semi span to a primary separation close to the leading edge and a secondary separation near the wing mid semi span. The shape of the wing apex determines the location of the chord plane where the separation begins. The solutions have not been compared with experimental data due to the nonavailability of the data at the time of writing. Sensitivity of the flow field to mesh refinement has been investigated and it has been found that the characteristics of the flow field remains the same even after mesh refinement.

Introduction

In the last few years there has been a renewed interest in hypersonic flow research. Large scale research programs have been initiated for both reentry vehicles and hypersonic transport aircraft. Current interest in developing advanced space planes like the NASP, HERMES, SANGER and the superConcorde has brought forth the problem of understanding the leeside vortical flow over these vehicles when they travel at hypersonic speeds. The vortical structures are characterized by shock waves, separated flow and shear layers and have important consequences on heating and local effects like a shock wave or shear layer impinging on a configuration detail, e.g. a flap or other protuberance. The vortex phenomena have been found responsible for intense local heating over the lee meridian of the delta wings. Due to the near vacuum pressures on the leeward side of the wing, the effectiveness of the control surfaces and lift performance are least affected by these vortical structures. The flow features of this low-pressure region, its structure and interactions are of primary interest. The effect of high angle of attack on the vortex structure in hypersonic flow has been investigated by Rizzi, et.al [1]. The effect of varying angle of attack on the vortex structure in hypersonic flow has been investigated by Srinivasan, et.al [2]. One of the objectives of the current research effort is to investigate the effect of varying the chord Reynolds number on the separation characteristics on the wing leeside. In the past we have analyzed the flow past a blunt edged delta wing with a sharp apex at a high chord Reynolds number. We decided to investigate the flow characteristics as the Reynolds number varies for the blunt apex delta wing.

Current understanding of the flow field over blunt delta wings at high Mach numbers is based on experimental data. However, due to limitations in instrumentation at very high speeds, detailed information of the flow field is not available. Numerical algorithms are used to simulate and resolve the flow field details. The computational results can be validated by comparing them to the available experimental data.

Numerical investigation of complex three dimensional flow requires enormous computer resources. The exponential growth of computer speed and storage capacity as well as algorithm sophistication has allowed application of advanced numerical methods to solve practical design problems in fluid mechanics and aerodynamics. The flow field around the delta wing is quite complex and requires a considerable computational effort to resolve the flow features accurately. The strong bow shock, separating shear layers, recirculating flow and cross flow shocks and their interactions form a complex flow structure. To resolve all these, the fully elliptic form of the governing partial differential equations must be solved. The objective of this paper is to compute the flow around the blunt edged delta wing and to get a better understanding of the flow features on the leeward side of the wing at three different chord Reynolds numbers.

Outline of Paper

The subject of the paper is the numerical simulation of a hypersonic vortical flow on the leeside of a blunt delta wing. The results to be presented are for flow at $M_\infty = 8.7$ and $\alpha = 30°$ incidence past a thick delta wing of 70° sweep and constant leading edge radius; a non-conical wing. The computations have been performed by solving the full Navier-Stokes equations to evaluate the inviscid and viscous mechanisms of leeside vortex formation, for three different chord Reynolds number, viz. 5.625×10^5, 5.625×10^6, and 10.0×10^6.

Numerical Method

Several numerical schemes have been developed over the years to solve the fully elliptic three dimensional Navier-Stokes equations. The results presented here have been obtained using a finite-volume code developed at FFA, that originated for solving transonic flows [3], but has been extensively modified and adapted for hypersonic flows, by Bengt Winzell at SAAB[4].

Space Discretization: Inviscid Terms

The scheme for the inviscid terms is a cell-centered finite volume discretization in space that is fully conservative. The flux at a cell face is computed as the flux of the average of the two state variables to the left and right of the face. An artificial smoothing term that is a nonlinear blend of second and fourth differences stabilizes this discretization.

Space Discretization: Viscous Terms

Cell-Vertex Method

The viscous terms are computed by a 3-point cell-vertex scheme using an auxillary staggered grid. The vertices S of the staggered grid are given by the cell centers of the main grid where the state variables are located (Fig. 1). The vertices of the main grid N are then taken as the cell centers of the staggered grid. Derivatives at the vertices, say $N1$, are computed from the gradient theorem applied to the cell defined by the surrounding staggered-grid points, $S1$ to $S4$. The viscous tensor can then be obtained and the viscous and thermal fluxes computed at the vertices N. They are averaged over the cell face to give the numerical flux at the center of the face.

Time Integration

The standard explicit four stage Runge Kutta time integration with local time stepping has been used to advance the solution in time.

Boundary Conditions

Appropriate boundary conditions are required on the wing surface, at the inflow, outflow, farfield and symmetry boundaries. The usual no-slip boundary conditions have been enforced on the wing. The condition for pressure on the wing surface has been set by second-order extrapolation from the field values and the wall considered to be isothermal at a constant temperature of $T_w = 300°K$. The farfield has been assumed to be inviscid and treated by either assigning or extrapolating the locally one dimensional Riemann invariants. The trailing edge and its tip are thick with sharp corners and all the variables in the computation have been extrapolated at the outflow boundary from the interior, which in effect makes the wing infinitely long. No problems were encountered with this type of boundary condition. The boundary conditions on the symmetry plane are straightforward. The flows on either side of the symmetry plane are mirror images of each other. The flow variables in the ghost cell have been assigned the value of the corresponding interior cell with the exception of the velocity normal to the symmetry plane, which has been set to the negative of the interior cell value.

Grid Structure

The computational grid for the delta wing is shown in fig. 2. The C-O grid around the delta wing has been generated using transfinite interpolation, with appropriate grid clustering in the boundary layer, around the leading edge and the apex. The details of the grid structure in the symmetry plane, the crossflow plane and on the wing surface are displayed in figs. 2a, 2b and 2c, respectively. A grid size of 41 x 97 x 193 in the streamwise, normal and spanwise directions, respectively has been used. The grids in the streamwise direction extends only up to the trailing edge. The trailing edge wake has not been simulated in the computations. There is a polar singular line at the apex of the wing due to the choice of C-O grid topology. The presence of this polar singular line at the apex caused some convergence problems for the Navier-Stokes solver.

Discussion of the Results

INRIA and GAMNI organized a workshop on "Hypersonic Flows for Reentry Problems" in January 1991 to investigate the suitability of computational techniques for hypersonic flow simulation. The calculations of laminar flow over a 70° swept blunt delta wing at 30° angle of attack, Mach number of 8.7, chord Re of 5.625×10^5, wall temperature of $300K$ and free stream temperature of $55K$ were carried out as part of the workshop. Further calculations were made for higher Reynolds numbers to understand how the flow structures change with increasing Reynolds number and they are presented here.

The model is a blunt nosed, blunt leading edged delta wing with a leading edge sweep of 70°. It has a chord length of $0.25m$. The surface geometry cannot be defined by an analytic expression. The geometry definition, coordinates and the surface mesh were obtained from MBB Germany. The leading edge radius is constant throughout, resulting in a non-conical wing. A detailed diagram of the wing geometry is shown in fig. 3.

The computations were performed in four steps in order to limit the computational time. In the first step, the computations were started on the coarsest grid of 21 × 25 × 25, obtained by taking every fourth point of the finest mesh in the spanwise and normal direction and every other point in the streamwise direction. A converged solution was obtained on this grid. The solutions from the coarse mesh were interpolated onto the medium grid of 21 × 49 × 49. Again a converged solution was obtained on this medium mesh. The computations were performed then on the fine mesh of 41 × 97 × 97 as the third step to convergence. Finally the converged solution was obtained on a superfine mesh of 41 × 97 × 193. This was typically reached after 7500 time steps.

Symmetry Plane

The Mach contours in the symmetry plane, at $\alpha = 30°$, and Reynolds number= 5.625×10^5, 5.625×10^6 and 10.0×10^6 are shown in fig. 4. As expected in hypersonic flows, the strong bow shock on the windward side and the flow expansion on the upper surface are clearly visible. The expansion on the leeward side is strong which drives the static pressure to near vacuum conditions.

Cross flow Plane

Figure 5 presents isoMach contours of the flowfield at the 50% chord section for the three cases. In all the three cases, the flow negotiates the leading edge and separates just past the maximum expansion forming a shear layer. This shear layer extends inboard to where it meets the crossflow shock over the wing and then turns downward towards the wing surface where a small embedded shock turns it once again outboard towards the leading edge. It finally rolls up to form a vortical layer. Between the wing and the shear layer a secondary vortex develops from the boundary layer separating from the wing. However, for the low Reynolds number case, the shear layer separates farther inboards of the leading edge. As the Reynolds number increases, both the crossflow shock and the small embedded shock are more pronounced. The direction of the flow motion in this plane is given by velocity vectors and their integrated path lines (i.e. in-plane streamlines) as displayed in Fig. 6. In each case the vortex is clearly identified. The flow features for the two high Reynolds number cases are more or less similar. As the Reynolds number increases, the flow tends to separate and form a shear layer, closer to the leading edge.

Upper Surface

The skin friction lines on the upper surface are shown in fig. 7a, for all the three cases. It is clear that for the low Reynolds number case there is one separation line, whereas for the two higher Reynolds number cases, the flow separates near the leading edge and reattaches and there is a secondary separation close to the mid span of the wing. As the Reynolds number increases, the shear layer tends to separate closer to the leading edge. This is clearly seen in the surface skin friction plot. In the case of the highest Reynolds number, the secondary separation line is curved and moves a little outboards. The blow up of the skin friction lines near the apex region is dispalyed in fig. 7b. The shape of the wing apex has some influence on the location of the chord plane where the separation begins. The y component of skin friction on the wing at 50% chord for the three cases is displayed in fig. 7c. The values of Cf_y show trends that agree with that of the velocity vector plots. As the flow accelerates past the leading edge, the velocity gradient normal to the wing increases, thereby increasing the value of Cf_y. However, at the leading edge there is a sudden drop in the Cf_y and it becomes negative as the flow moves inboard on the leeward surface. The Cf_y value then goes to zero as the flow passes over the separated region and then to a positive value in the vicinity of the reverse flow close to the symmetry plane. As the Reynolds number increases, the Cf_y decreases as seen in the figure.

Since the flow features at 80% chord station are similar to the ones at 50% chord, they have not been shown in the paper. There are no significant differences at these two stations.

The Stanton number on the upper surface of the wing is shown in fig. 8, for the three cases. As expected, the highest heat transfers occur around the leading edge and the apex, which is clearly displayed in the figure by the concentration of the contour lines. The Stanton number on the wing at 50% chord position for all the three cases is displayed in fig. 8b. The negative values indicate that on both the upper and lower surface, heat is being transferred from the surroundings to the wing. As expected, the Stanton number plots follow the same trend as observed in the skin friction plots, in that as the Reynolds number increases, the Stanton number decreases.

Lastly, the Cp distribution on the upper surface of the wing for the three cases is shown in fig. 9a. The Cp values are extremely low and near vacuum. Fig. 9b. shows that Cp values at 50% chord position. Since the static pressure on the upper surface is very near vacuum, we do not see any variation. In order to notice the variation, the Cp on the upper surface has been plotted separately in fig. 9c. As the Reynolds number increases, the expansion on the upper surface is stronger and thereby the pressure drops to a lower value which is indicated in the line plots. Even though we see a difference between the value of Cp on the windward side and at the leading edge for the three cases, the differences on the leeward side are minimal.

Mesh Refinement

In order to ascertain how dependent the flow features are on the mesh resolution, we present here a comparison of two results, one on the super fine mesh (discussed above) and the other on a fine mesh with half the number of points in circumferential direction of the super fine mesh for the case of Re = 5.65×10^5. The skin friction lines in Fig. 10a show that the flow structure of one primary separation remains unchanged and that its location is grid independent. Even after the grid refinement, there is no change in the location of the primary separation. The comparison of iso Mach contours shown in fig. 10b, in the 50% chord plane shows that there are no significant changes as the mesh is refined. A grid converged solution has therefore been obtained. However this grid refinement study has not been carried out for the higher Reynolds number cases.

Conclusions

Instead of the concentrated vortex usually found over a delta wing at transonic speed, the flow in hypersonic speed is dominated by a shear layer that separates just past the blunt leading edge and forms a more distributed vortical region over the wing. The behaviour of the shear layer variation with the chord Reynolds number has been investigated. The computations confirm the expected trends, namely that the Cfy decreases as the Reynolds number increases and the Stanton number follows the same trend. As the Reynolds number increases from 5.65×10^5 to 5.65×10^6, the shear layer separates closer to the leading edge and in turn gives rise to a secondary separation closer to the centre of the wing span. A further increase in the Reynolds number from 5.65×10^6 to 10.0×10^6 does not change the flow structure significantly leading to the conclusion that the variation in the Reynolds number beyond a limit does not affect the flow features. The shape of the wing apex determines the location of the chord plane where the separation begins.

Acknowledgments

The two higher Reynolds number flow cases were computed on the SX-2 supercomputer, and we are grateful to the NEC Corporation for granting us the computer time and to G. Van der Velde for his valuable assistance.

References

1. Rizzi, A., Murman, E.M., Eliasson, P. and Lee, M.K., " Calculation of Hypersonic Leeside Vortices Over Blunt Delta Wings ", Proc. AGARD Symposium on Vortex Flow Aerodynamics, Netherlands, 1-4 Oct 1990.

2. Srinivasan, S., Eliasson, P. and Rizzi, A., "Navier Stokes Computations of Hypersonic Flow Past a Blunt Edge Delta Wing at Several Angles of Attack ", AIAA 91-1698 22nd Fluid Dynamics, Plasma Dynamics and Lasers Conference, Hawaii, 24-27 June 1991.

3. Müller, B. and Rizzi, A., " Navier-Stokes Calculations of Transonic Vortices over a Round Leading Edge Delta Wing ", **Intl J. Numerical Methods Fluid Mechanics**, Vol 9, 1989, pp 943-962.

4. Winzell, B., " Validation of a Navier-Stokes Code for Hypersonic Flow ", SAAB Report TUKL-HERM 89:08,

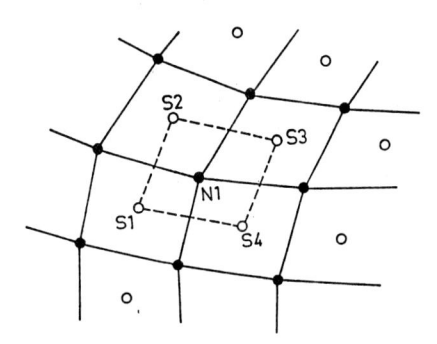

Fig. 1 Staggered Mesh for viscous flux calculation

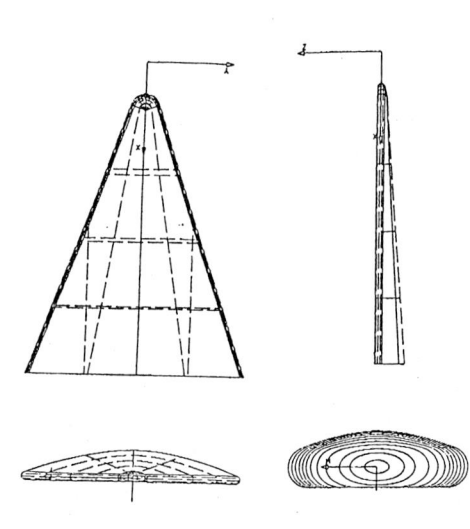

Fig. 3 Geometry delta wing model

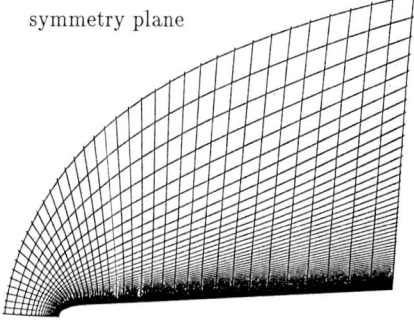

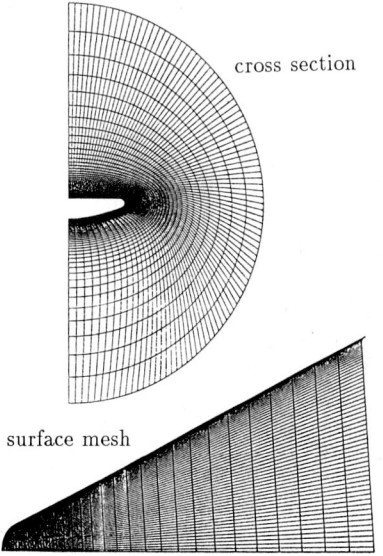

Fig. 2 C-O Mesh topology 41x97x193

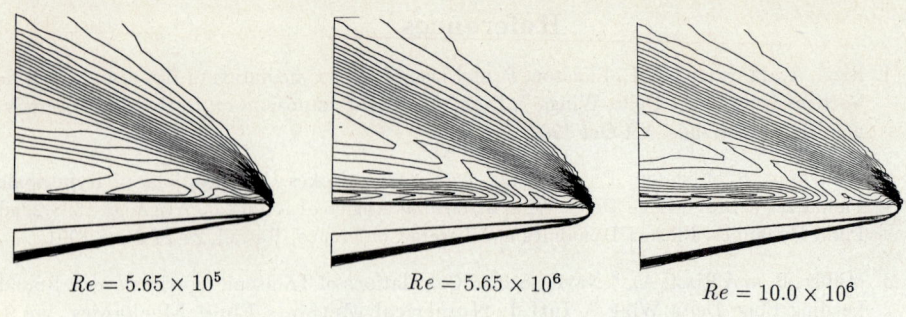

Fig. 4 IsoMach contours on symmetry plane

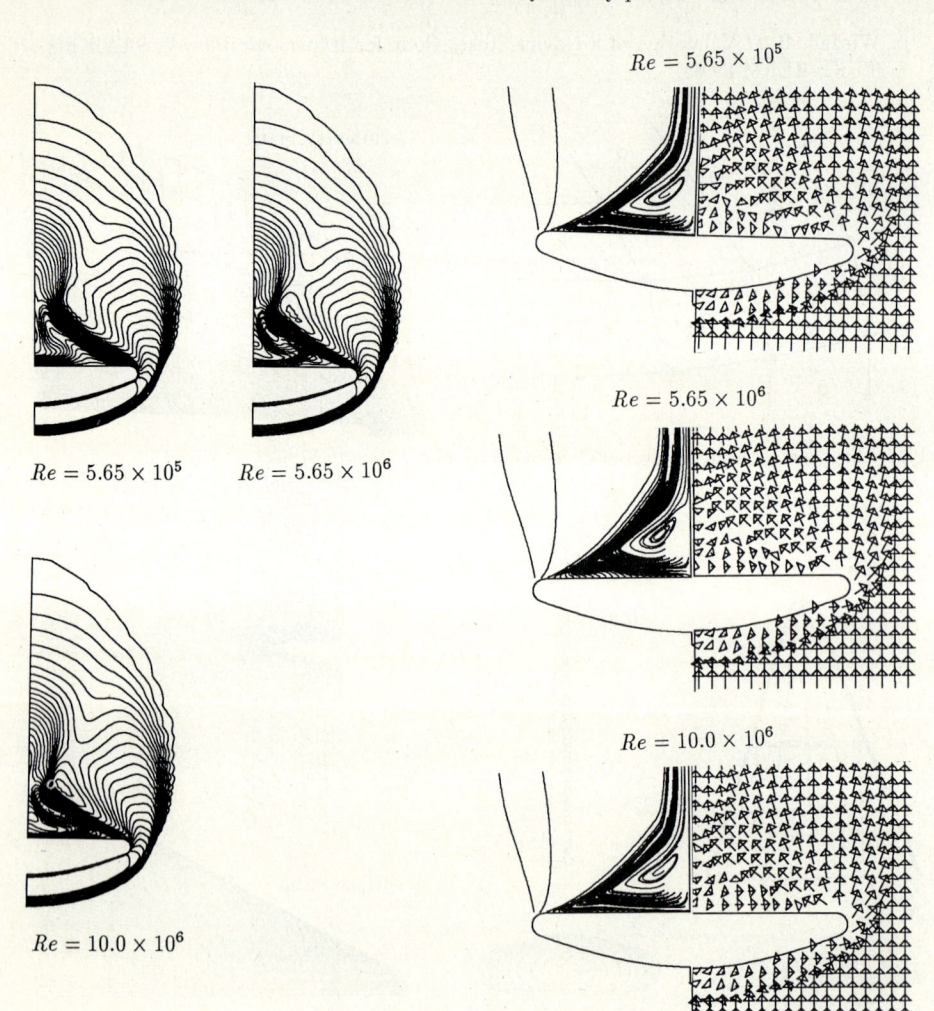

Fig. 5 IsoMach contours at 50% chord plane

Fig. 6 Velocity vectors and inplane streamlines, 50% chord plane

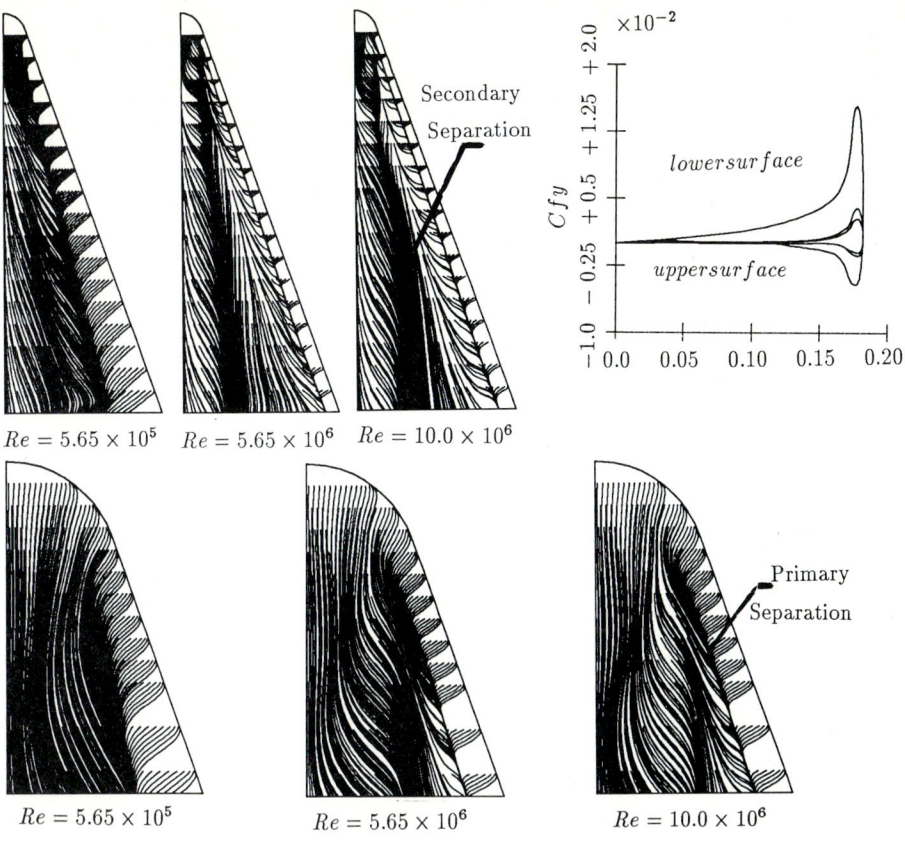

Fig. 7 a) Upper surface skin friction lines, b) Blow up in the nose region
c) Cfy in the 50% chord plane

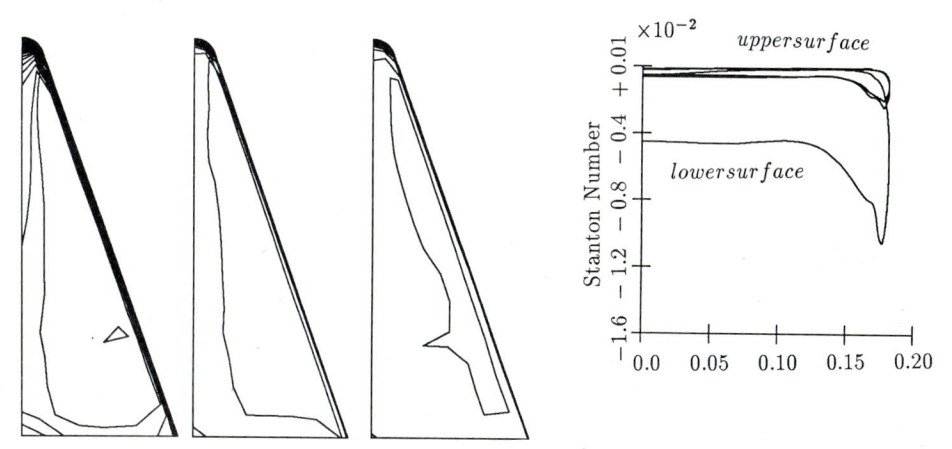

Fig. 8 a) Upper surface Stanton number contours, b) Stanton number distribution, 50% chord plane

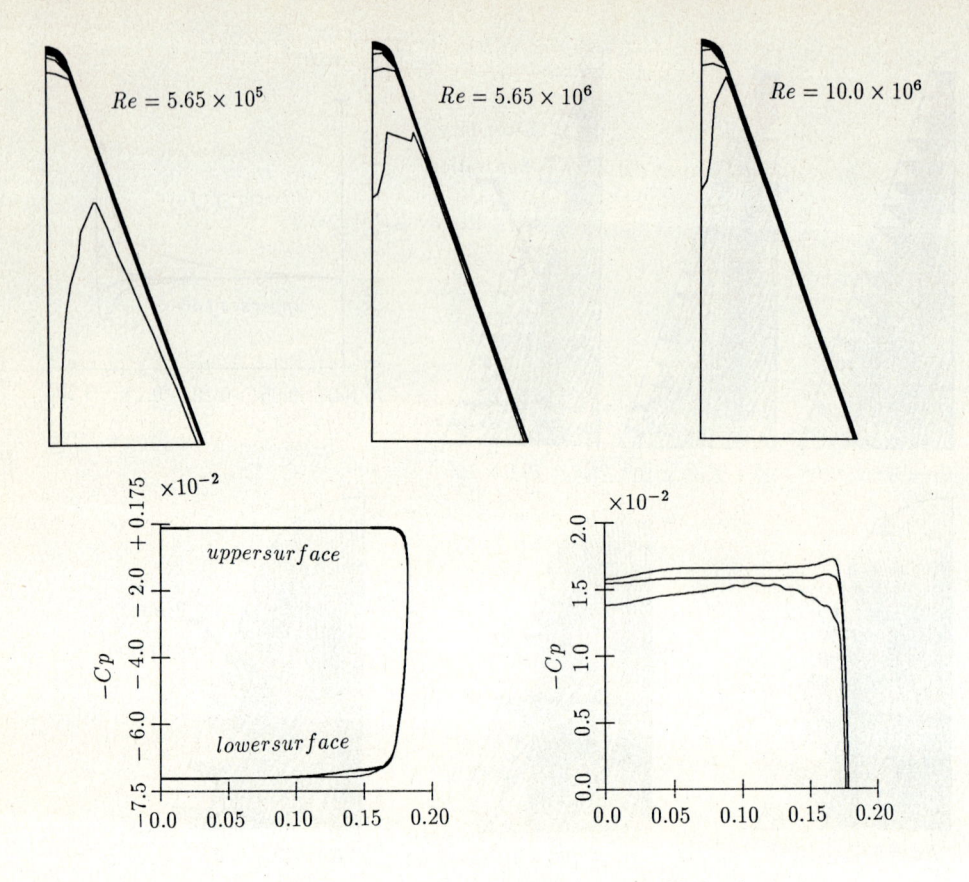

Fig. 9 a) Upper surface Cp contours, b) Cp distribution at 50% chord plane
i) upper+lower surface ii) upper surface

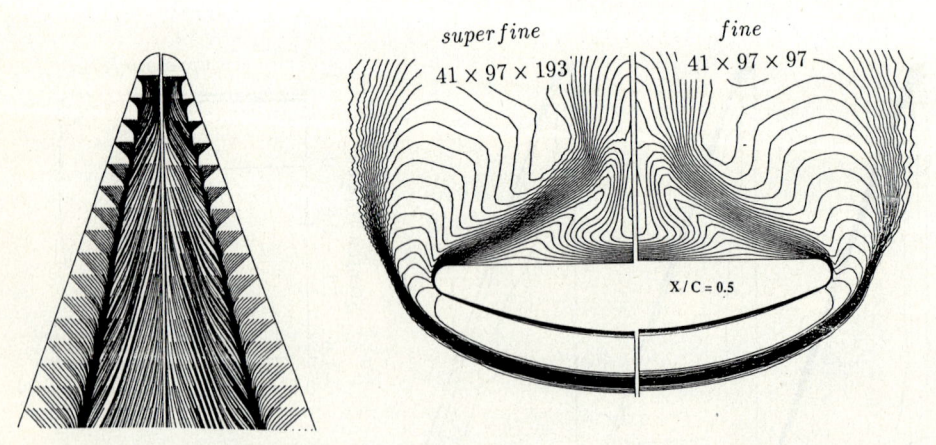

a) Upper surface Skin friction lines   b)   Mach number contours at 50% chord plane

Fig. 10 Mesh refinement effects

STUDY INTO THE LIMITS OF AN EULER EQUATION METHOD
APPLIED TO LEADING-EDGE VORTEX FLOW[*]

by

J.I. van den Berg, H.W.M. Hoeijmakers and H.A. Sytsma
National Aerospace Laboratory, NLR
Anthony Fokkerweg 2, 1059 CM Amsterdam
The Netherlands

SUMMARY

A steady-flow Euler method is applied to the subsonic leading-edge vortex flow about a 65-deg sharp-edged cropped delta wing at high incidence. Above a critical value of the incidence the numerical procedure fails to fully converge, the solution at successive iterations exhibiting a behaviour similar to the one observed during vortex breakdown. The occurrence of this "solution breakdown" indicates the limits of the domain of applicability of the steady-flow Euler method for the case of subsonic leading-edge vortex flow. Analysis of the solution at incidences just below the critical value reveals that a bubble-type of flow feature develops within the vortex core.

INTRODUCTION

For sharp-edged slender delta wings the formation of the leading-edge vortex above the upper surface of the wing as well as the influence of the flow over the upper surface of the wing and consequently the lift and moment coefficients are only slightly dependent on Reynolds number. This suggests that the flow can be described using an inviscid flow model. The application of Euler methods to vortical flow is rather attractive because these methods allow for rotational flow everywhere in the flow field, describe stretching and convection of vorticity and therefore "capture" vortical flow regions as integral part of the discrete flow solution.
The generation of vorticity, for the present sub-critical flow conditions through the process of flow separation only, requires a Kutta condition or a viscous-flow model. However, it is well-accepted now that at a sharp edge artificial dissipation present in the current numerical procedures to solve the Euler equations causes the flow to separate at the edge, thereby implicitly satisfying the Kutta condition.
In the past few years the structure of the flow field above (sharp-edged) delta-wing configurations has been studied using computational methods based on the Euler equations ([1]-[6]).

Aside from the separation at the sharp leading edge, which results in the leading-edge vortex, a secondary separation occurs at a point on a smooth part of the upper wing surface underneath the leading-edge vortex. This type of separation is not captured automatically by an Euler method like the one used in the present investigation.

Increasing the incidence leads to an increase in the strength of the leading-edge vortex. It has been observed in experiments that above a critical value of the incidence the internal structure of the vortex changes drama-

---

[*] This investigation has been carried out under contract with the Netherlands Agency for Aerospace Programs (NIVR) for the Netherlands Ministry of Defence

tically, resulting in an abrupt increase in the cross-sectional size of the vortex core. Downstream of the region where the phenomenon of vortex breakdown (burst) occurs the flow might recover, but usually the flow becomes unsteady and it loses its well-ordered structure. Vortex breakdown marks the end of the favourable effects induced by leading-edge vortex flow and for example determines the maximum attainable lift coefficient. Understanding the mechanisms of the onset of vortex breakdown is of paramount interest to the design of fighter-like aircraft and missiles operating in the high-angle-of-attack flight regime.

Two types of vortex breakdown have been observed in experiments, i.e. vortex breakdown of bubble- and of spiral-type. The former manifests itself as a more or less axially symmetric closed bubble with reversed flow situated near the center of the vortex core, the latter manifests itself as a spiralling of the central region of the vortex core in a sense opposite to the sense of the circulation within the vortex core.

Usually leading-edge vortex breakdown starts in the wake, where the vortex is strongest and where there is a rapid increase of the static pressure along the vortex core, progressing forward with increasing incidence. A possible cause of the burst of the vortex is the low total pressure at the center of the vortex core, due to viscous-flow losses within the vortex core. The pressure rise at the trailing-edge or in the near wake can then easily bring the flow to stagnation, leading to the bubble-type of breakdown. An alternative possibility is that instabilities within the (inviscid part of the) vortex core cause the onset of vortex breakdown.

The phenomenon of vortex breakdown has first been identified in experiments, by Peckham and Atkinson [7], Elle [8] and others. It has been studied in detail in both experimental (e.g. [9]) and computational investigations using Euler methods (e.g. [3], [4]) and Navier-Stokes methods (e.g. [10]). Until now the mechanisms leading to vortex breakdown are still unclear, see e.g. Hall [11].

In this paper results are presented from an Euler method applied to the subsonic ($M_\infty = 0.50$) high-angle-of-attack flow about an isolated 65-deg cropped delta-wing with sharp leading edges. Fig. 1, taken from [14], provides an overview, in the (M_∞, α) plane, of the experimentally found main flow features for a configuration featuring the same delta wing. In the wind-tunnel investigation vortex breakdown occurs at incidences above 25 deg for the present free-stream Mach number. Therefore in the numerical investigation solutions have been considered for incidences of 20 deg and higher, this in order to determine the highest ("critical") incidence for which the numerical procedure produces a converged (steady-state) solution. The structure of the vortex core above the wing and in the near wake will be analyzed for two solutions, the one for $\alpha = 20$ deg and the one for the critical incidence.

OUTLINE OF COMPUTATIONAL METHOD USED

The Euler method, as developed at NLR [12], solves the time-dependent Euler equations employing the fully conservative scheme of Jameson et al. [13]. The five equations for the conservation of mass, momentum in each of the three space directions and energy are discretized using a cell-centered central-difference scheme. Second- and fourth-order artificial dissipation terms are added to the discretized equations. The fourth-order term is required to provide the background dissipation for suppressing the tendency for odd-even point decoupling of the numerical solution. In regions with large gradients, e.g. near shocks, the second-order dissipative term is required to damp pre- and post-shock oscillations. For the present case of sub-critical flow the second-order term has been switched off.

To obtain a steady-state solution, integration in time is carried out by a four-stage Runge-Kutta scheme in which the dissipative terms are evaluated at the first step only. Convergence to steady state is accelerated by using local time-stepping. In the present investigation also enthalpy damping and implicit residual averaging has been utilized.
At the surface of the wing the boundary condition of zero normal velocity is applied, combined with a second-order accurate extrapolation of the pressure towards the solid-wall. At the outer boundary, the boundary conditions are based on one-dimensional Riemann invariants.

GRID

An O-O type topology grid is used to discretize the starboard half-space around the wing configuration. The grid has dimensions 144×38×28 (153,126 cells), i.e. on both the upper and lower side of the wing 72 cells in chordwise and 38 cells in spanwise direction, while there are 28 cells between the wing surface and the outer boundary. The grid is symmetric with respect to the horizontal plane of symmetry of the wing, quasi-conical on the wing surface (see Fig. 2), while grid lines are clustered near the apex, leading- and trailing-edge. Adjacent to the surface of the configuration, where the gradients in the flow are expected to be large, the cell-stretching ratio in the direction normal to the surface remains close to unity. The grid has a singular line which runs along the x-axis from the apex in upstream direction towards the outer boundary. The outer boundary of the computational domain is formed by the surface of a sphere with center at $x/c_R = 0.7$, $y/s = 0.0$, $z = 0.0$ and with radius $3c_R$, with c_R the root chord length and s the semi-span.

RESULTS

Solution breakdown
Computations have been carried out for a series of incidences, starting at 20 deg and increasing the incidence in small increments, keeping all the other parameters identical. It turned out that fully converged solutions could be obtained for incidences up to 21.25 deg, although the number of iterations required increased significantly with incidence. For incidences of 22.5 deg and higher the numerical procedure did neither really diverge nor converge: it resulted in a phenomenon termed "solution breakdown", characterized by the situation in which the solution in the region around the forward part of the wing appears to have reached its final state, while above the aft part of the wing the solution persists in changing substantially during the iteration process without reaching something like a final state. Since the numerical procedure is not time-accurate these solutions have to be discarded.

The influence of artificial dissipation on solution breakdown has been studied in some detail. It appears that adding more artificial viscosity to the solution procedure, in the form of the second-order term, results in solution breakdown at an incidence somewhat lower than 20 deg.
The influence of grid parameters (grid-point distribution, mesh size) on solution breakdown has also been considered in some detail. There is a

clear effect of the topology of the grid, but a less clear-cut effect of grid refinement on the critical value of the incidence.

Analysis of the solution at incidences close to the critical value
In the following the solution at $\alpha = 20$ deg and the one at 21.25 deg are analyzed in some detail. Both solutions are fully converged and steady-state values have been reached for the force and moment coefficients, for the surface pressure distribution as well as for the flow-field quantities.

Fig. 3 presents for $\alpha = 20$ deg the upper-surface isobar pattern and for the cross-flow planes $x/c_R = 0.60$ and 0.95 the distributions of the static pressure, the total-pressure loss and the chordwise velocity component. In Fig. 4 the corresponding results are shown for the solution at $\alpha = 21.25$ deg. In inviscid flow simulations total-pressure losses should occur in shock waves only. However, in the results of Euler methods total-pressure losses also arise due to artificial viscosity terms of the numerical algorithm, discretization errors, implementation of approximate boundary conditions, etc. For the present type of applications, these numerically induced total-pressure losses are largest in regions with large gradients, specifically in regions with concentrated vorticity such as, free shear layers and in particular vortex cores. This implies that distributions of total-pressure loss can be used to identify vortical flow regions.

In the results presented in Figs. 3 and 4 the upper-surface isobar pattern features an elongated region of low static pressure, which forms the footprint of the leading-edge vortex. Since the region with low pressure extends up to the apex it indicates that the flow separates all along the leading edge. The minimum pressure occurs very near the apex, typical for subsonic flow. In chordwise direction the influence of the presence of the trailing edge increases, that of the singularity at the apex decreases, resulting in a gradual rise of the pressure at the surface. Remark that the isobar pattern shows that on the aft part of the wing for $\alpha = 21.25$ deg the region in the footprint where the pressure increases most rapidly is more forward than for $\alpha = 20$ deg.

The results in the cross-flow plane at $x/c_R = 0.6$ also show that a well-developed vortex has formed above the wing, for both $\alpha = 20$ and 21.25 deg. This is indicated by the compactness of the region with low pressure and high total-pressure loss as well as by the high value of the chordwise velocity component just above the center of the vortex core. The vortex sheet leaving the leading edge is clearly visible in the plot for the axial velocity component. The results in the cross-flow plane further downstream, at $x/c_R = 0.95$, show that for $\alpha = 20$ deg the solution has the same structure as in the plane further upstream. However, for $\alpha = 21.25$ deg the leading-edge vortex has clearly undergone a dramatic change in its structure between the two cross-flow planes.
In the upstream plane, near the center of the vortex core, the flow is jet-like, i.e. the velocity component along the vortex axis has values exceeding free-stream. At the more downstream section the flow near the center of the vortex core is wake-like, i.e. it has retarded and has even reversed its direction.

Analysis of the solution at further sections has shown that a 'bubble' with reversed flow forms near the center of the vortex core. The bubble starts at $x/c_R = 0.85$ and terminates at $x/c_R = 1.15$. At $x/c_R = 0.95$ the bubble is evident as a rather significant increase of the static pressure at the center of the core, i.e. C_p increases from a value of -4.0 for $\alpha = 20$ deg to a value of -1.8 for $\alpha = 21.25$ deg.

At the section $x/c_R = 0.95$ the isobar of -1.0 surrounds a comparable cross-sectional area for both incidences, indicating that primarily the internal structure is affected by the phenomenon. During the process of the change of the structure the loss in total-pressure in the vortex core remains at the same level, but the cross-sectional area in which the largest losses occur blows up, filling the entire bubble region.

Fig. 5 presents, again for the sections at $x/c_R = 0.6$ and 0.95, the projection on the cross-flow plane of the vorticity vector, obtained from the solution for $\alpha = 20$ deg. Shown is part of plane in the region of the vortex core. Fig. 6 shows the corresponding results just before solution breakdown, i.e. for $\alpha = 21.25$ deg. The cross-flow-plane component of the vorticity is directly related to the distribution of the chordwise velocity component, while the chordwise component of the vorticity is directly related to the distribution of the velocity components in the cross-flow plane, see e.g. [15]. For the solutions considered here the chordwise component of the vorticity is always positive, i.e. it is directed in streamwise direction. At $\alpha = 20$ deg the cross-flow-plane vorticity vector turns around the center of the vortex core in the same sense as the circulation of the core and this all along the vortex axis, i.e. from the apex into the wake. In the bubble the cross-flow-plane vorticity reverses its direction, it now spirals in opposite sense around the center. Directly associated with this switch in sign, the chordwise velocity component near the center of the core decreases abruptly and becomes even negative.

Crocco's relation states that, in steady inviscid flow and in absence of total-pressure variation, the vorticity vector and the velocity vector are aligned. For the present solutions apparently the angle between the two vectors is zero everywhere, except in the bubble where it is 180 deg. The mechanism for the change-over in the direction of the cross-flow-plane vorticity vector is not known. However, the stability of the vortex might be greatly affected by this phenomenon, possibly causing solution breakdown at a slightly higher incidence.

It remains to be investigated whether or not the bubble contains closed streamlines. In case streamlines are closed it must be expected that viscous effects might greatly modify the flow feature, invalidating the present inviscid bubble model.

Solution in the near wake
The development of the flow just upstream and just downstream of the trailing edge (the near wake) is investigated next, again for $\alpha = 20.0$ and 21.25 deg. Hummel [16] showed that, at least for incompressible flow conditions, a complex mushroom-shaped vortex structure develops in the near wake. The constitutive elements of this flow feature are the leading-edge and the so-called trailing-edge vortex. The latter vortex collects the wake vorticity which chordwise component is of sign opposite to that of the vorticity contained in the leading-edge vortex. It starts to form immediately downstream of the trailing edge. Although the sign (negative) of the circulation of the trailing-edge vortex is the same as that of the vortex resulting from the smooth-surface secondary separation, it is a different vortex. This has been confirmed in inviscid flow simulations (e.g. see [15] and [17]).

For $\alpha = 20$ deg Fig. 7a presents the total-pressure losses and the chordwise component of the vorticity in five consecutive cross-flow planes, i.e. at $x/c_R = 0.95, 1.0, 1.025, 1.05$ and 1.10. In Fig. 7b the corresponding results are shown obtained from the solution for $\alpha = 21.25$ deg.

The solution for α = 20 deg indeed indicates the formation of a wake with two regions of increased total-pressure loss, indicative for regions with vortical flow as is confirmed by the plots giving isolines of the chordwise vorticity component. Two regions with chordwise vorticity, of high magnitude but different sign, can be distinguished. The region with positive chordwise vorticity and with the maximum total-pressure loss can be identified as the continuation of the leading-edge vortex into the wake. Viewed in downstream direction this vortex moves outboard and upward. Due to the coarseness of the grid in the wake, the total-pressure loss increases and is spread over a larger cross-sectional area. The region with negative chordwise vorticity identifies the trailing-edge vortex. Its formation and subsequent interaction with the leading-edge vortex is clearly visualized.

For the slightly higher incidence of 21.25 deg, in the sections close to the trailing edge, the larger region with high total-pressure loss identifies the bubble in the vortex core. Note that the chordwise component of the vorticity reaches not as high values as for α = 20 deg, though the circulation of the core will have about the same value. For cross-flow planes further downstream the cross-sectional area of the leading-edge vortex remains about equal, while the region near the center where the total-pressure loss is highest shrinks. Simultaneously the magnitude of the chordwise vorticity increases, both indicating the recovery of the leading-edge vortex. At x/c_R = 1.15 the bubble with reversed flow terminates. Also indicated in the cross-flow planes in the near wake is a region with negative chordwise vorticity corresponding with the trailing-edge vortex. Its formation is apparently not much affected by the change of the internal structure of the leading-edge vortex.

Behaviour of the solution along the vortex axis
For α = 20 and 21.25 deg Fig. 8 presents, as a function of the chordwise coordinate, the coordinates of the point in the cross-flow plane where the total-pressure loss attains it maximum, i.e. the trajectory of the center of the vortex core. Also shown are the value of the static and total pressure coefficient along the trajectory.

For such a small increase in incidence the location of the vortex axis does not change significantly, neither in lateral nor in vertical direction.
Most of the scatter in the plot is attributed to the post-processing procedure to find the trajectory of the vortex center. For α = 20 deg upstream of the trailing edge, for α = 21.25 deg upstream of the bubble (x/c_R<0.85), the total-pressure loss increases and the pressure decreases slightly when the incidence is raised from 20 to 21.25 deg. Near the trailing-edge (α = 20 deg) and in the bubble (α = 21.25 deg) the total-pressure loss increases to a somewhat higher level, while the static pressure increases sharply.
In retrospect, re-examining Figs. 7a and 8, it appears that also for α = 20 deg a (small) bubble develops in the vortex core as it passes through the near wake.

CONCLUDING REMARKS

(i) Solutions obtained with an Euler method for the steady subsonic flow about a 65-deg cropped sharp-edged delta wing have been analyzed. Converged solutions could be obtained for incidences up to 21.25 deg. For higher incidences the numerical procedure no longer converges and "solution breakdown" is encountered.
(ii) When approaching the critical incidence the vortex core experiences a dramatic change in its internal structure, a bubble with a strongly reduced

chordwise velocity component forms within the vortex core. Associated with the bubble is a change-over in the direction of the cross-flow-plane vorticity vector, as can be explained in terms of inviscid flow properties.
(iii) Investigation of the near wake has shown that the leading-edge vortex recovers in the wake and that, similar to lower incidences, a trailing-edge vortex develops. The interaction of the two vortices leads to the formation of a mushroom-shaped vortex system.
(iv) It is unclear if there is a direct relation between the appearance of the bubble in the Euler solution and the onset of vortex breakdown in the wind tunnel (at a significantly higher α), but there appears to be some similarity. The influence of grid parameters and parameters of the numerical procedure on the solution needs further investigation.

REFERENCES

[1] Murman, E.M., Rizzi, A.: Applications of Euler Equations to Sharp-Edged Delta Wings with Leading Edge Vortices. AGARD CP412, Paper 15 (1986).
[2] Longo, J.M.A.: The Role of the Numerical Dissipation on the Computational Euler-Equations-Solutions for Vortical Flows. AIAA Paper 89-2232 (1989).
[3] Hitzel, S.M.: Wing Vortex-Flows up into Vortex Breakdown. A Numerical Simulation. AIAA Paper 89-2232 (1989).
[4] O'Neill, P.J., Barnett, R.M., Louie, C.M.: Numerical Simulation of Leading-Edge Vortex Breakdown using an Euler Code. AIAA Paper 89-2189 (1989).
[5] Raj, P., Sikora, J.S., Keen, J.M.: Free-Vortex Flow Simulation using a Three-Dimensional Euler Aerodynamic Method. J. of Aircraft, Vol. 25, No. 2 (1988).
[6] Wardlaw Jr, A.B., Davies, S.F.: Euler Solutions for Delta Wings. AIAA Paper 89-3398 (1989).
[7] Peckham, P.D., Atkinson, S.A.: Preliminary Results of Low Speed Wind Tunnel Tests on a Gothic Wing of Aspect Ratio 1.0, ARC Rep. CP-508 (1957).
[8] Elle, B.J.: An Investigation at Low Speed of the Flow near the Apex of Thin Delta Wings with Sharp Leading Edges. ARC Rep. R&M. No. 3176 (1958).
[9] Lambourne, N.C., Bryer, D.W.: The Bursting of Leading-Edge Vortices- Some Observations and Discussion of the Phenomenon. ARC Rep. R&M. No. 3282 (1961).
[10] Ekaterinaris, J.A., Schiff, L.B.: Numerical Simulation of the Effects of Variation of Angle of Attack and Sweep Angle on Vortex Breakdown over Delta Wings. AIAA-90-3000-CP (1990).
[11] Hall, M.G.: Vortex Breakdown. Annual Review of Fluid Mechanics, Vol. 4, pp. 195-218 (1972).
[12] Boerstoel, J.W.: Progress Report of the Development of a System for the Numerical Simulation of Euler flows, with Results of Preliminary 3D Propeller-Slipstream/Exhaust-Jet Calculations. NLR TR 88008 U (1988).
[13] Jameson, A., Schmidt, W., Turkel, E.: Numerical Solution of Euler Equations by Finite Volume Methods Using Runge-Kutta Time-Stepping Scheme. AIAA Paper 81-1259 (1981).
[14] Elsenaar, A., Hoeijmakers, H.W.M.: An Experimental Study of the Flow Over a Sharp-Edged Delta Wing at Subsonic and Transonic Speeds. AGARD CP-494, Paper 15 (1991).
[15] Hoeijmakers, H.W.M.: Computational Aerodynamics of Ordered Vortex Flows. NLR TR 88088 U (1989).
[16] Hummel, D.: On the vortex Formation over a Slender Wing at Large Angles of Incidence. AGARD CP-247, Paper 15 (1979).
[17] Hoeijmakers, H.W.M., Jacobs, J.M.J.W., Berg, J.I. van den: Numerical Simulation of Vortical Flow Over a Delta Wing at Subsonic and Transonic Speeds. ICAS Paper 90-3.3.3 (1990).

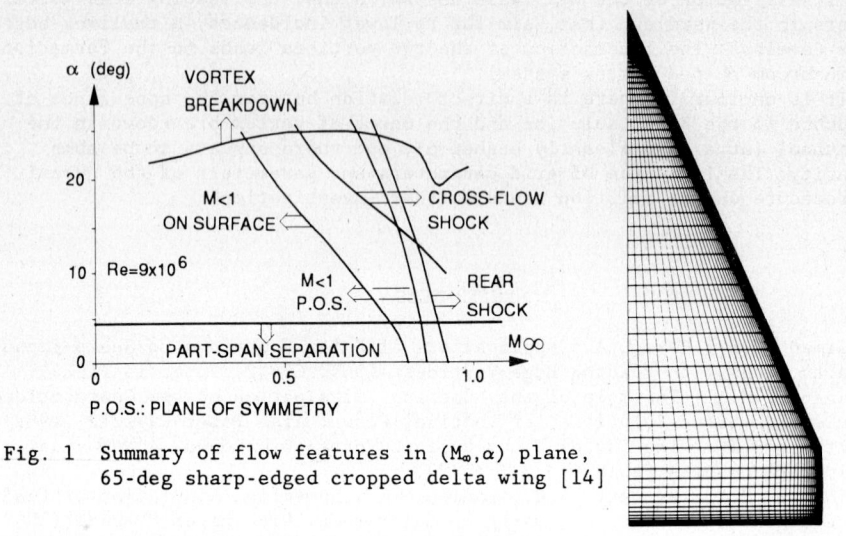

Fig. 1 Summary of flow features in (M_∞, α) plane, 65-deg sharp-edged cropped delta wing [14]

Fig. 2 Grid on upper wing surface (72 × 38).

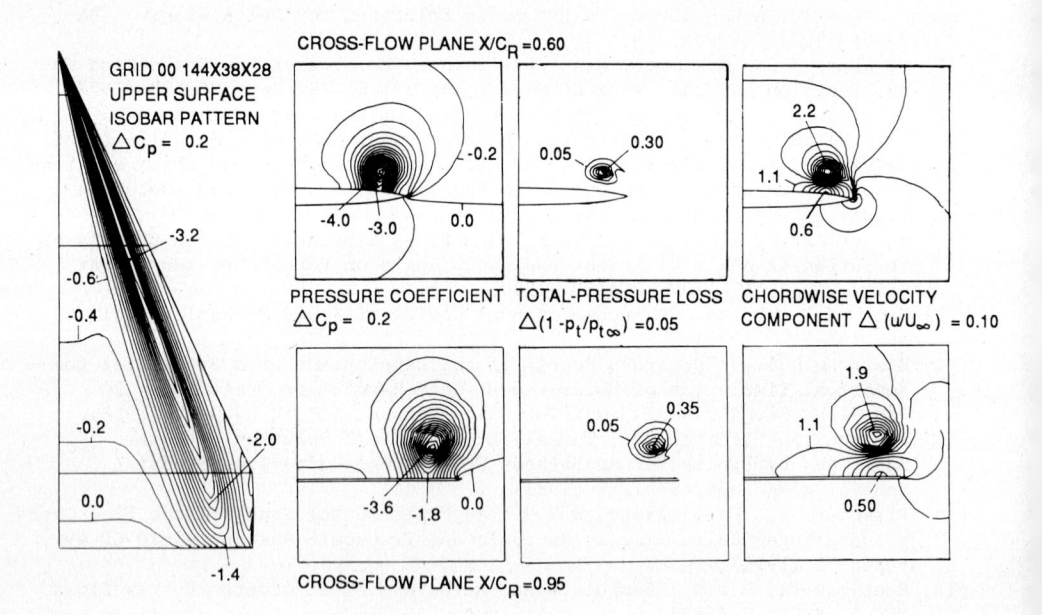

Fig. 3 Euler solution for 65-deg sharp-edged delta wing $M_\infty = 0.50$, $\alpha = 20$ deg.

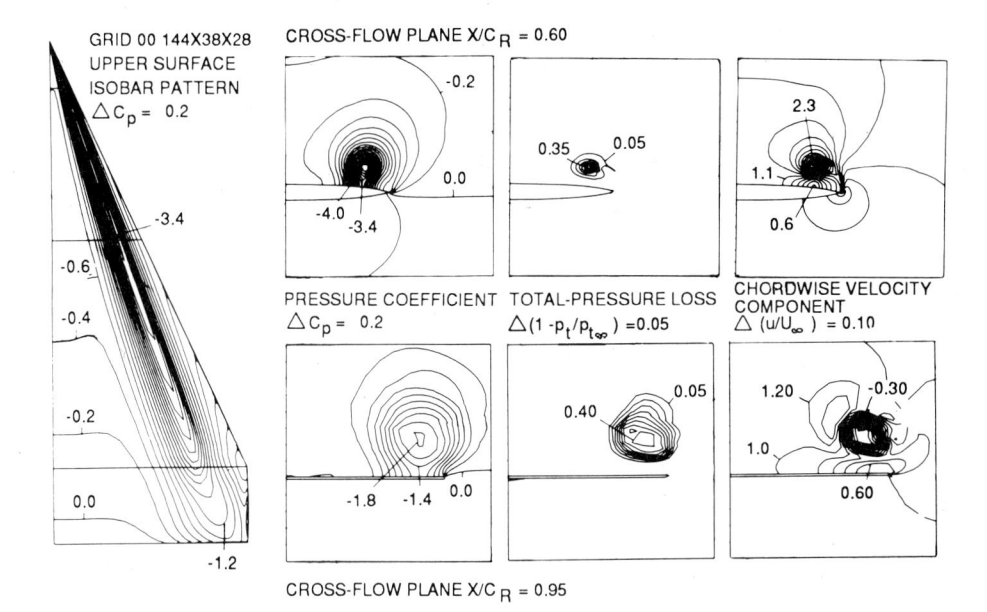

Fig. 4 Euler solution for 65-deg sharp-edged delta wing
$M_\infty = 0.50$, $\alpha = 21.25$ deg.

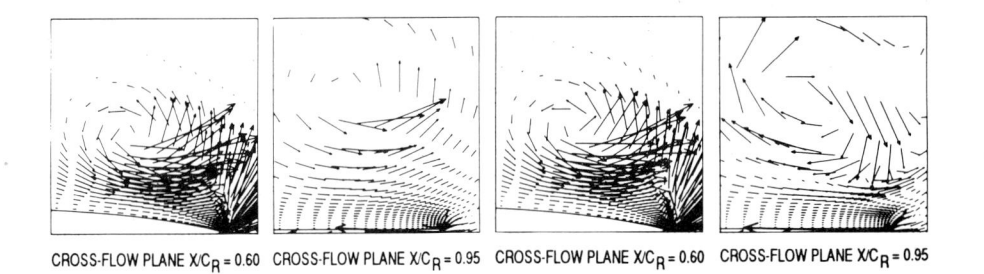

Fig. 5
Cross-flow-plane vorticity vector
$\vec{e}_x \times (\vec{\omega} \times \vec{e}_x)(c_R/U_\infty)$
$M_\infty = 0.50$, $\alpha = 20$ deg.

Fig. 6
Cross-flow-plane vorticity vector
$\vec{e}_x \times (\vec{\omega} \times \vec{e}_x)(c_R/U_\infty)$
$M_\infty = 0.50$, $\alpha = 21.25$ deg.

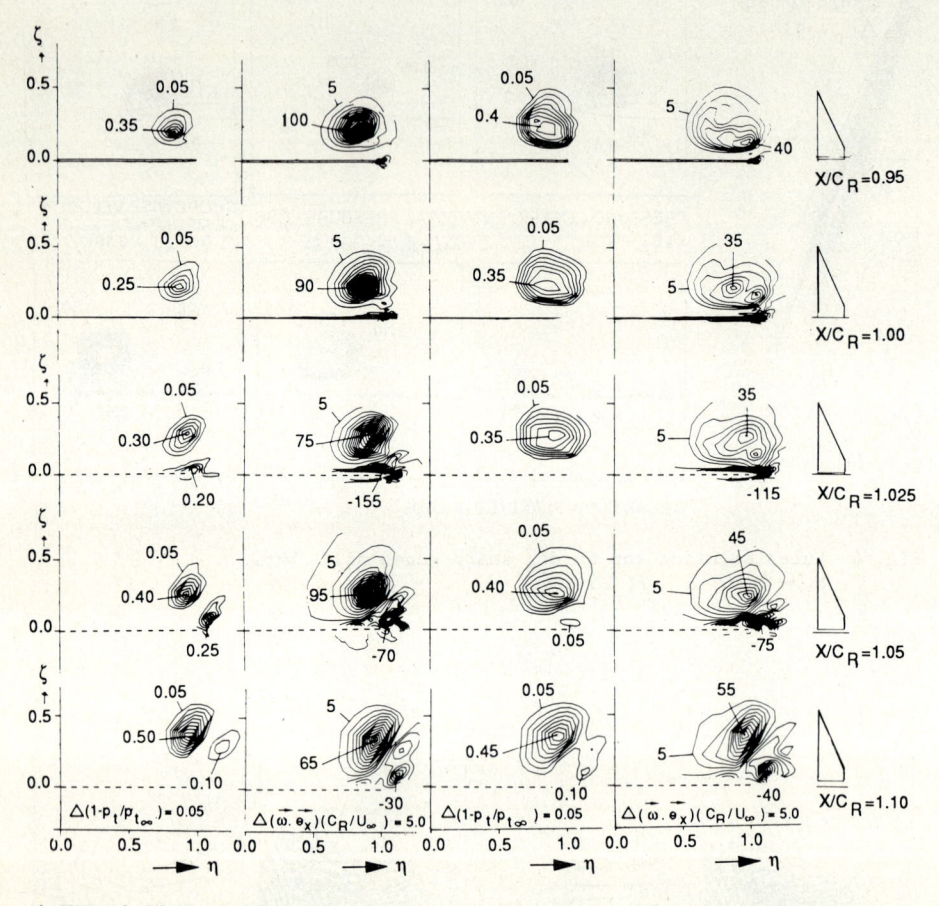

a) $M_\infty = 0.50$, $\alpha = 20$ deg.

b) $M_\infty = 0.50$, $\alpha = 21.25$ deg.

Fig. 7 Distribution of total-pressure and chordwise vorticity $(\vec{\omega}\cdot\vec{e}_x)(c_R/U_\infty)$ in the near wake.

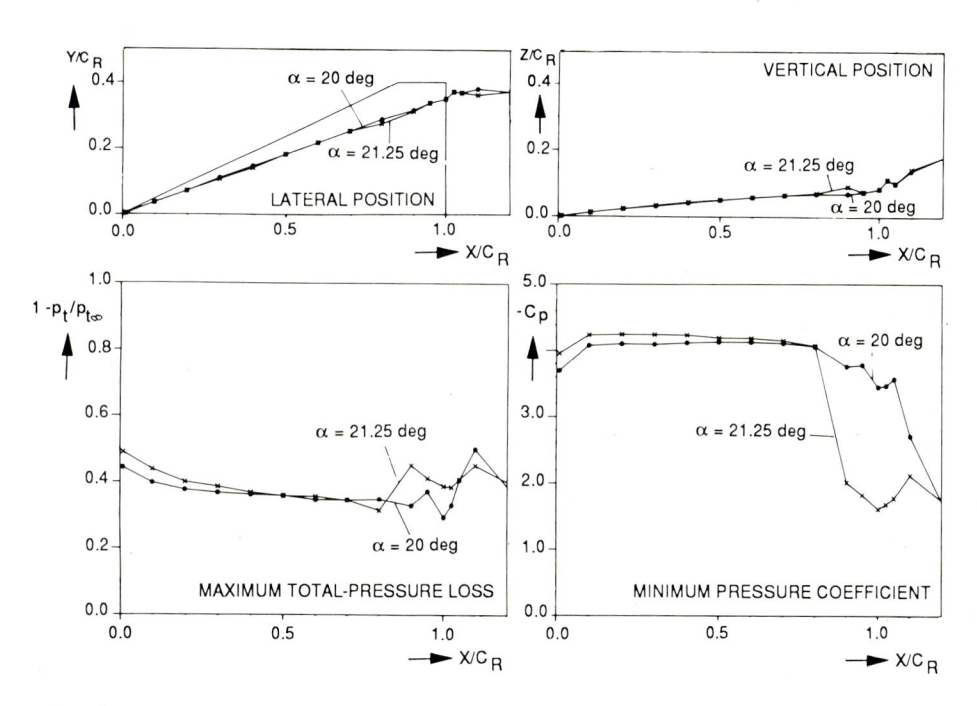

Fig. 8 Vortex core trajectory, total-pressure loss and pressure along vortex trajectory.

EXPLICIT MULTIGRID SMOOTHING FOR MULTIDIMENSIONAL UPWINDING OF THE EULER EQUATIONS

L. A. Catalano, P. De Palma, M. Napolitano
Politecnico di Bari, via Re David 200, **70125** Bari, Italy

SUMMARY

This paper provides a method for solving nonlinear scalar advection problems and the conservation-law Euler equations. A recently developed genuinely multi-dimensional upwind fluctuation splitting discretization is combined with a multi-stage Runge–Kutta scheme with optimal coefficients, accelerated by a multigrid strategy. Numerical results are presented to demonstrate the merits of the proposed approach.

INTRODUCTION

In the last few years, one of the most interesting and fruitful areas of research in Computational Fluid Dynamics has been the development of upwind methods for the Euler and Navier–Stokes equations. For the case of one space dimension, methods based on the solution of Riemann problems, proposed for the first time by Godunov [1], have achieved a remarkable level of accuracy at a reasonable computational cost, see, e.g., [2,3]. Unfortunately, the extension to more than one space dimensions has been based mostly on directional splitting, in the absence of a truly multi-dimensional Riemann solver. Therefore, to date, upwind methods are really satisfactory only when discontinuities are aligned with the mesh. In order to overcome such a difficulty, two genuinely multi-dimensional upwind methods have been proposed [4,5], which aim at resolving discontinuous flows with minimum grid dependency. In particular, the interest of the authors is to develop a conservative, genuinely multi-dimensional, upwind scheme on an unstructured adaptive triangular (in 2D) mesh, as well as an efficient and easy-to-vectorize explicit integration scheme, accelerated by a multigrid strategy.

Concerning the spatial discretization, the so-called fluctuation splitting (FS) approach, proposed in [6], is chosen as a suitable framework for implementing genuinely multi-dimensional upwind methods.

Concerning the time-marching integration, a very efficient multigrid smoother has been recently developed, by providing a general procedure for optimizing the coefficients and time steps of multi-stage Runge–Kutta schemes [7]. Furthermore, such a procedure has already been succesfully applied to solve nonlinear advection problems on a finite volume mesh [8].

The aim of this paper is twofold. Firstly, the approach proposed in [7] is extended to the FS discretization to provide an efficient and accurate solution method for nonlinear scalar advection problems on triangular grids. Secondly and more importantly, such a generalized solution procedure is applied to compute steady solutions to the Euler equations using the multi-dimensional upwinding proposed in [5] and implemented in [9].

SCALAR ADVECTION

Fluctuation splitting space discretization

A linear advection equation is considered at first:

$$\frac{\partial u}{\partial t} = -\mathbf{a} \cdot \nabla u = \text{Res}(u), \qquad \mathbf{a} = (a,b), \tag{1}$$

with prescribed initial conditions and Dirichlet conditions at the inflow boundaries. A triangular mesh is used to discretize the computational domain. The *flux balance over each triangle*, T, is defined as the *fluctuation*, Φ_T, namely,

$$\Phi_T = -\iint_T \mathbf{a} \cdot \nabla u \, dS = \oint_{\partial T} u\, \mathbf{a} \cdot \mathbf{n} \, ds, \tag{2}$$

$\mathbf{n}$ being the *inward* unit vector normal to the boundary ∂T of each triangle. The unknowns are located at the cell vertices and u is assumed to vary linearly over each triangle. After some algebra, one obtains:

$$\Phi_T = -\frac{1}{2}\sum_{j=1}^{3} u_j \, \mathbf{a} \cdot \mathbf{n}_j \, l_j = -\sum_{j=1}^{3} k_j u_j, \qquad k_j = \frac{1}{2} \mathbf{a} \cdot \mathbf{n}_j \, l_j, \tag{3}$$

l_j being the length of the side j (the one opposed to the vertex j).

The fluctuations are then used to construct $\text{Res}(u)$ at each cell vertex v, as:

$$\text{Res}(u)_v = \frac{1}{S_v} \sum_T \alpha_v^T \, \Phi_T. \tag{4}$$

In eq. (4), the summation is extended over all the triangles having the vertex v in common, the area S_v is a suitable scale factor (see, e.g., [6]) and α_v^T are the distribution coefficients, namely, the coefficients which *send* the correct amounts of the fluctuations Φ_T to the node v. Obviously, in order to ensure conservation, for every triangle T, the following condition must be satisfied:

$$\sum_{j=1}^{3} \alpha_j^T = 1. \tag{5}$$

In this work, the choice of these distribution coefficients is based on upwinding. If only one of the parameters k_j is positive, namely, if the flow is entering T through one of its sides only, the fluctuation is obviously sent to the downstream node. Otherwise, when two inflow sides are present, different ways of distributing Φ_T over the downstream nodes are possible. In [6], these coefficients are chosen so as to satisfy at least one of the two important properties called positivity and linearity preservation. Here, the choice is made according to the *N-scheme*, namely, the generalization of Roe's monotone scheme [10], due to Deconinck et al. [6]; for example, for $k_3 < 0$, one has:

$$\text{Res}(u)_1^T = -k_1 (u_1 - u_3) \tag{6a}$$

$$\text{Res}(u)_2^T = -k_2 (u_2 - u_3). \tag{6b}$$

This scheme satisfies only the positivity property.

The FS approach can be applied to the more general nonlinear equation

$$\frac{\partial u}{\partial t} = -\nabla \cdot \mathbf{F}, \qquad \mathbf{F}(u) = (f(u), g(u)), \tag{7}$$

if a proper linearization of the flux function is performed, as follows. For every triangle, one has:

$$\iint_T \nabla \cdot \mathbf{F}\, dS = \iint_T \mathbf{a}(u) \cdot \nabla u\, dS = \nabla u \cdot \iint_T \mathbf{a}(u)\, dS, \tag{8}$$

∇u being a constant for linearly varying u. Furthermore, by defining an appropriate average convection speed $\overline{\mathbf{a}}$, satisfying the so-called U property [6], namely,

$$\overline{\mathbf{a}} = \frac{1}{S_T} \iint_T \mathbf{a}(u)\, dS, \qquad S_T = \iint_T dS, \tag{9}$$

one finally obtains

$$\iint_T \nabla \cdot \mathbf{F}\, dS = \overline{\mathbf{a}} \cdot \nabla u\, S_T. \tag{10}$$

In the present paper, $\mathbf{a} = \left(\dfrac{\partial f}{\partial u}, \dfrac{\partial g}{\partial u}\right)$ is assumed to be a linear function of u, so that the integral in eq. (9) is computed exactly as

$$\overline{\mathbf{a}} = \frac{1}{3}\sum_{j=1}^{3} \mathbf{a}(u_j) = \mathbf{a}\left(\frac{1}{3}\sum_{j=1}^{3} u_j\right) = \mathbf{a}(\overline{u}). \tag{11}$$

Time marching integration procedure (explicit smoother)

In the present section, an optimal Runge–Kutta scheme is developed for the FS approach, along the lines described in [7] and [8]. A *diamond grid*, as the one shown in figure 1, obtained from a uniform Cartesian grid ($\Delta x = \Delta y = h$), is used for simplicity.

The FS N-scheme, described in the previous section, is used for approximating the residual of the linear advection equation (1) ($a \geq b \geq 0$). The resulting discrete equation is formally written as:

$$\left(\frac{\partial u}{\partial t}\right)_{i,j} = \left(L^h(u)\right)_{i,j}, \tag{12}$$

where L^h is the discrete spatial operator defined on the grid h and applied at point (i,j). The design of the smoother can be performed on the simpler ordinary differential equation

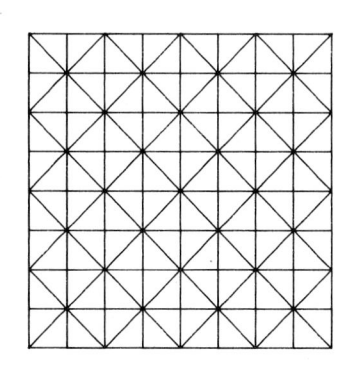

Figure 1: Cartesian diamond grid

$$\frac{du}{dt} = \lambda u, \quad \lambda = \frac{a}{h} z\left(\beta_x, \beta_y, R\right). \tag{13}$$

In eq. (13), $\lambda \in \mathcal{C}$ is the Fourier transform of L^h, β_x and β_y are the spatial wave numbers in the two coordinate directions and $R = b/a$. The function $z(\beta_x, \beta_y, R)$ takes a different form, depending on the shape of the grid; here, for a vertex with four concurrent triangles,

$$z = \left(1 - \frac{R}{2}\right) e^{-i\beta_x} + \frac{R}{2} e^{-i\beta_y} - 1, \qquad (14a)$$

whereas, for a vertex with eight concurrent triangles,

$$z = \left(1 - \frac{R}{2}\right) e^{-i\beta_x} + R e^{-i(\beta_x + \beta_y)} + \frac{R}{2} e^{-i\beta_y} - (R+1). \qquad (14b)$$

The time step is nondimensionalized using the larger component of the convection speed, a, to provide a single CFL number, $\nu = a\Delta t/h$. Equation (13) is thus rewritten as

$$\frac{du}{dt} = \frac{\nu z}{\Delta t} u . \qquad (15)$$

A multi-stage Runge–Kutta scheme is used to discretize the time derivative in eq. (15), as follows:

$$\begin{aligned} u^{(0)} &= u^\ell & (16a) \\ u^{(k)} &= u^{(0)} + c_k \nu z\, u^{(k-1)}, \quad k = 1, ..., n & (16b) \\ u^{\ell+1} &= u^{(n)}. & (16c) \end{aligned}$$

The resulting amplification factor is the following polynomial of degree n

$$P_n(z) = 1 + c_n \nu z (1 + c_{n-1} \nu z (... (1 + c_1 \nu z))). \qquad (17)$$

The use of a Runge–Kutta-type scheme in combination with a multigrid cycle suggests the choice of the n parameters c_k, $k = 1, ..., n-1$, and of ν, so as to minimize the smoothing factor σ, namely, the maximum amplification factor for a high frequency error component [7]:

$$\sigma = \max |P_n(\beta_x, \beta_y, R)|, \qquad |\underline{\beta}| = \max(|\beta_x|, |\beta_y|) \quad \left\{\frac{\pi}{2} \leq |\underline{\beta}| \leq \pi\right\}.$$

It is noteworthy that, where the advection velocity is aligned with the grid, P_n equals 1 for some waves and the standard minimization procedure would set the entire amplification factor field equal to one (see [11], for details). As suggested in [11], this difficulty is removed by excluding the boundary between the high and low frequency ranges, $\beta_y = \pi/2$, from the domain of definition of σ. Of course, the resulting schemes are not fully optimized. The coefficients of the three-stage smoothers, for vertices with four and eight triangles, are given in tables 1 and 2, respectively, for 12 different values of R.

Multigrid acceleration

The smoother described above is designed to work effectively in conjunction with a multigrid method. Here, the well-known FAS FMG scheme of Brandt [12] is employed;

the main features of the MG cycle, namely, the grid transfer operators, are briefly outlined. At each grid point (i,j) of the fine mesh h, the defect $d_{i,j}^h = -Res(u)_{i,j}$ is computed and collected on the coarser grid $H=2h$, by applying the classical full-weighting collection operator. Since nested iteration makes the scheme converge sufficiently at all levels, no restriction of the current solution is needed. The coarse-grid correction is then transferred back to the fine grid-points by means of standard bilinear interpolation. In the present work, a non-optimized V-cycle is applied, with one pre- and one post-relaxation at all levels, each relaxation step being performed using the aforementioned three-stage Runge–Kutta smoother.

Table 1: *optimal coefficients for the 3-stage FS N-scheme (4 triangles)*

	0	5	9	14
$R/50$	19	23	28	33
	38	42	46	50
	.222	.226	.229	.234
$c_1\nu$	.238	.242	.247	.252
	.257	.262	.266	.271
	.600	.613	.625	.640
$c_2\nu$	.656	.670	.688	.706
	.725	.740	.755	.770
	1.500	1.528	1.553	1.588
ν	1.627	1.662	1.708	1.757
	1.808	1.851	1.894	1.937

Table 2: *optimal coefficients for the 3-stage FS N-scheme (8 triangles)*

	0	5	9	14
$R/50$	19	23	28	33
	38	42	46	50
	.222	.215	.210	.204
$c_1\nu$	.198	.194	.189	.183
	.180	.174	.168	.162
	.600	.588	.579	.568
$c_2\nu$	.556	.545	.531	.515
	.499	.486	.473	.459
	1.500	1.476	1.460	1.446
ν	1.430	1.415	1.392	1.365
	1.331	1.317	1.296	1.269

Numerical results

The proposed approach has been applied to solve the nonlinear advection equation

$$\frac{\partial u}{\partial t} + \frac{\partial}{\partial x}\left(\frac{u^2}{2}\right) + \frac{\partial}{\partial y}(u) = 0,$$

in a unit square, with initial conditions $u(x,y) = 0$ and boundary conditions $u(x=0,y) = 1.5$, $u(x=1,y) = -.5$, $u(x,y=0) = 1.5 - 2x$: the exact solution is a converging compression fan creating an oblique shock, impinging on the top-right corner $(x=1, y=1)$. Results have been obtained using *diamond grids* with 129×129, 65×65 and 33×33 nodes, respectively. The numerical solution on the coarsest grid is given in figure 2: the numerical and exact $u(x)$ profiles at $y = .75$ and $y = .25$ are given as symbols and solid lines, respectively.

The major interest being the efficiency of the multigrid optimal Runge–Kutta scheme in combination with the FS discretization, convergence histories are presented in figure 3: the logarithm of the L^1-norm of the residual is plotted versus the work; one *work unit* is defined as *one single-stage residual* calculation on the finest grid; the work required by the nested iteration has been accounted for. In figure 3, the solid lines show the single and multigrid results for the finest 129×129 grid, whereas the broken and dotted lines refer to the intermediate 65×65 grid and to the coarsest 33×33 grid, respectively. The multigrid curves are the steeper ones in all cases and their convergence rate is almost independent of the grid. Finally, the convergence history for the multigrid scheme, using nonoptimal coefficients for the smoother ($c_1 = 1/3$, $c_2 = 1/2$, $\nu = .7$) and the finest 129×129 grid, is also shown in figure 3 as circles. The overall effectiveness of the proposed approach is clearly demonstrated. Incidentally, all solutions converge to machine zero without any numerical difficulty.

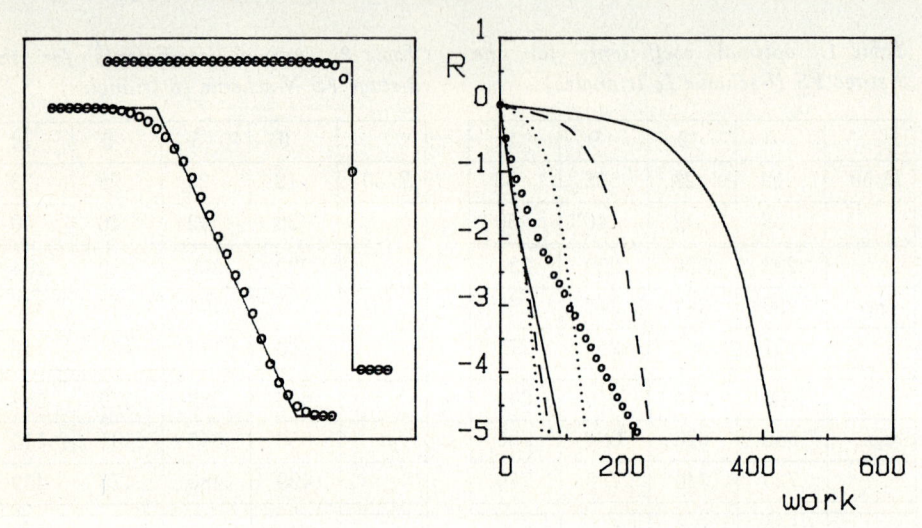

Figure 2: cuts of the solution at y=.25 and y=.75 for the oblique shock problem

Figure 3: convergence histories for the oblique shock problem

EULER EQUATIONS

The two genuinely multi-dimensional upwind methods proposed so far for the conservation-law Euler equations [4,5] are both based on some form of signal pattern recognition. The four-wave model of Deconinck et al. [4] selects four characteristic compatibility equations leading to an optimal decoupling of the Euler system into a set of four 2D advection equations for two acoustic, one entropy and one shear waves. Unfortunately, this approach, which is the most suitable to be combined with the present FS multigrid scheme, to date has given good results only for supersonic flows. Therefore, in this paper, Roe's decomposition method [5] is employed, which allows to recast the Euler equations into six scalar wave equations, as described below.

Roe's decomposition method

The quasi linear form of the Euler equations is considered:

$$\frac{\partial U}{\partial t} + A\frac{\partial U}{\partial x} + B\frac{\partial U}{\partial y} = 0. \tag{18}$$

A simple wave solution of eq. (18) is any solution U depending only on a scalar parameter $v = v(x, y, t)$, for which,

$$\nabla U = \frac{dU}{dv} \nabla v, \qquad (19)$$

where,

$$\nabla v = |\nabla v| \, \mathbf{n}. \qquad (20)$$

For a simple wave solution, eq. (18) can be rewritten as:

$$\frac{\partial v}{\partial t} \frac{dU}{dv} + (An_x + Bn_y) \, |\nabla v| \, \frac{dU}{dv} = 0. \qquad (21)$$

Equation (21) shows that $\frac{dU}{dv}$ is the right eigenvector of the matrix $An_x + Bn_y$ with corresponding eigenvalue λ, namely,

$$(An_x + Bn_y) \, r = \lambda r, \qquad r = \frac{dU}{dv}. \qquad (22a,b)$$

Equation (21), together with eqs. (20) and (22), provides the following advection equation for the scalar parameter v:

$$\frac{\partial v}{\partial t} + \lambda \frac{\partial v}{\partial n} = 0. \qquad (23)$$

Equation (23) has a wavelike solution of the form:

$$v_n^\alpha = v_n^\alpha (xn_x + yn_y - \lambda_n^\alpha t) = v_n^\alpha(q), \qquad (24)$$

where α denotes the wave type (acoustic, entropy or shear wave) and $\mathbf{n}$ is the direction of propagation. Equation (22b), using eq. (24), can be rewritten in vector form as

$$\nabla U_n^\alpha = r_n^\alpha \nabla v_n^\alpha = \left(r \frac{dv}{dq} \mathbf{n} \right)_n^\alpha. \qquad (25)$$

Roe's decomposition method [5] is based on the fact that any perturbation of the primitive variables $\tilde{U}$ can be reconstructed by superposition of the various wave contributions, as:

$$\nabla \tilde{U} = \sum_\alpha \nabla \tilde{U}_n^\alpha = \sum_\alpha (\tilde{r}k\mathbf{n})_n^\alpha. \qquad (26)$$

In eq. (26), $k_n^\alpha = dv_n^\alpha / dq$ is defined as the strength of the wave α travelling in the direction $\mathbf{n}$. Equation (26) constitutes an algebraic system whose unknowns are the parameters describing the decomposition, namely, the strengths and directions of propagation of the waves. Since eight components of the primitive variable gradients appear as known parameters in the right hand side of eq. (26), eight degrees of freedom are available for choosing a wave model.

Roe [5] proposed two different models allowing to solve eq. (26), analytically, with the following choices of parameters. One entropy wave with unknown direction and intensity (two unknowns). Four acoustic waves orthogonal to each other, with unknown direction and strengths (five parameters). The choice of the remaining parameter distinguishes the two models. Model A: the last unknown is the strength of the local vorticity. Model B: the last unknown is the intensity of a shear wave travelling perpendicularly to the streamline. In [9], a variant to model B has been proposed, model C: the last unknown is the intensity of a shear wave travelling in the direction of the pressure gradient. Such a model produces monotone results, a feature not enjoyed by both models A and B.

Linearization and scalar advection

A conservative linearization of the Euler equations is available [13], which allows to evaluate analytically the fluctuation Φ_T as:

$$-\iint_T \frac{\partial U}{\partial t} dS = -\Phi_T = \iint_T \left[\frac{\partial F(U)}{\partial x} + \frac{\partial G(U)}{\partial y} \right] dS = \left[\overline{A} \, \overline{\frac{\partial U}{\partial x}} + \overline{B} \, \overline{\frac{\partial U}{\partial y}} \right] S_T. \quad (27)$$

$\overline{A}$ and $\overline{B}$ are appropriate average Jacobians whereas $\overline{\partial U/\partial x}$ and $\overline{\partial U/\partial y}$ are average derivatives of the conservative variables. All of them can be computed analytically if Roe's parameter vector $W = \sqrt{\rho}(1, u, v, H)$ is used as the independent variable, varying linearly over the triangles, see [13] for details. Also, the average transformation matrix from primitive to conservative variables, $\overline{M}$, is obtained analytically so as to determine the average gradient $\nabla \tilde{U} = \overline{M}^{-1} \nabla U$ of the primitive variables, which must be used in eq. (26) to calculate the eight parameters of any of the three models described previously; equation (26) thus becomes:

$$\overline{\frac{\partial \tilde{U}}{\partial x}} = \sum_\alpha (\tilde{r} k n_x)_n^\alpha, \quad \overline{\frac{\partial \tilde{U}}{\partial y}} = \sum_\alpha (\tilde{r} k n_y)_n^\alpha. \quad (28a,b)$$

Furthermore, using the aforementioned matrix $\overline{M}$, eq. (27) can be rewritten as

$$-\Phi_T = \overline{M} \left[\overline{\tilde{A}} \, \overline{\frac{\partial \tilde{U}}{\partial x}} + \overline{\tilde{B}} \, \overline{\frac{\partial \tilde{U}}{\partial y}} \right] S_T, \quad (29)$$

where $\overline{\tilde{A}}$ and $\overline{\tilde{B}}$ are the average Jacobians in primitive variables. Equations (27-29), together with eq. (22a), finally provide the following expression

$$\iint_T \frac{\partial U}{\partial t} dS = \Phi_T = -S_T \sum_\alpha (r \, k \, \lambda)_n^\alpha = \sum_\alpha \Phi_T^\alpha, \quad (30)$$

where the global fluctuation is expressed as the sum of the six wave contributions and, by definition, $\Phi_T^\alpha = S_T (r \, k \, \lambda)_n^\alpha$ is the fluctuation due to a single wave.

Numerical methods and results

Thanks to the decomposition and the linearization described above, the FS approach, together with the previously discussed time integration and multigrid method are applied to each scalar component of the Euler system, as given in eq. (30), using a single CFL number, based on the maximum convection speed.

The proposed methodology has been tested versus the well known shock-reflection problem, the oblique shock impinging upon a flat plate at a 29 degree angle. The numerical solutions obtained using model C, on a 145×49 uniform diamond grid, are given in figure 4 as the Mach number distributions along the wall ($y=0$) and at $y=.25$. The present results are monotonic and typical of a good first-order-accurate, genuinely multi-dimensional upwind method.

Finally, convergence histories are provided in figure 5 as the logarithm of the L^1-norm of the density residual versus the work, one work unit being again defined as *one single-stage residual* calculation on the finest grid. The solid lines refer to the single- and multi-grid solutions; the multigrid curve obtained using the nonoptimal coefficients, $c_1 = 1/3, c_2 = 1/2, \nu = .7$, is also given as broken lines, for comparison. Figure 5 clearly demonstrates the efficiency of the multigrid scheme in conjunction with the optimal smoother. It is noteworthy that in the present Euler calculations the residuals drop about 9 orders of magnitude, without freezing the directions of propagation. This is a striking improvement with respect to most previous calculations employing genuinely multi-dimensional upwind methods [9,14]. In this respect, work is still in progress, see [15].

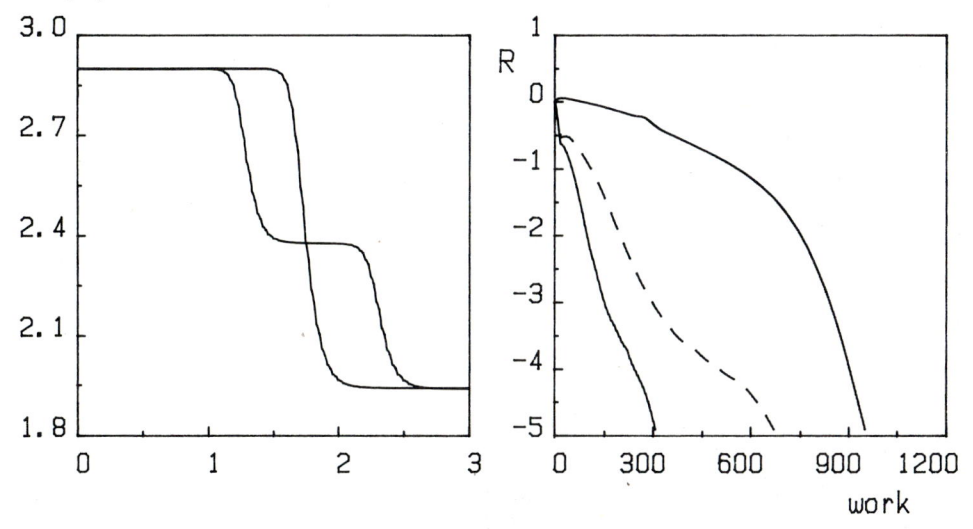

Figure 4: Mach number distributions at $y=0$ and $y=.25$ for the oblique shock reflection problem

Figure 5: convergence histories for the oblique shock reflection problem

CONCLUSION

An explicit multigrid strategy developed for nonlinear scalar advection problems using a finite volume approach has been extended successfully to the fluctuation splitting space discretization and to the Euler equations, using Roe's six-wave decomposition model.

ACKNOWLEDGEMENTS

The present research was carried out as part of the Brite/Euram Contract no. AERO-0003-C (Area 5: aeronautics), issued by the European Economic Community and coordinated by prof. H. Deconinck.

REFERENCES

[1] S. K. Godunov, "A finite-difference method for the numerical computation of discontinuous solutions of the equations of fluid dynamics", *Mat. Sbornik*, 47, 1959, pp. 357-393.

[2] P. L. Roe, "Characteristic-based schemes for the Euler equations", *Ann. Rev. Fluid Mech.*, 18, 1986, pp. 337-365.

[3] A. Harten, S. Osher, "Uniformly high-order accurate nonoscillatory schemes, I.", *SIAM J. Numer. Anal.*, 24, 1987, pp. 279-309.

[4] H. Deconinck, C. Hirsch, J. Peuteman, "Characteristic Decomposition Methods for the Multidimensional Euler Equations", *Lecture Notes in Physics*, 264, Springer Verlag, 1986, pp. 216-221.

[5] P. L. Roe, "Discrete models for the numerical analysis of time-dependent multidimensional gas dynamics", *J. Comp. Phys.*, 63, 1986, pp. 458-476.

[6] R. Struijs, P. L. Roe, H. Deconinck, "Fluctuation splitting schemes for the 2D Euler equations", *VKI-LS* 1991-01, von Karman Institute, Belgium, 1991.

[7] L. A. Catalano, H. Deconinck, "Two-dimensional optimization of smoothing properties of multi-stage schemes applied to hyperbolic equations", *Multigrid Methods: Special Topics and Applications II*, GMD-Studien Nr. 189, GMD St. Augustin, Germany, 1991, pp. 43-55.

[8] L. A. Catalano, M. Napolitano, H. Deconinck, "Optimal multi-stage schemes for multigrid smoothing of two-dimensional advection operators", submitted to *Communications in Applied Numerical Methods*.

[9] P. De Palma, H. Deconinck, R. Struijs, "Investigation of Roe's 2D wave decomposition models for the Euler equations", *Technical Note* 172, von Karman Institute, Belgium, 1990.

[10] P. L. Roe, "Linear advection schemes on triangular meshes", Cranfield Institute of Technology, CoA Report No. 8720, Cranfield, Bedford, U. K., November 1987.

[11] L. A. Catalano, H. Deconinck, "Two-dimensional optimization of smoothing properties of multistage schemes applied to hyperbolic equations", *Technical Note* 173, von Karman Institute, Belgium, 1990.

[12] A. Brandt, "Guide to multigrid development", *Lecture Notes in Mathematics* 960, Springer Verlag, Berlin, 1982, pp. 220-312.

[13] P. L. Roe, R. Struijs, H. Deconinck, "A conservative linearization of the multidimensional Euler equations", to appear on *J. Comp. Phys.*.

[14] R. Struijs, H. Deconinck, P. De Palma, P. L. Roe, K. G. Powell, "Progress on Multidimensional Upwind Euler Solver for Unstructured Grids", AIAA P 91-1550 CP, Honolulu, 1991.

[15] L. A. Catalano, P. De Palma, G. Pascazio, "A multi-dimensional solution adaptive multigrid solver for the Euler equations", submitted to the 13th International Conference on Numerical Methods in Fluid Dynamics, Rome, July 1992.

TWO-DIMENSIONAL LAGRANGIAN FLOW SIMULATION USING FAST, QUADTREE-BASED ADAPTIVE MULTIGRID SOLVER

C. Gáspár, J. Józsa
Water Resources Research Centre (VITUKI)
Kvassay Jenő út 1, H-1095 Budapest, Hungary

SUMMARY

A new method for solving the unsteady, two-dimensional Navier-Stokes equations of incompressible flow is proposed. The method applies the usual discrete vortex approach. At each time step, the problem is split into a pure advection part, a pure diffusion part as well as a Poisson equation for the stream function. The advection is solved in the usual Lagrangian way by moving the vortex particles along their trajectories. The advection-free subproblems are treated on an unstructured grid generated by the very efficient quadtree algorithm at each time step. The grid generation is controlled by the vortex particles, thus, the grid can follow the density of the particles from time step to time step. Appropriate difference schemes are defined and a multigrid technique is also developed in the quadtree context. As a demonstrative example, an application to the well-known Kelvin-Helmholtz instability of a free shear layer is also presented.

INTRODUCTION

The accurate and at the same time efficient simulation of natural and industrial flow phenomena still represents a challenging task in computational fluid mechanics. In general, the simulation requires the solution of the nonlinear transport of momentum or that of vorticity. As is well known, the transport can usually be divided to advection and molecular or eddy diffusion representing different time and space scales. Moreover, in vortex simulation, besides the vorticity transport, the velocity field has to be updated via a Poisson equation.

There are a lot of unsteady flow problems which contain relatively small, strongly advective evolving regions with rapid change in the flow variables, surrounded by regions with much smoother variation. Let us mention here just three typical cases. First, consider the problem of the numerical prediction of thunderstorms. As pointed out by Lilly [1], to predict the highly advective motion of these atmospheric elements over broad areas it is necessary to use refined resolution but it is enough to confine it to only a few percent of the total region. He mentions the adaptive coordinate streching and the adaptive refined grid windowing as possible solution techniques but seeks much more flexible remedy.

Another example is the evolution of free shear layers known as the Kelvin-Helmholtz instability. In this case two streams with different velocities merge resulting in intensive mixing in the high velocity gradient i.e. high vorticity region. As described e.g. by Ho and Huerre [2] the evolving flow field often shows, more or less coherent, organized two-dimensional character. According to Lesieur [3], in general the presence of an inflexion point in the velocity profile may give birth to transient, quasi two-dimensional vortices seen in flow visualisations. These rolling vortices dominate then the shear layer, which is thus locally characterized by steep gradients and high unsteadiness.

Large scale travelling eddies can be observed also in lakes. Satellite pictures show large horizontal structures and irregularities which are in contrast to the smooth pattern produced by traditional numerical flow models. Analysing the problem, Svensson [4] concludes that these discrete, transient, large eddies may have great influence on lake hydrodynamics and transport. The prediction of their advection, however, needs a numerical model with extremely low numerical diffusion.

In all the above mentioned cases, for realistic simulation, the accurate approximation of the advection is of primary importance. Although a lot of effort have been made to develop Eulerian techniques used fixed space discretisation, in this system the easy handling of the diffusion and the Poisson equation could hardly be extended to that of the advection. To accurately treat the advection, or, more precisely, to eliminate the need to explicitly solve the advective derivatives, the Lagrangian methods seem to be really promising. However, unlike in Eulerian system, here it is the approximation of the diffusion and the Poisson equation that poses numerical difficulties. As the advection is usually represented by scattered points or particles moving along the characteristic lines, the diffusion and the Poisson equation has to be interpreted on an often very irregular and unstructured set of points. In order to overcome this difficulty, a grid-free class of Lagrangian techniques, the so-called *particle methods* (see e.g. [5]) calculate the diffusion and update the velocity field by direct interaction of the computer particles. The required amount of operation is, however, $O(N^2)$ per timestep (N is the number of the particles), which is rather expensive.

Another remedy is the use of the Lagrangian methods which are completely free from conventionally structured grid. The advection-free part of the flow problem is approximated by finite difference or finite volume methods defined on special grids, which are, nevertheless, structured in an unusual way and updated in every timestep. The key to efficiency is to conceive a *rapid grid generation* and then a *fast solution* of the problem *on this irregular grid*. In our paper such a method is presented. The tracking of moving points is coupled with a fast, automatically generated quadtree grid on which the advection-free problem is solved by multigrid techniques. The main features of the method are shown in two-dimensional vortex simulation: the local resolution of the continuously updated grid is controlled by the density of the moving point vortices, initially distributed over the high vorticity regions. It will be seen how naturally this adaptivity suits the character of the Kelvin-Helmholtz instability in planar free shear layers.

GOVERNING EQUATIONS AND SOLUTION TECHNIQUES

Consider the Navier-Stokes equations of the unsteady, incompressible flow:

$$\text{div } \mathbf{u} = 0$$
$$\frac{\partial \mathbf{u}}{\partial t} + (\mathbf{u} \cdot \nabla)\mathbf{u} - \nu \cdot \Delta \mathbf{u} = 0 \quad (1)$$

which is valid in the flow domain Ω and for every t>0. In the important two-dimensional special case the equations can be simplified by introducing the stream function, ψ, and the vorticity function, ω:

$$\mathbf{u} = (\frac{\partial \psi}{\partial y}, -\frac{\partial \psi}{\partial x}), \qquad \omega := \text{rot } \mathbf{u}. \tag{2}$$

This leads to the well-known vorticity-stream function formulation:

$$\frac{\partial \omega}{\partial t} + \mathbf{u} \cdot \text{grad } \omega - \nu \cdot \Delta \omega = 0 \tag{3}$$

$$\Delta \psi = -\omega. \tag{4}$$

Eq.(3) is formally a convective diffusion problem for the function ω, which is, of course, nonlinear because the velocity function $\mathbf{u}$ depends also on ω. Eq.(4) is a simple Poisson equation for every $t>0$.

To avoid the numerical (false) diffusion which is generally appears in solving (3) by a traditional finite difference or finite element technique, it is usual to apply partly or fully Lagrangian methods which eliminate the advective term in the left-hand side of (3). Let us discretize (3)-(4) with respect to the time variable by the equidistant time steps $0=t_0<t_1<\ldots$ After a usual operator splitting procedure, we obtain the following semi-discretized subproblems (τ denotes the length of the time step):

$$\frac{\omega^{n+1/2} - \omega^n}{\tau} + \mathbf{u}^n \cdot \text{grad } \omega^n = 0 \tag{5}$$

$$\frac{\omega^{n+1} - \omega^{n+1/2}}{\tau} - \nu \cdot \Delta \omega^{n+1} = 0 \tag{6}$$

$$\Delta \psi^{n+1} = -\omega^{n+1} \tag{7}$$

$$\mathbf{u}^{n+1} = (\frac{\partial \psi^{n+1}}{\partial y}, -\frac{\partial \psi^{n+1}}{\partial x}). \tag{8}$$

At each time step, (5) is a pure advection problem; (6) is a diffusion problem, while (7) remains a Poisson equation. After solving (5)-(7), the velocity field has to be recomputed by the explicite expression (8).

In the conventional Lagrangian vortex methods, the vorticity field is represented by a set of discrete vortices in the following form:

$$\omega(x,y) = \sum_{k=1}^{N} \omega_k \cdot \delta(x-x_k) \cdot \delta(y-y_k) \tag{9}$$

where δ is the Dirac distribution, (x_k, y_k) is the actual position of the kth vortex, ω_k is its strength. According to the Lagrangian approach, the vortices are transported by the velocity field $\mathbf{u}$, thus, their new position can be obtained by the time integration of the system of ordinary differential equations

$$\frac{d}{dt}(x_k, y_k) = \mathbf{u}(x_k, y_k) \qquad (k=1,\ldots,N) \tag{10}$$

between t_n and t_{n+1} (ω_k remains unchanged).

If the flow is inviscid, the subproblem (6) i.e. the diffusion of the vortices is not present. Otherwise, it can be carried out either by a

Monte-Carlo approach (random walk method, see [6]), or by a deterministic method. In most of the latter methods, the advection-free subproblem (6) is treated in an Eulerian system. Using a fixed grid, some interpolation is required from the Lagrangian system into the Eulerian one and vice versa (which, unfortunately, generates more or less numerical diffusion).

The next and perhaps most important step is to create an efficient solution for the Poisson equation (7). The traditional and most natural way is to use again a fixed Eulerian system. Since (7) must be solved at each time level, only really fast solvers should be applied, such as the Fast Fourier Transform (if applicable) or, rather, multigrid solvers in which the computational cost is proportional to the *first* power of the number of unknowns (gridpoints) only. It should be emphasized, however, that the use of a fixed Eulerian grid is far from being the most economical, since it cannot exploit the special properties of the solution of (7) (namely, that the support of the source term ω is relatively small compared with the flow domain). In the vicinity of the particles, much finer grid would be preferable then far from them, because the stream function ψ exhibits rapid changes close to the particles only: elsewhere the flow is near irrotational, thus, the stream function is much smoother, which could allow much coarser grid in such subregions. Otherwise, the number of the gridpoints becomes unnecessarily large, which makes the solution procedure slow even if a fast solver is applied.

In order to construct a more efficient solver to the subproblem (7), a possible way is to apply *grid-free* techniques e.g. the so-called *multipole method* (see [7],[8]). This method is based on an extremely efficient approximate evaluation of finite sums of logarithmic potentials which appear in the solution of (7). Denoting by N the number of the vortex particles, up to a prescribed accuracy, the computational cost of the multipole method is $O(N)$ only, while the direct summation of these potentials would need $O(N^2)$ operations.

It is also possible, of course, to apply some grid-dependent technique, but in this case the structure of the grid should follow the density of the particles as described above. This can be performed either by a curvilinear grid generation or by a local refinement technique.

We shall apply a different approach, namely, the *unstructured grid generation*. The resulting grid, however, does have a nice structure which is always controlled by the vortex particles. The computational cost of the grid generation is quite moderate, thus, it is not a hard task to repeat the procedure at each time step. The crucial points are as follows:

- to derive proper schemes on the unstructured grid;
- to speed up the solution of the discrete equations as much as possible by defining a multigrid technique which fits the unstructured grid context (see also [9]).

UNSTRUCTURED GRID GENERATION USING THE QUADTREE ALGORITHM CONTROLLED BY THE VORTEX PARTICLES

Now we briefly outline the quadtree-based grid generation technique: for details, see e.g. [10]. First, suppose that a bounded domain Ω and a finite set of points contained in Ω

$$S := \{(x_k, y_k) \in \Omega : k=1, \ldots, N\}$$

are given. For simplicity, assume that Ω is the unit square. The quadtree alghorithm produces a sequence of subsets of Ω and S defined by the following recursive procedure:

1. Check a pre-defined condition list. If at least one of these conditions is fulfilled, the procedure ends. Otherwise, continue from Step 2.

2. Divide Ω into four congruent subsquares $\Omega_1, \ldots, \Omega_4$, and redistribute the points of S among these subsquares. Thus, we obtain the disjoint subsets $S_1, \ldots, S_4$: S_j consists of the points of S which are contained in Ω_j: $S_j := S \cap \Omega_j$ ($j=1, \ldots, 4$).

3. Repeat the procedure for each (Ω_j, S_j), $j=1,2,3,4$ from Step 1.

The condition list in Step 1 controls the subdivision process and has to ensure the finiteness of the algorithm. It should contain the following condition types: the subdivision level has reached a prescribed maximal level; the number of the elements of the actual subset S_k has decreased below a prescribed number etc.

Written in a formal, PASCAL-like language, the main features of the algorithm can easily be seen:

procedure build($\Omega, S, level$);
var finished: boolean;
begin
  checklist($\Omega, S, level, finished$);
  if not finished then begin
    ...{determine $\Omega_1, \Omega_2, \Omega_3, \Omega_4, S_1, S_2, S_3, S_4$}
    build($\Omega_1, S_1, level+1$);
    build($\Omega_2, S_2, level+1$);
    build($\Omega_3, S_3, level+1$);
    build($\Omega_4, S_4, level+1$);
  end;
end;

The procedure generates a subsquare (cell-) system: the cells which are not subdivided further, form the desired computational grid. The cell system can be represented by a tree-like graph in a natural way: the cells correspond to the points of the graph, the branches mean the subdivisions, Ω is the root element. Thus, the grid itself is represented by the *leaves* of the tree.

Remark: It is clear that the algorithm can be defined in arbitrary dimension in a quite similar way (*octrees* in 3D).

In contrast to the structured e.g. elliptic grid generation, the computational cost of the quadtree algorithm is low: it can be esimated by $O(N \cdot \log N)$, or even by $O(N)$ if the maximal subdivision level is bounded, independently of N.

We shall say that two cells are *neighbours* if none of them contains the other but have common boundary point(s). The neighbours are called *corner neighbours* if the common boundary point is a corner only, *face neighbours*, otherwise.

In most of applications it can be necessary to complete the subdivision procedure by extra subdivisions in order to ensure that the maximal ratio of the cell sizes of the face neighbours is at most 2. This can easily be built in the recursive definition of the quadtree grid and makes the definition of the difference schemes much simpler, and minimizes the discretization error, as well. Observe that unlike the uniform grids, the finding of the neighbours is not obvious but can always be performed quickly using the graph representation (see again [8]).

The obtained grid is generally a non-equidistant, non-uniform grid, the density of the cell follows the density of the control point set. Fig. 1 shows a typical quadtree-generated grid generated by 10 points; the maximal number of the points contained in a cell was 1.

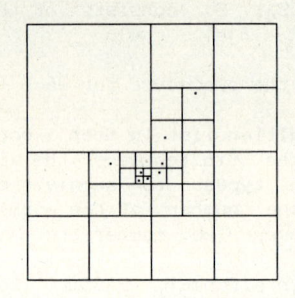

In our present application, the role of the control points can be played partly by the vortex particles. Nevertheless, it is also possible to locate additional (fixed) control points on the subregions which require fine resolution e.g. along the boundary of Ω (especially, if the actual flow domain does not coincide with the unit square).

Fig. 1. A typical quadtree-grid generated by 10 points

DISCRETIZATION AND MULTIGRID SOLUTION

Having obtained the quadtree-generated grid, the next problem is to create proper schemes on it in such a way that it is consistent with the Laplace operator.

The use of a finite element approach is always possible by creating a triangularization over the cell system, but there exist simpler techniques as well. Using finite differences, it is desirable to choose the gridpoints to be either the *corners* of the cells, or the *centers* of them. In the first case the data structure does not fit the grid structure well, but the discretization is quite straightforward. Due to the fact that there are no abrupt changes in cell sizes (the ratio of the adjacent cell sizes is at most 2), the standard - not necessarily central - difference schemes can be used to approximate the second derivatives with respect to x and also to y: the only exceptional case is shown in Fig. 2. Using Taylor series expansion, straightforward calculations show that in this latter case

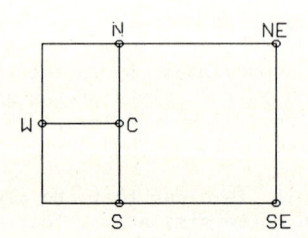

Fig. 2. Adjacent cells with different sizes: the stencil of the difference scheme

$$(\Delta u)_C = \frac{2}{3h^2} \cdot \left[u_N + u_W + u_S + \frac{u_N + u_{NE} + u_S + u_{SE}}{4} - 4 \cdot u_C \right] + O(h) \tag{11}$$

for every sufficiently smooth function u. Should u be harmonic in E, the above scheme becomes of order 2.

In the second case (*cell-centered* schemes) the data structure of the gridpoints coincides exactly with the quadtree graph (i.e. each gridpoint corresponds to exactly one cell), but the construction of the proper schemes as well as the estimation of the discretization error is less obvious. However, thanks to the absence of abrupt changes in cell sizes again, we always have only few different cases.

(a) The central cell is not smaller then its face neighbours (see Fig.3a). The scheme may then be based on the central cell and its face neighbours only. In the case of Fig.3a the following scheme can be derived:

$$(\Delta u)_C = \frac{1}{21h^2} \cdot (20u_N + 20u_S + 24u_E + 16u_{NW} + 16u_{SW} - 96u_C) + O(h). \tag{12}$$

(b) The central cell has a bigger face neighbour (see Fig.3b). Now a corner neighbour is also needed (due to the presence of the second mixed derivative): it is easy to see that

$$(\Delta u)_C = \frac{1}{21h^2} \cdot (8u_E + 6u_{SE} + 18u_N + 18u_W + 16u_S - 66u_C) + O(h). \tag{13}$$

All the other cases can be treated in a similar way.

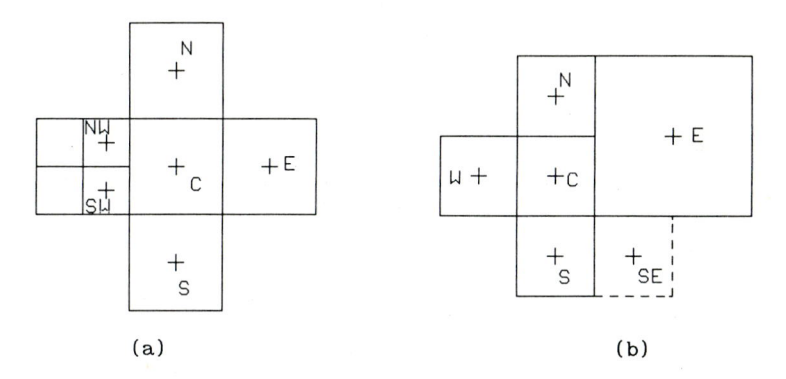

Fig.3. The stencils of cell-centered schemes

Finally, we note that the integrated schemes (*control volume* approach) can be derived in the quadtree context in a comfortable way. Integrating Δu over the central cell, Green's formula implies that

$$\int_C \Delta u \, d\Omega = \int_{\partial C} \frac{\partial u}{\partial n} \, d\Gamma. \tag{14}$$

Thus, the fluxes $\partial u/\partial n$ have to be approximated by using the central values. Based on some interpolation from the neighbouring cell values, it is possible to create simple schemes which are of order 1/2, 3/2 or even higher (with respect to the discrete H^1-norm). For details, see [11].

Remark: The schemes also call for the fast finding of the neighbouring cells. However, using the graph representation, it is a straightforward and not time consuming task as pointed out earlier.

To speed up the solution of the discrete equations, we used a multigrid technique in the quadtree context. Recall that the following essential steps are required (for more details of the multigrids, see e.g. [12]):

1. to define a sequence of nested grids ("fine" and "coarse" grids);
2. to define appropriate restriction and prolongation operators between the adjacent grids;
3. to define a robust smoother (iteration scheme) in each grid, consistent with the original problem (which is able to reduce the high-frequency components of the error).

We define the grid sequence as follows. Let L be the maximal level of subdivision and denote by G_L the quadtree-generated grid describe above. Then G_L will be considered the finest grid. Now let us omit the cells at the level L and replace them by their "parents" at the level L-1, that is, let us omit the highest-level subdivision procedure. Let the coarser grid G_{L-1} consist of the obtained cells which are not subdivided, and so on. It is clear that the representing graph of G_{L-1} can be obtained from that of G_L by cutting all the subgraphs starting from the level L-1 etc. Thus, we have a grid sequence G_L, $G_{L-1},\ldots,G_1$, each of them is coarser than the previous one.

Remark: It is obvious that each grid G_k has the property that the ratio of the adjacent cell sizes is at most 2. Thus, the derivation of the difference schemes is quite similar on each grid. It is also clear that during performing the quadtree algorithm, we obtain all the cells contained in all grids: that is, in order to construct the grid sequence, no additional computational work is required. Finally, it should be pointed out that in contrast to the conventional multigrid techniques, each grid contains cells *with different cell sizes*, in general.

The restriction operator can be defined by a simple averaging technique. If a cell of the coarser grid is contained also in the finer grid, the associated value remains unchanged: otherwise, it is defined as the average of the values corresponding to the subcells contained in it. The simplest prolongation operator is the piecewise constant prolongation; the value attached to the parent cell is transferred to the contained subcells without modification. Weighting methods can also be applied. As a smoothing procedure, the Seidel iteration generated by the above defined schemes can be used. Now all the usual tricks of the multigrid technique can be implemented without difficulty, such as coarse grid correction, full multigrid cycle etc.

Remark: As pointed out earlier, if a fixed grid is used in solving the diffusion subproblem (6), the interpolation procedures between the Lagrangian and Eulerian systems can generate some numerical diffusion. This can be minimized (or even eliminated) by using moving grids (i.e. Lagrangian approach also for the diffusion subproblem): unfortunately, the moving grids are necessarily be distorted sooner or later, which would require again an interpolation into a regular grid. A good compromise is to use the same quadtree-generated non-uniform grid applied in solving the Poisson equation (7). For applying the unstructured grids to convective diffusion problem, see [13].

APPLICATION TO THE KELVIN-HELMHOLTZ INSTABILITY PROBLEM

Finally, we illustrate the proposed method by applying it to the well-known Kelvin-Helmholtz instability problem. We have considered an initially horizontal planar flow in the unit square which has different velocities in the upper part (y>0.5) and the lower part (y<0.5). We assumed the velocity to be zero in the lower part. The velocity shear implies a vorticity distribution which is concentrated on the interface line y=0.5. This interface is unstable and rolls up in time in a characteristic way (see [14]).

In treating this problem, we applied the above-described Lagrangian method. using initially 800 particles. The diffusion subproblem (6) was simulated by random walk. The Poisson equation (7) was supplied with Dirichlet boundary conditions. At each time step we generated the cell system by the quadtree algorithm using max. 6 subdivision levels. We used cell-centered schemes and Seidel iteration as a smoother in the multigrid procedure. In Fig.4 the actual grid and the distribution of the vortex particles can be seen in three different stages showing the time evolution of the free shear layer.

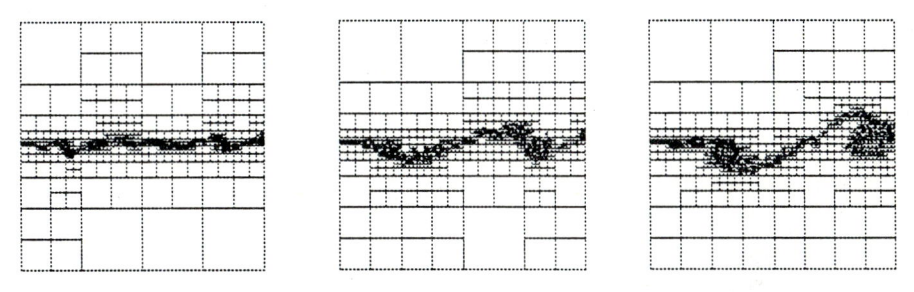

Fig.4. Time evolution of the free shar layer after 0.04, 0.08 and 0.11 time units. Upper velocity: 10, lower velocity: 0, diffusion coefficient: 10^{-4} (in arbitrary units).

REFERENCES

[1] Lilly,D.K.: Numerical prediction of thunderstorms - has its time come? Q. J. R. Meteorol. Soc. **116**, 779-798 (1990).

[2] Ho,C.H., Huerre,P.: Perturbed free shear layers. Ann. Rev. Fluid Mech. **16**, 365-424 (1984).

[3] Lesieur,M.: Structures cohérentes et turbulence en écoulement libre. Derniers progres dans la connaissance en turbulence, La Houille Blanche, No. 7-8: 569-572 (1987).

[4] Svensson,U.: Turbulence lakes. In: Physical processes in lakes (Ed. by J.Virta), 5-11, NHP-Report No. 16 (1986).

[5] Raviart,P.A.: Particle numerical models in fluid dynamics. In: Numerical methods for fluid dynamics II. (Ed. by K.W.Morton, M.G.Baines), 231-253 (1986).

[6] Kinzelbach,W.: Groundwater modelling. Elsevier, Amsterdam, 1986.

[7] Carrier,J., Greengard,L., Rokhlin,V.: A Fast Adaptive Multipole Algorithm for Particle Simulations. SIAM J. Sci. Stat. Comput. Vol. 9 No. 4, pp 669-686, July, 1988.

[8] van Dommelen,L. Rundensteiner,E.A.: Fast, Adaptive Summation of Points Forces in the Two-Dimensional Poisson Equation. J. of Comput. Physics **83**, 126-147 (1989)

[9] Gáspár,C., Józsa,J: A Coupled Lagrangian Particle Tracking and Quadtree-based Adaptive Multigrid Method with Application to Shear Layer Evolution. Proc. of the XXIV. IAHR Congress, 9-13 Sept 1991, Madrid, Spain.

[10] Cheng,J.H., Finnigan,P.M., Hathaway,A.F., Kela,A., Schroeder,W,J.: Quadtree/Octree Meshing with Adaptive Analysis. In: Numerical Grid Generation in Computational Fluid Mechanics '88 (Ed. by S.Sengupta, J.Hauser, P.R.Eiseman, J.F.Thompson), 633-642. Pineridge Press, Swansea, UK, 1988.

[11] Ewing,R.E., Lazarov,R.D., Vassilevski,P.S.: Local Refinement Techniques for Elliptic Problems on Cell-Centered Grids. I. Error Anlysis. Math. Comp. **56**, 437-462 (1991).

[12] Hackbusch,W.: Multi-Grid Methods and Applications. Springer-Verlag, Berlin, Heidelberg, New York, Tokyo, 1985.

[13] Gáspár,C., Józsa,J., Simbierowicz,P.: Lagrangian Modelling of the Convective Diffusion Problem Using Unstructured Grids and Multigrid Technique. Proc. of the First Int. Conf. on Water Pollution, 3-5 Sept 1991, Southampton, UK.

[14] Baker,G.R.: The "Cloud in Cell" Technique Applied to th Roll Up of Vortex Sheets. J. of Comput. Physics **31**, 76-95 (1979).

Multi-D Upwinding and Multigridding for Steady Euler Flow Computations

B. Koren, P.W. Hemker

Center for Mathematics and Computer Science
P.O. Box 4079, 1009 AB Amsterdam, The Netherlands

Abstract

Multi-dimensional upwind discretizations for the steady Euler equations are studied, with the emphasis on both a good accuracy and a good efficiency. The discretizations consist of a one-dimensional Riemann solver with locally rotated left and right cell face states, the rotation angle depending on the local flow solution. First, on the basis of a linear, scalar model equation, a study is made of the accuracy and stability properties of these schemes. Next the extension is made to the steady Euler equations. It is shown that for Euler flows, an appropriate local rotation angle can be found by maximizing a Riemann invariant along the middle subpath of the wave path in state space. For the steady, two-dimensional Euler equations, numerical results are presented for some supersonic test cases with either oblique contact discontinuity or oblique shock wave.

Note: This work was supported by the European Space Agency (ESA), through Avions Marcel Dassault - Bréguet Aviation (AMD-BA).

1. Introduction

Many upwind schemes used in multi-dimensional (multi-D) flow computations are based on the application of some one-dimensional (1-D) shock capturing scheme in a grid-aligned manner. Despite the rigorous mathematics involved in these 1-D upwind schemes, in most multi-D flow computations, the underlying 1-D upwind results are just superposed without rigorous mathematical justification. Besides this inconsistency in methodology, the grid-alignment (i.e. grid-dependency) is also inconsistent with the upwind principle that discretizations should be dependent on the solution only. It seems most natural nowadays to no longer ignore the multi-D nature of a multi-D flow in the upwind scheme itself. Various grid-decoupled multi-D upwind discretizations have been published already and many more are still in development, a development which is at a rapid pace. The emphasis in most of this research clearly lies on a good accuracy.

In the present paper, for the steady Euler equations in a cell-centered finite volume context we present: multi-D upwind methods with some optimal balance between accuracy and efficiency. The steady equations are solved directly (so not through an unsteady form). For good efficiency we rely on nonlinear multigrid (multigrid-Newton) iteration. As the smoothing technique in the multigrid iteration we apply point Gauss-Seidel relaxation, using the exact derivative matrices (exact Newton). The latter requires the cell face fluxes to be continuously differentiable. If for a multi-D scheme the multigrid solver does not meet our standards we rely on defect correction iteration as the most outer iteration, just as in e.g. [1]. The multi-D upwind schemes to be considered here are very simple schemes. They use neither decoupling of the Euler equations as in [2,3], nor rotated fluxes as in [4,5]. The schemes are based on *rotated left and right cell face states solely*. Per cell face, just as with grid-aligned 1-D upwind schemes, only a single numerical flux is computed: the one normal to the cell face. The only difference between grid-aligned 1-D upwind schemes and the present multi-D upwind schemes is that whereas in the first schemes the left and right cell face states are computed from a *solution-independent, 1-D* subset of the local multi-D solution, in the present multi-D upwind schemes, these states are computed from a *solution-dependent, multi-D* subset. The numerical flux function to be applied should allow a good resolution of both oblique shock waves and oblique contact discontinuities, which makes flux difference splitting schemes to be preferred above flux splitting schemes. Given the good experience with Osher's scheme [6] in combination with nonlinear multigrid [7], we apply this flux difference splitting scheme.

2. First-order accurate upwind schemes for a model equation

For sake of simplicity, in the present paper we still restrict ourselves to first-order accurate, multi-D upwind schemes. We make an analysis of such schemes for the linear, scalar, 2-D model equation

$$a\frac{\partial u}{\partial x} + b\frac{\partial u}{\partial y} = 0, \quad 0 \leq \theta \equiv atan\left(\frac{b}{a}\right) \leq \frac{\pi}{2}, \tag{1}$$

where θ is the angle between the characteristic direction and the x-axis. Discretization of the model equation on a square, cell-centered finite volume grid yields

$$a(u_{i+\frac{1}{2},j} - u_{i-\frac{1}{2},j}) + b(u_{i,j+\frac{1}{2}} - u_{i,j-\frac{1}{2}}) = 0, \tag{2}$$

where the half-integer indices refer to the cell faces between the (full-integer indexed) cell centers. Taking into account the positive signs of a and b, various first-order upwind choices

$$\begin{pmatrix} u_{i+\frac{1}{2},j} \\ u_{i,j+\frac{1}{2}} \end{pmatrix} = \begin{pmatrix} L_h(u_{i,j}, u_{i,j-1}, u_{i+1,j-1}) \\ L_v(u_{i,j}, u_{i-1,j}, u_{i-1,j+1}) \end{pmatrix}, \quad 0 \leq \theta \leq \frac{\pi}{2}, \tag{3}$$

can be made, which (with similar choices for $u_{i-\frac{1}{2},j}$ and $u_{i,j-\frac{1}{2}}$) lead to the general 6-point compact stencil

$$\begin{bmatrix} -\alpha_{i-1,j+1} & \cdot & \cdot \\ -\alpha_{i-1,j} & \alpha_{i,j} & \cdot \\ -\alpha_{i-1,j-1} & -\alpha_{i,j-1} & -\alpha_{i+1,j-1} \end{bmatrix}, \quad 0 \leq \theta \leq \frac{\pi}{2}. \tag{4}$$

A general stability requirement imposed to these schemes is that they satisfy the positive coefficients rule [8]. (Schemes satisfying this rule do not allow unstable, oscillatory solutions.)

Definition 1
Consider the general stencil (4). In this paper, schemes satisfying the positive coefficients rule

$$\alpha_{i,j} \geq 0, \quad \alpha_{i\pm m, j\pm n} \geq 0, \quad \forall m,n \in \mathbb{N}, \tag{5}$$

are called *positive*.

Applying truncated Taylor series expansions to the discrete equation corresponding to (4) and transforming to characteristic coordinates, one derives a modified equation of the general form

$$\frac{\partial u}{\partial s} - \frac{h}{2}\left[\mu_{ss}\frac{\partial^2 u}{\partial s^2} + \mu_{sn}\frac{\partial^2 u}{\partial s\partial n} + \mu_{nn}\frac{\partial^2 u}{\partial n^2}\right] = O(h^2). \tag{6}$$

In the present paper, we prefer multi-D upwind schemes with both $\mu_{sn} = 0$ and $\mu_{nn} = 0$; discretizations which guarantee a low crosswind diffusion for $\frac{\partial u}{\partial s} = r(s,n)$, with $r(s,n)$ arbitrary.

Definition 2
Consider the general modified equation (6). In this paper, $(\mu_{sn}, \mu_{nn})^T$ is called the *crosswind diffusion* and schemes for which $(\mu_{sn}, \mu_{nn})^T = 0$ are called *zero-crosswind diffusion schemes*.

2.1. The standard, grid-aligned scheme
With the standard, first-order accurate, grid-aligned 1-D upwind scheme, given the positive signs of a and b, for the cell face states occurring in (2) one takes

$$\begin{pmatrix} u_{i+\frac{1}{2},j} \\ u_{i,j+\frac{1}{2}} \end{pmatrix} = \begin{pmatrix} u_{i,j} \\ u_{i,j} \end{pmatrix}, \quad 0 \leq \theta \leq \frac{\pi}{2}, \tag{7}$$

with similar choices for $u_{i-\frac{1}{2},j}$ and $u_{i,j-\frac{1}{2}}$. The scheme is positive, which clearly appears from its stencil

$$\begin{bmatrix} \cdot & \cdot & \cdot \\ -a & a+b & \cdot \\ \cdot & -b & \cdot \end{bmatrix}, \quad 0 \leq \theta \leq \frac{\pi}{2}. \tag{8}$$

Further it is continuously differentiable, which allows the application of Newton iteration. However, applying the modified equation approach, one finds the poor accuracy properties given in Fig. 1a. It appears that zero-crosswind diffusion occurs only in case of $\theta = 0$ or $\theta = \frac{\pi}{2}$, i.e. in case of grid-alignment of the characteristic direction. The weak spot of scheme (7) is of course that it only uses two off-diagonal points. In constructing multi-D upwind schemes it is self-evident to add at least $u_{i-1,j-1}$ to the scheme. In the following we give two first-order accurate, multi-D upwind schemes in which this has been done.

2.2. A positive, continuously differentiable scheme
The present scheme is derived in [9]. (In another way and in another context, it is also derived in [10].) For the cell face states it takes:

$$\begin{pmatrix} u_{i+\frac{1}{2},j} \\ u_{i,j+\frac{1}{2}} \end{pmatrix} = \frac{1}{a+b} \begin{pmatrix} (a+\frac{1}{2}b)u_{i,j} + \frac{1}{2}bu_{i,j-1} \\ (b+\frac{1}{2}a)u_{i,j} + \frac{1}{2}au_{i-1,j} \end{pmatrix}, \quad 0 \leq \theta \leq \frac{\pi}{2}, \tag{9}$$

leading to the stencil

$$\frac{1}{a+b} \begin{bmatrix} \cdot & \cdot & \cdot \\ -a^2 & a^2+ab+b^2 & \cdot \\ -ab & -b^2 & \cdot \end{bmatrix}, \quad 0 \leq \theta \leq \frac{\pi}{2}, \tag{10}$$

and the diffusion coefficients given in Fig. 1b. The scheme has been constructed such that $\mu_{sn} = 0$ over the complete range of θ considered. Notice that compared to scheme (7), also μ_{nn} is smaller. Moreover, in [9] it is shown that this scheme still allows a more efficient smoothing of point Gauss-Seidel relaxation than scheme (7). All this makes that it is a more appropriate candidate for our multigrid purposes than scheme (7).

2.3. A zero-crosswind diffusion scheme
Following the modified equation approach, in [9] the derivation is also given of the 6-point compact scheme

$$\begin{pmatrix} u_{i+\frac{1}{2},j} \\ u_{i,j+\frac{1}{2}} \end{pmatrix} = \begin{pmatrix} u_{i,j} \\ \frac{1}{2}(1+\frac{b}{a})u_{i-1,j} + \frac{1}{2}(1-\frac{b}{a})u_{i-1,j+1} \end{pmatrix}, \quad 0 \leq \theta \leq \frac{\pi}{4}, \tag{11a}$$

$$\begin{pmatrix} u_{i+\frac{1}{2},j} \\ u_{i,j+\frac{1}{2}} \end{pmatrix} = \begin{pmatrix} \frac{1}{2}(1+\frac{a}{b})u_{i,j-1} + \frac{1}{2}(1-\frac{a}{b})u_{i+1,j-1} \\ u_{i,j} \end{pmatrix}, \quad \frac{\pi}{4} \leq \theta \leq \frac{\pi}{2}, \tag{11b}$$

leading to the stencils

$$\begin{bmatrix} \frac{1}{2}b(1-\frac{b}{a}) & \cdot & \cdot \\ b(\frac{b}{a}-\frac{a}{b}) & a & \cdot \\ -\frac{1}{2}b(1+\frac{b}{a}) & \cdot & \cdot \end{bmatrix}, \quad 0 \leq \theta \leq \frac{\pi}{4}, \tag{12a}$$

$$\begin{bmatrix} \cdot & \cdot & \cdot \\ \cdot & b & \cdot \\ -\frac{1}{2}a(1+\frac{a}{b}) & a(\frac{a}{b}-\frac{b}{a}) & \frac{1}{2}a(1-\frac{a}{b}) \end{bmatrix}, \quad \frac{\pi}{4} \leq \theta \leq \frac{\pi}{2}, \tag{12b}$$

and the diffusion coefficients given in Fig. 1c. This scheme has been constructed such that it has $(\mu_{sn}, \mu_{nn})^T = 0$ over the complete range of θ considered. Concerning the solution of the corresponding discretized equations, due their non-positivity, no such efficient smoother exists as for the equations belonging to scheme (9). To solve the present equations we rely on defect correction iteration with as 'working horse' scheme in the inner multigrid iteration: scheme (9). In [9] it is shown that this is a satisfactory approach.

3. The foregoing schemes for the Euler equations
We proceed by extending the foregoing model schemes to the steady, 2-D Euler equations. The extension is straightforward. Because in the model equation the characteristic information is coming from the left, for the Eulerian numerical flux function the components of the left cell face states are computed in the same way as the cell face states for the model equation. For the right cell face states, we simply take the *point symmetric* counterpart of the left states. In this way, in case of all characteristic information coming from the right (supersonic flow from the right), one also has the proper discretization. Also notice that in this way an n-point scheme for the model equation becomes a $(2 \times n - 1)$-point scheme for the Euler equations. In mathematical formulae, the procedure reads as follows. Consider as general relation for the left cell face states to be substituted into the numerical flux function:

$$\begin{pmatrix} q_{i+\frac{1}{2},j} \\ q_{i,j+\frac{1}{2}} \end{pmatrix}^l = \begin{pmatrix} L_h(q_{i,j}, q_{i,j-1}, q_{i+1,j-1}) \\ L_v(q_{i,j}, q_{i-1,j}, q_{i-1,j+1}) \end{pmatrix}, \quad 0 \leq \theta \leq \frac{\pi}{2}, \tag{13}$$

then the right cell face states simply obey:

$$\begin{pmatrix} q_{i+\frac{1}{2},j} \\ q_{i,j+\frac{1}{2}} \end{pmatrix}^r = \begin{pmatrix} L_h(q_{i+1,j}, q_{i+1,j+1}, q_{i,j+1}) \\ L_v(q_{i,j+1}, q_{i+1,j+1}, q_{i+1,j}) \end{pmatrix}, \quad 0 \le \theta \le \frac{\pi}{2}. \tag{14}$$

3.1. The standard, grid-aligned scheme

For sake of completeness we also make the extension for scheme (7). For this scheme we get the known result

$$\begin{pmatrix} q_{i+\frac{1}{2},j} \\ q_{i,j+\frac{1}{2}} \end{pmatrix}^l = \begin{pmatrix} q_{i,j} \\ q_{i,j} \end{pmatrix}, \quad 0 \le \theta \le \frac{\pi}{2}, \tag{15}$$

$$\begin{pmatrix} q_{i+\frac{1}{2},j} \\ q_{i,j+\frac{1}{2}} \end{pmatrix}^r = \begin{pmatrix} q_{i+1,j} \\ q_{i,j+1} \end{pmatrix}, \quad 0 \le \theta \le \frac{\pi}{2}. \tag{16}$$

3.2. The positive, continuously differentiable scheme

Positive, continuously differentiable scheme (9) as derived for model equation (1), is applied as

$$\begin{pmatrix} q_{i+\frac{1}{2},j} \\ q_{i,j+\frac{1}{2}} \end{pmatrix}^l = \frac{1}{1+\tan\theta} \begin{pmatrix} (1+\frac{1}{2}\tan\theta)q_{i,j} + \frac{1}{2}\tan\theta q_{i,j-1} \\ (\frac{1}{2}+\tan\theta)q_{i,j} + \frac{1}{2}q_{i-1,j} \end{pmatrix}, \quad 0 \le \theta \le \frac{\pi}{2}, \tag{17}$$

$$\begin{pmatrix} q_{i+\frac{1}{2},j} \\ q_{i,j+\frac{1}{2}} \end{pmatrix}^r = \frac{1}{1+\tan\theta} \begin{pmatrix} (1+\frac{1}{2}\tan\theta)q_{i+1,j} + \frac{1}{2}\tan\theta q_{i+1,j+1} \\ (\frac{1}{2}+\tan\theta)q_{i,j+1} + \frac{1}{2}q_{i+1,j+1} \end{pmatrix}, \quad 0 \le \theta \le \frac{\pi}{2}. \tag{18}$$

3.3. The zero-crosswind diffusion scheme

Similarly, zero-crosswind diffusion scheme (11) is applied as

$$\begin{pmatrix} q_{i+\frac{1}{2},j} \\ q_{i,j+\frac{1}{2}} \end{pmatrix}^l = \begin{pmatrix} q_{i,j} \\ \frac{1}{2}(1+\tan\theta)q_{i-1,j} + \frac{1}{2}(1-\tan\theta)q_{i-1,j+1} \end{pmatrix}, \quad 0 \le \theta \le \frac{\pi}{4}, \tag{19a}$$

$$\begin{pmatrix} q_{i+\frac{1}{2},j} \\ q_{i,j+\frac{1}{2}} \end{pmatrix}^l = \begin{pmatrix} \frac{1}{2}(1+\frac{1}{\tan\theta})q_{i,j-1} + \frac{1}{2}(1-\frac{1}{\tan\theta})q_{i+1,j-1} \\ q_{i,j} \end{pmatrix}, \quad \frac{\pi}{4} \le \theta \le \frac{\pi}{2}, \tag{19b}$$

and

$$\begin{pmatrix} q_{i+\frac{1}{2},j} \\ q_{i,j+\frac{1}{2}} \end{pmatrix}^r = \begin{pmatrix} q_{i+1,j} \\ \frac{1}{2}(1+\tan\theta)q_{i+1,j+1} + \frac{1}{2}(1-\tan\theta)q_{i+1,j} \end{pmatrix}, \quad 0 \le \theta \le \frac{\pi}{4}, \tag{20a}$$

$$\begin{pmatrix} q_{i+\frac{1}{2},j} \\ q_{i,j+\frac{1}{2}} \end{pmatrix}^r = \begin{pmatrix} \frac{1}{2}(1+\frac{1}{\tan\theta})q_{i+1,j+1} + \frac{1}{2}(1-\frac{1}{\tan\theta})q_{i,j+1} \\ q_{i,j+1} \end{pmatrix}, \quad \frac{\pi}{4} \le \theta \le \frac{\pi}{2}. \tag{20b}$$

4. Rotation angle for the Euler equations

Per cell face we have to select a single rotation angle from the local, multi-D solution. Here we present a technique which looks at all cell faces at either the local flow angle or a local shock wave angle. The technique considers a wave path in state space. As the wave path we consider the one of the P-variant of Osher's scheme [7]. For the steady 2-D Euler equations and a perfect gas, with $c \equiv \sqrt{\gamma \frac{p}{\rho}}$ and $z \equiv \ln\left(\frac{p}{\rho^\gamma}\right)$, the P-variant's wave path is shown in Fig. 2. For the determination of the angle θ, the wave path states $q_0 \equiv (u_0, v_0, c_0, z_0)^T$ and $q_1 \equiv (u_1, v_1, c_1, z_1)^T$ are taken dependent on θ as

$$q_0(\theta) = T(\theta)q^l, \quad q_1(\theta) = T(\theta)q^r, \tag{21a}$$

with

$$T(\theta) = \begin{pmatrix} \cos\theta & \sin\theta & 0 & 0 \\ -\sin\theta & \cos\theta & 0 & 0 \\ 0 & 0 & 1 & 0 \\ 0 & 0 & 0 & 1 \end{pmatrix}, \tag{21b}$$

and with the states q^l and q^r given. At the cell faces $i+\frac{1}{2}, j$ and $i, j+\frac{1}{2}$, for q^l and q^r we take e.g.

$$\begin{pmatrix} q_{i+\frac{1}{2},j} \\ q_{i,j+\frac{1}{2}} \end{pmatrix}^l = \begin{pmatrix} q_{i,j} \\ q_{i,j} \end{pmatrix}, \tag{22a}$$

$$\begin{pmatrix} q_{i+\frac{1}{2},j} \\ q_{i,j+\frac{1}{2}} \end{pmatrix}^r = \begin{pmatrix} q_{i+1,j} \\ q_{i,j+1} \end{pmatrix}. \tag{22b}$$

Then, a suitable rotation angle can be found by maximizing one of the two Riemann invariants along the middle subpath of the P-variant's wave path in state space; i.e. either the invariant velocity component

$$u_{\frac{1}{2}}(\theta) = \frac{1}{1+\alpha}\left[\left(u_1(\theta) - \frac{2}{\gamma-1}c_1\right) + \alpha\left(u_0(\theta) + \frac{2}{\gamma-1}c_0\right)\right], \tag{23a}$$

or the invariant pressure

$$p_{\frac{1}{2}}(\theta) = \left[\frac{\gamma-1}{2(1+\alpha)\sqrt{\gamma e^{\frac{z_0}{\gamma}}}}\left(u_0(\theta) - u_1(\theta) + \frac{2}{\gamma-1}(c_0+c_1)\right)\right]^{\frac{2\gamma}{\gamma-1}}, \tag{23b}$$

where in both: $\alpha \equiv e^{\frac{z_1-z_0}{2\gamma}}$.

Theorem 1
The orientation of a contact discontinuity follows from the maximization of the velocity component $u_{\frac{1}{2}}(\theta)$ along the P-variant's wave path in state space.

Proof
From (23a), it follows with (21a)-(21b):

$$\frac{du_{\frac{1}{2}}}{d\theta} = \frac{1}{1+\alpha}\left[-\sin\theta\, u^r + \cos\theta\, v^r + \alpha(-\sin\theta\, u^l + \cos\theta\, v^l)\right], \tag{24}$$

from which it is found that $u_{\frac{1}{2}}$ is maximal for

$$\tan\theta = \frac{\alpha v^l + v^r}{\alpha u^l + u^r}, \tag{25}$$

which is a biased relation for the orientation of a contact discontinuity. □

Remark 1
For the isentropic case, $\alpha = 1$, (25) simplifies to the known centered relation

$$\tan\theta = \frac{v^l + v^r}{u^l + u^r}. \tag{26}$$

Remark 2
The physical meaning of (25) for $\alpha \neq 1$ can be explained for the isobaric case $p^l = p^r$. For this case, (25) simplifies to the (still biased) relation

$$\tan\theta = \frac{\sqrt{\rho^l}v^l + \sqrt{\rho^r}v^r}{\sqrt{\rho^l}u^l + \sqrt{\rho^r}u^r}. \tag{27}$$

From (27) it follows that of the two states q^l and q^r, the state with the higher density has a stronger weight in the determination of the angle θ. To our opinion this is physically more proper than the absence of any such weight in the commonly used centered relation (26).

Theorem 2
The orientation of the normal at a shock wave follows from the maximization of the pressure $p_{\frac{1}{2}}(\theta)$ along the P-variant's wave path in state space.

Proof
From (23b), it follows with (21a)-(21b):

$$\frac{dp_{\frac{1}{2}}}{d\theta} = \frac{1}{1+\alpha}\sqrt{\frac{\gamma p_{\frac{1}{2}}^{\frac{\gamma+1}{\gamma}}}{e^{\frac{s_0}{\gamma}}}}\left(-\sin\theta(u^l - u^r) + \cos\theta(v^l - v^r)\right), \tag{28}$$

from which it is found that $p_{\frac{1}{2}}$ is maximal for

$$\tan\theta_n = \frac{v^l - v^r}{u^l - u^r}. \tag{29}$$

This is a relation for the orientation of the normal at a shock wave, because - given the θ-independence of p_0 and p_1 - the maximization of $p_{\frac{1}{2}}$ is identical to the maximization of either $p_{\frac{1}{2}} - p_0$ or $p_{\frac{1}{2}} - p_1$; the underlying, natural quantities for finding the orientation of the normal at a shock wave. □

Corollary
The shock wave angle itself, i.e. the rotation angle θ, satisfies

$$\tan\theta = \frac{u^l - u^r}{v^r - v^l}. \tag{30}$$

Remark 3
In contrast with (25), the result (30) is known. It directly follows from the jump relation which states that the tangential velocity components at the up- and downstream side of a shock wave are equal.

Remark 4
As opposed to (25), (30) contains differences, which makes it sensitive to noise and non-uniqueness. In principle, as a remedy against this, a blended formula like that proposed in [5], can also be constructed on the basis of (25) and (30).

Remark 5
Notice that by taking the shock wave angle as rotation angle, the upwinding is not done normal to the shock wave, cf. e.g. Davis [4], but - instead - along the shock wave; i.e. along the (merged) characteristics.

5. Numerical results

For the steady, 2-D Euler equations and a perfect gas with $\gamma = 1.4$, numerical experiments are performed for some supersonic, unit square flows with either oblique contact discontinuity or oblique shock wave. First, flows with contact discontinuity are considered for the flow angles $\theta = 0.1\pi, 0.2\pi, 0.3\pi$ and 0.4π (Fig. 3a). Next, flows with shock wave are considered for the shock wave angles $\theta = \frac{\pi}{4}$ and $\theta = \frac{\pi}{8}$ (Fig. 3b). All these flows are computed on a uniform 32×32-grid. In all cases - for simplicity - at each of the four boundaries, the exact solution is imposed (overspecification). Further, in all cases, standard nonlinear multigrid method (FAS) is applied; with a 2×2-grid as the coarsest grid, with V-cycles and with a single pre- and post-relaxation sweep per level. We remark that with scheme (17)-(18) to be locally linearized in the inner multigrid iteration, one has 4×4 derivative matrices containing contributions which originate from the solution-dependent rotation angle. In all cases we take as the initial solution: the solution with $q = q^L$ (the exact q^L's from Figs. 3a and 3b) uniformly constant over the complete domain.

5.1. Flows with contact discontinuity

First, in Fig. 4, reference results are given for the present four test cases; results obtained by the first-order, grid-aligned 1-D upwind scheme (15)-(16). In Fig. 4a, we plotted on top of each other: the enthalpy $(e + \frac{p}{\rho})$ distributions for $\theta = 0.1\pi, 0.2\pi, 0.3\pi$ and 0.4π. The iso-enthalpy values considered in these and all following enthalpy distributions are: $1.1, 1.2, 1.3, \ldots, 1.9$. Because of the severe smearing of the scheme, hardly any distinction can be made between the four solutions. (Notice that the layers along $x = 1$ and $y = 1$ in Fig. 4a, and also in the following enthalpy graphs, are only due to the overspecification.) The convergence histories belonging to the standard, grid-aligned 1-D upwind scheme, are given in Fig. 4b.

In Fig. 5, results are given as obtained by multi-D scheme (17)-(18), with as the rotation angle: the local flow angle according to (25). Though more accurate than the grid-aligned reference distributions in

Fig. 4a, the present enthalpy distributions (Fig. 5a) are still insufficiently accurate. Though not as very fast as the reference convergence in Fig. 4b, the present scheme's multigrid convergence (Fig. 5b) is still very good.

In Fig. 6a we give the enthalpy distributions for zero-crosswind diffusion scheme (19)-(20), as obtained after 10 defect correction cycles (with a single nonlinear multigrid cycle per defect correction cycle), and with also (25) for the rotation angle considered at each cell face. All these four enthalpy distributions appear to be almost free of crosswind diffusion. Although in principle the non-positivity of the scheme allows solutions with spurious oscillations, the distributions in Fig. 6a are still monotone. For comparison, in Fig. 6b we still give the distributions obtained by a higher-order accurate, grid-aligned 1-D upwind scheme: the $\kappa = \frac{1}{3}$-scheme [11]. All four enthalpy distributions of first-order accurate scheme (19)-(20) appear to be even less diffused than those of the higher-order accurate $\kappa = \frac{1}{3}$-scheme !

5.2. Flows with shock wave

Reference results obtained by the first-order, grid-aligned 1-D upwind scheme (15)-(16) (Mach number distributions) are given in Fig. 7a. Here solutions are also plotted on top of each other. The iso-Mach number values shown are: (i) $1.50, 1.55, 1.60, \ldots, 1.95$ for the case with $\theta = \frac{\pi}{4}, M^L = 2$, and (ii) $3.30, 3.35, 3.40, \ldots, 3.95$ for the case with $\theta = \frac{\pi}{8}, M^L = 4$. (Similar to the flows with contact discontinuity, the layers along $x = 1$ and $y = 1$ are caused by the overspecification.) In Fig. 7b we give the Mach number distributions as obtained after two defect correction cycles with zero-crosswind diffusion scheme (19)-(20). The rotation angle considered here is the shock wave angle according to (30). After two defect correction cycles, the solution seems to be free of crosswind diffusion, but for $\theta = \frac{\pi}{4}$ it has become non-monotone. Construction of a compact multi-D limiter might be useful. For further comparisons, in Fig. 7c we still give the distributions obtained by the limited $\kappa = \frac{1}{3}$-scheme from [12]. The monotone solution as obtained for $\theta = \frac{\pi}{8}$ by the first-order accurate zero-crosswind diffusion scheme appears to be less diffused than that of the higher-order accurate $\kappa = \frac{1}{3}$-scheme.

6. Conclusions

Multi-D upwinding through a 1-D Riemann solver with a local, solution-dependent rotation of the left and right Riemann states, allows to keep the number of flux computations per cell face equal to one only. Good efficiency is further guaranteed through nonlinear multigrid iteration and defect correction iteration. The accuracy and efficiency of the multi-D results are promising. For flows with contact discontinuities, the performance of nonlinear multigrid with point Gauss-Seidel relaxation is (still) good when one applies the positive, continuously differentiable, 7-point compact scheme. Also for flows with contact discontinuities, the solutions obtained by the zero-crosswind diffusion, 9-point compact scheme appear to be nearly free of any crosswind diffusion. Moreover, their computation by means of defect correction iteration (with the positive, continuously differentiable, 7-point compact scheme as the approximate scheme) is efficient. The zero-crosswind diffusion scheme seems to be well-suited for an accurate and efficient computation of e.g. vortex flows.

The numerical techniques presented in this paper do not require any tuning of parameters. Further, they can be carried over to 3-D and extended to non-Cartesian grids.

References

1. HEMKER, P.W.: Defect correction and higher order schemes for the multi grid solution of the steady Euler equations, *Lecture Notes in Mathematics*, **1228** (Springer, Berlin, 1986) pp. 149-165.

2. HIRSCH, CH., LACOR, C., DECONINCK, H.: Convection algorithms based on a diagonalization procedure for the multidimensional Euler equations, *AIAA paper 87-1163* (1987).

3. ROE, P.L.: Discrete models for the numerical analysis of time-dependent multidimensional gas dynamics, *J. Comput. Phys.*, **63** (1986) pp. 458-476.

4. DAVIS, S.F.: A rotationally biased upwind difference scheme for the Euler equations, *J. Comput. Phys.*, **56** (1984) pp. 65-92.

5. LEVY, D.W., POWELL, K.G., VAN LEER, B.: An implementation of a grid-independent upwind scheme for the Euler equations, *AIAA paper 89-1931* (1989).

6. OSHER, S., SOLOMON, F.: Upwind difference schemes for hyperbolic systems of conservation laws, *Math. Comput.*, **38** (1982) pp. 339-374.

7. HEMKER, P.W., SPEKREIJSE, S.P.: Multiple grid and Osher's scheme for the efficient solution of the steady Euler equations, *Appl. Numer. Math.*, **2** (1986) pp. 475-493.

8. PATANKAR, S.V.: *Numerical Heat Transfer and Fluid Flow* (Hemisphere, New York, 1980).

9. KOREN, B.: Low-diffusion rotated upwind schemes, multigrid and defect correction for steady, multi-dimensional Euler flows, *International Series of Numerical Mathematics*, **98** (Birkhäuser, Basel, 1991) pp. 265-276.

10. LAYTON, W.: On the principal axes of diffusion in difference schemes for 2D transport problems, *J. Comput. Phys.*, **90** (1990) pp. 336-347.

11. VAN LEER, B.: Upwind-difference methods for aerodynamic problems governed by the Euler equations, *Lectures in Applied Mathematics*, **22** (Amer. Math. Soc., Providence, RI, 1985) pp. 327-336.

12. KOREN, B.: Upwind discretization of the steady Navier-Stokes equations, *Int. J. Numer. Meth. Fluids*, **11** (1990) pp. 99-117.

a. Standard, grid-aligned scheme (7).

b. Positive, continuously differentiable scheme (9).

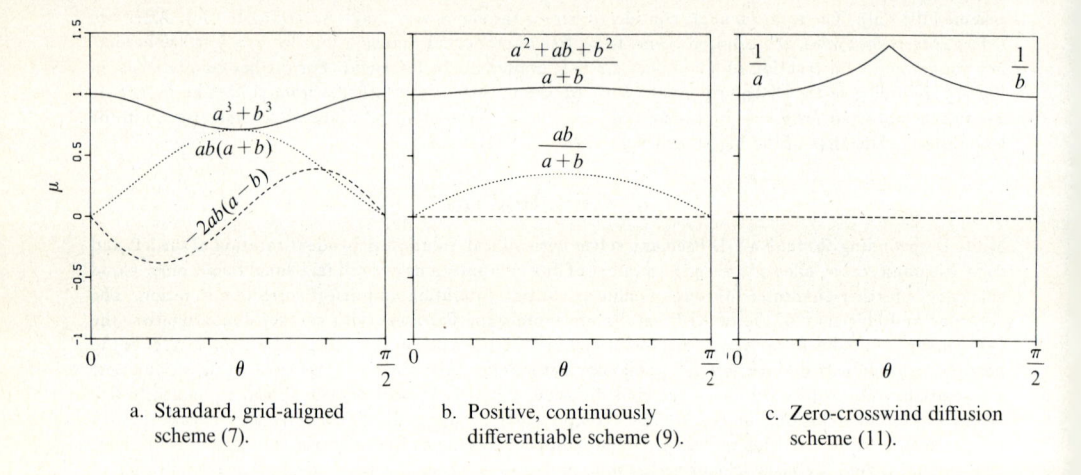

c. Zero-crosswind diffusion scheme (11).

Fig. 1. Diffusion coefficients modified equations (μ_{ss}: ——, μ_{sn}: ------, μ_{nn}: ....).

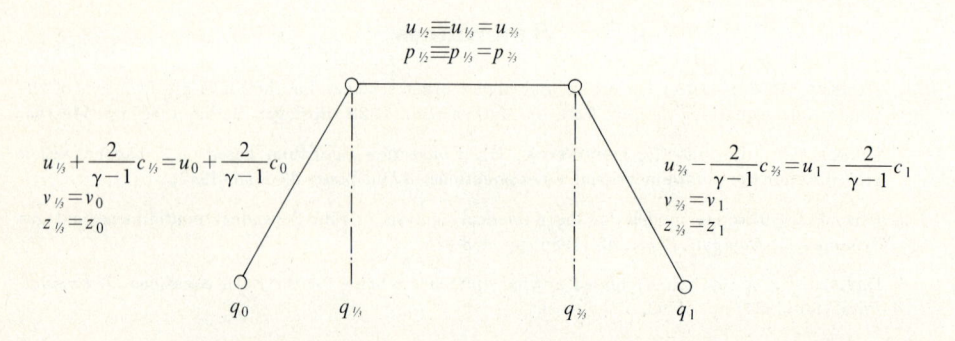

Fig. 2. Wave path in state space according to P-variant Osher scheme (perfect gas).

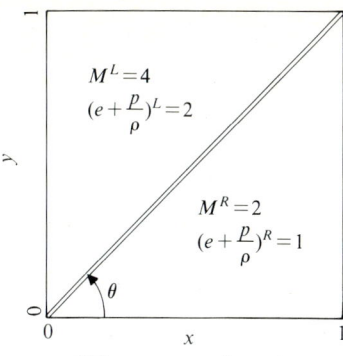

a. Oblique contact discontinuity
($\theta = 0.1\pi, 0.2\pi, 0.3\pi, 0.4\pi$, $p = 1$).

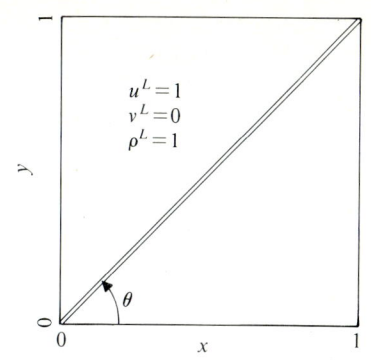

b. Oblique shock wave
($\theta = \dfrac{\pi}{4}, M^L = 2$ and $\theta = \dfrac{\pi}{8}, M^L = 4$).

Fig. 3. Test cases to be considered on unit square.

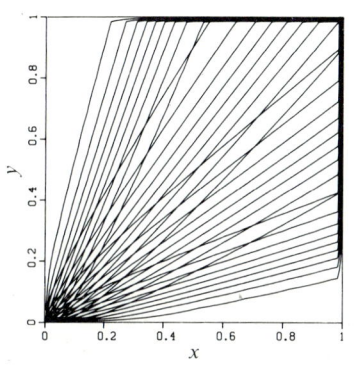

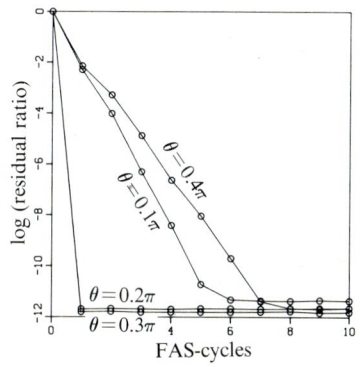

a. Enthalpy distributions.

b. Multigrid convergence histories.

Fig. 4. Results standard, grid-aligned scheme (15)-(16), flows with contact discontinuity.

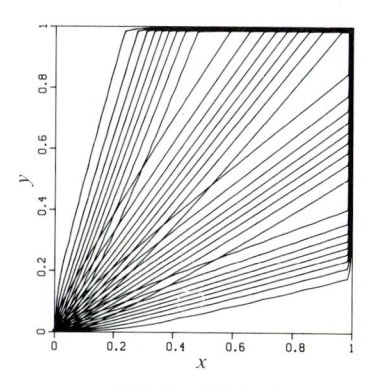

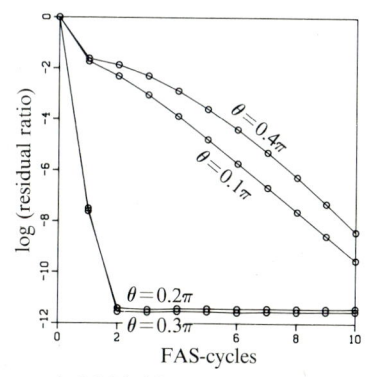

a. Enthalpy distributions.

b. Multigrid convergence histories.

Fig. 5. Results positive, continuously differentiable scheme (17)-(18), flows with contact discontinuity.

a. Zero-crosswind diffusion scheme (19)-(20).

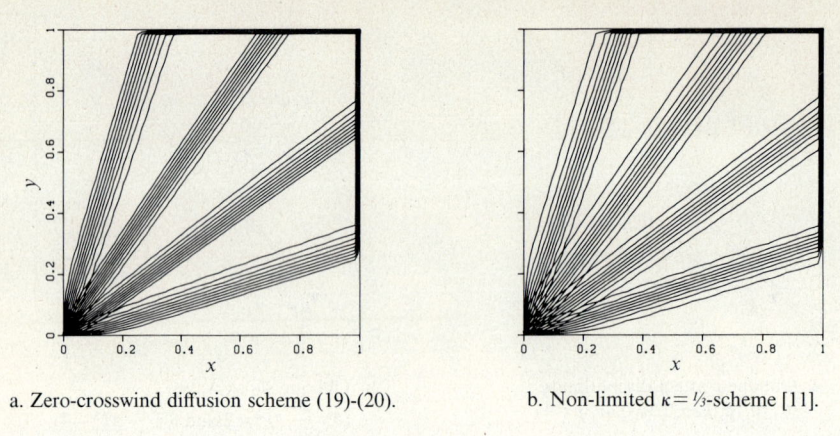

b. Non-limited $\kappa = \frac{1}{3}$-scheme [11].

Fig. 6. Enthalpy distributions, flows with contact discontinuity.

a. Standard, grid-aligned upwind scheme (15)-(16).

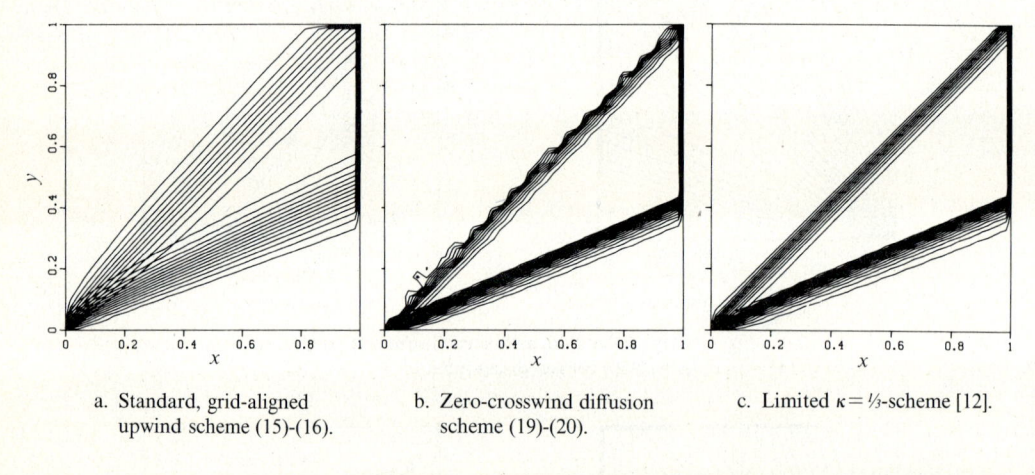

b. Zero-crosswind diffusion scheme (19)-(20).

c. Limited $\kappa = \frac{1}{3}$-scheme [12].

Fig. 7. Mach number distributions, flows with shock wave.

A MULTIGRID METHOD FOR A DISCRETIZATION OF THE INCOMPRESSIBLE NAVIER-STOKES EQUATIONS IN GENERAL COORDINATES

C.W. Oosterlee, P. Wesseling
Delft University of Technology
Faculty of Math. & Inf.
Mekelweg 4, 2628 CD Delft

SUMMARY

A discretization for the incompressible Navier-Stokes equations in general coordinates is derived. A staggered grid arrangement is used for the unknowns: contravariant flux-components and pressure.
The performance of the nonlinear multigrid algorithm, in particular the performance of two smoothing methods SCGS and SCGS/LS is investigated.

INTRODUCTION

For the computation of flows in complex geometries boundary-fitted coordinates are popular nowadays. An arbitrarily shaped physical domain is mapped onto a computational rectangular block, causing a transformation of the equations. The equations considered are the incompressible Navier-Stokes equations. In developing numerical methods for the incompressible Navier-Stokes equations in curvilinear coordinates some choices need to be made:

(i) a staggered or non-staggered grid

(ii) the type of velocity unknowns: Cartesian, contravariant or other

(iii) the iterative solver

In this work a staggered grid is used with contravariant flux components as unknowns. The solution method chosen is a multigrid method. Tensor notation proves indispensable for formulating physical conservation laws in general coordinates. It is found that discretization accuracy depends strongly on the way the geometric quantities (base vectors, metric tensor, Christoffel symbols) are discretized. It is assumed that the coordinate mapping is given only in terms of the physical coordinates of the points in the computational grid. The geometric quantities require differentiation of the mapping once or twice. The resulting work and inaccuracy which occurs if certain rules are not followed has led many authors to consider non-coordinate-invariant discretizations. However, this usually compromises stability and generality of the discretization. It seems attractive to use invariant discretizations of invariant physical laws, provided discretization accuracy can be maintained, and this is the approach that we follow. Desirable properties are: constant solutions are exact solutions of the discrete equations, and no loss of accuracy in the presence of grid discontinuities. These aims can be met by discretizing the geometric quantities in a special way, and by using fluxes rather than velocities as unknowns. If a coordinate invariant discretization of the equations is formulated in Gibbs' vector notation as in [7] explicit occurrence of Christoffel symbols is avoided. Instead other geometric quantities appear implicitly defining approximations of Christoffel symbols.
A nonlinear multigrid method is used to solve the equations. The behaviour of two smoothing methods, Symmetric Coupled Gauss-Seidel (SCGS) proposed in [11] and a variant (SCGS/LS) (Line Solver)) [6], [8] is compared for discretizations in curvilinear coordinates.

THE INCOMPRESSIBLE NAVIER-STOKES EQUATIONS IN GENERAL COORDINATES

The notation used will be tensor notation ([1], [9]). In general curvilinear coordinates the incompressible Navier-Stokes equations are given by:

$$U^\alpha_{,\alpha} = 0 \tag{1}$$

$$T^{\alpha\beta}_{,\beta} = (\rho U^\alpha U^\beta)_{,\beta} + (g^{\alpha\beta} p)_{,\beta} - \tau^{\alpha\beta}_{,\beta} = \rho F^\alpha \tag{2}$$

where ρ is the fluid density and $\tau^{\alpha\beta}$ represents the deviatoric stress tensor given by

$$\tau^{\alpha\beta} = \mu(g^{\alpha\gamma} U^\beta_{,\gamma} + g^{\gamma\beta} U^\alpha_{,\gamma}) \tag{3}$$

with μ the viscosity coefficient. The contravariant metric tensor $g^{\alpha\beta}$ is defined as:

$$g^{\alpha\beta} = \boldsymbol{a}^{(\alpha)} \cdot \boldsymbol{a}^{(\beta)}$$

with $\boldsymbol{a}^{(\alpha)} = \frac{\partial \xi^\alpha}{\partial \boldsymbol{x}}$ the contravariant base vectors.

The equations are transformed, because the arbitrarily shaped domain Ω is mapped onto a rectangular block G, resulting in a boundary fitted grid as shown in Figure 1; $\boldsymbol{x}$ are Cartesian coordinates, $\boldsymbol{\xi}$ boundary conforming curvilinear coordinates.

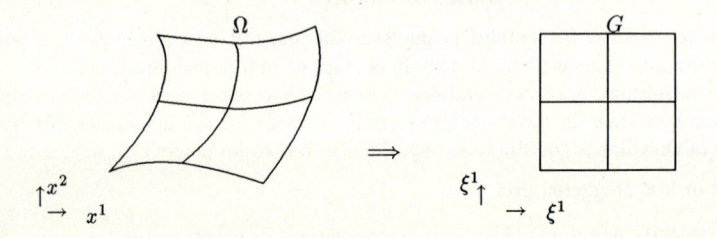

Figure 1: The mapping of a physical domain Ω onto a rectangular block G

In order to get accurate discretizations on non-smooth grids, some requirements should be met [10]:

(i) The geometric identity $\oint_S a^{(\alpha)}_\beta \, dS_{(\alpha)} = 0$ (coming from the application of the divergence theorem to a constant vector field) should be satisfied numerically. This requirement imposes rules on the covariant and contravariant base vectors.

(ii) Uniform flow fields should satisfy the discrete equations exactly. From this requirement the use of $V^\alpha = \sqrt{g} U^\alpha$, a relative contravariant tensor of weight one is found to be preferred as primary unknown, instead of U^α, although this by itself is not sufficient to meet this requirement; we will not go further into this here. Details can be found in [10], [4].

The covariant derivative of a relative contravariant tensor of weight one is defined by

$$V^\alpha_{,\beta} = \frac{\partial V^\alpha}{\partial \xi^\beta} + \left\{ {\alpha \atop \beta\gamma} \right\} V^\gamma - \left\{ {\gamma \atop \gamma\beta} \right\} V^\alpha.$$

The equations are discretized with a finite volume method on a **staggered** grid arrangement, i.e. pressure unknowns are placed at the cel centers, the flux unknowns V^1 are located at the centers of ξ^2 faces, the V^2 unknowns at the centers of ξ^1-faces. The total number of variables linked together in a momentum equation is 19 (13 flux unknowns, 6 pressure unknowns).

THE MULTIGRID SOLUTION ALGORITHM

The discretized equations are solved with the standard nonlinear multigrid algorithm [3], [2], [12]. Details of this algorithm are presented in [5]. Two smoothing methods are investigated for the discretization in curvilinear coordinates; the Symmetric Coupled Gauss-Seidel scheme (SCGS), proposed in [11], where the Navier-Stokes equations on staggered Cartesian grids were solved, and the line variant SCGS/LS proposed in [6], [8], which solves pressure unknowns along a line simultaneously. SCGS/LS is not well known yet and will be explained briefly: SCGS/LS is based on the SCGS smoother. For this smoother the five discretized equations per cell $(2 \times V^1 + 2 \times V^2$ momentum equations + continuity) are:

$$(A_c^1)_{i-1/2,j} V^1_{i-1/2,j} = F^1_{i-1/2,j}$$
$$(A_c^1)_{i+1/2,j} V^1_{i+1/2,j} = F^1_{i+1/2,j}$$
$$(A_c^2)_{i,j-1/2} V^2_{i,j-1/2} = F^2_{i,j-1/2}$$
$$(A_c^2)_{i,j+1/2} V^2_{i,j+1/2} = F^2_{i,j+1/2}$$

and

$$(V^1_{i+1/2,j} - V^1_{i-1/2,j})/\delta\xi^1 + (V^2_{i,j+1/2} - V^2_{i,j-1/2})/\delta\xi^2 = 0$$

where for example

$$\begin{aligned}
F^1_{i+1/2,j} = {} & A^1_{sw} V^1_{i-1/2,j-1} + A^1_s V^1_{i+1/2,j-1} + A^1_{se} V^1_{i+3/2,j-1} + \\
& A^1_w V^1_{i-1/2,j} + A^1_e V^1_{i+3/2,j} + \\
& A^1_{nw} V^1_{i-1/2,j+1} + A^1_n V^1_{i+1/2,j+1} + A^1_{ne} V^1_{i+3/2,j+1} + \\
& A^2_{sw} V^2_{i,j-1/2} + A^2_{se} V^2_{i+1,j-1/2} + \\
& A^2_{nw} V^2_{i,j+1/2} + A^2_{ne} V^2_{i+1,j+1/2} + \\
& A^3_{sw} p_{i,j-1} + A^3_{se} p_{i+1,j-1} + A^3_w P_{ij} + \\
& A^3_e p_{i+1,j} + A^3_{nw} p_{i,j+1} + A^3_{ne} p_{i+1,j+1} \;.
\end{aligned}$$

These equations are rewritten for solving them along a ξ^1-line in terms of corrections, indicated by primes and residuals, as follows:

$$\begin{aligned}
(A_c^1)_{i-1/2,j}(V^1)'_{i-1/2,j} - (A^3_w)_{i-1/2,j} p'_{i-1,j} - (A^3_e)_{i-1/2,j} p'_{i,j} &= R^1_{i-1/2,j} \quad (4)\\
&= F^1_{i-1/2,j} - (A_c^1)_{i-1/2,j} V^1_{i-1/2,j} \\
(A_c^1)_{i+1/2,j}(V^1)'_{i+1/2,j} - (A^3_w)_{i+1/2,j} p'_{i,j} - (A^3_e)_{i+1/2,j} p'_{i+1,j} &= R^1_{i+1/2,j} \\
&= F^1_{i+1/2,j} - (A_c^1)_{i+1/2,j} V^1_{i+1/2,j} \\
(A_c^2)_{i,j-1/2}(V^2)'_{i,j-1/2} - (A^3_n)_{i,j-1/2} p'_{i,j} &= R^2_{i,j-1/2} \\
&= F^2_{i,j-1/2} - (A_c^2)_{i,j-1/2} V^2_{i,j-1/2} \\
(A_c^2)_{i,j+1/2}(V^2)'_{i,j+1/2} - (A^3_s)_{i,j+1/2} p'_{i,j} &= R^2_{i,j+1/2} \\
&= F^2_{i,j+1/2} - (A_c^2)_{i,j+1/2} V^2_{i,j+1/2} \\
(V^{1'}_{i+1/2,j} - V^{1'}_{i-1/2,j})/\delta\xi^1 + (V^{2'}_{i,j+1/2} - V^{2'}_{i,j-1/2})/\delta\xi^2 &= R^3_{i,j} \\
&= (V^1_{i-1/2,j} - V^1_{i+1/2,j})/\delta\xi^1 + \\
&\quad (V^2_{i,j-1/2} - V^2_{i,j+1/2})/\delta\xi^2
\end{aligned}$$

(The equations for solving along a ξ^2-line are found in a similar way).

Solving the set of equations (4) gives us a tri-diagonal system for pressures along the line:

$$(P_w)_{i-1,j} p'_{i-1,j} + (P_c)_{i,j} p'_{i,j} + (P_e)_{i+1,j} p'_{i+1,j} = R^p_{i,j}$$

where

$$(P_w)_{i-1,j} = (-\frac{A^3_w}{A^1_c})_{i-1/2,j} \cdot \frac{1}{\delta \xi^1}$$

$$(P_e)_{i+1,j} = (\frac{A^3_e}{A^1_c})_{i+1/2,j} \cdot \frac{1}{\delta \xi^1}$$

$$(P_c)_{i,j} = (-\frac{A^3_e}{A^1_c})_{i-1/2,j} \cdot \frac{1}{\delta \xi^1} + (\frac{A^3_w}{A^1_c})_{i+1/2,j} \cdot \frac{1}{\delta \xi^1} - (\frac{A^3_n}{A^2_c})_{i,j-1/2} \cdot \frac{1}{\delta \xi^2} + (\frac{A^3_s}{A^2_c})_{i,j+1/2} \cdot \frac{1}{\delta \xi^2}$$

$$R^p_{i,j} = R^3_{i,j} + (\frac{R^1}{A^1_c})_{i-1/2,j} \cdot \frac{1}{\delta \xi^1} - (\frac{R^1}{A^1_c})_{i+1/2,j} \cdot \frac{1}{\delta \xi^1} + (\frac{R^2}{A^2_c})_{i,j-1/2} \cdot \frac{1}{\delta \xi^2} - (\frac{R^2}{A^2_c})_{i,j+1/2} \cdot \frac{1}{\delta \xi^2}.$$

After solving the tridiagonal system the corrections for p are added to the current solution. The corrections for V^1 and V^2 are found from (4):

$$V^{1'}_{i-1/2,j} = [R^1_{i-1/2,j} + (A^3_w)_{i-1/2,j} p'_{i-1,j} + (A^3_e)_{i-1/2,j} p'_{i,j}]/(A^1_c)_{i-1/2,j}$$

$$V^{1'}_{i+1/2,j} = [R^1_{i+1/2,j} + (A^3_w)_{i+1/2,j} p'_{ij} + (A^3_e)_{i+1/2,j} p'_{i+1,j}]/(A^1_c)_{i+1/2,j}$$

$$V^{2'}_{i,j-1/2} = [R^2_{i,j-1/2} + (A^3_n)_{i,j-1/2} p'_{i,j}]/(A^2_c)_{i,j-1/2}$$

$$V^{2'}_{i,j+1/2} = [R^2_{i,j+1/2} + (A^3_s)_{i,j+1/2} p'_{ij}]/(A^2_c)_{i,j+1/2}.$$

We can move line by line in a forward or backward direction. The velocities are updated twice, while the pressures are updated once.

Underrelaxation is implemented as in [11] by modifying the main diagonal coefficients:

$$(A^1_c)_{i\pm 1/2,j} := (A^1_c)_{i\pm 1/2,j}/\alpha_1$$
$$(A^2_c)_{i,j\pm 1/2} := (A^2_c)_{i,j\pm 1/2}/\alpha_2.$$

TEST PROBLEMS, SOME RESULTS

To investigate the behaviour of SCGS and SCGS/LS in the multigrid method some driven cavity problems are calculated. The first problem is the classical problem: An equidistant grid in a square cavity. Then: a non-equidistant grid in a square cavity, an equidistant grid in a skewed cavity (skew angle 63^0), and a non-equidistant grid in an L-shaped cavity. Figures 2a to 2d show the last three cavities. Boundary conditions for the first three cavities are: $u = 0$ on boundary I, II and IV, $u = -1, v = 0$ on III. The L-shaped cavity causes extra problems, namely boundary II consists of two parts as does IV. On only half of II ($y = 1$) the tangential velocity $u = -1$ must be prescribed.

All problems are computed with $Re = 100$ and $Re = 1000$.

Average reduction factors r are compared, defined as

$$r = \left(\frac{\|res\|_{nit}}{\|res\|_0}\right)^{1/nit}$$

i.e. the 2-norm of the residual after nit iterations divided by the 2-norm of the starting residual. The algorithm starts with nested iteration; the starting vector is the zero solution on the coarsest ($= 2 \times 2$) grid.

The number of pre-smoothing iterations is 1 as is the number of post-smoothing iterations. The

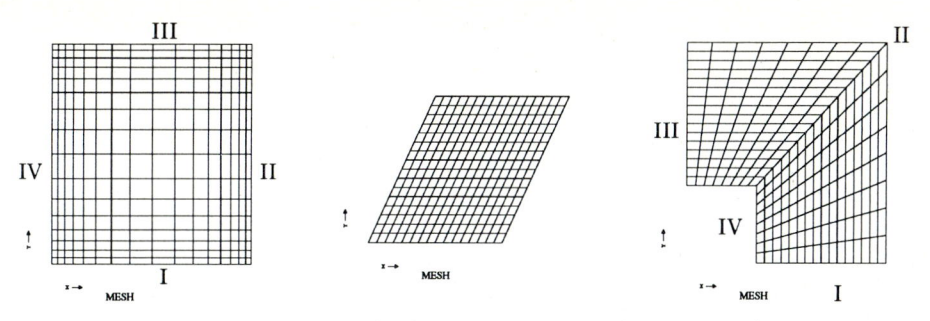

Figure 2: The different cavities, 16 × 16 grid

results given are the reduction factors for the W-cycle, which was more efficient then the V-cycle closely followed by the F-cycle.

The smoothers: For different Reynolds numbers different values of the underrelaxation factors α_k are needed. However for both smoothing methods the optimal α_k's are the same:

$$\text{For} \quad \begin{array}{l} Re = 100 \quad : \quad \alpha_k = 0.7 \\ Re = 1000 \quad : \quad \alpha_k = 0.3 \end{array}.$$

For both Reynolds numbers the smoothing iterations are performed in alternating directions, a sweep in the ξ^1-direction is followed by a sweep in the ξ^2-direction.

Table 1 presents the reduction factors r for the first three problems for $Re = 100$, $nit = 10$. Table 2 shows r for $Re = 1000$, $nit = 20$. Finally table 3 gives r for the L-shaped cavity, $nit = 20$. Figure 3 shows streamlines for the skew cavity for $Re = 100$ and $Re = 1000$ (64 × 64 grid). In figure 4 the streamlines for the L-shaped cavity are presented for the 64 × 64 grid.

Table 1: Average reduction factors for the different cavity problems, $Re = 100$.

number of levels	grid	sq. cav. eq. grid		sq. cav. neq. grid		skew. cav. eq. grid	
		SCGS	SCGS/LS	SCGS	SCGS/LS	SCGS	SCGS/LS
4	16 x 16	.208	.220	.155	.209	.234	.185
5	32 x 32	.153	.200	.227	.270	.176	.174
6	64 x 64	.147	.153	.242	.336	.152	.130
7	128 x 128	.142	.169	.241	.520	.150	.146

For the square cavity with the non-equidistant grid r_{SCGS} seems better than $r_{SCGS/LS}$. This is due to very good reduction in the first iterations. The convergence factor however is the same for both smoothing methods:

$$\rho \approx 0.55$$

Some ideas to overcome problems with stretched cells are currently under investigation.

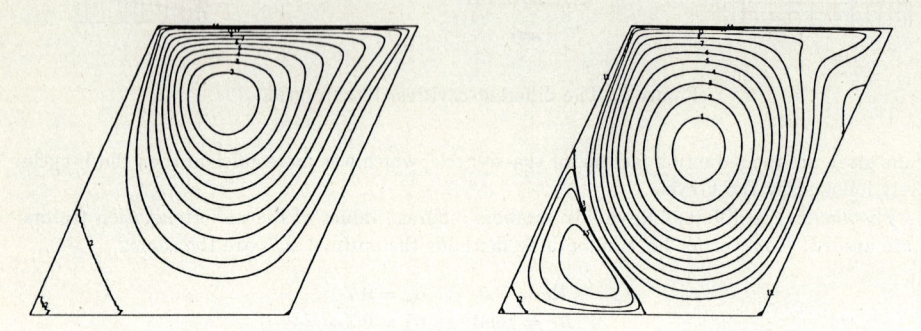

Figure 3: Streamlines for the skew cavity, a) $Re = 100$, b) $Re = 1000$, 64×64 grid.

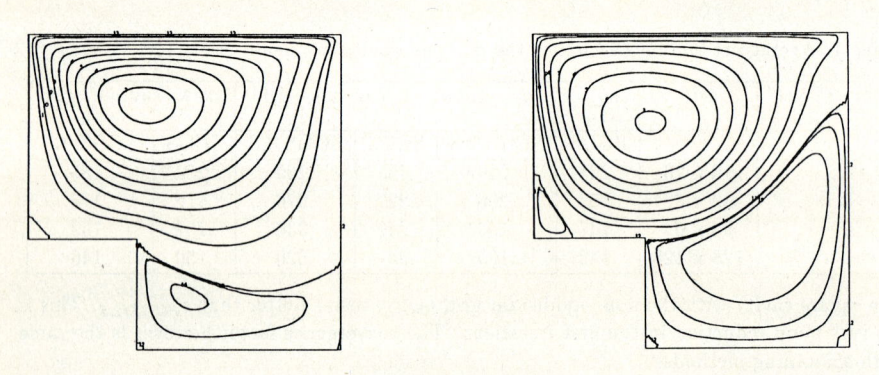

Figure 4: Streamlines for the L-shaped cavity, a) $Re = 100$, b) $Re = 1000$, 64×64 grid.

Table 2: Average reduction factors for the different cavity problems, $Re = 1000$.

number of levels	grid	sq. cav. eq. grid		sq. cav. neq. grid		skew. cav. eq. grid	
		SCGS	SCGS/LS	SCGS	SCGS/LS	SCGS	SCGS/LS
4	16 x 16	.516	.466	.405	.459	.495	.565
5	32 x 32	.565	.495	.445	.459	.540	.540
6	64 x 64	.564	.569	.530	.499	.601	.585
7	128 x 128	.517	.558	.456	.519	.602	.607

For $Re = 1000$ there is not much difference for the three test problems. All reduction factors are level-independent. For the non-equidistant grid of problem 2 r is not worse than r of problem 1. This is due to upwind discretization in large parts of the domain.

Table 3: Average reduction factors for the L-shaped cavity, $Re = 100$ and $Re = 1000$.

number of levels	grid	$Re = 100$		$Re = 1000$	
		SCGS	SCGS/LS	SCGS	SCGS/LS
4	16 x 16	.334	.311	.608	.637
5	32 x 32	.402	.370	.686	.687
6	64 x 64	.411	.410	.767	.688

CONCLUSION

An invariant formulation of the incompressible Navier-Stokes equations in general coordinates has been presented in which Christoffel symbols occur. The discretization of the invariant formulation shows good results for many geometries and fairly non-uniform grids. A level independent convergence rate has been found for the test problems solving with a multigrid solution algorithm. Reduction factors for rectangular and some more complex geometries do not seem to differ much. Reduction factors for domains in which stretched cells occur are worse than for domains with square cells. This is a well known fact when a Gauss-Seidel type of smoothing method is being used. At the moment we are attempting to improve these reduction factors. The behaviour of the two smoothing methods SCGS and SCGS/LS does not differ much. It seems that SCGS is more robust in the variation of underrelaxation factors, especially for low Reynolds flows and in the choice of smoothing directions.

REFERENCES

[1] R. Aris. *Vectors, tensors and the basic equations of fluid mechanics.* Prentice-Hall, Inc., Englewood Cliffs, N.J., 1962.

[2] A. Brandt. *Guide to multigrid development.* In: W. Hackbusch and U. Trottenberg (eds.) Multigrid methods, 960: 220-312, Springer Verlag, Berlin, 1982

[3] W. Hackbusch. *Multi-grid methods and applications.* Springer-Verlag, Berlin, 1985.

[4] A.E. Mynett, P. Wesseling, A. Segal and C.G.M. Kassels *The ISNaS incompressible Navier-Stokes solver: invariant discretization* Appl. Scient. Research 48: 175-191, 1991

[5] C.W. Oosterlee and P. Wesseling. *A multigrid method for an invariant formulation of the incompressible Navier-Stokes equations in general coordinates.* Report 91-12, Faculty of Technical Mathematics and Informatics, Delft University of Technology, Delft, 1991.

[6] J.I. Rollet, D.F. Mayers, and T.M. Shah. *Analysis and Application of an efficient line solver based upon the Symmetric Coupled Gauss-Seidel scheme.* Report 88/10, Oxford Univ. Comp. Lab., 1985.

[7] M. Rosenfeld, D. Kwak, and M. Vinokur. *A Solution Method for the Unsteady and Incompressible Navier-Stokes Equations in Generalized Coordinate Systems.* AIAA Paper AIAA-88-0718, 1988.

[8] T.M. Shah. *Analysis of a Multigrid method.* Master's thesis, Oxford Univ. Comp. Lab., 1987.

[9] I.S. Sokolnikoff. *Tensor analysis.* John Wiley & Sons, Inc., Englewood Cliffs, N.J., 1964.

[10] J.J.I.M. Van Kan, C.W. Oosterlee, A. Segal, and P. Wesseling. *Discretization of the incompressible Navier-Stokes equations in general coordinates using contravariant velocity components.* Report 91-09, Faculty of Technical Mathematics and Informatics, Delft University of Technology, Delft, 1991.

[11] S.P. Vanka. Block-implicit multigrid solution of Navier-Stokes equations in primitive variables. *J. Comp. Phys.*, 65:138–158, 1986.

[12] P. Wesseling. Multigrid methods in computational fluid dynamics. *Z. angew. Math. Mech.*, 70:T337–T348, 1990.

VISCOUS HYPERSONIC FLOWS

Numerical Solution of Viscous Hypersonic Flows in Chemical Non Equilibrium

V. Bellucci and F. Grasso
Department of Mechanics and Aeronautics
University of Rome "La Sapienza"
Via Eudossiana, 18 – 00184 – Rome, Italy

ABSTRACT

A high order scheme for the solution of viscous hypersonic flows with "real gas effects" has been developed. The scheme is based on an upwind biased total variation diminishing formulation to solve the species, momentum and energy conservation equations in the presence of non equilibrium chemistry. The approach has been applied first to assess the effects of different models for thermodynamic properties, transport phenomena and chemical kinetics on the solution of the flow over a 10° wedge at 8100 m/s and 61 km of altitude. Then, the effects of chemical non equilibrium versus an equilibrium assumption are evaluated by computing the flow about a double ellipse at $M_\infty = 25$, 30° of incidence and an altitude of 75 km.

INTRODUCTION

The modeling of hypersonic flows has intrinsic difficulties due to uncertainties in the description of transport and chemical kinetics mechanisms and lack of high temperature data. The purpose of this work is to develop a high order scheme in the presence of "real gas effects" and to analyse the effects of the physical submodels on the numerical solution of viscous hypersonic flows in chemical non equilibrium.

Two approaches are in general possible in selecting the formulation of the governing equations for a multicomponent mixture. The first one solves for the continuity equation and $N - 1$ species continuity equations (where N is the total number of species) [3,12,13,16,18,19]. In the second approach N species conservation equations are solved [8,15,17,20]. From a physical point of view the two approaches are equivalent. However, from the numerical point of view the computed solution may be affected by the selected formulation. In particular, the first approach may lead to some inaccuracies due to round off errors and even to instabilities during the first stage of the computation. In the present work the second approach has been followed, and all species conservation equations are solved. A sensitivity analysis of the flow over a two-dimensional wedge is first discussed; then, results of equilibrium and non equilibrium flows around a double ellipse at high Mach number are presented.

GOVERNING EQUATIONS

The governing equations are the species, momentum and energy conservation equations

$$\frac{\partial}{\partial t}\int_V \mathbf{W}\,dV + \int_S [(\mathbf{F}-\mathbf{F}_v)n_x + (\mathbf{G}-\mathbf{G}_v)n_y]\,dS = \int_V \mathbf{H}\,dV.$$

The vector $\mathbf{W}$, the inviscid flux $(\mathbf{F},\mathbf{G})$ and the viscous flux $(\mathbf{F}_v,\mathbf{G}_v)$ are

$$\mathbf{W} = \begin{bmatrix} \rho_q \\ \rho u \\ \rho v \\ \rho E \end{bmatrix} \qquad \mathbf{F} = \begin{bmatrix} \rho_q u \\ \rho u^2 + p \\ \rho u v \\ \rho u H \end{bmatrix} \qquad \mathbf{G} = \begin{bmatrix} \rho_q v \\ \rho u v \\ \rho v^2 + p \\ \rho v H \end{bmatrix}$$

$$\mathbf{F}_v = \begin{bmatrix} -\rho_q u_q \\ \sigma_{xx} \\ \sigma_{xy} \\ u\sigma_{xx} + v\sigma_{xy} - Q_x \end{bmatrix} \qquad \mathbf{G}_v = \begin{bmatrix} -\rho_q v_q \\ \sigma_{yx} \\ \sigma_{yy} \\ u\sigma_{yx} + v\sigma_{yy} - Q_y \end{bmatrix} \qquad \mathbf{H} = \begin{bmatrix} \dot{w}_q \\ 0 \\ 0 \\ 0 \end{bmatrix}$$

where ρ_q, $\dot{w}_q$, u_q and v_q are respectively the q-th species density, the reaction rate and the x- and y-component of the diffusion velocity. Furthermore

$$\boldsymbol{\sigma} = \mu\left[\text{grad } \mathbf{u} + (\text{grad } \mathbf{u})^T\right] - \frac{2}{3}\mu(\text{div } \mathbf{u})\mathbf{U}$$

$$\mathbf{Q} = -\eta\,\text{grad } T + \sum_q h_q \rho_q \mathbf{u}_q$$

$$E = e + \frac{u^2+v^2}{2} = \sum_q Y_q e_q + \frac{u^2+v^2}{2}$$

$$h_q = e_q + R_q T$$

$$H = E + \frac{p}{\rho}$$

$$\rho = \sum_q \rho_q$$

where Y_q and R_q are respectively the mass fraction and the gas constant of the species q, and $\mathbf{U}$ is the unit tensor.

Thermodynamic Relations

In the present work air is assumed to be a multicomponent mixture of perfect gases, and the sensitivity of the solution upon different thermodynamic relations is studied.

Model 1 (RT1)

The first model assumes that the translational and rotational energy modes are fully excited and for the vibrational mode the harmonic oscillator expression is used [1,2,12,16,17,18]. For atomic species one has

$$e_q = \frac{3}{2} R_q T + \Delta h_q^\circ \quad .$$

For diatomic species

$$e_q = \frac{5}{2} R_q T + \frac{R_q \theta_q^v}{\exp\left(\theta_q^v/T\right) - 1} + \Delta h_q^o$$

where θ_q^v and Δh_q^o are respectively the characteristic vibrational temperature and the enthalpy of formation of the species q.

Model 2 (RT2)

In the second model a polynomial fit of temperature for the specific heat coefficients [3,4,5,19] leads to the following expression for the species internal energy

$$e_q = R_q \sum_{k=1}^{5} \frac{A_{q,k}}{k} T^k - R_q T + \Delta h_q^o \ .$$

Transport Coefficients

In general, the transport properties are evaluated either by using Eucken approximation together with Wilke's rule, or by using expressions obtained from kinetic theory. The effects of the two approaches are studied.

Model 1 (CT1)

The derivation of the transport coefficients is based on the Chapman-Enskog theory and on an extension of Yos' formula [2,4,5]. This approach has a broad range of validity, even though it is computationally expensive and with some uncertainties in the evaluation of the collision integrals.

The mixture viscosity is defined as

$$\mu = \sum_q \frac{m_q \gamma_q}{\sum_r \gamma_r \Delta_{q,r}^{(2)}(T)}$$

where m_q and $\Delta_{q,r}^{(2)}$ are respectively the mass of the species q and the modified collision integral, and

$$\gamma_q = \frac{\rho_q}{\rho W_q} \ .$$

The thermal conductivity η is expressed as

$$\eta = \eta_{tr} + \eta_{int}$$

$$\eta_{tr} = \frac{15}{8} \kappa \sum_q \frac{\gamma_q}{\sum_r a_{q,r} \gamma_r \Delta_{q,r}^{(2)}(T)}$$

$$a_{q,r} = 1 + \frac{\left[1 - (m_q/m_r)\right]\left[0.45 - 2.54\,(m_q/m_r)\right]}{\left[1 + (m_q/m_r)\right]^2}$$

$$\eta_{int} = \kappa \sum_q \frac{\gamma_q\,(C_{v,q}/R_q - 3/2)}{\sum_r \gamma_r \Delta_{q,r}^{(1)}(T)}$$

where κ is the Boltzmann constant.

The binary diffusion coefficient $(D_{q,r})$ and the diffusion coefficient of the species q in the mixture (D_q) are given by

$$D_{q,r} = \frac{\kappa T}{p \Delta_{q,r}^{(1)}(T)}$$

$$D_q = \frac{\gamma_{tot}^2 W_q (1 - W_q \gamma_q)}{\sum_{\substack{r \\ r \neq q}} \gamma_r / D_{q,r}}.$$

The modified collision integrals $\Delta_{q,r}^{(k)}$ are defined as fits of temperature as proposed in Ref.[5].

Model 2 (CT2)

In the present model, for temperature lower than 1000 K, Sutherland's law is used to evaluate the mixture viscosity. The thermal conductivity is obtained by assuming a constant Prandtl number, defined with the frozen specific heat $(Pr = 0.71)$.

For temperature greater than 1000 K, the viscosity of the species q is calculated from curve fits [2,3,15]

$$\mu_q = 0.1 \left[\exp C_{\mu_q} \right] T^{[A_{\mu_q} \ln T + B_{\mu_q}]}.$$

The thermal conductivity is obtained by Eucken's formula

$$\eta_q = \mu_q \left(C_{v,q} + \frac{9}{4} R_q \right).$$

Wilke's mixing rule yields the mixture viscosity and thermal conductivity. The diffusion coefficient is computed by assuming a constant Lewis number $(Le = 1.4)$.

Chemistry Model

In the present work, neglecting ionization phenomena, five major species (O, N, NO, O_2, N_2) are assumed and the sensitivity of the solution upon the reaction mechanism is studied.

Model 1 (CC1)

In the first model the selected reaction mechanism is that proposed by Park [11]

$$\begin{aligned} O_2 + M &= O + O + M \\ NO + M &= N + O + M \\ N_2 + M &= N + N + M \\ N_2 + O &= NO + N \\ NO + O &= O_2 + N \end{aligned}$$

where M is anyone of the five species.

According to Park, the forward reaction rate constant is expressed by

$$k_{f,r} = C_{f,r} T^{n_{f,r}} \exp(-E_{f,r} / \kappa T)$$

and the backward reaction rate constant is defined via the equilibrium constant.

Model 2 (CC2)

The second model is a simplification of Park's model by neglecting the dissociation reactions of N_2 and NO, under the assumption that they are slower than the two exchange reactions [12].

Model 3 (CC3)

The third model is that of Blottner [4], that differs from Park's only in the evaluation of the reaction rates.

NUMERICAL SCHEME

The numerical scheme is based on a finite volume approach. The discretized form of the conservation laws applied to the (i,j) cell of volume $V_{i,j}$ is

$$\frac{\partial \mathbf{W}_{i,j}}{\partial t} V_{i,j} + \sum_{\beta=1}^{4} \left[(\mathbf{F} - \mathbf{F}_v) n_x + (\mathbf{G} - \mathbf{G}_v) n_y \right]_\beta S_\beta = \mathbf{H}_{i,j} V_{i,j} .$$

The discretization of the inviscid flux is obtained by an approximate Riemann solver [6]. At a cell face $\beta = 2$, shared by the adjacent cells (i,j) and $(i+1,j)$ one has (see Fig. 1)

$$(\mathbf{F} n_x + \mathbf{G} n_y)_{\beta=2} = \frac{1}{2} \left[(\mathbf{F}_{i,j} + \mathbf{F}_{i+1,j}) n_x + (\mathbf{G}_{i,j} + \mathbf{G}_{i+1,j}) n_y + \mathbf{R}_{i+1/2,j} \, \mathbf{\Phi}_{i+1/2,j} \right]$$

where $\mathbf{R}$ is the right eigenvector matrix of the normal flux jacobian and is defined as

$$\mathbf{R} = \begin{bmatrix} \delta_{q,r} & 0 & Y_q & Y_q \\ u & -c n_y & u + c n_x & u - c n_x \\ v & c n_x & v + c n_y & v - c n_y \\ \dfrac{\mathbf{u} \cdot \mathbf{u}}{2} - \dfrac{\chi_r}{K} & (-u n_y + v n_x) c & H + c u_n & H - c u_n \end{bmatrix}$$

where K and χ_q are the pressure derivatives and c is the frozen speed of sound.

The vector $\mathbf{\Phi}_{i+1/2,j}$ ensures that the scheme satisfies second order TVD properties. In the present work, Yee's formulas [6] for ideal gases have been extended to account for real gas effects, and a *minmod* function is selected as limiter for its better computational efficiency and speed of convergence. The values at the interfaces are calculated by using a generalization of Roe's averaging to account for chemical non equilibrium [7,8].

The discretization of the viscous flux is symmetric.

The time integration is performed by a three stage Runge-Kutta point implicit algorithm [13]:

$$\mathbf{W}_{i,j}^{(0)} = \mathbf{W}_{i,j}^n$$

$$\mathbf{P}_{i,j}^{(k)} \left(\mathbf{W}_{i,j}^{(k)} - \mathbf{W}_{i,j}^{(0)} \right) = \alpha_k \Delta t \left\{ -\frac{1}{V} \sum_{\beta=1}^{4} \left[\left(\mathbf{F}^{(k-1)} - \mathbf{F}_v^{(0)} \right) n_x + \left(\mathbf{G}^{(k-1)} - \mathbf{G}_v^{(0)} \right) n_y \right]_\beta S_\beta + \mathbf{H}_{i,j}^{(k-1)} \right\}$$

$$\mathbf{W}_{i,j}^{n+1} = \mathbf{W}_{i,j}^{(3)}$$

where

$$\mathbf{P}^{(k)} = \mathbf{I} - \alpha_k \Delta t \left(\frac{\partial \mathbf{H}}{\partial \mathbf{W}}\right)^{(k-1)}$$

is a precondition matrix introduced to reduce the stiffness due to finite rate chemistry. For computational efficiency, a partial jacobian of the source term $(\partial \mathbf{H}/\partial \mathbf{W})$ is used (the dependency of $\mathbf{H}$ on ρu, ρv, ρE is neglected), without affecting the accuracy of the steady-state solution.

RESULTS

The flow over a 10° wedge at free stream velocity of 8100 m/s, an altitude of 61 km and a wall temperature of 1200 K has been simulated.

This test case has been investigated by other authors [3,9] and shows a small degree of nonequilibrium in the boundary layer. The test case is a "simple" one considering the simplicity of the geometry. However, it contains all of the relevant features of complex hypersonic flows, and has been selected to study the sensitivity of the solution upon the physical submodels.

All computations have been obtained on a 176×48 grid with normal mesh spacing ranging from $\Delta Y/L = 0.0015$ to $\Delta Y/L = 0.069$, and cell aspect ratios that vary between 0.5 and 13. Referring to Fig. 2, the boundary conditions imposed are: symmetry conditions on Γ_1; no slip, diabatic, non catalytic conditions on Γ_2; first order extrapolation conditions on Γ_3; free stream on Γ_4.

In Figs. 3–4 the effects of the transport mechanism and chemical kinetics on the solution are shown. The computed results are compared versus the reference test case solution at $x = 3.5$ m [3]. The results show that the model based on Eucken's approximation yields a lower peak temperature within the boundary layer, due to a lower level of viscosity (approximately 7% for $T = O(6000 \text{ K})$).

The effects of eliminating N_2 and NO dissociation reactions are shown in Figs. 5–6, where the results obtained with the first (**CC1**) and second (**CC2**) Park model are reported. The figures show that the wall values of atomic oxygen and nitric oxide mass fractions are affected the most.

The sensitivity of the solution upon the thermodynamic relationships is shown in Figs. 7–8. The differences in the peak temperature and mass fractions are essentially negligible. Oxygen and nitric oxide mass fractions are affected the most at the wall, most likely on account of streamwise history effects.

The second test case corresponds to the flow over a double ellipse at Mach number $M_\infty = 25$ and angle of attack $\alpha = 30°$. The freestream conditions correspond to an altitude of 75 km and a wall temperature of $T_w = 1500$ K. The computations have been performed on a 176×74 grid with normal mesh spacing ranging from $\Delta Y/L = 0.1\,10^{-5}$ to $\Delta Y/L = 0.02$, and cell aspect ratio varying between $0.9\,10^{-3}$ and 1800. The selected model for the thermodynamic relationships, transport mechanism and chemical kinetics corresponds to **RT2**, **CT1** and **CC1** and contains the least approximations. For this test case two solutions are reported, corresponding to equilibrium and non equilibrium chemistry (the equilibrium solution is obtained by multiplying the reaction rate constants by the factor 10^6).

The distribution of temperature along the stagnation line (Fig. 11) shows a peak temperature of 14851 K for the non equilibrium case and 5715 K for the equilibrium case. This is due to the freezing of chemical kinetics through the shock in the non equilibrium case. The non equilibrium shock detachment distance is about 1.02 cm and the equilibrium one 0.44 cm, as inferred from Figs. 9–11. As a consequence the stagnation heat loss for the nonequilibrium case is smaller, as observed from the Stanton number distribution (Fig. 14). Little differences are found on the pressure and on the skin friction coefficient (Figs. 12–13).

CONCLUSIONS

A high order scheme for the solution of viscous hypersonic flows with "real gas effects" has been developed. The scheme is based on an upwind biased total variation diminishing formulation. All species, momentum and energy conservation equations are solved under the assumption of thermal equilibrium and chemical non equilibrium.

The modeling of hypersonic flows has intrinsic difficulties due to uncertainties in the description of transport and chemical kinetics mechanisms and lack of high temperature data. The effects of the available models for thermodynamic relationships, molecular transport and chemistry have been assessed by performing a sensitivity analysis. Computed results for the flow over a 10° wedge at 8100 m/s indicate that: wall values of atomic oxygen and nitric oxide mass fractions are affected the most; the solution is more sensitive to transport and chemical kinetics mechanisms rather than to thermodynamic relationships.

The effects of equilibrium and non equilibrium chemistry have also been assessed, by computing the flow about a double ellipse at $M_\infty = 25$; 30° of incidence and an altitude of 75 km. The results of the non equilibrium case show that chemistry freezes at the bow shock and a peak temperature of 14851 K is obtained along the stagnation line. A smaller heat loss is predicted and little effects on the pressure and skin friction coefficient distributions are observed.

REFERENCES

[1] Vincenti W.G., Kruger C.H. Jr., *Introduction to Physical Gas Dynamics*, John Wiley and Sons, Inc., New York, 1965.

[2] Lee J.H., *Thermal Design of Aeroassisted Orbital Transfer Vehicles*, Nelson H.F. ed., Volume 96 in Progress in Astronautics and Aeronautics, AIAA, 1985.

[3] Prabhu D.K., Tannehill J.C., Marvin J.G., AIAA J., Vol. 26, 1988.

[4] Gupta R.N., Yos J.M., Thompson R.A., Lee K.P., NASA Reference Publication 1232, 1990.

[5] Gnoffo P.A., Gupta R.N., Shinn J.L., NASA Technical Paper 2867, 1989.

[6] Yee H.C., NASA Technical Memorandum 101088, 1989.

[7] Roe P.L., J. Comp. Phys. 43, 1981.

[8] Liu Y., Vinokur M., AIAA–89–0201.

[9] Candler G., AIAA–89–0312.

[10] Park C., AIAA–85–0247.

[11] Park C., AIAA–89–1740.

[12] Park C., Yoon S., AIAA-89-0685.
[13] Bussing T.R.A., Murman E.M., AIAA J., Vol. 26, 1988.
[14] Hornung H.G., J. Fluid Mechanics 53, 1972.
[15] Hollanders H., *et al.*, ONERA.
[16] Palmer G., AIAA-89-1701.
[17] Grossman B., Cinnella P., Garrett J., AIAA-89-1653.
[18] Desideri J.A., Glinsky N., Hettena E., Comp. and Fluids, Vol. 18, n. 2, pp. 151-182, 1990.
[19] Shuen J.S., Yoon S., AIAA J., Vol. 27, 1989.
[20] Yee H.C., Shinn J.L., AIAA J., Vol. 27, 1989.

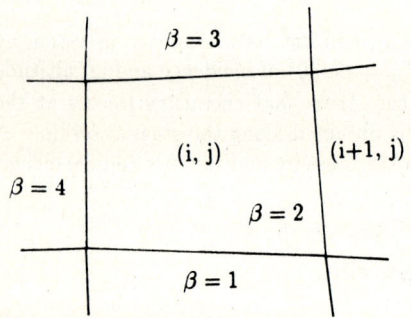

Fig. 1 – Computational cell.

Fig. 2 – Computational domain for wedge test case.

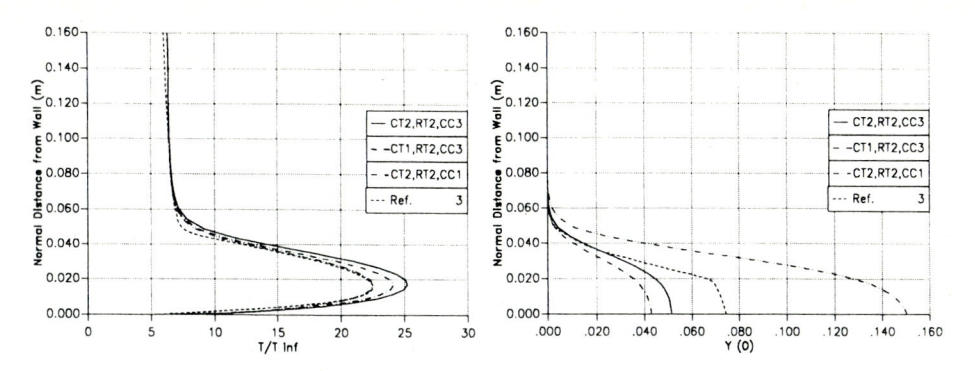

Fig. 3 – T/Tinf vs η.

Fig. 4 – Oxygen mass fraction vs η.

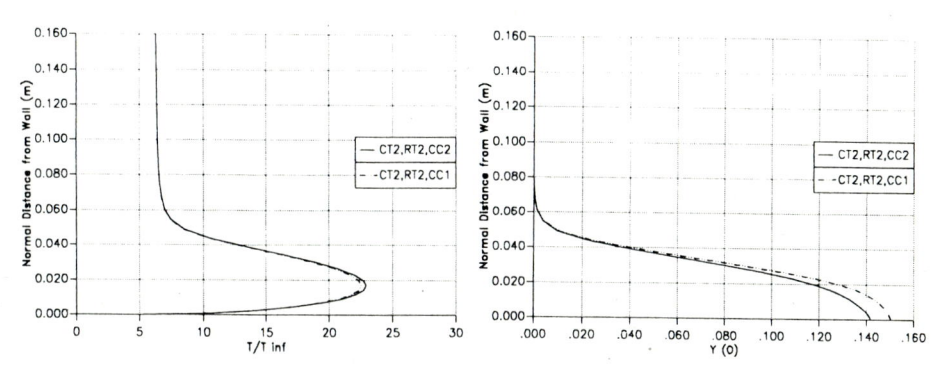

Fig. 5 – T/Tinf vs η.

Fig. 6 – Oxygen mass fraction vs η.

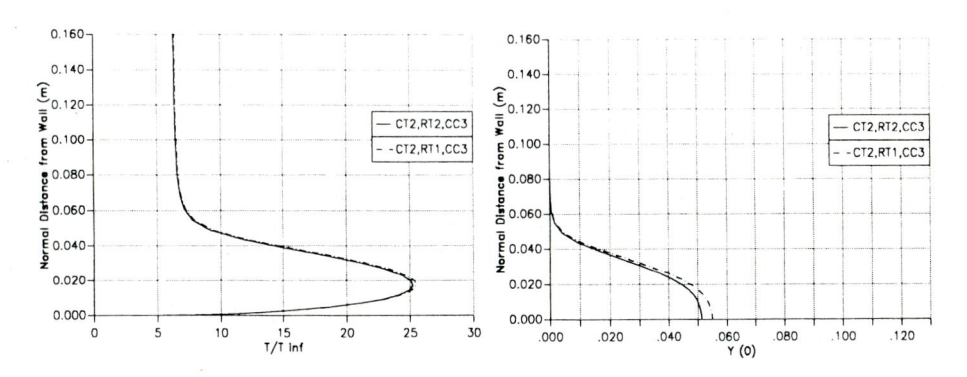

Fig. 7 – T/Tinf vs η.

Fig. 8 – Oxygen mass fraction vs η.

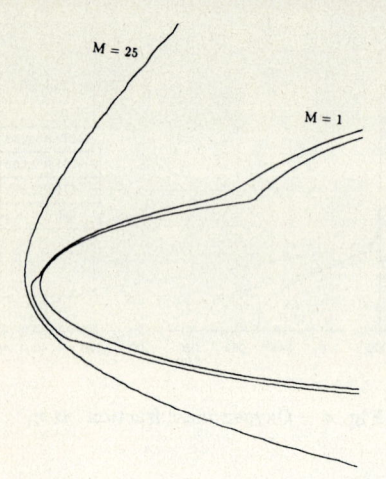

Fig. 9 – Iso-Mach (with non equilibrium chemistry).

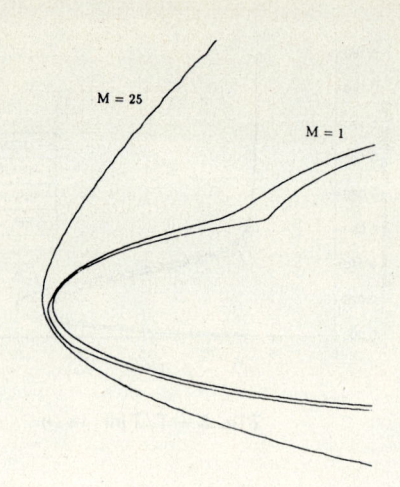

Fig. 10 – Iso-Mach (with equilibrium chemistry).

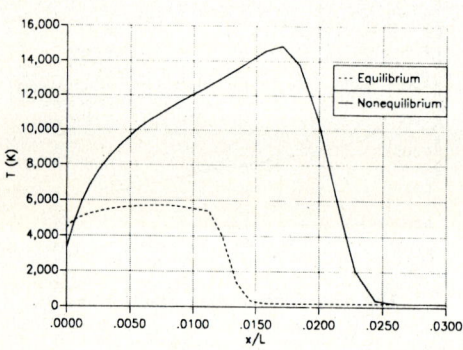

Fig. 11 – T distribution along stagnation line.

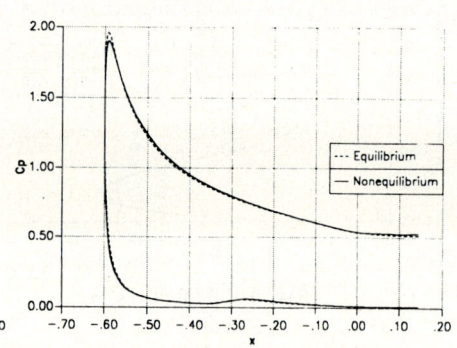

Fig. 12 – Pressure coefficient vs x.

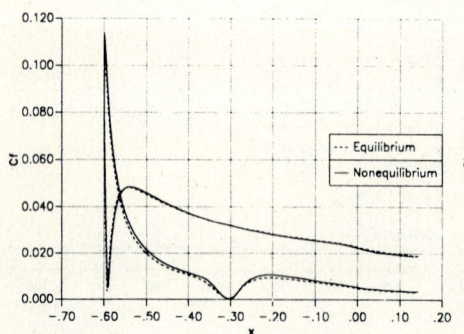

Fig. 13 – Skin friction coefficient vs x.

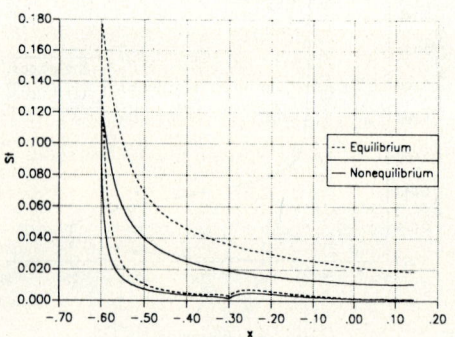

Fig. 14 – Stanton number vs x.

Influence of numerical diffusion in high temperature flow *

M. Fey and R. Jeltsch
Seminar für Angewandte Mathematik
ETH Zürich
CH-8092 Zürich, Switzerland

Abstract

In high temperature flow it is necessary to introduce new physical phenomena to the governing equations. Chemical reactions and vibrational excitation of the molecules lead to inhomogeneous Euler equations with a source term and an additional equation of conservation of mass for each species. From the mathematical point of view we only get additional contact discontinuities for the different species. From the numerical perspective the treatment of the fluxes of partial densities has a large influence on the results. For a given shock-capturing scheme we discuss three methods to compute the fluxes of partial densities for the same total density flux. We compare the numerical diffusion for the different fluxes. In a two-dimensional testcase we illustrate the advantages and disadvantages of these schemes. The third one shows a good resolution of the strong gradients in the mass fractions for this special testcase.

1 Introduction

For a class of flow problems such as the reentry of a space vehicle or the conditions in a shock tube the numerical simulation becomes important. From the need to calculate the flow field around a double ellipse at an altitude of 70 km during the reentry, several different experiments were made concerning the numerical diffusion of the scheme. A more or less academic problem is the stagnation point temperature for an inviscid flow. This is a very sensitive value in the flow field which depends strongly on the composition of the gas. Due to the strong density and temperature gradients near the stagnation point, numerical diffusion becomes important. For small-scaled bodies these gradients are infinite so that most of the numerical schemes do not calculate the correct temperature as it was shown in [1]. This problem is an interesting example to examine the influence of numerical diffusion. We will explain our investigations for this testcase. First we present two ways to introduce chemical reactions into a standard Euler-solver. The first one is a straightforward procedure

---

*This research has been supported by the research and developement program in the Hermes preparatory phase under contract RDANE 19/87 Step 3

with a large amount of numerical diffusion. The second one based on the physical background of contact surfaces does not show the correct behavior in the steady state solution. The combination of both ideas then leads to an improved scheme with reduced numerical diffusion.

2 The governing equations

The equations describing the one-dimensional inviscid flow with chemical reactions are the Euler equations in the form

$$U_t + F(U)_x = S(U) \tag{1}$$

where

$$U = \begin{pmatrix} \varrho_1 \\ \vdots \\ \varrho_N \\ m \\ E \end{pmatrix}, \quad F(U) = \begin{pmatrix} \varrho_1 u \\ \vdots \\ \varrho_N u \\ mu + p \\ u(E+p) \end{pmatrix}, \quad S(U) = \begin{pmatrix} s_1(\varrho_1, \ldots, \varrho_N, T) \\ \vdots \\ s_N(\varrho_1, \ldots, \varrho_N, T) \\ 0 \\ 0 \end{pmatrix}. \tag{2}$$

Here $\varrho_i, i = 1, \ldots, N$ denote the partial densities, m is the momentum and E the total energy including the formation enthalpy of the different species, p is the pressure, u the velocity and T the temperature of the gas. In the case of chemical non-equilibrium the eigenvalues of the Jacobian matrix of F(U) are $u - c, u, \ldots, u, u + c$. c is the frozen speed of sound given by $c^2 = \gamma p/\varrho$ where ϱ is the total density and γ is the ratio of the heat capacities for constant pressure and constant volume. In comparison to a pure gas there are additional contact discontinuities due to the different species all moving with velocity u. Equation (1) is equivalent to

$$\hat{U}_t + F(\hat{U})_x = \hat{S}(\hat{U}) \tag{3}$$

with

$$\hat{U} = \begin{pmatrix} \varrho_1 \\ \vdots \\ \varrho_N \\ m \\ \hat{E} \end{pmatrix}, \quad \hat{S}(\hat{U}) = \begin{pmatrix} s_1(\varrho_1, \ldots, \varrho_N, T) \\ \vdots \\ s_N(\varrho_1, \ldots, \varrho_N, T) \\ 0 \\ -\sum_{i=1}^{N} h_i^0 s_i \end{pmatrix}. \tag{4}$$

$\hat{E}$ is the total energy without the heat of formation, i.e.

$$E = \sum_{j=1}^{N} \varrho_j(e_j(T) + h_j^0) + \varrho \frac{u^2}{2} = \hat{E} + \sum_{j=1}^{N} \varrho_j h_j^0 \tag{5}$$

where $e_j(T)$ is the internal energy per unit mass of species j and h_j^0 the formation enthalpy.

3 The numerical scheme

The main underlying scheme is a finite volume method with the Van Leer flux vector splitting [3] to evaluate the fluxes at the cell interfaces. Starting from a discrete function U_i^n at time $n\Delta t$ at the point $i\Delta x$ we solve the inhomogeneous equations by an operator splitting approach. We first integrate the ordinary differential equation

$$\frac{\partial U}{\partial t} = S(U) \qquad (6)$$

for an intermediate value $U_i^{n+1/2}$. In this step we use formulation (1) to avoid the source term in the energy equation. In the next step we solve the homogeneous Euler equation. We adapt the standard solver to chemical reacting flow by the following procedure. First we rewrite the energy flux $F^E(U)$ to separate the heat of formation:

$$\begin{aligned} F^E(U) = u(E+p) &= u(\hat{E}+p) + \sum_{j=1}^N \varrho_j u h_j^0 \\ &= F^E(\hat{U}) + \sum_{j=1}^N h_j^0 F^{\varrho_j}(\hat{U}). \end{aligned} \qquad (7)$$

So we get the total energy flux as the standard energy flux plus the sum of the formation enthalpy of each species times their density flux $F^{\varrho_j}(\hat{U})$. Since there are no reactions in this step the mass fractions Y^j in each cell are constant. For this step we reduce the system to

$$W_t + \hat{F}(W)_x = 0$$

with

$$W = W(U) = \begin{pmatrix} \sum_{j=1}^N \varrho_j \\ m \\ E - \sum_{j=1}^N \varrho_j h_j^0 \end{pmatrix} = \begin{pmatrix} \varrho \\ m \\ \hat{E} \end{pmatrix}, \quad \hat{F}(W) = \begin{pmatrix} m \\ mu + p \\ u(\hat{E}+p) \end{pmatrix}$$

by summing over all partial densities and removing the heat of formation. We update the values of W with the Van Leer flux vector splitting method for real gas. We have

$$W_i^{n+1} = W_i^n - \frac{\Delta t}{\Delta x}\left(\hat{F}_{i+1/2} - \hat{F}_{i-1/2}\right) \qquad (8)$$

with

$$\hat{F}_{i+1/2} = \hat{F}_{i+1/2}(W_i^n, W_{i+1}^n) = F^+(W_i^n) + F^-(W_{i+1}^n).$$

The fluxes are

$$F^+(W) = \begin{cases} \hat{F}(W) & \text{if} \quad u > c \\ 0 & \text{if} \quad -c \geq u \end{cases}, \quad F^-(W) = \begin{cases} 0 & \text{if} \quad u > c \\ \hat{F}(W) & \text{if} \quad -c \geq u \end{cases},$$

and if $|u| \leq c$ then

$$\begin{aligned} F^{\varrho\pm}(W) &= \pm\frac{\varrho}{4c}(u \pm c)^2, \\ F^{m\pm}(W) &= \frac{\varrho}{4\gamma c}(u \pm c)^2 (2c \pm (\gamma-1)u), \\ F^{\hat{E}\pm}(W) &= \pm\frac{\varrho}{4c}(u \pm c)^2 \left[\frac{(2c \pm (\gamma-1)u)^2}{2(\gamma^2-1)} + \frac{c^2}{\gamma}\left(\frac{\bar{\gamma}}{\bar{\gamma}-1} - \frac{\gamma}{\gamma-1}\right)\right]. \end{aligned}$$

$\bar{\gamma}$ is an averaged value of $\gamma = \gamma(T)$ over a temperature region.

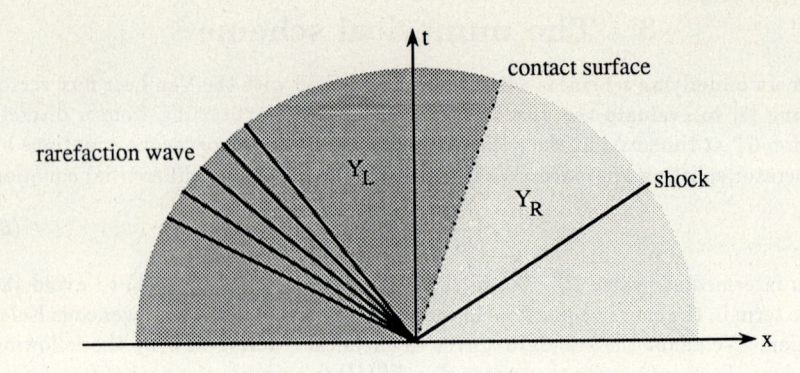

Figure 1: Separation of different mass fractions by the contact surface

With these properties we can construct the fluxes of the partial densities in a straightforward manner by setting

$$F^{\varrho j}_{i+1/2} = Y_i^j \hat{F}^{\varrho+}(W_i^n) + Y_{i+1}^j \hat{F}^{\varrho-}(W_{i+1}^n). \tag{9}$$

In the second approach we take a closer look at the physics of a contact surface. Different compositions of the gas are separated by a contact discontinuity and remain separated. From the direction of this wave we can derive the movement of the different mass fractions (Fig. 1). The propagation of this wave depends directly on the sign of the total mass flux. If this flux is positive then the contact surface moves to the right and vice versa. With this idea we can construct a second flux for the partial densities setting

$$F^{\varrho j}_{i+1/2} = \begin{cases} Y_i^j \hat{F}^{\varrho}_{i+1/2} & \text{if} \quad \hat{F}^{\varrho}_{i+1/2} > 0 \\ Y_{i+1}^j \hat{F}^{\varrho}_{i+1/2} & \text{if} \quad \hat{F}^{\varrho}_{i+1/2} \leq 0 \end{cases}. \tag{10}$$

With the fluxes of the partial densities in (9) or (10),

$$F^m_{i+1/2} = \hat{F}^m_{i+1/2}$$

for the momentum, and

$$F^E_{i+1/2} = \hat{F}^E_{i+1/2} + \sum_{j+1}^{N} h_j^0 F^{\varrho j}_{i+1/2} \tag{11}$$

for the energy we update the quantities in U in the same manner as in (8) using the intermediate state $U_i^{n+1/2}$ as the initial value.

$$U_i^{n+1} = U_i^{n+1/2} - \frac{\Delta t}{\Delta x}(F_{i+1/2} - F_{i-1/2}). \tag{12}$$

Both fluxes with (9) and (10) are consistent in the sense that $F_{i+1/2}(U, U) = F(U)$. We now compare these fluxes. To keep the notation as simple as possible we set:

$$\begin{aligned} F_i^{\pm} &:= \hat{F}^{\varrho\pm}(W(U_i^{n+1/2})) \\ F_{i+1/2} &:= \hat{F}^{\varrho}_{i+1/2} \\ G^j_{i+1/2} &:= F^{\varrho j}_{i+1/2} \quad \text{from (9)} \\ H^j_{i+1/2} &:= F^{\varrho j}_{i+1/2} \quad \text{from (10)} \end{aligned}$$

We look at the difference

$$(H^j_{i+1/2} - H^j_{i-1/2}) - (G^j_{i+1/2} - G^j_{i-1/2})$$

$$= \begin{cases} Y^j_i(F^+_i + F^-_{i+1}) - Y^j_{i-1}(F^+_{i-1} + F^-_i) & \text{if } F_{i+1/2} > 0, F_{i-1/2} > 0 \\ Y^j_i(F^+_i + F^-_{i+1}) - Y^j_i(F^+_{i-1} + F^-_i) & \text{if } F_{i+1/2} > 0, F_{i-1/2} \leq 0 \\ Y^j_{i+1}(F^+_i + F^-_{i+1}) - Y^j_{i-1}(F^+_{i-1} + F^-_i) & \text{if } F_{i+1/2} \leq 0, F_{i-1/2} > 0 \\ Y^j_{i+1}(F^+_i + F^-_{i+1}) - Y^j_i(F^+_{i-1} + F^-_i) & \text{if } F_{i+1/2} \leq 0, F_{i-1/2} \leq 0 \end{cases}$$

$$- \left(Y^j_i F^+_i + Y^j_{i+1} F^-_{i+1} - Y^j_{i-1} F^+_{i-1} - Y^j_i F^-_i \right)$$

$$= \begin{cases} (-F^-_{i+1})(Y^j_{i+1} - Y^j_i) + (-F^-_i)(Y^j_{i-1} - Y^j_i) & \text{if } F_{i+1/2} > 0, F_{i-1/2} > 0 \\ (-F^-_{i+1})(Y^j_{i+1} - Y^j_i) + F^+_i(Y^j_{i-1} - Y^j_i) & \text{if } F_{i+1/2} > 0, F_{i-1/2} \leq 0 \\ F^+_{i+1}(Y^j_{i+1} - Y^j_i) + (-F^-_i)(Y^j_{i-1} - Y^j_i) & \text{if } F_{i+1/2} \leq 0, F_{i-1/2} > 0 \\ F^+_{i+1}(Y^j_{i+1} - Y^j_i) + F^+_i(Y^j_{i-1} - Y^j_i) & \text{if } F_{i+1/2} \leq 0, F_{i-1/2} \leq 0. \end{cases}$$

For subsonic flow, i.e. $|u| < c$, there exists a constant $K \in \mathbb{R}$ such that

$$F^+ \geq K > 0 \quad \text{and}$$
$$-F^- \geq K > 0.$$

Then the fluxes in (9) and (10) can be compared yielding

$$G^j_{i+1/2} - G^j_{i-1/2} \approx H^j_{i+1/2} - H^j_{i-1/2} - K(Y^j_{i+1} - 2Y^j_i + Y^j_{i-1})$$

and the additional numerical diffusion can be estimated by the last term. In this estimate only the fluxes in the partial densities are needed.

For the numerical tests we changed the geometry of the Antibes testcase from an ellipse to a cylinder. This does not change the values in the stagnation point, it only removes additional complications due to the non-symmetric mesh and the curved streamline ending at the stagnation point. For the cylinder we can calculate a solution by integrating an ODE system for the partial densities and the velocity along the streamline. Figures 2 – 4 show the solutions for different scalings of the body. The size of the body is proportional to the inverse of v_y, the derivative of the y-velocity component in the y-direction. We can see from the plots of density, temperature, and mass fraction of nitrogen atoms Y^N that the chemical reactions move more and more to the stagnation point when the body becomes smaller and smaller.

Test calculations with these two fluxes show an unacceptable property of the flux in (10). From the energy equation we get

$$\frac{\partial E}{\partial t} + \frac{\partial}{\partial x}(u(E+p)) = \frac{\partial E}{\partial t} + \frac{\partial}{\partial x}(u\varrho H) = 0. \tag{13}$$

In the steady state we have

$$0 = \frac{\partial}{\partial x}(u\varrho H) = \varrho u \frac{\partial}{\partial x} H + H \frac{\partial}{\partial x} \varrho u. \tag{14}$$

Due to the conservation of mass the second term vanishes and thus H is constant along streamlines. Fig. 5 shows the total enthalpy H along the stagnation point streamline. For the flux in (10) the enthalpy increases drastically near the stagnation point in contrast to the standard Van Leer flux (9). This behavior can be explained by the following: The mass fractions Y^j are independent in different cells as they used to be due to the contact surface. The same holds for the heat of formation because in (11) the same flux of the partial densities is used. This contradicts equation (14) and the fact that the total enthalpy which includes the formation enthalpy is constant.

Therefore we define a third scheme. For the fluxes of the partial densities we keep the formula (10), but for the energy flux (11) we use the $F^{\varrho j}_{i+1/2}$ defined in (9):

$$F^{\varrho j}_{i+1/2} = \begin{cases} Y^j_i \hat{F}^{\varrho}_{i+1/2} & \text{if} \quad \hat{F}^{\varrho}_{i+1/2} > 0 \\ Y^j_{i+1} \hat{F}^{\varrho}_{i+1/2} & \text{if} \quad \hat{F}^{\varrho}_{i+1/2} \leq 0 \end{cases}$$

$$F^E_{i+1/2} = \hat{F}^E_{i+1/2} + \sum_{j+1}^{N} h^0_j \left[Y^j_i \hat{F}^{\varrho+}(W^n_i) + Y^j_{i+1} \hat{F}^{\varrho-}(W^n_{i+1}) \right]. \tag{15}$$

The resulting scheme is still consistent in the above sense. We now have a combination of both properties. The mass fractions are now nearly independent of the neighboring values and the numerical diffusion is reduced as shown above. On the other hand we introduce a mechanism to control the heat of formation over the cell interfaces in the energy flux.

The last sequence of figures shows the values of temperature, density, and mass fraction of nitrogen atoms for the three schemes with two different meshes. The size of the cell in front of the stagnation point is 0.5 mm and 0.2 mm, respectively.

4 Conclusions and Remarks

In the case of chemical reacting flow it is necessary to reduce the numerical diffusion. A comparison of the results seems to show that the combined scheme (15) gives a better approximation of the mass fractions than the standard Van Leer scheme (9). The investigations in this article are not restricted to Van Leer's scheme, for some other methods like TVD or ENO it is also possible to replace the standard flux of partial densities by the one in (10).

From useful discussions with colleagues it seems that for some special Godunov schemes an increase of the total enthalpy in the steady state solution also results from the contact surface of the tangential velocity in a two dimensional flow. This contact surface has the same properties as those mentioned above.

References

[1] Proceedings of the Antibes Workshop Part I (1990), to appear

[2] K.E. Brenan, S.L. Campbell, L.R. Petzold, *Numerical Solution of Initial-Value Problems in Differential-Algebraic Equations,*
North-Holland, New York, Amsterdam, London (1989)

[3] B. VAN LEER, *Flux-Vector Splitting for the Euler Equations,*
Lect. Notes in Phys. 170 (1982), pp 507-512

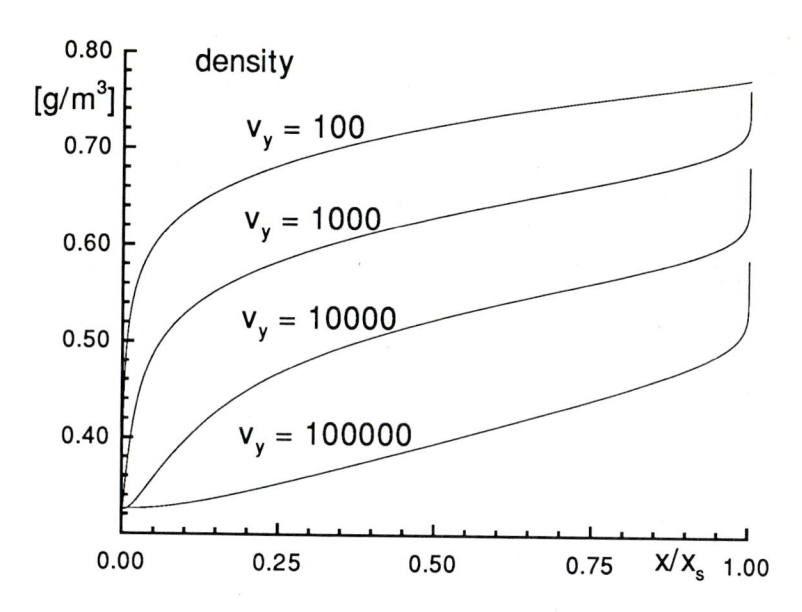

Figure 2: Density along the streamline with the ODE Solver DDASSL [2]

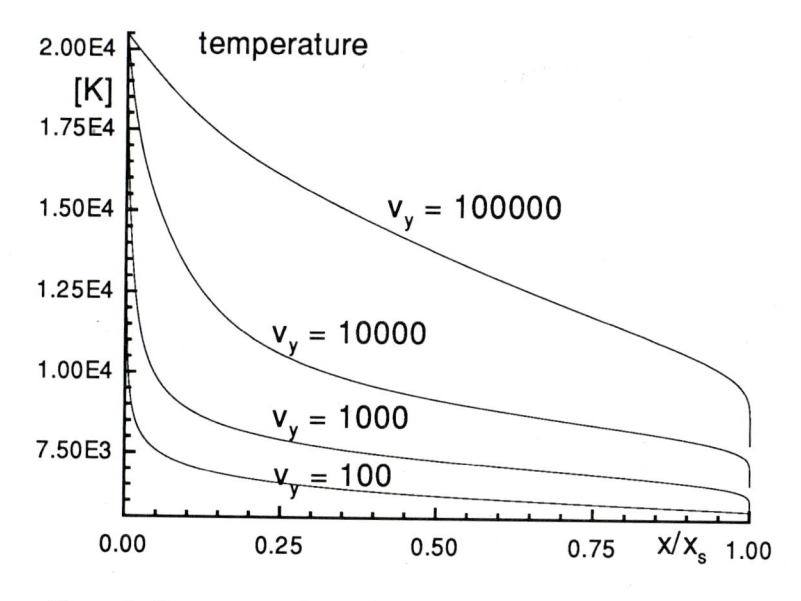

Figure 3: Temperature along the streamline with the ODE Solver DDASSL

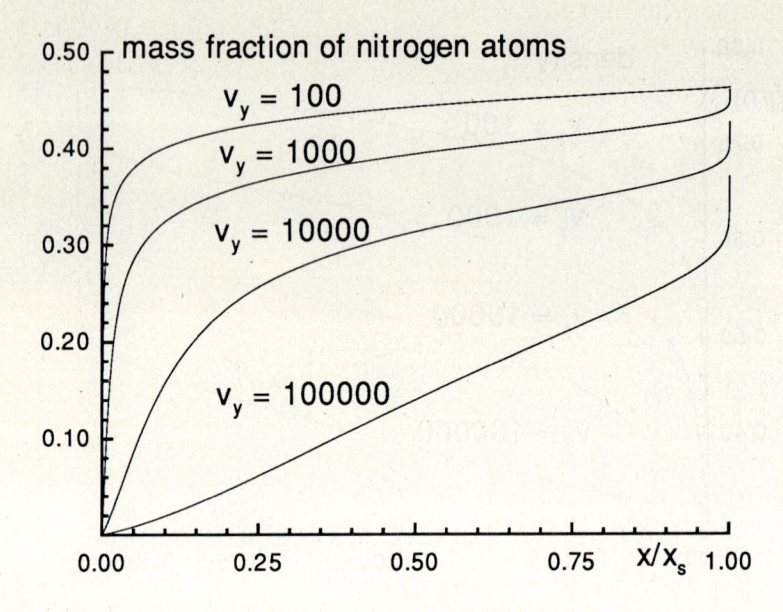

Figure 4: Mass fraction of nitrogen atoms along the streamline with the ODE Solver DDASSL

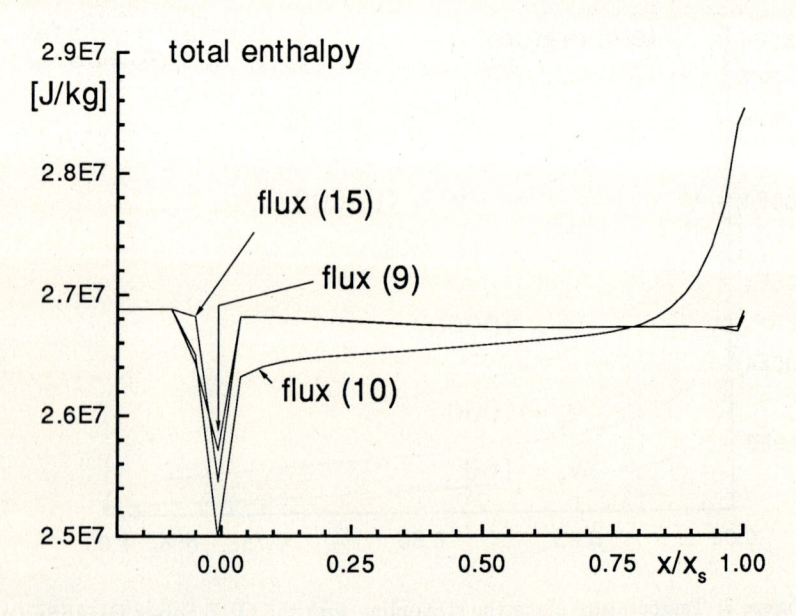

Figure 5: Total enthalpy along the streamline with different fluxes of partial densities

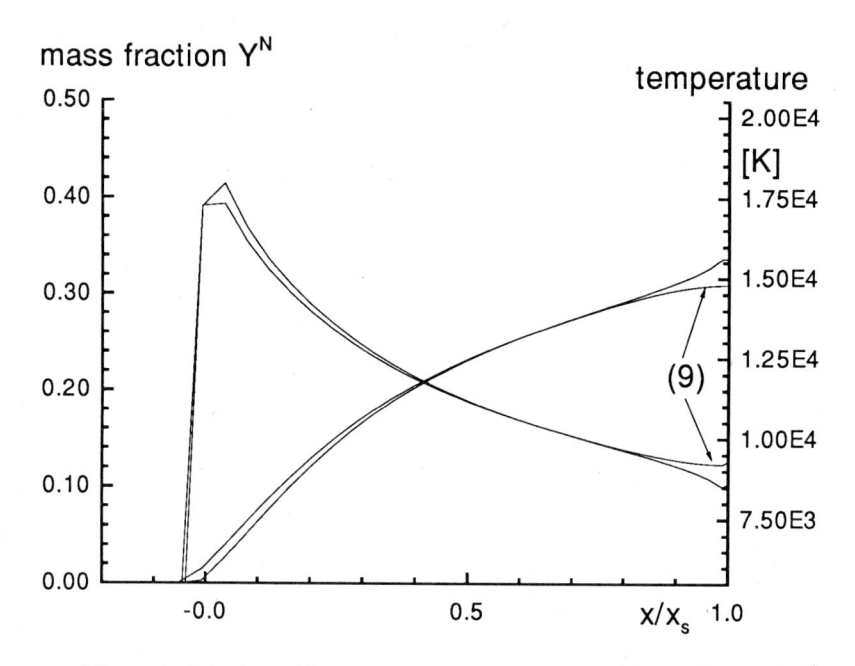

Figure 6: Solution with 0.5 mm cell size and fluxes (9) and (15).

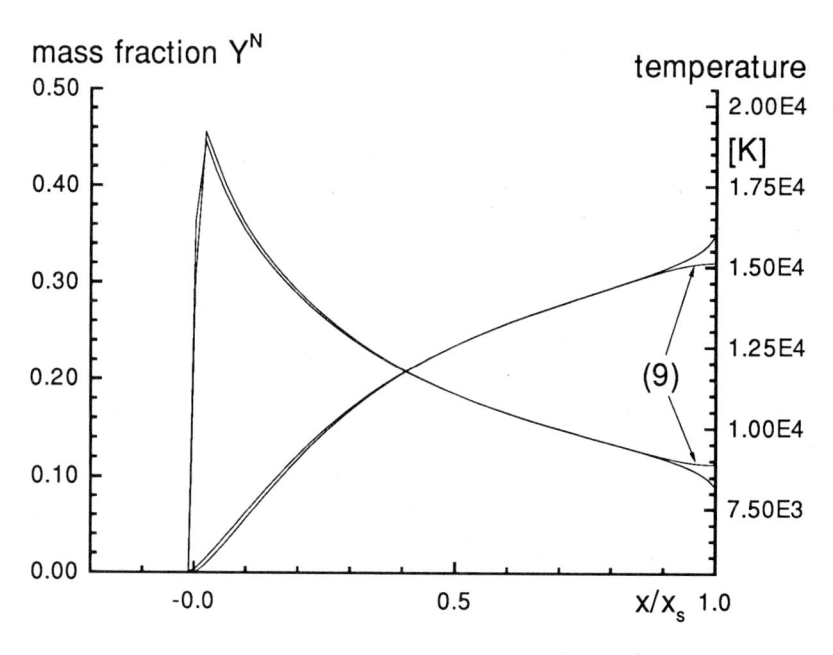

Figure 7: Solution with 0.2 mm cell size and fluxes (9) and (15).

ROBUST COMPUTATION OF 3D VISCOUS HYPERSONIC FLOW PROBLEMS

W. Schröder and G. Hartmann
Messerschmitt-Bölkow-Blohm GmbH
Space Systems Group, P.O.Box 80 11 69
8000 Munich 80, Germany

SUMMARY

A symmetric TVD relaxation method to solve the three-dimensional Navier-Stokes equations for equilibrium real gas flows is presented. The coefficient matrix is based on a linearization with respect to density, velocity and total enthalpy which reduces the work to compute the elements of the flux matrices and provides a more robust algorithm. A subgrid procedure for the viscous layer around body surfaces is introduced to focus the relaxation on the part of the computational domain in which generally a less benign convergence behavior is encountered and, hence, to save computer time. Results are presented for some laminar and turbulent viscous hypersonic flows.

INTRODUCTION

Since the late 1980s the investigation of hypersonic flowfields has seen a resurgence due to the conceptual design studies of several space crafts, as, e.g., the National Aerospace Plane NASP in the USA, the European reentry vehicle HERMES or the space transportation system SAENGER in Germany. To analyze the aerothermodynamics of these vehicles comprehensive experimental and numerical studies are being conducted. Among other things viscous flow features such as heat loads and separation regions have to be determined, i.e., to study such flowfields computationally the Navier-Stokes equations have to be solved.

In recent years relaxation methods have been proven extremely successful in solving viscous flow problems [1,2,3,4]. Their applications range from incompressible steady and unsteady flows to highly compressible flows. It turns out, however, that especially in hypersonic flows the efficiency and the robustness of transonic flow computations cannot be achieved. This is primarily caused through dramatic changes of the flowfield in shock waves and/or expansion regimes. Thus it is still necessary to improve the convergence and the versatile applicability of these methods. This is particularly true when viscous equilibrium real gas flows over "real" configurations are investigated.

The purpose of the paper is threefold. First, we present an ample description of our relaxation method to solve three-dimensional viscous hypersonic equilibrium real gas flow problems. Second, several ideas which ameliorate the efficiency and the robustness of relaxation solvers for the Navier-Stokes equations are briefly discussed. To be more precise, we introduce a linearization of the inviscid and the viscous fluxes with respect to density, velocity, and total enthalpy such that the flux matrices are easier to compute and better suited for hypersonic flows. When solutions of the Navier-Stokes equations are sought the convergence in the immediate wall region is generally poorer than in the rest of the computational domain. Therefore we implement a subgrid procedure that concentrates most of the iteration steps only on this portion of the computational domain and, hence, saves computer time. Third, to validate our solver it is applied to several two-dimensional and three-dimensional, laminar and turbulent test problems. The quality of the results is discussed by comparing them with other numerical and/or experimental data. Subsequently, the flowfield over a complete two-stage space system is computed and analyzed.

PROBLEM DEFINITION

Let ρ denote the density, u, v, w the Cartesian velocity components and ε, p the specific internal energy and the pressure. Then, neglecting body forces and heat sources, the thin-layer approximation to the convervative, nondimensional form of the time-dependent 3D Navier-Stokes equations of a compressible fluid (TLNS) can be written in a generalized, body-fitted coordinate system $\xi = \xi(x, y, z), \eta = \eta(x, y, z), \zeta = \zeta(x, y, z)$

$$Q_t + E_\xi + F_\eta + G_\zeta - Re_\infty^{-1}(E_{v,\xi} + F_{v,\eta} + G_{v,\zeta}) = 0 \tag{1}$$

where $Q = J^{-1}(\rho, \rho u, \rho v, \rho w, e)^T$ using $e = \rho(\varepsilon + 0.5(u^2 + v^2 + w^2))$ is the vector of conservative variables, $J = \partial(\xi, \eta, \zeta)/\partial(x, y, z)$ is the metric Jacobian, E, F, G represent the inviscid fluxes, E_v, F_v, G_v correspond to the simplified viscous fluxes and $Re_\infty = \rho_\infty u_\infty L/\mu_\infty$ is the Reynolds number defined by the reference values. To close the system of equations we use the equation of state $p = p(\rho, \varepsilon)$, which is valid for equilibrium real gas flows. The bulk viscosity is determined via Stokes hypothesis $\lambda + 2/3\mu = 0$ to formulate the simplified form of the shear stress. Unless otherwise stated, the molecular viscosity is evaluated using Sutherland's law

$$\frac{\mu}{\mu_\infty} = (\frac{\bar{T}}{\bar{T}_\infty})^{\frac{3}{2}} \frac{\bar{T}_\infty + \bar{C}}{\bar{T} + \bar{C}} \tag{2}$$

where $\mu_\infty, \bar{T}_\infty$ are the freestream vicosity and the freestream temperature, respectively, and $\bar{C}$ is Sutherland's constant for air.

The inflow and outflow boundary conditions are determined by the characteristics of the 1D Euler equations. The no-slip condition and either adiabatic or isothermal conditions are imposed on solid walls and undisturbed flow is assumed for the far field.

NUMERICAL METHOD

General Discrete Problem

The time and space discretization are separated to assure a steady state independent of the time step. The numerical scheme is implicit in time and employs Newton's method to compute the solution at each new time step. More precisely, after discretizing the TLNS equations in time, expanding in Δq, not in ΔQ, where q corresponds to $q = J^{-1}(\rho, u, v, w, H = (e + p)/\rho)^T$, and after linearizing we get [5]

$$Q_q \frac{\Delta q^{n+1,s}}{\Delta t} + rL^{n+1,s} = -\{\frac{Q^{n+1,s} - Q^n}{\Delta t} + rR^{n+1,s} + (1-r)R^n\} \tag{3}$$

$$R \equiv E_\xi + F_\eta + G_\zeta - Re_\infty^{-1}(E_{v,\xi} + F_{v,\eta} + G_{v,\zeta})$$

$$L \equiv (A\Delta q)_\xi + (B\Delta q)_\eta + (C\Delta q)_\zeta -$$
$$Re_\infty^{-1}((A_v\Delta q)_\xi + (B_v\Delta q)_\eta + (C_v\Delta q)_\zeta).$$

The indices n, s define the time and the iteration level, respectively. Setting $r = 1$ yields an $O(\Delta t)$ and $r = 1/2$ an $O(\Delta t^2)$ scheme. The quantities $Z = A, B, C$ and $Z_v = A_v, B_v, C_v$ are the flux matrices of the inviscid and viscous fluxes. We use the linearization with respect to q instead of Q to ensure efficiency and robustness of the scheme also for hypersonic flows. In such flows density changes dramatically throughout the flowfield due to strong shock waves and expansions, and sometimes during the convergence phase it even approaches zero. This leads to stiff matrices due to the reciprocal occurrence of the density in the elements. In any case low density values impair the robustness and the maximum possible time step which generally deteriorates the convergence behavior of the method. Therefore we try to reduce the number of terms containing the factor ρ^{-1} in the flux matrices by expanding the fluxes in Δq. Furthermore, using such an expansion the matrices Z_v are less costly to compute than $\partial(E_v, F_v, G_v)/\partial Q$, i.e., Z_v contains more zero elements than $\partial(E_v, F_v, G_v)/\partial Q$ and as such less operations per cell volume have to be conducted. Using $\omega = \xi, \eta, \zeta$ the matrices Q_q, Z, Z_v read

$$\frac{\partial Q}{\partial q} = \begin{pmatrix} 1 & 0 & 0 & 0 & 0 \\ u & \rho & 0 & 0 & 0 \\ v & 0 & \rho & 0 & 0 \\ w & 0 & 0 & \rho & 0 \\ H - p_\rho & -p_u & -p_v & -p_w & \rho - p_H \end{pmatrix}$$

$$Z = \begin{pmatrix} \vartheta_\omega & \rho\omega_x & \rho\omega_y & \rho\omega_z & 0 \\ u\vartheta_\omega + p_\rho\omega_x & \rho\vartheta_\omega + (\rho u + p_u)\omega_x & \rho u\omega_y + p_v\omega_x & \rho u\omega_z + p_w\omega_x & \rho H\omega_x \\ v\vartheta_\omega + p_\rho\omega_y & \rho v\omega_x + p_u\omega_y & \rho\vartheta_\omega + (\rho v + p_v)\omega_y & \rho v\omega_z + p_w\omega_y & \rho H\omega_y \\ w\vartheta_\omega + p_\rho\omega_z & \rho w\omega_x + p_u\omega_z & \rho w\omega_y + p_v\omega_z & \rho\vartheta_\omega + (\rho w + p_w)\omega_z & \rho H\omega_z \\ H\vartheta_\omega & \rho H\omega_x & \rho H\omega_y & \rho H\omega_z & \rho\vartheta_\omega \end{pmatrix}$$

$$Z_v = J \begin{pmatrix} 0 & 0 & 0 & 0 & 0 \\ 0 & \mu\alpha_{2,\omega}\frac{\partial\bullet}{\partial\omega} & \mu\beta_{2,\omega}\frac{\partial\bullet}{\partial\omega} & \mu\gamma_{2,\omega}\frac{\partial\bullet}{\partial\omega} & 0 \\ 0 & \mu\alpha_{3,\omega}\frac{\partial\bullet}{\partial\omega} & \mu\beta_{3,\omega}\frac{\partial\bullet}{\partial\omega} & \mu\gamma_{3,\omega}\frac{\partial\bullet}{\partial\omega} & 0 \\ 0 & \mu\alpha_{4,\omega}\frac{\partial\bullet}{\partial\omega} & \mu\beta_{4,\omega}\frac{\partial\bullet}{\partial\omega} & \mu\gamma_{4,\omega}\frac{\partial\bullet}{\partial\omega} & 0 \\ \chi\frac{\partial}{\partial\omega}(T_\rho\bullet) & \mu(\varphi_2\bullet + \phi_\alpha\frac{\partial\bullet}{\partial\omega}) + \chi\frac{\partial}{\partial\omega}(T_u\bullet) & \mu(\varphi_3\bullet + \phi_\beta\frac{\partial\bullet}{\partial\omega}) + \chi\frac{\partial}{\partial\omega}(T_v\bullet) & \mu(\varphi_4\bullet + \phi_\gamma\frac{\partial\bullet}{\partial\omega}) + \chi\frac{\partial}{\partial\omega}(T_w\bullet) & \chi\frac{\partial}{\partial\omega}(T_H\bullet) \end{pmatrix}$$

where we have introduced the cell-face normals $\omega_x, \omega_y, \omega_z$ and

$$\begin{aligned}
\vartheta_\omega &= u\omega_x + v\omega_y + w\omega_z & \alpha_{2,\omega} &= \omega_x^2 4/3 + \omega_y^2 + \omega_z^2 \\
\Gamma &= \tfrac{1}{\rho}\tfrac{\partial p}{\partial \varepsilon}|_{\rho=const} & \beta_{3,\omega} &= \omega_y^2 4/3 + \omega_x^2 + \omega_z^2 \\
p_\rho &= (\tfrac{\partial p}{\partial \rho}|_{\varepsilon=const} + p\Gamma/\rho)/(1+\Gamma) & \gamma_{4,\omega} &= \omega_z^2 4/3 + \omega_x^2 + \omega_y^2 \\
p_u &= -\Gamma\rho u/(1+\Gamma) & \alpha_{3,\omega} &= \beta_{2,\omega} = \omega_x\omega_y/3 \\
p_v &= -\Gamma\rho v/(1+\Gamma) & \alpha_{4,\omega} &= \gamma_{2,\omega} = \omega_x\omega_z/3 \\
p_w &= -\Gamma\rho w/(1+\Gamma) & \beta_{4,\omega} &= \gamma_{3,\omega} = \omega_y\omega_z/3 \\
p_H &= \Gamma\rho/(1+\Gamma) & \varphi_i &= \alpha_{i,\omega}u_\omega + \beta_{i,\omega}v_\omega + \gamma_{i,\omega}w_\omega \\
T_\rho &= \tfrac{\partial T}{\partial \rho}|_{\varepsilon=const} + \tfrac{\partial T}{\partial \varepsilon}|_{\rho=const}(\tfrac{p}{\rho} - p_\rho) & \phi_\delta &= u\delta_{2,\omega} + v\delta_{3,\omega} + w\delta_{4,\omega} \\
T_{u,v,w,H} &= \tfrac{p_{u,v,w,H}}{\rho\Gamma}\tfrac{\partial T}{\partial \varepsilon}|_{\rho=const} & \chi &= \kappa|\nabla\omega|^2 \tfrac{\kappa_\rho T_\infty}{u_\infty^2 \mu_\infty}.
\end{aligned}$$

The TLNS equations are solved by a finite volume method, i.e., flow variables are defined at cell centers and coordinates at cell vertices. Using a general notation for the spatial discretization the discrete problem reads

$$Q_q \frac{\Delta q^{n+1,s}}{\Delta t} + r L_\delta^{n+1,s} = -\left\{ \frac{Q^{n+1,s} - Q^n}{\Delta t} + r R_\delta^{n+1,s} + (1-r)R_\delta^n \right\} \tag{4}$$

$$R_\delta \equiv \delta_\xi E + \delta_\eta F + \delta_\zeta G - Re_\infty^{-1}(\delta_\xi E_v + \delta_\eta F_v + \delta_\zeta G_v)$$

$$L_\delta \equiv \delta_\xi(A\Delta q) + \delta_\eta(B\Delta q) + \delta_\zeta(C\Delta q) - Re_\infty^{-1}(\delta_\xi(A_v\Delta q) + \delta_\eta(B_v\Delta q) + \delta_\zeta(C_v\Delta q)).$$

The symbol $\delta_\omega M$ with $M = E, F, G$, $\omega = \xi, \eta, \zeta$, and $m = i, j, k$ is understood as $\delta_\omega M = M_{m+\frac{1}{2}} - M_{m-\frac{1}{2}}$ where the constant subscripts along the ω-line are dropped. Upon convergence the spatial accuracy of the solution of the approximate Navier-Stokes equations (4) is determined only by the discretization of the right-hand side, i.e., by the R_δ-expression. Therefore the spatial approximation of the L_δ-term on the left-hand side will be chosen under the condition to develop an efficient algorithm to invert the solution matrix in every time step.

Spatial Discretization of the Explicit Operator

The viscous terms are expressed by central differencing. The derivatives of the inviscid fluxes are approximated by a symmetric TVD scheme [6,7]. At the cell interface $\Gamma_{m+\frac{1}{2}}$ the general flux $M_{m+\frac{1}{2}}$ is formulated as

$$M_{m+\frac{1}{2}} = \frac{1}{2}\{M(Q_{m+1},\Gamma_{m+\frac{1}{2}}) + M(Q_m,\Gamma_{m+\frac{1}{2}}) -$$
$$(\varphi^{-1}R^N\psi(\Lambda^N))_{m+\frac{1}{2}}(\alpha^N_{m+\frac{1}{2}} - S(\alpha^N_{m-\frac{1}{2}},\alpha^N_{m+\frac{1}{2}},\alpha^N_{m+\frac{3}{2}}))\} \quad (5)$$

where a four argument minmod limiter

$$S(\alpha^N_{m-\frac{1}{2}},\alpha^N_{m+\frac{1}{2}},\alpha^N_{m+\frac{3}{2}}) = minmod(\alpha^N_{m-\frac{1}{2}},\alpha^N_{m+\frac{1}{2}},\alpha^N_{m+\frac{3}{2}},\frac{1}{2}(\alpha^N_{m-\frac{1}{2}}+\alpha^N_{m+\frac{1}{2}})) \quad (6)$$

is applied to the difference of the Riemann invariants

$$\alpha^N_{m+\frac{1}{2}} = \varphi(R^N_{m+\frac{1}{2}})^{-1}(Q_{m+1} - Q_m). \quad (7)$$

To accomplish a sensitive limiting the Riemann invariants are scaled by $\varphi = c_R^2/\rho$ where the square of the sound velocity c_R^2 is defined by

$$c_R^2 = \frac{\partial p}{\partial \rho}|_{\varepsilon=const} + \frac{p}{\rho^2}\frac{\partial p}{\partial \varepsilon}|_{\rho=const} = \frac{\partial p}{\partial \rho}|_{\varepsilon=const} + \Gamma(H - \varepsilon - \frac{1}{2}(u^2+v^2+w^2)). \quad (8)$$

The entropy correction function $\psi(f)$

$$\psi(f) = \begin{cases} |f| & |f| \geq \epsilon \\ (f^2+\epsilon^2)/2\epsilon & |f| < \epsilon \end{cases}$$

with a small positive parameter ϵ is used, the matrices $R^N, (R^N)^{-1}$ are the right and left eigenvector matrices of the flux matrices $N = \partial M/\partial Q$ and $\Lambda^N_{m+\frac{1}{2}}$ represents the eigenvalue matrix

$$\Lambda^N_{m+\frac{1}{2}} = diag(\vartheta_\omega, \vartheta_\omega, \vartheta_\omega, \vartheta_\omega + c_R|\nabla\omega|, \vartheta_\omega - c_R|\nabla\omega|) \quad (9)$$

where ϑ_ω is the contravariant velocity. Unless otherwise stated, the variables at cell interfaces are computed as Roe-averages [8]. This discretization results in a second-order accurate approximation of the steady state operator provided the mesh and the solution are smooth enough.

Spatial Discretization of the Implicit Operator

The overall approximation of the implicit operator is first-order accurate in space. Central differences are applied to the second derivatives. The terms of the flux matrices $Z = A, B, C$ are again formulated as a balance across a cell $\delta_\omega(Z\Delta q) = (Z\Delta q)_{m+\frac{1}{2}} - (Z\Delta q)_{m-\frac{1}{2}}$ and $(Z\Delta q)_{m+\frac{1}{2}}$ is given by

$$(Z\Delta q)_{m+\frac{1}{2}} = \frac{1}{2}\{Z(Q_{m+1},\Gamma_{m+\frac{1}{2}})\Delta q_{m+1} + Z(Q_m,\Gamma_{m+\frac{1}{2}})\Delta q_m -$$
$$(Q_q X^Z \psi(\Lambda^Z)(X^Z)^{-1})_{m+\frac{1}{2}}(\Delta q_{m+1} - \Delta q_m)\} \quad (10)$$

where Λ^Z corresponds to Λ^N and $X^Z, (X^Z)^{-1}$ are the right and left eigenvector matrices of Z

$$X^Z = \begin{pmatrix} \tilde{\omega}_x & \tilde{\omega}_y & \tilde{\omega}_z & 1 & 1 \\ 0 & -\tilde{\omega}_z & \tilde{\omega}_y & \tilde{\omega}_x c_R/\rho & -\tilde{\omega}_x c_R/\rho \\ \tilde{\omega}_z & 0 & -\tilde{\omega}_x & \tilde{\omega}_y c_R/\rho & -\tilde{\omega}_y c_R/\rho \\ -\tilde{\omega}_y & \tilde{\omega}_x & 0 & \tilde{\omega}_z c_R/\rho & -\tilde{\omega}_z c_R/\rho \\ -\tilde{\omega}_x c_\varepsilon + v\tilde{\omega}_z & -\tilde{\omega}_y c_\varepsilon - u\tilde{\omega}_z & -\tilde{\omega}_z c_\varepsilon + u\tilde{\omega}_y & \frac{c_R}{\rho}(c_R+\vartheta_\omega) & \frac{c_R}{\rho}(c_R-\vartheta_\omega) \\ -w\tilde{\omega}_y & +w\tilde{\omega}_x & -v\tilde{\omega}_x & & \end{pmatrix}$$

$$(X^Z)^{-1} = \begin{pmatrix} \tilde{\omega}_x \sigma & \tilde{\omega}_x \sigma \frac{\rho u}{c_R^2} & \tilde{\omega}_z + \tilde{\omega}_x \sigma \frac{\rho v}{c_R^2} & -\tilde{\omega}_y + \tilde{\omega}_x \sigma \frac{\rho w}{c_R^2} & -\tilde{\omega}_x \sigma \frac{\rho}{c_R^2} \\ \tilde{\omega}_y \sigma & -\tilde{\omega}_z + \tilde{\omega}_y \sigma \frac{\rho u}{c_R^2} & \tilde{\omega}_y \sigma \frac{\rho v}{c_R^2} & \tilde{\omega}_x + \tilde{\omega}_y \sigma \frac{\rho w}{c_R^2} & -\tilde{\omega}_y \sigma \frac{\rho}{c_R^2} \\ \tilde{\omega}_z \sigma & \tilde{\omega}_y + \tilde{\omega}_z \sigma \frac{\rho u}{c_R^2} & -\tilde{\omega}_x + \tilde{\omega}_z \sigma \frac{\rho v}{c_R^2} & \tilde{\omega}_z \sigma \frac{\rho w}{c_R^2} & -\tilde{\omega}_z \sigma \frac{\rho}{c_R^2} \\ \frac{1}{2(1+\Gamma)} & \frac{\rho}{2c_R}(\tilde{\omega}_x - \frac{u\sigma}{c_R}) & \frac{\rho}{2c_R}(\tilde{\omega}_y - \frac{v\sigma}{c_R}) & \frac{\rho}{2c_R}(\tilde{\omega}_z - \frac{w\sigma}{c_R}) & \frac{\rho}{2c_R^2}\sigma \\ \frac{1}{2(1+\Gamma)} & -\frac{\rho}{2c_R}(\tilde{\omega}_x + \frac{u\sigma}{c_R}) & -\frac{\rho}{2c_R}(\tilde{\omega}_y + \frac{v\sigma}{c_R}) & -\frac{\rho}{2c_R}(\tilde{\omega}_z + \frac{w\sigma}{c_R}) & \frac{\rho}{2c_R^2}\sigma \end{pmatrix}$$

with

$$\sigma = \Gamma/(1+\Gamma) \qquad \tilde{\omega}_{x,y,z} = \omega_{x,y,z}/|\nabla \omega|$$
$$c_\varepsilon = c_R^2 \left(\frac{\partial p}{\partial \varepsilon}\right)^{-1}_{\rho=const} \qquad \tilde{\vartheta}_\omega = \vartheta_\omega/|\nabla \omega| .$$

To reduce the computational costs and to strengthen the main diagonal of the solution matrix we generally neglect the difference $Z(Q_m, \Gamma_{m+\frac{1}{2}}) - Z(Q_m, \Gamma_{m-\frac{1}{2}})$ and use only the maximum eigenvalue to determine $\psi_{max}^Z = \psi(\Lambda_{max}^Z)$ such that Eq. (10) can be rewritten [5]

$$(Z\Delta q)_{m+\frac{1}{2}} = \frac{1}{2}\{Z(Q_{m+1}, \Gamma_{m+\frac{1}{2}})\Delta q_{m+1} - (Q_q I \psi_{max}^Z)_{m+\frac{1}{2}}(\Delta q_{m+1} - \Delta q_m)\} \qquad (11)$$

where I is the identity matrix. Arithmetic averaging is applied to evaluate the variables at cell interfaces. To compute $p = p(\rho, \varepsilon)$, $T = T(\rho, \varepsilon)$ etc. we incorporate the vectorized fitting routines used in [9].

Algorithm

The discretization described in the preceding subsections results in a block-pentadiagonal or a block-heptadiagonal system of equations depending on the dimensionality of the flow problem. To discuss the solution method it is appropriate to rewrite the discrete problem in a more compact form. Omitting the indices n, s of Eq. (4) and using the notation $(T_m + T_{m+n})\Delta q = T_m \Delta q_m + T_{m+n} \Delta q_{m+n}$ we obtain

$$(T_{ijk} + T_{i+1jk} + T_{ij+1k} + T_{ijk+1} + T_{i-1jk} + T_{ij-1k} + T_{ijk-1})\Delta q = RHS_{ijk}, \qquad (12)$$

where the matrices T_{mnl} contain all the coefficients of the cell volume (mnl) and RHS_{ijk} represents the right-hand side of Eq. (4). Since we want a method with good convergence properties and one that is easy to vectorize and with an appropriate structure for concurrent computers we use a symmetric point Gauss-Seidel relaxation scheme with red-black ordering for j=const. planes to obtain an approximate solution to Eq. (12).

$$\begin{aligned} \Delta q_{ijk} &= RHS_{ijk} - (T_{i-1jk} + T_{ij-1k} + T_{ijk-1})\Delta q & \text{(red: ik)} \\ \Delta q_{ijk}^{n+1,s} &= RHS_{ijk} - (T_{i-1jk} + T_{ij-1k} + T_{ijk-1})\Delta q & \\ & \quad - (T_{i+1jk} + T_{ij+1k} + T_{ijk+1})\Delta q^{n+1,s} & \text{(black: ik)} . \end{aligned} \qquad (13)$$

Subsequently, the vector of solution $Q^{n+1,s+1}$ is updated via

$$Q^{n+1,s+1} = Q^{n+1,s} + Q_q \Delta q^{n+1,s}. \qquad (14)$$

The reciprocal value of the maximum residual, i.e., the right-hand side of Eq. (4), controls the time step. In the beginning of the computation the time step is increased with decreasing maximum residual. If, however, the time step is greater than $\Delta t_{max} = 1$ then Δt is set $\Delta t = \Delta t_{max}$. Generally a much faster convergence can be observed in the inviscid part of the flow than within the viscous layer around a body. In this viscous part the density and temperature distribution still change while the outer flowfield can be considered already converged. Therefore we apply a subgrid procedure [5], i.e., we decompose the mesh into two parts. The first part contains the complete grid, whereas the subgrid comprises the part of the flowfield in which mainly viscous effects prevail. First the relaxation is performed on all cells, then, when the residual has dropped, say, four orders of magnitude at 50 percent of the grid points, we switch to the subgrid procedure. That means, the solution vector is iterated in every time step only on the subgrid whereas on the outer cells it is altered just in every fifth sweep.

In the iteration from level n to level $n + 1$ the boundary conditions are treated explicitly. That is, the boundary values are updated only after the values at interior points have been iterated. As a consequence the solution vector Q has all its components at the same order of accuracy only when the iterations have converged.

RESULTS

We will present results for several 2D and 3D hypersonic flows. In all flow problems only steady state solutions are sought, i.e., the parameter in Eq. (4) is set $r = 1$ and $n + 1, s \rightarrow n$. The impact of the q-linearization and the subgrid procedure on the convergence rate has been shown in [4] for a similar method and will not be discussed here.

A two-dimensional perfect gas flow at a freestream Mach number $M_\infty = 14.1$ and a freestream Reynolds number $Re_\infty/m = 6 \times 10^6$ over a $15°$ ramp is investigated. The freestream temperature is $\bar{T}_\infty = 57K$ and the temperature at the wall is $\bar{T}_w = 290K$. The Mach number and the density contours in Figs. 1a,b illustrate the flowfield (shocks, boundary layer, entropy layer, separation bubble) for an assumed laminar flow. However, the state of the flow, laminar or turbulent, is not known a priori. Therefore we perform several computations for laminar and turbulent [10] flows and flows undergoing transition on a (251×59)-cell mesh for the flow and the crossflow direction. In Fig. 1c the results of the Stanton number distribution are compared with experimental data of [11]. Assuming laminar flow the size of the separation region at the compression corner is in close agreement with the experiment whereas the level after the reattachment $(X/L \cong 1.2)$ is much better predicted when the flow is considered turbulent. The flow seems to undergo transition after the compression since the best overall agreement between numerical and experimental data is achieved when a laminar-turbulent transition point is fixed at $(X/L = 1.3)$.

To validate the present method of solution for equilibrium real gas flows the flow over a hyperbola at $M_\infty = 10, Re_\infty/m = 1.6 \times 10^5, \bar{T}_\infty = 220K$, at a freestream density $\bar{\rho}_\infty = 7.7 \times 10^{-4} kg/m^3$, and at a freestream pressure $\bar{p}_\infty = 48.67 N/m^2$ is computed. The mesh that is clustered near the body surface consists of 150×60 cells along and normal to the wall. In Fig. 2 the results of a perfect and a real gas flow are contrasted (Fig. 2a) and compared with an Euler-boundary-layer solution [9] (Fig. 2b). Due to the higher density values in real gas flows the bow shock is much closer to the body than in the perfect gas solution. The comparison of the shock locations on the symmetry line $x_s/r(perf., Nav.Stok.) = 0.556$ and $x_s/r(real, Nav.Stok.) = 0.321$ with $x_s/r(perf., [9]) = 0.560$ and $x_s/r(real, [9]) = 0.314$, where $r = 0.015m$ defines the nose radius, evidences the good agreement between the present solution and the results of [9] for perfect and real gas flows. When the variables at the cell interfaces are computed by simple arithmetic instead of Roe averaging we get essentially the same results, i.e., the convergence history doesn't change and the shock location, the pressure contours etc. are alike.

Next a 3D laminar flow over a double ellipsoid [11] at $M_\infty = 8.15, Re_\infty/m = 1.67 \times 10^7, \bar{T}_\infty = 56K, \bar{T}_w = 288K$, at an angle of attack $\alpha = 30°$ is considered. The flowfield is resolved by 100 crossflow planes with 63 cells almost equally spaced in the azimuthal direction and 53 cells normal to the body clustered near the wall using a minimum step size 10^{-4}. Using the skin-friction coefficient and the Stanton number distribution in the symmetry plane we compare our results with numerical and experimental results of [11,12] (Fig. 3). Especially the agreement with Monnoyer's boundary-layer solution is excellent. Unlike Hänel we do not detect the experimentally confirmed very weak separation at $X/L \cong 0.9$ (Fig. 3a). Satisfactory correspondence is achieved for the Stanton number results on the windward and on the leeward side (Fig. 3b).

Results for the laminar hypersonic flow over a delta wing [12] at $M_\infty = 8.7, Re_\infty/m = 2.25 \times 10^6, \bar{T}_\infty = 55K, \bar{T}_w = 300K$, and $\alpha = 30°$ are presented in Fig. 4. The mesh contains 64 crossflow planes using 102 cells in the azimuthal direction and 54 cells normal to the body with an axially varying minimum step size from 10^{-5} to 10^{-4}. Our results for the skin-friction coefficient and the pressure coefficient in the symmetry plane (Figs. 4a,b) are very close to Monnoyer's boundary-layer solution [12] and Menne's Navier-Stokes data [13]. Fig. 4c shows Mach number contours in the symmetry plane. The bow shock is sharply captured and no separation occurs in mainstream direction. The Mach number contours in the crossflow plane $X/C = 0.8$ display a crossflow shock wave and a crossflow separation on the leeward side (Fig. 4d). The fact that there is only one leeward vortex pair is evidenced by the wall-streamlines in Fig. 4e. The total lift and drag coefficients and their parts due to pressure "P" and friction "F" of the present computation "SH", of Menne [13], and of Riedelbauch [12], all of which are determined on a fine and a coarse mesh, are compiled in Table 1.

Table 1: Total lift and drag coefficients for delta-wing flow with grid refinement.

	grid	C_L	C_{LP}	C_{LF}	C_D	C_{DP}	C_{DF}	C_L/C_D
SH	coarse	0.5328	0.5365	-0.00363	0.4094	0.4003	0.00907	1.301
	fine	0.5329	0.5364	-0.00349	0.4094	0.4003	0.00911	1.302
[13]	coarse	0.5338	0.5369	-0.00306	0.4087	0.4007	0.00802	1.306
	fine	0.5344	0.5377	-0.00339	0.4098	0.4014	0.00844	1.303
[12]	coarse	0.5409	0.5447	-0.00372	0.4149	0.4066	0.00834	1.303
	fine	0.5409	0.5447	-0.00374	0.4151	0.4067	0.00846	1.303

Preliminary results for a laminar hypersonic perfect gas flow over a two-stage system at $M_\infty = 6, Re_\infty/m = 4.2 \times 10^5, \bar{T}_\infty = 229K, \alpha = 0°$, and at a relative angle and a maximum gap width between the upper and the lower stage $\Delta\alpha = 0°, \Delta z = 2.24m$ are discussed. Adiabatic body surfaces are assumed. We use a two-block mesh the outer of which contains $51 \times 139 \times 90$ normal $\times$ axial $\times$ azimuthal cells. The inner $(40 \times 139 \times 30)$-cell mesh resolves the gap between the upper and the lower stage (Fig. 5a). The minimum normal step size varies from 10^{-4} to 10^{-3}. For lack of storage capacity we apply a block-marching technique similar to the one described in [13] to conduct the computation. Looking at the density contours and the quasi-streamlines within the gap between the upper and the lower stage the intricacy of the flowfield becomes very evident (Fig. 5). The bow shocks of the upper and the lower stage are sufficiently captured, the shock-shock interaction and the shock-boundary-layer interaction are adequately resolved (Figs. 5a,b,c,d). The shock wave of the upper stage impinges upon the upper surface of the lower stage and causes a massive separation region. The shock is reflected and generates another flow separation, in flow and in crossflow direction, on the lower side of the upper stage (Figs. 5e,f,g). The analysis of such flowfields is extremely important to understand what will happen when the upper and the lower stage separate.

CONCLUSION

The symmetric TVD relaxation method for equilibrium real gas flows has been discussed. Several laminar and turbulent, 2D and 3D Navier-Stokes problems were computed to show the robustness, quality, and versatile applicability of the method when hypersonic flows over complex geometries are numerically investigated. The results are in satisfactory to good agreement with other computational and/or experimental data.

REFERENCES

1. J. L. Thomas and R. W. Walters, *AIAA Paper* 85-1501 (1985).
2. S. Yoon and A. Jameson, *AIAA Paper* 87-600 (1987).
3. W. Schröder and D. Hänel, *Comp. & Fluids* **15**(3), 313 (1987).
4. W. Schröder and H. B. Keller, *J. Comp. Phys.* **91**(1), 197 (1990).
5. W. Schröder and G. Hartmann, to be published in *Comp. & Fluids*.
6. H. C. Yee, *J. Comp. Phys.* **68**, 151 (1987).
7. M. Pfitzner, W. Schröder, S. Menne and C. Weiland, *Int. Conf. on Hypersonic Aerodyn.*, Manchester, UK (1989).
8. M. S. Liou, B. van Leer and J. S. Shuen, *J. Comp. Phys.* **87**, 1 (1990).
9. Ch. Mundt, M. Pfitzner and M. A. Schmatz, *Notes on Numerical Fluid Mechanics* **29**, 419 (1990).
10. B. S. Baldwin and H. Lomax, *AIAA Paper* 78-257 (1978).
11. INRIA, GAMNI/SMAI, *Workshop on Hypersonic Flows for Reentry Problems*. Antibes, France (1990).
12. INRIA, GAMNI/SMAI, *Workshop on Hypersonic Flows for Reentry Problems*. Antibes, France (1991).
13. S. Menne, *First Europ. Symp. on Aerodyn. for Space Vehicles*, Noordwijk, Netherlands (1991).

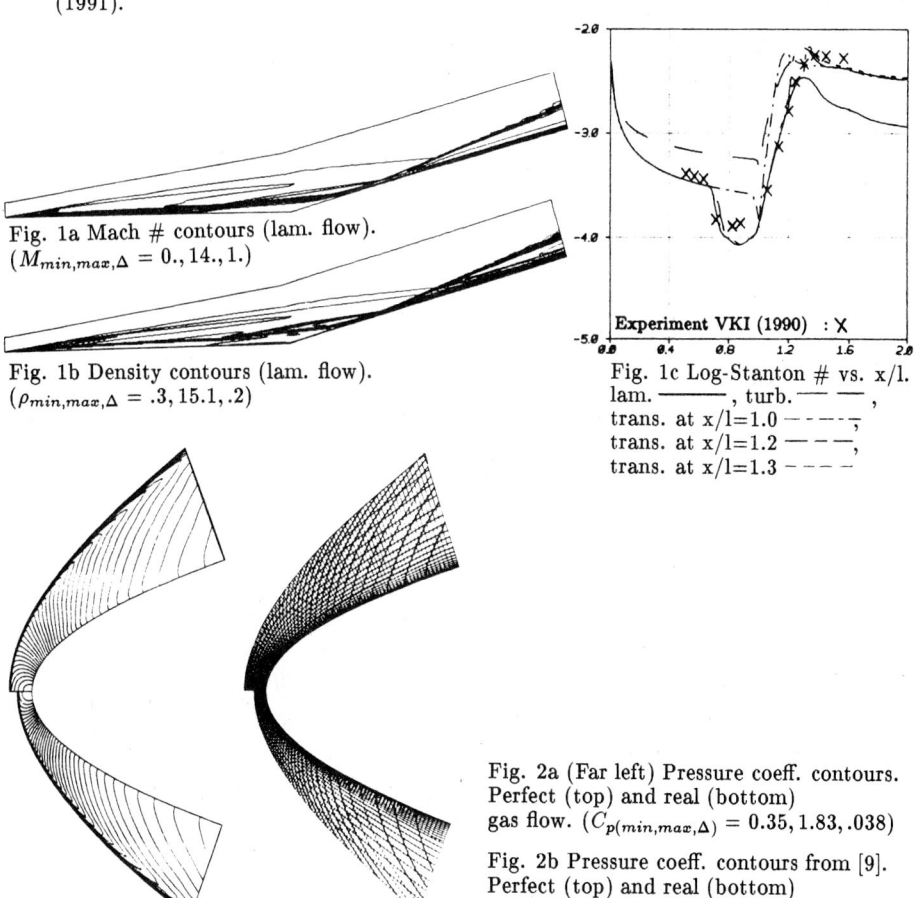

Fig. 1a Mach # contours (lam. flow). $(M_{min,max,\Delta} = 0., 14., 1.)$

Fig. 1b Density contours (lam. flow). $(\rho_{min,max,\Delta} = .3, 15.1, .2)$

Fig. 1c Log-Stanton # vs. x/l. lam. ———, turb. —— —, trans. at x/l=1.0 — - - -, trans. at x/l=1.2 — — —, trans. at x/l=1.3 - - - -

Fig. 2a (Far left) Pressure coeff. contours. Perfect (top) and real (bottom) gas flow. $(C_{p(min,max,\Delta)} = 0.35, 1.83, .038)$

Fig. 2b Pressure coeff. contours from [9]. Perfect (top) and real (bottom) gas flow. $(C_{p(min,max,\Delta)} = 0.35, 1.83, .038)$

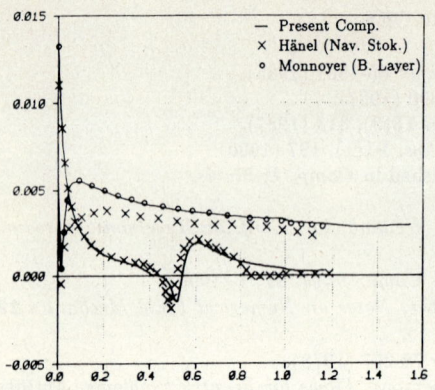

Fig. 3a Skin-friction coeff. vs. x/l in the sym. plane.

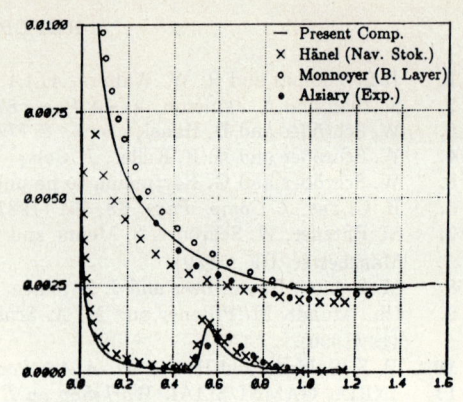

Fig. 3b Stanton # vs. x/l in the sym. plane.

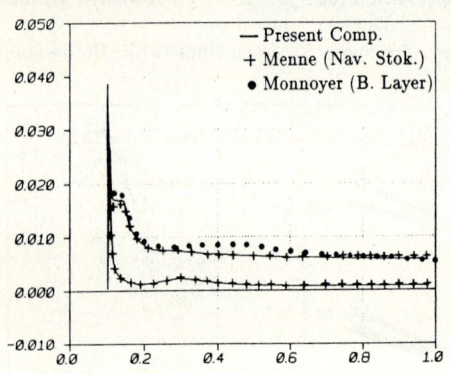

Fig. 4a Skin-friction coeff. vs. x/l in the sym. plane.

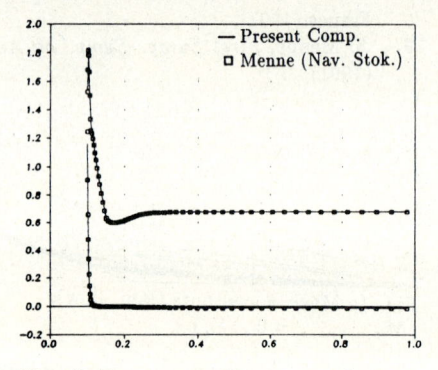

Fig. 4b Pressure coefficient vs. x/l in the sym. plane.

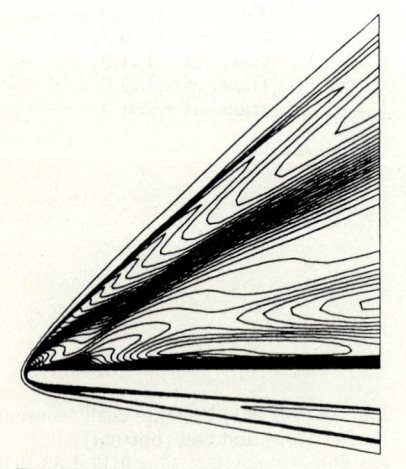

Fig. 4c Mach # contours. ($M_{min,max,\Delta} = 0, 8.6, .2$)

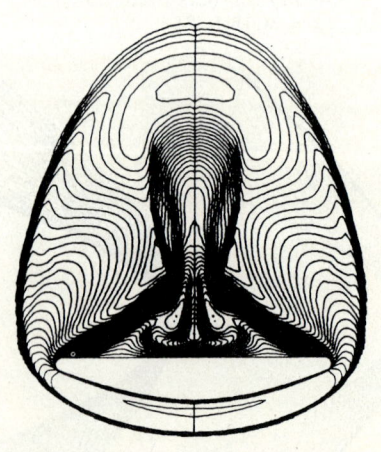

Fig. 4d Mach # contours. ($M_{min,max,\Delta} = 0, 8.6, .2$)

Fig. 4e Wall-streamlines. (leeward side)

Fig. 5a Density contours in the sym. plane. ($\rho_{min,max},\Delta = .15, 7.75, .1$)

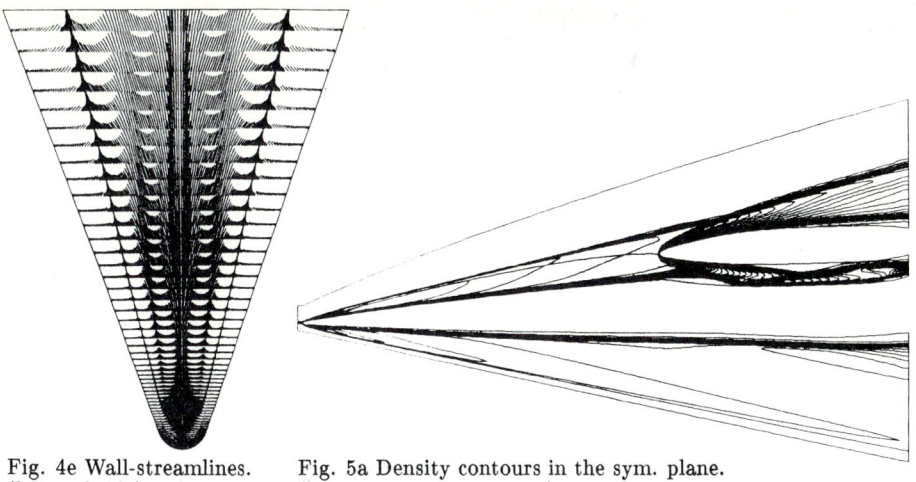

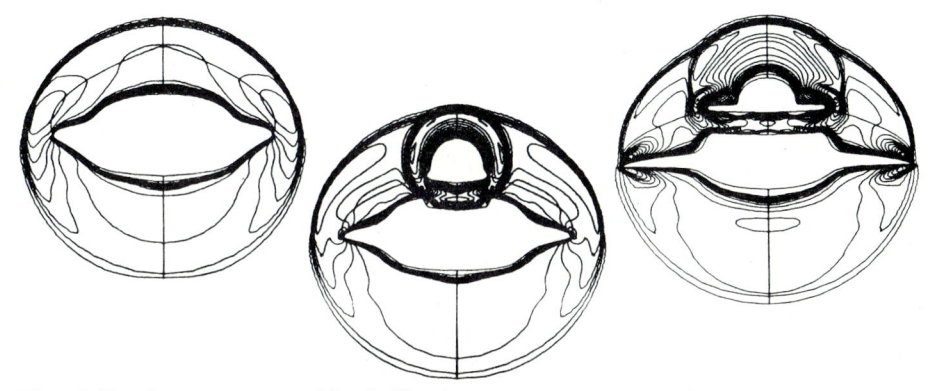

Fig. 5b Density contours at x=40.2m. ($\rho_{min,max},\Delta = .15, 7.75, .1$)

Fig. 5c Density contours at x=53.7m. ($\rho_{min,max},\Delta = .15, 7.75, .1$)

Fig. 5d Density contours at x=67.1m. ($\rho_{min,max},\Delta = .15, 7.75, .1$)

Fig. 5e Quasi-streamlines at x=56.3m

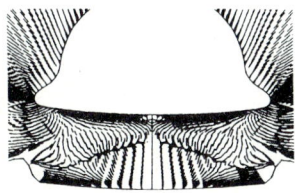

Fig. 5f Quasi-streamlines at x=59.0m

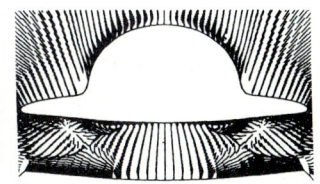

Fig. 5g Quasi-streamlines at x=63.3m

Comparison of thin layer Navier-Stokes and coupled Euler / boundary layer calculations of non-equilibrium hypersonic flows

S. Wüthrich[1], F. Perrel[2], M.L. Sawley[1] and A. Lafon[3]

[1] *IMHEF, Ecole Polytechnique Fédérale de Lausanne, ME - Ecublens, CH-1015 Lausanne, Switzerland*
[2] *CERFACS, 42 Av. Coriolis, F-31057 Toulouse, France*
[3] *CERT, 2 Av. Belin, F-31055 Toulouse, France*

SUMMARY

A comparison of thin layer Navier-Stokes and coupled Euler / boundary layer methods is presented. For the Navier-Stokes method, the inviscid fluxes are discretized using a Total Variation Diminishing (TVD) scheme, while a second-order central difference scheme is applied to the viscous terms. A fully-implicit, fully-coupled time discretization is employed. The external bow shock is captured by the scheme. For the coupled Euler / boundary layer method, the boundary layer equations, correct to second order in $Re^{-1/2}$, are applied in the inner (viscous) region and the Euler equations including displacement effects are solved for the outer (inviscid) region. The external inviscid bow shock is determined by a shock-fitting procedure. The results of calculations for flow over a hyperboloid at Mach = 25.5 show excellent agreement for the wall coefficients and the normal profiles using the two methods.

1. INTRODUCTION

For conditions relevant to hypersonic re-entry into the earth's atmosphere, strong shock waves occur in the flow region, with the air behind the bow shock in chemical non-equilibrium. The numerical simulation of the flow by means of the full Navier-Stokes equations requires an excessive amount of computer time and memory storage for present-day capabilities. This imposes severe limitations on the simulation of such flows. On the other hand, straightforward first-order boundary layer calculations do not take into account several physical features that may play a dominant role in the flow. There is therefore a need to develop relatively simple, accurate and efficient computational methods, especially for their use as a design tool for hypersonic vehicles. This paper presents a comparison of two different methods that have been employed to calculate hypersonic flows in chemical non-equilibrium: a thin layer Navier-Stokes method and a coupled Euler / boundary layer method.

The hypersonic flows considered here, having a relatively high Reynolds number, present a preferential direction (tangential to the body) along which the viscous gradients are much smaller than the normal gradients. The thin layer Navier-Stokes equations are thus obtained by neglecting the tangential viscous gradients, while retaining all the inviscid terms (unlike the boundary layer approach). These equations allow the computation of mixed subsonic-supersonic flows as well as moderately separated flows.

Coupled methods are based on asymptotic matching theory, as described by Van Dyke [1]. In this approach, the Navier-Stokes equations are expanded for both an inner (viscous) region and an outer (inviscid) region in terms of a small parameter, $Re^{-1/2}$, where Re is a characteristic Reynolds number. To first order in $Re^{-1/2}$, the classical Prandtl boundary layer equations are obtained for the inner region, while flow in the outer region is governed by the usual Euler equations. For some flow situations, however, such as thick boundary layers with an external rotational flow, it is found that a suitable matching between the inner and outer region can only be obtained by a second-order expansion of the Navier-Stokes equations in the two regions. The inclusion of second-order terms in $Re^{-1/2}$ accounts for several physical features such as displacement effects on the inviscid flow and the influence of longitudinal and transverse surface curvature, external vorticity and normal pressure gradients in the boundary layer region [2,3]. Thus the coupled Euler / boundary-layer method can be applied to flows of interest for which the boundary layer concept is valid.

2. PHYSICAL MODELLING

In order to compare results obtained using the two different numerical methods, the same physical model has been chosen for both calculations. The gas is considered to be a mixture of five species (N_2, O_2, NO, N and O), each species being assumed to behave as a perfect gas. The pressure and enthalpy of the mixture are then given by

$$p = \rho R T \sum_s \frac{Y_s}{M_s} \quad , \quad h = \sum_s Y_s h_s(T) \quad , \tag{1}$$

where Y_s is the mass fraction ($Y_s = \rho_s / \rho$), h_s the enthalpy and M_s the molar mass of species s. The enthalpy of each of the constituent species is written as a polynomial function of temperature, based on the fitting of theoretical data valid for temperatures up to 25,000 K. The diffusion fluxes are calculated assuming Fick's law with a constant Lewis number (Le = 1.2) for all the species. The viscosity coefficient for each species is obtained using the curve fits to experimental data given by Blottner [4], and the coefficient of thermal conductivity for each species determined using the Eucken relation. The viscosity and thermal conductivity of the mixture are calculated using the Wilke mixing rule.

For the calculation of non-equilibrium chemistry, a chemical reaction scheme consisting of seventeen reactions is employed. The forward reaction rates k_f depend on the temperature and are calculated from a modified Arrhenius equation, while the equilibrium constants $K_{c,k}$ are obtained using a fourth-order polynomial:

$$k_f = B_k \, T^{\alpha_k} \, e^{-E_k/T} \quad , \quad K_{c,k}(T) = \exp\{A_1 + A_2 Z + A_3 Z^2 + A_4 Z^3 + A_5 Z^4\} \quad , \tag{2}$$

where $Z = (10^4 / T)$. The coefficients A_i, i=1...5, determined by fitting experimental spectroscopic data, as well as the coefficients B_k, α_k and E_k, are as given by Park [5].

3. THIN LAYER NAVIER-STOKES METHOD

3.1 Basic Equations

The unsteady thin layer Navier-Stokes equations for a mixture of reacting gases can be written in a cartesian (X,Y) coordinate system as

$$\frac{\partial}{\partial t} \mathbf{w} + \frac{\partial}{\partial X} \mathbf{F} + \frac{\partial}{\partial Y} \mathbf{G} = \frac{\partial}{\partial Y} \mathbf{G}_v + \mathbf{S} \quad . \tag{3}$$

Using a general body-fitted coordinates system (ξ,η) (where ξ is the longitudinal and η the normal direction), this set of equations has the following form:

$$\frac{1}{J}\frac{\partial}{\partial t}\mathbf{w} + \frac{\partial}{\partial \xi}\widehat{\mathbf{F}} + \frac{\partial}{\partial \eta}\widehat{\mathbf{G}} = \frac{\partial}{\partial \eta}\widehat{\mathbf{G}}_v + \frac{1}{J}\mathbf{S} \quad , \tag{4}$$

where the planar case is considered here for sake of simplicity. In Eq. (4), $J = \xi_X \eta_Y - \xi_Y \eta_X$ is the Jacobian of the coordinate transformation $(X,Y) \rightarrow (\xi,\eta)$, $\mathbf{w}$ represents the vector of variables (species densities, momentum, total energy density), $\widehat{G}_V$ denotes the viscous terms, **S** the reactive source terms, and $\widehat{F}, \widehat{G}$ the convective fluxes. These eight-component vectors are given by

$$w = \begin{pmatrix} \rho Y_s \\ \rho U \\ \rho V \\ \rho E \end{pmatrix}, \quad F = \begin{pmatrix} \rho Y_s U \\ \rho U^2 + p \\ \rho UV \\ U(\rho E + p) \end{pmatrix}, \quad G = \begin{pmatrix} \rho Y_s V \\ \rho UV \\ \rho V^2 + p \\ V(\rho E + p) \end{pmatrix}, \quad \widehat{G}_V = \frac{\mu}{J} \begin{pmatrix} \alpha_4 \frac{Le}{Pr} \frac{\partial Y_s}{\partial \eta} \\ (gv)_6 \\ (gv)_7 \\ (gv)_8 \end{pmatrix}, \quad S = \begin{pmatrix} \omega_s \\ 0 \\ 0 \\ 0 \end{pmatrix} \quad (5)$$

with

$$\widehat{F} = \frac{1}{J}\left(\xi_X F + \xi_Y G\right) \; ; \quad \widehat{G} = \frac{1}{J}\left(\eta_X F + \eta_Y G\right) , \quad (6)$$

and

$$(gv)_6 = \alpha_1 \frac{\partial U}{\partial \eta} + \alpha_3 \frac{\partial V}{\partial \eta} \; ; \quad (gv)_7 = \alpha_3 \frac{\partial U}{\partial \eta} + \alpha_2 \frac{\partial V}{\partial \eta} \; ;$$

$$(gv)_8 = \frac{1}{2}\alpha_1 \frac{\partial U^2}{\partial \eta} + \frac{1}{2}\alpha_2 \frac{\partial V^2}{\partial \eta} + \alpha_3 \frac{\partial UV}{\partial \eta} + \frac{\alpha_4}{Pr}\left[(1 - Le)\, C_p f \frac{\partial T}{\partial \eta} + Le \frac{\partial h}{\partial \eta}\right] .$$

Here α_1 to α_4 are metric coefficients, defined by

$$\alpha_1 = \frac{4}{3}\eta_X^2 + \eta_Y^2 \; ; \quad \alpha_2 = \frac{4}{3}\eta_Y^2 + \eta_X^2 \; ; \quad \alpha_3 = \frac{1}{3}\eta_X \eta_Y \; ; \quad \alpha_4 = \eta_X^2 + \eta_Y^2 .$$

The total energy E is defined as $E = e + (U^2 + V^2)$, where e is the internal energy. The boundary conditions at the four boundaries of the computational domain are: 1) the freestream condition at the inflow boundary (chosen at a sufficient distance such that the detached bow shock lies entirely within the computational domain); 2) a symmetry condition along the symmetry line; 3) a zero-order extrapolation (numerical boundary condition) of the variables at the outflow boundary where the flow is essentially supersonic; and 4) along the wall $U = V = 0$, $T = T_w$, $\partial p/\partial \eta = 0$, while the species mass fractions are given by either a fully catalytic or non-catalytic wall assumption.

3.2 Numerical method

Spatial discretization

For the computation of hypersonic viscous reacting flows, an attractive approach is to use a Total Variation Diminishing (TVD) type scheme for the discretization of the inviscid fluxes, and a second-order central difference scheme for the viscous terms, with the source term evaluated at the centre of the computational cell.

Indeed, TVD schemes have proved to give oscillation-free solutions (with the sharp capture of discontinuities) for the Euler equations, (see [6] for a review on TVD schemes). Among the class of TVD type schemes, the symmetric Yee-Roe-Davis scalar scheme [6] has been chosen. This scheme is second-order accurate in space except at extrema points of the solution where the limiting procedure involved in the scheme to ensure the TVD property reduces the accuracy to first order. The scalar scheme is applied to the hyperbolic system of Euler equations via Roe's approach [7]. This approach allows the definition at each point of a local system of characteristic directions such that the original hyperbolic system is transformed into a system of uncoupled scalar equations; the scalar TVD scheme is applied to each of these scalar characteristic equations [6]. The expansion of Roe's approach for a mixture of gases as proposed by Montagne and Vinokur [8] is followed here. A detailed formulation can be found in [9,10].

Numerical procedure

A time-marching procedure is used to obtain the steady-state solution. For viscous reacting flows, several characteristic times (for convection, diffusion and chemical processes) are involved. The stability criteria imposed by an explicit time discretization can thus be quite severe and an implicit time discretization is preferred. A fully-implicit, fully-coupled method, which has proved to enhance convergence on fine meshes for viscous flows [6], is used. A linearized implicit form is derived from the resulting set of nonlinear equations (see [10] for a precise formulation). The steady-state solution is obtained by iteratively solving this linear system using a line relaxation alogorithm with backward and forward sweeping of the computational domain [9]. A uniform time step, based on a CFL condition, is used since it has been shown to enhance the convergence rate (compared to a local time step) for reactive flow computations [10].

4. COUPLED EULER / BOUNDARY LAYER METHOD

4.1 Basic equations

Inner region

Using an (x,y) body-oriented coordinate system, with x denoting the tangential direction and y the normal direction, the boundary layer equations correct to second order in $Re^{-1/2}$ can be written as [11]:

global continuity equation:

$$\frac{\partial}{\partial x}(R\rho u) + \frac{\partial}{\partial y}(HR\rho v) = 0 , \qquad (7)$$

species continuity equations:

$$\rho \frac{u}{H}\frac{\partial Y_s}{\partial x} + \rho v \frac{\partial Y_s}{\partial y} = \omega_s + \frac{1}{R}\frac{\partial}{\partial y}\left(R\mu \frac{Le}{Pr}\frac{\partial Y_s}{\partial y}\right) + \kappa\mu\frac{Le}{Pr}\frac{\partial Y_s}{\partial y} , \qquad (8)$$

x-momentum equation:

$$\rho\left(\frac{u}{H}\frac{\partial u}{\partial x} + v\frac{\partial u}{\partial y} + \kappa u v\right) = -\frac{1}{H}\frac{\partial p}{\partial x} + \frac{1}{R}\frac{\partial}{\partial y}\left(R\mu\frac{\partial u}{\partial y}\right) + \kappa\left(\mu\frac{\partial u}{\partial y} - u\frac{\partial \mu}{\partial y}\right) , \qquad (9)$$

y-momentum equation:

$$\rho \kappa u^2 = \frac{\partial p}{\partial y} , \qquad (10)$$

energy equation:

$$\rho\frac{u}{H}\frac{\partial}{\partial x}\left(h + \frac{u^2}{2}\right) + \rho v \frac{\partial}{\partial y}\left(h + \frac{u^2}{2}\right) = \frac{1}{R}\frac{\partial}{\partial y}\left\{R\left(\mu u \frac{\partial u}{\partial y} - q_y\right)\right\} - \kappa\left\{u\frac{\partial}{\partial y}(\mu u) + q_y\right\} , \qquad (11)$$

where the local heat flux in the normal direction is given by

$$q_y = -\lambda \frac{\partial T}{\partial y} - \frac{\mu}{\rho} \frac{Le}{Pr} \sum_s h_s \frac{\partial Y_s}{\partial y} \quad .$$

In the above equations, $H = 1 + \kappa(x) y$ and $R = [\, r(x) + y \cos \theta(x) \,]^m$, where $\kappa(x)$ is the inverse longitudinal surface radius of curvature, $\theta(x)$ is the angle between the wall surface and the freestream direction, and $m = 1$ for axisymmetric flow. Equations (7) - (11) are valid if the local inverse radius of curvature is of order less than or equal to $Re^{-1/2}$ that is, $O(1/\kappa) \leq O(\delta)$, where δ is the boundary layer thickness. It can be noted that if the longitudinal curvature effects are of $O(Re^{-1/2})$, the above equations to $O(1)$ reduce to the classical Prandtl boundary layer equations. The same boundary conditions are applied at the wall as for the thin layer Navier-Stokes method.

Outer region

The flow in the outer (inviscid) region is governed by the Euler equations, which are given by Eq. (3) and (5) with the viscous term $\widehat{G}_v$ set to zero [12].

It can be shown that displacement effects, associated with terms of second order in $Re^{-1/2}$, can be incorporated into the inviscid calculation through the concept of an "equivalent source" at the wall [13]. This equivalent source can be written in the form of a normal component of velocity at the wall that is related to the first-order mass flow at the wall by

$$\rho(0)\, V(0) = \frac{\partial}{\partial x} \left(\rho_{ext}\, u_{ext}\, \delta_1 \right) \quad , \tag{12}$$

where δ_1 is the displacement thickness given by

$$\delta_1 = \int_0^\delta \left(1 - \frac{\rho\, u}{\rho_{ext}\, u_{ext}} \right) dy \quad .$$

For first-order calculations, the normal velocity at the wall is set to zero in the absence of displacement effects. The outer edge of the inviscid region is determined by the bow shock, calculated with a shock-fitting procedure. The flow at the outflow boundaries is assumed to be supersonic, with first-order extrapolation providing the necessary conditions. It should be noted that no viscous terms appear in the Euler equations to second order in $Re^{-1/2}$, these being of third or higher order [11].

4.2 Numerical method

Inner region

The second-order boundary layer equations (7) - (11) being parabolic in nature are solved using a space-marching, fully-implicit, finite difference scheme. First-order spatial differences are used for the convective terms, while the diffusive terms are discretized by second-order differences. A decoupled, iterative method is employed using a primitive variable formulation, without the use of any similarity transformation. The global continuity and x-momentum equations are first solved simultaneously, with the y-momentum equation used to update the pressure. The species continuity equations are then resolved simultaneously, followed by the energy equation. To account for coupling between equations, iterations are performed at each station along the flow direction until the desired level of convergence is obtained. Only three species continuity equations are resolved, the mass fractions of other two species being determined from the resolution of the continuity equation for elemental nitrogen (which contains no source term) and the conservation of species. The remaining chemical source terms are linearized with respect to iteration number, and treated in a semi-implicit fashion. This has been found to provide an efficient and robust method over the entire hypersonic range [14,15].

Outer region

The Euler equations are discretized using a space-centered finite volume scheme. Both second- and fourth-order artificial dissipation terms are added, to avoid spurious oscillations in the vicinity of discontinuities and suppress odd / even oscillations. The external bow shock is determined using a shock-fitting procedure, with any internal shocks being captured by the numerical scheme. The time integration of the discretized Euler equations is carried out using an explicit fourth-order Runge-Kutta scheme. The incorporation of non-equilibrium chemical effects is implemented without using special procedures to treat the chemical source term [12].

Coupling procedure

The matching procedure adopted is as follows. First the first-order Euler equations are solved for the flow quantities in the outer inviscid region. The calculated tangential velocity at the wall, together with the stagnation point values of pressure and mass fractions, are then used to solve the first-order boundary layer equations. From the results of the first-order boundary layer calculations, the equivalent source values at the wall required as a boundary condition for the second-order Euler calculation are determined, as well as the boundary layer thickness. The interface between inner and outer regions is chosen to be at a distance δ_u from the wall, where δ_u corresponds to the position where the tangential velocity of the first-order boundary layer solution equals 99% of its limiting outer value. Following the second-order Euler calculation, a computational mesh for the new inner region defined by the first-order boundary layer thickness is then constructed to perform the second-order boundary layer calculation. To undertake this calculation, the required values of the pressure, temperature, tangential velocity and the species mass fractions at the interface are obtained by interpolation.

The above-described coupling procedure requires both first- and second-order boundary layer and Euler calculations to be performed in a consecutive manner. This necessitates a greater computation time than a first-order calculation alone. It should be noted, however, that the second-order Euler calculation is initiated using the first-order result, leading to a computation time for both Euler calculations that is less than twice that for the first-order calculation. In addition, each of the boundary layer calculations require significantly less computation time than an Euler calculation to obtain a comparable degree of accuracy.

5. RESULTS

Comparison of results obtained by the two above-described methods have been carried out for various flow conditions. The results presented here have been obtained for a non-equilibrium axisymmetric flow at $M_\infty = 25.5$ over a hyperboloid [16]. This geometry enables the re-entry flow along the windward symmetry plane of the American space shuttle to be modelled, via the concept of an "equivalent axisymmetric body" first introduced by Cooke [17]. The nose radius of the hyperboloid is 1.276 m and the half angle is 40.75°. Calculations have been undertaken up to a distance of 20 m from the stagnation point, which corresponds to slightly smaller than the distance from the nose of the shuttle to the hinge of the body flap. The chosen Mach number corresponds to an altitude of 74.89 km on the STS-2 flight re-entry trajectory, and to the following freestream conditions:

$$p_\infty = 2.170 \text{ Pa}, \quad \rho_\infty = 3.815 \times 10^{-5} \text{ kg m}^{-3}, \quad Re_\infty = 2.7 \times 10^4 \text{ m}^{-1} \ .$$

Calculations have been performed assuming a constant wall temperature of 1050 K along the entire body surface. This value corresponds to the temperature measured at a distance of 10 m from the nose of the shuttle [16]. Both fully catalytic and non-catalytic wall conditions have been considered.

The Navier-Stokes calculations employed an analytical 150x100 mesh stretched in both the tangential and normal directions. An equidistant mesh of 100x50 points was chosen for the Euler calculations, while a 200x100 mesh stretched in both directions was used for the boundary layer calculations. About 40% of the Navier-Stokes mesh points were used in the viscous boundary layer region, and 10 to 20% to capture the shock and in the freestream. Due to the use of an especially fine mesh in the vicinity of the wall, the computer time required to solve the Navier-Stokes equations was about six times more than for the coupled Euler / boundary layer calculation.

Figure 1 shows convergence plots for the different methods using the fully-catalytic wall condition. Figure 1(a) presents the residual as a function of the number of iterations for the Navier-Stokes calculation. The corresponding plots for the two Euler calculations (first- and second-order) are shown in Fig. 1(b), and in Fig. 1(c) for the two boundary-layer calculations (first- and second-order) at a given streamwise station midway along the body.

Figure 2 shows plots of the surface pressure coefficient C_p, Stanton number St, and skin friction coefficient C_f defined as follows:

$$C_p = \frac{p_w - p_\infty}{\frac{1}{2}\rho_\infty u_\infty^2} \ , \quad St = \frac{q_{y,w}}{\rho_\infty u_\infty c_{p\infty} (T_{0\infty} - T_w)} \quad \text{and} \quad C_f = \frac{\mu_w \left[\frac{\partial u}{\partial y}\right]_w}{\frac{1}{2}\rho_\infty u_\infty^2} \ ,$$

where $T_{0\infty}$ is the total enthalpy divided by the freestream specific heat. The non-radiative component of the wall heat flux, deduced from the surface temperature flight data using a simple transient analysis [18], are also presented in Fig. 2.

A comparison of the results presented in Fig. 2 shows that there is little difference between the pressure coefficient calculated using the two methods. With the exception of the nose region, the pressure coefficient is constant in the remaining, essentially-conical, region of the body. The difference observed in the Stanton number is less than 10% for a fully catalytic wall and less than 7% for a non-catalytic wall. The agreement between the results of the skin friction is excellent, despite a slight discrepancy in the slope of the coefficients which may be due to a different mesh refinement in the vicinity of the wall surface. The maximum skin friction is calculated to occur very close to the stagnation point, at x = 0.35 m; the difference between the two calculations at this location is less than 1% for the fully catalytic wall and about 3% for the non-catalytic wall.

The profiles of the temperature, pressure and mass fraction of species NO for a catalytic wall, calculated for a streamwise location x = 10 m, are shown in Fig. 3(a). The mass fraction of species N_2, N and O for a catalytic wall are presented in Fig. 3(b) and for a non-catalytic wall in Fig. 3(c).

A difference is observed in the shock position calculated by the two methods, with the shock-capturing procedure of the Navier-Stokes method smearing out the shock over three mesh points. With the exception of the shock region, the profiles obtained by the two methods show excellent agreement. The catalytic wall condition results in the recombination of molecular nitrogen and oxygen for the chosen wall temperature, while this effect is not observed for the non-catalytic wall.

On the stagnation line, the maximum temperature of about 14,000 K is reached just behind the shock and results in a strong dissociation of the molecules. The temperature decreases to 7,200 K at the edge of the boundary layer (viscous-inviscid interface). The shock stand-off distance, defined at the maximum of the density gradient, is calculated by the Navier-Stokes calculation to be about 0.091 m, whereas the Euler calculation gives a stand-off distance of 0.089 m.

6. CONCLUSIONS

The comparison of thin layer Navier-Stokes and coupled Euler / boundary layer methods to model non-equilibrium hypersonic flows has been presented. Given identical physical models, the two different approximations to the full Navier-Stokes equations have been demonstrated to lead to very similar flow solutions, despite the different numerical procedures adopted. Excellent agreement has been found for the surface coefficients, as well as for the computed normal profiles. Slight discrepencies have been observed in the shock region, associated with the different techniques used to determine the shock position. The coupled Euler / boundary layer method has been found to be computationally less expensive, and enabled a better resolution of the viscous region due to the larger number of points used. However, while such a coupled method can only be applied in flow regions of weak viscous-inviscid interactions, the thin layer Navier-Stokes method has a wider domain of applicablility.

7. REFERENCES

[1] M. Van Dyke, *Perturbation methods in fluid mechanics*, Academic Press, New York (1964).
[2] M. Van Dyke, *Higher approximation in boundary layer theory. Part 1*, J. Fluid Mech., **14**, 161 (1962).
[3] M. Van Dyke, *Higher-order boundary layer theory*, Ann. Rev. Fluid Mech., **1**, 265 (1969).
[4] F.G. Blottner, M. Johnson and M. Ellis, *Chemically reacting viscous flow program for multicomponent gas mixture*, Technical Report Sc-RR-70-754 Sandia Laboratories (1970).
[5] C. Park, *On convergence of computation of chemically reacting flows*, AIAA Paper 85-0247, 23rd Aerospace Science Meeting, Reno (1985).
[6] H.C. Yee, *A class of high resolution explicit and implicit shock capturing methods*, VKI lecture series 1989-04 (1989).
[7] P. Roe, *Approximate Riemann solvers, parameters vectors and differences schemes*, J. Comp. Phys. **43**, 357-372 (1981).
[8] J.L. Montagne, H.C. Yee, G.H. Klopfer and M. Vinokur, *Hypersonic blunt body computations including real gas effects*, NASA TM 100074 (1988).
[9] A. Lafon, *Calcul d'écoulements visqueux hypersoniques*, CERT-DERAT, 32/5005.22 (1990).
[10] F. Perrel, *Simulation numérique d'écoulements hypersoniques visqueux en déséquilibre chimique*, Ph.D. thesis (1991), to appear.
[11] R. Grundmann, *Boundary layer equations and methods of solution*, VKI Lecture series 1987-02 (1987).
[12] J.B. Vos and C.M. Bergman, *Chemical equilibrium and non-equilibrium inviscid flow simulation using an explicit scheme*, Comp. Phys. Comm., **65** (1991).
[13] M.J. Lighthill, *On displacement thickness*, J. Fluid Mech. **4**, 383 (1958).
[14] M.L. Sawley and S. Wüthrich, *Non-equilibrium hypersonic flow simulations using a coupled Euler / boundary layer method*, Proc. of the First European Symposium on Aerothermaodynamics for Space Vehicles, ESA SP-318, p. 387 (1991).
[15] M.L. Sawley and S. Wüthrich, *Non-equilibrium hypersonic flow simulations using the second-order boundary layer equations*, Comp. Meth. Appl. Mech. Eng., **89** (1991), to appear.
[16] J.L. Shinn, J.N. Moss and A.L. Simmonds, *Viscous shock-layer heating analysis for the shuttle windward symmetry plane with surface finite catalytic recombination rates*, AIAA Progress in Astronautics and Aeronautics, **85**, 149 (1983).
[17] J.C. Cooke, *An axially symmetric analogue for general three-dimensional boundary layers*, British A.R.C., R. & M. No 3200 (1961).
[18] S.D. Williams, *Columbia: the first five flights entry heating data series*, NASA-CR-171820 (1984).

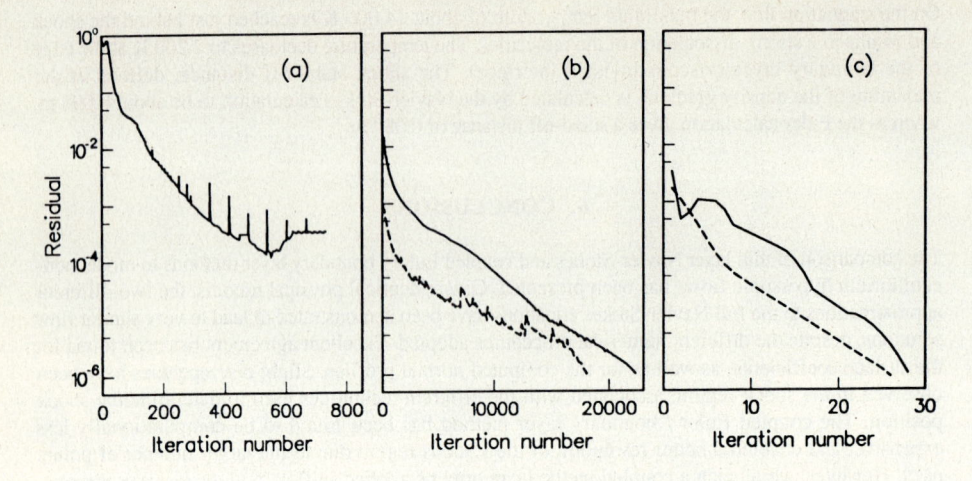

Fig. 1. Plots of residual for (a) Navier-Stokes, (b) Euler (——— first-order; − − − second-order) and (c) the boundary layer calculations (——— first-order; − − − second-order).

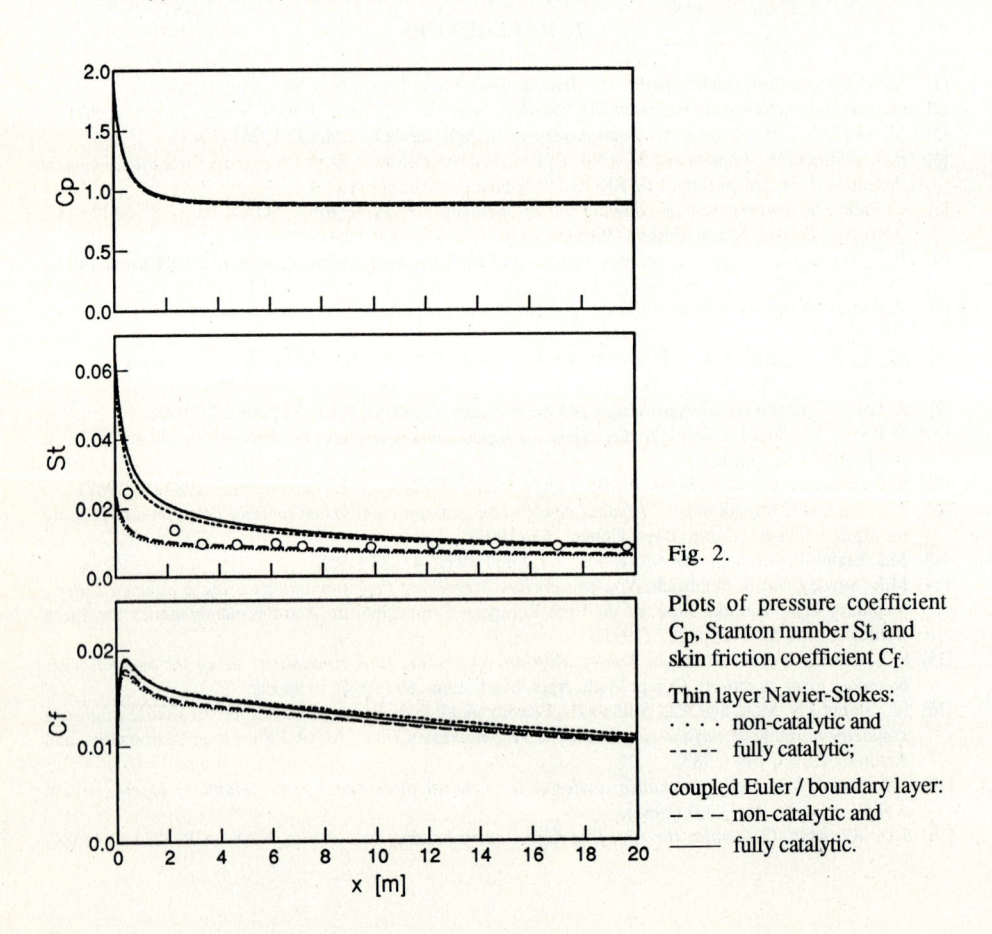

Fig. 2. Plots of pressure coefficient C_p, Stanton number St, and skin friction coefficient C_f.

Thin layer Navier-Stokes:
· · · · non-catalytic and
- - - - - fully catalytic;

coupled Euler / boundary layer:
− − − non-catalytic and
——— fully catalytic.

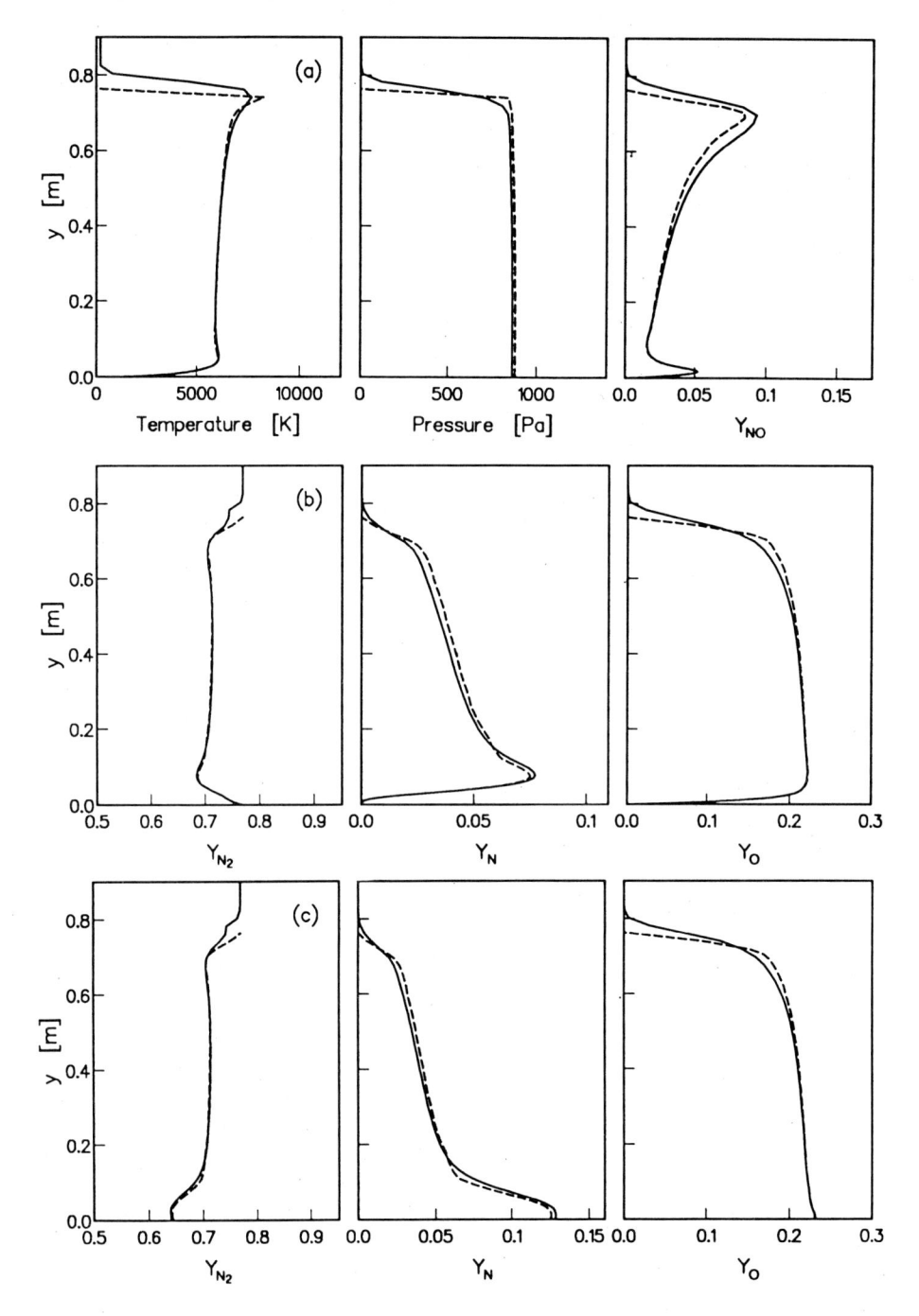

Fig. 3. Profiles, calculated at x=10 m, of temperature, pressure and mass fraction of species NO for a catalytic wall (a); mass fraction of species N_2, N and O for a catalytic wall (b), and for a non-catalytic wall (c). (—— thin layer Navier-Stokes, - - - coupled Euler / boundary layer)

MULTI BLOCK TECHNIQUES FOR VISCOUS COMPRESSIBLE FLOWS

A Multi Block Flow Solver for Viscous Compressible Flows

C.M. Bergman[1] and J.B. Vos[2]

[1]C2M2, *The Royal Institute of Technology, S-100 44 Stockholm, Sweden*
[2]*IMHEF, Ecole Polytechnique Fédérale de Lausanne, CH-1015 Lausanne, Switzerland*

SUMMARY

A 2 dimensional Compressible Navier Stokes flow solver has been developed. The Navier Stokes equations are discretized in space following the method of Jameson using central differencing in space with added artificial dissipation. For the computation of the viscous fluxes, gradients of the velocities and temperature are evaluated using the gradient theorem on auxiliary cells. The resulting system of ODE's is integrated in time using the explicit Runge Kutta procedure. To facilitate the grid generation for complex geometries and to permit an efficient use of parallel computers a multi block structure has been adopted yielding the possibility to split the computational domain into multiple patched blocks. The code was validated comparing calculated results for the flow over a flat plate with Blasius theory, and comparing the results for the flow around a NACA 0012 profile with other computations. Comparing single and multi block computations showed that the multi block structure does not affect the convergence of the numerical procedure or the calculated results.

1. INTRODUCTION

The task of calculating the flow around supersonic and hypersonic aircraft or re-entry spacecraft has attracted considerable attention during the last few years. In Europe, projects as HERMES, HOTOL and SAENGER have stimulated the research in developing numerical tools to simulate these flows. In these research projects numerical simulation has become as important as experimental simulation since experimental simulation of high Mach number high enthalpy re-entry flows is not possible. Today, typically, the Euler equations are solved to predict integrated coefficients such as lift, drag, and momentum. If the flow is attached it is possible to couple inviscid flow solvers to boundary layer solvers in order to calculate the heat flux to the wall. This procedure does not work in separated flow regions, where one has to use the full Navier Stokes equations. For hypersonic flows, it is necessary to include the effects of dissociating air into the solver making the cost of the solution of the 3D Navier Stokes equations prohibitive with the present state-of-the-art computer technology. For the near future, only parallel computers seem to provide the necessary computational power and memory to solve the 3D Navier Stokes equations for aerodynamic design. For these computers, a multi block approach provides a natural means to use them efficiently [1]. For example each block can be sent to a different processor, and a time step is made in parallel in each block. Only periodically information has to be exchaged between blocks, giving high parallel efficiencies.

Presently IMHEF and C2M2 carry out a joint project on the development of a family of multi block flow solvers using structured grids. The principal interest of IMHEF in these solvers is to be able to calculate the flow around complex geometries by cutting the computational domain in patched blocks which can be meshed independently of each other. The principal interest of C2M2 in this project is to develop and explore algorithms for parallel computers.

2. GOVERNING EQUATIONS

The multi block flow solver described here solves the 2D Navier Stokes equations. In cartesian coordinates these equations can be written as

$$\frac{\partial}{\partial t}(w) + \frac{\partial}{\partial x}(f_q - f_v) + \frac{\partial}{\partial y}(g_q - g_v) = 0 \qquad (1)$$

where the state vector w and the convective flux vectors are given by

$$w = \begin{pmatrix} \rho \\ \rho u \\ \rho v \\ \rho E \end{pmatrix}, \quad f_q = \begin{pmatrix} \rho u \\ \rho u^2 + p \\ \rho u v \\ u(\rho E + p) \end{pmatrix}, \quad g_q = \begin{pmatrix} \rho v \\ \rho v u \\ \rho v^2 + p \\ v(\rho E + p) \end{pmatrix}. \qquad (2)$$

In these equations ρ denotes the density, u and v the cartesian velocity components, E the total energy and p the pressure. The viscous fluxes are

$$f_v = \begin{pmatrix} 0 \\ \tau_{xx} \\ \tau_{xy} \\ (\tau U)_x - kq_x \end{pmatrix}, \quad g_v = \begin{pmatrix} 0 \\ \tau_{yx} \\ \tau_{yy} \\ (\tau U)_y - kq_y \end{pmatrix} \qquad (3)$$

where U is the velocity vector. The shear stress tensor τ is given by

$$\tau_{xx} = \frac{2}{3}\mu\left(2\frac{\partial u}{\partial x} - \frac{\partial v}{\partial y}\right) \quad \tau_{xy} = \tau_{yx} = \mu\left(\frac{\partial v}{\partial x} - \frac{\partial u}{\partial y}\right) \quad \tau_{yy} = \frac{2}{3}\mu\left(-\frac{\partial u}{\partial x} + 2\frac{\partial v}{\partial y}\right). \qquad (4)$$

The frictional heating in the energy equation is calculated from

$$(\tau U)_x = \tau_{xx} u + \tau_{xy} v \qquad (\tau U)_y = \tau_{yx} u + \tau_{yy} v \qquad (5)$$

and the conductive heat transport equals

$$q_x = -k\frac{\partial T}{\partial x} \qquad q_y = -k\frac{\partial T}{\partial y}. \qquad (6)$$

The equations are non-dimensionalized as in [2] yielding that the constant viscosity can be computed as $\mu = \sqrt{\gamma}M/Re$, where M is the Mach number, Re the Reynold number and γ the ratio between specific heats, equal to 1.4. Assuming a constant Prandtl number (for air Pr = 0.72), the heat conductivity can then be found by $k=\gamma\mu/(\gamma-1)Pr$. The system of equations is completed by specifying the relation between the pressure p and the state vector w. For a caloric perfect gas, this equation states

$$p = (\gamma - 1)\rho\left(E - \frac{1}{2}(u^2 + v^2)\right) \qquad (7)$$

and, after some algebra, the temperature can be found as $T = p/\rho$.

3. NUMERICAL METHOD.

3.1 Finite Volume Method.

Following the concept of the finite volume method, Equation (1) is integrated over a domain Ω with boundary $\partial\Omega$. Then by using the theorem of Gauss one obtains

$$\int_{\Omega} \frac{\partial}{\partial t} w \, dS + \int_{\partial\Omega} H \cdot n \, ds = 0 \qquad (8)$$

where $H = (f_q - f_v, g_q - g_v)$ is the flux tensor, see Equation (2), and n the normal at the boundary of Ω, pointing in the outward direction. Assume that the domain Ω is a Finite Volume cell with index (i,j), and consider $w_{i,j}$ as an approximation to the state vector in this cell. Then Equation (8) reduces to

$$\frac{d}{dt}(S_{i,j} w_{i,j}) + Q_{i,j} = 0 \qquad (9)$$

where $S_{i,j}$ is the volume of the cell (i,j), and where $Q_{i,j}$ represents the net flux leaving and entering the cell through the surface, calculated from

$$Q_{i,j} = h_{i+1/2, j} + h_{i-1/2, j} + h_{i, j+1/2} + h_{i, j-1/2} \qquad (10)$$

with

$$h_{i-1/2,j} = \int_{i-1/2,j} H \cdot n \, ds = H_{i-1/2,j} \cdot \int_{i-1/2,j} n \, ds \, . \qquad (11)$$

The convective part of the flux tensor, $H_{i-1/2,j}$ is calculated from Equation (2) using the average of the state vectors having the surface at $i-1/2,j$ in common. On a equispaced cartesian grid this results in a centered scheme which is second order accurate in space. For the computation of the viscous fluxes in Equation (3), gradients of the velocities and temperature are evaluated using the gradient theorem on auxiliary cells.

3.2 Adaptive Dissipation.

The Finite Volume scheme is augmented by the addition of a second order artificial viscosity term to be used near discontinuities, and of a fourth order dissipation term to suppress odd/even oscillations allowed for by centered schemes. The dissipation terms are formed from the second and fourth order differences of the state vector multiplied by a scaling factor and a weight, the latter usually referred to as a switch [3]. This switch is constructed from the absolute value of the normalized second difference of the pressure, implying that the second order dissipation term is small except in regions of large pressure gradients, such as in the neighborhood of a shock wave or a stagnation point. The fourth order difference is used everywhere except in regions where the second order dissipation is strong in order to prevent the generation of oscillations in these regions.

After addition of the dissipative terms, the following numerical scheme results,

$$\frac{d}{dt}(S_{i,j} w_{i,j}) + Q_{i,j} - D_{i,j} = 0 \qquad (12)$$

where $D_{i,j}$ is the dissipation operator. The conservative form of the discretized equations is preserved by, analogous to the convective fluxes, introducing dissipative fluxes for each equation.

The operator $D_{i,j}$ is then split as

$$D_{i,j} = d_{i+1/2, j} - d_{i-1/2, j} + d_{i, j+1/2} - d_{i, j-1/2} \qquad (13)$$

where each dissipative flux is evaluated at the same location as the corresponding convective flux.

3.3 Explicit time integration.
Equation (12) is integrated in time using the five stage explicit Runge Kutta scheme [3]. This type of scheme is normally used for their high accuracy in time, but here properties as stability and damping are of greater interest. Furthermore the explicit integration scheme does not require a matrix inversion which is difficult to parallelize, as for example is the case for implicit schemes.

3.4 Boundary conditions.
Boundary conditions need to be prescribed at all the sides of the calculation domain. At solid walls, the no-slip condition is imposed, and the pressure at the wall is obtained by linear extrapolation from the interior field to the wall. At the outflow and farfield boundary three option exist: the state vector w can be extrapolated, set to the free stream conditions or found using 1D Riemann invariants described for example in [4].

4. STRUCTURE OF THE MULTI BLOCK SOLVER

4.1 Introduction.
From a mathematical and numerical point of view, there is no difference between a single block and a multi block solver, since the equations and the numerical method are exactly the same. It is on the level of the incorporation of the boundary conditions that there appears a difference since a block connectivity boundary condition must be introduced. Explicit schemes do not require special solution procedures for periodic and block connectivity boundary conditions, making these schemes very suitable for multi block calculations, this compared to the more complex and cumbersome implicit schemes. The step from a single block to a multi block solver is small, and mainly a matter of defining a good and efficient data structure. Here the data structure is that of a shared memory machine. The state vector and coordinates are stored in global arrays for all blocks. The computations are performed block by block, copying back and forth to the global arrays. This means that storage for the temporary arrays, like fluxes, differences and scaling factors, is allocated only for the largest block.

Before proceeding, proper definitions of the computational space shown in Figure 1 are introduced. The basic idea is that it should be possible to connect any point on the boundary of a block to any other point on any side of any block. Here this connection is done by means of patched multi blocks.

4.2 Definition of the computational domain.
In Figure 1 the computational domain is shown. The following definitions are introduced:

1) **Multiple Blocks.** Several local curvilinear coordinate systems that are connected together, resulting from a decomposition of the physical domain.
2) **Block.** One local curvilinear coordinate system with I and J as running indices.
3) **Side.** Represents the boundary of the block, numbered from 1 to 4.
4) **Window.** On each side several boundary conditions can be imposed. The definition of the line segment where a particular boundary condition can be imposed is called a window.

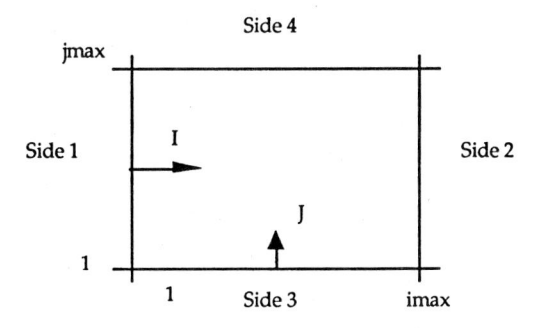

Figure 1: Single Block Computational Domain

4.3 Boundary arrays and ghost cells

The implementation of the boundary conditions has been uncoupled from the algorithm used to update the interior cells. The link between the boundary conditions and the algorithm for the interior cells is made using boundary arrays and 'ghost cells'. In general, finite volume flow solvers use a certain number of ghost cells which are used in enforcing the boundary conditions. Here, two layers of ghost cells are used at each side of the calculation domain, and the values of the state vector w and pressure p at these cells are obtained from the boundary arrays. These boundary arrays are filled in boundary condition routines in such a way that the boundary conditions are satisfied when using the ghost cells. Since the difference stencils for the convective fluxes and the artificial dissipation fluxes are not the same, separate boundary arrays have been defined for each stencil (respectively bw and bd). For the evaluation of the viscous fluxes the boundary array bw can also be used. In Table 1 the ghost cells which are needed for a given boundary array are indicated with an x.

Table 1: Ghost cells

Array	Ghost cells		Interior cells	
	-1	0	1	2
bw		x		
bd	x	x		
ba	x	x		

In the Table 1, ba is the boundary array for the calculation of the switch in the artificial dissipation. For <u>each window</u>, the following information is provided as input to the solver:

-Boundary condition type
-Window definition (start and end indices)
-Adjacent block number
-Side number on connecting adjacent block
-Index direction on adjacent block
-Starting address on adjacent block

To minimize the communication between the blocks, boundary arrays addressing adjacent blocks, are updated only once per five Runge Kutta stages. Boundary conditions using local arrays are updated in the two first stages. Using this structure a time step can be made in each block independent of the other blocks, and only after completion of a time step data is exchanged between the blocks.

5. VALIDATION

5.1 Flow over a flat plate

As validation case the flow over a flat plate was computed and compared with the Blasius theory [5]. The Mach number was set to 0.2, the Reynolds number to 200 and the Prandtl number to 0.72. Free-stream density and pressure were set both to 1. The computational domain is shown in Figure 2.

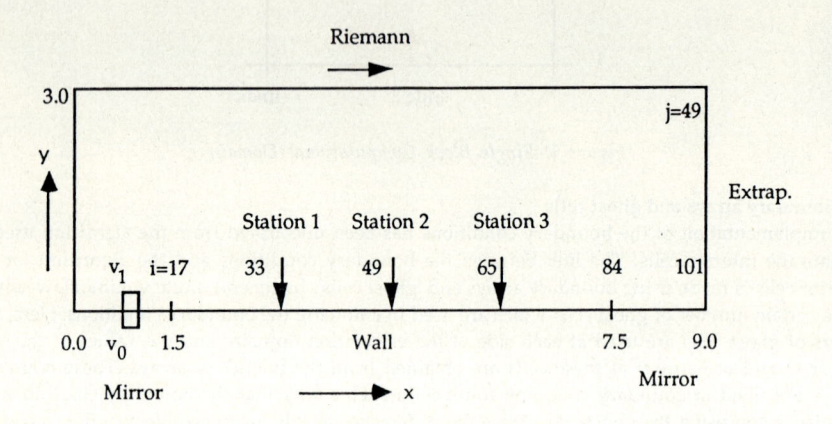

Figure 2: Computational domain, flow over a flat plate.

The simulation of this case required special attention to the incorporation of the boundary conditions. First of all the flat plate can not be connected directly to inviscid free-stream conditions as Riemann Invariants [5], due to the important viscous effects at the leading edge of the flat plate. Therefore the lower side of the computational domain in Figure 2 was divided into three segments, the middle segment being the wall, and at the first and last segment the velocity was mirrored by setting $v_0=-v_1$ to produce a symmetry condition. At the outflow all state variables were extrapolated linearly.

As can be seen in Figure 2, a 101x49 grid was used. A hyperbolic stretching function was applied in the direction normal to the wall to resolve the boundary layer. After 3000 iterations, using local time-steps, the L2 norm of the residual dropped 3 decades to a level of $0.3*10^{-6}$. Comparing the results after 3000 steps with those after 6000 steps showed hardly and difference in results, and it was concluded that 3000 time steps was sufficient for this flow problem to reach steady state. It was observed that in the boundary layer the smallest viscous timestep was a factor of 1000 smaller than the convective time step. The simulation was run on a single processor of the CRAY-2 at the computing center of EPFL taking 248 CPU seconds. Calculated boundary layer profiles were compared with Blasius theory at three stations, $x_{plate}=1.5$, $x_{plate}=3.0$ and $x_{plate}=4.5$, and the results are shown in Figures 3a), 3b), and 4a). As can be seen in these figures, the calculated u-velocity profiles compare well with the results from the Blasius theory [5]. At station 1 and 3 a small difference is visible, which is due to the close presence of the inflow and outflow boundaries which affect the solution. At station 2, see Figure 3b), at the largest distance from the boundaries, the profiles match very well. A comparison of the skin friction coefficient was also made, see Figure 4b). The calculatedskin friction is about 20 % below that obtained from the Blasius theory showing that it is not possible ro resolve properly the flow at the leading edge of the plate using this grid system. The poor resolution at the leading edge also affects the solution downstream which explains the difference in skin friction.

Exactly the same calculation was made using respectively 2 and 8 blocks. In the first case two blocks where created by cutting the flow domain in the I-direction at the middle. In the second case, the flow domain was divided 4 times in the I-direction and twice in the J-direction. It was found that the calculated results where numerically the same up to the third digit.

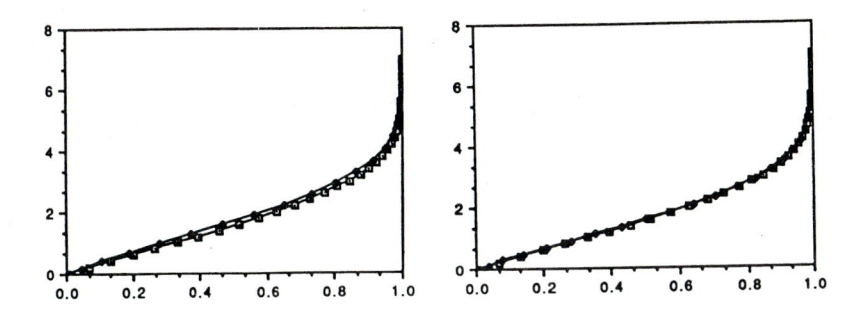

Figure 3: Boundary layer profiles a)station 1, b) station 2

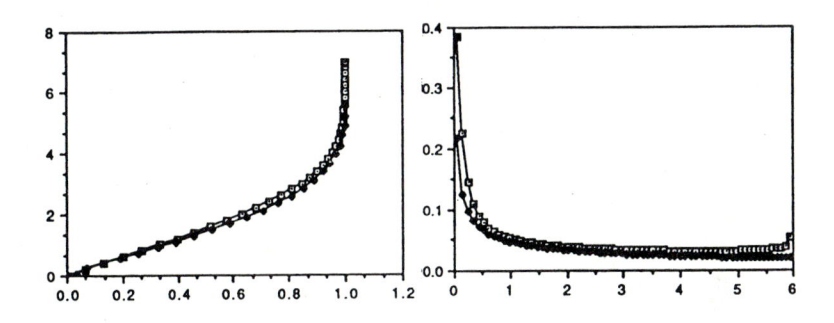

Figure 4: a) Boundary layer profile station 3, b) c_f versus x_{plate}

5.2 NACA 0012 Profile

For further validation the flow around a NACA 0012 profile was computed using three different grid systems. This case was taken from the workshop on viscous compressible flows [6]. In Table 2 the input for the computation is given.

Table 2: Test case for NACA 0012 Profile

Mach	0.8	CFLVIS	0.8
p0	1.0	EPS	10^{-6}
ρ0	1.0	DKI2/DKJ2	1.0
Alpha	0.0	DK4	0.03
Re	500	NRUNGK	5
CFL	2.0	TALFA	1/4, 1/6, 3/8, 1/2

Since the equations are in non-dimensional form the free-stream pressure, p0, and density, ρ0, were set to 1.0. As for the flat plate calculation the viscous time step was smaller than the convective during the complete time integration process. The stability limit for the former, CFLVIS, did therefore determine the convergence and the maximum was found to be 0.8. For the convective time step a value of 2.0 was set which is well within the stability limit. The computation was, as for the flat plate, made using a local timestep to accelerate the convergence. The values of the second order dissipation coefficients, DKI2/DKJ2 and the fourth order coefficient, DK4, where set to their standard values. To reach a grid independent solution three different grid systems were used. The results obtained using the coarse grid of 65x17 points were symmetric, and showed some expected flow features, but they did not compare well with the results found in [6], see also Figure 5a which shows the iso Mach contours. The medium grid with 129x33 grid points, see Figure 5b), had enough resolution to produce the same results as found in [6]. Using the fine grid of 257x65 point a grid converged solution was found. This grid, as can be seen in Figure 6a), was split into 8 blocks of 33x65 points giving a solution that was, within round-off, the same as for the single block. As can be seen in the Figure 6a) the iso-mach contours are smooth across block interfaces.

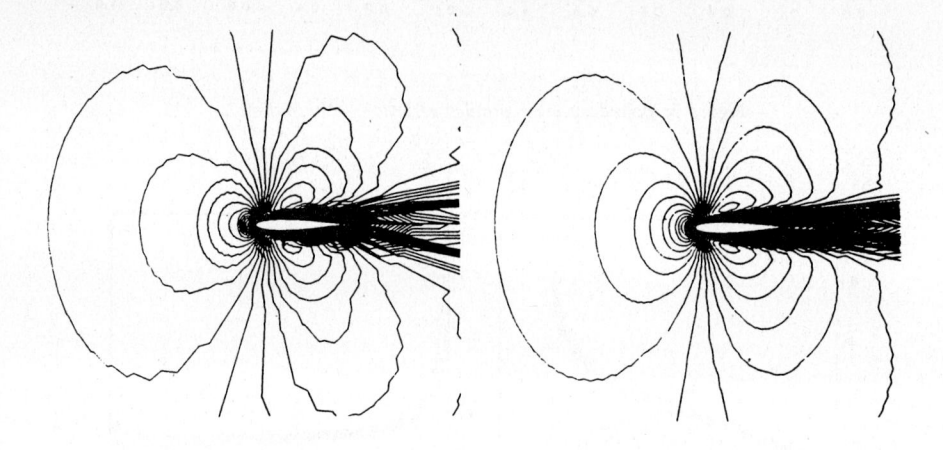

Figure 5: Iso-Mach lines, ΔM=0.025, a) 65x17 grid, b) 129x33 grid

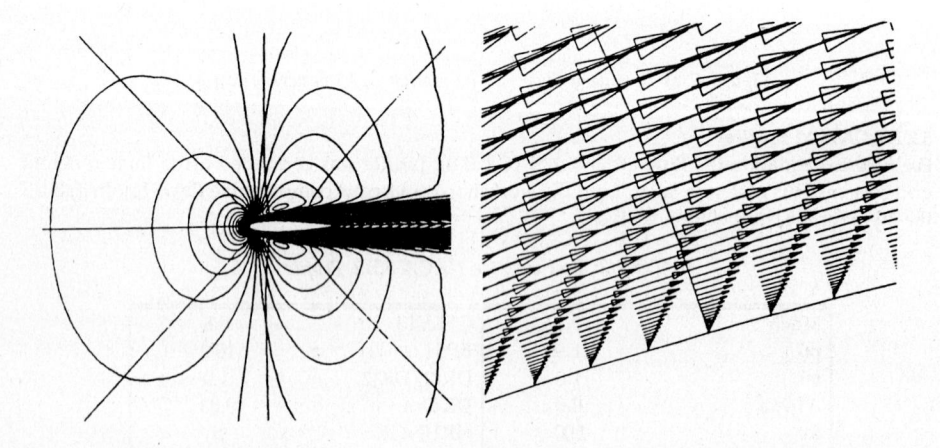

Figure 6: a) Iso-Mach lines, ΔM=0.025, 8x33x65 grid b) Boundary layer velocity vectors

In Table 3 it can be seen that the increase of the number of iterations with the size of the problem is not linear as should be expected. The reason for this is that the initial residual level is lower for finer meshes, and thus the number of decades to convergence is smaller. A converged solution was, however, for all cases reached.

Table 3: Test runs for NACA 0012. CPU and Mflops on the CRAY-2

Mesh	# iterations	CPU (Seconds)	Mflops
65x17	2890	58.7	59
129x33	7570	465	78
257x65	13650	2324	112
8x33x65	13640	4803	54

Since the compiler vectorizes only over I-direction, the computational performance, measured in Mflops, degrades for smaller block sizes. This can be avoided by hand-collapsing the loops giving the compiler the possibility to vectorize over all interior points. As seen in Table 3 the convergence to steady-state is not affected by the division into multiple blocks. The blocking strategy chosen, divisions in the direction around the profile, was done such that the profile and the boundary would be connected, since the dominant transport is in this direction. In Figure 6b) the velocity vectors in the boundary layer are plotted. The cell center values are interpolated to the vertex of the computational cells. As can be seen, the number of points in the boundary layer is at least 30 which is more than sufficient to resolve it. Furthermore it can be seen that the vectors are contiguous over the block boundaries.

6. CONCLUSIONS

A first step in validating a Multi block Compressible Navier Stokes Solver has been presented, using single and multi block meshes. The computed velocity profiles over a flat plate correspond well with Blasius theory. A NACA 0012 profile has also been computed and the obtained iso-contours compare well with results found in the literature. The finest mesh was split into 8 blocks and the results where found almost identical to those obtained in single block mode. The computations using an explicit scheme were found to be very expensive and for steady state solutions convergence acceleration techniques, as for example Multi Grid, should be considered.

8. REFERENCES

[1] Bergman C.M. and Vos J.B. , "Parallelization of CFD Codes", Second World Congress on Computational Mechanics, August 27-31, 1990, Stuttgart, FRG

[2] Muller, B. and Rizzi, A.,"Navier Stokes Calculations of Transonic Vorticies over a Round Leading Edge Delta Wing", Int. J. Num. Met. in Fluid Mechanics, Vol. 9, 1989, pp 943-962

[3] Jameson, A., "Numerical Solution of the Euler Equations for Compressible Inviscid Fluids", In: "Numerical Methods for the Euler Equations of Fluid Dynamics", Ed: Angrand, F., et al SIAM, Philidelphia, 1985

[4] Bergman C.M., "Development of Numerical Techniques for Inviscid Hypersonic Flows Around Re-entry Vehicles", Ph.D. Thesis, INP Toulouse, June 20, 1990

[5] Schlichting H., "Boundary-Layer Theory", McGraw-Hill Series in Mechanical Engineering, Seventh Edition

[6] Bristeau M.O., Glowinski R., Periaux J., Viviand H. eds., Notes on Numerical Fluid Mechanics, Vieweg, Vol. 16, 1985

A Three Dimensional Multigrid Multiblock Multistage Time Stepping Scheme for the Navier-Stokes Equations

Alaa Elmiligui & Frank Cannizzaro, Old Dominion University
N. Duane Melson, NASA, Langley Research Center
E. von Lavante, University of Essen.

SUMMARY

A general multiblock method for the solution of the three-dimensional, unsteady, compressible, thin-layer Navier-Stokes equations has been developed. The convective and pressure terms are spatially discretized using Roe's flux differencing technique while the viscous terms are centrally differenced. An explicit Runge-Kutta method is used to advance the solution in time. Local time stepping, adaptive implicit residual smoothing, and the Full Approximation Storage (FAS) multigrid scheme are added to the explicit time stepping scheme to accelerate convergence to steady state. Results for three-dimensional test cases are presented and discussed.

INTRODUCTION

In recent years progress has been made in obtaining solutions of the Navier-Stokes equations for three-dimensional flow. Many methods, both explicit and implicit, have been developed for iteratively solving these equations. The aim of the present work was to develop a finite-volume scheme suitable for efficient computations using block structured grids. Upwind schemes were used due to their high degree of reliability in viscous flow computations and their superior shock-capturing capability [1].

In the present work, an explicit multistage time-stepping scheme is used to construct the algorithm for solving the unsteady, compressible, Navier-Stokes equations. The basic algorithm for the inviscid terms in the Navier-Stokes equations is based on Roe's flux differencing [2]. This type of upwind scheme is simplified by linearizing the Riemann problem between two cell interfaces about a state defined by Roe's averaging procedure. The present discretization employs the finite volume approach, with the state variables at the cell interface determined by the MUSCL interpolation using the so-called κ scheme [3]. The viscous and the heat transfer terms in the equations are centrally differenced.

A multiblock strategy is used to allow greater geometric flexibility. The solution domain is divided into multiple zones (blocks) and the grid for each zone is then generated. In the present multiblock implementation the interface between blocks requires C° continuity. The type of boundary conditions used on each of the faces of the blocks is provided through an input file at run time rather than the traditional method of hard-coding different grid topologies.

GOVERNING EQUATIONS.

The three-dimensional, time-dependent, compressible, thin-layer, Navier-Stokes equations (neglecting body forces and heat source terms) are written in general curvilinear coordinates (ξ,η,ζ) in nondimensional strong conservative form as:

$$\frac{\partial Q}{\partial t} + \frac{\partial (F - F_v)}{\partial \xi} + \frac{\partial (G - G_v)}{\partial \eta} + \frac{\partial (H - H_v)}{\partial \zeta} = 0 \qquad (1)$$

where,

$$Q = \frac{1}{J}\begin{bmatrix} \rho \\ \rho u \\ \rho v \\ \rho w \\ e \end{bmatrix} ; \quad F^l = \begin{bmatrix} \rho U^l \\ \rho u U^l + l_x p \\ \rho v U^l + l_y p \\ \rho w U^l + l_z p \\ (e + p) U^l \end{bmatrix} ; \quad F_v^l = \frac{M_\infty}{Re}\frac{\mu}{J}\begin{bmatrix} 0 \\ u_l \phi_l + l_x \psi_l \\ v_l \phi_l + l_y \psi_l \\ w_l \phi_l + l_z \psi_l \\ \left\{\left(\frac{q^2}{2}\right)_l + \frac{T_l}{Pr(\gamma-1)}\right\}\phi_l + U^l \psi_l \end{bmatrix} \qquad (2)$$

$l = \xi, \eta$ and ζ where, $F^\xi = F$; $F^\eta = G$; $F^\zeta = H$

$$U^l = ul_x + vl_y + wl_z \; ; \; q^2 = u^2 + v^2 + w^2 \; ; \; Re = \frac{\rho_\infty^* q_\infty^* L^*}{\mu_\infty^*} \qquad (3)$$

$$\phi_l = l_x^2 + l_y^2 + l_z^2 \; ; \; \psi_l = \frac{1}{3}[u_l l_x + v_l l_y + w_l l_z]$$

ρ, u, v, w, p, e, and T are the density, the three Cartesian velocity components, the static pressure, the total energy, and the temperature respectively. U^l is the contravariant velocity component, J is the Jacobian of transformation, and μ is the molecular viscosity (simple power law is used to determine μ). Pr is the Prandtl number; Re and M_∞ are the reference Reynolds number and Mach number respectively.

The above equations are normalized with a reference length L^*, and the free stream values of density ρ_∞^*, speed of sound a_∞^*, and molecular viscosity μ_∞^*. The total energy is normalized by $\rho_\infty^*(a_\infty^*)^2$. For an ideal gas, the pressure is given by:

$$p = (\gamma - 1)\left[e - \frac{\rho q^2}{2}\right]. \qquad (4)$$

COMPUTATIONAL ALGORITHM

The present algorithm employs a finite volume approach with the state variables at the cell interface determined by the MUSCL interpolation. The convective and pressure terms are upwind differenced using Roe's flux differencing [2]. The viscous and heat flux terms are discretized using a second order accurate central differencing operator. The spatial derivatives across a cell interface, for example the ξ-direction, are written as a flux balance such that:

$$\frac{\partial (F - F_v)_i}{\partial \xi} = (F - F_v)_{i+\frac{1}{2}} - (F - F_v)_{i-\frac{1}{2}}. \qquad (5)$$

The inviscid interface flux, F, is evaluated as:

$$F_{i+\frac{1}{2}} = \frac{1}{2}\left[F(Q_R) + F(Q_L) - S_\xi |\Lambda_\xi| S_\xi^{-1}(Q_R - Q_L)\right]. \qquad (6)$$

$F(Q_R)$ and $F(Q_L)$ are the inviscid flux vectors computed from the left and right states; S_ξ and Λ_ξ are the eigenvector and eigenvalue matrices of the flux Jacobian matrix $\frac{\partial F}{\partial Q}$, evaluated using the Roe averaged flow variables given in [1].

The cell interface fluxes are obtained by the extrapolation of the state variables (primitive variables) based on a locally one-dimensional model of wave interactions normal to the cell interface. The extrapolation of the state variables to the cell interface are obtained from:

$$(Q_l)_{i+\frac{1}{2}} = Q_i + \frac{1}{4}[(1-\kappa)\nabla q_i + (1+\kappa)\Delta q_i]$$
$$(Q_l)_{i+\frac{1}{2}} = Q_i - \frac{1}{4}[(1-\kappa)\nabla q_{i+1} + (1-\kappa)\Delta q_{i+1}]$$
(7)

where ∇ and Δ are the backward and forward differences respectively as expressed with a minmod limiter [4]. The value of the parameter κ determines the spatial accuracy of the scheme; $\kappa = -1$ is the second order fully upwind scheme; $\kappa = 0$ is the upwind biased second order Fromm scheme; $\kappa = 1/3$ is the third order upwind biased scheme and $\kappa = 1$ is the second order central difference scheme. In this study $\kappa = 0$ and $\kappa = 1/3$ were used.

MODIFIED RUNGE-KUTTA METHODS

Modified Runge-Kutta methods with standard coefficients have been successful with central difference spatial discretization [5]. They have not worked as well with upwind differencing. The standard coefficients have been modified successfully by the authors [3] to achieve better performance with upwind differencing. To advance the solution from time level n to n+1 using k stages the scheme takes the following form:

$$\frac{\partial Q}{\partial t} = -RHS$$
$$Q^0 = Q^n$$
$$Q^1 = Q^n - \alpha_1 \Delta t RHS(Q^0)$$
$$Q^2 = Q^n - \alpha_2 \Delta t RHS(Q^1)$$
$$\vdots$$
$$Q^k = Q^n - \alpha_k \Delta t RHS(Q^{k-1})$$
$$Q^{n+1} = Q^k.$$
(8)

The modified Runge-Kutta requires two time levels to be stored in memory which is a crucial consideration in three dimensional computations. The solution obtained for the steady-state calculations is independent of the time step. The time step is calculated from,

$$\frac{1}{\Delta t} \geq \frac{1}{\Delta t_\xi} + \frac{1}{\Delta t_\eta} + \frac{1}{\Delta t_\zeta} + \frac{1}{\Delta t_\xi^v} + \frac{1}{\Delta t_\eta^v} + \frac{1}{\Delta t_\zeta^v}$$
(9)

where, $\dfrac{1}{\Delta t_l} \geq \lambda_l = |U^l| + a\sqrt{\phi_l^2}$

and $\dfrac{1}{\Delta t_l^v} \geq \lambda_l^v = \dfrac{\mu}{\rho}\left[\phi_l\left\{max\left(\dfrac{4}{3}, \dfrac{\gamma}{Pr}\right)\right\} + \dfrac{1}{3}\{|l_x l_y| + |l_x l_z| + |l_y l_z|\}\right].$

The first three terms on the right hand side of equation (9) are due to the stability limitation on the inviscid flux, while the last three terms are due to stability limitations on the viscous flux. Adding the viscous limitation, $\frac{1}{\Delta t_l^v}$, makes the scheme more robust on fine grids and in boundary layer type flows [6].

In steady-state calculations, it was desirable to remove the stiffness from the governing set of equations, and accelerate convergence. Hence, local time stepping, implicit residual smoothing and the full approximation storage (FAS) multigrid scheme were added to the explicit multistage time stepping scheme.

ADAPTIVE IMPLICIT RESIDUAL SMOOTHING

Implicit residual smoothing extends the stability limit and improves the damping properties of the multistage time-stepping scheme [5]. The authors decided to use adaptive implicit residual smoothing, which is only activated in a coordinate direction where it is needed, in contrast to constant scalar residual smoothing which is activated in all three coordinate directions equally. The adaptive implicit residual smoothing of [7] has been extended to three dimensions in the following form:

$$(1 - \beta_\xi \nabla_\xi \Delta_\xi)(1 - \beta_\eta \nabla_\eta \Delta_\eta)(1 - \beta_\zeta \nabla_\zeta \Delta_\zeta) RHS^* = RHS$$

where, $\beta_\xi = max \left\{ \frac{1}{4} \left[\left(\frac{CFL}{CFL^*} \frac{\lambda_\xi}{(\lambda_\xi + \frac{1}{16}(\lambda_\eta + \lambda_\zeta))} \right)^2 - 1 \right], 0 \right\}$ (10)

β_η and β_ζ have a similar formulation. The β_l are the adaptive residual smoothing coefficients which are a function of the grid aspect ratio and the spectral radii λ_l. $\frac{CFL}{CFL^*}$ is the ratio of the CFL number of the smoothed scheme to that of the basic explicit scheme. $CFL \approx 3.25$ gave the best rate of convergence for the test cases presented in this study. Increasing the ratio of $\frac{CFL}{CFL^*} > 2.0$ caused the high frequency damping of the scheme to vanish which is detrimental to multigrid convergence. The smoothing operator is applied after every stage of the Runge-Kutta time stepping scheme.

MULTIGRID METHOD

The basis for multigrid is the use of successively more coarse grids to calculate corrections to the solution of a set of partial differential equations on a fine mesh. These corrections reduce the low frequency components of error in the fine grid solution. Since the coarse grid contains significantly fewer points than the fine grid, less work is required to perform a computation there than on the fine grid.

The standard Full Approximation Storage multigrid scheme is used in this study. The FAS restriction operator used in the current work has two forms. One form is used to restrict the dependent variables; i.e. the flow quantities $\rho, \rho u, \rho v, \rho w,$ and e. For these, the volume weighted average of the values of the function at the midcells of the eight fine grid cells is used to set the cell value on the coarse grid. The other form of the restriction operator is for the restriction of residuals. A simple summation of the residuals over the eight fine grid cells is used to give the forcing function for the coarse grid cell. The restriction operations are performed for all interior points. At the inflow/outflow and block interface boundaries, only the values of the functions are restricted, with no residual restriction. The residual values are frozen to the fine grid values and are not updated on the coarse grids. On wall surfaces, the same boundary conditions are used for all the grids. At the interfaces, values from coincident cells in adjacent blocks are used to set the values in the ghost cells.

The description of the FAS scheme and a practical approach to the coding of multigrid in a multiblock environment is presented by the authors in reference [8].

MULTIBLOCK STRATEGY

With the progress achieved in numerical algorithms and also in the field of computer technology, computing flow fields around complex three-dimensional configurations is now possible. The generation of acceptable body-fitted grids is often quite a burden. Multiblock topologies offer the potential to reduce this complexity. The trade-off is the overhead required to transfer data between neighboring blocks. In the present multiblock implementation, C^o continuity of the grid is used because it avoids the necessity of spatial interpolation of the data when loading the ghost cells at block interfaces.

There are two possible strategies for the implementation of multigrid with a multiblock grid structure: (1) multigrid inside of multiblock and (2) multiblock inside of multigrid. The first strategy implies that for each successive block, one or more multigrid cycles is performed before the next block is considered. The drawback of this scheme is that the coarse grids for each of the blocks cannot communicate with each other except through the fine grids. This greatly reduces the benefits of multigrid. In the latter strategy each block is updated before proceeding to the next grid level. This has the advantage of allowing communication between the blocks at all grid levels, which reduces the time lag of information flow. The authors chose the multiblock inside of multigrid strategy, which can identically reproduce the convergence history of a single block calculation using an explicit algorithm.

BOUNDARY CONDITIONS

Standard boundary conditions were used for inflow/outflow, solid wall, and symmetry plane boundaries. For an interface boundary condition two ghost cells were used, which were set equal to the latest values in the coincident interior cells in the adjacent block. The updates of the interface ghost cells are performed before each iteration in a given block. The iteration on each block can then proceed without the need for further information from adjacent blocks.

In the original implementation of the multiblock strategy by the authors [1], a homogeneous boundary condition was used on each block face. This has been extended to allow multiple boundary conditions per face. The particular type of boundary condition, on each segment of each face of a particular block in the computational domain, is specified explicitly depending on the case. The multiple boundary conditions per face increased significantly the flexibility of the code to handle complex boundaries around three-dimensional geometries.

RESULTS

The present scheme was first tested using a single block to compute the inviscid flow through a rectangular channel configuration, with two compression ramps forming a compression corner about which the channel was symmetric. The back and bottom walls of the channel have converging ramps, each at 9.0°. A supersonic inlet flow of Mach = 3.17 was used for the test case, producing three shock surfaces. Two of the shock surfaces are two-dimensional wedge-flow shocks, which can be verified using two-dimensional analysis based on the Mach number normal to the leading edge of the wedge. The third surface is formed where the two wedge shocks coalesce to form a three-dimensional flow region which is shaped like a cone. This can be seen in Figure 1, where Mach line contours are shown on the back and bottom walls, and on the exit plane of the channel. The positions of the wedge shocks are shown on the back and bottom walls by the regions of highly concentrated Mach lines perpendicular to the flow direction. Also, on these two walls the edges of the cone shaped surface can be seen. On the exit plane, four flow regions are present. In the upper right corner is free stream, which is one-dimensional flow. From the middle of the plane to the lower left corner, where the flow is three-dimensional, the bottom of the cone surface, a partial disc, can be seen. The two wedge shock planes can be seen in the upper left corner and the lower right corner of the exit plane. Note that since the geometry of the channel is symmetric about the compression corner (the one joining the back and bottom walls to the exit plane of the channel), the flow field should be and is symmetric about this corner.

The above calculation was repeated using eight-blocks. The eight block grid was obtained by dividing the domain of the single block grid in half in each of the three coordinate directions. Results of the multiblock calculation are shown in Figure 2 and identically reproduce the results obtained with the single block calculation.

The convergence histories for the single-block and the eight-block calculations are shown in Figure 3. Notice that the convergence rate for the explicit scheme shows little degradation for the multiple block calculation. This is due primarily to the choice of the multiblock-inside-of-multigrid strategy which allows communication between the coarse grids in the multigrid scheme.

To verify the viscous capability of the method, flow over a flat plate was computed. The freestream Mach number was 0.5 and the Reynolds number was 5000. The normalized minimum spacing in the normal direction to the wall was 10^{-3}. The computed velocity profile in the boundary layer, and the skin friction coefficient along the plate are shown in Figure 4. Comparison between the computed results and Blasius boundary layer solution shows good agreement. The flow over the flat plate was recomputed after activating the adaptive implicit residual smoothing. Comparison of the convergence history using local time stepping, and adaptive implicit residual smoothing with results produced without the smoothing is shown in Figure 5.

The authors were interested in studying the flow field of a jet impinging normally on a flat plate in ambient air. This case is of particular interest as it contains some of the physics encountered by V/STOL aircraft in hover. The jet Mach number was 0.5 and the Reynolds number was 1×10^5. The jet exit was 5 jet diameters away from the ground plate and the plate extended radially outwards 15 jet diameters. Residual smoothing and multigrid were not used in this test case. The flow properties were normalized with respect to the jet inlet conditions.

The numerical simulation started with the jet entering the computational domain and interacting with quiescent air. The solution was then advanced in time using a nondimensional global time step. The time step was based on the stability criteria of the scheme. Snap shots, at different time steps, of the Mach number and the stagnation pressure contours are shown in Figure 6. A starting vortex develops as the free jet propagates into the computational domain as shown in Figure 6a. The free jet impinges on the ground plate and is deflected to a wall jet as shown in figure 6b. The stagnation region creates a favorable pressure gradient which causes the wall jet to accelerate rapidly as it departs the stagnation region, Figure 6c. The primary vortex, located on top of the wall jet, produces a local unsteady adverse pressure gradient which causes the flow to separate, Figure 6d. The wall jet with the primary vortex located on top moves with the separation point radially outward as shown in Figure 6e. The different flow fields that identify this type of flow, namely the free jet, the stagnation region, the wall jet and the moving separation, were predicted accurately.

All of the test cases presented in this paper have been run with same source code; only inputs to the program have been changed.

CONCLUSIONS

A three-dimensional multigrid multiblock multistage time stepping scheme for the Navier-Stokes equations has been developed. The multiblock structure provides complete topological and geometric flexibility. The only requirement is the C° continuity of the grid across block interfaces. The accelerating techniques (local time stepping, implicit residual smoothing and multigrid procedure) are effective in accelerating the rate of convergence to steady-state.

The present work demonstrates the feasibility of the multiblock multigrid multistage time stepping scheme for accurately predicting complex three-dimensional compressible flows. The method has been tested on a number of diverse three-dimensional test cases with increasingly complex flow characteristics. Supersonic inviscid flow through a simple duct was calculated to verify the multiblock structure. Viscous flow over a flat plate was computed to verify the viscous results and show the effectiveness of the adaptive implicit residual smoothing on accelerating convergence to steady-state. Unsteady viscous flow resulting from a jet impinging on a ground plate was calculated, which demonstrated the capability of the developed scheme to perform time-dependent calculations.

ACKNOWLEDGMENT

The work of Mr. Elmiligui and Mr. Cannizzaro was supported by NASA Langley Research Center Grant No. NAG1-633 monitored by Mr. Manuel D. Salas.

REFERENCES

1. Cannizzaro, F. E., Elmiligui, Alaa, and Melson, N. D., von Lavante, E., "A Multiblock Multigrid Method for the Solution of the Three-Dimensional Euler Equations." AIAA Paper 90-0105 (1990).
2. Roe, P., L., "Characteristic Based Schemes for the Euler Equations", A numerical Review of Fluid Mechanics (1986), pp. 337–365.
3. von Lavante, E., Elmiligui, A., Cannizzaro, F. E., and Warda, H., "Simple Explicit Upwind Schemes For Solving Compressible Flows", Proceedings of the Eighth GAMM-Conference on Numerical Methods in Fluid Mechanics, NNFM vol 29 (1989), pp. 291–301.
4. Kuruvila, G., and Salas, M. D. "Three Dimensional Simulation of Vortex Breakdown.", NASA TM 102664 (1990).
5. Jameson, A., Schmidt, W., and Turkel, E., "Numerical Solutions of the Euler Equations by Finite-Volume Methods Using Runge-Kutta Time-Stepping Schemes.", AIAA Paper 81-1259 (1981).
6. Vatsa, V., N., Turkel, E., and Abolhassani, J., S. "Extension of Multigrid Methodology to Supersonic/Hypersonic 3–D Viscous Flows", Fifth Copper Mountain Conference on Multigrid Methods, March 31–April 5, 1991.
7. Swanson, R. C., Turkel, E., and White, J. A. "An Effective Multigrid Method for High Speed Flows.", Fifth Copper Mountain Conference on Multigrid Methods, March 31–April 5, 1991.
8. Melson, N. D., Cannizzaro, F. E., and Elmiligui, A. "A Multigrid Multiblock Programming Strategy", Fifth Copper Mountain Conference on Multigrid Methods, March 31–April 5, 1991.

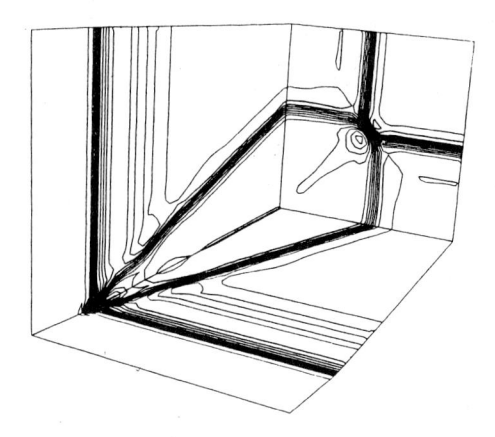

Figure 1: Mach Contours from single-block calculation for flow through a $9°$ compression corner (inlet Mach number of 3.17)

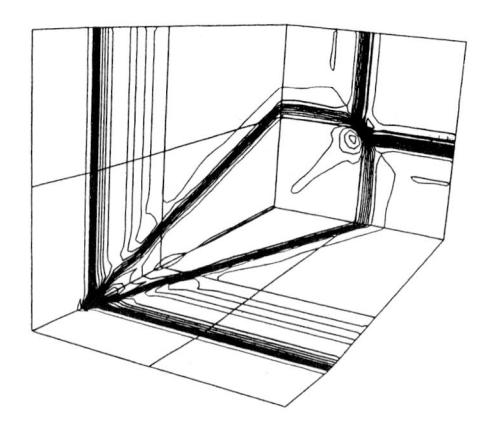

Figure 2: Mach Contours from eight-block calculation for flow through a $9°$ compression corner (inlet Mach number of 3.17).

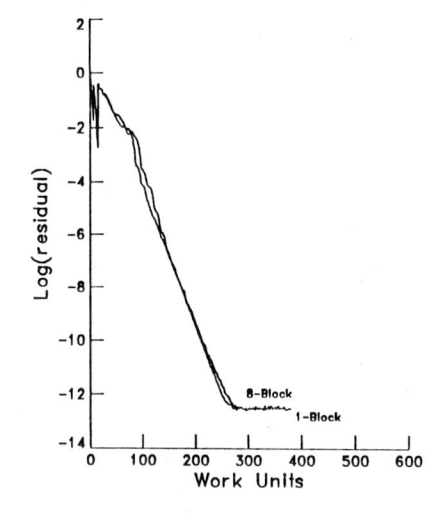

Figure 3: Comparison of convergence histories of single and eight block calculations of Mach contours for corner flow (inlet Mach number of 3.17).

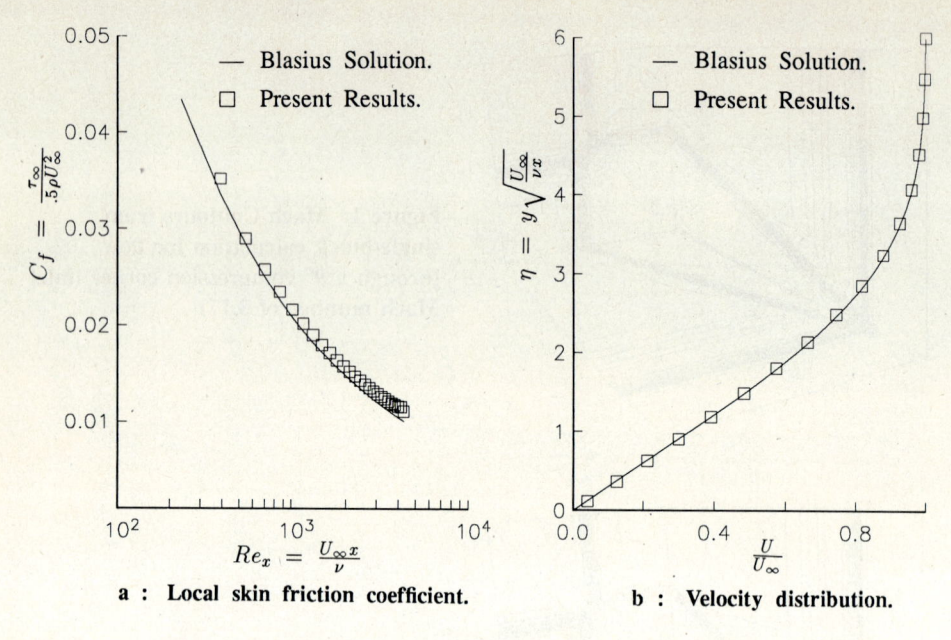

Figure 4: Laminar flow over a flat plate. ($M_\infty = 0.5$, $Re_\infty = 5000$.)

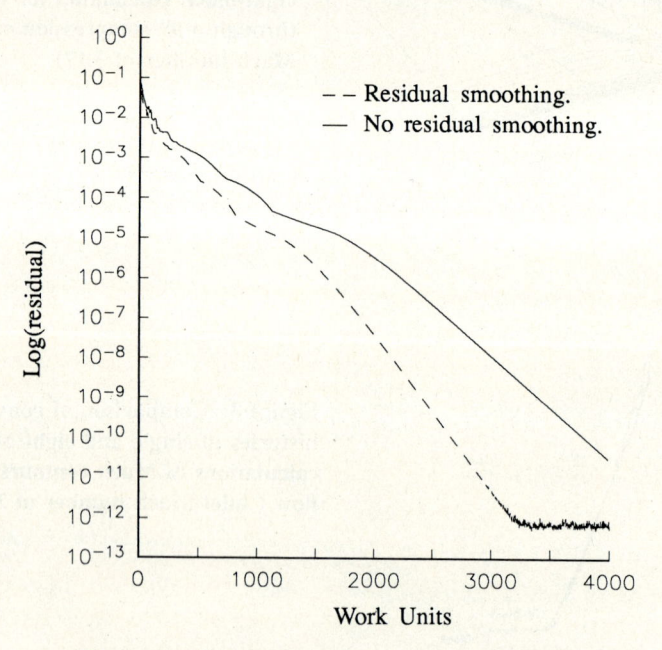

a : Local skin friction coefficient.

b : Velocity distribution.

Figure 5: Comparison of convergence history for a Laminar flow over a flat plate. ($M_\infty = 0.5$, $Re_\infty = 5000$.)

$M_{jet} = 0.5$

(a)

(b)

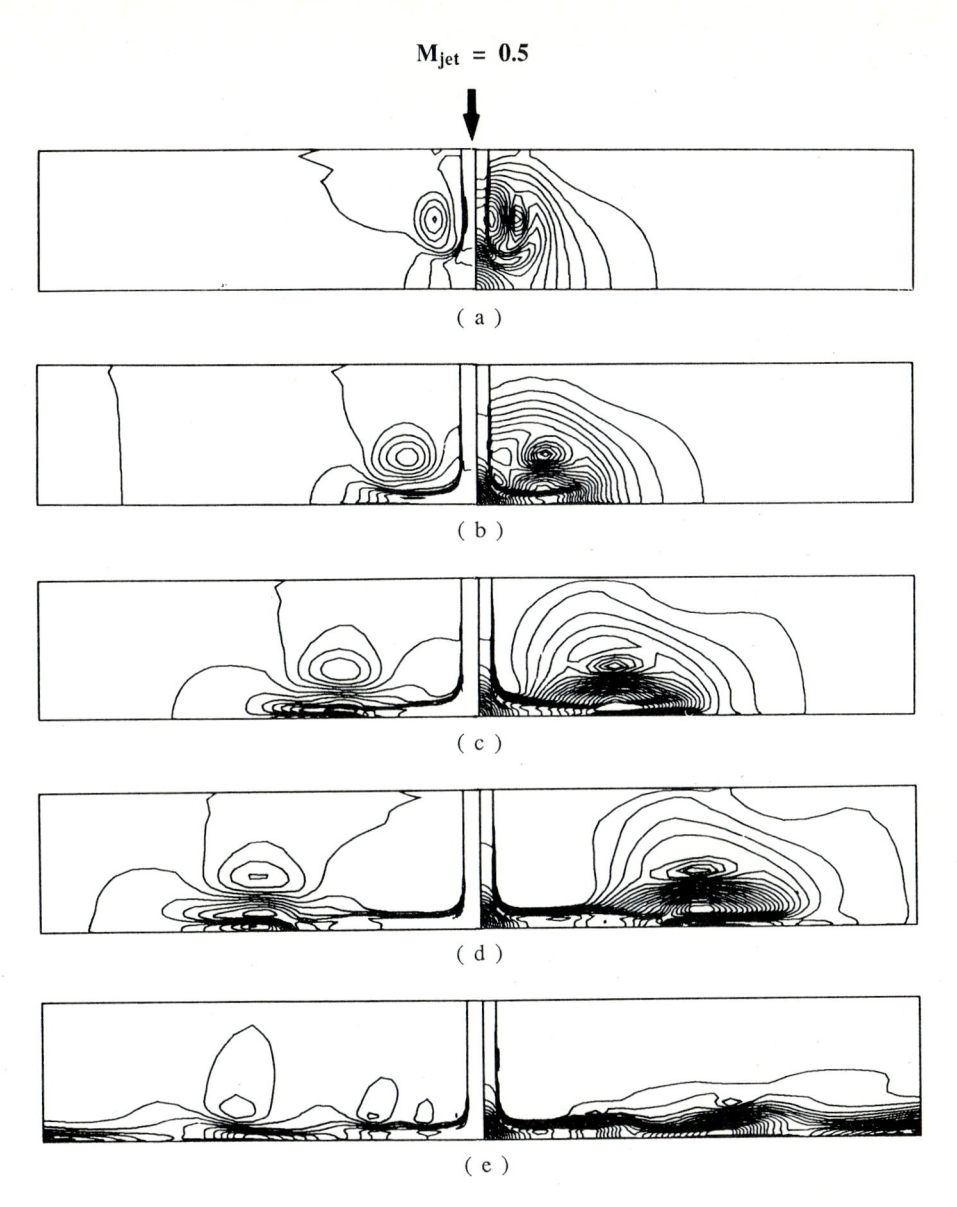

(c)

(d)

(e)

Stagnation Pressure Contours   **Mach Number Contours**

Figure 6: Jet impingement on a ground plate.
($M_\infty = 0.0$, $M_{jet} = 0.5$, $P_{jet}/P_\infty = 1.0$, $T_{jet}/T_\infty = 1.0$, $Re = 100,000$).

Numerical Solutions of Compressible Viscous Flows Using Multidomain Techniques

M. Fatica, F. Grasso
Department of Mechanics and Aeronautics
University of Rome "La Sapienza"
Via Eudossiana, 18 – 00184 – Rome, Italy

ABSTRACT

A multidomain technique has been developed for the solution of viscous transonic and supersonic flows. The approach is suitable for parallel architectures and allows operator adaptation by introducing viscous and inviscid subdomains. Results are presented for flows around NACA 0012 and RAE 2822 airfoils, and for a shock wave/laminar boundary layer interaction.

INTRODUCTION

In the present work, viscous two-dimensional transonic and supersonic flows are solved using a multidomain technique.

One of the critical aspects of computational fluid dynamics is related to the disparity of characteristic scales in the different regions of the flow: regions where the behavior is essentially inviscid, boundary layers, shock waves, geometric singularities, etc. Moreover, the solution of engineering problems on modern computers with parallel architecture needs a partitioning of computer instruction and data on the available processors.

A "natural" methodology to resolve the intrinsic difficulties of computational fluid dynamics and for an efficient utilization of parallel architectures is the domain decomposition. With the use of such a technique, some of the main problems, which CFD encounters in simulating flows over complex configurations, can be overcome. In particular: in each subdomain grids can be independently generated (this is very useful especially for three-dimensional complex geometries); different numerical schemes can in principle be employed in the different regions of the flow (for example accurate schemes, which are computationally expensive, can be used only in the subdomains where high order schemes are needed); different governing equations can be used in the different regions (for example the Navier-Stokes equations can be solved only in the regions of the flow where the viscous effects are important, whereas the Euler equations, or even the potential ones, can be solved in the other zones, with an obvious gain in CPU time); the use of multidomain decomposition allows the execution in parallel of a complex aerodynamic problem, whereby the solution on all domains is obtained concurrently by reducing a large scale problem to several (depending on the number of available processors) small-scale partitions of the same problem.

In general, there are two alternatives for subdividing the computational domain: the first one introduces domains which can "overlay" with an arbitrary orientation; the second one subdivides the computational domain into "patched" zones that share common boundaries without overlapping.

The main advantage of "overlaid" subdomain decomposition is the flexibility in grid generation. However, difficulties may arise in ensuring conservation; moreover, the effects of the size of overlaid zones on the accuracy and convergence rate are difficult to predict. Berger has derived a conservative flux scheme [1] for 2D problems, which, however, cannot be easily extended to 3-D flows. Moon and Liou [2] have introduced a constraint on the conservative quantities such as mass, momentum and total energy.

"Patched" subdomain decomposition introduces some constraints on the grid generation. However, it makes it simple to ensure a conservative treatment of block interfaces. Rai [3] has ensured global conservation by enforcing surface fluxes and the continuity of the dependent variables at the interfaces. Yadlin and Caughey [4] have developed a patched block approach for the solution of the Euler equations by implementing a diagonal implicit multigrid approach. At the (subdomain) interfaces they have imposed the continuity of the dependent variables. More recently [5] they have reported an extension of the algorithm to the solution of the Navier Stokes equations. In particular, they have exploited the advantage of operator adaptation by solving the Navier Stokes equations in the near wall domains and in the wake region, while solving the Euler equations in the far field.

In the present work a multidomain technique is developed for the solution of viscous two-dimensional transonic and supersonic flows. A "patched" subdomain decomposition with continuous grids is introduced that ensures global conservation and it yields no distortion of discontinuities (such as shocks and slip surfaces) that cross (grid/subdomain) interfaces. Results of flows over NACA 0012 and RAE 2822 airfoils and a shock wave/boundary layer interaction problem are discussed to demonstrate the properties of the method mainly for operator adaptation and parallel applications.

NUMERICAL ALGORITHM

The governing equations solved are the two-dimensional compressible Navier-Stokes equations in conservation form.

$$\frac{\partial}{\partial t} \int_S \mathbf{W} \, dS = - \oint_{\partial S} (\mathbf{F}_E - \mathbf{F}_V) \cdot \underline{n} \, ds \tag{1}$$

where $\mathbf{W}$, $\mathbf{F}_E$ and $\mathbf{F}_V$ are respectively the vector unknown, the inviscid and viscous fluxes, defined as:

$$\mathbf{W} = [\rho, \rho \underline{u}, \rho E]^T$$

$$\mathbf{F}_E = \left[\rho \underline{u}, \rho \underline{u}\underline{u} + p \underline{\underline{U}}, \rho \underline{u}\left(E + \frac{p}{\rho}\right)\right]^T$$

$$\mathbf{F}_V = \left[0, \underline{\underline{\sigma}}, -\left(\underline{q} - \underline{u} \cdot \underline{\underline{\sigma}}\right)\right]^T$$

and

$$p = (\gamma - 1)\rho(E - \underline{u} \cdot \underline{u}/2)$$
$$\underline{\underline{\sigma}} = \mu\left(\nabla \underline{u} + \nabla \underline{u}^T\right) - \frac{2}{3}\mu \nabla \cdot \underline{u}\,\underline{\underline{U}}$$
$$\underline{q} = -\lambda \nabla T.$$

The basic algorithm for solving the system of governing equations is based on a finite volume approach with a cell centered formulation. A system of ordinary differential equations is obtained by using the method of lines, which allows separation of space and time discretization.

Approximating surface and boundary integrals by means of the mean value theorem and mid-point rule, the governing equations are cast in the following discretized form

$$\frac{d\mathbf{W}_{i,j}}{dt} S_{i,j} + \sum_{\beta=1,4} \left(\mathbf{F}_{E,num} - \mathbf{F}_{V,num}\right) \cdot \underline{n}\, ds = 0 \tag{2}$$

where β indicates the cell face, and the subscript num stands for numerical.

In the present work, two different approaches are used to evaluate the numerical inviscid flux ($\mathbf{F}_{E,num}$): the first one is based on a central difference formula; and the second one on an upwind discretization. In both cases, $\mathbf{F}_{E,num}$ is formally defined as

$$(\mathbf{F}_{E,num} \cdot \underline{n})_{i+1/2,j} = \frac{1}{2}\left(\mathbf{F}_{E_{i,j}} + \mathbf{F}_{E_{i+1,j}}\right) \cdot \underline{n}_{i+1/2,j} + D_{i+1/2,j}. \tag{3}$$

The term $D_{i+1/2,j}$ represents the adaptive dissipation flux, needed when a central difference formula is used. Such a term is an adaptive blending of first- and third-order derivatives, necessary to prevent oscillations and to ensure stability and convergence to steady state, and the model of Jameson [6] is used.

Viceversa, when the upwind discretization formula is used, the term $D_{i+1/2,j}$ represents the numerical antidiffusive flux that makes the scheme upwind biased and ensures the TVD property. Following Harten and Yee [7] its expression is obtained by characteristic decomposition along the normal direction to cell face [8], as a function of the differences of the characteristic variables, and the right eigenvector matrix (defined in terms of the covariant and contravariant velocity components), with the use of minmod limiter for the antidiffusive flux.

The viscous flux depends on the gradient of the primitive variables (velocity and temperature). The gradient is numerically evaluated by applying Gauss theorem to a computational cell whose vertices are the two grid nodes (I,J) and $(I,J-1)$ and the centers of the two adjacent cell (i,j) and $(i+1,j)$. Hence, the discretized flux contribution at cell face $i+1/2,j$ is a function of grid and cell center values, i.e.

$$\mathbf{F}_{V,num} = \mathbf{g}\left(W_{I,J}; W_{I,J-1}; W_{i,j}; W_{i+1,j}\right) \tag{4}$$

where the vertex values of the variables are obtained by bilinear interpolation of the cell center values.

The time integration is performed by a three- and five-stage Runge Kutta algorithm with only one evaluation of viscous contribution.

Parallel Domain Decomposition and Data Structure

Domain decomposition is a natural methodology for an efficient utilization of parallel computers [9], and the instruction set and the data can easily be partitioned on the available processors and memories. Moreover, the technique allows for operator adaptation and mesh refinement, and it is suitable for complex geometries. In the present work the technique has been implemented to exploit its properties for parallel applications and operator adaptation. The computational domain is partitioned in patched-type subdomains, called blocks in logical space. However, a fictitious layer of cells is introduced at the boundaries (as in the single domain approach) for an efficient treatment of the boundary conditions, thus making all blocks overlaid. The partitioning in blocks has been obtained starting from the single domain with the constraint of grid continouities.

In a multiblock approach the computation is carried out separately on each block (henceforth its natural use for parallel applications). However, for a correct evolution of the solution, block adjacency relationships must be defined. In the present work, we have introduced a connection matrix C (a square matrix $N_B \times N_B$, where N_B is the number of subdivision of the computational domain), whose element $C_{i,j}$ yields the correlation between the i- and j-th domain. For two-dimensional applications each block has four boundary faces, which have been numbered as indicated in Fig. 1. Then, $C_{i,j}$ is equal to the number representing the face shared by the two blocks; if i and j share no boundaries, $C_{i,j}$ is set equal to zero (the elements of the principal diagonal are meaningless and are set equal to zero). Moreover, a block is not allowed to share the same face with more than one block. Hence, in two-dimensions the number of non zero elements on each row is at most four, and $C_{i,j} \neq C_{i,\ell} \; \forall j \neq \ell$. To handle efficiently the boundary conditions, we have also assumed that each face can only be of one type (either wall, or wake or farfield, etc.).

Domain decomposition is well suited for operator adaptation. This amounts to either numerical operator adaptation (for example use higher order schemes only on those subdomains where greater accuracy is required), or physical operator adaptation (for example solve for the Navier Stokes equations only on those subdomains where viscous effects are important). In the present work, the subdomain decomposition technique has been developed for physical operator adaptation. For flows around airfoils the near wall and wake regions are the ones where viscous effects are important, while the flow can reasonably be assumed inviscid in the outer part. Such a consideration naturally allows a decomposition in "viscous" and "euler" subdomains corresponding to the type of equations solved. At present, the interface between the two regions has been set *a priori* at a distance of the order of one chord to avoid the uncertainties in the boundary conditions when such interface is a function of the solution. Referring to Fig. 2, the connection matrix is constructed as a block structured matrix, where the diagonal blocks represent the connection matrices for the inner and outer subdomains, and the off-diagonal ones yield the adjacency relationships between the "viscous" and the "inviscid" blocks. All the topological and physical properties of the subdomains can be obtained by scanning the rows and columns of the connection matrix. However, to increase the efficiency of the algorithm, rather than scanning the connection matrix at each computational cycle, an incidence matrix L (a rectangular matrix $N_B \times 4$) is constructed, where $L_{n,\ell}$ yields the number of the block that shares face ℓ with block

n. To identify blocks in the wake, $L_{n,\ell}$ assumes a negative value.

Then, for each block the data structure requires: the dimension (in logical space) NX_B and NY_B; the metric variables X_B (coordinates) and S_B (surfaces); the field variables $\mathbf{W}_B$, p_B, T_B etc.; the parameters that characterize geometrical topology of the flows (wall, wake etc.) and physical topology (viscous or inviscid). Observe that all variables are of two types: local and global. The local variables are variables of the block. The global variables are defined as one dimensional arrays that are addressed by defining entry points corresponding to each block, and are the only variables that allow exchange of information between blocks.

The total number and the dimension of blocks depend upon their topological properties. For example, the operation count of viscous blocks is about three times that of inviscid ones. Therefore, computational efficiency and load-balance require either that the dimension of a viscous block should be approximately one third the inviscid one, or sinchronization of the tasks.

The computational strategy here devised has been implemented in a computer code that runs on a shared memory parallel computer (IBM 3090/600). On such computer three levels of parallelization are possible: autotasking; microtasking; and macrotasking. With autotasking, parallelization is done automatically by the compiler at do-loop level. With microtasking, the user decides which do-loops and instruction set are to be executed in parallel. Multitasking amounts to perform in parallel large tasks (at subroutine levels), amounting to coarse-grain parallelism, thus reducing communication overhead. The domain decomposition here developed is a natural candidate for multitasking, and the computation is performed concurrently on all-blocks by defining a set of tasks equal to the number of blocks.

Boundary Conditions and Interface Treatment

The partitioning of the computational domain in subdomains introduces internal boundaries corresponding to the interfaces between adjacent blocks. Therefore, the use of a multidomain technique requires boundary conditions along the true boundaries and at block interfaces. The continuity of the variables is enforced [10] along the boundaries shared by two adjacent blocks. The introduction of a fictitious layer of cells surrounding the blocks allows to enforce the interface conditions by injecting the solution of the underlying block onto the overlaying one. However, only one fictitious layer of cells is not sufficient to evaluate the flux at the interface. Indeed, the dependency of the adaptive dissipation flux [6] upon the third derivatives of the variables requires at least two fictitious layers of cells. Likewise, the upwind biased TVD scheme [7,8] needs more than one layer to maintain second order accuracy. In the present work, for an efficient exploitation of the parallelism, all interfaces and boundaries are treated asynchronously. Morevoer, to reduce the storage requirement the problem has been solved by adding to the data structure two interface matrices, that yield for each block the first derivatives of the variables needed for a correct evaluation of the inviscid numerical flux.

RESULTS

The validity of the proposed approach has been assessed by simulating viscous transonic flows around NACA 0012 and RAE 2822 airfoils, corresponding to some of the test cases reported in the Viscous Transonic Airfoil Workshop [11]. The methodology has also been applied to compute a shock wave/laminar boundary layer interaction.

NACA 0012

The first test case corresponds to the flow around a NACA 0012 airfoil at a Mach number $M_\infty = .55$, an angle of attack $\alpha = 8.34°$ and Reynolds number $Re = 9\,10^6$. The conditions are such that the flow is turbulent, and the algebraic turbulence model of Baldwin and Lomax has been used, with transition set at 5% of the chord, as prescribed in the workshop. The flow shows a supersonic zone on the upper surface with separation at the foot of the shock. The rather large angle of attack (nearly equal to stall) causes a rapid expansion immediately followed by a shock. The latter interacts (weakly) with the boundary layer inducing a small separation at the interaction site. For this test case four computations have been performed: 1) single domain; 2) three subdomains; 3) six subdomains; 4) three viscous subdomains and three inviscid subdomains.

The first computation (SD) has been performed on a C-type grid of 176×48 cells. In the second computation (SDD3) the grid of the reference test case SD has been partitioned in three blocks of 24×48, 128×48 and 24×48 cells, where the first and third block are in the wake region, and the second one is the block around the airfoil. In the third computation (SDD6), six blocks have been generated by partitioning the original C-grid in two strips (each of 24 cells in the normal direction) and subsequently subdividing the latter in three blocks, thus obtaining three inner blocks of 24×24, 128×24 and 24×24 cells, and three outer blocks of 24×24, 128×24 and 24×24 cells. In the fourth computation (SDD6VE) the six blocks have been generated as described before. However, the Navier Stokes Equations have been solved on the three inner blocks and the Euler equations on the outer ones, so as to assess the ability of the approach for operator adaptation.

The results of the four computations are reported in Figs. 3 and 4, where the distribution of the pressure and skin friction coefficients vs. x/c are reported. No differences in the pressure distribution is noticeable, however some discrepancies in the skin friction distribution are observed in the proximity of the foot of the shock. The computed lift coefficients for the four computations show differences of the order of 3% with respect to the experimental one. For all cases, the drag counts are overpredicted due to deficiencies in the turbulence model as also found by other authors [11]. Comparing the CPU times of the SDD6 and SDD6VE computations, one observes a gain in efficiency of about 13% when three viscous blocks plus three inviscid ones are used, and practically no differences in the computed solution.

RAE 2822

The second test case corresponds to the flow around a RAE 2822 airfoil at a Mach number $M_\infty = .73$, an angle of attack $\alpha = 2.79°$, Reynolds number $Re = 6.5\,10^6$, and transition set at 3% of the chord.

Two computations have been performed: 1) single domain with 176×48 cells; 2) four subdomains of 24×48, 64×48, 64×48, and 24×48 cells. Fig. 5 shows the computed pressure vs. x/c. The computed results obtained with four (all viscous) subdomains (SDD4) compare well with the single domain solution (SDD). The disagreement with the experimental distribution of Cook, Firmin and McDonald [12] has to be ascribed to deficiencies in the turbulence model as concluded in Refs. [4,6].

Shock Wave/Boundary Layer Interaction

A shock wave/boundary layer interaction problem (SWBLI) has been computed to assess the ability of the method to resolve highly interacting flows. The test case corresponds to a shock wave that impinges on a laminar boundary layer that develops on a flat plate. The angle of the impinging shock and its distance from the leading edge are $\theta = 32.6°$ and $x_{sh} = .16$ ft; the Mach number is $M_\infty = 2$ and the Reynolds number $Re_{x_{sh}} = .296\ 10^6$. The computation has been carried out with the upwind biased TVD algorithm. The domain has been partitioned in four subdomains of 4×48, 46×48, 35×48, and 15×48 cells. The first block corresponds to the free stream region before the plate leading edge. The second and fourth block are respectively located upstream and downstream of the interaction. The third block is located in the region of interaction. Such a partitioning has been selected to check if impinging, separation and reattachment shocks are properly transmitted across block interfaces.

The wall pressure and skin friction coefficients are shown in Figs. 7–8. The measurements of Hakkinen [13] and the results of Mac Cormack [14] are also reported. The comparison indicates that the present approach yields overall good prediction of the flow. The extent of the recirculation region is in good agreement with the data of Hakkinen, even though a slightly slower recompression of the flow is predicted. The computed results also indicate that the complex wave pattern (incident, separation and reattachment shocks) is not altered by the decomposition technique.

CONCLUSIONS

A multidomain technique has been developed for the solution of viscous transonic and supersonic flows. The technique uses a patched domain decomposition with continuous grids, it ensures global conservation and it does not produce distortions of discontinuities (such as shocks and slip surfaces) that cross interfaces.

Domain decomposition has been here exploited for operator adaptation and for an efficient utilization of parallel computers. In general, operator adaptation amounts to either numerical or physical operator adaptation. The former one amounts to use different schemes in different blocks depending upon the degree of accuracy required. The latter one amounts to adapt the governing equations in the different blocks according to the controlling local physical phenomena. The methodology has been developed for physical operator adaptation by introducing "viscous subdomains", in which the full Navier Stokes equations have been solved, and "inviscid subdomains" where the Euler equations are solved. The computed solution is found to be as accurate as the single domain one, indicating the validity of the technique, particularly as far as the interface treatment is concerned. Comparison of the results with those obtained by partitioning only in viscous subdomains indicate a gain in computational efficiency of about 15%, and practically no deterioration of the accuracy. The technique uses either an upwind biased total variation diminishing discretization or a central one with adaptive dissipation. However, the two algorithms have not yet been implemented for numerical operator adaptation. The subdomain decomposition technique has been structured for parallel applications on shared memory computers at multitasking level. However, the efficiency of the method has not been fully established, and an analysis of the optimization of the partitioning and of the speed-up versus serial execution is required.

REFERENCES

[1] Berger, M., "On Conservation at Grid Interfaces", SIAM J. Numer. Anal., Vol. 24, October 1987.

[2] Moon, Y.J., and Liou, M.S., "Conservative Treatment of Boundary Interfaces for Overlaid Grids and Multi-Level Grid Adaptations", AIAA Paper 89–1990–CP.

[3] Rai, M.A., "A Conservative Treatment of Zonal Boundaries for Euler Equation Calculations", Journal of Computational Physics, 62 (1986).

[4] Yadlin, Y., and Caughey, D.A., "Block Multigrid Implicit Solution of the Euler Equations of Compressible Fluid Flow", AIAA Journal, Vol. 29, pp. 712–719, 1991.

[5] Yadlin, Y., Tysinger, T.L., and Caughey, D.A., "Parallel Block Multigrid Solution of the Compressible Navier Stokes Equations", AIAA 10th CFD, Conference Proceedings, pp. 965–966, 1991.

[6] Jameson, A., "Transonic Flow Calculations", MAE Report # 1651.

[7] Yee, H.C., Warming, R.F., and Harten, A., "Implicit TVD Schemes for Steady State Calculations", Journal of Computational Physics, Vol. 57, pp. 327–360, 1985.

[8] Bassi, F., Grasso, F., and Savini, M., "Finite Volume TVD Runge Kutta Scheme for Navier Stokes Computations", Lecture Notes in Physics, Vol. 323, pp. 131–136, 1989.

[9] Gropp, W.D., and Keyes, D.E., "Domain Decomposition Methods in Computational Fluid Dynamics", *to appear as a 1991 ICASE Technical Report*.

[10] Quarteroni, A., "Domain Decomposition Methods for Systems of Conservation Laws: Spectral Collocation Approximation", ICASE Report, N. 89–5, 1989.

[11] Holst, T.L., "Viscous Transonic Airfoil Workshop Compendium of Results", AIAA Paper 87–1460, January 1987.

[12] Cook, P.H., Mc Donald, M.A., and Firmin, M.C.P., "Aerofoil RAE 2822 Pressure Distributions and Boundary Layer and Wake Measurement", AGARD Advisory Report N. 138, May 1979.

[13] Hakkinen, R.J., et al., "The Interaction of an Oblique Shock Wave with a Laminar Boundary Layer", NASA Memorandum, 1959.

[14] MacCormack, R.W., "A Numerical Method for Solving the Equations of Compressible Viscous Flow", AIAA Journal, Vol. 20, pp. 1275–1281, 1982.

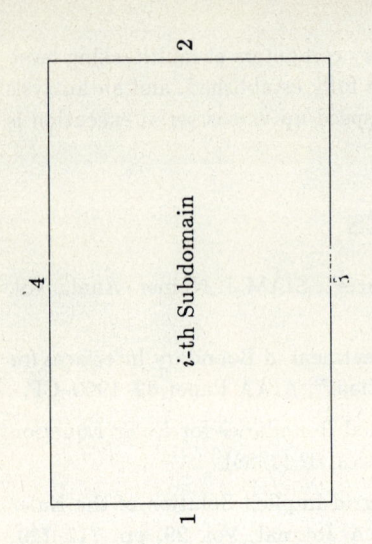

Fig. 1 – Numbering of subdomain faces (for adjacency relationship).

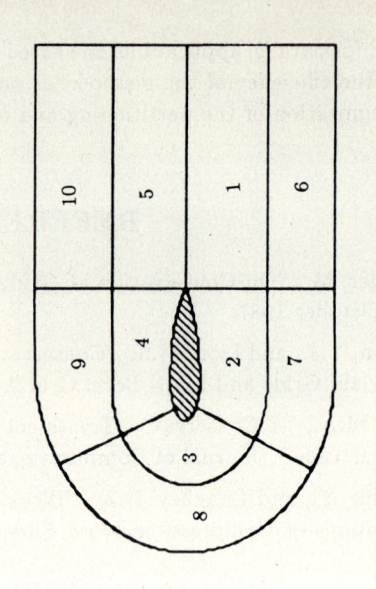

Fig. 2 – Typical partitioning for airfoil computations.

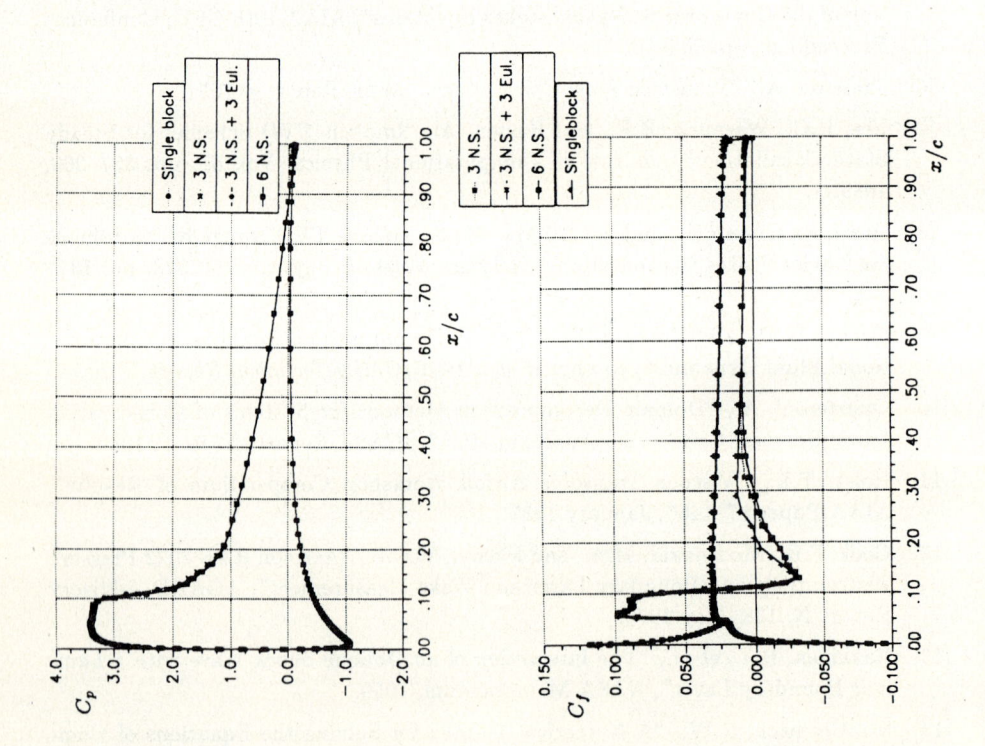

Fig. 3 – NACA 0012: pressure coefficient vs x/c.

Fig. 4 – NACA 0012: skin friction coefficient vs x/c.

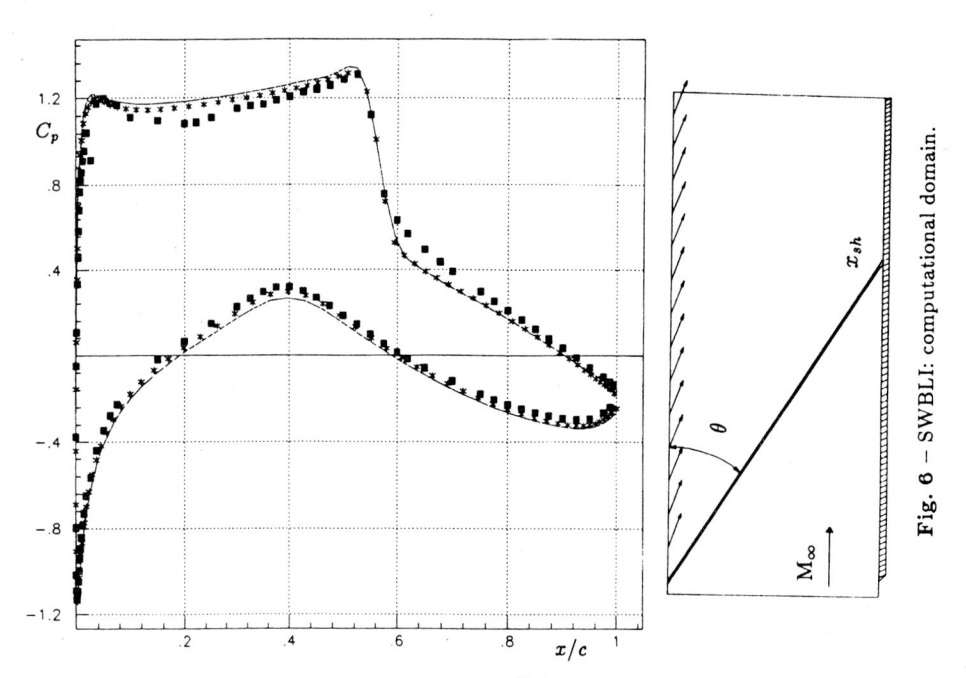

Fig. 5 – RAE 2822: pressure coefficient vs x/c. ■ Ref.[12]; * SDD; — SDD4

Fig. 6 – SWBLI: computational domain.

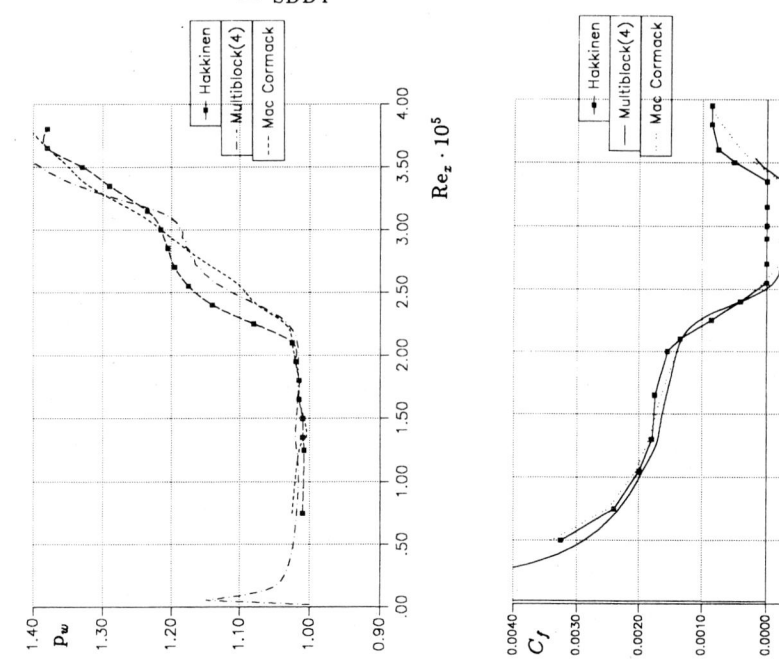

Fig. 7 – SWBLI: wall pressure vs $Re_z \cdot 10^5$.

Fig. 8 – SWBLI: skin friction vs $Re_z \cdot 10^5$.

INCOMPRESSIBLE NAVIER STOKES METHODS

THREE-DIMENSIONAL NAVIER-STOKES COMPUTATIONS OF SPATIALLY DEVELOPING GÖRTLER VORTICES

Alessandro Bottaro
Institut de Machines Hydrauliques et de Mécanique des Fluides (IMHEF)
Ecole Polytechnique Fédérale de Lausanne (EPFL)
CH-1015 Lausanne, Switzerland

SUMMARY

The Görtler flow is studied through three-dimensional incompressible Navier-Stokes computations, that solve for the spatially developing flow. By the use of inflow-outflow boundary conditions (as opposed to the temporal approach) we allow in a natural way for the development of the boundary layer, we preserve the *convective* nature of the Görtler instability and we do not impose a priori streamwise wavelengths (such as the wavelength of the secondary instability). Furthermore, we consider more than one vortex pair in the cross-section so that vortex interaction mechanisms are not overlooked. This is likely to be an important requirement in the selection of the secondary instability mode. The results computed in the primary instability regime are in very good agreement with experiments. To trigger the secondary instability we introduce the equivalent of a "vibrating ribbon". The secondary instability starts as a sinuous motion of the low speed streaks. The spanwise movement of the streaks is a feature that the Görtler flow shares with shear layer structures in near-wall turbulence.

INTRODUCTION

Longitudinal vortices developing on curved boundary layers are the object of intense investigations. They are found in a variety of situations: turbine blades, heat exchangers and airfoils, to name just a few, and affect significantly heat transfer and friction factors. Understanding the formation and breakdown of streamwise vortices is of importance in determining the performance and efficiency of the application in question. Furthermore, Görtler-like vortices constitute an essential ingredient of coherent structures in shear turbulence. As such, a study focussing on the breakdown and mutual interaction of an array of vortices could elucidate some of the underlying processes governing the laminar-to-turbulent transition. A numerical study of Görtler vortices can employ the temporal approach [1, 2] or the spatial approach [3]. Both approaches have their advantages and drawbacks. In the temporal approach the streamwise direction, x, is periodic and a source term, representing an average pressure gradient, drives the flow. This means that the boundary layer develops in time (instead of space) and a *convection velocity* must be adopted to match experimental and numerical results. This velocity cannot be determined a priori, and is generally different in the laminar, transitional and turbulent regime [1]. Finally, it is questionable whether a uniform convection velocity throughout the cross-section could properly match streamwise advection to temporal development. This is because different regions in the cross-section may act on very different time scales (e.g., at the onset of the secondary instability). The temporal approach presents the advantage that open (inlet-outlet) boundary conditions do not need to be specified, and the computational domain in x can be taken relatively short. In particular, when looking only at the primary Görtler instability, the flow can be computed on the two-dimensional cross-section. The parallel flow assumption invoked in the temporal approach has been disputed by Hall [4]. The alternative of spatially developing simulations requires the adoption of suitable open boundary conditions. These conditions should:

(i) not be the cause of a numerical resonance between inlet and outlet (see Huerre and Monkewitz [5] for a discussion of the problem),

(ii) not produce appreciable upstream reflection of outgoing waves in the neighborhood of the exit boundary.

The Görtler flow presents no separation near the outlet section. Additionally, in the transitional and turbulent regime the flow near the channel exit is quasi-parallel.

These circumstances are such that the specification of an outflow boundary condition presenting little upstream reflection is not exceedingly difficult [6]. Also, because of the *convective* nature of the primary instability, an inlet perturbation must be provided to drive the vortex development. If the constant inlet perturbation is of sufficiently large amplitude it will overrule the (eventual) perturbation generated at the entrance of the channel by the inlet-outlet resonance mechanism, making this latter - undesirable - phenomenon not operational. The additional drawback of a spatial simulation lies in the large number of grid points required in the x direction to resolve accurately the streamwise scales (e.g., those of the secondary instability). The simulation discussed here is sufficiently resolved to capture the details of the wavy secondary instability that develops on top of the Görtler vortices.

Computational domain, mesh system and boundary conditions

The geometry of the channel being studied is sketched in figure 1. The cross-section of the channel has an aspect ratio equal to three, the inner radius of curvature is taken equal to 4.5 times the width a of the channel and the angular opening is $2\pi/7$; a is used as length scale. A channel like the one sketched in figure 1 has been built at IMHEF and hot wire measurements in air of the streamwise velocity component have been carried out by O.J.E. Matsson and P.H. Alfredsson of The Royal Institute of Technology in Stockholm, Sweden. Görtler vortices occur in a thin boundary layer near the concave wall so that, for computational economy, the calculation domain is suitably reduced. The domain over which we discretize the equations is shown shaded in figure 1. The computational length in the spanwise direction is one sixth of the total spanwise length of the duct, and in the normal direction the length is equal to half the channel width. The grid used employs 256 streamwise points and 64^2 points in the cross section. The mesh is graded in the normal direction with a concentration of points near the concave wall, and uniform in the other two directions. To match the experimental environment suitable boundary conditions must be applied. The inlet conditions are Dirichlet conditions on the streamwise speed u, normal speed v and spanwise speed w:

$$u = U,$$
$$v = \varepsilon U \sin(\alpha z),$$
$$w = \varepsilon U \cos(\alpha z)/(\alpha r),$$

for $0 \leq z \leq z_{max}$, with ε varying in a stepwise fashion from 0 at the inner boundary $r = (r_{out} + r_{in})/2$ to 0.01 near the outer boundary $r = r_{out}$; α is the spanwise wavenumber, $\alpha = 6\pi/z_{max}$.

Such inlet conditions satisfy continuity and are such that a low amplitude wave (composed of three full cycles) describes the secondary motion in a neighborhood of the concave wall. At $r = r_{out}$ no-slip conditions are used; at $r = (r_{out} + r_{in})/2$ a uniform "free stream" velocity prevails. However, since we are dealing with duct flow (and not a canonical boundary layer flow for which the free stream speed is fixed) the outer layer velocity increases with the thickening of the boundary layers. Such free stream velocity is not known a priori[1]. On the inner boundary of the computational domain we have chosen to specify stress-free conditions:

$$\partial u/\partial r - u/r = 0,$$
$$v = 0,$$
$$\partial w/\partial r = 0.$$

The conditions on the spanwise direction are periodic. The choice has been dictated by the need to avoid an overspecification of spanwise grid points to properly resolve flow

---
[1] Note that this is not the case for the bulk velocity U which is constant.

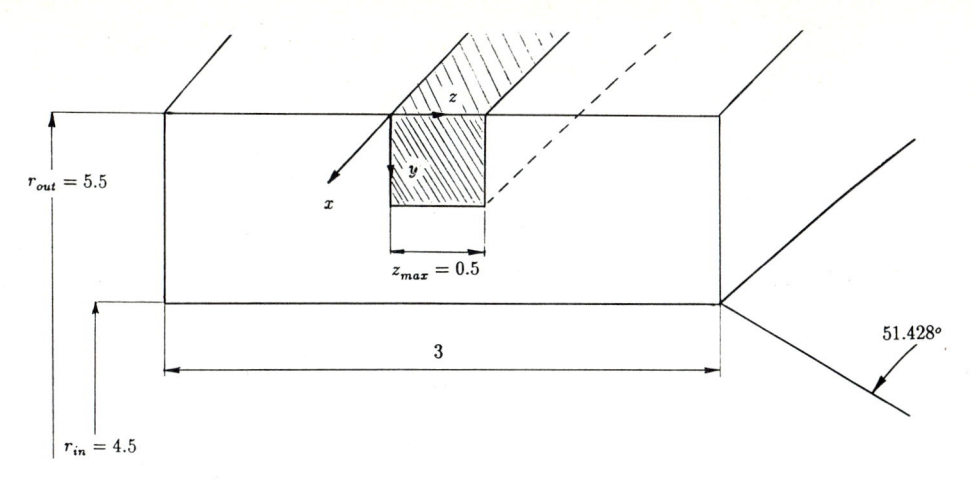

Figure 1: Computational domain (shaded) in the duct and geometrical nomenclature.

structures (Ekman vortices) induced by the side walls. The study of such wall vortices is beyond the scope of the present investigation. At the outlet cross-section convective boundary conditions have been adopted. They take the form:

$$\partial u/\partial t + c\,\partial u/\partial x = 0,$$
$$\partial v/\partial t + c\,\partial v/\partial x = 0,$$
$$\partial w/\partial t + c\,\partial w/\partial x = 0.$$

The quantity c is a phase speed and has been taken to be constant and equal to the bulk velocity U. Different choices of c are possible but tests have shown that the results are essentially unaffected (see, e.g., Pauley, Moin and Reynolds [7]). The above equations have been treated explicitely, to first order in time and space, like in [6].

The control parameter in the simulation is the Reynolds number, defined as

$$Re = aU/\nu,$$

where the bulk velocity U is, hereafter, used as velocity scale. We consider the case $Re = 10,000$. For flat plate boundary layer this translates to $0 \leq Re_x \leq 49,400$ from the beginning of the boundary layer to the end. The Blasius momentum thickness δ_2 is then

$$0 \leq \delta_2 = 0.664\,x\,Re_x^{-1/2} \leq 0.015,$$

while the displacement thickness δ^* is

$$0 \leq \delta^* = 1.721\,x\,Re_x^{-1/2} \leq 0.038,$$

which gives a displacement-thickness-based Reynolds number $0 \leq Re_{\delta^*} = U\delta^*/\nu \leq 380$. The upper limit is superior to the experimentally measured value for the onset of transition in Blasius flow which is of approximately 300. The linear stability value for the onset of Tollmien-Schlichting (TS) waves in plane boundary layer flow is about 400. Therefore, TS waves are stable in the regime where Görtler vortices develop. In the presence of curvature the boundary layer grows differently from the plane case, so that the values of maximum momentum and displacement thicknesses should be somewhat adjusted. However, based on these values, we can estimate the Görtler number to be

$$0 \leq G\ddot{o} = \frac{U\delta_2}{\nu}\left(\frac{\delta_2}{r_{out}}\right)^{1/2} \leq 7.7.$$

The value of 7.7, although approximate, is considerably larger than any stability limit for the onset of streamwise vortices available in the literature. It has been found experimentally [8] that the Görtler number for the onset of the secondary instability is more or less independent of primary (spanwise) wavelength and is as low as 7.1^2. We conclude that, in the parameter range chosen, Görtler vortices grow out of the low amplitude steady forcing provided at the inlet. Furthermore, somewhere near the exit of the channel, a wavy type of secondary instability should begin. This instability is reported to be of shear-type and to persist when the centrifugal mechanism is removed [9]. These conjectures are confirmed by the experimental results obtained at IMHEF.

NUMERICAL TECHNIQUE

We have considered the incompressible Navier-Stokes equations expressed in cylindrical polar coordinates, and have discretized them by a finite volume technique on a staggered mesh. The spatial scheme adopts a second order central difference operator to treat total (convective and diffusive) fluxes. Solutions are obtained by fully implicit time marching. Typically, three internal iterations are performed at each time step to ensure that the maximum pressure and velocity residuals decrease to acceptable levels (10^{-5} is the level required for each physical quantity). A time step dt equal to 0.005 is used. It is sufficiently small to provide confidence in the time accuracy of the numerical procedure: four time steps are needed to advance a fluid structure (in an average sense) of about one streamwise mesh width at speed U. The *pressure coupling* is treated with the SIMPLER pressure correction technique described by Patankar [10]. The computational domain is scanned with radial and azimuthal *zebra* sweeps which do not eliminate the (inevitable) recursion, but allow vectorization of all inner loops. The resulting tridiagonal systems of linear equations are solved by the use of the Thomas algorithm. The computer used for the simulations is the Cray-2 at EPFL.

PRIMARY INSTABILITY RESULTS

The inlet velocity field drives a vortical flow with three pairs of vortices in the cross-section of the channel. The presence of *three* vortex pairs is related to the *convective* nature of the instability. A different stationary inlet forcing generates a different downstream flow. The steady results obtained are displayed in terms of streamline velocity isolines in figure 2 and secondary motion in figure 3. The results in figure 2 should be compared to the hot wire measurements in air carried out by Matsson and Alfredsson and displayed in figure 4. In both figures the dimensionless spacing between isolines is 0.1. The agreement is good until $\theta \approx 40°$. Downstream of this measurement station the background noise present in the experiments is responsible for the amplification of a secondary instability mode which manifests itself as large amplitude oscillations of the low speed streaks. From the numerical point of view such an event needs to be somehow triggered (unless one has an "infinite" amount of computer time available to let the round-off error start the instability). We will address this point in the following section.

The numerical results (physically significant until $\theta \approx 40°$) show the progressive deflection of the streamwise velocity in the proximity of the concave wall boundary

---

[2] We must, however, remark that the onset of the secondary instability ought to be a function of the curvature ratio, which is absent in the definition of δ_2 provided above.

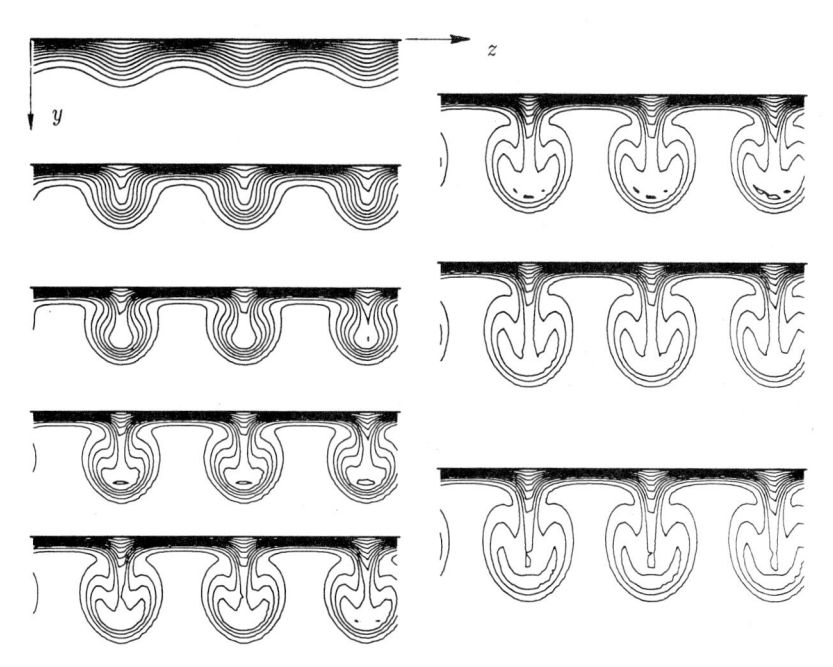

Figure 2: Streamwise velocity isocontours for, from top to bottom and left to right, $\theta = 21.6°, 25.6°, 29.7°, 33.7°, 37.8°, 41.8°, 45.9°$ and $49.9°$.

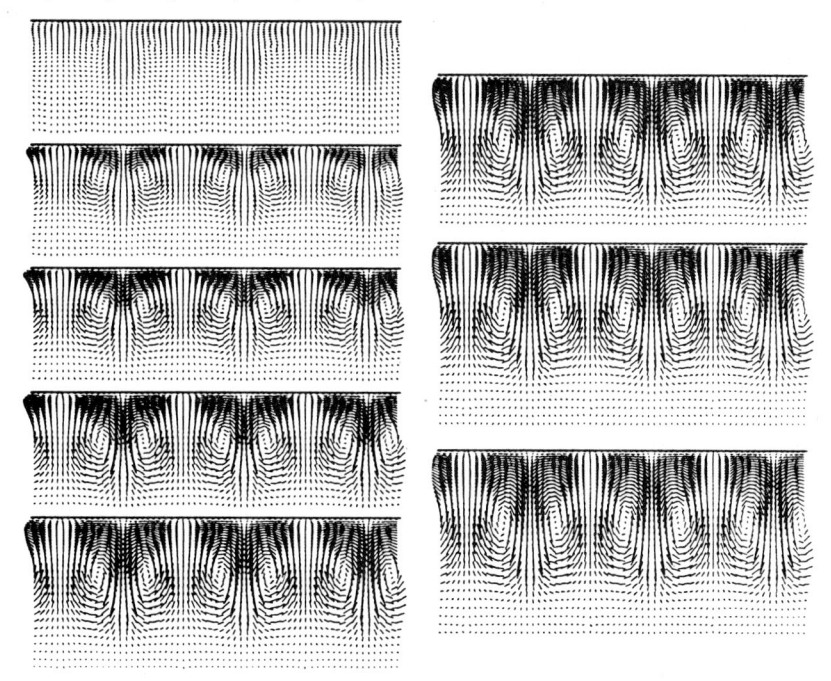

Figure 3: Secondary velocity flow at the same streamwise locations as figure 2.

$\theta = 20°$

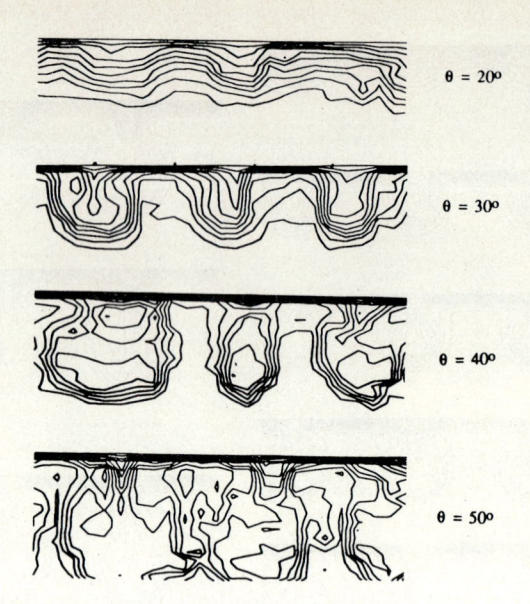

$\theta = 30°$

$\theta = 40°$

$\theta = 50°$

Figure 4: Hot wire measurements of the streamwise velocity.

layer to generate narrow low velocity regions. In correspondence of these low velocity regions the fluid moves away from the wall. These *ejections* become more pronounced at large θ where a mushroom-shaped structure has formed. The vortices have initially a square-shaped form; with increasing streamwise distance the center of the vortices moves away from the wall and the vortices become more rectangular. Spanwise wavenumbers are the same in experiments and simulations because special tripping devices were placed near the inlet section of the experimental apparatus to match the wavelength with the one imposed in the calculations. The spanwise wavelength is also approximately the same to the one occurring naturally in the apparatus of Swearingen and Blackwelder [11]. In the present work, however, the vortices develop much faster that in [11] because of the higher curvature ratio. This points out the limitation of Blasius-related quantities (and in particulat the Görtler number based on the Blasius momentum thickness) as similarity parameters.

The boundary layer thicknesses are in very good agreement between experiment and simulation (cf. figures 2 and 4). In figure 5 we have plotted displacement and momentum thicknesses in the low and high speed regions, and, as a reference, the Blasius flow solutions. These plots have the same trends as those available in the literature (see, e.g., [1, 11]). At $x = 3.3$ ($\theta = 34.4°$) the displacement thickness is maximum in the low speed region. It is here that Swearingen and Blackwelder [11] notice the beginning of the sinuous motion of the low speed streaks, signalling the beginning of the secondary instability. From figure 2 one can deduce that strongly inflectional streamwise velocity profiles form, both in the normal and in the spanwise directions. It is argued that a Rayleigh-type of instability is responsible for the breakdown of the longitudinal vortices.

SECONDARY INSTABILITY RESULTS AND CONCLUSIONS

Experiments with air carried out at IMHEF show that the frequency of the secondary instability is approximately 27 $[Hz]$ at $Re = 20000$. To trigger the secondary instability numerically we introduced at the inlet of the channel the equivalent of a "vibrating ribbon". This consists in adding, at $r = 5.486$ and for $0 \leq z \leq z_{max}$, a small time dependent perturbation of the form

$$\sigma sin(2\pi t/T).$$

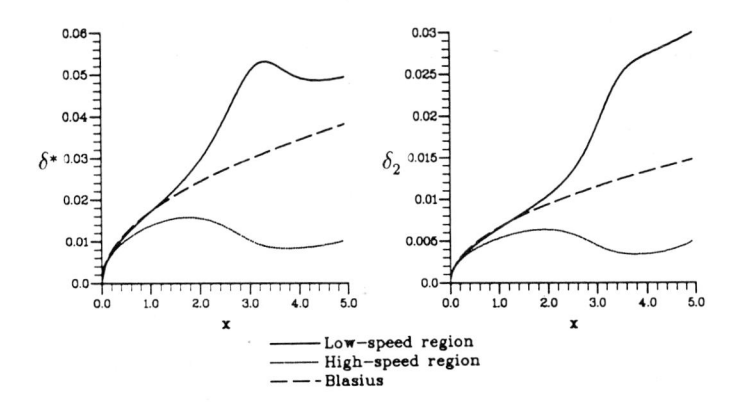

Figure 5: Displacement thickness and momentum thickness as function of $x = r_{out}\theta$.

to all three components of the inlet velocity. We chose $\sigma = 10^{-4}$ and $T = 0.32$. Obviously such a periodic forcing excites a mode of perturbation which depends on σ and T, and such a mode is not necessarily the most unstable one. One can also argue that the same is true for the experiments, where random perturbations on the flow at the entrance of the channel might amplify different types of waves. In particular, one should notice that in the experiments of Swearingen and Blackwelder [11] sinuous and varicose waves could coexist. While we should not exclude that the choice of sinuous or varicose mode could be caused by factors such as the spanwise wavenumber of the vortex pair, it should be remembered that the latter is selected by inlet perturbations (because of the *convective* nature of the instability).

Calculations show that the spanwise component of the velocity, w, is the first one to experience very small ("infinitesimal") oscillations in time; they start in the upwash region of each vortex pair and amplify downstream. Hence, the stem of the "mushroom", where significant spanwise inflectional velocity profiles of the basic flow occur, start oscillating in z. For fixed y and z, and at different streamwise stations x, the oscillations are out of phase, indicating that a finite wavespeed c_r ($c_r \neq 0$) propagates the inlet perturbation. We can expand in a triple Fourier series the small perturbation to the basic (nearly parallel) Görtler flow computed in the previous section as:

$$\phi(x,y,z,t) = Real\{\hat{\phi}(y,z)\,exp[i\alpha x + i\beta z - i\alpha ct]\},$$

where $\hat{\phi} = (\hat{u}, \hat{v}, \hat{w}, \hat{p})$ are complex eigenfunctions and $\phi = (u', v', w', p')$ represents the infinitesimal perturbation of the initial linear stage of the secondary instability; α and β are real streamwise and spanwise wavenumbers (for oblique waves $\alpha, \beta \neq 0$) and $c = c_r + ic_i$ is a wavespeed. The wave grows in time like $exp(\alpha c_i t)$, $\alpha c_i > 0$. At the beginning of the time oscillations the dimensionless period is approximately equal to 1.15^3, which indicates that $\alpha c_r \approx 2\pi/1.15 = 5.464$. Also, through the phase lag in the time oscillations at different streamwise stations we can estimate $c_r = \Delta x/\Delta t \approx 0.897$. The wavenumber in x is therefore $\alpha \approx 6.091$ which means that at the onset of the secondary instability a long wave is established with streamwise wavelength $\lambda = 2\pi/\alpha \approx 0.969$. Notice that this value of λ has been obtained just by inspection of the time signals. Isoline plots of u, v, and w in spanwise-streamwise planes fail to reveal the presence of a

---

[3]For flow of air this corresponds to a frequency of $1.54[Hz]$.

wave, signalling that $\hat{\phi} \to 0$. As the wave amplifies in x and t new modes are excited and may overrule the first unstable mode which was characterized by a very small growth rate αc_i. As a consequence we notice that the frequency of the oscillations increases. Ultimately, non-linearities start playing a significant role once the amplification of the wave is no more "infinitesimal". We remark, however, based on past experience, that linear theory should be able to provide accurate answers until quite large values of ϕ. It is found that oscillations in time of significant amplitude are characterized by a sinuous motion of the low speed streaks. Eventually, tiny secondary vortices develop in an inner layer near the concave wall and interact with the primary Görtler vortices. A detailed description of the mechanisms at play is reserved for a future communication.

Acknowledgements

I would like to thank O.J.E. Matsson and P.H. Alfredsson for permission to reproduce their experimental results and for several helpful discussions.

References

[1] W. Liu and J. Domaradzki. Direct numerical simulation of transition to turbulence in Görtler flow. *AIAA Paper*, 90-0114, 1990.

[2] D.S. Park. *The primary and secondary instabilities of Görtler flow*. PhD thesis, University of Southern California, 1990.

[3] A. Bottaro. Spatially developing flow in curved duct. In C. Taylor, J.H. Chin, and G.M. Homsy, editors, *Numerical Methods in Laminar and Turbulent Flow*, volume VII, pages 696–706. Pineridge Press, 1991.

[4] P. Hall. The linear development of Görtler vortices in growing boundary layers. *J. Fluid Mech.*, 130:41–58, 1983.

[5] P. Huerre and P.A. Monkewitz. Local and global instabilities in spatially developing flows. *Annual Review of Fluid Mechanics*, 22:473–537, 1990.

[6] A. Bottaro. Note on open boundary conditions for elliptic flows. *Num. Heat Transfer B*, 18:243–256, 1990.

[7] L.L. Pauley, P. Moin, and W.C. Reynolds. The structure of two-dimensional separation. *J. Fluid Mech.*, 220:397–412, 1990.

[8] R. Y. Myose and R. F. Blackwelder. Controlling the spacing of streamwise vortices on concave walls. *AIAA Journal*, 29:1901–1905, 1991.

[9] A.S. Sabry and J.T.C. Liu. Longitudinal vorticity elements in boundary layers: nonlinear development from initial Görtler vortices as a prototype problem. *J. Fluid Mech.*, 231:615–664, 1991.

[10] S.V. Patankar. *Numerical Heat Transfer and Fluid Flow*. McGraw-Hill, New York, 1980.

[11] J.D. Swearingen and R.F. Blackwelder. The growth and breakdown of streamwise vortices in the presence of a wall. *J. Fluid Mech.*, 182:255–290, 1987.

NEW FULLY COUPLED SOLUTIONS OF THE NAVIER-STOKES EQUATIONS

G.B. Deng, M. Ferry, J. Piquet, P. Queutey & M. Visonneau
CFD Group, LHN-URA1217 CNRS, ENSM
1, Rue de la Noe, 44072 Nantes Cedex, France

SUMMARY

Three fully-coupled methods for the solution of incompressible Navier-Stokes equations are investigated. They share in common a fully implicit time discretization of momentum equations, the standard linearization of convective terms, a cell-centered collocated grid approach and a block-nanodiagonal structure of the matrix of nodal unknowns. The Methods differ by the interpolation for the flux reconstruction problem, by the basis iterative method for the fully coupled system and by the acceleration means that control the global efficiency of the procedures, the performances of which are discussed.

1. INTRODUCTION

The computation of turbulent incompressible viscous flows past threedimensional geometries like a shiplike hull, a fuselage or an afterbody is nowadays possible using the Reynolds-averaged Navier-Stokes equations (RANSE). However, the main limitations of present calculations lie primarily in the lack of grid resolution of results, and, subsidiary, in difficulties associated with turbulence modelling. While the latter problem is not discussed here, the former is often connected to the lack of robustness of used solvers in that the convergence slows down on fine grids, forbidding a grid independent rate of convergence.

The difficulty of obtaining a solution of incompressible flow equations results from the lack of a pressure time-derivative term in the continuity equation. Several methods have been suggested to overcome this problem and they can be distinguished by the way the incompressibility constraint is enforced. Apart from methods, such as the pseudo compressibility method, in which the pressure-velocity coupling is simulated with a suitable modification of the continuity equation, the most commonly used methods follow the so-called seggregated or *Poisson pressure equation approach* where the pressure velocity coupling is solved iteratively, updating successively the velocity variables in the momentum equations and the pressure in a pressure equation, in such a way that solenoidality is satisfied at convergence [1][2]. Efforts for improving the robustness of the pressure solver, which is the key part of the method in that it partly controls its convergence rate, bring significant benefits for the overall procedure [3] because the increased stiffness of the threedimensional pressure matrix is the most immediate consequence of an increased grid clustering. However, the main drawback of all these methods where the pressure velocity coupling is not enforced at each iteration lies in the slowing-down of convergence when the number of grid points or when the clustering ratios over curvilinear grids increase.

In the following, we only consider the so-called *fully-coupled methods* where the continuity equation is satisfied identically at each iteration. The first attempt to solve (simultaneously) the momentum equations and the continuity equation in a coupled way has been the so-called SIVA algorithm [4]. More recent methods follow the boundary layer practice in which the flow domain is swept from upstream to downstream, implicit differencing of momentum being used for marching stability. Upstream influence through the pressure field has to be accounted for, and this is done introducing some form of forward differencing for the streamwise pressure gradient which allows departure-free behavior. Such methods have been mainly developped in the framework of the partially parabolic approximation, for instance in [5].

Unfortunately, in [4][5], the coupling between dependent variables is performed only in small subdomains (a cell volume or cells with the same given longitudinal station and girth). In such cases, the resulting matrices are easy to handle but poor convergence rates are obtained, especially on fine grids, because of the weak coupling between subdomains. The situation can be in some respects improved with multigrid methods [6][7], although the fully coupled approach [6] does not appear to bring significant improvements with respect to standard Poisson-based methods [8][9].

It appears therefore that the coupling between the *solenoidal* velocity field and the pressure has to be performed over the whole domain, in spite of the increased complexity of the algebraic system. Velocity and pressure fields are then simultaneously updated in a linear sense, iterations being performed only to solve for the non linearity. In contrast to [6][7] where the continuity equation is retained in its primitive form, the present work follows [10] in the use of an approximate Poisson equation but it departs from it in the choice of coupled dependent variables as well as in the selected iterative method.

In the following, significant convergence rates for fully coupled methods are presented on several two or three dimensional, steady or unsteady problems, including geometries described by a curvilinear body-fitted coordinate system. Other significant aspects of the proposed methods include the use of accelerating techniques such as multigrid or preconditioned conjugate gradient techniques for the solution of the fully coupled system, as a means to overcome the poor convergence rates resulting from too local a coupling.

2. FORMULATION

The full RANSE (2.1) for the mean velocity field U and for the mean pressure deviation P

$$\nabla . U = 0, \quad \frac{\partial U}{\partial t} + (U.\nabla)U + \nabla P = Re^{-1}\nabla^2 U - \nabla . \overline{uu} \tag{1}$$

need a closure assumption for the Reynolds stresses $-\overline{uu}$ which is taken under the isotropic form (2.2)

$$\overline{uu} = \frac{2}{3}k\mathbf{1} - \nu_T [\nabla U + \nabla^T U] \tag{2}$$

where the turbulent viscosity ν_T is specified by the Baldwin & Lomax model.

Equations (1, 2) which classically involve the cartesian velocity components U,V,W are partially transformed from the cartesian rectilinear coordinates x, y, z in the physical space, to the curvilinear coordinates λ,η,ζ in the so-called computational space. The transport equations for the mean momentum can be written under the following convective form (3) of the equation for the generic variable Φ [3]:

$$g^{11}\Phi_{\lambda\lambda} + g^{22}\Phi_{\eta\eta} + g^{33}\Phi_{\zeta\zeta} = 2A_\phi\Phi_\zeta + 2B_\phi\Phi_\eta + 2C_\phi\Phi_\lambda + R_\phi\Phi_t + S_\phi . \tag{3}$$

The index ϕ refers to any of the advected quantities, namely U, V, W; the indices λ, η, ζ refer to the partial derivatives. For the sake of brevity, the values of convective coefficients A_ϕ, B_ϕ, C_ϕ, R_ϕ and source terms S_ϕ - which include pressure gradients and second order cross derivatives of Φ - are omitted here (see [3] for their detailed values).

3. THE NUMERICS

3.1. *The discretization of the momentum equations and the coupled systems*

All the methods to be discussed in the following share in common the use of a cell-centered collocated layout of unknowns, a fully implicit backward Euler time discretization, and a convective formulation of momentum equations with the standard linearization technique $\{[U.\nabla] U\}^{(n+1)} \rightarrow [U^{(n)}.\nabla] U^{(n+1)}$, where t^{n+1} is the unknown fictitious time step. Also, they lead in both cases to a block-nanodiagonal system of unknowns (due to their lexicographic ordering) where the size of the blocks (3*3 or 5*5) depends on the choice of the coupled primitive variables. Methods to be discussed are classified according to : (i) the numerical scheme for advection equations, (ii) the flux reconstruction method, ie the closure which provides the interpolated values needed to fulfill the continuity equation, (iii) the choice of the coupled primitive unknowns. The corresponding information relative to the three tested methods is gathered in Table 1

Table 1

Method	Advection equations	Reconstruction	Coupled primitive unknowns
A [11]	linear+exponential shape function for U, V on the eight triangular parts of the quadrilateral control volume	resulting from the closure (see [12])	U, V, P block 3*3
B [13]	centered scheme	seven point stencil "CPI" approach	U, V, P block 3*3
C [14]	9 point uniexponential scheme [15] in 2D, 7 point multiexponential [15] scheme in 3D	Rhie & Chow (see [16])	U, V, $\hat{U}$, $\hat{V}$, P block 5*5

Among the three mentionned classification criteria, the central issue is the method used for the flux reconstruction. It is possible to discretize the momentum equation written on a closure volume different from the control volume. In method A, the closure volumes are four quadrilateral subcells of the control volume, each of them defining one half part of two fluxes [11]. In method B, the closure volume surrounds the control volume interface centers following the practice [12]. In both cases the stencil for nodal unknowns involves the eight neighbours of point C, the center of the control volume. The other alternative, used in method C, is to approximate such an interpolation formula by weighting the terms which do not involve the pressure gradient (interpolation [16] in the computational plane), ie the coefficients of pressure gradients and the so-called pseudovelocity components, while discretizing the pressure gradient. In this case, the interpolation [16] connects each flux to twelve velocity points so that the continuity equation involves twenty neighbours of point C. In order to avoid such a large stencil, the pseudo velocity components $\hat{U}, \hat{V}$ are retained as primitive unknowns. They gather all contributions of the momentum equations but pressure gradients, as indicated by the following projection form written at the center of the control volume :

$$U_C = \hat{U}_C - C_{nn}^U R_{eff} [P_x]_C \; ; \; V_C = \hat{V}_C - C_{nn}^U R_{eff} [P_y]_C . \tag{4}$$

Once the pressure gradients are discretized, a suitable normalisation of influence coefficients leads to

$$\hat{U}_C - \sum C_{nb}^U U_{nb} = C_{nn}^U [eU_C^{n-1} - S_U] \; ; \; \hat{V}_C - \sum C_{nb}^U V_{nb} = C_{nn}^U [eV_C^{n-1} - S_V] , \tag{5}$$

S_U and S_V are the source terms S_ϕ where pressure gradients P_x, P_y have been omitted ; e is the inverse of the time step τ times R_{eff}. The 5*5 block nano-diagonal system is defined by the discretized form of 4, by (5) and by the continuity constraint which leads to a Poisson-like equation for pressure (6) once the flux reconstruction problem is solved by means of the interpolation [16].

3.2. *The linear fully-coupled systems (formulation and preconditionning)*

The algebraic fully-coupled system can be organized in different ways depending on the ordering of unknowns. In both cases, the system can be written as

$$A_C X_C + \sum_{nb} A_{nb} X_{nb} = S_C . \tag{6}$$

For methods A and B the vector unknown is $X \equiv \| U, V, P \|^T$, while, for the method C, $X \equiv \| \hat{U}, \hat{V}, U, V, P \|^T$. With a lexicographic ordering of unknowns, the linearized twodimensional equations give rise to a block Nb*Nb (Nb = 3 or 5) nanodiagonal matrix :
$A = \| ...0...A_{SW} \; A_S \; A_{SE}...0...A_W \; A_C \; A_E...0...A_{NW} \; A_N \; A_{NE}...0... \|$ such that $AX=b$ (7). With a preconditioning matrix M and the residual matrix R ($A = M - R$), the iterative method can be written $X^{(k+1)} = M^{-1} [S + R \; X^{(k)}]$ and its efficiency as a means to solve (7) depends on the choice of M. A Block-point Jacobi method $M = \text{diag}(A) = \| ..0..A_C..0.. \|$ is a vectorizable preconditionner which needs only the inversion of the Nb*Nb submatrix coefficients at point C. Other block-point iterative methods (SOR, Zebra) can be as easily applied. Block line iterative methods retain the Nb*Nb-block-tridiagonal matrices :
$\| ...0...A_W \; A_C \; A_E...0... \|$ or $\| ...0...A_S, \; A_C, \; A_N...0... \|$ of discrete points along a line λ (W, C, E) or η (S, C, N). M is then efficiently solved using a block version of the Thomas algorithm. However, such methods require $3*Nb^2$ working matrices in order to store the factorisations and they have been found less efficient than ILU decomposition methods, especially when the grid is stretched in more than one direction.

The Block-ILU decomposition is such that $M = LD^{-1}U$ where D is a Nb*Nb-block diagonal matrix which is determined according to $\text{diag}A = \text{diag}LD^{-1}U$. The simplest way to construct L and U is to take :
$L = \| ...0...A_{SW} \; A_S \; 0...0...A_W \; A_C \; 0...0... \|$;
$U = \| ...0...0 \; A_C \; A_E...0...0 \; A_N \; A_{NE}...0... \|$ where A_{SE} and A_{NW} have been put to zero in L and U, respectively, in order to allow the vectorization to be performed along the diagonal line $\xi + \eta = $ const.

To fully specify the methods A, B and C, it remains to indicate (Table 2) the basis iterative method used for the solution of the fully coupled system as well as the acceleration techniques that allow an efficient calculation procedure.

Table 2

Method	Basis Iterative Method	Acceleration Techniques
A	Alternate line underrelaxed (.8) Jacobi method	Linear Black-box Multigrid method
B	Jacobi Method	Preconditionned Conjugate gradients CGS [17] or CGStab [18]
C	Block - ILU decomposition [14]	Preconditionned Conjugate Gradients CGS or CGStab

3.3. Acceleration Techniques.

For elliptic problems, the convergence rate of classical basis iterative methods, like Jacobi or Gauss-Seidel methods, is approximately proportional to N^{-2}, where N is the number of unknowns per direction. When applied to the solution of the linearized algebraic system resulting from the discretization of the unsteady or steady incompressible Navier-Stokes equations, the behaviour of a block-iterative method is about the same. As a result, the number of iterations increases dramatically with the number of grid points. For this reason, convergence acceleration, like multigrid and conjugate gradient, is needed, both for steady and for unsteady flow calculations, especially in the three-dimensional case. When a basis iterative method is used as a smoother or a preconditionner, the convergent rate of a multigrid method (independent of grid size) is higher than that of a conjugate gradient method (approximately proportional to N^{-1} according to our experience).

This has been confirmed by results obtained with method A where the convergence is accelerated by a black box linear multigrid method. Let the linear system on the fine grid be $L_f U = S_f$, and define the residual over the fine grid by $r_f = S_f - L_f U_f = L_f (U-U_f)$ if U is the converged solution of the linear system. We introduce the restriction operator R which transfers residuals from the fine grid to the coarse grid and the prolongation (coarse to fine) operator P. The multigrid procedure generates a coarse grid operator L_c such that $L_c U = S_c = R \, r_f$ with $L_c = R \, L_f \, P$. Once the relaxation operator for the approximate inversion of L_f is selected, the multigrid method is fixed by the choice of the operators P and R and by the type of cycles used. A linear interpolation is taken for the prolongation operator and a full weighting operator for the restriction. The alternate line Jacobi method, used as a smoother, is underrelaxed (about 0.8, the underrelaxation factor needs to be decreased when the Reynolds number increases). The performances of this linear multigrid method will be discussed in § 4.

In spite of its lower convergent rate, the conjugate gradient method has been used for the following reasons: (i) It is independent of grid configuration and easier to implement than MG methods. (ii) the efficiency of a multigrid method depends strongly on the control parameters, on interpolation and restriction operators, on the construction of coarse grid operators, on the choice of the smoother and of its relaxation factor, boundary condition treatment, influence of grid aspect ratio, etc. (iii) The multigrid efficiency may slow down over clustered grids. (iv) Unlike the multigrid method, the conjugate gradient method is nearly independent of eventual control parameters.

For a non-symmetric linear system, the CGS algorithm [17] is found to be very efficient. However, when applied to a non-linear problem with conventional (non-Newton) linearization, the computational effort is prohibitive because the CGS method does not satisfy a minimization property so that a non-monotone decrease of the error is found, even for an unsteady simulation where the time change remains small. A variant of CGS, the so-called "CGStab" algorithm [18] is used since it allows a smoother convergence and a better efficiency for time dependent problems. The efficiency of a CG method depends on the condition number of the matrix problem. Linearized algebraic equations resulting from the discretization of the Navier-Stokes equations are usually ill-conditionned, especially when large grid aspect ratios are used, making the preconditioning necessary. Table 2 indicates the preconditioners defining methods B and C. Resulting tests of both methods are now presented.

4. RESULTS

The classical 2D square driven cavity problem $0 \le x \le 1$; $0 \le y \le 1$ with $U(x,1) = 1$ is first studied. Multigrid convergence curves for the non linear operator at Re = 100 are obtained for a regular 129*129 finest grid (fig.1). It is seen that seven orders of residual reduction are obtained in about 2ms per grid point on a Siemens VP200. The convergence is found "better than N" because of the improved vectorisation properties when a large number of grid points per direction is involved. It may be mentionned here that a further reduction of a factor about two would result from the use of a FAS method instead of the presently used linear multigrid method. The efficiency of the multigrid method decreases when the Reynolds number increases as indicated by fig.2 where the convergence history is shown, together with the centerline velocity profiles $V(x,1/2)$ and $U(1/2,y)$, at Re = 10000. The converged solution to a given level of residual, which looks in good agreement with that of [19], is obtained with a cpu time about five times higher than for Re = 100.

The efficiency of accelerated fully coupled methods can be discussed from fig.3, obtained for the square driven cavity at Re = 400. The evolution with grid refinement of the computational effort σ (cpu per grid point on VP200) is considered. In both cases, a residual reduction of six orders in magnitude is achieved. The increase of σ with the number of points N per direction is quicker than N^2 for a seggregated (PISO-like) method (although the curve is not drawn for high values of N since this level of convergence is no more obtained in a reasonable cpu time on a fine grid). In contrast, the fully coupled method C gives an asymptotic behavior σ = O(N), with CG acceleration. For the sake of comparison, optimal MG efficiency is demonstrated for method A.

In order to qualify the different tested discretizations, Tables 3 and 4 gather significant values of minimal and maximal centerline velocities at Re = 400 and 1000, respectively.

Table 3. Square driven cavity ; Re = 400

Method	Grid	Umin	Vmin	Vmax
Ghia, Ghia & Shin [19]	129*129	-.3273	-.4499	.3020
A	25*25	-.3253	-.4351	.2953
A	49*49	-.3298	-.4485	.3023
A	129*129	-.3290	-.4531	.3036
A	25*25 §	-.3346	-.4491	.3079
A	49*49 §	-.3272	-.4491	.3012
B ("CPI" approach [13])	25*25	-.297	-.416	.272
B ("CPI" approach [13])	49*49	-.320	-.444	.296
Finite analytic [20]	49*49	-.2960	-.4051	.2662
upwind scheme	49*49	-.2240	-.3462	.2101
centered scheme (staggered grid)	49*49	-.310	-.429	.285
Multiexponential scheme [15]	49*49	-.2771	-.3842	.247
Uniexponential scheme [15]	49*49	-.294	-.406	.264

Table 4 : Square driven cavity ; Re = 1000

Method	Grid	Umin	Vmin	Vmax
Ghia, Ghia & Shin [19]	129*129	-.3829	-.5155	.3709
Bruneau et Al [21]	256*256	-.3764	-.5208	.3665
A	257*257	-.3870	-.5251	.3754
A	129*129	-.3830	-.5201	.3716
A	65*65	-.3721	-.5078	.3609
B ("CPI" approach [13])	66*66	-.3750	-.507	.364
centered scheme (staggered grid)	65*65	-.357	-.485	.344
Finite Analytic [20]	129*129	-.3689	-.5037	.3553
Multiexponential scheme [15]	129*129	-.3460	-.4858	.3330

It is apparent, from the case Re = 400, that both methods A and B perform correctly with respect to the target values [19]. In particular, the sensitivity of method A with respect to the grid size is very low ; also, results § are obtained with a randomized grid such that x(i,j) = (i-1)/(i$_{max}$-1) + .3*A(I,J) ; y(i,j) = (i-1)/(j$_{max}$-1) + .3*B(I,J) where A and B are matrices of random numbers between -1 and 1. The accuracy of corresponding results do not appear to be seriously affected by the lack of regularity of the grid. Method B is also found to give results very similar to those of method A. However, the closure scheme of the former method is far simpler than that of the latter in the three-dimensional case ; for this reason, only method B has been further developed. The improvement provided by methods A and B with respect to the upwind, finite analytic or even multiexponential or centered schemes demonstrates, if necessary, the controlling influence of the flux closure on the global accuracy of the scheme. Table 4 confirms these trends for Re = 1000. In particular, the method B appears to give improved results with respect to more standard schemes, even with a grid 66*66 and it still performs almost as well as method A.

The cubic lid-driven cavity flow is considered now because it provides an interesting benchmark test case for three-dimensional methods. The physics of this simple geometry is complex since the flow undergoes significant transverse motions, like Taylor-Görtler-like (TGL) vortices, endwall vortices and, for a given high enough Reynolds number, stronger unsteady effects than in the 2D case. Also a series of interesting experiments have been conducted on this type of geometry (e.g; [22]). To make evident the global differences with the two-dimensional case, the velocity profiles in the symmetry plane z = 1/2 are compared with their two-dimensional counterparts in a steady case, for Re = 400 (figs.4). The "CPI" approach (method B) has been used on two grids, with 16 (fig.4a) and 25 points (fig.4b) in the z direction (the driving face is y = 1 where the flow is along the x-axis). Obtained velocity profiles appear less steep in the 2D case than in the 3D case because of the global divergence of the flow away from the symmetry plane. Symbols "o" indicate the two-dimensional results of Ghia et Al [19] while 2D and 3D pseudo spectral results [23] are indicated by symbols "+" and "Δ", respectively. The significantly different V velocity profiles are believed to be due to the use of a regularized velocity profile U along the edges of plane y = 1, in order to avoid discontinuous boundary conditions which would destabilize spectral-type calculations. Method B has been also successfully used to compute the 3:1:1 parallelopipedic cavity at a Reynolds number of 3200. For this case, the highly unsteady flow behavior has been studied up to t = 100 on a 64^3 grid for the half cavity space [13], and transverse motions with three or four pairs of TGL vortices have been found.

The fully-coupled method C, which has been seen to use the classical flux closure [16], is specific in the choice of the dependent variables which lead to 5*5 blocks. The convergence behavior of the method has been tested on several benchmark cases among which the afterbody 3 [24] has been selected. This axisymmetric case is interesting since it involves a moderately complex geometry with a body-fitted coordinate system and a turbulent flow (Re = 5.9 10^6) with a small separation bubble, slightly upstream of the aft point. Moreover, some pressure and velocity experimental data [24] are available for comparison.

The calculation uses a grid consisting in 90 axial stations, the first of which is at midbody x/L = .5 and 50 radial points which are clustered near the wall, far field conditions being imposed one body length away from the symmetry axis. A standard Baldwin Lomax turbulence model is used. The fully-coupled procedure using the Block-ILU preconditionned CGStab method described in §3. The pressure coefficient on the body and along its wake is presented in fig.5. The good agreement with experimental data, which can be noticed, is confirmed by comparisons over the longitudinal velocity profile at a station located in the small separation region close to the aft.

Results are obtained in convergence conditions which are compared in fig.6 with similar results using a PISO-like seggregated method with the same grid, inlet conditions and turbulence model. Because of the occurrence of turbulent separation, a PISO-like method leads to a saturation of (non linear) residuals when their magnitude has been reduced by about 3.5 orders. In contrast, the fully coupled method can be driven to more than seven orders of residual reduction. It is seen that the convergence behavior produced by Cgstab is more regular than that produced by CGS. Taken as a whole, such results imply that for a residual reduction of three orders in magnitude for which both methods still converge, the fully coupled method is about eight times quicker than the seggregated method. Such results establish the robustness of the fully coupled algorithm.

5.CONCLUSION

The superiority of the fully-coupled method for the iterative calculation of steady flow problems has been demonstrated on several characteristic problems. This superiority comes from the following aspects. (i) With a fully-coupled method, the iterative procedure accounting for the non linearity becomes distinct from the problems connected to the resolution of the linear system ; this makes easier the understanding of difficulties specific of the linearization. (ii) Because the correction step of seggregated methods like "PISO" is suppressed, the sensitivity of such methods to initial conditions is strongly reduced. Underrelaxation factors become closer from one since their value is determined by constraints provided by the linearization rather than by the correction step of seggregated methods. (iii) For steady flow problems, the strength of CGS methods provides high rates of convergence and allows quasi-infinite time steps as well. (iv) For unsteady flow problems, the limitations over the size of the time step arise exclusively from the physics of the problem (presence of characteristic periods of unsteadiness). (v) The evolution of the computational effort is strongly improved when the number of grid points increases. (vi) There is still a serious potential for improving the global methods through the use of better preconditionners for the coupled system, may be by a combination of MG and CG techniques.

Among the drawbacks of the method, the use of conjugate gradient methods has to be paid by a serious storage penalty ; the required two or threedimensional arrays for the CG method have to be added to the Nb^2 needed two or threedimensional arrays required to store the fully- coupled system. It is however considered that this storage penalty is a very weak drawback with respect to the increased efficiency and robustness of the fully coupled method.

Acknowledgments. Authors gratefully acknowledge partial financial support of DRET through Contract 89-117. Computations have been performed on the Cray2 (CCVR) with cpu provided by the Scientific Committee of CCVR and on the VP200 (CIRCE) with cpu provided by the DS of SPI.

6. REFERENCES

[1] Patankar, S.V. "Numerical Heat Transfer and Fluid Flow", Mc. Graw Hill, New York, 1980.
[2] Issa, R.I. "Solution of the Implicitly Discretized Fluid Flows Equations by Operator-Splitting", J. Comp. Phys. **62**, N^0 1, pp.40-65 (1986).
[3] Piquet, J. & Visonneau, M. "Computation of the Flow past Shiplike Hulls", Comp. & Fluids **19**, 183-215 (1991).
[4] Caretto, L.S., Curr, R.M. & Spalding, D.B. "A Calculation Procedure for Heat, mass and momentum transfer in three-dimensional parabolic flows", Int. J. Heat Mass Transfer **15**, pp.1878-1806 (1972).
[5] Hoekstra, M. "Recent Developments in a Ship Stern Flow Prediction Code", Proc. 5th. Symp. Num. Ship Hydrodynamics, Hiroshima, Mori, K.H. Ed. (1989) pp.87-101.
[6] Vanka, S.P. "Block-implicit multigrid solution of Navier-Stokes Equations in primitive variables", J. Comp. Phys. **65**, pp. 138-158 (1986).
[7] Bruneau, C.H. & Jouron, C. "An Efficient Scheme for Solving Steady Incompressible Navier-Stokes Equations", J. Comp. Phys. **89**, pp. 389-413 (1990).
[8] Arakawa, Ch., Demuren, A.O., Rodi, W. & Schönung, B. "Application of Multigrid Methods forr the coupled and decoupled solution of the Incompressible Navier-Stokes Equations", Proc. 7th. GAMM-Conf. Numerical Methods in Fluid Mechanics, Deville, M. Ed.; Notes Num. Fluid Mechanics **20**, 1-7 (1987) Vieweg Verlag.
[9] Linden, J., Lonsdale, G.L, Steckel, B. & Stüben, K. "Multigrid for the Steady-state Incompressible Navier-Stokes Equations : A survey", GMD Arbeitspapier 322 (1988).
[10] Schneider, G.E. & Zedan, M. "A Coupled Modified Strongly Implicit Procedure for the Numerical Solution of Coupled Continuum Problem", AIAA Paper 84-1743 (1984).
[11] Ferry, M. & Piquet, J. "A New Fully-coupled Method for the Solution of Navier-Stokes Equations", Proc. 3rd. European Multigrid Conference, Bonn (1991) GMD Studien **189**, Trottenberg Ed.
[12] Schneider, G.E. & Raw, M.J. "Control Volume Finite Element Method for Heat Transfer and Fluid Flow using collocated Variables", Num. Heat Transfer **11**, 363-400 (1987).
[13] Deng, G.B., Piquet, J., Queutey, M. & Visonneau, M. "A Fully Implicit and Fully Coupled Approach for the Simulation of Three-Dimensional Unsteady Incompressible Flows", Proc. GAMM Workshop on Num. Sim. 3D Incomp. Unst. Visc. Lam. Int/Ext. Flows. Deville, M., Lé, T.H. & Morchoisne, Y. Eds., Num. Notes in Fluid Mech.*To appear* (1991).
[14] Deng, G.B., Piquet, J. & Visonneau, M. "Viscous Flow Computations using a Fully-coupled Technique", Proc. 2nd Int. Coll. Viscous Fluid Dyn. in Ship & Ocean techn., Osaka, (1991).
[15] Deng, G.B., Piquet, J., Queutey, P. & Visonneau, M., "Three-dimensional full Navier-Stokes Solvers for incompressible Flows past arbitrary geometries", Int. Journ. Num. Methods in Eng. **31**, 1427-1451 (1991).
[16] Rhie, C.M. & Chow, W.L. "Numerical Study of the Turbulent Flow past an Airfoil with Trailing Edge Separation", AIAA Journ. **21**, 1525-1532 (1983).
[17] Sonneveld, P. "CGS : a fast Lanczos-type solver for nonsymmetric linear systems, SIAM J. Sci. Statist. Comput. **10**, pp.36-52 (1989).
[18] Van der Vorst, H.A. & Sonneveld, P. "CGSTAB : A more smoothly converging variant of CG-S", preprint (1990).

[19] Ghia, U., Ghia, K.N. & Shin, C.T. "High-Re Solution for Incompressible Flow using the Navier-Stokes Equations and multigrid method", J. Comp. Phys. **48**, 387-411 (1982).
[20] Chen, C.J. & Chen, H.J. "Finite Analytic Numerical Method for Unsteady two-dimensional Navier-Stokes Equations", J. Comp. Phys. **53**, 209-226 (1984).
[21] Bruneau, C.H. & Jouron, C. "An Efficient Scheme for Solving Steady Incompressible Navier-Stokes Equations and Multigrid Method", J. Comp. Phys. **48**, 389-413 (1990).
[22] Prasad, A.K. & Koseff, J.R. "Reynolds Number and End-wall Effects on a Lid-driven cavity flow", Phys. Fluids A **1(2)**, 208-218 (1989).
[23] Ku, H.C., Hirsh, R.S. & Taylor, T.D. "A Pseudospectral Method for the Solution of the Three Dimensional Incompressible Navier-Stokes Equations", J. Comp. Phys. **70**, 439-462 (1987).
[24] Huang, T.T., Wang, H.T., Santelli, N. & Groves, N.C. "Propeller/Stern/ Boundary Layer Interaction on Axisymmetric Bodies : Theory and Experiment", DTNSRDC Report 76-0113 (1976).

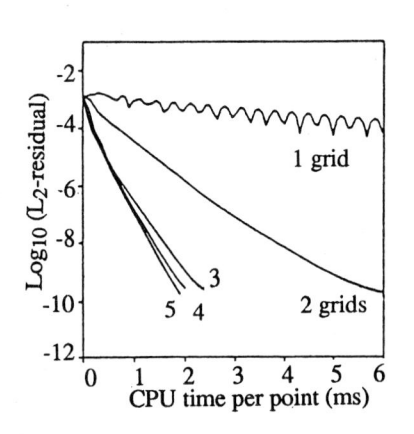

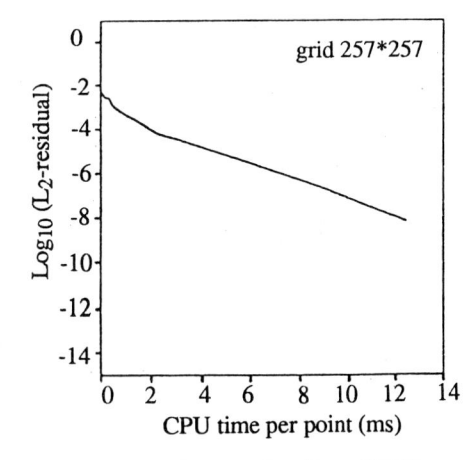

Fig.1. Multigrid convergence for Method A (1, 2, 3, 4, 5 grids). Driven cavity, Re=100. Finest grid : 129*129.

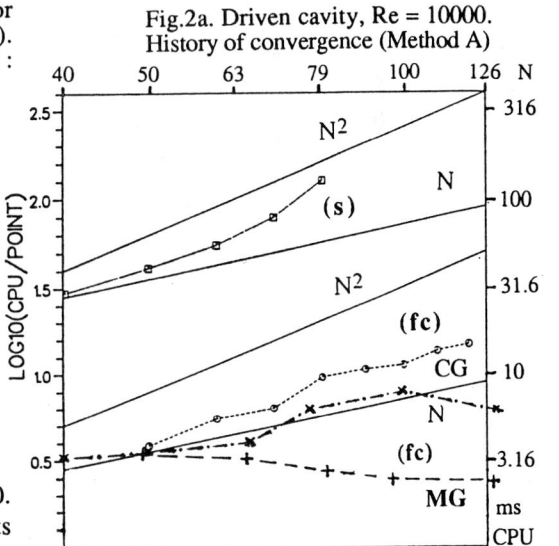

Fig.2a. Driven cavity, Re = 10000. History of convergence (Method A)

Fig.2b. Driven cavity, Re = 10000. —, results of Method A. ◻, Results [19].

Fig.3. Driven Cavity. Re = 400. Evolution of the Computational effort with the number N of points per direction. ◻, seggregated (Piso-like) method. — + — , fully coupled MG method A ; - ·x - ·, fully coupled CPI method B ; ···◌···, fully coupled BILU-PCG method C. ——, evolution of the cpu per point as N or N^2.

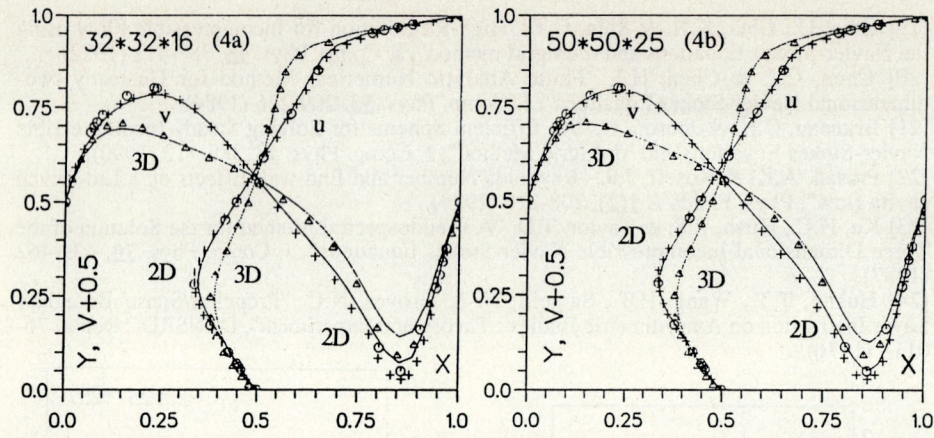

Fig.4. Comparison of the centerline velocity profiles u(0.5, y) and v(x, 0.5) for the square driven cavity and the cubic cavity in the plane z = 1/2, at Re = 400. O, 2D results [19]. +, 2D results [23]. Δ, 3D results [23]. ——, present calculations with method B. Half domain calculation

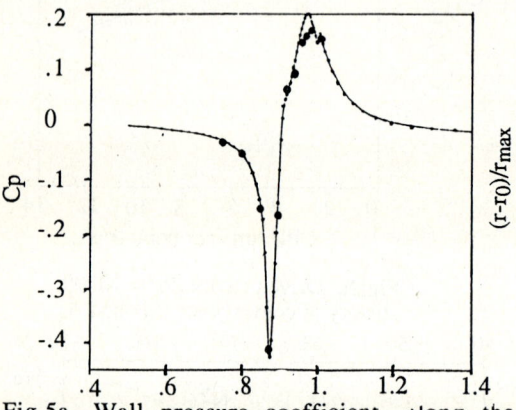

Fig.5a. Wall pressure coefficient along the Afterbody 3 as a function of x/L. ●, Experimental data [24]. ——, present calculations with method C.

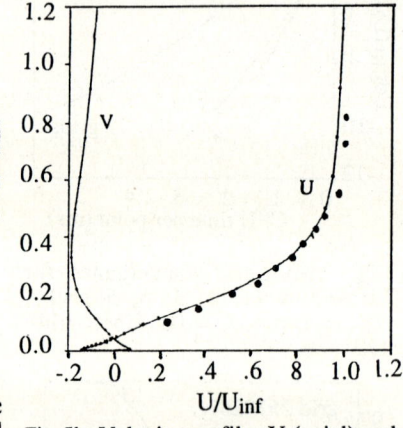

Fig.5b. Velocity profiles U (axial) and V (radial) for x/L = .934 on the afterbody 3. ●, Experimental data [24]. ——, present calculations with method C. •, grid points.

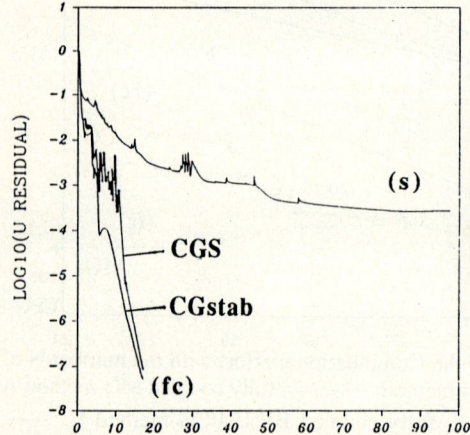

Fig.6. Convergence history for the non linear residuals. Comparison between the seggregated piso (s) and the Block-ILU-PCG fully coupled method (fc). The computational effort is in mseconds per point on VP200. Initial residuals are normalised to 1.

Evaluation of the Artificial Compressibility Method for the Solution of the Incompressible Navier-Stokes Equations

Yves P. Marx
Institut de Machines Hydrauliques et de Mécanique des Fluides
Ecole Polytechnique Fédérale de Lausanne
CH-1015 Lausanne, Switzerland.

Introduction

The numerical solution of the Navier-Stokes equations for an incompressible fluid still represents a computational challenge. The main reason is due to the lack of an equation for the pressure which makes the coupling between the velocity field and the pressure field a bit mysterious (see the abstract of Gresho and Sani [1] for a description of the mystery). In the late sixties and the early seventies several pressure-velocity coupling procedures were proposed [2, 3, 4]. Surprisingly, the underlying philosophy of each procedure remained unchanged since their introduction, and almost all the methods employed today for solving the incompressible Navier-Stokes equations used one of these coupling procedures. Furthermore, the question "which pressure-velocity coupling procedure is the most suited for the numerical simulation of incompressible flows" is still debated between the antagonist churches. The purpose of the present work is to contribute to the debate by a *quantitative* evaluation of the artificial compressibility method. Before discussing in details the investigations performed, the numerical method will be described briefly.

Numerical Method

The artificial compressibility method belongs to the family of preconditioning techniques that modify the original system to remove its stiffness. In the artificial compressibility method, this is achieved by forcing the pressure waves to propagate at a finite speed. In the present study, the simplest form, where only the continuity equation is altered, has been employed. The transformed continuity equation used for the resolution is

$$\frac{\partial P}{\partial t} + \text{div}(c^2\,\vec{u}) = 0,$$

where c^2 represents the artificial compressibility constant. The numerical method employed to solve the modified flow equations is then simply an extension to the incompressible case of standard compressible techniques. Symbolically, the pseudo-unsteady equations are solved with an implicit upwind method of the form

$$(I + \Delta t L)\delta U = -\Delta t R^n, \quad U^{n+1} = U^n + \delta U,$$

where L is a spatial operator and R is the residual of the steady equations.

The residuals R are evaluated following the MUSCL approach [5]; the reconstruction is performed with the κ scheme and Roe's Riemann solver is employed to calculate the fluxes at the cell faces. Two implicit operators L were tested to accelerate the convergence to the steady state :
  • a non-factored first-order upwind approximation of the Jacobian of the residual R,
  • a diagonal (in the sense of Chaussee and Pulliam [6]) ADI approximation of the previous operator.
The non-factorized operator was designed to be diagonally dominant and a multigrid scheme that combines zebra-relaxation smoothing with semi-coarsening, was developed to solve efficiently the linear implicit system [7].

Numerical Investigations

Influence of the artificial compressibility parameter

Since its introduction by Chorin [2], the importance of the artificial compressibility parameter c^2 has been greatly elucidated either by numerical experiments or by theoretical analyses [8, 9, 10]. Here, we will show by numerical experiments, that the parameter c^2 has an admissible range which is not too narrow for an implicit scheme. The experiments also suggest that the theoretical optimal value given by Turkel [9] leads effectively to near optimal convergence rates. The classical lid driven cavity problem, for Reynolds number varying from Re= 100 to Re= 10000, will serve as test case to illustrate this feature of the method.

In figures (1–4), the shape of the function, $N = F_{c^2}(\text{CFL})$, where N is the number of iterations needed to reduce the residual by four orders, are represented for the factored implicit scheme. As stated, for the different Reynolds numbers and for the different meshes used, the value recommended by Turkel,

$c^2 = 3\,q_{\max} = 3\,(u^2 + v^2)_{\max}$, gives almost the best convergence rate. The shape of the function F_{c^2} is typical of ADI type scheme, for which an optimal CFL number exists (Fourier analyses shows that the amplification factor presents a minimum for some CFL number). It can be noted that for $0.5 \leq c^2 \leq 2$ (the velocity has been non-dimensionalized by the velocity of the moving wall), the number of iterations N is roughly the same. N increases significantly only for $c^2 \gg 2$ and $c^2 \ll 0.5$. The maximum CFL number indicated in figures (1–4) is not always the maximum value that can be used. Actually, the maximum CFL number increases as a function of c^2, fig.(5). Thus, from robustness considerations, it may be preferable to select a c^2 higher than the optimal value. Numerical experiments have also shown that the use of a variable — in time — c^2, linked to the maximum velocity, can also stabilize an otherwise diverging calculation.

Similar tests were repeated with the non-factored scheme, fig.(6–9). For the non-factored scheme, it has been observed that it is preferable to use a large CFL number (CFL $\geq 10^{+3}$) and to under-relax the increments, $U^{n+1} = U^n + \alpha\,\delta U$. As a consequence, for the non-factored scheme, the number of iterations N needed to reduce the residual by four orders, will be given as a function of the under-relaxation parameter α. Again, the "optimal" value, $c^2 = 3\,q_{\max}$, appears to be a good choice. In contrast to the results obtained with the factored scheme, there is no clear plateau, where the convergence rates are optimals. The best rates are obtained for under-relaxation parameters close to the highest values that give convergence. Whether the non-factored scheme is less sensitive to the artificial compressibility parameter c^2 than the factored scheme is unclear. Theoretically [9, 11] it should be less sensitive, but since the non-factored operator L is not the exact Jacobian of the residual R, the implicit method is not equivalent to a Newton method for large CFL numbers. Therefore, the above cited results are not really applicable, and to explain the dependence of the non-factored scheme to the artificial compressibility parameter c^2, a more careful analyses of the damping properties of the implicit scheme has to be made. In all the calculation performed so far, we have not experienced any problem with $c^2 = 3\,q_{\max}$.

As opposed to classical preconditioning methods, the solution calculated with the artificial compressibility method depends slightly on c^2. This dependence results from the upwinding which introduces an artificial dissipation scaled by the eigenvalues of the Jacobian matrix, i.e. scaled by u, $u \pm \sqrt{u^2 + c^2}$. Therefore, the influence of c^2 on the accuracy must also be studied. Assuming that the solution calculated on a 257×257 grid is the exact solution, the distribution of the error for v across the center of the cavity for several c^2 is given in figures (10–11). It can be noted that in the admissible range, the error differences obtained with different c^2, are of one order less than the absolute error. Furthermore, the accuracy of the method appears to be of second order. To assess more precisely the accuracy of the scheme, the regularized driven cavity problem proposed by Shih et al. [12] — for which the exact solution is known — has been used. For the "optimal" c^2, the order of the method is given in table (1). From Taylor expansions, a change on the c^2 should modify only the level of the error but not the order of the scheme. This has been verified numerically, fig.(12–13). It can be observed that a change on the c^2 has only a minor effect on the accuracy of the results, especially on sufficiently fine grids, fig.(13). The accuracy of the method has also been evaluated on non-uniform meshes. For Shih's test case, it was observed that improved results were obtained with stretched meshes, fig.(14). Reverse observations are found with an uniform mesh that has been distorted randomly. In that case, the order of the scheme reduces to 0.8.

The artificial compressibility parameter appears to have more influence on the convergence rate of the method than on the accuracy of the results. However, it has been shown that a value of the parameter leading to almost optimal convergence rate can be calculated automatically by inspection of the velocity field. In consequence, the necessity to specify an artificial parameter cannot be considered as a drawback of the artificial compressibility method.

Influence of the boundary conditions

The importance of the boundary conditions cannot be over-stressed. Unfortunately, they are often discussed in few lines, although they usually determine the accuracy and the convergence rate of the method. The first boundary condition examined is the solid wall condition.

Mathematically, there is no condition to be imposed on the pressure when the velocities are prescribed. This is also the case when a staggered grid is employed. However, for a co-located finite-volume method, the pressure at the wall is needed to calculate the normal momentum flux. How this pressure should be evaluated is not clear. Several techniques were tested. Near the wall, three different topologies of the grid were considered, fig.(15), and for each topology, different methods for calculating the pressure at the wall were studied. For the topology I, the pressure at the wall was calculated by enforcing either the continuity or the momentum equation on the half mesh adjacent to the wall. Since the κ scheme cannot be used to evaluate the left state at the first cell, two interpolation procedures were tested; a linear interpolation, $q_L = (q_1 + q_2)/2$, or a zero-order extrapolation, $q_L = q_1$. The conditions used with the topology I are thus,

type 1 : Pressure determined from the continuity equation, linear interpolation for the velocity and the pressure.
type 2 : Pressure determined from the continuity equation, linear interpolation for the velocity and zero-order extrapolation for the pressure.
type 3 : Pressure determined from the continuity equation, linear interpolation for the pressure and zero-order extrapolation for the velocity.
type 4 : Pressure determined from the momentum equation, linear interpolation for the velocity and the pressure.

For the topology II, four distinct techniques were also tested,
type 1 : Pressure calculated from a linear extrapolation, $P_w = (3\,P_1 - P_2)/2$.
type 2 : Pressure calculated from a zero-order extrapolation, $P_w = P_1$.
type 3 : Pressure calculated from the solution of the characteristic equation.
type 4 : Pressure calculated from the solution of the normal momentum equation.

The boundary conditions used with the topology III are technically more complicated.
type 1 : The velocity at the phantom cell is calculated in such a way that the κ scheme gives $u_w = 0$. The pressure is linearly extrapolated. The wall values used for the computation of the flux, are interpolated using the κ scheme.
type 2 : The velocity at the phantom cell is calculated by reflection and a zero wall velocity is imposed in the computation of the fluxes. The pressure is treated similarly as in the type 1 method.
type 3 : The velocity is calculated as in the type 1 method. The pressure is calculated by enforcing that the incoming Riemann invariant at the wall, is equal to the outgoing one. This is "equivalent" to enforce a zero wall velocity.
type 4 : A reflection technique is used to calculate the velocity at the phantom cell. The pressure at the wall is calculated such that the Roe scheme gives no mass flux through the wall.

All these techniques were tested on three one-dimensional problems. In all the tests, the computational domain was limited at both ends by solid walls. Source terms were also added to the equations. The source terms were chosen to enforce specific solutions. The accuracies found for each procedure are reported in the tables (2–4). The order indicated in these tables is the *worst* order obtained for the three problems. An inspection of the tables (2–4) shows clearly that the method used to calculate the pressure at the wall has a larger influence on the accuracy of the results for the grid topology I than for the other topologies. All the computations were done using an implicit scheme. But, as with the non-factored scheme, it was found necessary to under-relax the increments. The numerical experiments have shown that this under-relaxation coefficient is more stringent, and also more problem-dependent for the topology I than for the topologies II and III. These experiments indicate unequivocally that the topology I should be rejected in favor of the topologies II or III. The mathematical reason is unclear, especially for the type 1 method. In this case, there are as many discrete equations (continuity and momentum) as unknowns. Furthermore, if a centered scheme is employed, it is easy to verify that the boundary condition obtained for the derived discrete Poisson equation for the pressure, is the normal momentum equation, which is known to be the correct condition [1]. For this topology, one can demonstrate that the enforcement of the continuity equation gives *the* correct pressure condition. Figures (16,17) illustrates this point. The first figure corresponds to a case where the pressure was determined at one wall by enforcing the continuity equation. At the other wall, the pressure was specified. In this case, as shown in fig.(16), at convergence, the discrete continuity equation is automatically satisfied at the other wall. This indicates that the computations should not converge when the momentum equation is used to calculate the pressure at the wall. This has been found effectively, fig.(17). However, the level of the convergence remains below the level of the truncation error and the scheme is still of second order.

The differences between the results obtained on the topologies II and III are more subtle. Actually, it should be possible to employ very similar treatments in both cases. The numerical implementation of the boundary conditions for the topology III are however more cumbersome, and difficulties have been experienced with boundary conditions slightly different from the one described here. Therefore, the best topology appears to be topology II. For this topology, the simplest procedure that gives accurate results is to extrapolate linearly the pressure. This procedure however has the worst convergence rate, fig.(18). The best rate is obtained when the characteristic equation is employed, fig.(19). If the use of the characteristic equation normal to the boundary is easy in the one-dimensional case, its extension to the three-dimensional case is tedious. A good compromise is to extrapolate the outgoing Riemann invariant, $P + (\vec{u}.\vec{n} + a)\vec{u}.\vec{n}$, where $\vec{n}$ is the outward normal and a the pseudo speed of sound. The multidimensional results presented so far were obtained with this procedure. Its effect on the convergence is illustrated in fig.(20). Whereas the zero-order extrapolation on the pressure leads to an accuracy of order 2 for the velocities and of order 1.5 for the pressure, the other treatments lead all to second order accuracy for both velocities and pressure. These results are in good agreement with the numerical studies performed on the one-dimensional inviscid problems. For viscous flows, two boundary conditions were imposed at the wall. The conditions described above are imposed for the inviscid part of the operator, For the viscous part, $\partial u_n / \partial n = 0$ is used. This condition is equivalent to impose the continuity equation at the wall.

The second boundary that has been examined is the open boundary condition. The backward facing step problem was chosen as test case. For laminar flows, a Poiseuille flow sets up behind the back-flow region. Then, if the computational boundary is located sufficiently far from the back-flow region, in accordance with the parabolic nature of the flow, it should be possible to extrapolate linearly all the variables. With this procedure, accurate solutions were found in the back-flow region. However, it was impossible to converge the scheme to machine accuracy. Whereas to extrapolate all the variables agrees well with the parabolic nature of the steady solution, this condition violates the well-posed conditions imposed by the hyperbolic character of the pseudo-unsteady equations, since at the boundary there is one incoming characteristic. The procedure was then modified, and two boundary conditions were imposed. One for the inviscid part of the operator and one for the viscous part. Dirichlet conditions were imposed for the inviscid part. Since an upwind scheme is employed, this is equivalent to impose the incoming Riemann invariant. For the viscous part, the velocities were extrapolated linearly. This corresponds to neglect the stream-wise diffusion, which is correct for Poiseuille flow. With this procedure, accurate results were obtained and the residuals could be converged to machine zero. To neglect the stream-wise diffusion may be questionable in the general situation, the procedure was nevertheless found to give accurate results when the computational boundary crosses the back-flow region. This finding is illustrated in fig.(21). Two computations were performed. The first was made with the boundary located far from the back-flow region. The solution obtained on a section crossing the back-flow was then stored and used as Dirichlet condition for the second calculation. Comparisons between the two solutions show that the computed fields obtained on the short domain were almost identical to the fields computed on the large domain. An analyses explaining why this boundary condition gives accurate solutions, can be found in [13].

The previous computations have shown that the imposition of the incoming Riemann invariants leads to accurate solutions and to good convergence. Unfortunately, the procedure cannot be used for more complex flows, for which the Riemann invariants are rarely known in advance. The boundary conditions that should be imposed will then depend on the available data. The previous investigations indicate however, that the boundary conditions should be derived from hyperbolic considerations.

In the treatment of a wall and of an open boundary, two distincts boundary conditions were imposed for the advection and the diffusion part of the Navier-Stokes operator It is thought that this procedure has some advantages. The procedure achieves the "decoupling" the advection and the diffusion operators (separate approximations appropriate for the inviscid and the viscous terms were already used at an internal point). Thus, when the Reynolds number tends to infinity, the Navier-Stokes solver tends smoothly to an Euler solver. This "decoupling" simplify also the analyses of the boundary conditions since they are derived for pure hyperbolic or pure parabolic initial value problems. It is then easier to derive boundary conditions that bounds the energy growth.

Other investigations

The numerical scheme presented in this paper was also tested against well-established solutions, i.e. driven cavity and backward facing step solutions, for several Reynolds numbers. Excellent agreements were observed. On the backward facing step problem, detailed comparisons were performed with results obtained using a pure centered scheme on a MAC staggered grid. On sufficiently fine grids, identical solutions were computed. However, it was observed, that the present method gives slightly better solutions on coarser meshes. The comparison was repeated on a three-dimensional curved channel with solid walls on top and bottom. Periodic conditions are applied on the span-wise and stream-wise directions. The flow is then driven by an imposed pressure gradient, (for a more complete description of the problem see [14]). The differences between the unstable analytic stream-wise velocity and the computed velocity, for respectively the upwind and the centered method, are given in fig.(22,23). The velocity differences were plotted on two adjacent computational domains. It can be observed that identical shifted pattern are obtained with the two methods. Since the flow is periodic in the span-wise direction, the solutions calculated by the two methods are thus identical.

This final test confirms that accurate solutions can be computed with the artificial compressibility method. The efficiency of the method then depends on the iterative procedure employed; explicit, fully implicit or ADI methods. To assess precisely the efficiency of the iterative procedure, further investigations are needed.

Acknowledgements

The author is grateful to Mr. M. Blomjous for having performed most of the calculations presented in this paper, and to Dr. A. Bottaro for having enabled, by running his MAC code, the detailed comparisons between the upwind and the centered schemes.

References

[1] P.M. Gresho and R.L. Sani. On pressure boundary conditions for the incompressible Navier-Stokes equation. *Int. J. Numer. Methods Fluids*, 7:1111–1145, 1987.

[2] A.J. Chorin. A numerical method for solving incompressible viscous flow problems. *J. of Comput. Physics*, 2:12–26, 1967.

[3] A.J. Chorin. Numerical solution of the Navier-Stokes equations. *Mathematics of Computation*, 22:745–762, 1968.

[4] S.V. Patankar and D.B. Spalding. A calculation procedure for heat, mass and momentum transfer in three dimensional parabolic flows. *Int. J. Heat Mass Transfer*, 15:1787–1806, 1972.

[5] B. Van Leer. Towards the ultimate conservative difference scheme IV. A new approach to numerical convection. *J. of Comput. Physics*, 23:276–298, 1977.

[6] D. S. Chaussee and T. H. Pulliam. A diagonal form of an implicit approximate factorization algorithm with application to a two dimensional inlet. *AIAA Paper*, 80-0067, 1980.

[7] Y.P. Marx. Multigrid solution of the advection-diffusion equation with variable coefficients. In *5th Copper Mountain Conference*, april 1991.

[8] J.L.C. Chang and D. Kwak. On the method of pseudo compressiblility for numerically solving incompressible flows. *AIAA Paper*, 84-0252, 1984.

[9] E. Turkel. Preconditioned methods for solving the incompressible and low speed compressible equations. *J. of Comput. Physics*, 72:277–298, 1984.

[10] J. Feng and C.L. Merkle. Evaluation of preconditioning methods for time-marching systems. *AIAA Paper*, 90-0016, 1990.

[11] D. Choi and C.L. Merkle. Application of time-iterative schemes to incompressible flows. *AIAA Journal*, 23(10):1518–1524, 1985.

[12] T.M. Shih, C.H. Tan, and B.C. Hwang. Effects of grid staggering on numerical schemes. *Int. J. Heat Mass Transfer*, 9:193–212, 1989.

[13] B. Gustafsson. The Euler and Navier-Stokes equations, wellposedness, stability and composite grids. *VKI-LS*, 1991-01, 1991.

[14] W. H. Finlay, J. B. Keller, and Ferziger J. H. Instability and transition in curved channel flow. Technical Report TF-30, Stanford University, 1987.

Table 1: Order of the method, Shih et al. problem.

	accuracy		
Re	u	v	p
10	2	2	1.75
100	2	2	2.10
500	2.3	2.3	2.3
1000	2.5	2.5	2.5

Table 2: Order of the method, grid topology I.

	accuracy	
type	u	p
1	2.5	2.
2	1.5	1.
3	1.5	1.
4	2.	2.

Table 3: Order of the method, grid topology II.

	accuracy	
type	u	p
1	2.5	2.5
2	1.5	1.5
3	2.5	2.5
4	2.5	2.5

Table 4: Order of the method, grid topology III.

	accuracy	
type	u	p
1	2.4	2.5
2	2.5	2.5
3	2.5	2.5
4	2.5	2.

Figure 1: Re=100

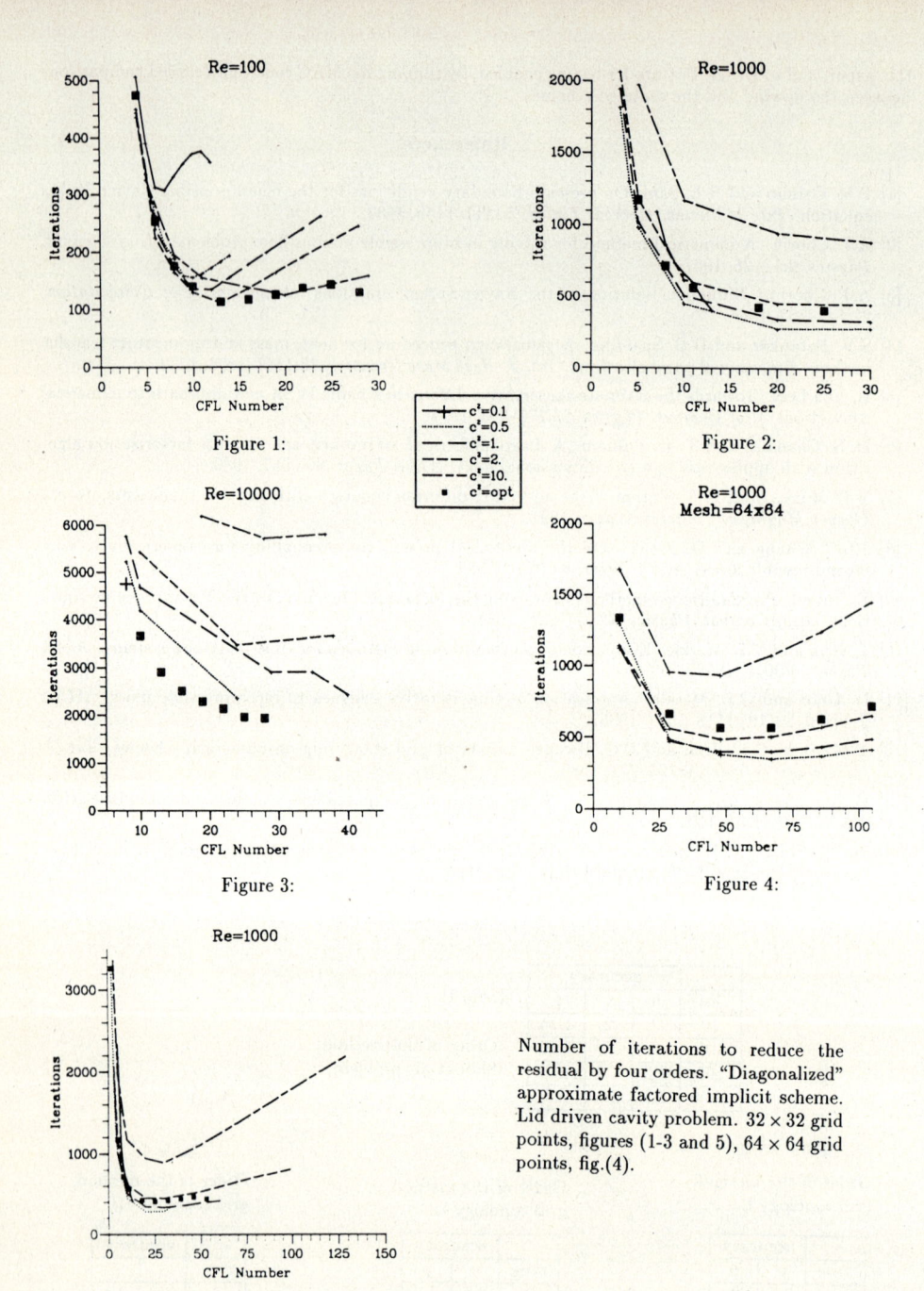

Figure 2: Re=1000

Figure 3: Re=10000

Figure 4: Re=1000, Mesh=64x64

Figure 5: Re=1000

Legend: $c^z=0.1$, $c^z=0.5$, $c^z=1$, $c^z=2$, $c^z=10$, $c^z=\mathrm{opt}$

Number of iterations to reduce the residual by four orders. "Diagonalized" approximate factored implicit scheme. Lid driven cavity problem. 32×32 grid points, figures (1-3 and 5), 64×64 grid points, fig.(4).

Figure 6:

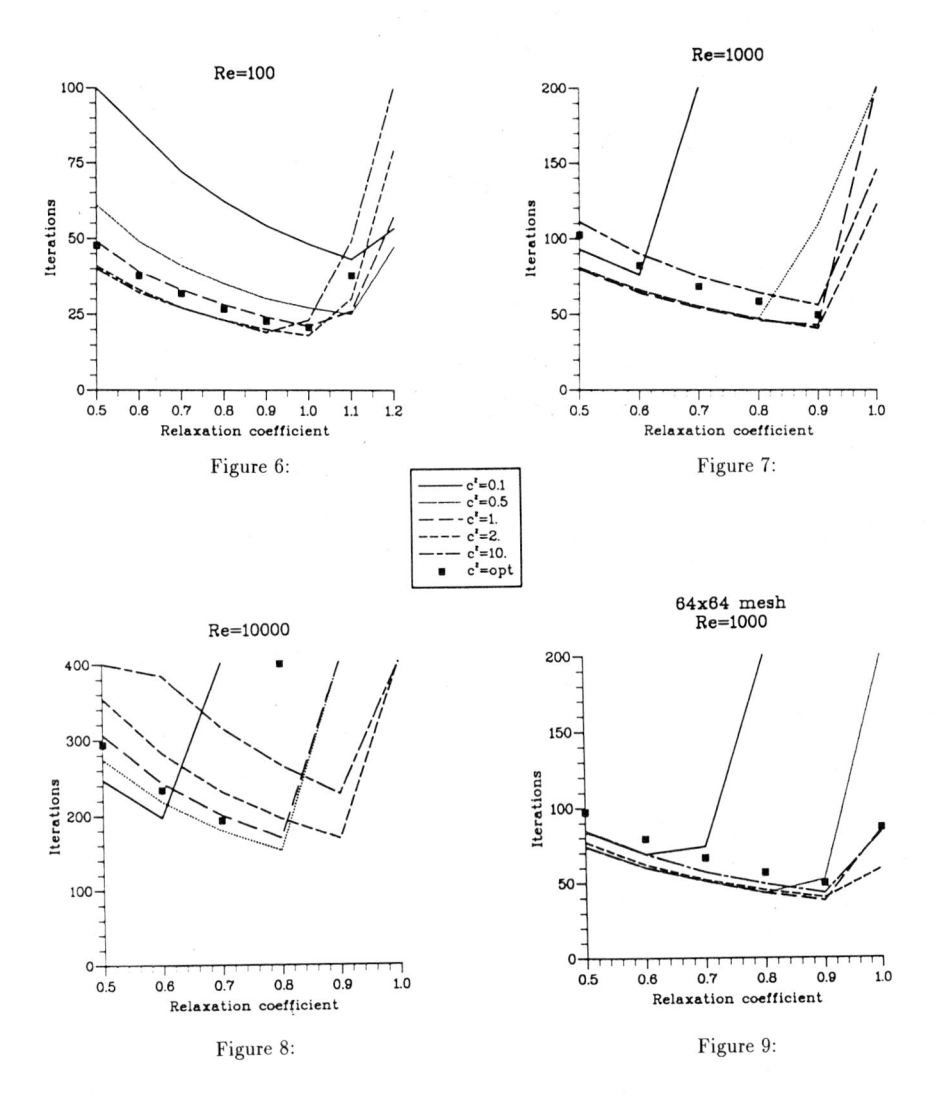

Figure 7:

Legend: $c'=0.1$, $c'=0.5$, $c'=1.$, $c'=2.$, $c'=10.$, ■ $c'=\text{opt}$

Figure 8:

Figure 9:

Number of iterations to reduce the residual by four orders. Non-factored implicit scheme. Lid driven cavity problem. 32×32 grid points, figures (6-8), 64×64 grid points, fig.(9).

207

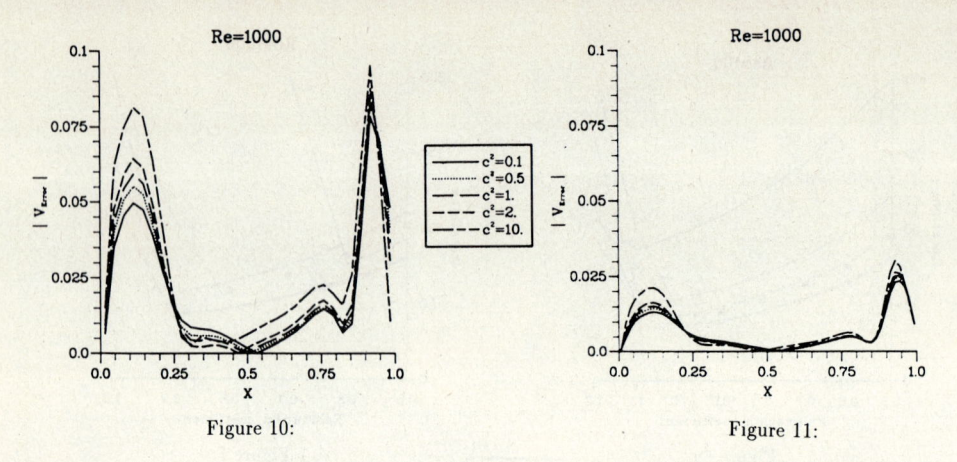

Figure 10:

Figure 11:

Error distribution across the center of the cavity. 32×32 grid points, fig.(10). 64×64 grid points, fig.(11).

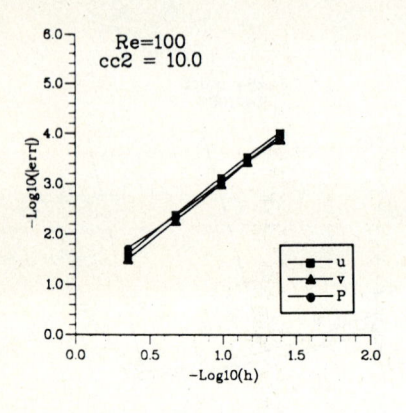

Figure 12:

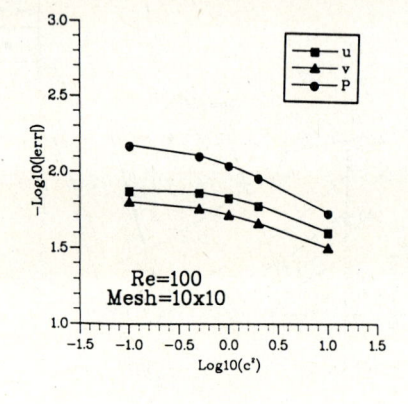

Figure 13:

L_2 norm of the error as a function of the grid size ($c^2 = 10$), fig.(12); as a function of the artificial compressibility parameter, fig.(13). Shih et al. problem.

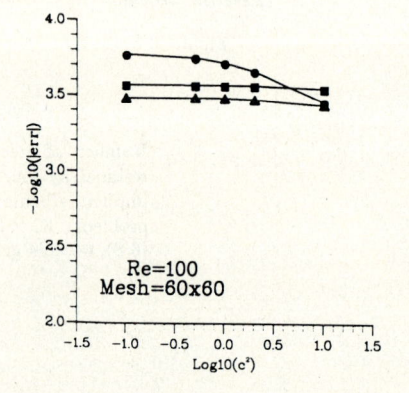

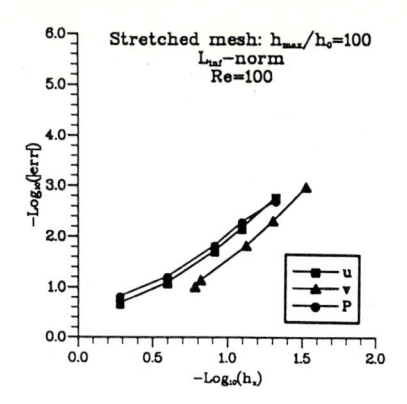

Figure 14:
Evolution of the error for a stretched mesh. Shih et al. problem.

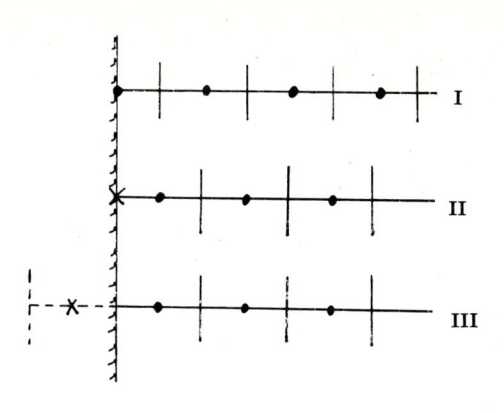

Figure 15:
Grid topologies tested near a wall

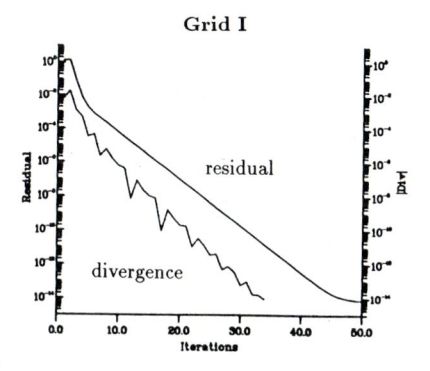

Figure 16:
Evolution of the residual and of the divergence of the velocity near the wall.

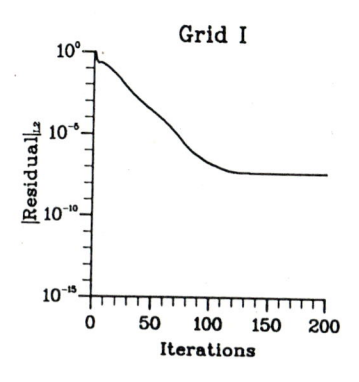

Figure 17:
Convergence history. Pressure calculated from the momentum equation.

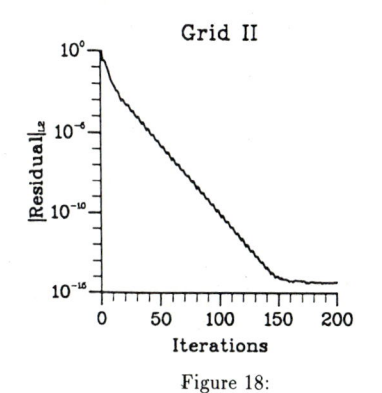

Figure 18:
Convergence history. Pressure calculated by linear extrapolation.

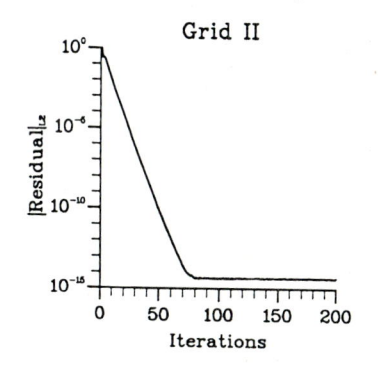

Figure 19:
Convergence history. Pressure calculated from the characteristic equation.

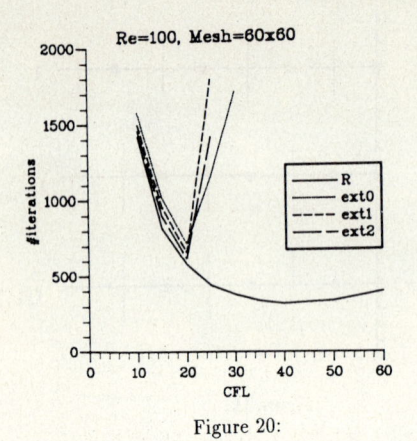

Figure 20:

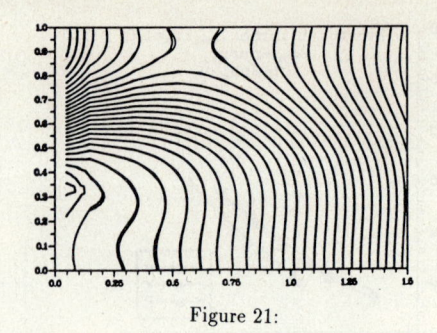

Figure 21:

Constant pressure contours behind the step.
Small domain : ....
Large domain : ———

Number of iteration to reduce the residual by four orders. R = extrapolation on the Riemann invariants. ext0 = zero-order extrapolation. ext1 = linear extrapolation. ext2 = quadratic extrapolation. Lid driven cavity problem. 32×32 mesh.

Figure 22:

Figure 23:

Stream-wise velocity perturbation in a cross section of the three-dimensional curved channel. Upwind scheme, 60×60 grid points, fig.(22). Centered scheme, 30×30 grid points, fig.(23).

VISCOUS SUPERSONIC FLOWS

THREE-DIMENSIONAL THIN-LAYER AND SPACE-MARCHING NAVIER-STOKES COMPUTATIONS USING AN IMPLICIT MUSCL APPROACH: COMPARISON WITH EXPERIMENTS AND EULER COMPUTATIONS.

M. Borrel, P. D'Espiney and C. Jouet
ONERA - BP 72 - 92322 Châtillon Cedex (FRANCE)

Abstract

A space-marching and a thin-layer Navier-Stokes unsteady method is presented for supersonic laminar flows. The numerical scheme is based on an upwind implicit scheme for the convective and pressure terms and a centered approximation for the viscous terms. Results on vortex flows around a fuselage with a lenticular section, are compared with available experimental data and with Euler results. To obtain the right vortex pattern, it was found essential to take into account the viscous effects and to use a sufficiently refined grid.

I. Introduction

Three-dimensional supersonic separated/vortex flows around sharp bodies have frequently received significant attention for being related to most of supersonic and hypersonic aerodynamic problems. Numerically, this problem is very difficult due to predominant viscous effects. However the computation costs can generally be reduced by using some approximation of the Navier-Stokes equations: namely the parabolized approximation (PNS) and the thin-layer approximation (TLNS).

The laminar, non-dimensional, time-dependent compressible Navier-Stokes equations, expressing the conservation of mass, momentum, and energy for an ideal gas, are discretized in conservation law form and Cartesian coordinates using a finite volume method on structured grids. The viscous terms are discretized in only two grid directions. An implicit upwind differencing for the inviscid terms is used. In fact, we incorporated into the 3D Euler code FLU3C [1] the differencing of the viscous terms which corresponds classically to second-order centered differences [2],[3]. The numerical algorithm used is compatible with the space-marching strategy included in the code.

The objective of this paper is both to validate this newly developed PNS/TLNS code and to provide a better understanding of some problems associated with complex vortex flows.

II. Numerical Method

Explicit scheme

Briefly, the governing equations can be written in conservation law form:

$$\partial W/\partial t + \partial(f - f^v)/\partial x + \partial(g - g^v)/\partial y + \partial(h - h^v)/\partial z = 0 \qquad (1)$$

where $W = (\rho, \rho u, \rho v, \rho w, e)^t$ are the conserved quantities, $F = (f, g, h)^t$ represents the inviscid fluxes and $F^v = (f^v, g^v, h^v)^t$ the viscous fluxes. It can be noticed that $F = F(W)$ is a function of W, whereas F^v is a function of the primitives variables $\tilde{W} = (\rho, u, v, w, T)^t$ and of their gradients. Typically, the viscous fluxes can be written in matrix form:

$$f^v = N_x \partial \tilde{W}/\partial x + M_y \partial \tilde{W}/\partial y + L_z \partial \tilde{W}/\partial z. \qquad (2)$$

The finite volume technique used in the code FLU3C leads to discretize the variables W at the nodes of a ijk-structured grid, and to define control volumes as the barycentric cells surrounding each node. The time incremental evolution of W is first computed through the explicit formula:

$$\Delta W_{ijk}^{\exp} = W_{ijk}^{n+1} - W_{ijk}^n = -(\Delta t_\Omega / vol(\Omega)) \Sigma_{l=1}^6 |\partial \Omega_l| (F_l - F_l^v) \cdot \vec{n_l} \qquad (3)$$

where $\vec{n_l}$ represents the outwards unit normal to the control interface $\partial \Omega_l$, Δt_Ω the local time step, F_l and F_l^v the numerical fluxes. The values of the inviscid numerical fluxes are obtained using the Roe flux formula associated with the MUSCL approach:

$$F_l = 1/2 \ [\ F(W_l^+) + F(W_l^-) - (RDiag \ \psi(\lambda) R^{-1})^* \ (W_l^+ - W_l^-) \ ] \tag{4}$$

where $W_l^\pm$ correspond to values of W defined at both sides of the control interface from the neighbour values, incremented with slopes defined along each grid direction with the van Albada limiter applied on primitive variables; R and R^{-1} are the matrices of right and left eigenvectors of the Jacobian of fluxes with respect to the state W^* which represents the Roe averaging of W_l^+ and W_l^-; $Diag(\lambda)$ corresponds to the diagonal matrix of the eigenvalues λ and $\psi(\lambda)$ the Harten entropy correction. This correction has been applied uniformly in all the computational domain.

The values of the viscous fluxes are obtained through an evaluation of the gradients $\nabla \tilde{W}$ by an averaging process over a staggered cell, centered at the control interfaces and using the Green formula, as presented for example in [2] or [3]. What we have called the TLNS approximation consists in neglecting all the viscous terms along the k-grid direction.

Implicit operator

In order to accelerate the convergence towards the steady state, an implicit phase has been added. It corresponds to a Steger-Warming type linearization for only the inviscid first order terms. The viscous terms are treated following a technique introduced by Coakley [4] based on an evaluation of the spectral radius of the viscous matrix. Then, an ADI approximate factorization algorithm is used in two grid directions, coupled with a relaxation algorithm in the other direction.

In each of the two i,j grid directions, the implicit operator is derived from an evaluation of the numerical fluxes at the time step t^{n+1} in (3). For the inviscid fluxes we have:

$$F_l^{n+1} = F_l + A_l^+ \Delta W_l + A_l^- \Delta W_{l+1} \ , \quad A_l^\pm = (RDiag \ [\lambda \pm \psi(\lambda)] R^{-1})^* \tag{5}$$

where $W_l = W_{ijk}$ and $W_{l+1} = W_{adjacent \ cell}$.

For the viscous fluxes, we drop in (2) the cross grid derivatives and we take an estimation of the spectral radius of the remaining viscous matrix:

$$(F_l^v)^{n+1} = F_l^v + (\nu/\delta l) \times I \times (\Delta W_{l+1} - \Delta W_l) \tag{6}$$

with $\nu = \rho^{-1} max(\ 4/3 \ \mu, \gamma \mu/Pr \ )$, $\delta l = vol(\Omega_{ijk})/|\partial \Omega_l|$.

The unfactored implicit scheme, based on three time levels, can be written in "delta" form:

$$[I + \beta(\ \delta_i \Phi_i + \delta_j \Phi_j + \delta_k \Phi_k \ )] \ \Delta W^n = \alpha \Delta W^{exp} + (1-\alpha) \ \Delta W^{n-1} = RHS \tag{7}$$

where: $\delta_i \Phi_i = \sigma_{i+½} \Phi(F-F^v)_{i+½}^{n+1} - \sigma_{i-½} \Phi(F-F^v)_{i-½}^{n+1}$, $\sigma_{i \pm ½} = \Delta t / \delta l_{i \pm ½}$, (i = i, j or k) ,

$\Phi(F-F^v) = (F-F^v)^{n+1} - (F-F^v)^{exp}$.

The second order of accuracy in time is obtained with $\alpha = \beta = 2/3$. For robustness reasons, we have taken $\alpha = 2/3$ and $\beta = 1$ in the computations presented.

Following [3], the 2D approximate factorized implicit operator used consists of:

$$[I + r^{-1} \beta \ \delta_i \Phi_i \ ] \ [I + r^{-1} \beta \ \delta_j \Phi_j \ ] \ \Delta W^{n+1} = r^{-1} \ RHS \tag{8}$$

where r represents the spectral radius of the k-factor: $r = 1 + \beta \sigma_k (\vec{V}.\vec{k} + c)$.

Boundary conditions

All boundary conditions are treated both implicitly and explicitly. The use of an upwind scheme simplify the inflow/outflow boundary treatment of characteristic type, even in Navier-Stokes computations. Special attention must be paid to solid wall boundaries: the unknowns being located at the boundary (nodal treatment), the explicit treatment consists, as usual, of no slip velocity, zero normal pressure gradient and adiabaticity (zero normal temperature gradient) or prescribed temperature (isothermal wall). This treatment is applied after each timestep with fictitious nodes defined by extrapolation. The implicit wall treatment is as follow: if the index of the boundary is denoted by $L + 1$, the implicit operator (8) is modify using:

$$\Delta W_{L+1} = P \ D \ P^{-1} \ \Delta W_L \tag{9}$$

where P represents the Jacobian of the transformation from W to $\tilde{W}$, and D the diagonal matrix $Diag \ (1,0,0,0,\delta)$, $\delta = 1$ or 0 according the wall adiabatic or isothermal.

Space marching (PNS)

The "plane by plane" strategy included in the code FLU3C consists in iterating in time in each grid plane, until convergence, the three dimensional unsteady algorithm with extrapolated values in the next plane. For viscous computations, the Vigneron's correction to inviscid pressure terms in the fluxes along the marching direction has been implemented [5]. This correction returns to retain, along that direction, only a fraction $\omega \ p$ of pressure inside the subsonic boundary layer:

$$\omega = 1 \quad \text{if} \quad M_k^2 \geq 1/(2-\gamma) \ , \quad \omega = M_k^2/(1+(\gamma-1)M_k^2) \quad \text{if} \quad 0 \leq M_k^2 < 1/(2-\gamma) \tag{10}$$

where M_k represents the normal Mach number. In order to discard gradients in the marching direction, we use a zero order extrapolation between the current plane and the next one. In the computations presented, the initialization has always been done with the inviscid free stream conditions, applied on a fictitious plane. The treatment (10) is unconditionally stable in space, and so allows us to set the distribution of the nodes along the marching direction before the computation.

III. Results and Discussion

The three flow cases presented here were all obtained on a lenticular section fuselage, characterized by sharp edges and a boattail. An experimental data base has been obtained in different wind tunnels at ONERA and has already been used for comparison with inviscid computational methods [6]. For cases 1 and 2 ($M_\infty=2$) only TLNS computations have been achieved, for case 3 ($M_\infty=4.5$) TLNS and PNS computations have been carried out. The Prandtl number has been taken at the constant value $Pr = .72$.

At $M_\infty=2$, two models of different length (L=360mm and 1200mm) have been tested in wind tunnels, leading to the values: 2.10^6 and 18.10^6, for the Reynolds number based on the longitudinal body length. The smallest model is used for oil flow visualisations and external flowfield measurements in two cross sections (X/L =0.571 and 0.915) with a five-hole pressure probe. In fact, natural transition and triggered transition by mean of a carborandum strip do not change cross flow organisation in sections X/L = 0.571 and 0.915. Surface pressure measurements are only available with the longest test model at $M_\infty=2$ and 4.5, and $Re_\infty=18.10^6$, with natural transition of the boundary layer.

An O-H grid structure has been adopted for all the computations presented (Fig.1). Except near the nose where the density of the nodes is higher, the grid planes are equally distributed along the axis. The grids are built "plane by plane", with an algebraic refinement of about 30 nodes inside the boundary layer region. For viscous computations, the grid spacing near the wall is $\delta y /L \approx 2.5 \ 10^{-4}$; the number of the nodes in the radial direction is 85. For Euler computations, on cases 1 and 2, a coarser grid has been used: 43 nodes in circumferential direction, 28 nodes in radial direction, and 43 mesh planes in axial direction. Typically, with a 43x85x45 grid, at 700 iterations the residual value has decreased five orders of magnitude for a TLNS computation at a

CFL number of 15; the computation requires about two hours and half on a Cray-2 computer. To obtain the same convergence level, we need about six times less computer time for a space-marching computation (PNS or Euler with the same grid).

Case 1 - Mach 2, $\alpha=10$ °

A TLNS computation for $Re_\infty=3.10^6$ was first carried out on the forward part of the grid with two circumferential node densities : a coarse one with 43 nodes equally distributed and a fine one with 91 nodes among which 61 at the leeward side (Fig.1c). A special attention has been paid to the node distribution near the sharp edge in order to get meshes of the same size. As shown in Fig.2, two co-rotating vortices are predicted with the fine grid, like in experiment, but only one with the coarse grid.

Afterwards, the computation was continued on the backward part of the fine grid. From the computed skin friction lines and the cross-velocity plot (not shown here) in two cross sections, we propose in Fig.3a, a sketch interpretation of the structure with two co-rotating vortices which develops at the leeward side. The vortex F1 appears near the nose on smooth surface (ogive-cylinder problem), afterwards, the sharp-edge vortex F3 (delta-wing problem) also appears; these two vortices develop separately before coalescing, leading to a single vortex structure beyond section X/L = 0.8 for experiment, and X/L=0.571 for computations. The predicted secondary contrarotating vortex F2 is not clearly put in evidence by experiment, due to a lack of precision of the velocity measurements. In the backward part of the fuselage, except in the vortex area, the skin friction lines agree well with experiment (Fig.3b/c). The comparison of the total pressure in the two sections (Fig.4) shows that the level and the shape of the TLNS contours better agree with the experiment than the Euler ones. Nevertheless the two co-rotating vortices are too close to each other and a more refined grid would probably be needed to describe more precisely this vortex structure. This is still more true with the computation for $Re_\infty=18.10^6$. Indeed, the surface pressure measurements, only available for this Reynolds number, show a well separated vortex pair at section X/L=0.571. The predicted leeward surface pressure does not agree exactly with experiment: the flow is perhaps no more laminar, even for this upstream section X/L=0.253. In Fig.5, leeward surface pressure measurements for two cross sections near the nose are compared with TLNS (for two Reynolds numbers) and Euler results. A rather good agreement is obtained with TLNS ($Re_\infty=18.10^6$), although the vortex F1 is stronger and starting more upstream than in experiment.

Case 2 - Mach 2, $\alpha=20$ °

The computed leeward skin friction lines (Fig.6) give the main features of the experimental flow. In that case, we have a classical vortex structure with a well-defined pair of contrarotating vortices F1 and F2; a third contrarotating vortex F3 is also captured and disappears in the boattail region before section X/L=0.915. The accuracy of flowfield measurements is not high enough to describe in detail the secondary vortex. Nevertheless a good agreement is obtained between experiment and TLNS computations concerning the location and the size of the main and secondary vortices. Crossflow streamlines at X/L=0.571 (Fig.7) show that for both Euler and TLNS computations, the position of the main vortex is well predicted, but only TLNS computations give a satisfactory level and shape of total pressure lines (Fig.8). Comparison of surface pressure measurements, available only for $Re_\infty=18.10^6$, with Euler and TLNS computations is presented in Fig.9. As pointed out in a basic experiment on a supersonic vortex flow around an ogive-cylinder [7], for high angles of attack, very similar separated regions are obtained for laminar and turbulent flows: this can explain the quite good agreement between experiment and TLNS results on this case.

Case 3 - Mach 4.5, $\alpha=10$ and 15 °

Euler and PNS computations were performed on the same 43x85x66 grid. For $\alpha=10°$, a TLNS computation has also been carried out on the forward part. Differences between PNS and TLNS results are small if we look at the pressure profiles and at the skin friction lines, particularly concerning the position of the separation line (Fig.10). But the values of the skin friction coefficients are about 30% different. In the flowfield, the position of the bow shock is slightly farther from the wall in TLNS than in PNS computations.

When the incidence increases to 15 degrees, the flow features remain qualitatively the same. On the skin friction lines (Fig.11a), we can see that the flow, on the forward part, runs round the lateral edge without separating, and then separates leading to a mainly single vortex structure; the separation line comes near the lateral edge and then follows it. A second separation line, linked to the reattachement line and a secondary contrarotating vortex, can be seen in the rear part. As shown in Fig.11b/c, the vortex structure for the Euler computation results from the presence of a cross shock on the leeward-side, in contrast to the viscous computation.

Comparisons with experiment are made on the pressure profiles for the two angles of attack (Fig.12). The experiment was performed for a Reynolds number of 18.10^6. The PNS computation, although not carried out with the right Reynolds number, gives a better comparison at the leeward side than the Euler computation. However, it can be noticed that, the higher the Mach number or the angle of attack is, the more important the contribution of the windward side is, and, therefore, the less important the effect of the leeward vortex structure is, especially for the global coefficients.

IV. Conclusion

Approximate Navier-Stokes and Euler solutions are presented for complex vortex flows and are compared with experiments. These comparisons have stressed some well-known facts: Euler computations do not allow a right description of the vortex structure in contrast to PNS and TLNS computations; for viscous computations on some difficult vortex structures, a grid refinement is needed not only in the normal wall direction but also in the circumferential direction; the PNS strategy allows to decrease the computational costs but does not work for too low Mach numbers. To go further, we need more accurate measurements both on the body and in the flowfield, and a better understanding and modelling of the physics (turbulence, transition).

Acknowledgements : This work has been financially supported by the *Direction des Recherches, Etudes et Techniques* of The French Ministry of Defense. The authors would like also to thank Dr. J.L. Montagné, presently of DCN, for his fruitful contribution to the development of the method.

References

[1] M. Borrel, J.L. Montagné, J. Diet, P. Guillen, J. Lordon: *Upwind Scheme for Supersonic Flows around Tactical Missiles* ; La Recherche Aérospatiale, 1988-2.
[2] E. Chaput, F. Dubois, D. Lemaire, G. Moules, J.L. Vaudescal: *FLU3PNS: a Three-Dimensional Thin Layer and Parabolized NS Solver Using the MUSCL Upwind Scheme* ; AIAA paper 91-0728.
[3] S.R. Chakravarthy: *High Resolution Upwind Formulation for the NS Eqs.* ; VKI Lec. 1988-05.
[4] T.J.Coakley: *Implicit Upwind Methods for the Compressible NS Eqs.* ; AIAA J. vol.23-3,1983.
[5] Y.C. Vigneron, J.V. Rakich, J.C. Tannehill: *Calculations of Supersonic Viscous Flows over Delta Wings with Sharp Subsonic Leading Edges* ; AIAA Paper 78-1137.
[6] P. d'Espiney: *Comparaison de différentes méthodes de calcul appliquées à un fuselage de section lenticulaire* ; AGARD-CP-493, April 1990.
[7] D. Pagan, P. Molton: *Basic Experiment on a Supersonic Vortex Flow around a Missile Body* ; AIAA 29th Aerospace Science Meeting, January 7-10 1991.

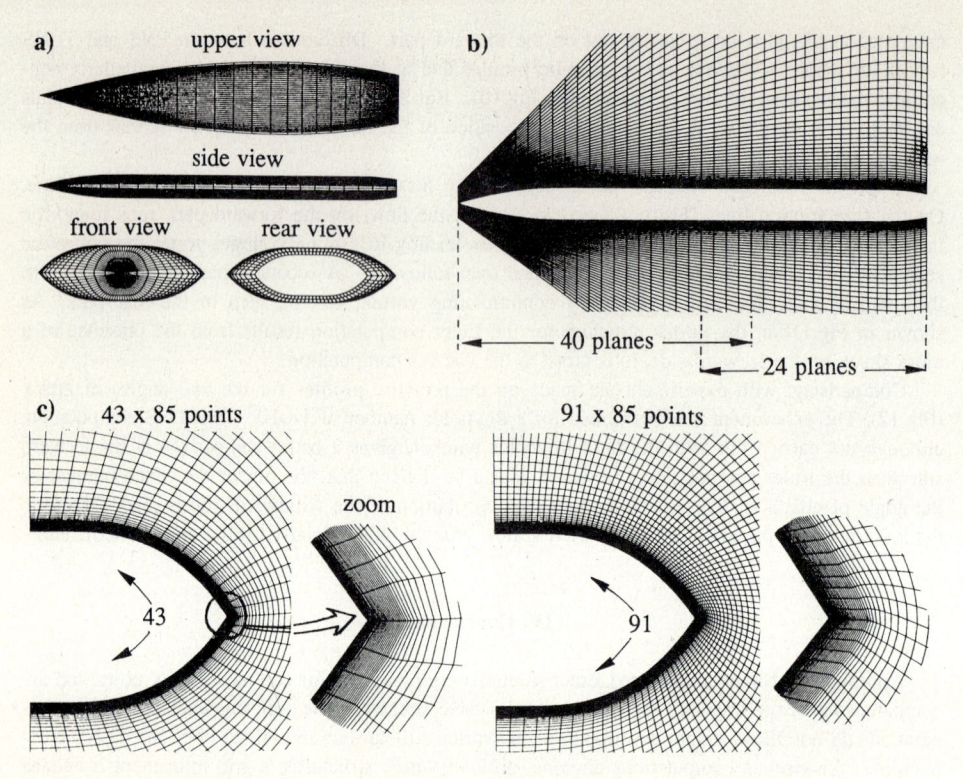

Fig. 1 : a) body geometry ; b) and c) computational grids for $M_\infty = 2$, $\alpha = 10^0$.

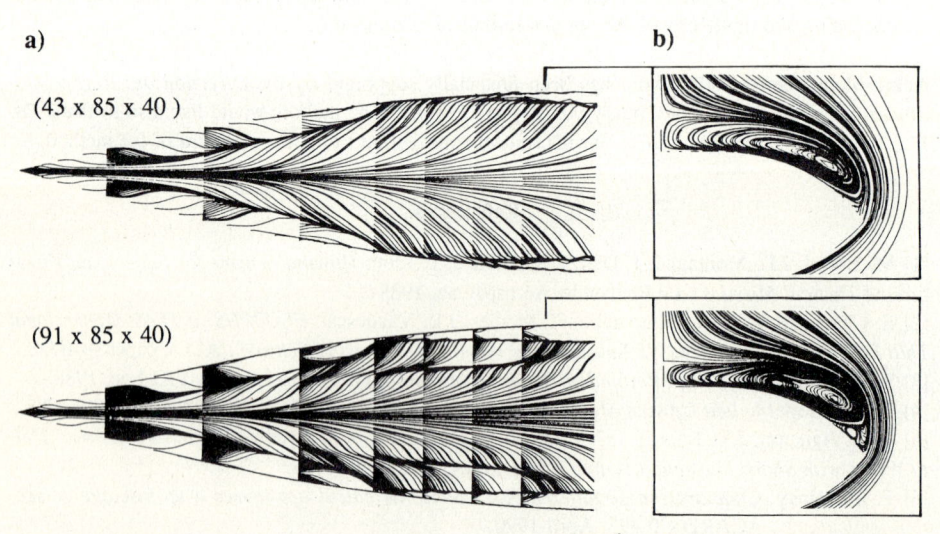

Fig. 2 Grid effects for $M_\infty = 2$, $\alpha = 10^0$, ($Re_L = 3.10^6$)
    a) leeward skin friction lines ; b) cross flow streamlines.

b) experiment - oil flow visualization

c) TLNS skin friction lines

a)

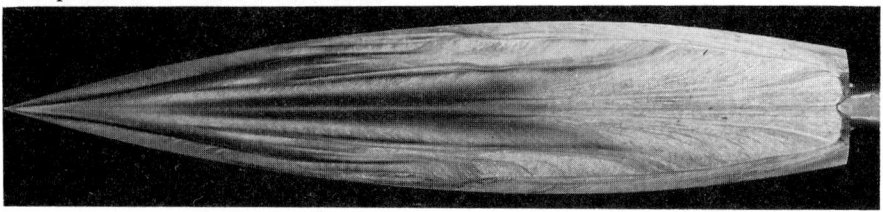

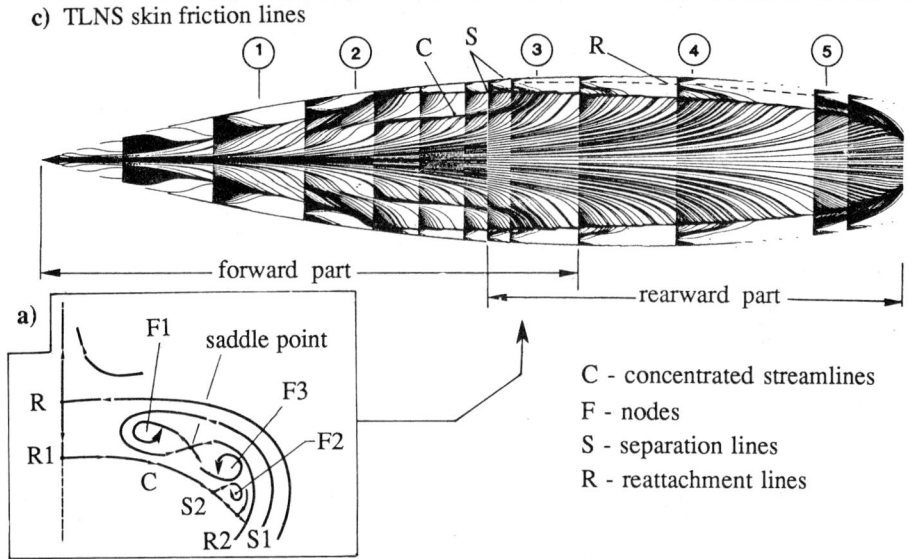

C - concentrated streamlines
F - nodes
S - separation lines
R - reattachment lines

Fig. 3 : comparison experiment / TLNS - $M_\infty = 2$, $\alpha = 10^0$, $Re_L = 3 \cdot 10^6$
 a) crossflow organization at $x/L \cong 0.571$
 b) - c) leeward skin friction patterns.

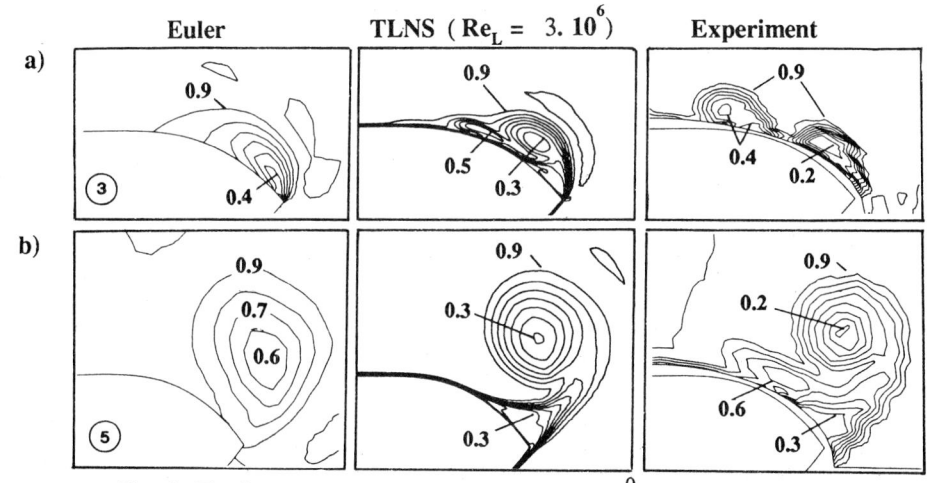

Fig. 4 Total pressure contours - $M_\infty = 2$, $\alpha = 10^0$ - (Euler / TLNS / Exp.)
 a) section $x/L = 0.571$; b) section $x/L = 0.915$

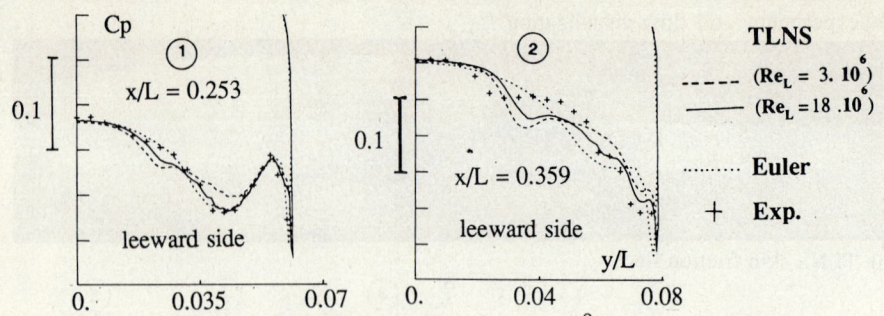

Fig. 5 Pressure coefficient distribution - $M_\infty = 2$, $\alpha = 10°$ – (Euler / TLNS / Exp.)

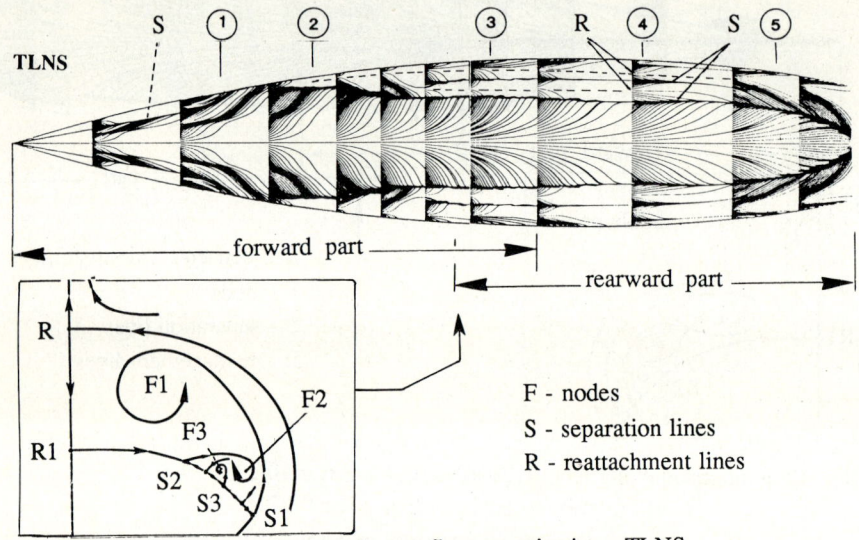

Fig. 6 Leeward skin friction lines and cross flow organization - TLNS
- $M_\infty = 2$, $\alpha = 20°$, $Re_L = 3.10^6$

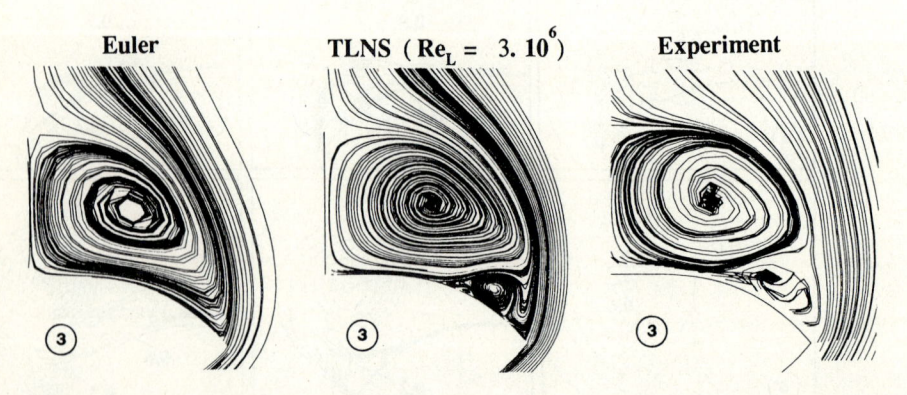

F - nodes
S - separation lines
R - reattachment lines

Fig. 7 Crossflow streamlines - $M_\infty = 2$, $\alpha = 20°$, section $x/L = 0.571$

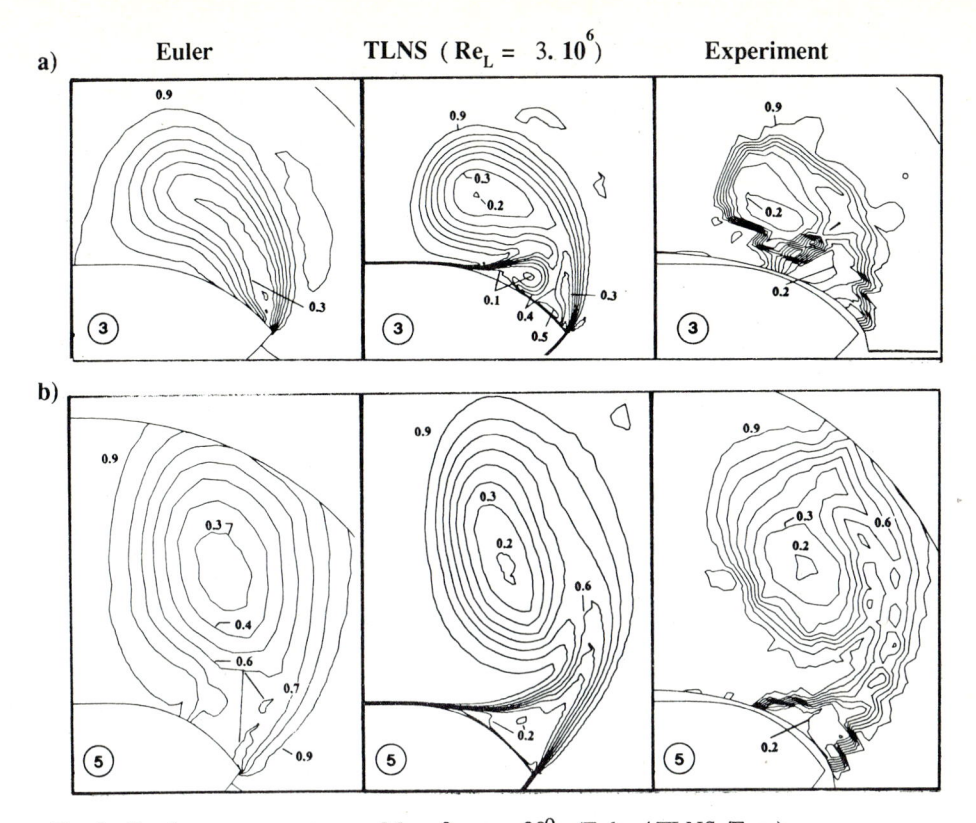

Fig. 8 Total pressure contours - $M_\infty = 2$, $\alpha = 20°$ – (Euler / TLNS /Exp.)
   a) section x/L = 0.571 ; b) section x/L = 0.915

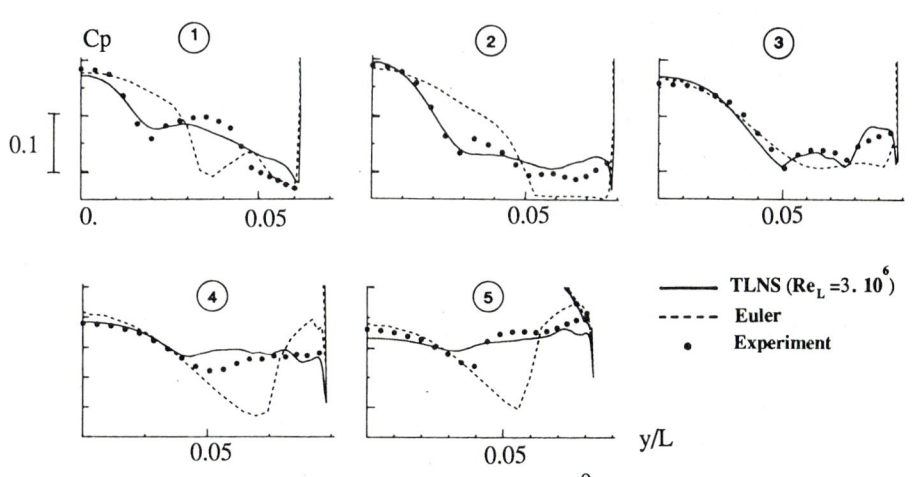

Fig. 9 Leeward pressure distributions - $M_\infty = 2$, $\alpha = 20°$
   (sections x/L = 0.253, 0.359, 0.571, 0.759, 0.915)

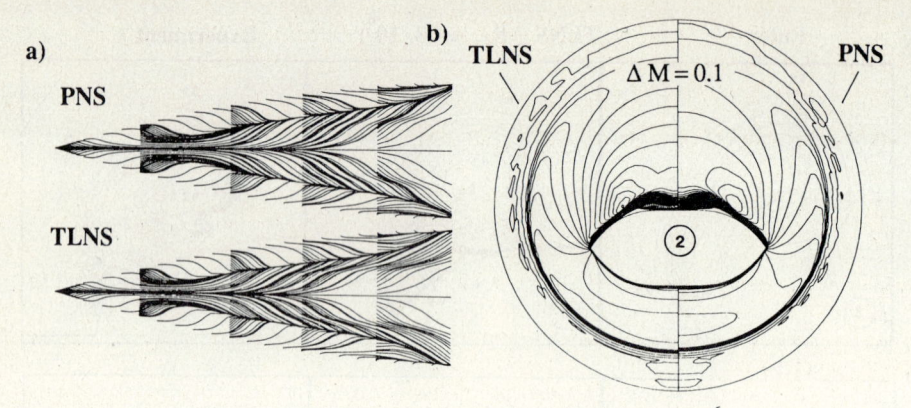

Fig. 10 Comparison TLNS / PNS - $M_\infty = 4.5$, $\alpha = 10^0$, $Re_L = 3.10^6$
a) leeward skin friction lines ; b) iso-Mach contours

a) leeward skin friction lines

b) crossflow streamlines

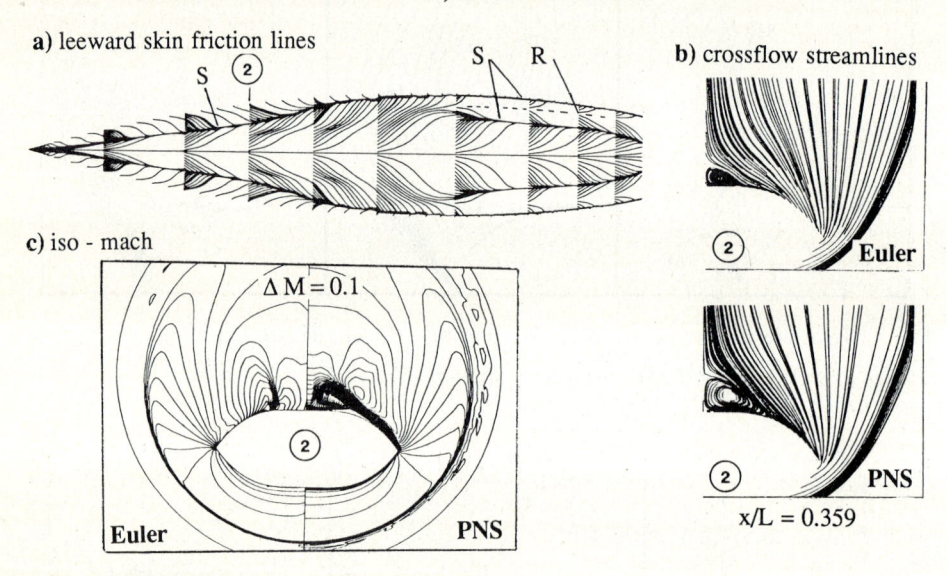

c) iso - mach

Fig. 11 Comparison Euler / PNS - $M_\infty = 4.5$, $\alpha = 15^0$, $Re_L = 3.10^6$

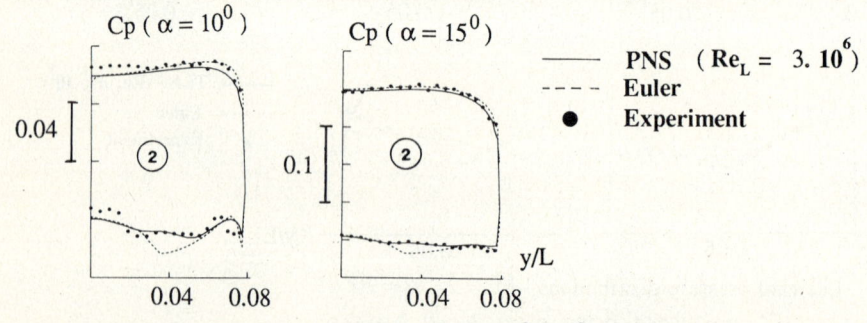

Fig. 12 Pressure distributions - $M_\infty = 4.5$, $\alpha = 10^0, 15^0$ – (PNS / Euler / Exp.)

HIGH SPEED FLOWS OVER COMPRESSION RAMPS

Pénélope Leyland, Roland Richter, Tristan Neve.

IMHEF, Ecole Polytechnique Fédérale de Lausanne, ME - Ecublens, CH-1015 Lausanne, Switzerland

SUMMARY

Numerical simulations of high speed inviscid and viscous flows compression corner geometries (wedge, ramp..) using three different approaches to calculate the numerical flux are presented. An assessment of the discontinuity capturing properties, coupled with the requirement of minimising the numerical dissipation, is given. The three schemes considered consisted of an explicit in time, centred scheme with additional artificial dissipation for stability and shock-capturing, an implicit in time upwind approximate Riemann Solver, and an implicit in time centred scheme coupled with a Harten-Yee TVD limiting term for discontinuity capturing. The governing equations are discretised in conservative form, allowing an equivalent cell vertex P1 Galerkin finite volume formulation on unstructured finite element meshes.

1. INTRODUCTION

High speed flows over compression corner geometries provide valuable predictions of realistic flow phenomena in aerothermodynamics. Shock-shock interactions, shock-boundary layer interactions, multiple shock reflexions, together with possible separation-recirculation secondary shocks, and rarefaction waves, may all be present simultaneously within these flows. Numerical simulation of these phenomena require high resolution schemes for discontinuity capturing, without the introduction of excessive numerical dissipation, particularly within the boundary layer.

The governing equations are the full Navier-Stokes system of equations. However, the detailed study of artificial dissipation via numerical viscosity matrices is first developed for the Euler equations. This sub-system of equations provide valuable information about the discontinuity capturing properties of the underlying numerical schemes. Academic test problems, as for example flows over a ramp, a wedge or an inlet configuration, are used to tune the coefficients of the artificial dissipation model. It is important to validate firstly for flow set-ups without the shock-boundary layer interactions, since these are of a highly complex physical nature, and a difficult challenge for numerical schemes. In this context, the mesh adaptation efficiency can be clearly established. Meshes are automatically adapted to the solution process by refinement and local enrichment, barycentric smoothing and stretching grid lines along strong gradients. This provides a robust shock capturing quality capability, using an optimally reduced number of grid points.

The paper is divided into three parts: in the first part details of the numerical flux functions used are presented, the second part discusses the Euler computations on wedges, and the third part deals with full Navier-Stokes calculations over compression ramps. In this last study the three numerical flux functions used are compared. Ramp test cases provide a challenge for numerical schemes due to the very complex flow features (shock-shock interactions, secondary separation shock formations, lambda-boundary-layer shock interactions). Existing experimental data for the test cases calculated here [1],[2], as well as simulations found in literature provide solid guide-lines.

The use of a finite element spatial discretisation allows freedom for the mesh adaptivity possibilities, and reduces the number of overall mesh points for a given accuracy. Boundary conditions are implemented exactly as part of the formulation, or imposed using a strong formulation, avoiding the need of extrapolation procedures. This has been incorporated into a finite volume type scheme by using a cell vertex P1 Galerkin formulation, which provides an exact formulation for the discretised terms (see e.g. [3], [4], [5]...).

All the test cases considered in this paper correspond to laminar, steady state flows. Time accuracy is thus not necessary to be superior to one. Second order space accuracy for the two implicit TVD schemes are obtained via the Harten-Yee entropy fix, rather than MUSCL limitation using upwind triangles [7]. The explicit in time centred scheme is theoretically of second order [8].

2. GOVERNING EQUATIONS

The Full 2D Navier-Stokes equations are written in conservative form:-

$$\int_{V(t)} \left\{ \frac{\partial W}{\partial t} + \text{div } N(W) \right\} dV(t) = 0 , \forall\ V(t) \in \Omega .\tag{2.1}$$

Ω is the discretisation domain, $V(t)$ is a control volume within Ω, for each time t.

Where

$$W = (\rho, \rho u, \rho v, \rho E)^T$$

and the continuous fluxes may be decomposed as

$$N(W) = F(W) + G(W) + R(W) + S(W) ,\tag{2.2}$$

which in non-dimensional form reads for the inviscid fluxes

$$F(W) = \begin{pmatrix} \rho u \\ \rho u^2 + p \\ \rho uv \\ u(\rho E + p) \end{pmatrix} , \qquad G(W) = \begin{pmatrix} \rho v \\ \rho uv \\ \rho v^2 + p \\ v(\rho E + p) \end{pmatrix}$$

and the viscous fluxes

$$R(W) = \frac{1}{Re} \begin{bmatrix} 0 \\ \frac{4}{3}\frac{\partial u}{\partial y} - \frac{2}{3}\frac{\partial v}{\partial x} = \tau_{xx} \\ \left(\frac{\partial u}{\partial y} + \frac{\partial v}{\partial x}\right) = \tau_{xy} \\ \tau_{xx} u + \tau_{xy} v + \frac{\gamma}{Pr}\frac{\partial e}{\partial x} \end{bmatrix} , \quad S(W) = \frac{1}{Re} \begin{bmatrix} 0 \\ \left(\frac{\partial u}{\partial y} + \frac{\partial v}{\partial x}\right) = \tau_{xy} \\ \frac{4}{3}\frac{\partial u}{\partial x} - \frac{2}{3}\frac{\partial v}{\partial y} = \tau_{yy} \\ \tau_{xy} u + \tau_{yy} v + \frac{\gamma}{Pr}\frac{\partial e}{\partial y} \end{bmatrix} .$$

Re represents the Reynolds number, p the static pressure, $p = f(\rho, e)$, and e the internal energy. The upstream conditions here correspond to perfect gas conditions thus total energy and pressure are related by the state equation

$$p = (\gamma - 1) \rho e = (\gamma - 1) (\rho E - \frac{1}{2} \rho (u^2 + v^2))$$

and

$$\rho E = \rho e + \frac{1}{2} \rho (u^2+v^2).$$

The Euler equations are by setting Re = ∞, and form a hyperbolic system of conservation laws:-

$$\frac{\partial W(x,t)}{\partial t} + \frac{\partial F(W(x,t))}{\partial x} + \frac{\partial G(W(x,t))}{\partial y} = 0, \quad (x,t) \in \mathbf{R}^n \times (0,T). \quad (2.3)$$

For Navier-Stokes calculations the viscosity and thermal conductivity μ and λ coefficients are taken to be functions only of temperature. Sutherland's law corrected for the low temperatures [9] of the experimental set-ups (wall temperature ≈ 290°K, upstream temperature ≈ 50°K, maximum temperatures ≤ 500 °K), and a constant Prandtl number Pr =0.72, are taken.

3. NUMERICAL VISCOSITY MATRICES

3.1. DISCRETISATION.

The system (2.1) is discretised over the spatial domain Ω_h by a P1-triangulation $\{\mathcal{T}_h\}$. A cell vertex approximation is adopted over the dual control volumes, which consists of constructing barycentric subdivision of the triangulation, by joining the barycentres of each triangle.

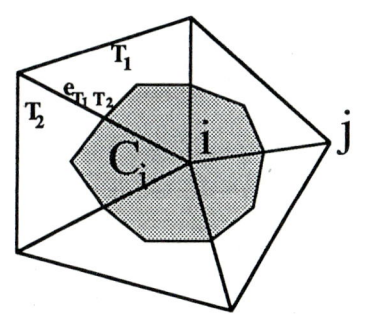

Figure 1. Cell vertex control volume.

These cells provide again a covering of Ω_h,

$$\Omega_h = \bigcup_{i=1}^{ns} C_i.$$

In n dimensions, the boundary of a dual cell C_i is made up of all the dual cells of all the simplexes of dimension (n-1) which contain i as vertex. Let K(i) denote the set of neighbouring nodes (ns) of $\{i\}$, then the boundary of C_i decomposes as

$$\partial C_i = \bigcup_{j \in K(i)} \partial C_{ij}.$$

3.2 NUMERICAL SCHEMES

Weak solutions of (2.3) admit discontinuities, i.e. shocks, rarefaction fans and contact discontinuities may be present within the physical solution. Among the set of admissible weak solutions of a system of conservation laws, the physical solution is selected as the one which satisfies the supplementary "Entropy Condition ", [10], [11]. In order to capture precisely the various discontinuities, the numerical simulation of such systems is often based on the extension of solving the 1D Rieman problems at the cell interfaces, ∂C_{ij}:-

$$\frac{\partial W(x,t)}{\partial t} + \vec{\nabla}.\mathcal{F}(W(x,t)) = 0 \text{ in direction } \vec{n}(x,t) \text{ at time } t$$

$$W(x,t) = W_L \quad \text{for } \vec{x}.\vec{n} < 0 \tag{3.1}$$

$$W(x,t) = W_R \quad \text{for } \vec{x}.\vec{n} > 0 .$$

Where L and R stand for the left and right states on each side of a possible discontinuity. For the finite element type space discretisation considered here this may be illustrated as follows

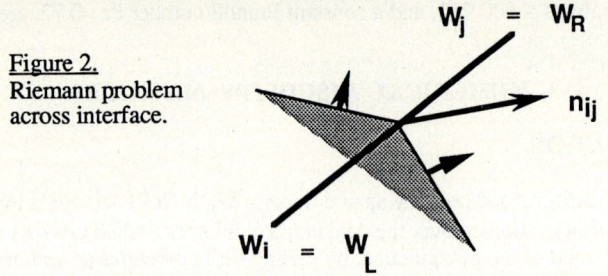

Figure 2.
Riemann problem across interface.

For the complete Navier-Stokes equations, following (2.1), the discretised viscous fluxes, R and S, are approximated by a P1 finite element approximation over the underlying triangulation to the dual mesh, whereas the inviscid fluxes F and G are discretised on the dual cell vertex control volume mesh.

From a general discretised form of (2.1),

$$\int_{\Omega_h} \left\{ \frac{W_h^{n+1} - W_h^n}{\Delta t_h^n} + \text{div } N(W_h^{n+1}) \right\}.\varphi_h dx = 0; \text{ for } \varphi_h \in \mathcal{V}_h \equiv \left\{ v_h \in C^0(\Omega); v_{h_T} \in P_1(T) \, \forall T \in \mathcal{T}_h \right\}$$

a numerical flux function can be defined as

$$\Phi_{ij}(W_i, W_j) = \int_{\partial C_{ij}} \vec{\mathcal{F}}(W).\vec{n}_{ij} . \tag{3.2}$$

For any numerical scheme this may be decomposed [12], [13] as an implicit part plus an explicit part, and a dissipative part which is either intrinsic or explicitly added,

$$\Phi_{ij}(W_i, W_j) = \Phi_{ij}^{impl} + \Phi_{ij}^{expl} + D_{ij}^{diss} . \tag{3.3}$$

In this paper, three different numerical fluxes are tested for the simulation of flows over compression ramps and wedges. These geometries provoke all the difficulties in the physical solution as mentioned above. The first scheme consists of an explicit in time, second order space accurate centred scheme with added artificial dissipation (RK5). The second and third schemes are implicit in time. The explicit numerical flux function for the second one is a two-stage centred scheme, with Harten-Yee-Davis TVD flux limiting (ICTVD), [14], for the third scheme an upwind approximate Riemann solver using Osher's Flux is employed, with again TVD flux limiting for second order accuracy (IUARS). The time integration for the first scheme is performed via a Runge-Kutta multi-(5)-stage scheme, which has been proven to be optimal (second order in time), [15], whereas for the implicit schemes an unfactorised matrix inversion is performed [16] (first order in time). Only convergence to a steady state solution is considered here.

The numerical flux function for the first scheme (RK5) can be written as

$$\Phi_{1ij}^{expl}(W_i, W_j) = \frac{1}{2}\{\mathcal{F}(W_i) + \mathcal{F}(W_j)\} - d_{ij}^{diss}. \quad (3.4)$$

Where d^{diss}_{ij} represents a second order plus a fourth order artificial dissipation term [8], to limit spurious oscillations in vicinity of shocks, and damp out numerical oscillations respectively. The standard construction of this flux consists of a first order plus third order differencing at the surfaces of the control volume, the latter obtained for finite element meshes by the construction of fictitious points i*, j* as in [15].

$$d_{ij}^{diss}(W_i, W_j) = \Theta_{ij}\alpha_2\chi_{ij}(W_i - W_j) - \Theta_{ij}\alpha_4\chi_{ij}(W_i^* - 3W_i + 3W_j - W_j^*). \quad (3.5)$$

The coefficients within the product ($\Theta_{ij} \alpha \chi_{ij}$) represent respectively a scaling factor, a coefficient (which both act globally) and a sensor based on a second order difference of either pressure or entropy, which localises shocks

$$\alpha_4 = \max(0, \varepsilon_4 - \alpha_2), \quad \chi_{ij} = \max(\chi_i, \chi_j), \quad \chi_i = \left|\frac{p_{i^*} - 2p_i + p_j}{p_{i^*} + 2p_i + p_j}\right|. \quad (3.6)$$

This original set of dissipation sensors can be improved by taking into account the natural upwinding nature of the flow in the vicinity of discontinuities. Equation (2.3) may then be written as

$$\frac{\partial W(x,t)}{\partial t} + A(W)\frac{\partial W(x,t)}{\partial x} + B(W)\frac{\partial W(x,t)}{\partial y} = 0, \quad (3.7)$$

The characteristic matrix $H(n_x, n_y, W) = [A.n_x + B n_y]$ is diagonalisable as $H = T \Lambda T^{-1}$ where T is an inversible matrix. (3.7) can thus be considered as a system of 1D -problems, each of which can be written in a local characteristics form:-

$$\frac{\partial U}{\partial t} + T\Lambda T^{-1}\frac{\partial U}{\partial X_k} = 0; W = T^{-1}U,$$

A matrix dissipation flux can be obtained by scaling the coefficients α_2 and α_4 to the largest eigenvalues of the Jacobiens A and B, [17],

$$\alpha_k = \frac{1}{2}\{\lambda_i + \lambda_j\}\alpha_k, \quad k = 1, 2; \quad \lambda_i = (\vec{u}.\vec{n} + c)_i \otimes \text{Area } C_i$$

which leads to a TVD numerical viscosity matrix

$$d_{ij}^{(2)} \approx \frac{1}{2}T|\Lambda|T^{-1}(W_i - W_j).$$

A similar expression for $d^{(4)}_{ij}$ can be obtained following formulae (3.5), (3.6). In general T is taken to be the matrix of right eigenvectors.

The dissipation sensors (3.6), are taken to be functions of either the pressure or entropy, depending on the application considered. For Euler calculations of external flows, it is custom to employ the above pressure sensor, which well localises shock waves. However for Navier-Stokes simulations of external flows, an entropy sensor similar to the above together with a boundary layer scaling to limit its action within the boundary layer is often preferred [18]. For the ramp flows considered here, there is a uniform entropy gradient within the boundary layer (see Figure 11), and the corner shock comes down to the angle; a pressure sensor is thus employed, giving a good localised dissipation without perturbing the near wall layer.

For the second scheme, (ICTVD), Φ^{expl}_{ij} is given by a predictor-corrector S^α_β scheme with a TVD viscosity matrix similar to the one used to attain second order accuracy for the third scheme, (see below). The combination of the two-stage scheme with TVD flux limiting gives a numerical viscosity matrix following [19], [20], and adapted to triangulations [14].

$$\Phi^{diss}_{TVD} = \Sigma_{T \in V(T(i))} \left\{ \chi_{ij} \int_{eTiTk} \text{diag} A_{ik} \cdot \overrightarrow{\mathcal{F}(u^n)} \cdot \vec{n} \, ds + (1 - \chi_{ij}) \int_{eTiTk} \overrightarrow{\mathcal{F}(u^*)} \cdot \vec{n} \, ds \right\}$$

following the notations of Figure 1, and $V(T(i))$ represents the neighbouring triangles around $T(i)$.

The third scheme is based on an upwind approximate Riemann solver, (IUARS), and uses the Osher flux

$$\Phi^{expl}(W_L, W_R) = \frac{1}{2} \left\{ F(W_L) + F(W_R) - \int_{W_L}^{W_R} |A_{LR}(W)| \, dW \right\}.$$

The integral represents an integration path along the set of k-waves.

For this scheme, second order accuracy is obtained using the Harten-Yee formulation where

$$\Phi^{Num}_{ij}(W_i, W_j) = \Phi^{expl}_{ij} + \Phi^{diss}_{TVD}, \qquad \Phi^{diss}_{TVD} = - \Psi(A_{ij}) \, T \, |\widetilde{\Lambda}| \, T^{-1} (W_i - W_j).$$
$$ 2$$

$\widetilde{\Lambda} = \Lambda (1 - Q(\text{diag}\widetilde{A}_{ij/2}))$, where $\widetilde{A}$ is the modified flux Jacobien taking into account the entropy correction, (term $\Psi(A_{ij/2})$. The limiter $Q(\lambda) = Q(\widetilde{\lambda} + \text{limiter})$, and the min mod limiter is chosen.

4. EULER CALCULATIONS

4.1. NUMERICAL METHOD.

The first scheme is applied to the calculation of supersonic flows over a ramp using various angles. For inviscid flows the solution of this problem can be calculated analytically. A second problem of a standard wedge geometry with multiple shock reflexions and an expansion fan is calculated. The capturing of shock waves having a large angle with respect to the coordinate direction of the underlying mesh is imprecise and leads to smearing of the wave over several mesh cells. Shocks waves can be more accurately captured when the grid is aligned to them. Mesh adaptation can be used to improve the shock capturing properties in this way. The meshes are automatically adapted to the converging solution by local enrichment, barycentric smoothing and directional stretching to align the grid lines across strong gradients. This results in a Multi-level scheme in space. An optimum number of auto-refined remeshings is found to be between 3 to 6 levels. The solution is P1-interpolated between different grids.

The parameters controlling the mesh adaption is based on the finite element criterion of the equidistribution of the local error via local enrichment, (reduction of h), [21]. For the P1 approximation considered here this reads

$$| U - U_h | \leq \varepsilon (h^2), \text{ where } | \lozenge | \text{ represents the } H_1 \text{ semi-norm.}$$

From the continuity equation, this can be estimated by the gradient of the density variable $\nabla \rho_h$. The directional stretching is performed by a barycentric spring analogy, [21], for which the stretching force is proportional to the gradient of the density. From discontinuous solutions of the continuity equation, $\rho_t + (\rho u)_x = 0$, changes in gradient of the density approximates the error with directionality,

$$\delta (\nabla \rho) \approx \varepsilon (h^2) \frac{\partial^2 \rho}{\partial x_i\, \partial x_j} \; .$$

This strategy is integrated into the solution procedure giving the global Multi-Level Scheme as discussed above.

4.2. RESULTS.

First a Mach 3 inviscid flow over a 24° compression ramp is simulated to test directional stretching to align the grid to the shock. The following series of refinements plus stretching is obtained yielding almost the correct analytical solution, see Figure 3.

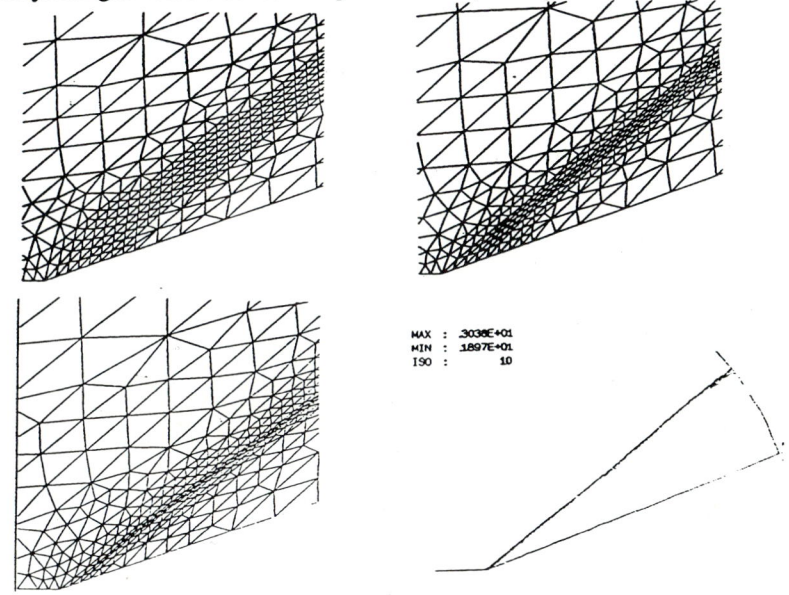

Figures 3: Grid alignment capacities.

Next a standard supersonic (Mach 2) wedge configuration is simulated to assess the advantages of combining local enrichment and grid alignment over local refinements.

A first series of calculations consists of only creating the multi-levels by local refinement. Altogether, six levels are created, $T_1 \subset T_2 \subset ... \subset T_6$, the final triangulation having 5769 nodes and 11322 triangles. Total CPU time to obtain a residual of $10^{-8} = 235$ on one processor of a Cray YMP, with aperformance of 118 Mflops. The results are given by Figures 4-5.

The second calculation is performed on a series of grids combining local refinement and grid alignment by stretching. The final level contains 5777 nodes and 11340 elements. The results are even more sharp than for the first series, as illustrated below by Figures 6-7.

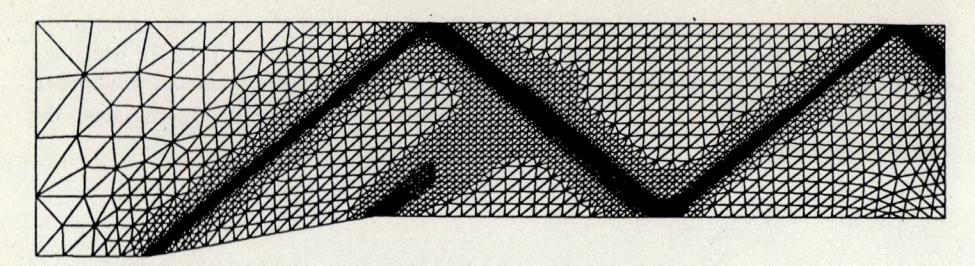

Figure 4: Final level grid for supersonic wedge, local enrichment only.

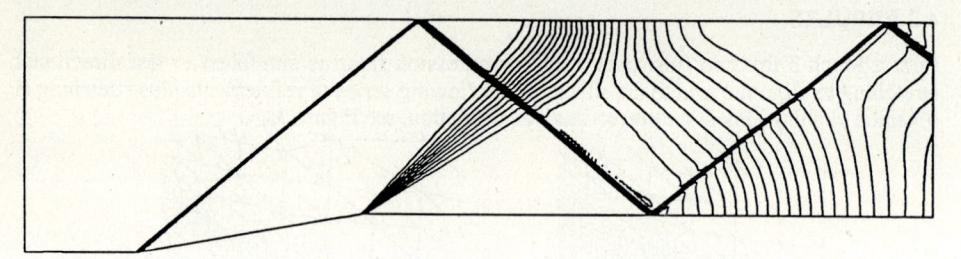

Figure 5: Iso Cp lines for grid system constructed using local enrichment only.

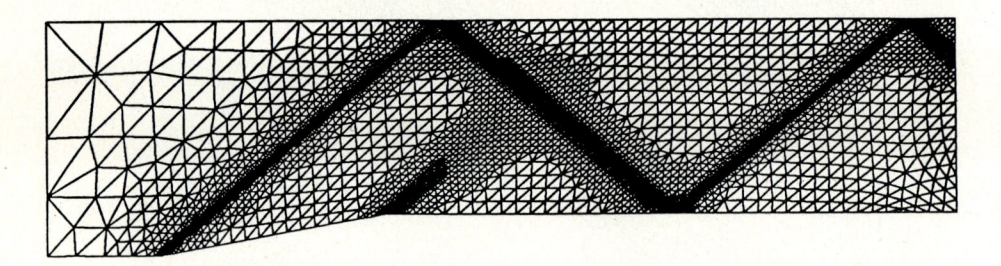

Figure 6: Final level grid for supersonic wedge, (stretching and refinement).

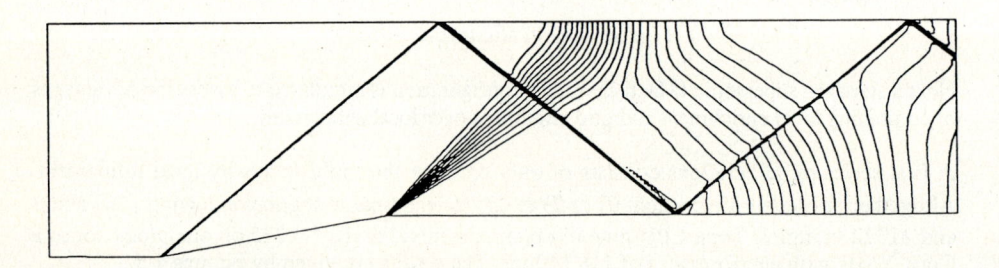

Figure 7: Iso Cp lines for grid system of Figure 6.

230

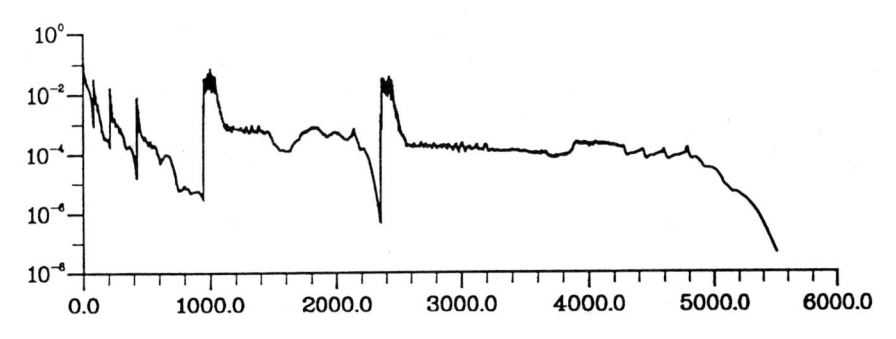

Figure 8: Convergence History for Multi-Level calculation

The cost of the second calculation for the same residual as before took 390 secs. An equivalent solution on a structured grid necessitated 10 times more points for the same accuracy, taking about nine times more CPU time than the unstructured solution. The following Figure shows a blow-up of the cross section along the middle line of the wedge around the first shock,

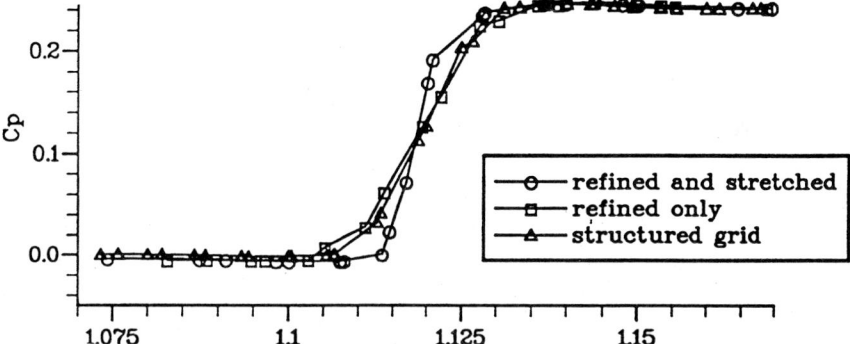

As can be clearly seen, the refined and stretched grid system gave the sharpest resolution. Also, on inspection of the iso-lines, (Figures 5 et 7), the expansion fan and the shock reflexions are captured optimally.

5. NAVIER STOKES CALCULATIONS

The three schemes discussed in section 3 are now applied to hypersonic compression ramp test cases. In this case the Navier-Stokes solver should be able to capture precisely not only the internal and external shocks, and contact discontinuities, but also shock-shock interactions, and boundary layer - shock interactions. Moreover an accurate description of the boundary layer properties, and separation / recirculation phenomena should be obtained. All these features should, of course, be resolved without excessive numerical dissipation in order to represent as close as possible the real viscous effects within the flow field.

5.1. SCHEMES

The above schemes give the numerical flux functions for the convective fluxes, the viscous terms are discretised by a P1 Galerkin approximation on the triangular elements. For the RK5 scheme, the dissipation is based on characteristic scaling, combined with a sensor, which for these ramp flows is chosen to be taken over the pressure variable (3.6), as a high entropy gradient is present near the wall boundaries, (see Figure 11). The other two schemes ICTVD and IUARS are unfactorised implicit schemes in time, where the full Navier-Stokes matrix has to be inverted. The automatic mesh adaption scheme during the calculation is more difficult than in the inviscid case; indeed, the leadind edge shock moves in towards the ramp during the time integration process, which requires a refinement /derefinement algorithm to render such a strategy efficient. As a consequence, only static mesh adaption by local enrichment is performed.

5.2. RESULTS

The two test cases considered are chosen because of the existence of experimental data, and results of other calculations in the literature. The first one corresponds to a high Reynold's number flow, where separation is present, and a secondary separation shock is visible. There is a strong reattachement zone. The second corresponds to a more viscous flow field, with a complex leading edge shock impinging on the corner shock giving a complex flow field downstream. The main features of such flows are illustrated by the following Figure.

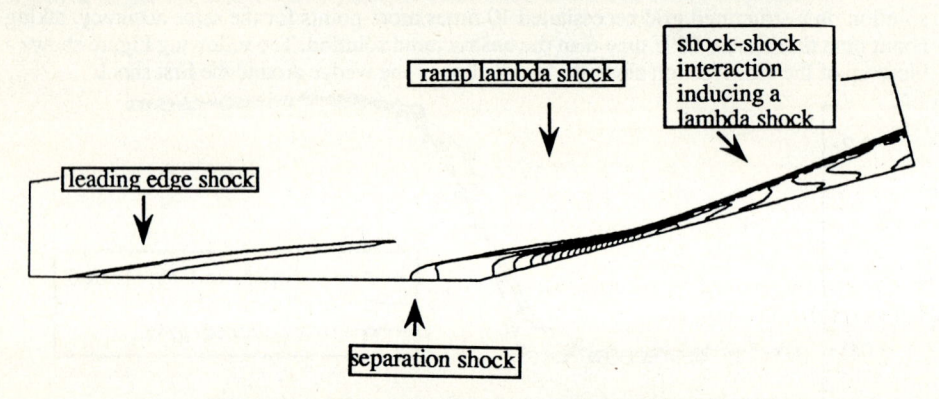

Iso Cp lines, Mach 11.68 test case (see below), RK5

The upstream flow conditions are: for the first case, wedge angle $\beta = 15°$; Mach $=11.68$; $Re_{/m} = 5.58.10^5$; $T_{inf} = 64.44°K$; $P_{inf} = 19.763$ Pa.; $T_{wall} = 296.67°K$. The length of the flat plate in front of the corner = 0.442 m, and the total length of the ramp is taken to be 0.9012 m in order to study the flow field after reattachement. For the second case, the upstream flow conditions are: again the wedge angle $\beta = 15°$, but now Mach $=14.1$; $Re_{/m} = 2.37 \ 10^5$; $T_{inf} = 72.2°K$; $P_{inf} = 19.$ Pa.; $T_{wall} = 296.67°K$. The length of the flat plate in front of the corner = 0.439 m, and the total length of the ramp is taken to be 0.88 m. There is no longer a strong separation in front of the ramp corner, however there is a strong interaction of the leading edge shock of the flat plate with the ramp's corner shock, giving a characteristic expansion at their intersection as is clearly illustrated by the Figures (15), (16).

The evaluation of the different schemes for these applications showed similar results for the representation of the flow field. The RK5 scheme proved to be very precise in the separation shock region for the Mach 11.68 case, but more oscillatory than the implicit TVD schemes in the ramp shock-boundary layer interaction regions. For the second test case (Mach number 14.1), the qualititative study of the simulated flow fields by the three schemes gave similar results, the upwind approximate Riemann solver being more dissipative, and smeared out at the rear expansion fan. The results are compared with the experimental data, and a reasonable agreement was found, see Figures (14),(17). For the Mach 11.68 case the TVD schemes underestimated the separation extent, but gave closer agreement in the reattachement region, whereas the RK5 scheme seemed to suffer from unadapted numerical dissipation in this region.

EFFICIENCY AND CONVERGENCE

All codes are optimised by enhancing the vectorisation using colouring techniques, and data reorganisation [16]. Auto-tasking on a two processor CRAY YMP was efficient upto 1.89 processor, and 3.23 on a four processor CRAY 2. The performance of all codes on a single processor of a YMP were 120Mflops.

TEST CASE	MACH 11.1, $\beta = 15°$	MACH 14.1, $\beta = 15°$
Number of mesh points	12085	13445
RK5	CPU = 4500 secs for convergence at 9500 iterations	CPU = 3572 secs for convergence at 8000 iterations
IUARS	CPU = 1766 secs for 2500 iterations	CPU = 1199 secs for 2500 iterations
ICTVD	CPU = 1452 secs for 2500 iterations	CPU = 1095 secs for 2500 iterations

Typical convergence histories are given below for the two types of schemes RK5 + mesh adaption (two-level static adaption), and the implicit schemes.

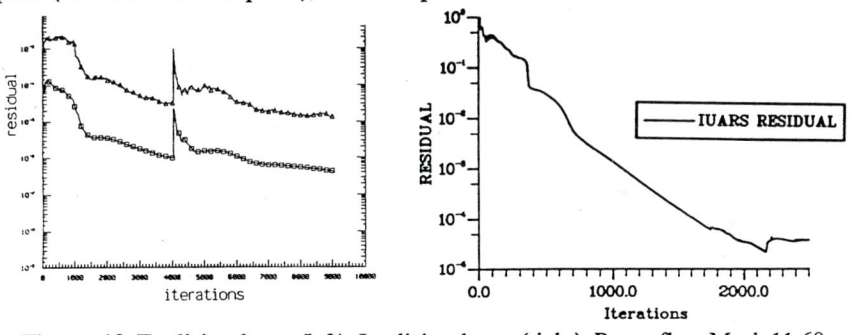

Figures 10. Explicit scheme (left), Implicit scheme (right), Ramp flow Mach 11.68

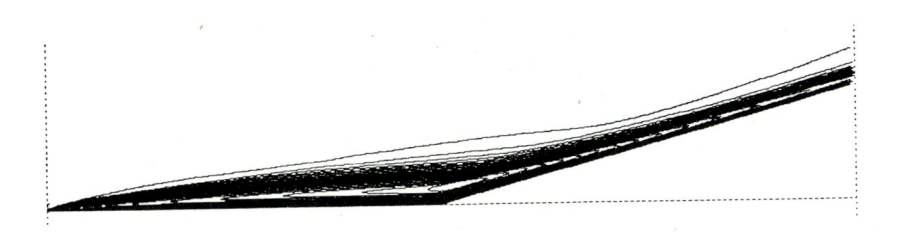

Figure 11. Entropy contours for Mach 14.1 flow, illustrating high entropy gradient within boundary layer.

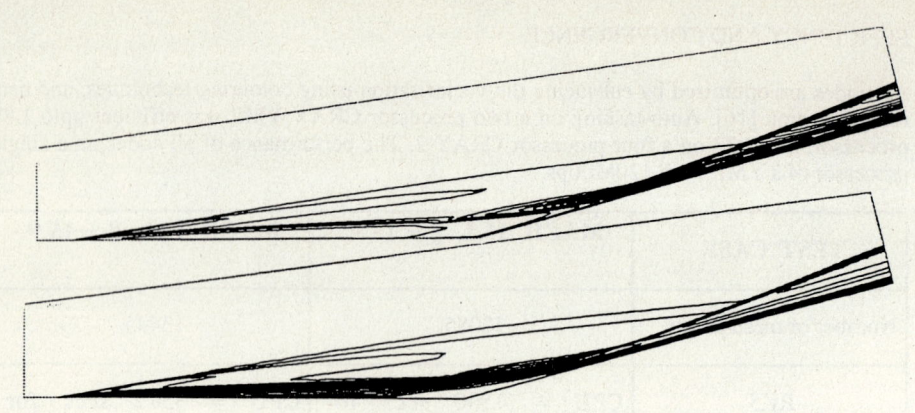

Figures 12. Mach 11.68 test case; Iso-density lines (top), Mach isolines (bottom), RK5 scheme.

Figures 13. Mach 11.68 test case; Iso-Cp lines , ICTVD scheme.

Figures 14. Mach 11.68 test case; CF Number at wall.
Comparaison RK5, ICTVD, IUARS, Holden & Moselle

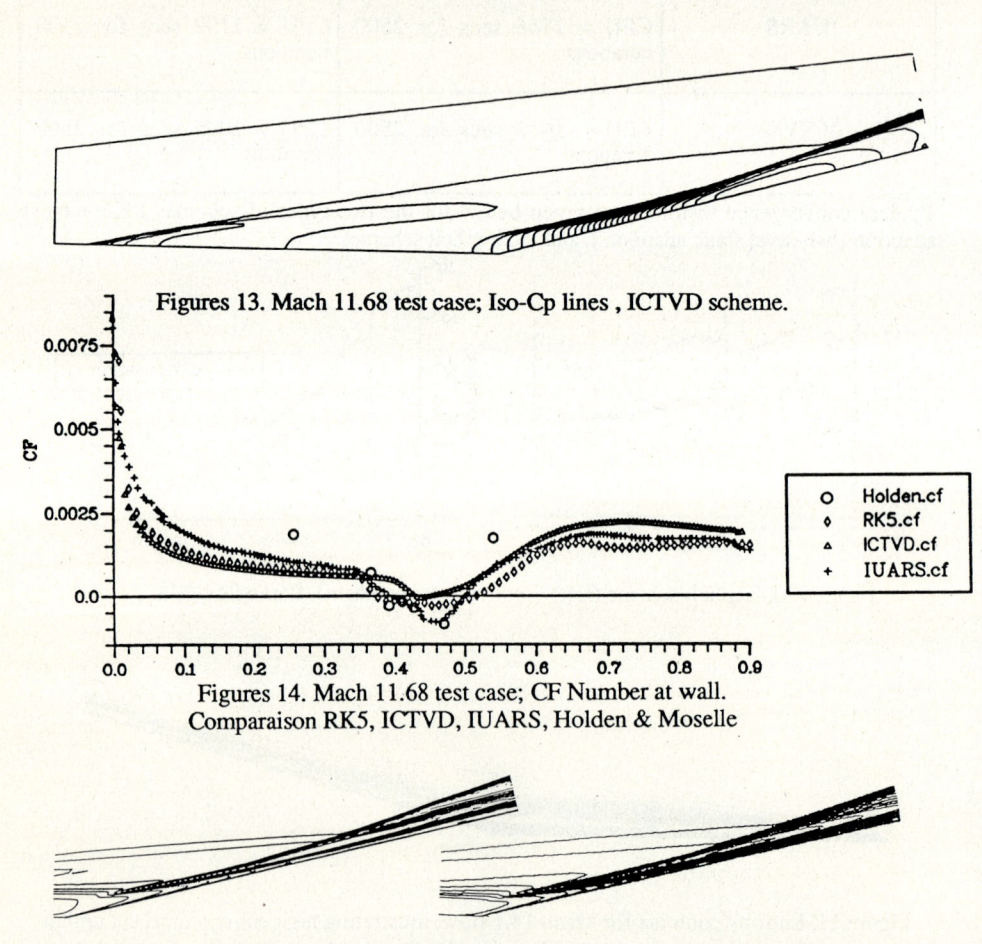

Figures 15. Mach 14.1 test case; Iso-Density lines , RK5 scheme.(left), ICTVD(right)

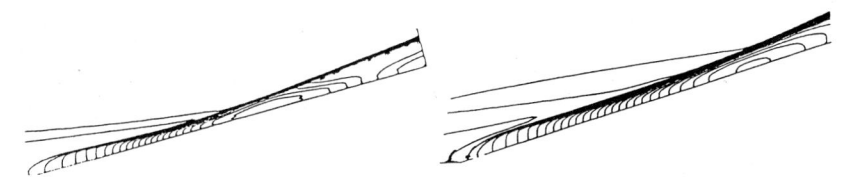

Figures 16. Mach 14.1 test case; Iso-Cp lines, RK5 scheme.(left), IUARS (right)

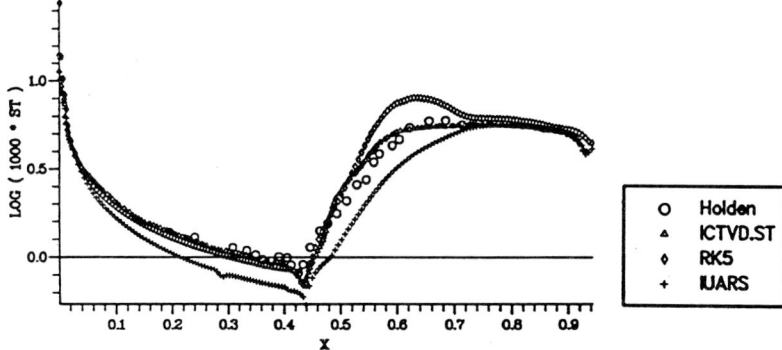

Figures 17. Mach 14.1 test case; Wall Stanton Number, Comparaison RK5, ICTVD, IUARS, Holden & Moselle

6. CONCLUSIONS

Three different codes using either imbedded numerical or explicit artificial dissipation have been applied to the calculation of the flow over wedges and compression ramps. It has been shown that via careful tuning of the numerical scheme, second order accuracy at steady state has been obtained, yielding precise solutions compatible with complex shock-shock formations and interactions. Combining Mesh adaptivity and Multi-Level Algorithms renders the explicit centred scheme competitive to the faster converging Implicit Schemes based on TVD limiting.

ACKNOWLEDGEMENTS

This work was undertaken at the Institut de Machines Hydrauliques et de Mécanique des Fluides at the Ecole Polytechnique Fédérale de Lausanne within Professor Ryhming's CFD group. The authors would like to thank Prof.I.L.Ryhming for his permanent interest and valuable comments on this work, and for providing the stimulating environment amidst the hypersonics group at the Institute. Discussions with Helen Yee have been extremely profitable, and the authors would also like to thank her for her interest. The calculations were performed on the CRAY 2 at EPFL, and on the CRAY YMP at the ETHZ at Zürich.

7. REFERENCES

[1] Holden, M.S., Moselle, J.R. Theoretical and Experimental Studies of the Shock Wave - Boundary Layer Interaction on Compression Surfaces in Hypersonic Flow. ARL 70-0002, Jan 1970, & CALSPAN Pept. AF-2410-A-1, Buffalo, NY, (October 1969).

[2] Holden, M.S., A study of Flow Separation in Regions of Shock-Wave-Boundary Layer Interactions in Hypersonic Low. AIAA 78-1169, July 1978.
[3] Dervieux, A. VKI Lecture Series, (1984).
[4] Stoufflet, B., Periaux, J. Fezoui, F.,Dervieux, A. Numerical Simulation of 3D Hypersonic Euler Flows Around Space Vehicles using adapted finite elements. AIAA -87-0560.
[5] Périaux, J. Finite Element Simulations of Three-Dimensional Hypersonic Reacting Flows around Hermes, Second joint Europe / US short course in Hypersonics, Colorado Springs, (January 1989).
[6] Baba, K., Tabata, M., On a conservative upwind finite element scheme for convective diffusion equations, RAIRO Numerical Analysis, Vol. 15, 3-25, 1981.
[7] Mallet, M., Periaux, J, Perrier, P, Stoufflet, B. Flow modelisation and computational methodologies for the aerothermal design of hypersonic vehicles: Application to the European Hermes; AIAA 88-2628.
[8] A. Jameson , T.J.Baker, and N.P.Weatherill. Calculation of inviscid transsonic flow over a complete aircraft. AIAA 86-0103. January 1986.
[9] Proceedings of Workshop on Hypersonic Flows for Reentry Problems. Part II. (to appear)
[10] Lax, P., Weak solutions of nonlinear hyperbolic equations and their numerical computation. Comm.Pure.Appl. Math., 7. 159-193. (1954).
[11] Oleinik, O., Discontinuous solutions of nonlinear differential equations. Usp.Mat.Nauk., 12. 3-73, (1957).
[12] Lerat. A. Difference methods for hyperbolic problems with emphasis on space-centred approximations. Von Karman Institute Lecture Series 1990-03.
[13] Yee H. A Class of High Resolution Explicit and Implicit Shock Capturing Methods. Nasa Tech. Mem.no.101088. Feb. 1989.
[14] Leyland, P., Vos, J.B., Real Gas Calculations for Viscous Hypersonic Flows on 2D unstructured meshes. Proc.First European Symposium on Aerothermodynamics for Space Vehicules.Estec, May 1991. Publ.ESA SP-318.
[15] Richter, R., Leyland, P., Mesh Adaptaion for 2D transsonic Euler Flows on Unstructured Meshes. in Numerical Grid Generation in CFD, ed. Arcilla et al. Elsevier Sc.Pub.(North Holland), 1991.
[16] Angrand, F, J.Erhel, Leyland, P. Fully vectorised implicit scheme for 2-D viscous hypersonic flow using adaptive finite element methods.7th GAMM International Conference on Computing methods, ed.Computational Mechanics.(1989).
[17] Swansen, R.C., Turkel, E., On Central-Difference and Upwind Schemes. ICASE Report No.90-44 (June 1990).
[18] Martinelli L. Thesis. Princeton University. (October 1987).
[19] Harten, A. High Resolution Schemes for Hyperbolic Conservation Laws. Journ. Comp. Phys. 49.357-393 (1983)..
[20] Arminjon, P., Dervieux, A. Construction of TVD-like Artificial Viscosities on 2-Dimensional arbitrary FEM Grids. RR-INRIA no. 1111.October 1989.
[21] Peraire, J., Vahdati, M., Morgan K., Zienkiewicz, O.C., Adaptive Remeshing for Compressible Flow Computations. J.Comp.Phys., Vol 71, 449-466 (1987).

A simple triangular element for the compressible Navier-Stokes equations

Y. Secretan* H. Thomann [†]

Institute for Fluid Dynamics, ETH Zürich
Sonneggstr. 3, 8092 Zürich, Switzerland

Summary

We have developed a new, stable and simple element to integrate the compressible Navier-Stokes equations by the finite element method. This element is well suited for grid refinement. The non-linear system is solved by an iterative method. Coupled with a new diagonal preconditioning, this leads to an efficient algorithm to solve the system. The element converges quadratically for all the variables and has been tested against results of the GAMM Workshop on Compressible Navier-Stokes Flows, on regular grids as well as on adapted grids.

Introduction

The application of the Finite Element Method (FEM) to CFD problems enjoys more and more attention, owing to the ease with which complex geometries can be represented and with which boundary conditions can be enforced. The element by element construction of the discrete system allows the use of unstructured grids and of techniques to dynamically adapt the grid to the solution. In order to solve large problems, we are looking for resolution schemes which do not require matrices, but which are stable and fast. We also favor very simple triangular elements for both grid refinement and computer efficiency. We have applied the FEM to the laminar compressible viscous flow of a newtonian perfect gas.

Mathematical model

We write the compressible Navier-Stokes equations for the primitive variables (ρ, **v**, T) and a perfect gas in a dimensionless form as

$$\mathbf{v} \cdot \nabla \rho + \rho \, \nabla \cdot \mathbf{v} = 0 \tag{1.a}$$

$$\rho \mathbf{v} \cdot \nabla \mathbf{v} + \nabla p - \frac{1}{Re} \nabla \cdot \boldsymbol{\tau} = \mathbf{f} \tag{1.b}$$

$$\left(\frac{c_p - R}{E}\right) \rho \mathbf{v} \cdot \nabla T + p \, \nabla \cdot \mathbf{v} - \left(\frac{1}{Re \, Pr \, E}\right) \nabla \cdot (\lambda \, \nabla T)) - \frac{1}{Re} \boldsymbol{\tau} : \nabla \mathbf{v} = q \tag{1.c}$$

$$p = \frac{R}{E} \rho T \tag{1.d}$$

with

---

*Research Assistant
[†]Professor

$$Re = \frac{\rho_0 v_0 L_0}{\mu_0} \qquad Pr = \frac{\mu_0 c_{p0}}{\lambda_0} \qquad E = (\gamma - 1)Ma_0^2 \qquad R = \frac{\gamma - 1}{\gamma}$$

and where the subscript "0" denotes reference values. The viscosity coefficient μ and the heat transfer coefficient λ, are kept constant to allow for a comparison with the results of the GAMM Workshop of Nice (1985) [3]. The Prandtl number Pr is fixed at $Pr = 0.72$ and the ratio of specific heat of the fluid γ at $\gamma = 1.4$.

The variational formulation underlying our finite element formulation is obtained in a classical manner by the weighted residual method and Galerkin weighting functions. After integration by part of all the second order derivatives, one obtains a weak form which is the finite element equivalent to a central discretisation.

Geometrical discretisation

In the finite element method, the computing domain Ω_h is decomposed into a finite number of sub-domains Ω_e of simple geometrical shape – the finite elements. The weak form over the domain is then replaced with the sum over the elements of the elementary weak form:

$$\int_{\Omega_h} d\Omega_h = \sum_{\text{elements}} \int_{\Omega_e} d\Omega_e .$$

Elements with the same geometrical form and the same degree of geometrical approximation are images of the same element, the reference element, through a unique geometric transformation τ.

For computer efficiency, we selected first order Lagrange reference elements. The Jacobian matrix of the transformation τ is then constant over an element. This permits the analytical integration of the terms of the weak form and suppresses the loop over the quadrature points, leading to a better vectorization of the code.

The only elements fulfilling this condition are the linear triangle in 2D (Figure 1) and the linear tetrahedron in 3D. With these elements, it is possible to describe every geometrical shape and to have only one type of reference element, which tremendously simplifies the program structure. Moreover, triangles are well suited for dynamical mesh refinement.

Approximation functions

For the approximations, the weak form of (1) imposes a global continuity of type C^0 (actually H^1 but for all practical purposes C^0). In order to compute the derivatives of the weak form, the approximations on the reference element for the different variables must be complete at least to degree 1.

The approximations have to be chosen carefully [12]. Equal approximations lead to a checker-board pattern, reminiscent of the stability problems encountered for incompressible viscous flows. We thus selected approximations fulfilling the Babuška-Brezzi condition and settled on the MINI element, proposed by Arnold et al. [1] for the incompressible Stokes and Navier-Stokes equations. Adapted to compressible flows [11], this element has a linear basis for the density and the temperature. For the velocity components, the linear basis is enriched with a bubble function at the barycenter of the element (Figure 1). As the degrees of freedom of the bubble are not dependant on any

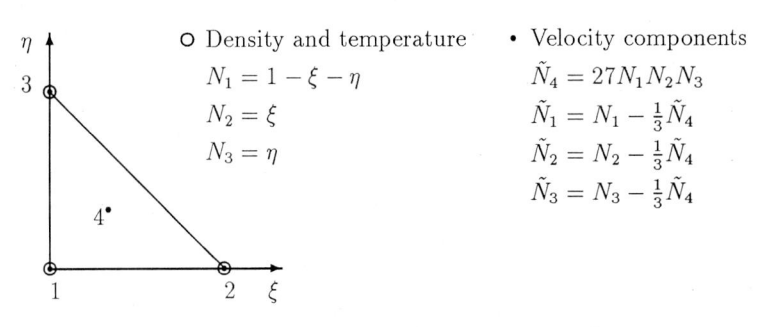

Figure 1: Reference element and approximation functions

other elements, they will be statically condensed.

For the compressible Stokes equations and after static condensation, on one side one obtains the discretisation of the fully linear element (i.e. without the bubble) and on the other side, a supplementary term. The effect of the bubble can be traced down to a dissipation operator acting on the pressure in the continuity equation and which has the form

$$\nabla \cdot (\boldsymbol{\kappa} \cdot \nabla p) \tag{2}$$

where $\boldsymbol{\kappa}$ is a symmetric tensor of the order

$$Re D \rho$$

and D the determinant of the Jacobian matrix. The bubble re-equilibrates the discrete sytem of equations, the continuity being the only equation without dissipation or diffusion.

For the Navier-Stokes equations, the convection also produces a term which can be associated with an anti-diffusive operator. But we believe, and extensive tests sustain this affirmation, that the resulting part of the discrete sytem associated with the continuity equation is badly conditioned, leading to poor convergence of the iterative methods.

However, the effect of the anti-diffusive operator can be simulated by relaxing (2) with

$$\omega \nabla \cdot (\boldsymbol{\kappa} \cdot \nabla p) \qquad \omega \in [0,1] \tag{3}$$

where the relaxation parameter ω varies linearly with the local pressure p as

$$\omega = \omega_0 + (1-\omega_0)\frac{p-p_0}{p_{max}-p_0} \qquad \omega \in [0,1] \tag{4}$$

p_0 being the reference pressure and where p_{max} is given by the isentropic relation

$$p_{max} = \frac{1}{\gamma Ma_0^2}\left(1+\frac{(\gamma-1)Ma_0^2}{2}\right)^{\frac{\gamma}{\gamma-1}}. \tag{5}$$

The remaining adjustable parameter ω_0 typically takes the value 0.3 for transonic flows. This element does not produce spurious pressure or density oscillations. It is very simple and inexpensive in terms of CPU time.

Resolution

The non-linear system resulting from the discretisation

$$K u = f \tag{6}$$

is solved by an inexact quasi-Newton-Raphson algorithm. Each linear sub-system [6]

$$Kt \, \Delta u = f - K u \tag{7}$$

is in turn solved by a preconditioned iterative method (GMRES [10]), where the product of the tangent matrix Kt times the unknown vector Δu is evaluated numerically [5,4]. GMRES is a Krylov subspace method and the dimension of the subspace influences the convergence. We generally used 25 vectors, but restarted the resolution up to 10 times to achieve a better convergence within a Newton-Raphson step. To initiate the resolution, we used non-preconditioned and partially converged Newton-Raphson steps. The resolution algorithm is then

1. 1 Newton-Raphson step without preconditioning but with 10 restarts;
2. 10 Newton-Raphson steps with preconditioning but without restart;
3. until convergence perform:
   1 Newton-Raphson step with preconditioning and with 10 restarts

The preconditioning we have developed is based on the matrix stability analysis for the linear sub-system. The error growth is given by

$$e_n = (I - P \, Kt)^n e_0 \tag{8}$$

where P is the pre-conditioning matrix and Kt the tangent matrix. The error e stays bounded if the condition

$$\|I - P \, Kt\| \leq 1 \qquad \forall \, \| \cdot \| \tag{9}$$

is fulfilled. Deliberately choosing for P a diagonal matrix, and for the matrix norm the infinity norm defined as

$$\|A\|_\infty = max_i \sum_j |a_{ij}| \tag{10}$$

and applying condition (9) to each row, we obtain for p_{ii}, the components of P,

$$p_{ii} \leq \frac{2}{\sum_j |kt_{ij}|} \, . \tag{11}$$

P is then a diagonal approximation of Kt^{-1} and is related to the local time step techniques used for explicit time marching schemes.

Convergence of the approximation

The convergence rate of the discrete system in function of the grid size has been controlled on regular grids of sizes ranging from 3×3 nodes up to 21×21 nodes. The exact solution used is

$$\rho = e^{-x-y} \qquad u = e^x \qquad v = e^y \qquad t = e^{x+y} \qquad x,y \in [0,1]$$

and generates a right hand side term in (1) which is incorporated in the discrete system as a source term. The boundary conditions are of Dirichlet type on the entire boundary and for all the variables.

The figure on the right shows in a log-log diagram the curves of the L_2 norme of the error against the grid size h, where p is the exponent of the expression

$$\| e \|_0 \leq c\, h^p \qquad (12)$$

which relates the global error to the grid size (c is independant of h). The convergence rate is quadratic for all the variables.

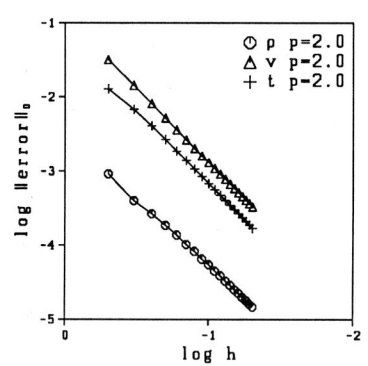

Error estimate

One strength of the FEM is the ability to deal with unstructured meshes, due to the element by element construction of the discrete system and to the local transformation between the real element and the reference element. This can be exploited to dynamically refine the grid in an attempt to distribute the approximation error more evenly on the grid. The approximation error, which is a local value and is a function of the grid size, has to be estimated with respect to known values [2]. We use an a priori error estimate based on the second order term of the Taylor serie extension of the function, and which is related to the approximation error for the linear element [6].

Introducing the symmetric tensor ε

$$\varepsilon = \frac{1}{2} \begin{bmatrix} \dfrac{\partial^2 \Phi}{\partial x^2}\Delta x^2 & \dfrac{\partial^2 \Phi}{\partial x \partial y}\Delta x \Delta y \\ \dfrac{\partial^2 \Phi}{\partial x \partial y}\Delta x \Delta y & \dfrac{\partial^2 \Phi}{\partial y^2}\Delta y^2 \end{bmatrix} \qquad (13)$$

we define the nodal error e as the norm of ε

$$e = \|\varepsilon\| = \sqrt{\sum \varepsilon_{ij}^2}\; . \qquad (14)$$

The error is thus based on second derivatives which are not directly available with a linear approximation. Since in the weak form the second derivatives are integrated by part, we also integrate ε by part and obtain for ε_{ij} the definition

$$\varepsilon_{ij} = \frac{1}{2}\, \frac{\int_\Omega N_{,xi}^T\, N_{,xj}\, \Phi\, d\Omega}{\int_\Omega N_{,xi}^T d\Omega\; \int_\Omega N_{,xj}^T d\Omega}\, \int_\Omega N^T d\Omega \qquad (15)$$

where N is the vector of the approximation functions (Figure 1), Φ the vector of the unknown used to compute the error and $,xi$ denotes differentiation with respect to x_i. The vector operations are done on a component basis.

For a multi-variable problem, one has to decide which variable to use. We included in our code the option of working on product of variables (for example the pressure) as well as on more than one variable at a time; in this case, the resulting error is the arithmetic mean of the errors.

Mesh refinement

The mesh refinement algorithm is straightforward [9,7]. The elements with all nodes having an error bigger than $0.2 \times e_{max}$ will have their sides halved, generating 4 triangles. The hanging nodes are captured with a bi-section of the neighbouring element (Figure 2). To eliminate the succesive bi-section of an angle, we follow the history of the refinement;

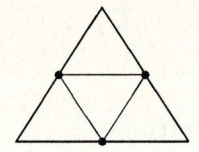

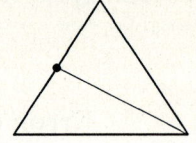

Figure 2: Quad-section and bi-section of a triangle

finally the grid is smoothed. The complete grid refinement algorithm is then:

1. Compute the error in each node (15);
2. Eliminate the bi-section of the previous refinement. This is done to prevent succesive bi-section of the same original angle; it obliges to follow the refinement history;
3. Divide in 4 the elements with all the nodal errors $e > 0.2 \times e_{max}$;
4. Loop until the number of created nodes is stable:
   - Divide in 4 the elements having one or several nodes with more than one degree of refinement of difference;
   - Divide in 4 the elements around a node with more than 9 neighbouring elements;
   - Divide in 4 the elements with 2 hanging nodes;
5. Divide in 2 the elements with one hanging node;
6. Move the new boundary nodes to the boundary. This point is dependant on the geometry of the problem;
7. Smooth the resulting grid by recentering a node with respect to the surrounding nodes;
8. Interpolate the values for the new nodes and set the new boundary conditions.

Comparison for the A5 case

The new element has been validated with the results of the GAMM Workshop on Compressible Navier-Stokes Flows hold in Nice in 1985 [3]. We present here a comparison for the case A5. This transonic flow over a NACA-0012 airfoil at 0° angle of attack,

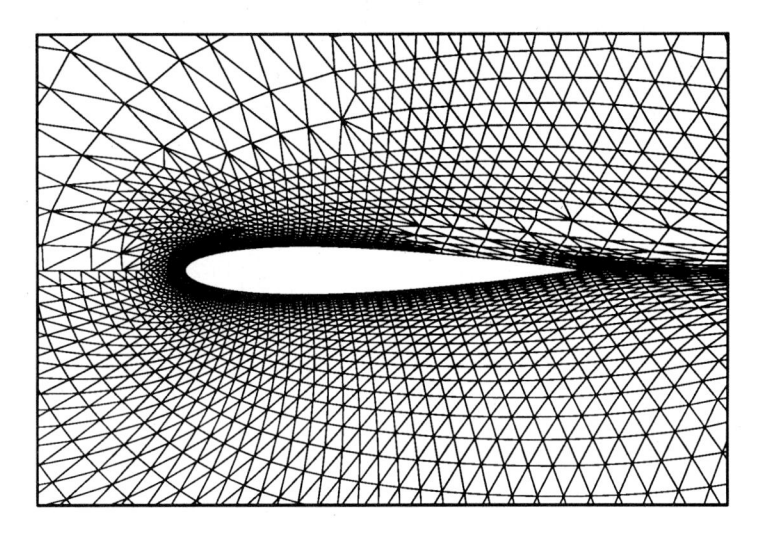

Figure 3: Case A5 — Top: Adapted grid – Bottom: Grid 121x41

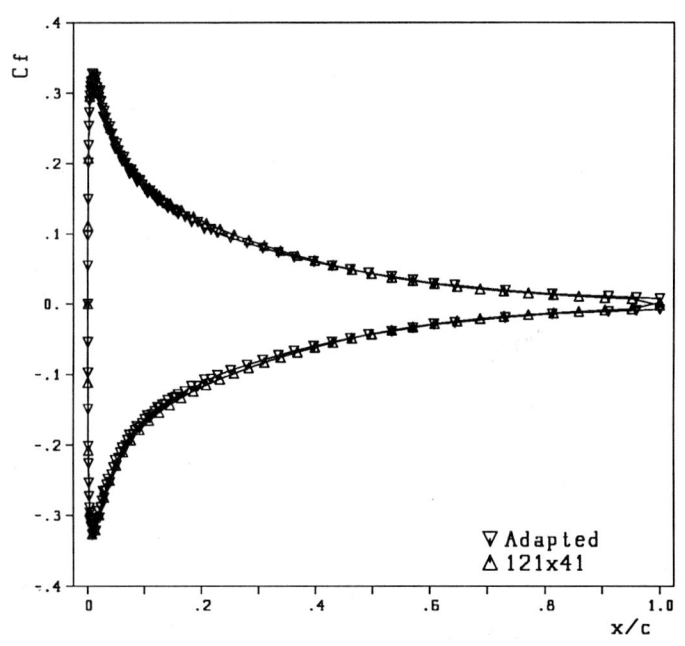

Figure 4: Cas-A5 – Friction coefficient Cf

defined by $Re = 500$, $Ma = 0.85$, $\gamma = 1.4$ and $Pr = 0.72$, has been solved on a structured grid of 121×41 nodes and on an adapted grid. Starting from a coarse grid of 31×11 nodes, we do four iterations of grid refinement with the error computed on all the variables. Figure 3 shows in the upper part, the adapted grid and in the lower part, the 121×41 grid. Figure 4 shows the very good agreement between both Cf curves. It also underlines the work of the grid refinement algorithm, above all in the leading edge region where the adapted grid is finer than the regular one. Due to the symmetry of the flow and of the grid, the momentum coefficient C_M and the lift coefficient C_L are null for both computations. Table 1 summarizes the results for both drag coefficients C_D and CPU-Time together with two results of the Workshop; the agreement is excellent. The most impressive difference is in the CPU-Time. For a difference of only 2.7% in C_D, the adapted computation required less than half the CPU-time.

Table 1: GAMM Workshop – Case A5

	Number of nodes	C_D	CPU-Time (s) Cray-YMP	$\|\cdot\|_0$
Grid 121x41	4940	0.223	115	$2.2 \; 10^{-7}$
Adapted Grid	3054	0.217	49	$7.2 \; 10^{-8}$
Secretan et al. [13]	5152	0.218	–	–
Müller et al. [8]	10595	0.220	–	–

Supersonic Flow

The grid refinement technique allows to adapt the grid and so to improve the solution. It is not limited to transonic flows. We applied this techique to the supersonic flow over a NACA-0012 airfoil at 10° angle of attack. The flow is defined by $Re = 1\,000$, $Ma = 2.0$, $\gamma = 1.4$ and $Pr = 0.72$. Starting with a regular grid of 61×21 nodes, we made first a global refinement, before applying four iterations of grid refinement. The error is computed, as in the previous case, on all the variables.

Figure 5 shows the resulting grid which counts 11 082 nodes and which is well adapted to the flow. Figure 6 shows the isolines of the dimensionless pressure; the distance

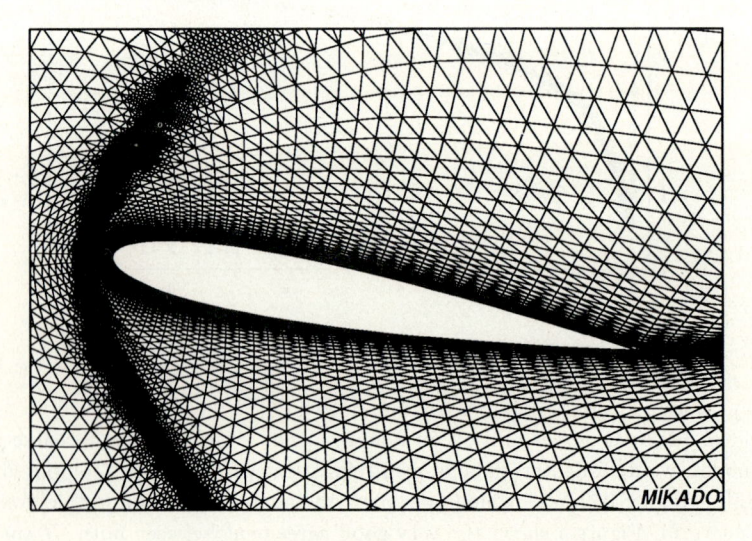

Figure 5: Supersonic flow – Adapted grid

between two isolines is 0.25, the maximum 5.82 and the minumum 0.59. The resolution of the shock is very good. This simulation required 310 s and 1.62 MWords on a Cray-YMP. The convergence in the $\|\cdot\|_0$ norme is of $1.3 \, 10^{-7}$. Over the four refinements and due to the shock, the maximum error could only be halved.

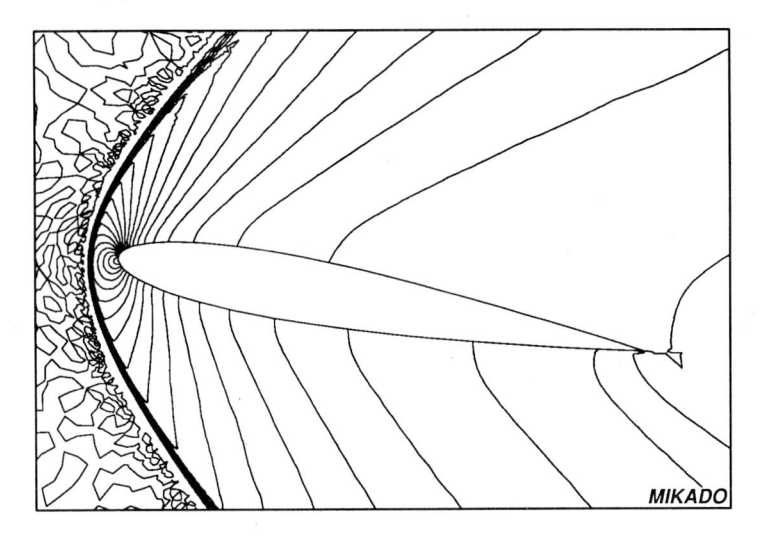

Figure 6: Supersonic flow – Iso-pressure

Conclusion

The element we propose is simple and efficient. Coupled with the grid refinement algorithm, it leads to a flexible scheme to integrate the compressible Navier-Stokes equations. GMRES proved to be robust and the diagonal preconditioning is well adapted for the purpose.

References

[1] D.N. Arnold, F. Brezzi, and M. Fortin. A stable finite element for the Stokes equations. *Calcolo*, 21, 1984.

[2] I. Babuška, O.C. Zienkiewicz, J. Gago, and E.R. de A. Oliveira, editors. *Accuracy estimates and adaptative refinements in finite element computations*. John Whiley & Sons, 1986.

[3] M.O. Bristeau, R. Glowinski, J. Periaux, and H. Viviand, editors. *Numerical Simulation of Compressible Navier-Stokes Flows*. Volume 18 of *Notes on Numerical Fluid Mechanics*, Vieweg, 1987. Proceedings of the GAMM-Workshop (Nice 1985).

[4] P. N. Brown. A local convergence theory for combined inexact-Newton/finite-difference projection methods. *SIAM J. Numer. Anal.*, 24:410–434, 1987.

[5] T.F. Chan and K.R. Jackson. Nonlineary preconditionned Krylov subspace methods for discrete Newton algorithms. *SIAM J. Sci. Stat. Comput.*, 5:533–542, 1984.

[6] G. Dhatt and G. Touzot. *Une présentation de la méthode des éléments finis. Collection Université de Compiègne*, Maloine S.A., deuxième édition, 1984.

[7] R. Löhner, K. Morgan, and O. C. Zienkiewicz. Adaptative grid refinement for the compressible Euler equations. In *Accuracy Estimates and Adaptive Refinements in Finite Element Computations*, pages 281–297, John Whiley & Sons, 1986.

[8] B. Müller, T. Berglind, and A. Rizzi. Implicit central difference simulation of compressible Navier-Stokes flow over a NACA-0012 airfoil. In *Numerical Simulation of Compressible Navier-Stokes Flows*, pages 183–200, Vieweg, 1987.

[9] B. Palmerio and A. Dervieux. Application of a FEM moving node adaptive method to accurate chock capuring. In *Numerical Grid Generation in CFD*, pages 425–436, 1986.

[10] Y. Saad and M. Schultz. GMRES: a generalized minimal residual algorithm for solving nonsymetric linear systems. *SIAM J. Sci. Stat. Comput.*, 7:856–869, 1986.

[11] Y. Secretan. *Contribution à la résolution des équations de Navier-Stokes compressibles par la méthode des éléments finis adaptatifs; Développement d'un élément simple*. PhD thesis, Eidgenössische Technische Hochschule, Zürich, Switzerland, 1991.

[12] Y. Secretan. *Modélisation par éléments finis des équations de Navier-Stokes compressibles appliquées aux écoulements externes*. Master's thesis, Université Laval, Québec, Canada, 1986.

[13] Y. Secretan, G. Dhatt, and D. Nguyen. Compressible viscous flow around a NACA-0012 airfoil. In *Numerical Simulation of Compressible Navier-Stokes Flows*, pages 219–236, Vieweg, 1987.

SEMIIMPLICIT SCHEMES FOR SOLVING THE NAVIER–STOKES EQUATIONS

E. von Lavante and J. Grönner
Universität - GH - Essen, FB 12 - Strömungsmaschinen -
Schützenbahn 70, 4300 Essen

SUMMARY

Different approaches for accelerating the convergence of explicit methods for solving the compressible Navier–Stokes equations were investigated. It was found that some of these, combined with the proper multi-grid procedure, can lead to efficient means of solving the Navier–Stokes equations for high speed flows. The findings will be demonstrated on several numerical examples.

INTRODUCTION

In recent time, several different numerical schemes for solving the Euler and Navier–Stokes equations were introduced and quickly gained popularity. The explicit methods in particular are in wide spread use due to their simplicity. Depending on the type of spatial discretization they use, one can loosely distinguish them as either central difference schemes or upwind schemes. Typical representatives of the central difference group of schemes are the prolific Runge–Kutta (R–K) time–stepping methods, based in principle on the approach introduced by Jameson, Schmidt and Turkel [1]. The upwind methods are more complex, as they are based on one of the several variations of the Riemann problem solver, but are usually more robust and reliable, especially when used in viscous and/or high Mach–number computations. Typical examples are the schemes developped by Rossow [2], Radespiel and Kroll [3] or von Lavante et al. [4]. In these schemes, high degree of efficiency was achieved by the use of multi–grid (MG) strategies, coupled with attemps to improve the critical high frequency damping. To this end, in Ref. [4], the coefficients of the R–K time–stepping procedure were optimized for good high frequency damping at relatively high CFL numbers. As a result, most two–dimensional and three–dimensional inviscid cases can be predicted rather efficiently. However, the achievable rates of convergence are very low when these schemes are applied to viscous flows.

In these cases, the well known implicit procedures such as the upwind scheme developped by Eberle et al. [5], seem to be more effective in overcoming the inherent stiffness of the resulting equations. In the context of the MG methods, the implicit formulations have been demonstrated to possess good smoothing properties. The explicit upwind methods are in this sense usually less than perfect.

The present work seeks to improve the convergence behaviour of explicit upwind schemes, applied to predictions of compressible viscous flows, by investigating the MG procedures combined with several simplified implicit procedures and other approaches for accelerating the convergence to steady state.

ALGORITHMS

The basic algorithm for the explicit part was the same as in Ref.[4]. The governing equations of motion were in this case the compressible Navier–Stokes equations in their thin shear layer

(TSL) form. They are in two–dimensional, body fitted coordinates, in non–dimensional strong conservation law form

$$\frac{\partial \hat{Q}}{\partial t} + \frac{\partial \hat{F}}{\partial \xi} + \frac{\partial (\hat{G} - \hat{G}_v)}{\partial \eta} = 0 \tag{1}$$

where

$$\hat{Q} = \frac{1}{J} \left\{ \begin{array}{c} \rho \\ \rho u \\ \rho v \\ e \end{array} \right\}, \quad \hat{F} = \frac{1}{J} \left\{ \begin{array}{c} \rho U \\ \rho u U + \xi_x p \\ \rho v U + \xi_y p \\ (e+p)U \end{array} \right\}, \quad \hat{G} = \frac{1}{J} \left\{ \begin{array}{c} \rho V \\ \rho u V + \eta_x p \\ \rho v V + \eta_y p \\ (e+p)V \end{array} \right\}$$

$$\hat{G}_v = \frac{1}{J} \left\{ \begin{array}{c} 0 \\ \mu \Phi_1 u_\eta + \mu \Phi_2 \eta_x \\ \mu \Phi_1 v_\eta + \mu \Phi_2 \eta_y \\ \frac{\gamma \kappa}{Pr} \Phi_1 e_\eta + \mu \left[\Phi_1 \left(\frac{u^2+v^2}{2} \right)_\eta + V \Phi_2 \right] \end{array} \right\}$$

μ is the viscosity devided by the Reynolds number Re_0, κ the conductivity, and Pr the Prandtl number.

$$U = \xi_x u + \xi_y v, \qquad V = \eta_x u + \eta_y v$$

$$\Phi_1 = \eta_x^2 + \eta_y^2, \qquad \Phi_2 = \frac{1}{3} \left(\eta_x u_\eta + \eta_y v_\eta \right)$$

$$J = \xi_x \eta_y - \xi_y \eta_x = \frac{1}{x_\xi y_\eta - x_\eta y_\xi}, \qquad \xi_x = y_\eta J$$

$$\xi_y = -x_\eta J, \qquad \eta_x = -y_\xi J, \qquad \eta_y = x_\xi J.$$

The scheme is based on Roe's flux difference splitting, as introduced in Ref. [6]. The present spatial discretization employed the finite volume approach, with the state variables (conservative variables) at the cell interfaces determined by the MUSCL interpolation (κ scheme). Best results were obtained for $\kappa = 0$, yielding very low dispersion errors. Details are given in, for example, Ref. [4], and will not be repeated here.

The temporal discretization followed the modified R–K method, as introduced in Ref. [1]. It is a multi–stage procedure for k–stages, where

$$\begin{array}{rcl} Q^1 & = & Q^n - \alpha_1 \Delta t^* R\left(Q^n\right) \\ Q^2 & = & Q^n - \alpha_2 \Delta t^* R\left(Q^1\right) \\ & \vdots & \\ Q^k & = & Q^n - \alpha_k \Delta t^* R\left(Q^{k-1}\right) \\ Q^{n+1} & = & Q^n - \Delta t^* R\left(Q^k\right) \end{array} \tag{2}$$

and R is the residual from eq. (1), $R = -Q_t$ and $\Delta t^* = \Delta J$.

The coefficients in the above R-K scheme were optimized in Ref. [4] for good high frequency damping, important in the MG schemes. The entire theory of the above optimization was based on the linear wave equation. It was therefore surprising that it worked rather well in the case of the two-dimensional and three-dimensional Euler equations, yielding fairly fast rates of convergence.

It should be stressed here that in MG procedures, the maximum achievable CFL number is almost irrelevant; important is good damping of high frequency errors (those close to $\beta = \pi$), which can occur at CFL numbers lower than their maximum. In the present explicit scheme, the R-K coefficients were designed for good damping in wide range of values up to the maximum. For the preferred case of $\kappa = 0$ (Fromm scheme), the maximum sustainable CFL number that also gave good damping in the MG was about 2.2, giving good convergence in a wide range of two-dimensional and three-dimensional simulations.

The Navier-Stokes equations are, however, a totally different story. Here, especially in the case of turbulent boundary layers, the requirement of good grid resolution in the viscous layers results in higly stretched grids with cells that have an extremely high aspect ratio. The system becomes very stiff, and the above R-K scheme looses its effectiveness in solving it. In combination with the MG procedures, the damping of the above scheme in the direction along the solid walls (in the longitudal direction of the alongated cells) is very weak and the MG will become unreliable. It was therefore decided to investigate several possibilities for enhancing the robustness of the present scheme. The most promising method was the addition of implicit damping to the basic explicit scheme. The main goal of the present work was to keep the simplicity of the original scheme while improving it. Therefore, the fully implicit versions of the flux difference splitting were not considered. Rather, simplified implicit operators were studied, giving rise to the notation of semiimplicit schemes. Out of the many possibilities, several of the more successfull choices will be discussed here.

The simplified implicit parts were constructed in the sense of implicit residual smoothing, to be applied after each stage of the R–K procedure. Again, it should be understood that improved damping properties are more important than a mere slight rise in the maximum CFL number.

SCHEME I.

This is a simple procedure introduced by Lerat [7]. In this case, the corresponding delta form of the governing equations in two dimensions is

$$\delta Q - \frac{(\Delta t^*)^2}{4} \left[(\rho_A^2 \delta Q_\xi)_\xi + (\rho_B^2 \delta Q_\eta)_\eta \right] = -\Delta t^* R \tag{3}$$

where ρ_A and ρ_b are the spectral radii of the Jacobian matrices A and B, corresponding to the flux vectors F and $(G - G_v)$, respectively. δQ represents here the change $Q^{n+1} - Q^n$.

The discretized version of this scheme is very simple, resulting in a scalar tri-diagonal system of equations. The coefficients of this system can be computed once per time step using Q^n and stored for the subsequent k stages of the R-K procedure.

This scheme worked relatively well for subsonic and transonic simulations, but displayed some problems for supersonic flows. This should be expected, since it is basically central difference operator with nonlinear coefficients, and is therefore inconsistant with the explicit upwind operator. In order to make it consistant and give it an upwind character, a slight modification was devised.

Considering, for the sake of clarity, only the ξ direction in eq. (3), its discrete version can be written as

$$\delta Q - \frac{(\Delta t^*)^2}{4} \left[\rho_{Ai+1/2}^2 (\delta Q_{i+1} - \delta Q_i) - \rho_{Ai-1/2}^2 (\delta Q_i - \delta Q_{i-1}) \right] \tag{4}$$

which is clearly a second order central difference approximation of the second derivative, consisting of the difference of two first order upwind differences. It can be easily made adjustable to the character of the flow field by introducing two switch parameters τ^+ and τ^- such that, in the case of supersonic flows, only one of the upwind differences will be in effect:

$$\delta Q - \frac{(\Delta t^*)^2}{4} \left[\rho_{Ai+1/2}^2 (\delta Q_{i+1} - \delta Q_i) \tau_{i+1/2}^- - \rho_{Ai-1/2}^2 (\delta Q_i - \delta Q_{i-1}) \tau_{i-1/2}^+ \right] \tag{5}$$

where, for example, $\tau^+ = 1$ when $\lambda_{A_{max}}^+ > 0$ and else $\tau^+ = 0$ or, similarly, $\tau^- = 1$ when $\lambda_{A_{min}}^- < 0$ and $\tau^- = 0$ elsewhere. This version of the Lerat scheme worked much better, giving effective damping for CFL numbers of up to 3.7.

SCHEME II.

Another possibility is obtained by replacing the Jacobian matrices A and B by their diagonal eigenvalue matrices Λ_A and Λ_B in the classical delta form of the governing equations:

$$\delta Q + \Delta t^* (\Lambda_A \delta Q)_\xi + \Delta t^* (\Lambda_B \delta Q)_\eta = -\Delta t^* R. \tag{6}$$

The diagonal matrices are then split into their non–negative and non–positive parts. The final form is obtained by replacing the split eigenvalue matrices by their spectral radii ρ_{A+}, ρ_{A-}, and upwind differencing these terms:

$$\left(I + \Delta t^* \left[(\rho_A^+)^b_\xi - (\rho_A^-)^f_\xi + (\rho_B^+)^b_\eta - (\rho_B^-)^f_\eta\right]\right) \delta Q = -\Delta t^* R. \tag{7}$$

This implicit residual smoothing was not as effective as scheme I. It worked only up to CFL=2.6, which certainly represents only a moderate rise over the explicit limit of 2.2. Besides, it was not as robust as scheme I., making scheme I. clearly a better choice.

OTHER SCHEMES.

Several investigators have also reported some success with simpler forms of implicit residual smoothing operators. These variations on the above theme were also tested in the present work. To name just few, they were:

a) Constant coefficient form; here, a simple second order central difference operator of the form $(I - \epsilon_\xi \delta^2)$ is used. The coefficient ϵ_ξ is a user selectable constant. Depending on the choice of ϵ, this smoothing procedure was more or less efective; mostly, it was not.

b) Variable coefficient form, introduced by Turkel et al. [8]. It is very similar to the central difference version of scheme I., except for some additional user selectable scaling constants. Its performance was comparable to the corresponding version of scheme I.

CONVERGENCE ACCELERATION

In order to speed-up the rate of convergence even more, some of the recently introduced convergence acceleration techniques by Cheung et al. [9] were tested. Generally, they performed rather poorly. They are relaxation procedures, applied to the change δQ^n:

$$Q^{n+1} = Q^n + \omega \delta Q \tag{8}$$

of the iterative process $\vec{Q}^{n+1} = \overline{\overline{G}} \vec{Q}^n$. The overrelaxation parameter ω is related to the iteration matrix $\overline{\overline{G}}$ through its spectral radius Π. Mostly, Π is not exactly known and has to be estimated. Several formulae can be found in the literature. In the present work, two such estimates were tested:

$$a) \ \Pi = \frac{\|\delta \vec{Q}^n\|}{\|\delta \vec{Q}^{n-1}\|} \qquad b) \ \Pi = \frac{\left(\delta \vec{Q}^{n-1}\right)^T \cdot \left(\delta \vec{Q}^n\right)}{\left(\delta \vec{Q}^{n-1}\right)^T \cdot \left(\delta \vec{Q}^{n-1}\right)}. \tag{9}$$

In Ref. [9], it is recommended to compute ω according to

$$\omega = \frac{2}{2 - \Pi} \tag{10}$$

and to apply it after every iteration. In present computations, this led to instabilities. The overrelaxation was therefore applied after every 4-16 iterations or cycles, and the factor ω was reduced using $\omega_{new} = (\omega_{original} - 1) c_r + 1$, with c_r selected between 0.5 and 0.25.

MULTIGRID

The MG procedure used in the present investigation was the standard FAS MG for solving nonlinear problems, described in detail in, for example, Ref. [4]. In order to improve its performance, it was applied in some cases only in one direction to improve the uniformity of the coarse grids. More discussion of this approach is offered below in the section RESULTS.

RESULTS

The above algorithms were tested on several configurations. Due to the space limitations, only some two-dimensional computations will be reported.

The basic test case was a supersonic diffusor with a weak oblique shock ($p_2/p_1 = 1.2$) generated by an upper wall inclined at $3°$. It causes boundary layer separation at the lower wall with a typical lambda shock and subsequent reattachment. The resulting pressure contours can be seen in Fig. 2 a). The extend of the separation zone critically depends on the accuracy of the computational scheme. The basic grid, shown in Fig. 1 at the top, consisted of 64x64 grid points. Since the Reynolds number was $3·10^5$, it was assumed that the flow remained laminar; the inflow Mach number was 2.0. The grid displayed relatively fine spacing, with first grid point at a distance corresponding to $y^+ = 1.0$.

Before comparing the relative speed of convergence of the different schemes, it should be stressed that the basis of these comparisons is an "honest" work unit, based on the total computer time of the various programs devided by the time of one iteration of the simple single grid code with no other overhead.

The single grid methods are compared with the best explicit MG calculation in Fig.3. The semi-implicit scheme I. performed at a CFL number of 3.5 even worse than the purely explicit method at a CFL = 2.2 because of its computational intensity. The fully implicit, Flux Vector Splitting code, developped by the authors, certainly converged faster at CFL=175. It is interesting to note that the MG scheme in its basic explicit form at CFL=2.2 (also shown as a) in Fig.4) was still better.

The different MG procedures are compared in Fig.4. As noted before, the MG with its usual coarsening in both directions was difficult to optimize for this case. Counting the grids from the fine to the coarse (finest = 1, coarsest = 4), it converged only for V cycles in the grid-sequence 1-2-1, with the iterations on each grid level distributed as 8-24-6 (Fig.4 a)). The reason becomes obvious after looking at Fig.1 a): The coarse grids still have very high cell aspect ratios and therefore poor damping in one direction, explaining the necessity of repeated application of smoothing. One possible remedy lies in coarsening only in direction normal to the wall, Fig.1 b). The cells become increasingly regular. In this case, the MG worked for a sequence 1-2-3 with the iterations 2-6-24-6-2. The FMG procedure, using coarsening in both directions, is shown as residuum c). All the MG schemes performed about the same, the one-directional MG mainly because of the larger amount of cells on the coarser grids.

The modified versions of the MG procedures are displayed in Fig.5, compared to the scheme in Fig.4 a). It easy to see that the convergence acceleration from Ref. [9] did not hurt but did not help much either. Additionally, it was very sensitive to the proper choice of the user selectable parameters and can not be therefore considered robust. The application of the implicit smoothing from scheme I. to the unidirectional MG brought a significant improvement in the comvergence behaviour, as now the iteration sequence 1-4-8-4-1, shown in Fig.5 as curve c), was somewhat faster than the other schemes.

As a next, more demanding, test case, the GAMM double nozzle was selected. The finest computational grid, consisting of 128x64 points, is shown in Fig.6. The resulting pressure contours and pressure and velocity along the nozzle axis for Re = 400 can be seen in Fig.7. They compare well with the results from the GAMM workshop. In this case, the MG method converged only with the semi-implicit scheme I., either with unidirectional coarsening or with the coarsening in both directions.

CONCLUSIONS

The explicit Flux Difference Splitting methods, combined with the R-K time stepping, are, without any further enhancement, not too effective in solving the Navier-Stokes equations. With the proper MG procedure, however, they could converge faster than the fully implicit methods. The semi-implicit operator, used in the present work, makes this scheme much more robust. The MG is less sensitive to the correct choice of coarsening and iterating strategies and works for a wider range of applications. The convergence acceleration techniques were usually much less effective and in general unreliable.

The question of improving the convergence of numerical schemes for the simulation of high speed flows needs obviously more attention.

Bibliography

[1] Jameson, A., Schmidt, W., and Turkel, E., "Numerical Solutions of the Euler Equations by Finite-Volume Methods Using Runge-Kutta Time-Stepping Schemes", AIAA Paper 81-1259, June 1981.

[2] Rossow, C.-C., "Flux-Balance Splitting - A New Approach for a Cell Vertex Upwind Scheme", Proceedings of the 12th International Conference on Numerical Methods in Fluid Dynamics, Oxford, England, July 1990.

[3] Radespiel, R., and Kroll, N., " A Multigrid Scheme with Semicoarsening for Accurate Computations of Viscous Flows", Proceedings of the 12th International Conference on Numerical Methods in Fluid Dynamics, Oxford, England, July 1990.

[4] von Lavante, E., El-Miligui, A., Cannizaro, F. and Warda, H.A., "Simple Explicit Upwind Schemes for Solving Compressible Flows", Proceedings ot the 8th GAMM Conference on Numerical Methods in Fluid Mechanics, Vieweg, 1990.

[5] Eberle, A., Schmatz, M.A., and Bissinger, N.C.,"Generalized Flux-Vectors for Hypersonic Shock-Capturing", AIAA Paper 90-0390, january 1990.

[6] Roe, P.L., " Approximate Riemann Solvers, Parameter Vectors, and Difference Schemes", Journal of Comp. Physics, Vol. 43, 1981, pp. 357-372.

[7] Lerat, A., " Implicit Methods of Second-Order Accuracy for the Euler Equations", AIAA Journal, Vol. 23, January 1985, pp. 33-40.

[8] Turkel, E., Swanson, R.C., Vatsa, V.N. and White, J.A.,"Multigrid for Hypersonic Viscous Two- and Three-Dimensional Flows", AIAA Paper presented at the 10th CFD Conference, Honolulu, Hawaii, June 24-26, 1991.

[9] Cheung, S., Cheer, A., Hafez, M. and Flores, J., "Convergence Acceleration of Viscous and Inviscid Hypersonic Flow Calculations", AIAA Journal, Vol.29, No. 8, August 1991, pp. 1214-1222.

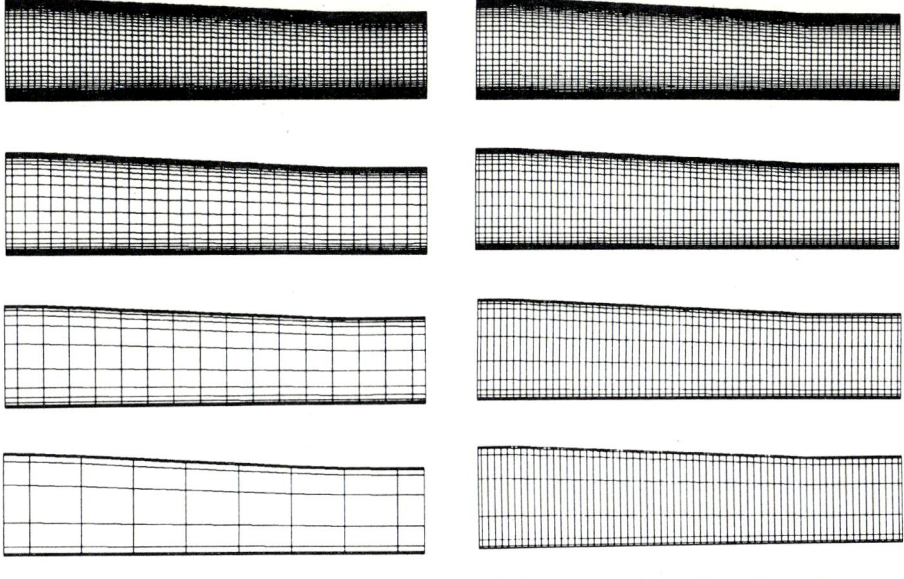

a) Coarsening in both directions

b) Coarsening in η- direction only

Fig. 1: Comparison of grid sequences for multi-grid predictions of supersonic diffusor flow.

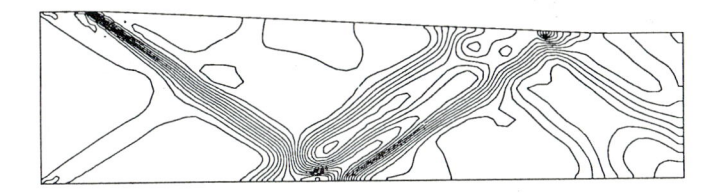

a) Finest Grid

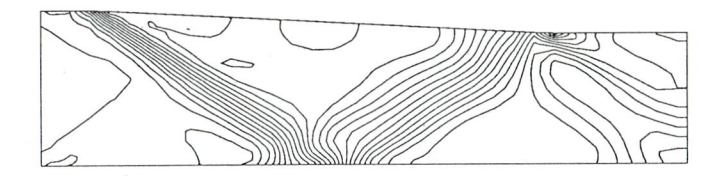

b) Second finest grid

Fig. 2: Resulting pressure contours in supersonic diffusor with inflow Mach-number of 2.0.

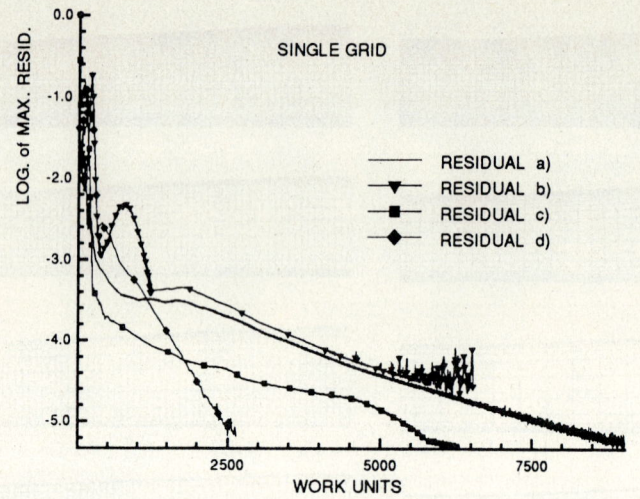

Fig 3.: Convergence history for supersonic diffusor; Comparison of single grid schemes:
  a) Single grid, explicit scheme, CFL = 2.2
  b) Single grid, semi-explicit scheme, CFL = 3.5
  c) Single grid, implicit FVS scheme, CFL = 175
  d) Best multi-grid in η-direction, explicit, CFL = 2.2

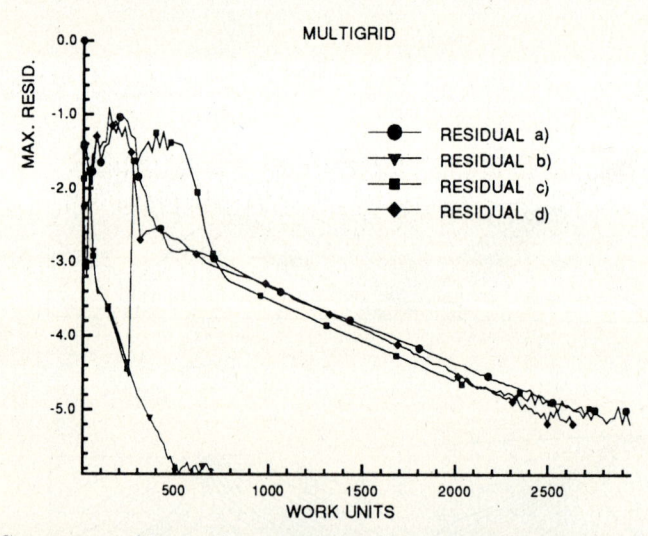

Fig. 4: Convergence history for supersonic diffusor; Comparison of basic multi-grid schemes:
  a) Multi-grid in ξ and η directions on grids 1-2, explicit, CFL = 2.2
  b) Multi-grid in ξ- and η- directions on coarse grids 2-3, explicit, CFL = 2.2
  c) Approximated full multi-grid, in ξ- and η- directions
  d) Multi-grid in η- direction only, V-cycle.

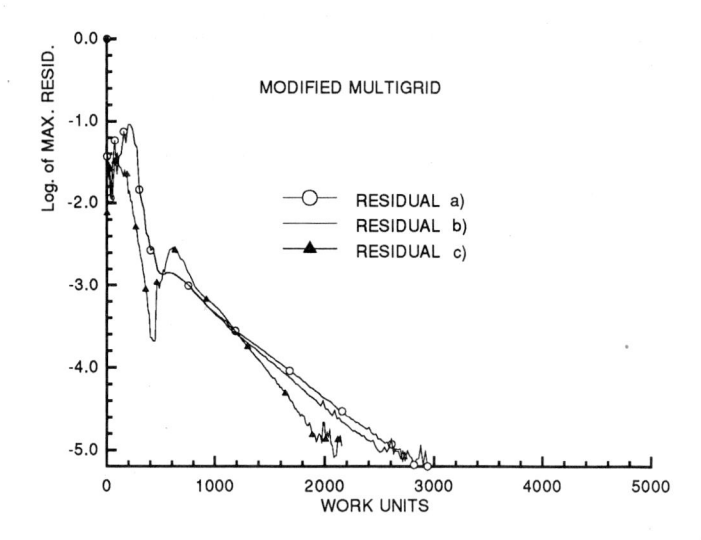

Fig. 5: Convergence history for supersonic diffusor; Comparison of modified multi-grid schemes:
   a) Basic explicit multi-grid scheme in ξ-η-direction
   b) Multi-grid with convergence acceleration
   c) Semi-implicit multi-grid scheme in ξ-direction

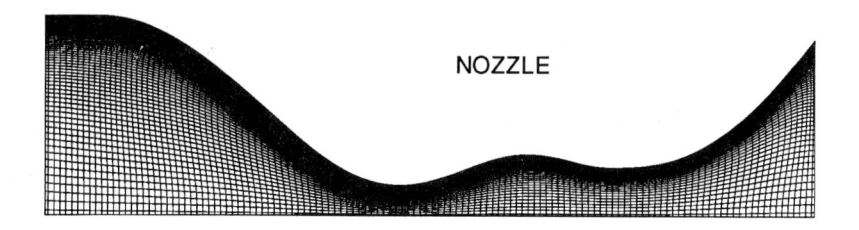

Fig. 6: Computational grid (finest) for GAMM-nozzle.

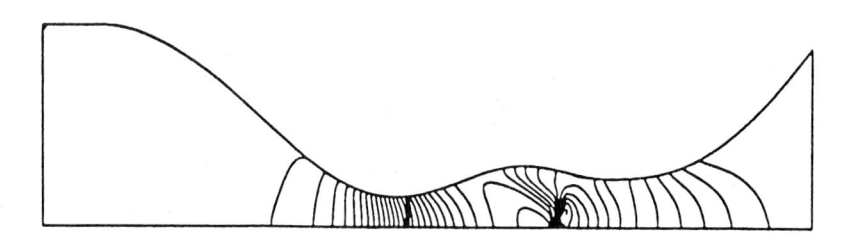

Fig. 7: Resulting pressure contours

INCOMPRESSIBLES FLOWS IN COMPLEX GEOMETRIES

TURBULENT FLOW SIMULATION OF FRANCIS WATER RUNNER WITH PSEUDO-COMPRESSIBILITY

Chuichi ARAKAWA, Yi QIAN, Masahiro SAMEJIMA
University of Tokyo, Faculty of Engineering
7-3-1 Hongo, Bunkyo-ku, Tokyo 113, Japan
Yuichi MATSUO   National Aerospace Laboratory, Tokyo, Japan
Takashi KUBOTA   Fuji Electric Co. Ltd., Kawasaki, Japan

Summary

The three-dimensional Navier-Stokes code with the pseudo-compressibility, the implicit formulation of finite difference, and the turbulence model has been developed for the Francis water runner. Two turbulence models are used; one is the so called Baldwin-Lomax zero-equation model and the other is the $k - \epsilon$ two-equation model. The viscous flow in the rotating field can be simulated well with these turbulence models, and their results agree with the experimental data and the Euler code in the design flow condition. Because these codes employing the wall function near the wall for the $k - \epsilon$ model do not require large CPU time, they can be used on the small computer like the so-called engineering work station.

INTRODUCTION

The three-dimensional Navier-Stokes code with the pseudo-compressibility, the implicit formulation of finite difference, and the turbulence model has been developed for the Francis water runner. The authors have already developed the 3-D turbulent flow calculation code for the advanced turboprop(ATP) with compressible flow scheme of the rotating blade, which enables one to predict the performance and the detailed flow including the location of shock successfully[1]. The Francis runner in the water turbine is employed in this paper as the next application of this code using the idea of the pseudo-compressibility. We take the implicit finite difference and approximate factorization method so as to get the comparatively large time step for the time marching [2]. The k-ϵ turbulence model, requiring two transport equations, is introduced for simulating the high Reynolds number flow, while the wall function is used to decrease the number of the grid points. The Baldwin-Lomax zero-equation model[3] is also introduced as a representative of a simple turbulence model in the flow problems. The TVD (Total Variations Diminishing) scheme is used in order to get converged results adding automatically a minimum artificial viscosity.

The computational model of the Francis runner shown in Fig.1 is the one provided by the organizer of the GAMM WORKSHOP LAUSANNE, 1989. It would be ideal to calculate the whole system of the Francis turbine consisting of a spiral-casing, a stay ring, guide vanes, a runner and draft tube. However, the domain of this analysis is limited to the water passage in the runner and its near region so that we need the pressure and velocity distributions at the inlet of the runner as boundary conditions.

These boundary conditions required in the calculation were provided by the organizer of the GAMM WORKSHOP and slightly modified so as to satisfy mass conservation. The report on this workshop will be published in future, and members in the group of the organizer have already presented some calculations, where they have examined the difference between mesh types used in the distributer and runner[4].

The results obtained with the Navier-Stokes calculations are compared with experimental data and with the results of Euler analysis [5], in order to show the necessity to employ the turbulence model. To save the computational time, we use the wall function for the $k - \epsilon$ turbulence model and the implicit scheme, which enables us to take not only the big supercomputer but also the desktop standard computer (e.g. workstation).

BASIC EQUATIONS

The unsteady, three-dimensional, incompressible flow with constant density is governed by the following equations written in the rotating Cartesian coordinates.

$$\frac{\partial Q}{\partial t} + \frac{\partial E}{\partial x} + \frac{\partial F}{\partial y} + \frac{\partial G}{\partial z} + H = Re^{-1}\left(\frac{\partial R}{\partial x} + \frac{\partial S}{\partial y} + \frac{\partial T}{\partial z}\right) \qquad (1)$$

$$Q = \begin{bmatrix} 0 \\ u \\ v \\ w \end{bmatrix}, \quad E = \begin{bmatrix} u \\ uu+p \\ uv \\ uw \end{bmatrix}, \quad F = \begin{bmatrix} v \\ vu \\ vv+p \\ vw \end{bmatrix}, \quad G = \begin{bmatrix} w \\ wu \\ wv \\ ww+p \end{bmatrix},$$

$$H = \begin{bmatrix} 0 \\ -\Omega^2 x - 2\Omega v \\ -\Omega^2 y + 2\Omega u \\ 0 \end{bmatrix}, \quad [R, S, T] = \begin{bmatrix} 0 & 0 & 0 \\ u_x & u_y & u_z \\ v_x & v_y & v_z \\ w_x & w_y & w_z \end{bmatrix}.$$

The direction of rotation is clockwise about the positive z-direction and Ω denotes the angular velocity. The notations p, u, v and w are pressure and relative velocity components respectively, and all variables are properly normalized. In order to introduce the pseudo-compressibility proposed by Chorin [6] and developed by Rogers. et al.[7], the continuity equation is modified by adding a time derivative of the pressure term, which allows for the pressure wave to propagate at some limited speed and disappear when the converged steady solution is achieved. General curvilinear coordinates (ξ, η, ζ) are introduced so as to accommodate fully three-dimensional geometries. After using the chain rule to transform the physical coordinates into curvilinear ones, we get the following form of the basic equations.

$$\frac{\partial \hat{Q}}{\partial t} + \frac{\partial \hat{E}}{\partial \xi} + \frac{\partial \hat{F}}{\partial \eta} + \frac{\partial \hat{G}}{\partial \zeta} + \hat{H} = Re^{-1}\left(\frac{\partial \hat{R}}{\partial \xi} + \frac{\partial \hat{S}}{\partial \eta} + \frac{\partial \hat{T}}{\partial \zeta}\right) \qquad (2)$$

where,

$$\hat{Q} = J^{-1}\begin{bmatrix} p^* \\ u \\ v \\ w \end{bmatrix}, \quad \hat{E} = J^{-1}\begin{bmatrix} \beta U \\ uU + \xi_x p^* \\ vU + \xi_y p^* \\ wU + \xi_z p^* \end{bmatrix}, \quad \hat{F} = J^{-1}\begin{bmatrix} \beta V \\ uV + \eta_x p^* \\ vV + \eta_y p^* \\ wV + \eta_z p^* \end{bmatrix},$$

$$\hat{G} = J^{-1} \begin{bmatrix} \beta W \\ uW + \zeta_x p^* \\ vW + \zeta_y p^* \\ wW + \zeta_z p^* \end{bmatrix}, \quad \hat{H} = J^{-1} \begin{bmatrix} 0 \\ -2\Omega v \\ 2\Omega u \\ 0 \end{bmatrix},$$

$$[\hat{R}, \hat{S}, \hat{T}] = (1+\nu_t)J^{-1} \times$$

$$\begin{bmatrix} 0 & 0 & 0 \\ (\xi_x\xi_x + \xi_y\xi_y + \xi_z\xi_z)u_\xi & (\eta_x\xi_x + \eta_y\xi_y + \eta_z\xi_z)u_\eta & (\zeta_x\xi_x + \zeta_y\xi_y + \zeta_z\xi_z)u_\eta \\ (\xi_x\eta_x + \xi_y\eta_y + \xi_z\eta_z)v_\xi & (\eta_x\eta_x + \eta_y\eta_y + \eta_z\eta_z)v_\eta & (\zeta_x\eta_x + \zeta_y\eta_y + \zeta_z\eta_z)v_\zeta \\ (\xi_x\zeta_x + \xi_y\zeta_y + \xi_z\zeta_z)w_\xi & (\eta_x\zeta_x + \eta_y\zeta_y + \eta_z\zeta_z)w\eta & (\zeta_x\zeta_x + \zeta_y\zeta_y + \zeta_z\zeta_z)w_\zeta \end{bmatrix}$$

(2b)

$$p^* = p - \frac{1}{2}\Omega^2(x^2 + y^2) + gz. \quad (3)$$

The parameter of pseudo-compressibility to control the speed of the pressure wave is denoted as β in the continuity equation, and is fixed to 1 in this paper. The notations U, V and W are contravariant velocities, and J is the Jacobian of the transformation in the metrics.

NUMERICAL SCHEME

The numerical algorithm to advance eq.(2) in time is an implicit, approximately factored, finite-difference scheme originally developed by Beam and Warming [2]. The basic equations are discretized in the conventional delta form with the use of Euler backward time differencing. For the spatial differencing of the explicit right-hand-side in the delta form, the upwind based high accuracy TVD scheme by Chakravarthy and Osher[8] is used without any explicit artificial dissipation terms being added. On the other hand, the implicit left-hand-side term relies on the diagonalized ADI method whereby the computational efficiency can be improved because the steady state solution is indifferent to implicit operators. The spatial difference utilizes an upwind flux-split technique. Each ADI operator is decomposed into the product of lower and upper bidiagonal matrices by using the diagonally dominant factorization, and the operator is efficiently inverted by the forward and backward sweeps [9]. The final algorithm results in the following discretized equations:

$$\begin{aligned}
&T_\xi\left[I + h\left(\nabla_\xi \Lambda_\xi^+ + \triangle_\xi \Lambda_\xi^- - Re^{-1}\bar{\delta}_\xi\nu_\xi I\right)\right]T_\xi^{-1} \\
&T_\eta\left[I + h\left(\nabla_\eta \Lambda_\eta^+ + \triangle_\eta \Lambda_\eta^- - Re^{-1}\bar{\delta}_\eta\nu_\eta I\right)\right]T_\eta^{-1} \\
&T_\zeta\left[I + h\left(\nabla_\zeta \Lambda_\zeta^+ + \triangle_\zeta \Lambda_\zeta^- - Re^{-1}\bar{\delta}_\zeta\nu_\zeta I\right)\right]T_\zeta^{-1} \\
&[I + hD^n](Q^{n+1} - Q^n) \\
&= -h\left[\hat{E}_{j+\frac{1}{2}}^n - \hat{E}_{j-\frac{1}{2}}^n + \hat{F}_{k+\frac{1}{2}}^n - \hat{F}_{k-\frac{1}{2}}^n + \hat{G}_{l+\frac{1}{2}}^n - \hat{G}_{l-\frac{1}{2}}^n\right. \\
&\left. - Re^{-1}\left(\bar{\delta}_\xi\hat{R} + \bar{\delta}_\eta\hat{S} + \bar{\delta}_\zeta\hat{T}\right) + \hat{H}\right]
\end{aligned} \quad (4)$$

where,

$$\frac{\partial \hat{E}}{\partial \hat{Q}} = T_\xi \Lambda_\xi T_\xi^{-1}, \cdots, \frac{\partial \hat{H}}{\partial \hat{Q}} = D, \nu_\kappa = 2(1+\nu_t)(\nabla\kappa \cdot \nabla\kappa)$$

and $h = \Delta t$ is the time step size. ∇u and $\triangle u$ denote the first-order backward and forward difference operators respectively, and I is the identity matrix.

For example, the numerical flux of ξ direction in the right hand side of Eq.(4) is defined as follows;

$$\hat{E}_{j+\frac{1}{2}} = \hat{h}_{j+\frac{1}{2}} - \frac{\phi}{2}[\tilde{d}f^-_{j+\frac{3}{2}}] - \frac{1-\phi}{2}[\tilde{d}f^-_{j+\frac{1}{2}}] + \frac{1-\phi}{2}[\tilde{d}f^+_{j+\frac{1}{2}}] + \frac{\phi}{2}[\tilde{d}f^+_{j-\frac{1}{2}}] \tag{5}$$

where,

$$\hat{h}_{j+\frac{1}{2}} = \frac{1}{2}[\hat{E}_j + \hat{E}_{j+1} - (df^+_{j+\frac{1}{2}} - df^-_{j-\frac{1}{2}})]$$
$$df^\pm_{j+\frac{1}{2}} = (J^{-1}T\Lambda^\pm T^{-1})_{j+\frac{1}{2}} \triangle_{j+\frac{1}{2}} JQ$$

$$\tilde{d}f^-_{j+\frac{3}{2}} = (J^{-1}T\Lambda^-)_{j+\frac{1}{2}} \mathrm{minmod}(T^{-1}_{j+\frac{1}{2}}\triangle_{j+\frac{1}{2}}JQ, bT^{-1}_{j+\frac{1}{2}})\triangle_{j+\frac{3}{2}} JQ$$
$$\tilde{d}f^-_{j+\frac{1}{2}} = (J^{-1}T\Lambda^-)_{j+\frac{1}{2}} \mathrm{minmod}(T^{-1}_{j+\frac{1}{2}}\triangle_{j+\frac{1}{2}}JQ, bT^{-1}_{j+\frac{1}{2}})\triangle_{j+\frac{1}{2}} JQ$$
$$\tilde{d}f^+_{j+\frac{1}{2}} = (J^{-1}T\Lambda^+)_{j+\frac{1}{2}} \mathrm{minmod}(T^{-1}_{j+\frac{1}{2}}\triangle_{j+\frac{1}{2}}JQ, bT^{-1}_{j+\frac{1}{2}})\triangle_{j-\frac{1}{2}} JQ$$
$$\tilde{d}f^+_{j-\frac{1}{2}} = (J^{-1}T\Lambda^+)_{j+\frac{1}{2}} \mathrm{minmod}(T^{-1}_{j+\frac{1}{2}}\triangle_{j-\frac{1}{2}}JQ, bT^{-1}_{j+\frac{1}{2}})\triangle_{j+\frac{1}{2}} JQ$$

with

$$\Lambda^\pm = \frac{1}{2}(\Lambda \pm |\Lambda|)$$
$$\triangle_{j+\frac{1}{2}} JQ = (JQ)_{j+1} - (JQ)_j$$
$$\mathrm{minmod}(x,y) = \mathrm{sign}(x) \cdot \max[0, \min(|x|, y \cdot \mathrm{sign}(x))] .$$

The cell interface values are determined by just simple averaging. The parameter ϕ is chosen to be equal to 1/3 which results in the third-order scheme. The compression parameter b is taken as the maximum allowable value of 4. The fluxes in the other directions can be obtained similarly. A detailed description of the TVD formulation is given by Chakravarthy et al.[8]. A spatially varying time step technique is used which results in a faster convergence of the solution.

As for the turbulence model, the standard $k - \epsilon$ model is employed to describe the viscous flow of high Reynolds number. A two-equation model such as the $k - \epsilon$ one is available for the complicated flow including the separation region. Further development of the code should result in a more powerful tool to predict the flow in off-design conditions as compared, e.g., to codes employing a simple zero-equation model such as Baldwin-Lomax [3]. The set of transport equations for the turbulent energy k and dissipation rate ϵ are :

$$\frac{\partial Q_t}{\partial t} + \frac{\partial E_{ti}}{\partial x_i} = Re^{-1} \frac{\partial R_t}{\partial x_i} + S_t \tag{6}$$

$$Q_t = \begin{bmatrix} k \\ \epsilon \end{bmatrix}, \ E_{ti} = \begin{bmatrix} ku_i \\ \epsilon u_i \end{bmatrix}, \ R_{ti} = \begin{bmatrix} \frac{\nu_t}{\sigma_k}\frac{\partial k}{\partial x_i} \\ \frac{\nu_t}{\sigma_\epsilon}\frac{\partial \epsilon}{\partial x_i} \end{bmatrix}, \ S_t = \begin{bmatrix} P - \epsilon \\ C_1 \frac{\epsilon}{k} P - C_2 \frac{\epsilon^2}{k} \end{bmatrix}.$$

where,

$$P = Re^{-1}\nu_t \left[\frac{\partial u_i}{\partial x_j}\left(\frac{\partial u_i}{\partial x_j} + \frac{\partial u_j}{\partial x_i}\right)\right]$$

$$\nu_t = C_\mu \frac{k^2}{\epsilon}$$

$$C_\mu = 0.09, \quad C_1 = 1.44, \quad C_2 = 1.92 \quad \sigma_k = 1.0, \quad \sigma_\epsilon = 1.3 .$$

The wall function is used in the near-wall region so as to reduce the number of grid points. In this method it is assumed that the production and dissipation of turbulent energy are in equilibrium and other terms can be neglected in the so-called logarithmic region near the wall. Finally the number of grid points is decreased to 20,000 in this paper from 200,000 in the ATP calculations where the low-Reynolds number model is used which requires the finer grid near the wall.

The Baldwin-Lomax model[3] is introduced in this paper to compare with the $k-\epsilon$ model. This model is called a zero-equation model because it does not need a transport equation to get the eddy viscosity. It is given in the inner layer with the function of the mixing length, and in the outer layer with some characteristic quantity of the wake and the intermittency factor. This model is very simple, but does not take the history of the turbulence quantities into account. Because of the restriction in memory size on small computer, we used the same grid system with both the turbulence models employed.

GEOMETRY AND BOUNDARY CONDITIONS

The flow is assumed to be cyclic around the runner and to be steady, so that the computational domain contains only one water passage in the runner as shown in Figs.2-4. The in- and out-flow boundaries are specified by the two measurement cross-sections, and the crown surface is smoothly attached to the rotating axis for simplicity (Fig.3). All of the boundary points are specified with an appropriate method, then interior points are distributed by using algebraic interpolation techniques. The H-mesh topology is adopted on the blade-to-blade plane in order to avoid leading edge singularities and to facilitate the periodic boundary condition. The mesh clustering near the surfaces is imposed to enhance the numerical accuracy for its explicitly specified boundary conditions. The mesh for this turbulent flow calculation is $65 \times 21 \times 21$. Fig.4 shows the computational domain mapped onto the z-const. plane.

As for boundary conditions, pressure and all the velocity components are given by the measured values, with some modifications in the velocity field at the inflow boundary. In a previous report[5], it was found that some modifications to the inlet values of the velocity were needed to conserve the mass flow mentioned as a catalogued value at the runner inlet, multiplying the constant value to the velocity component in the experiment. The axial symmetry requires that velocity components u and v are equal to zero and that other values are averaged from the surrounding points. On the crown, band and runner surfaces, impermeable conditions are specified as follows. The tangency condition is enforced by specifying the contravariant velocity $W = 0$ on the crown and band surfaces, and $V = 0$ on the runner surfaces. Other contravariant velocities and the pressure are obtained by linear extrapolation from interior points if necessary. The wall function is also applied on every surfaces as described before. At the outflow boundary, the derivatives of all the velocity components in the streamwise direction are assumed to be zero at first, and extrapolated velocities are corrected so as to conserve the mass flow rate.

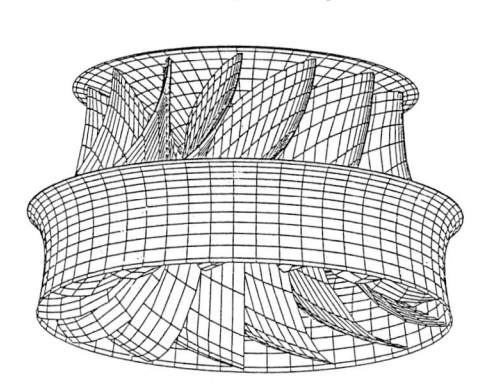

Fig.1 Computational model of Francis runner provided in GAMM workshop 1989.

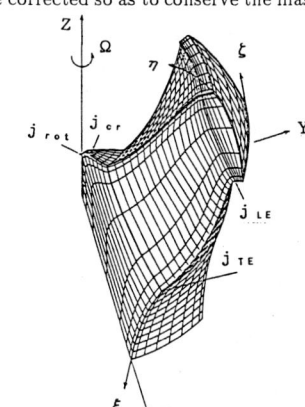

Fig.2 Computational domain.

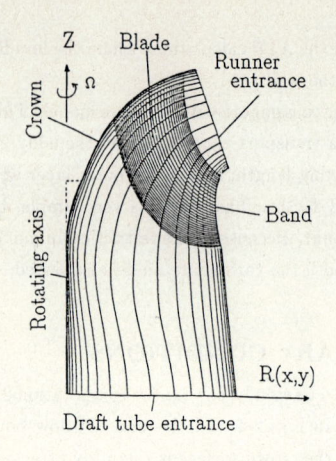

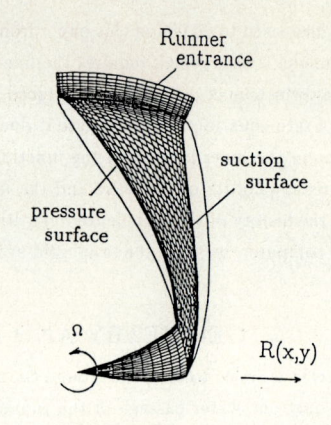

Fig.3 Computational domain (meridional surface).

Fig.4 Computational domain (z-const. surface).

RESULTS AND DISCUSSIONS

The results of Navier-Stokes calculations with the turbulence model are presented and compared with results from the Euler method[5]. These calculations are also compared with the experimental data provided by the GAMM organizer. Both of the calculations are carried out at the design point.

Fig.5 shows the vectors of relative velocities at the first grid point near the runner surface with the Navier-Stokes code and the $k-\epsilon$ model. A strong cross flow is observed near the leading-edge on the pressure surface and near the trailing-edge on the suction side. Comparing with the results of the Euler code shown in Fig.6, it is found that the secondary flows in the radial direction owing to the rotating effect are much stronger in the Navier-Stokes code than in the Euler code, which means that the former code reasonably predicts well the qualitative behavior of the flow field in the rotating blades. As for the comparison between the $k-\epsilon$ model with the Baldwin-Lomax model shown in Fig.7, there is not a big difference between the two codes, but a little smaller cross flow is observed in the region of trailing edge on the suction side in the Baldwin-Lomax model than in the $k-\epsilon$ model.

Pressure contours on both surfaces with the $k-\epsilon$ code are shown in Fig.8. There, a strong suction peak can be found near the corner between the leading edge and the band on the suction surface. The cavitation occurs at a similar place at first when the pressure level for the water-turbine system is down according to the explanation of the GAMM organizer. The difference among three codes is found to be small, although the suction peak is a little less strong in the Navier-Stokes codes.

Fig.9 shows surface pressure distributions calculated with two codes on three chord-wise sections. The experimental data are also plotted in the same figure. The sections 2,9,15 are the GAMM notation of wing sections; section 2 is the vicinity of the crown, section 9 is the middle of the runner, and section 15 is near the band. Both calculation results qualitatively agree with the experimental data which are plotted with two notations corresponding to experiments carried out at different times. The small difference between Navier-Stokes codes and Euler code appears near the leading edge on the suction surface in sections 15 and 9, because the strong pressure variation probably due to the curvature of the surface in the Euler code becomes very small in the Navier-Stokes codes. The boundary layer developing on the surface may play the role to keep the pressure peak small.

Fig.10 shows the absolute velocity and pressure distributions at the runner inlet and outlet. The numerical predictions coincide with the experimental data well, except for the tangential velocity at the inlet which was corrected to satisfy mass conservation. The difference among the three codes is so small that the effect of the viscosity is negligible for the macro-scale field of the velocity and pressure of the design condition. The difference of the velocity distribution between two turbulence models is caused mainly by the boundary layer development on the band and crown, where the first grid point is located a little too far from the surface owing to the limitation on the number of grid points. It means that the Baldwin-Lomax model may predict a larger thickness of the boundary layer. But the difference of the macro scale quantity like the torque is very small in this calculation.

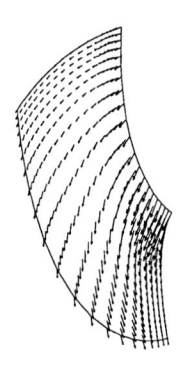

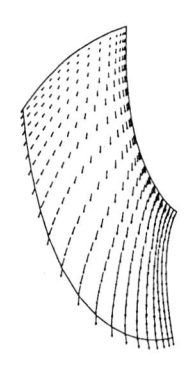

　　a. suction surface　　b. pressure surface
Fig.5 Velocity vector distribution at the first grid point near the runner surface by the $k - \epsilon$ code.

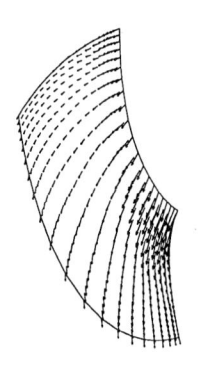

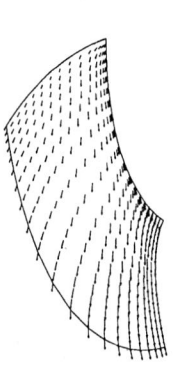

　　a. suction surface　　b. pressure surface
Fig.6 Velocity vector distribution at the first grid point near the runner surface by the Euler code.

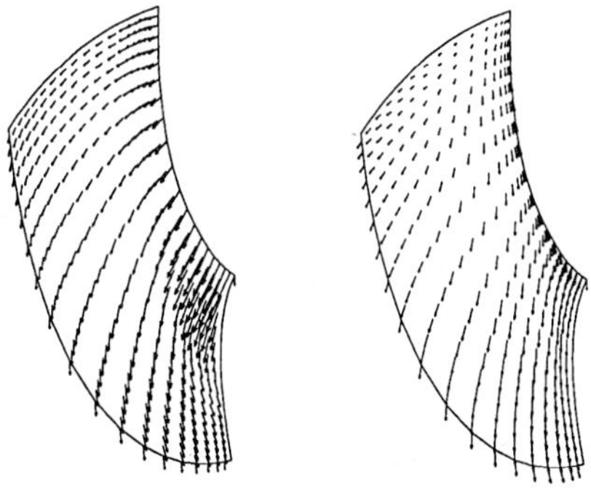

　　a. suction surface　　b. pressure surface
Fig.7 Velocity vector distribution at the first grid point near the runner surface by the Baldwin-Lomax code.

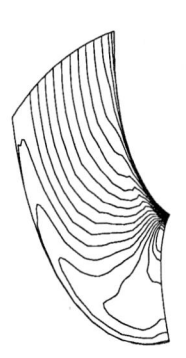

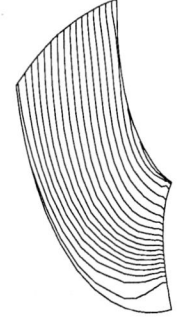

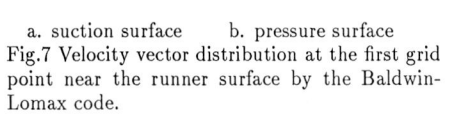

　　a. suction surface　　b. pressure surface
Fig.8 Pressure contours on the runner surface by the $k - \epsilon$ code.

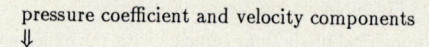

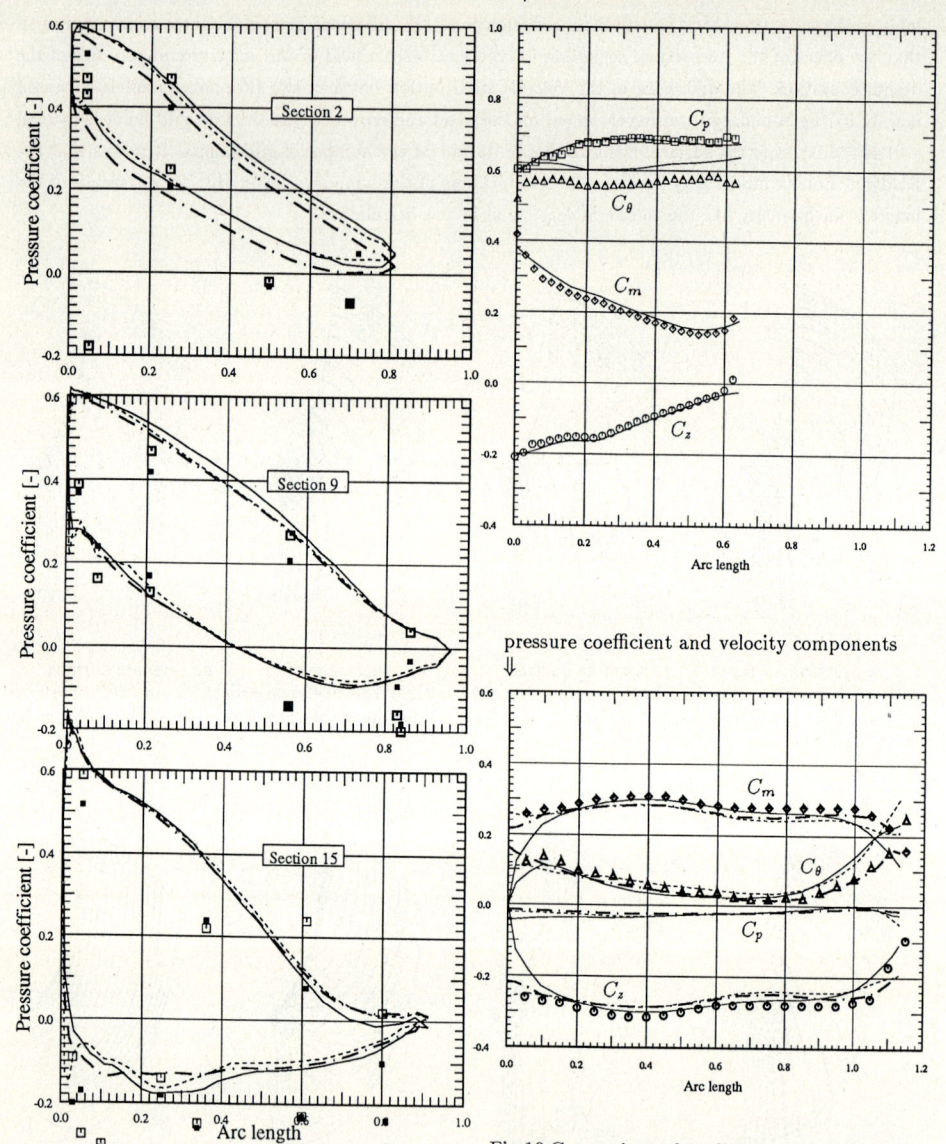

Fig.9 Pressure distributions on the wing sections of runner. (— · — : $k-\epsilon$ code, - - - : Euler code, ——— : Baldwin-Lomax code, ☐ ■ : experiments)

Fig.10 Comparison of predicted and measured distributions of pressure and velocity components at the inlet (above) and outlet (below). (— · — : $k-\epsilon$ code, - - - : Euler code, ——— : Baldwin-Lomax code, ◉ △ ◇ : experiments, C_p: pressure, C_z: axial velocity, C_θ : tangential velocity, C_m: meridional velocity)

The efficiency is 0.96 in both of two Navier-Stokes codes. The torque used in the prediction of the efficiency is derived with the integration of the pressure and the shear stress distributions on the surfaces. The total head for the runner including the loss is given with the experimental data. The efficiency drop of the Navier-Stokes code is due to the viscosity effect, while the drop is very small owing to the favorable flow pattern derived here which has no separations. For reference, the Reynolds number of the runner is about 10^6.

The comparisons between the $k - \epsilon$ and Baldwin-Lomax models are also carried out for the case of a two-dimensional aerofoil at moderate angle of attack. Fig.11 shows the pressure distributions on the aerofoil upper and lower surfaces with two turbulence models. They coincide with each other and agree well with the experimental data. We cannot conclude here which model is better in the aerofoil and Francis water turbine. The best model for the runner will be selected after computing the flow in the off-design operating conditions

Incidentally, the wall function used in this paper enables one to decrease the number of grid points as described before. If we employ the low-Reynolds-number $k - \epsilon$ model proposed by Jones-Launder [10], we need to put another 10 grid points near the wall. Fig.12 shows the pressure distribution on the 2-D aerofoil NACA-0012 calculated with two models, and both results coincide each other. However the number of grid points is increased from 157x31 to 157x41, and the number of iteration to get converged results is also increased from 1600 to 6000, because of the small time step required in the finer mesh. This fact means that the wall function used here makes it possible to predict the 3-D turbulent flow in a Francis runner even with the small computer.

The calculations have been carried out with the super engineering work station (TITAN) which contains a vector processor and whose speed is 16MIPS. The CPU time to get a converged result is about 20 hours for 3000 iterations in the Navier-Stokes code. It is found that the big super computer, whose speed is more than 1GFLOPS, is not indispensable to predict the flow of the Francis runner with the Navier-Stokes analysis. This fact implies that this code will be useful as a design tool even in the small office where supercomputers might not be available.

CONCLUSIONS

A three-dimensional incompressible Navier-Stokes code with the pseudo-compressibility, implicit finite difference formulation, a high accuracy TVD scheme and two turbulence models, has been developed and applied to the flow around the Francis runner. The simulation results with both turbulence models implemented show a reasonable comparison to the experimental data and to the Euler solution. The Navier-Stokes results show the proper influence of the viscosity, such as the secondary flow owing to the rotation. The difference between the two turbulence models is small in the design-flow condition used in this paper. In order to decrease the computational time, the implicit scheme and the wall function near the surface for the $k - \epsilon$ model are introduced efficiently and this fact makes this code capable of running efficiently even on small engineering work station.

ACKNOWLEDGEMENT

The authors would like to acknowledge Prof.I.L.Ryhming and Prof.P.Henry, the organizers of the GAMM WORKSHOP 1989, for providing the runner geometry, experimental data and discussions.

REFERENCES

[1] Matsuo,Y., Arakawa,C.,Saito,S. and Kobayashi,H., 1989, *Navier-Stokes Simulations around a Propfan Using Higher-order Upwind Scheme*, AIAA Paper 89-2699.

[2] Beam,R.F. and Warming,R.M., 1976, *An Implicit Finite-Difference Algorithm for Hyperbolic Systems in Conservation Laws,* Journal of Computational Physics, Vol.22, pp.87-110.

[3] Baldwin,B.S. and Lomax,H., 1978, *Thin Layer Approximation and Algebraic Model for Separated Turbulent Flows,* AIAA Paper 78-257.

[4] Neury,C. and Bottaro,A., 1990, *Influence of Mesh Type in the Simulation of Hydraulic Machines,* Proceedings, IAHR Symposium 1990, Belgrade Yugoslavia, Vol.1 C4, pp.1-13.

[5] Arakawa,C., Samejima,M., Matsuo,Y., and Kubota,T., 1990, *Numerical Simulation of Francis Runner Using the Pseudo-Compressibility,* Proceedings, IAHR Symposium 1990, Belgrade Yugoslavia, Vol.1 C1, pp.1-10.

[6] Chorin,A.J., 1967, *A Numerical Method for Solving Incompressible Viscous Flow Problems,* Journal of Computational Physics, Vol.2, pp.12-26.

[7] Rogers,S.E., Kwak,D. and Kiris,C., 1989, *Numerical Solutions of the Incompressible Navier-Stokes Equations for Steady-State and Time-Dependent Problems,* AIAA Paper 89-0463.

[8] Chakravarthy,S.R. and Oshers,S., 1985, *A New Class of High Accuracy TVD Schemes for Hyperbolic Conservations Laws,* AIAA Paper 85-0369.

[9] Obayashi,S. and Fujii,K., 1985, *Computation of Three - Dimensional Viscous Transonic Flows with the LU factored Scheme,* AIAA Paper 85-1510.

[10] Jones,W.P. and Launder,B.E., 1972, *The Prediction of Laminarization with a Two-Equation Model of Turbulence,* International Journal of Heat and Mass Transfer, Vol.15, pp.301-314.

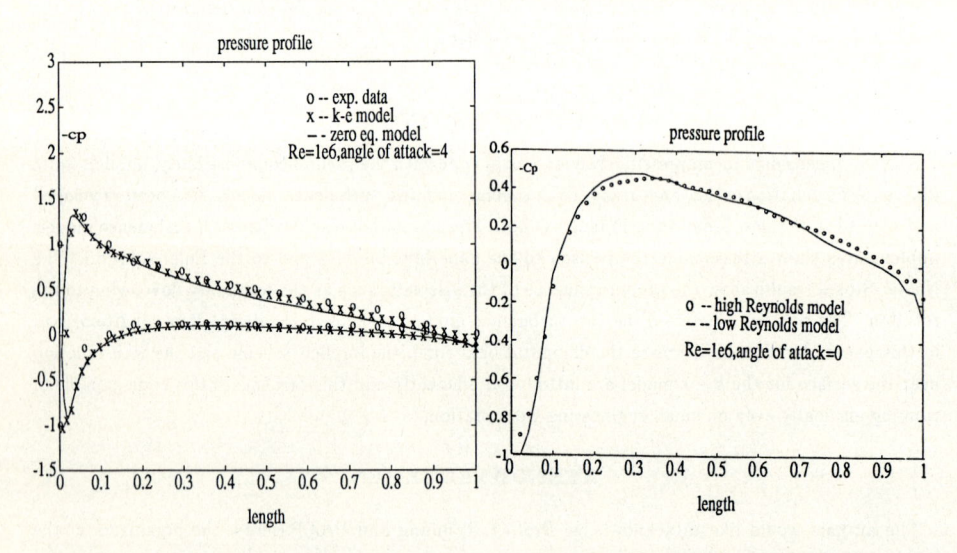

Fig.11 Pressure distributions of 2-D aerofoil with two turbulence models compared to experimental data.

Fig.12 Predicted pressure distributions on the suction surface of 2-D aerofoil with wall function and low Reynolds number model in $k - \epsilon$ code.

CALCULATION PROCEDURE FOR UNSTEADY INCOMPRESSIBLE 3D FLOWS IN ARBITRARILY SHAPED DOMAINS

A. Kost, L. Bai, N.K. Mitra, M. Fiebig
Institut für Thermo- und Fluiddynamik
Ruhr-Universität Bochum
Universitätsstr. 150, D-4630 Bochum, FRG

SUMMARY

A finite-volume method employing non-staggered variable arrangement and Cartesian velocity components is developed for the solution of the time-dependent three-dimensional incompressible Navier-Stokes equations on curvilinear boundary-fitted grids. The solution of the continuity equation is decoupled from the momentum equations by the SIMPLEC algorithm which enforces mass conservation by solving a pressure-correction equation. The computational scheme includes a 3D elliptic grid generation method for arbitrarily shaped domains. A number of 2D and 3D steady as well as unsteady flow examples have been computed and compared with other experimental and numerical results in order to validate the present code.

INTRODUCTION

In recent years, much effort has been put into the development of numerical procedures for the calculation of steady incompressible three-dimensional flows in complex geometries. Finite-volume methods employing Cartesian velocity components and non-staggered variable arrangement have been successfully used to obtain steady-state solutions of 3D Navier-Stokes equations on curvilinear boundary-fitted grids [1,2]. However, many interesting fluid flow phenomena are essentially time-dependent, e.g. turbomachinery flows. The calculation of three-dimensional unsteady flows in complex geometries needs efficient and accurate numerical methods to obtain realistic solutions with reasonable computational effort. The purpose of the present study is to develop and validate a solver for viscous incompressible unsteady flows in arbitrarily shaped domains.

Good accuracy of numerical solutions require sufficient temporal and spatial resolution. Hence, for the calculation of general flow situations with complex boundaries, curvilinear boundary-fitted grids with an easy control of the grid resolution and higher-order time discretization are used. Several 2D and 3D steady as well as unsteady flow computations demonstrate the flexibility and accuracy of the present computational procedure.

NUMERICAL PROCEDURE

Grid generation

A differential equation method has been used for the generation of complex 2D and 3D grids. The initial grid is generated on the basis of the concept of transfinite interpolation [3]. The differential equation method involves the solution of Poisson-type partial differential equations. The source terms are determined iteratively so that certain prescribed boundary conditions can be satisfied, e.g. grid intersection angles at the boundary or specified distribution of grid lines near the boundaries [4].

Governing equations and solution scheme

The equations governing the 3D laminar unsteady flow of an incompressible viscous fluid in generalized non-orthogonal coordinates using Cartesian velocity components are written in non-dimensional form:

$$\frac{\partial U_i}{\partial x_i} = 0 \qquad (1)$$

$$\frac{\partial u_k}{\partial t} + \frac{\partial}{\partial x_i} \{U_i u_k - \frac{1}{ReJ}(B_j^i \frac{\partial u_k}{\partial x_j} + \beta_j^i \omega_k^j)\} + \frac{\partial}{\partial x_j}(\beta_k^j p) = 0 \qquad (2)$$

where

$$U_i = u_j \beta_j^i \quad ; \quad B_j^i = \beta_m^i \beta_m^j \quad ; \quad \omega_k^j = \frac{\partial u_j}{\partial x_m} \beta_k^m . \qquad (3)$$

The Einstein summation convention is adopted. The curvilinear coordinates x_1, x_2 and x_3 and the Cartesian coordinates y_1, y_2 and y_3 are shown in Fig. 1. The coordinates y_1, y_2, y_3, the Cartesian velocity components u_1, u_2, u_3, time t and pressure p are non-dimensionalized by the appropriate scales for length, velocity, time and pressure, respectively. β_j^i is the cofactor of $\partial y_i/\partial x_j$ in the Jacobian J of the coordinate transformation $y_i = y_i(x_j)$ and Re denotes the Reynolds number. The equations are discretized by employing a finite-volume approach. The flow domain is subdivided into a finite number of control volumes (CV), see Fig. 1. All dependent variables are defined in the centre point P of the CV.

The convective fluxes are determined through the so-called "deferred-correction approach", which was first suggested by Khosla and Rubin [5]. According to this technique the convective flux I_C is split into an implicit part, expressed through the upwind differencing scheme (UDS) and an explicit part containing the difference between the central differencing scheme (CDS) and the UDS approximations.

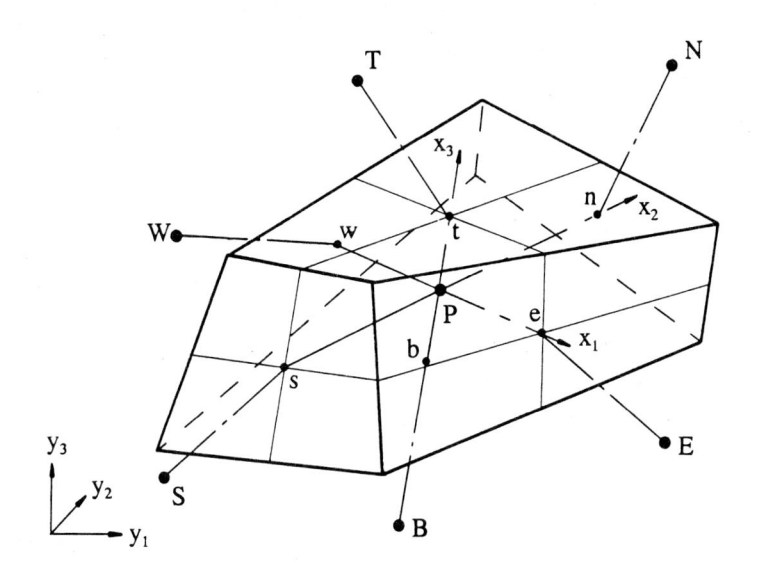

Fig. 1: Three-dimensional control volume and nomenclature

Thus:

$$I_C = I_C^{UDS} + \gamma (I_C^{CDS} - I_C^{UDS}).\tag{4}$$

The factor γ lies between 0 and 1. This technique is known to enhance the stability of the iterative solution algorithm [5].

The discretization of the time derivative in equation (2) has been chosen to be second-order accurate:

$$[\frac{\partial u_k}{\partial t}]^{n+1} \simeq \frac{1}{2\Delta t}(3u_k^{n+1} - 4u_k^n + u_k^{n-1}).\tag{5}$$

The resulting finite volume equation can then be written in a general form as:

$$a_P \Phi_P = \sum_{nb} a_{nb} \Phi_{nb} + S_C \Delta V \quad ; \quad nb = E, W, N, S, T, B$$

$$a_P = \sum_{nb} a_{nb} - S_P \Delta V \tag{6}$$

where Φ denotes the actual dependent variable and S_C and S_P are the two parts of the linearized source term $S = S_C + S_P \Phi_P$. ΔV denotes the CV volume. The coefficients a_{nb} contain the combined effects of convection and diffusion. The detailed expressions, which are similar to the expressions for steady flow, are beyond the scope of the present paper and can be found in the literature, e.g. [1].

Pressure-correction method

The velocity and pressure fields are interactively calculated with the SIMPLEC algorithm [6]. Due to the non-staggered variable arrangement the velocities at the CV faces have to be calculated from the adjacent CV-centred quantities. In order to avoid oscillations a special interpolation for the CV face mass fluxes [7] is employed. From the discretized u_1 momentum equation using relaxation factor α_u the CV-centred velocity u_P is expressed as follows:

$$u_P = \alpha_u \left[\frac{\sum_{nb} a_{nb} u_{nb} + S_C \Delta V}{a_P} + D_1^1 (p_e - p_w) \right] + (1-\alpha_u) u_P^o \qquad (7)$$

$$D_1^1 = -\frac{b_1^1}{a_P}$$

where one of the pressure difference terms has been taken out of the $S_C \Delta V$ term and b_1^1 represents the corresponding projection area. u_P^o denotes the value of the last iteration. Following the basic concept of the staggered variable arrangement [8], the interpolation for the velocities at the CV faces is given by:

$$u_e = \alpha_u \left[\overline{\left(\frac{\sum_{nb} a_{nb} u_{nb} + S_C \Delta V + a_P \frac{(1-\alpha_u)}{\alpha_u} u_P^o}{a_P} \right)}_e + \overline{(D_1^1)}_e (p_E - p_P) \right]. \qquad (8)$$

The overbars represent linear interpolation. According to the SIMPLEC algorithm the velocity corrections at the CV faces u_e' are related to the pressure corrections p_E' and p_P' as follows:

$$u_e' = -\overline{\left(\frac{b_1^1}{\frac{a_P}{\alpha_u} - \sum_{nb} a_{nb}} \right)}_e (p_E' - p_P'). \qquad (9)$$

The continuity equation in discrete form reads:

$$\sum_{nb} (F_{nb}^* + F_{nb}') = 0 \quad ; \quad \sum_{nb} F_{nb}^* = S_m \qquad (10)$$

thus,

$$\sum_{nb} F_{nb}' = -S_m \qquad (11)$$

where F_{nb}^* is the uncorrected mass flow, F_{nb}' denotes the mass flow correction and S_m is the actual mass imbalance. From these equations we get the pressure correction equation, which is expressed in form of the general equation (6). The equations are solved by using the strongly implicit procedure (SIP) of Stone [9]. Within each cycle of the SIMPLEC procedure only one iteration is performed for the momentum equations. The pressure-correction equation is iterated until the residual norm is reduced by a factor of 5 or for a maximum of 25 iterations.

APPLICATIONS

The calculation procedure presented above is applied to laminar flows in several complex 2D and 3D geometries. Emphasis is put on the calculation of unsteady flows, since the simulation of steady flows in complex geometries is well documented in the literature, e.g. [?]. For each variable, the sum of the absolute residuals in equation (6) is calculated and normalized by an appropriate quantity, typically the inlet mass or momentum flux. The calculations were declared to be converged when this sum has fallen below 10^{-3} for each variable.

Steady Laminar Flow in a 90° Bend of Square Cross Section

The flow in a square duct with 90° bend serves as an example for a strongly three-dimensional steady flow. The main flow in the streamwise direction is associated with strong secondary motion in the cross-sections arising from the centrifugal forces due to the curvature of the bend. Experimental and numerical study of this flow has been presented by Humphrey et al. [10] for a Reynolds number of 790 (based on the bulk velocity and hydraulic diameter of the duct). Their results show that perfect symmetry of the flow exists with respect to the vertical mid-plane ($x_3=0$, see Fig. 2a). For computational convenience only one symmetric half of the duct has been simulated. The velocity profile at the inlet of the duct corresponds to fully developed square-duct flow. The detailed geometrical parameters for this test case can be found in Fig. 2a which shows the 60x40x20 grid. Fig. 2b displays the calculated velocity vectors at channel mid-plane ($x_3=0$). Along the bend the peak in the velocity profile moves from the duct axis towards the outer wall. At $\vartheta=90°$ there are two peaks in the velocity profile. In Fig. 2c the computed streamwise velocities are compared with the experimental results of [10] for $\vartheta=60°$ and 90°. The reasonable agreement of the results gives confidence regarding the ability of the present method for calculation of complex 3D flows.

Unsteady Laminar Flow Around a Circular Cylinder in a Channel

The 2D flow around a circular cylinder is an example of confined bluff-body flow which is of certain practical interest. The flow is known to depend strongly on the Reynolds number and blockage ratio [11]. At higher Reynolds numbers the flow field is characterized by periodic vortex shedding. In the present study, the laminar vortex shedding in a channel with a built-in circular cylinder of diameter D is considered. Fig. 3a displays the geometry and the 100x50 non-orthogonal grid. Fig. 3b shows a close up of the grid around the cylinder. The channel has a total length of L/D=20 and a width-to-diameter ratio of 5. A fully developed parabolic velocity profile is prescribed at the channel inlet (3D upstream of the cylinder center). In the present case the vortex shedding is initiated by an asymmetric perturbation of the flow field which consists of shifting the center of the cylinder from the axis of symmetry for a few iteration steps

⟶ = 2.00

b) Velocity vectors at channel mid-plane ($x_3=0$)

a) Grid and geometrical details

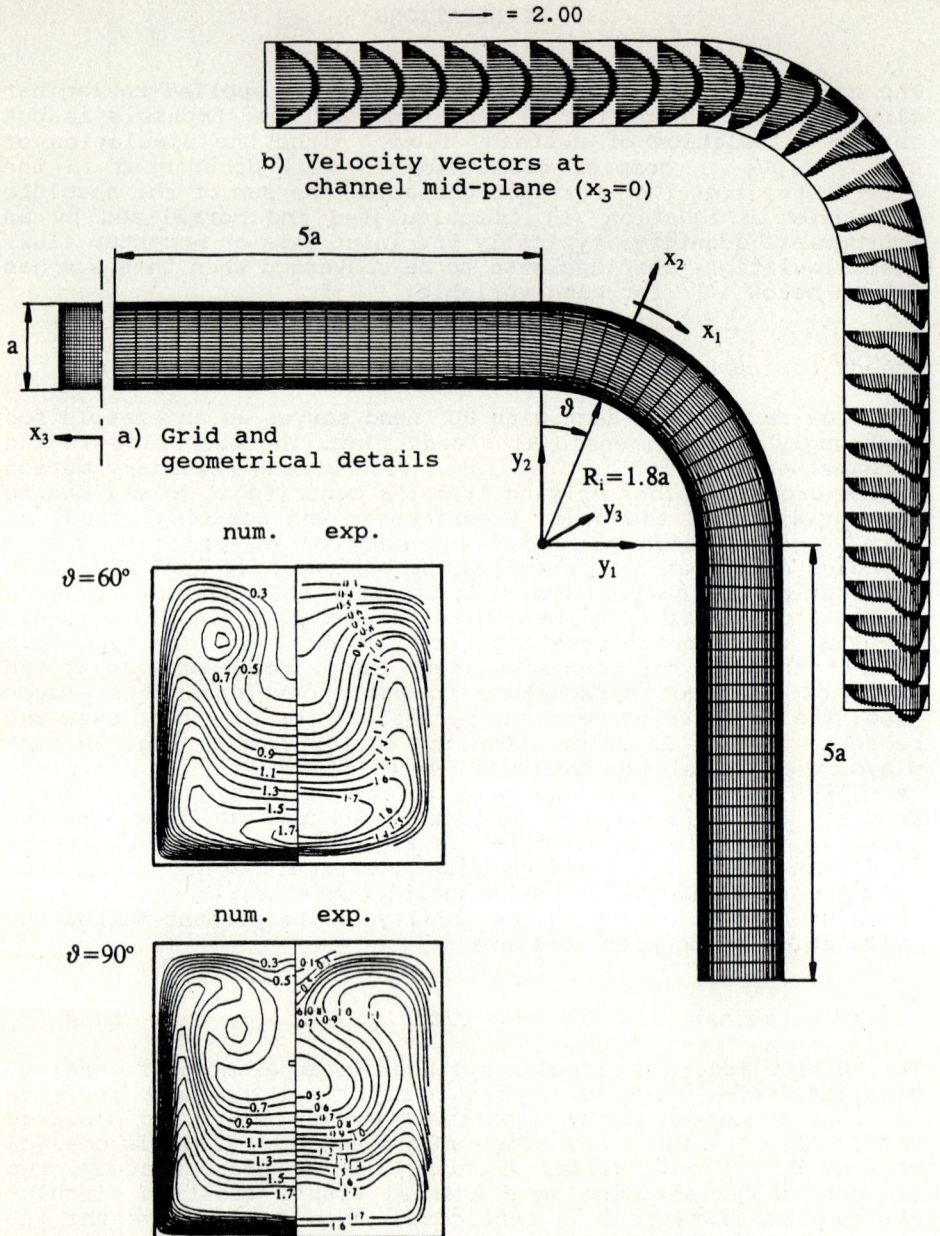

$R_i = 1.8a$

$\vartheta = 60°$

num.   exp.

$\vartheta = 90°$

num.   exp.

c) Comparison of computed and experimental [10] results

Fig. 2: Laminar flow in a 90° bend of square cross-section

at the beginning of the calculation. This flow has also been studied by Kiehm et al. [12] who found vortex shedding without imposing a perturbation. Possibly the round-off error triggered the vortex shedding. Fig. 3c displays instantaneous streamlines for one vortex shedding cycle. The periodic characteristics of the flow can be clearly seen.

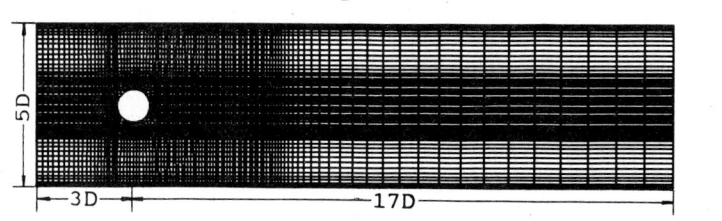

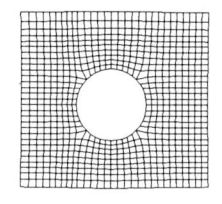

a) Grid and geometrical details

b) Close up of the grid

t=T/3

t=2T/3

t=T

c) Instantaneous streamlines for one vortex shedding cycle

Fig. 3: Laminar flow around a circular cylinder in a channel

a) Geometry and flow definitions

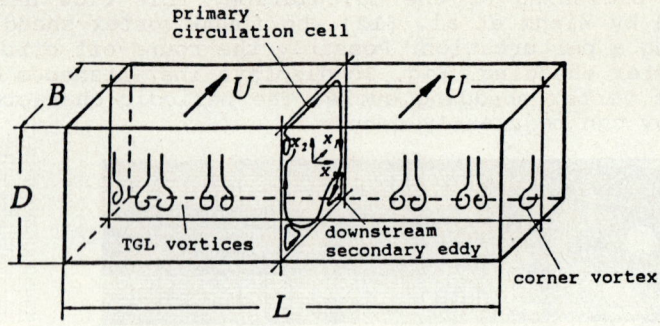

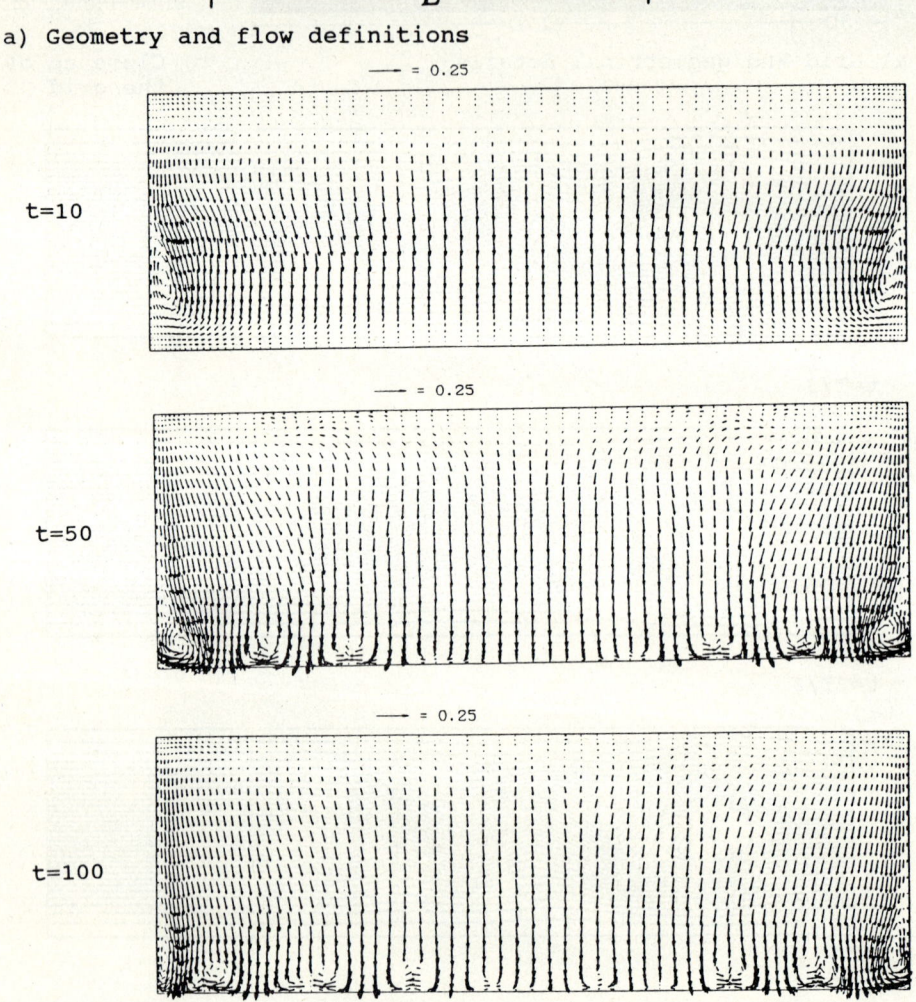

b) Time evolution of the flow field (x_1=4/15)

Fig. 4: Lid-driven cavity flow

3D Unsteady Laminar Flow in a Cavity

The 3D lid-driven cavity flow represents an example of a complex three-dimensional unsteady flow in a simple geometry and with straightforward boundary conditions. Due to the presence of the end-walls the structure of the 3D cavity flow exhibits substantial differences from two-dimensional solutions. Taylor-Görtler-like (TGL) longitudinal vortices have been experimentally observed and numerically investigated, e.g. [13]. In the present paper, the laminar flow in a lid-driven rectangular cavity is considered (see Fig. 4a for geometry and flow definitions). The flow is driven by the moving upper wall of the cavity. The Reynolds number considered (based on the speed of the upper wall and the cavity width) is taken equal to 3200. The cavity has a depth-to-width aspect ratio of $D/B=1$ and a lateral span-to-width aspect ratio of $L/B=3$. Due to computer hardware limitations only a 25x25x75 non-equidistant grid was used for this case which seems to be insufficient to resolve the TGL vortices appropriately. Fig. 4b shows the time evolution of the flow field at a plane near the downstream side-wall. The formation of two corner vortices and nine pairs of TGL vortices is indicated. This compares very well with other investigations, e.g. [13]. Further computational details and analysis of the flow structure can be found in [14].

CONCLUSIONS

A procedure has been presented for the calculation of three-dimensional unsteady incompressible flows in arbitrarily shaped domains. Curvilinear boundary-fitted grids were generated for complex 2D and 3D geometries. The finite-volume method employs non-staggered variable arrangement and Cartesian velocity components. The application examples presented have demonstrated the flexibility and efficiency of the code to handle general flow situations. However, further improvements of the code concerning convergence acceleration by multigrid methods, turbulence modeling and solution of energy equation are in progress. Once these features have been implemented in the code, this will be a powerful tool for calculating fluid flow and heat transfer problems of practical interest.

REFERENCES

[1] Perić, M.: "A finite volume method for the prediction of three-dimensional fluid flow in complex ducts", Ph.D. thesis, University of London (1985).

[2] Majumdar, S., Rodi, W. and Schönung, B.: "Calculation procedure for incompressible three-dimensional flows with complex boundaries", in: Notes on Numerical Fluid Mechanics, Vol. 25, Finite Approximations in Fluid Mechanics II, ed. by E.H. Hirschel, Braunschweig: Vieweg (1989), pp. 279-294.

[3] Thompson, J.F., Warsi, Z.U.A. and Mastin, C.W.: "Numerical grid generation. Foundations and applications", New York: North-Holland (1985).

[4] Hilgenstock, A.: "A fast method for elliptic generation of 3D grids with full boundary control", in: Numerical Grid Generation in Computational Fluid Mechanics, ed. by S. Sengupta et al., Swansea: Pineridge (1988), pp. 137-146.

[5] Khosla, P.K. and Rubin, S.G.: "A diagonally dominant second-order accurate implicit scheme", Comp. Fluids 2 (1974), pp. 207-209.

[6] Van Doormaal, J.P. and Raithby, G.D.: "Enhancement of the SIMPLE method for predicting incompressible fluid flows", Num. Heat Transfer 7 (1984), pp. 147-163.

[7] Rhie, C.M. and Chow, W.L.: "A numerical study of the turbulent flow past an isolated airfoil with trailing edge separation", AIAA J. 21 (1983), pp. 1525-1532.

[8] Patankar, S.V.: "Numerical heat transfer and fluid flow", New York: McGraw-Hill (1980).

[9] Stone, H.L.: "Iterative solution of implicit approximations of multidimensional partial differential equations", SIAM J. Num. Anal. 5 (1968), pp. 530-558.

[10] Humphrey, I.A.C., Taylor, A.M.K. and Whitelaw, I.H.: "Laminar flow in a square duct of strong curvature", J. Fluid Mech. 83 (1977), pp. 509-527.

[11] Shair, F.H., Grove, A.S., Peterson, E.E. and Acrivos, A.: "The effect of confining walls on the stability of the steady wake behind a circular cylinder", J. Fluid Mech. 17 (1965), pp. 546-550.

[12] Kiehm, P., Mitra, N.K. and Fiebig, M.: "Numerical investigation of two- and three-dimensional confined wakes behind a circular cylinder in a channel", AIAA paper 86-0035 (1986).

[13] Freitas, C.J. and Street, R.L.: "Non-linear transient phenomena in a complex recirculating flow: A numerical investigation", Int. J. Num. Meth. Fluids 8 (1988) pp. 769-802.

[14] Kost, A., Mitra, N.K. and Fiebig, M.: "Numerical simulation of three-dimensional unsteady flow in a cavity", to be published in: Proc. GAMM Workshop on "3D incompressible unsteady viscous laminar flows", Braunschweig: Vieweg.

A FULLY IMPLICIT 3-D EULER-SOLVER FOR ACCURATE AND FAST TURBOMACHINERY FLOW CALCULATION

S. Lecheler, H.-H. Frühauf
Institut für Raumfahrtsysteme, Universität Stuttgart,
Pfaffenwaldring 31, 7000 Stuttgart 80, Germany

Summary

A fully implicit three-dimensional Euler finite-difference code has been developed for the analysis of turbomachinery blade row flows and other internal flows. In the code development emphasis is laid on accuracy and fast convergence. The Euler equations are transformed to a body-fitted curvilinear coordinate system, rotating around the x-axis. Numerical boundary conditions are handled using characteristic concepts. The governing equations are discretized centrally and solved on H- and C-grids using the approximate factorization implicit finite-difference method of Beam and Warming. All boundary conditions are implemented implicitly. Typical Courant-numbers are around 45. An accurate and nearly monotone shock resolution is obtained by a nonlinear artificial dissipation model with a special treatment at the boundaries. At singular mesh points the accuracy is enhanced by a local finite-volume discretization. The code has been assessed for axial, mixed-type and radial blade rows. In this paper results will be presented for steady transonic flows. The selected configurations are a supersonic channel with 4% circular arc bump and the AGARD-WG-18 test cases E/CA-4 (low supersonic compressor cascade) and E/CO-4 (transonic compressor rotor).

1. Introduction

Many 3-D Euler codes have been developed up to now for internal and turbomachinery flow applications. Some difficulties were reported in the literature in finding accurate 3-D Euler solutions in complex flow situations such as complicated inviscid secondary flows or multishocked transonic flows in highly staggered blade rows. Difficulties were also found in computing subsonic potential flows through radial compressor rotors. Even though inviscid flows can be significantly different from viscous dominated real flows, the development of accurate Euler solvers is necessary, because they form a very important part of Navier-Stokes solvers. The aim of this paper is to describe the theory of a new 3-D Euler code very briefly. Emphasis is hereby on the presentation of important algorithmic details which enhance the accuracy and the convergence of the code.

2. Governing Equations

The governing equations are transformed to a Cartesian (x, y, z) coordinate system rotating with the angular velocity Ω around the x-axis. The transformation maintains the absolute Cartesian flow velocities in the relative system and introduces source terms in the y- and z-momentum equations. These equations are mapped to a general body-fitted (ξ, η, ζ) coordinate system using standard techniques:

$$\frac{\partial}{\partial \tau} \hat{Q} + \frac{\partial}{\partial \xi} \hat{E} + \frac{\partial}{\partial \eta} \hat{F} + \frac{\partial}{\partial \zeta} \hat{G} + \hat{H} = 0 \quad . \tag{1}$$

$\hat{Q}$ is the vector of conservative variables, $\hat{E}$, $\hat{F}$ and $\hat{G}$ are the flux vectors in ξ-, η- and ζ-direction and $\hat{H}$ is the rotational source term:

$$\hat{Q} = \frac{1}{J} \begin{bmatrix} \rho \\ \rho u \\ \rho v \\ \rho w \\ e_t \end{bmatrix} \quad \hat{E} = \frac{1}{J} \begin{bmatrix} \rho U' \\ \rho u U' + \xi_x p \\ \rho v U' + \xi_y p \\ \rho w U' + \xi_z p \\ e_t U' + p U \end{bmatrix} \quad \hat{H} = \frac{1}{J} \begin{bmatrix} 0 \\ 0 \\ -\Omega \rho w \\ \Omega \rho v \\ 0 \end{bmatrix}$$

with the absolute and relative contravariant velocity component

$$U = \xi_x u + \xi_y v + \xi_z w \qquad U' = U + \xi_t + \Omega\left(\xi_y z - \xi_z y\right) \qquad (2)$$

and the static pressure

$$p = (\kappa - 1)\left[e_t - \frac{1}{2}\rho\left(u^2 + v^2 + w^2\right)\right] \quad . \qquad (3)$$

The flux vectors $\hat{F}$ and $\hat{G}$ are obtained by changing ξ in $\hat{E}$, U and U' to η and ζ. ρ is density, u, v, w are the absolute Cartesian velocity components and e_t is the specific total energy per volume. The metric terms ξ_x, ξ_y etc. and the Jacobian J of the transformation are defined as usual, e.g. in [5]. This quantities are nondimensionalized by a reference density $\bar{\rho}_{ref}$ [kg/m³], a reference speed of sound $\bar{a}_{ref}$ [m/s] and a reference radius $\bar{r}_{ref}$ [m].

3. Boundary conditions

A particular choice or combination of boundary conditions can have a considerable influence on the accuracy and stability of the computational scheme. The most accurate numerical boundary conditions for the Euler equations are characteristic compatibility relations. They guarantee a low entropy error and an accurate convection of total pressure at the boundaries. Furthermore, numerical calculations showed that they improve the robustness of the procedure remarkably, especially if skewed grids or short inlet and outlet regions are used. We followed the concept of Chakravarthy [2], which couples the compatibility relations and the time differenced physical boundary conditions, leading to a boundary condition matrix formulation. A physical boundary condition B_i is written as $B_i = 0$ and a time derivation yields

$$\frac{\partial}{\partial \tau} B_i = \frac{\partial}{\partial \hat{Q}} B_i \frac{\partial}{\partial \tau} Q = J \frac{\partial}{\partial Q} B_i \frac{\partial}{\partial \tau} \hat{Q} = 0 \quad . \qquad (4)$$

The compatibility relations are linear combinations of the conservation equations. With $l_{\xi,j}$ as the j-th row eigenvector of the left eigenvector matrix L_ξ of the Jacobian-matrix $\hat{A} = \partial \hat{E}/\partial \hat{Q}$ one obtains, e.g., for a surface ξ=constant (t is the transposed of a vector)

$$(l_{\xi,j})^t \left(\frac{\partial}{\partial \tau} \hat{Q} + \frac{\partial}{\partial \xi} \hat{E} + \frac{\partial}{\partial \eta} \hat{F} + \frac{\partial}{\partial \zeta} \hat{G} + \hat{H}\right) = 0 \quad . \qquad (5)$$

The combination of these two equations (4) and (5) gives

$$M_1 \frac{\partial}{\partial \tau} \hat{Q} + M_2 \left(\frac{\partial}{\partial \xi} \hat{E} + \frac{\partial}{\partial \eta} \hat{F} + \frac{\partial}{\partial \zeta} \hat{G} + \hat{H}\right) = 0 \qquad (6)$$

where M_1 contains i physical boundary conditions and $j = 5 - i$ left eigenvectors. M_2 is similar to M_1, with zeros replacing the physical boundary conditions. Multiplying (6) by M_1^{-1} gives $M = M_1^{-1} M_2$ and the following equation results:

$$\frac{\partial}{\partial \tau} \hat{Q} + M \left(\frac{\partial}{\partial \xi} \hat{E} + \frac{\partial}{\partial \eta} \hat{F} + \frac{\partial}{\partial \zeta} \hat{G} + \hat{H}\right) = 0 \; . \qquad (7)$$

The matrix M is equal to the unit matrix I in the interior and at periodic boundaries, while at inlet, outlet and solid wall boundaries M represents the physical and numerical boundary conditions. As physical boundary conditions for subsonic flow we choose entropy $s = ln\,(\kappa\,p) - \kappa\,ln\,\rho$, total enthalpy $h_t = (e_t + p)/\rho$ and the flow directions v/u and w/u at inlet and static pressure p at outlet. For supersonic flow u is given additionally at inlet and no physical boundary condition at outlet. At solid walls the flow must be tangential (e.g. $U'=0$ for a surface ξ=constant).

The conservative flux formulation in the compatibility equations (5) does not converge for skewed meshes. In contrast, using a quasilinear formulation of the fluxes at the boundaries, e.g.

$$\frac{\partial}{\partial \xi} \hat{E} = \frac{\partial}{\partial \hat{Q}} \hat{E} \frac{\partial}{\partial \xi} \hat{Q} = \frac{\partial}{\partial \hat{Q}} \hat{E} \frac{\partial}{\partial Q} Q \frac{\partial}{\partial \xi} Q = \frac{1}{J} \hat{A} \frac{\partial}{\partial \xi} Q \quad , \tag{8}$$

the method is stable also for skewed grids. However, with this quasilinear formulation shocked flows can not be handled accurately. A more conservative and converging formulation is obtained by splitting the fluxes at the boundaries in a conservative and a quasilinear part [4] as follows:

$$\frac{\partial}{\partial \xi} \hat{E} = \frac{\partial}{\partial \xi} \hat{E}_1 + \frac{\partial}{\partial \xi} \hat{E}_2 = \frac{\partial}{\partial \xi} \hat{E}_1 + \frac{1}{J} \hat{A}_2 \frac{\partial}{\partial \xi} Q \quad , \tag{9}$$

with

$$\hat{E}_1 = \frac{1}{J} \begin{bmatrix} \rho U' \\ \rho u U' \\ \rho v U' \\ \rho w U' \\ e_t U' + p U \end{bmatrix} \qquad \hat{E}_2 = \frac{p}{J} \begin{bmatrix} 0 \\ \xi_x \\ \xi_y \\ \xi_z \\ 0 \end{bmatrix} \qquad \hat{A}_2 = \frac{\partial}{\partial \hat{Q}} \hat{E}_2 \quad . \tag{10}$$

Another possibility is a more complex splitting with a projection of the conservative equations onto the tangential plane and of the quasilinear equation into the normal direction [3].

For flows with locally high flow gradients which are not resolved adequately, e.g. around thin leading and trailing edges, non-physical local entropy layers occur. Due to the accurate convection property of the classical characteristic solid wall boundary condition, these local numerical errors are transported downstream, leading to an extended entropy layer and to false Mach number distributions. Therefore, we replaced the compatibility relation describing entropy transport in the direction tangential to the wall by the following expression

$$\frac{\partial s}{\partial n} = \begin{cases} 0 \\ \text{constant} \end{cases} \quad . \tag{11}$$

With this formulation, the entropy at the wall is extrapolated from the interior flow field. This is theoretically correct only for homentropic shock-free flows. Nevertheless, numerical calculations showed that reasonable accuracy can be obtained also for shocked flows and flows with locally high gradients.

4. Numerical method

The governing equations are solved on H- and C-grids using the implicit approximate factorization finite-difference method of Beam and Warming [1]. After implicit time differencing, linearization, factorization, with 2nd order central spatial differences δ and artificial dissipation terms $\mathcal{D}$ one obtains the discretized equation

$$\begin{aligned}&\left[I + \theta \Delta \tau_{im} M^n \left(\delta_\xi \hat{A}^n + \mathcal{D}^n_{im,\xi}\right)\right] \left[I + \theta \Delta \tau_{im} M^n \left(\delta_\eta \hat{B}^n + \mathcal{D}^n_{im,\eta}\right)\right] \\ &\left[I + \theta \Delta \tau_{im} M^n \left(\delta_\zeta \hat{C}^n + \hat{D}^n + \mathcal{D}^n_{im,\zeta}\right)\right] \Delta \hat{Q}^n \\ &= -\Delta \tau_{ex} M^n \left(\delta_\xi \hat{E}^n + \delta_\eta \hat{F}^n + \delta_\zeta \hat{G}^n + \hat{H}^n\right) + M^n \left(\mathcal{D}^n_{ex,\xi} + \mathcal{D}^n_{ex,\eta} + \mathcal{D}^n_{ex,\zeta}\right) \hat{Q}^n\end{aligned} \tag{12}$$

where $\hat{A}$, $\hat{B}$, $\hat{C}$ and $\hat{D}$ are the Jacobians of the flux vectors $\hat{E}$, $\hat{F}$ and $\hat{G}$ and of the rotational source term $\hat{H}$. θ is the usual parameter for different time discretizations in a general implicit scheme ($\theta = 1 \to$ Euler-implicit). For time asymptotic calculations a spatially variable time step, depending only on the Jacobian determinant J is used, which proved to be efficient [5]. Furthermore, by using a smaller time step on the implicit part of equation (12) the factorization error is reduced and larger explicit time steps can be used:

$$\Delta \tau_{ex} = \Delta \tau_{in} \frac{1}{1 + \sqrt{J}} \qquad \Delta \tau_{im} = \omega \, \Delta \tau_{ex} \quad , \tag{13}$$

where $\Delta \tau_{in}$ is an input value. With an optimum value of $\omega \approx 0.6$ the convergence speed can be doubled. The spatial discretization of the 3-D metric terms ξ_x, ξ_y etc. needs an additional averaging to satisfy conservation and uniform flow conditions [4,5].

5. Nonlinear dissipation

In contrast to upwind schemes, which provide enough internal damping, the central spatial discretization (and the 3-D factorization) needs some added damping terms for stability. Furthermore, an accurate and oscillation-free shock resolution can only be obtained by nonlinear damping terms. Because TVD-concepts are rather CPU-time consuming a blended nonlinear 2. and 4. order dissipation model similar to the one used in [6] is used. The explicit part, e.g. in ξ-direction, is:

$$\mathcal{D}_{ex,\xi} = \Delta_\xi^- \left[\left(\frac{\sigma_{\hat{A}} I}{J} \right)_{j+\frac{1}{2}} \left(\varepsilon^{(4)}_{ex,j+\frac{1}{2}} \Delta_\xi^+ \Delta_\xi^- \Delta_\xi^+ J_j - \varepsilon^{(2)}_{ex,j+\frac{1}{2}} \Delta_\xi^+ J_j \right) \right] \quad , \tag{14}$$

with the nonlinear coefficients:

$$\varepsilon^{(2)}_{ex,j+\frac{1}{2}} = \varepsilon^{(2)}_{ex,in} \Delta \tau_{ex,j+\frac{1}{2}} \, Max\,(\gamma_j, \gamma_{j+1}) \tag{15}$$

$$\varepsilon^{(4)}_{ex,j+\frac{1}{2}} = Max \left(0, \varepsilon^{(4)}_{ex,in} \Delta \tau_{ex,j+\frac{1}{2}} - \varepsilon^{(2)}_{ex,j+\frac{1}{2}} \right) \tag{16}$$

and the shock sensor and onedimensional spectral radius:

$$\gamma_j = \frac{|p_{j+1} - 2\,p_j + p_{j+1}|}{|p_{j+1} + 2\,p_j + p_{j+1}|} \qquad \sigma_{\hat{A}} = |U'| + a\,\sqrt{\xi_x^2 + \xi_y^2 + \xi_z^2} \quad . \tag{17}$$

The implicit artificial dissipation term, e.g. in ξ-direction, is expressed by:

$$\mathcal{D}_{im,\xi} = -\varepsilon_{im,in} \Delta_\xi^- \left(\frac{\sigma_{\hat{A}} I}{J} \right) \Delta_\xi^+ J \quad . \tag{18}$$

Δ^+ and Δ^- are first order foreward and backward difference formulas, the values at $j + 1/2$ are arithmetic mean values of j und $j + 1$ and $\varepsilon^{(4)}_{ex,in} \approx 0.04$, $\varepsilon^{(2)}_{ex,in} \approx 1.0$ and $\varepsilon_{im,in} \approx 0.25$ are optimum input values for stability, accuracy and convergence.

The characteristic formulation of the boundary conditions needs artificial damping terms at the boundaries. At periodic boundaries the same damping terms can be used as in the interior as well as in tangential directions of inlet, outlet and solid wall boundary surfaces. Modifications are necessary for the latter boundaries in the normal direction, where one-sided differences are used. Pulliam [6] investigated the stability properties of the 4th order damping terms for various difference stencils at the boundaries by an eigenvalue analysis, but with his suggested formulation we could not obtain good results for our characteristic boundary treatment. Therefore, we applied damping terms at and near the boundaries, which are consistent with the flux discretization and which are nearly conservative. At a boundary surface, e.g., $j = 1$ the fluxes in the normal direction are discretized by one-sided differences. This choice provides enough internal dissipation. Therefore no added viscosity terms are needed in the normal direction. For a neighbouring surface $j = 2$ we extrapolate the flow variable for $j - 2$ and use the same formulation (14) as in the interior flow field.

6. Implicit implementation of boundary conditions

Good convergence rates can only be obtained if all boundary conditions are implemented implicitly. Especially the implicit treatment of the periodicity condition is very important. The boundary elements have to be inserted into the block-tridiagonal matrices of the left-hand-side of equation (12). For periodic boundaries the following set of equations results, e.g., for the η-sweep:

$$\begin{bmatrix} \mathcal{M}_2 & \mathcal{N}_3 & 0 & \cdots & \mathcal{L}_{kx} \\ \mathcal{L}_2 & \mathcal{M}_3 & \mathcal{N}_4 & \cdots & 0 \\ \vdots & & \ddots & & \vdots \\ 0 & \cdots & \mathcal{L}_{jx-2} & \mathcal{M}_{kx-1} & \mathcal{N}_{kx} \\ \mathcal{N}_2 & \cdots & 0 & \mathcal{L}_{kx-1} & \mathcal{M}_{kx} \end{bmatrix} \begin{bmatrix} \Delta \hat{Q}_2 \\ \Delta \hat{Q}_3 \\ \vdots \\ \Delta \hat{Q}_{kx-1} \\ \Delta \hat{Q}_{kx} \end{bmatrix} = \begin{bmatrix} f_2 \\ f_3 \\ \vdots \\ f_{kx-1} \\ f_{kx} \end{bmatrix} \quad (19)$$

where $\mathcal{L}$, $\mathcal{M}$ and $\mathcal{N}$ denote tridiagonal elements, f is the known right-hand-side and $\Delta \hat{Q}$ is unknown. Deriving this system of equations, the relations

$$\mathcal{L}_1 \Delta \hat{Q}_1 = \mathcal{L}_{kx} \Delta \hat{Q}_{kx} \quad \text{and} \quad \mathcal{N}_{kx+1} \Delta \hat{Q}_{kx+1} = \mathcal{N}_2 \Delta \hat{Q}_2 \quad (20)$$

have been used. This is not trivial, since on surfaces of revolution periodicity is only valid in a cylindrical coordinate system. The terms $\mathcal{L}$, $\mathcal{N}$ and $\Delta \hat{Q}$ themselves, containing Cartesian velocity components, are therefore not periodic. Only the product of $\mathcal{L} \cdot \Delta \hat{Q}$ is periodic, because it represents a flux term. For inlet, outlet and solid wall boundaries, where one-sided differences of 2. order accuracy are used, the following set of equations applies:

$$\begin{bmatrix} \mathcal{M}_1 & \mathcal{N}_2 & \mathcal{O}_3 & 0 & 0 & \cdots & 0 \\ \mathcal{L}_1 & \mathcal{M}_2 & \mathcal{N}_3 & 0 & 0 & \cdots & 0 \\ 0 & \mathcal{L}_2 & \mathcal{M}_3 & \mathcal{N}_4 & 0 & \cdots & 0 \\ \vdots & & & \ddots & & & \vdots \\ 0 & \cdots & & \mathcal{L}_{kx-3} & \mathcal{M}_{kx-2} & \mathcal{N}_{kx-1} & 0 \\ 0 & \cdots & & 0 & \mathcal{L}_{kx-2} & \mathcal{M}_{kx-1} & \mathcal{N}_{kx} \\ 0 & \cdots & & 0 & \mathcal{O}_{kx-2} & \mathcal{L}_{kx-1} & \mathcal{M}_{kx} \end{bmatrix} \begin{bmatrix} \Delta \hat{Q}_1 \\ \Delta \hat{Q}_2 \\ \Delta \hat{Q}_3 \\ \vdots \\ \Delta \hat{Q}_{kx-2} \\ \Delta \hat{Q}_{kx-1} \\ \Delta \hat{Q}_{kx} \end{bmatrix} = \begin{bmatrix} f_1 \\ f_2 \\ f_3 \\ \vdots \\ f_{kx-2} \\ f_{kx-1} \\ f_{kx} \end{bmatrix} \quad (21)$$

Implicit boundary conditions are implemented for H- and C-type meshes. Coupling two periodic boundaries or two one-sided boundaries, the boundary conditions can be treated implicitly with system (19) and (21) respectively. For C-grids it is necessary to use a half pitch shifted domain for the η-sweep, coupling two solid wall boundaries (Fig. 1). Equations (19) and (21) are solved by two different lower-upper-decompositions.

7. Discretization at singular mesh points

At singular mesh points, e.g. at trailing edges, the central spatial discretization of the metric and flux terms is not unique. This leads to local inaccuracies, which can be significant in regions of steep flow gradients. An equivalent finite-volume discretization with a local flux box (Fig. 2) is used for the computation of flow quantities at these points. It can be shown [4] that this is equivalent to a finite-difference discretization with double the number of grid points. Time discretization has to be explicit in order not to destroy the tridiagonal structure of the left-hand-side matrices. The discretized equation at singular mesh points is:

$$\hat{Q}_{k,j,l}^{n+1} = \hat{Q}_{k,j,l}^n - \frac{1}{8} \Delta \tau_{k,j,l} \left[\left(\hat{E}_{j+1} - \hat{E}_{j-1} \right)_{k+\frac{1}{2}, l+\frac{1}{2}} + \left(\hat{E}_{j+1} - \hat{E}_{j-1} \right)_{k-\frac{1}{2}, l+\frac{1}{2}} \right.$$
$$\left. + \left(\hat{E}_{j+1} - \hat{E}_{j-1} \right)_{k+\frac{1}{2}, l-\frac{1}{2}} + \left(\hat{E}_{j+1} - \hat{E}_{j-1} \right)_{k-\frac{1}{2}, l-\frac{1}{2}} + \left(\hat{F}_{k+1} - \hat{F}_{k-1} \right)_{j+\frac{1}{2}, l+\frac{1}{2}} \right.$$

$$+ \left(\hat{F}_{k+1} - \hat{F}_{k-1}\right)_{j-\frac{1}{2},l+\frac{1}{2}} + \left(\hat{F}_{k+1} - \hat{F}_{k-1}\right)_{j+\frac{1}{2},l-\frac{1}{2}} + \left(\hat{F}_{k+1} - \hat{F}_{k-1}\right)_{j-\frac{1}{2},l-\frac{1}{2}} \quad (22)$$

$$+ \left(\hat{F}_{k_u} - \hat{F}_{k_o}\right)_{j+\frac{1}{2},l+\frac{1}{2}} + \left(\hat{F}_{k_u} - \hat{F}_{k_o}\right)_{j+\frac{1}{2},l-\frac{1}{2}} + \left(\hat{G}_{l+1} - \hat{G}_{l-1}\right)_{k+\frac{1}{2},j+\frac{1}{2}}$$

$$+ \left(\hat{G}_{l+1} - \hat{G}_{l-1}\right)_{k-\frac{1}{2},j+\frac{1}{2}} + \left(\hat{G}_{l+1} - \hat{G}_{l-1}\right)_{k+\frac{1}{2},j-\frac{1}{2}} + \left(\hat{G}_{l+1} - \hat{G}_{l-1}\right)_{k-\frac{1}{2},j-\frac{1}{2}} \Big] = 0$$

where the fluxes are averaged e.g. $\hat{E}_{k+\frac{1}{2}} = (\hat{E}_k + \hat{E}_{k+1})/2$.

8. 3-D C-grid generation

It is well known that the truncation error increases with the mesh obliqueness and the local stretching. Therefore the quality of the grid is very important for accuracy. The generation of proper three-dimensional meshes around complex geometries like turbomachinery bladings is not easy and, furthermore, time consuming. We developed a program for the generation of three-dimensional C- and H-type grids which can be used for axial, mixed-type and radial blade rows of turbomachines [4]. The blade-like mesh surface is generated by algebraic functions while for surfaces of revolution an elliptic differential equation is solved. Thereby a good control of mesh points at boundaries and a smooth and nearly orthogonal distribution in the interior is achieved.

9. Results

The 3-D Euler-code has been tested for axial, mixed type and radial turbine and compressor blade rows [4]. In this paper we present solutions for well-known benchmark configurations, in order to show the convergence and accuracy of the 3-D Euler solver developed. The selected examples are a supersonic channel with a 4% circular arc bump, the AGARD-WG-18 test cases E/CA-4 (DLR low supersonic compressor cascade) and E/CO-4 (DLR transonic compressor rotor). For all examples the input values for nonlinear dissipation are the ones already mentioned in section 5.

The implicit implementation of the periodicity condition improves the convergence rate by a factor of 7.5 in comparison to an explicit treatment (Fig. 3). Furthermore the residual is much smoother. An additional increase in the convergence rate can be obtained with a reduced implicit time step ($0.6 \leq \omega \leq 1$) in equation (13). The following table summarizes typical convergence-related quantities for transonic flows on coarse and fine grids. CN_{opt} is the maximum Courant-number for optimal convergence and Ma-convergence is achieved, when the local Mach number at leading edge is converged (Ma-convergence for an interior point needs about 20 % less iterations). A mesh and problem independent Courant number of about 45 can be obtained.

Example	Ma_{max}	Grid	CN_{opt}	Iter. until Ma-converg.	Iter. for one order decrease in L_2-Res
Compressor cascade	1.3	201x25-C	45	300	100
Compressor rotor	1.6	101x13x9-C	41	100	100
Compressor rotor	1.6	201x25x17-C	45	800	350

In Fig. 4 results are presented for the 2-D supersonic flow through a channel with a 4 % circular arc bump and an inflow Mach number 1.4. This example is selected to show the shock resolution of the central difference scheme with added nonlinear dissipation and the solution accuracy close to the wall using the classical characteristic solid wall boundary conditions. The grid lines of the 141x41 H-grid are not adapted to the shocks (Fig. 4c). To resolve the shock 3 to 4 grid points are necessary. The entropy jumps across the shocks at leading and trailing edges are well predicted (Fig. 4b) in comparison to the ones calculated by one-dimensional theory ($\Delta s_{theory,LE} = 0.0049$, $\Delta s_{theory,TE} = 0.0035$). However, there are still oscilllations in front of the shocks, which produce small overshoots in the contour Mach-number distribution (Fig. 4a). This is due to the fact that the simplifications used in the nonlinear dissipation model lead only to an approximately monotone scheme. The oscillations could be reduced with higher input values, leading, however, to incorrect entropy jumps.

In the second example the flow is computed through a low supersonic compressor cascade ($\beta_s = 138.51°$, $t/c = 0.621$, $c = 90mm$) with a very thin round leading edge (0.5 % c) and a wedge-type trailing edge, referred to AGARD test case E/CA-4. The inlet flow conditions are $M_1 = 1.03$ and $\beta_1 = 148.5°$. A fine (201x25-C) grid computation is performed. The small leading edge radius is resolved by 15 points. Fig. 5b shows the transonic flow. A strong supersonic expansion and recompression occurs at the thin leading edge (Figs. 5a,c). This complex chocked flow with locally very high flow gradients is computed to demonstrate the accuracy enhancement which can be obtained by characteristic boundary equations. Here, the entropy extrapolation (11) has to be used in order not to convect the numerical entropy errors, which are generated at the leading edge, further downstream. The entropy jump across the normal shock is accurately predicted. The entropy error at the leading edge is much smaller (Fig. 5d) in comparison to standard boundary conditions (Fig. 5e). A further reduction of this error can be obtained in our opinion only by an adequate mesh resolution in this region.

The DLR transonic axial compressor rotor (AGARD test case E/CO-4, N=20260 rpm, $\dot{m}$=17.3 kg/s) is computed with the 3-D Euler code. The original trailing edge is modified to a wedge-type trailing edge. Thereby the chord length has been extended by 4 % on average. A 2.8 % higher mass flow rate was selected to adapt the numerically computed shock system to the experimental shock locations. A fine mesh computation (201x25x17-C) was performed. At the suction surface of the blade the shock extends from hub to tip (Fig. 6a) with an axial shift towards the trailing edge in the spanwise direction. In Fig. 6b the computed relative Mach contours are compared to the experimental ones. The computed and measured flow field agrees qualitatively. The locations of the shocks are well predicted. Differences may be explained by the absence of any viscous corrections and tip clearance effects in the 3-D Euler code.

Acknowledgments

The authors acknowledge the support of Bundesministerium für Forschung und Technologie and BMW-Rolls-Royce.

References

[1] Beam R.M., Warming R.F., An Implicit Factored Scheme for the Compressible Navier-Stokes-Equations, AIAA Journal, Vol. 16, Apr. 1978

[2] Chakravarthy S.R., Euler Equations - Implicit Schemes and Implicit Boundary Conditions, AIAA Paper 82-0228, Jan. 1982

[3] Küster U., Boundary Procedures for the Euler-Equations, Notes on Numerical Fluid Mechanics, Vol .13, Vieweg, Sep. 1985

[4] Lecheler S., Ein Voll-implizites 3-D Euler-Verfahren zur genauen und schnellkonvergenten Strömungsberechnung in Schaufelreihen von Turbomaschinen, Doctoral Thesis, to be published in 1992

[5] Pulliam T.H., Efficient Solution Methods for the Navier-Stokes-Equations, VKI Lecture Series on Numerical Techniques for Viscous Flow Computation in Turbomachinery Bladings, Brussels, Belgium, Jan. 1986

[6] Pulliam T.H., Artificial Dissipation Methods for the Euler Equations, AIAA-Journal, Vol. 24, Dec. 1986

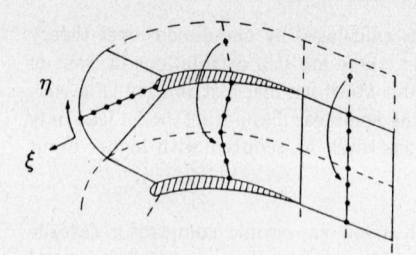

Fig. 1 Implicit implementation of periodicity condition

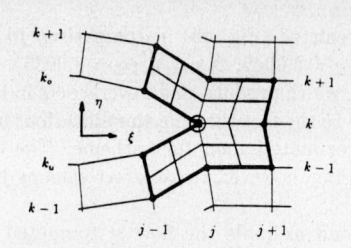

Fig. 2 Finite volume flux box for singular mesh points

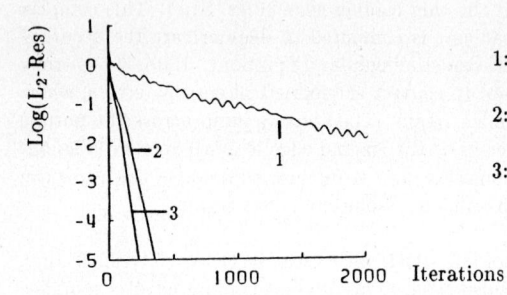

1: explicit periodicity condition and $\omega = 1.0$
2: implicit periodicity condition and $\omega = 1.0$
3: implicit periodicity condition and $\omega = 0.6$

Fig. 3 Convergence improvements due to implicit periodicity condition and reduction of implicit time step

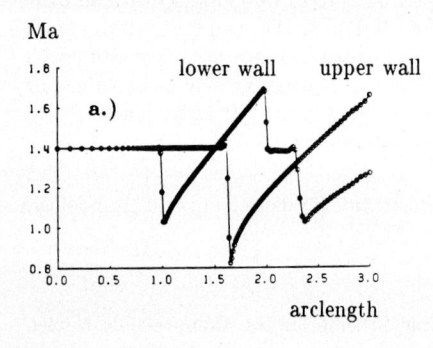

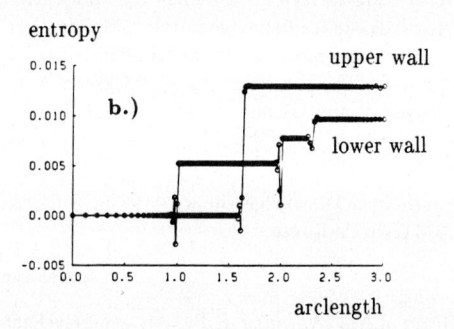

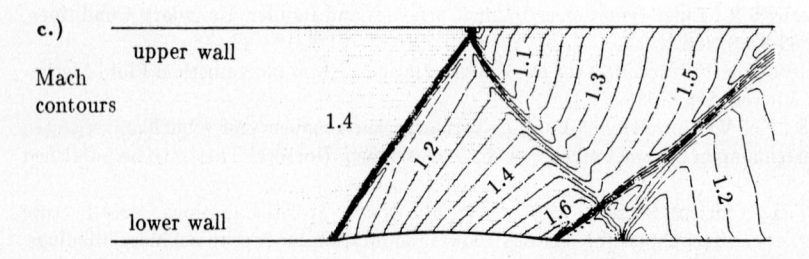

Fig. 4 Supersonic channel with 4 % circular arc

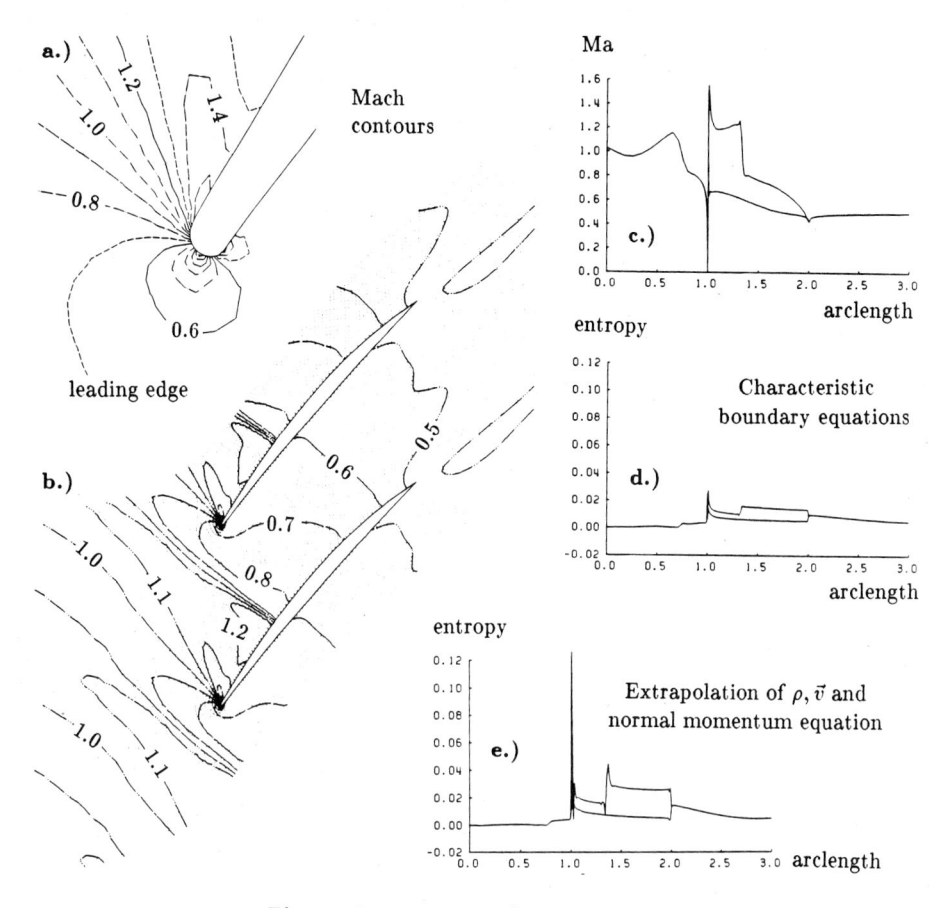

Fig. 5 Low supersonic compressor cascade

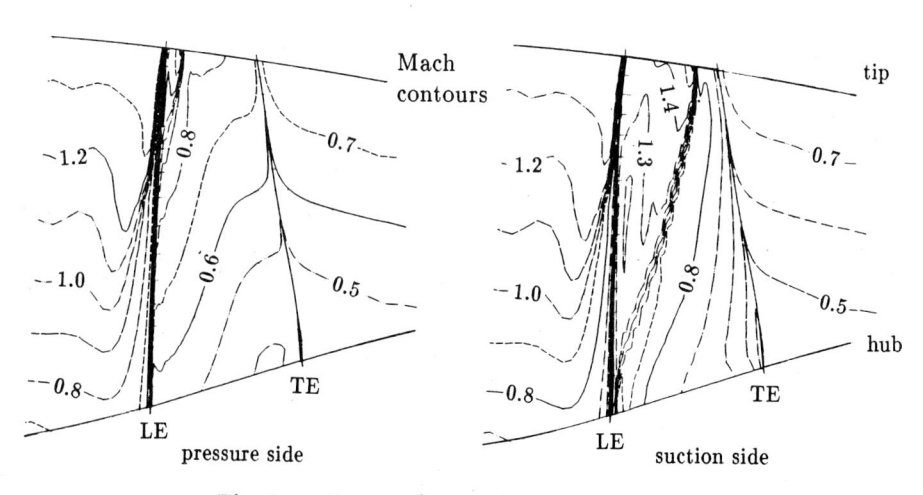

Fig. 6 a Transonic compressor rotor

89 % span

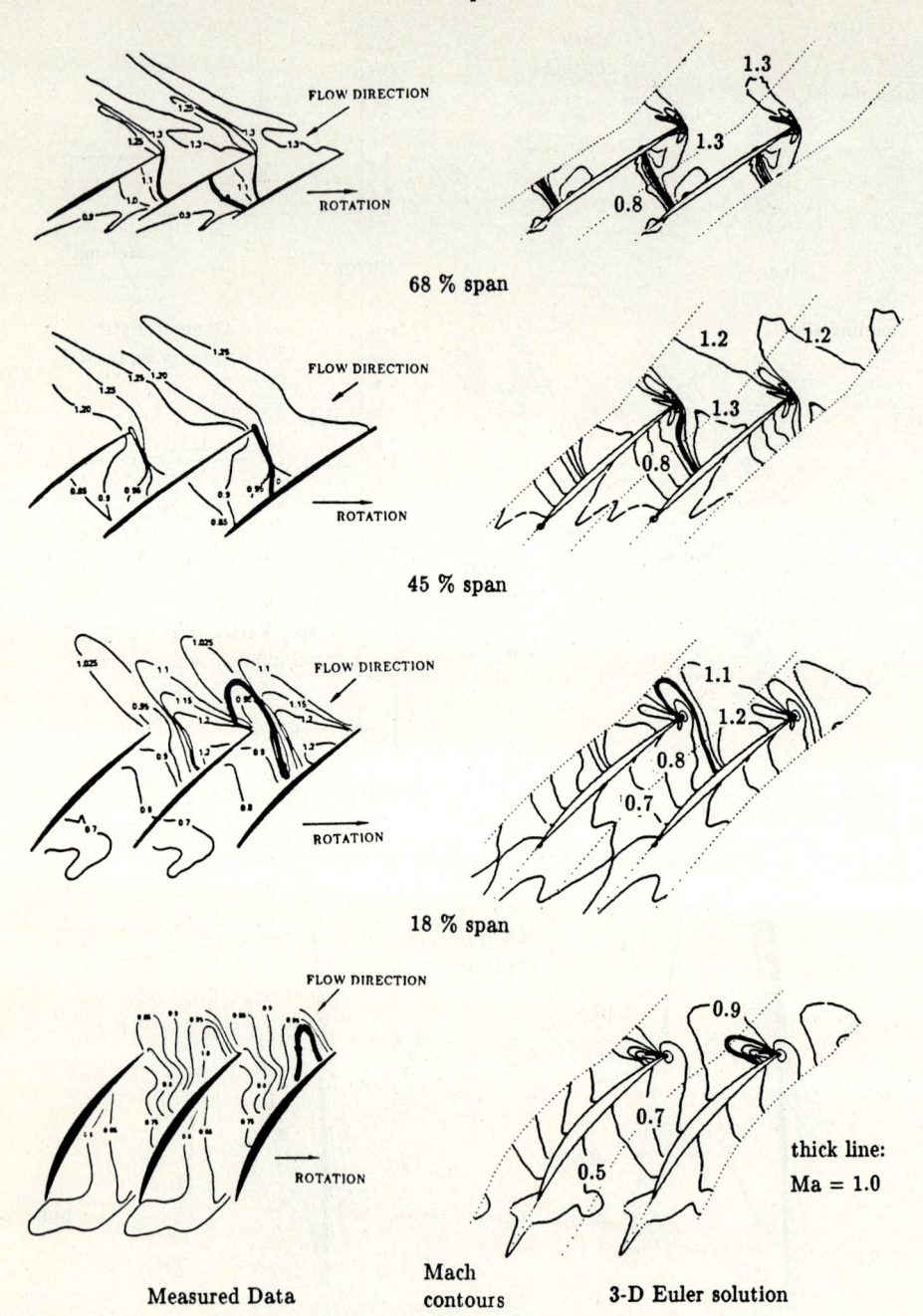

68 % span

45 % span

18 % span

Measured Data   Mach contours   3-D Euler solution

thick line: Ma = 1.0

Fig. 6 b   Transonic compressor rotor

GALERKIN/LEAST–SQUARES APPROXIMATIONS OF INCOMPRESSIBLE FLOW PROBLEMS

Gert Lube
Department of Mathematics, Magdeburg University of Technology,
PF 4120, O-3010 Magdeburg, Germany
Andreas Auge
Institute of Fluid Mechanics, Dresden University of Technology,
Mommsenstr. 13, O-8027 Dresden, Germany

Abstract

We consider stabilized mixed finite element methods for incompressible flow problems which do not require satisfaction of the Babuška-Brezzi condition and thus allow for arbitrary velocity-pressure interpolations. Least–squares forms of the Stokes or Navier–Stokes equations are added to the basic Galerkin discretization in order to stabilize the discrete problem without sacrificing accuracy of the solution. Stability and convergence results on non–uniform meshes are given in the whole range from diffusion to convection dominated situations. Numerical results in 2D are presented for low order velocity/pressure interpolation.

INTRODUCTION

Stabilization techniques in mixed finite element methods for incompressible flow problems prevent those numerical oscillations that might be generated by inappropriate combinations of velocity/pressure interpolation functions or by dominant convective terms. Basic ideas of such methods can be found in [1], [2], [3] for convection–diffusion problems and for more complicated problems as in case of Euler and Navier–Stokes equations.

INCOMPRESSIBLE NAVIER–STOKES PROBLEM

Let $\Omega \subset R^d, d \leq 3$ be the flow domain with a piecewise smooth boundary Γ. We consider the following velocity/pressure formulation of Navier–Stokes type equations governing steady incompressible flows

$$N(a, \hat{u}) = -\nu \Delta u + (a \circ \nabla)u + \nabla p = f \quad \text{in } \Omega \quad (1)$$
$$\nabla \circ u = 0 \quad \text{in } \Omega \quad (2)$$

where $\hat{u} = (u, p)$, u and p are velocity and pressure, and f is a given body force. ν is the inverse Reynolds number. We consider two cases:
case I: a: given vector field with $\nabla \circ a = 0$ (linearized Navier-Stokes problem) (LNS)
case II: $a = u$ (Navier-Stokes problem). (NS)

For simplicity we analyze only homogeneous Dirichlet boundary conditions

$$u = 0 \quad \text{on } \Gamma. \tag{3}$$

Combinations of essential and natural boundary conditions are possible [4].

By $W^{t,p}(D)$ we denote a Sobolev space with a corresponding norm or seminorm

$$\|\cdot\|_{t,p,D} \quad , \quad |\cdot|_{t,p,D} \quad .$$

Further let

$$X = V \times Q \quad , \quad V = W_0^{1,2}(\Omega)^d \quad , \quad Q = L_0^2(\Omega)$$

where V is the space of vector functions with square integrable generalized first order derivatives and zero boundary values and Q is the space of square integrable functions with vanishing mean value. For given

$$a \in H_{div}(\Omega) = \{v \in L^2(\Omega)^d : \nabla \circ v \in L^2(\Omega)\} \quad , \quad \nabla \circ a = 0$$

there exists a unique solution $\hat{u} = (u,p) \in X$ of (LNS). In case II there exists at least one solution $\hat{u} \in X$ of (NS) which is additionally unique for small data. The solution of (NS) is "essentially" finite [5].

GALERKIN/LEAST-SQUARES STABILIZATION

The finite element discretization is achieved by dividing Ω into regular elements K_j, $j = 1, \cdots, N$ with diameter h_j. Let $T_h = \{K_j\}$. We assume, for simplicity, $\bar{\Omega} = \cup \bar{K}_j$. With this discretization, we define the following conforming finite element interpolation function spaces for velocity and pressure

$$X_h = V_h \times Q_h \subset X = V \times Q \tag{4}$$
$$V_h = \{v \in V : v|_K \in P_l(K)^d, \forall K \in T_h\} \tag{5}$$
$$Q_h = \{q \in Q : q|_K \in P_k(K), \forall K \in T_h\} \tag{6}$$

with integers l, k, $l \geq 1$, $k \geq 0$. $P_s(K)$ denotes the set of (piecewise) polynomials of degree s. $(\cdot, \cdot)_D$ is the inner product in $L^2(D)$. Then the basic Galerkin discretization of (1) - (3) reads:

Find $\hat{u}_h = (u_h, p_h) \in X_h$ such that $\forall \hat{v}_h = (v_h, q_h) \in X_h$

$$B_G(\hat{u}_h, \hat{v}_h) = \nu(\nabla u_h, \nabla v_h) +$$
$$\frac{1}{2}\{((a \circ \nabla)u_h, v_h)_\Omega - ((a \circ \nabla)v_h, u_h)_\Omega\} - (p_h, \nabla \circ v_h)_\Omega = (f, v_h)_\Omega \tag{7}$$
$$(\nabla \circ u_h, q_h)_\Omega = 0. \tag{8}$$

Numerical oscillations in such mixed finite element methods might be generated by

i. inappropriate combinations of velocity/pressure interpolation functions which do not pass the inf–sup condition

$$\inf_{q_h \in Q_h} \sup_{v_h \in V_h} \frac{(q_h, \nabla \circ v_h)_\Omega}{\|q_h\|_Q |v_h|_V} \geq \gamma > 0 \qquad (9)$$

with a mesh–independent constant γ [6],

ii. the presence of dominant convective terms such that for the local Reynolds number the following inequality holds:

$$Re_j = \frac{\|a\| h_j}{2\nu} > 1 \quad . \qquad (10)$$

The idea of Galerkin/least–squares (GLS) methods consists of adding least–squares residuals of (1) and (2) to (7) and (8), respectively, so that the problem becomes:

Find $\hat{u} = (u_h, p_h) \in X_h$ such that $\forall \hat{v}_h = (v_h, q_h) \in X_h$

$$B_G(\hat{u}_h, \hat{v}_h) + \sum_{j=1}^{N} \delta_j \Big(N(a, \hat{u}_h) - f, N(a, \hat{v}_h) \Big)_{K_j} = (f, v_h)_\Omega \qquad (11)$$

$$(\nabla \circ u_h, q_h)_\Omega + \sum_{j=1}^{N} \tau_j (\nabla \circ u_h, \nabla \circ v_h)_{K_j} = 0 \qquad (12)$$

where δ_j and τ_j are parameters to be specified below. (11) and (12) are consistent with (1) to (3) in the sense that an exact solution still satisfies the stabilized formulation. With vanishing parameters we get again (7), (8). In the following we discuss the effect of the stabilization terms in the linearized problem (LNS).

CHOICE OF PARAMETERS AND A MODIFIED INF–SUP CONDITION

Simple low order pairs of velocity/pressure interpolation functions (as P1/P1, Q1/Q1 or Q1/P0) which are attractive from the computational point of view (with respect to unsteady 3D flow with adaptive mesh refinement and multigrid methods) do not pass the inf–sup condition (9). Using an idea of Verfürth, cf. [7], we can prove a modified inf–sup condition which guarantees the solvability of (11), (12) for arbitrary velocity/pressure interpolation. Let $\|\| \cdot \|\|$ be a norm on X_h defined by

$$\|\|\hat{u}_h\|\|^2 = \nu |u_h|_{1,2,\Omega}^2 + \nu \|p_h\|_{0,2,\Omega}^2 + \sum_{j=1}^{N} \tau_j \|\nabla \circ u_h\|_{0,2,K_j}^2$$

$$+ \sum_{j=1}^{N} \delta_j \| - \nu \Delta u_h + (a \circ \nabla) u_h + \nabla p_h \|_{0,2,K_j}^2 \qquad (13)$$

which characterizes the "stabilizing" effect of (11), (12) for $\delta_j, \tau_j > 0$. Then it holds for all $\hat{u}_h \in X_h$ [8]:

$$\sup_{\hat{v}_h \in X_h} \left\{ B_G(\hat{u}_h, \hat{v}_h) + \sum_{j=1}^{N} \delta_j \Big(N(a, \hat{u}_h), N(a, \hat{v}_h) \Big)_{K_j} + (\nabla \circ u_h, q_h)_\Omega \right.$$

$$\left. + \sum_{j=1}^{N} \tau_j (\nabla \circ u_h, \nabla \circ v_h)_{K_j} \right\} / |||\hat{v}_h||| \geq \gamma |||\hat{u}_h||| \quad (14)$$

if

$$\delta_j = \delta_0 h_j \|a\|^{-1} \zeta(Re_j) \quad (15)$$

$$\zeta(t) = \min\{1; \frac{1}{3}t\} \quad \text{(cf. fig. 1)} . \quad (16)$$

$\delta_0 > 0$ is a "tuning" parameter. In the diffusion dominated case (as Stokes flow) and

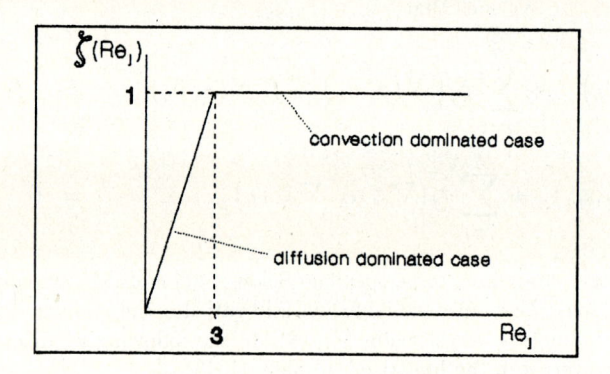

Figure 1: Definition of function $\zeta(Re_j)$

in the convection dominated case

$$\delta_j = O(h_j^2 \nu^{-1}) \quad \text{and} \quad \delta_j = O(h_j \|a\|^{-1}) \; ,$$

hold, respectively.

Furthermore, we choose (with an additional "tuning" parameter $\tau_0 > 0$)

$$\tau_j = \tau_0 \max\{\nu; h_j\} \quad (17)$$

in order to achieve best interpolation results with (15) to (17) in X_h with respect to the expression

$$|||\hat{u} - \pi\hat{u}|||^2 + \sum_{j=1}^{N} \delta_j^{-1} \|u - \pi u\|_{0,2,K_j}^2 \quad (18)$$

[9], [8]. $\pi : V \times [Q \cap W^{1,2}(\Omega)] \to X_h$ denotes an interpolation operator.

ERROR ANALYSIS FOR LINEARIZED PROBLEMS

For the linearized Navier–Stokes model (LNS) with

$$N(a, \hat{u}) = -\nu \Delta u + (a \circ \nabla) u + \nabla p \tag{19}$$

we find for sufficient smooth solutions $\hat{u} = (u, p)$ with

$$u \in V \cap W^{t+1,2}(\Omega)^d \quad , \quad p \in Q \cap W^{s+1,2}(\Omega) \tag{20}$$

and integers $1 \leq t \leq l$, $0 \leq s \leq k$, That there exists a unique discrete solution $\hat{u}_h = (u_h, p_h) \in X_h$ of (11), (12) such that

$$\begin{aligned}
|||\hat{u}_h - \hat{u}_h|||^2 &= \nu |u - u_h|_{1,2,\Omega}^2 + \nu \|p - p_h\|_{0,2,\Omega}^2 \\
&+ \sum_{j=1}^{N} \left(\delta_j \|N(a, \hat{u}_h) - f\|_{0,2,K_j}^2 + \tau_j \|\nabla \circ (u - u_h)\|_{0,2,K_j}^2 \right) \\
&\leq C_1 \sum h_j^{2t} (\nu + h_j) |u|_{t+1,2,K_j}^2 \\
&+ C_2 \sum h_j^{2s+1} \min\{1; \frac{h_j}{\nu}\} |p|_{s+1,2,K_j}^2 \quad .
\end{aligned} \tag{21}$$

For Stokes flow ($a = 0$) we recover the result of [10]. Note that (21) is valid on an arbitrary (regular) mesh and gives control of the discrete residuals corresponding to (1), (2). This could be exploited in adaptive mesh refinement methods [11]. L^2–error estimates for velocity follow via duality argumentation.

ERROR ANALYSIS FOR THE NAVIER–STOKES MODEL

Now let $a = u$ in (19). Furthermore, we add a penalty term $\alpha(p_h, q_h)_\Omega$ in (11), (12) for technical reasons, with

$$\alpha = \max_j \{h_j\}^{1+2z} \quad , \quad z = \max\{k; l\} \quad . \tag{22}$$

There exists at least one solution of (11), (12). Let now $\hat{u} \in X$ be a regular solution of (NS) for $\nu > 0$, i.e. the Frechet derivative of the operator $\mathcal{N} : X \to X^*$ corresponding to (1) – (3) with respect to $\hat{u}$ is an isomorphism of X. Then, there exists a regular solution $\hat{u}_h \in X_h$ of (11), (12) which is unique in a sufficiently small neighbourhood of $\hat{u}$. The following asymptotic error estimate holds:

$$\begin{aligned}
&\nu |u - u_h|_{1,2,\Omega}^2 + (\nu + \alpha) \|p - p_h\|_{0,2,\Omega}^2 \\
&\leq C_3(\nu) \alpha \|p\|_{0,2,\Omega}^2 + C_4(\nu) \inf_{v \in V_h} |u - v|_{1,2,\Omega}^2 + C_5(\nu) \inf_{q \in Q_h} \|p - q\|_{0,2,\Omega}^2 \quad . \tag{23}
\end{aligned}$$

For small data according to

$$\beta \nu^{-2}\|f\|_{0,2,\Omega} \le \omega < 1 \quad , \quad \beta = \sup_{u,v,w \in V} \frac{((u \circ \nabla)v, w)_\Omega}{|u|_V |v|_V |w|_V}$$

the stabilized error estimate (21) with $a = u_h$ holds.

NUMERICAL RESULTS

We present simple 2D examples with Lagrangian P1/P1-interpolation of velocity and pressure. For further 2D results and other low–order elements we refer to [4], [12]. Theoretical results for the linear model (LNS) on a quasi–uniform mesh (with $h_j \sim h$) and smooth solutions according to (20) with $t = 1$, $s = 0, 1$ are given in Table 1. As an

Table 1: Theoretical error estimates for low–order interpolation

	diffusion dominated case	convection dominated case
$\sqrt{\nu}\|u - u_h\|_{1,2}$	$O(\sqrt{\nu}h + \frac{h^{0.5+s}}{\sqrt{\nu}})$	$O(h^{0.5+s})$
$\sqrt{\nu}\|p - p_h\|_{0,2}$	$O(\sqrt{\nu}h + \frac{h^{0.5+s}}{\sqrt{\nu}})$	$O(h^{0.5+s})$
$\|N(a, \hat{u}_h) - f\|_{0,2}$	$O(\nu + h^s)$	$O(h^s)$
$\|\nabla \circ u_h\|_{0,2}$	$O(h + \frac{h^{s+1}}{\nu})$	$O(h^s)$

accuracy test we performed calculations for Stokes problem for the fully developed Poiseuille flow with Dirichlet outlet (a1) and with free outlet (a2) and for two body force problems (b,c) - cf. [13]. Averaged numerical convergence rates are given in Table 2.

Note that for P1/P1 interpolations the discrete Laplacian in (11), (12) vanishes. So we add a consistent approximation of the discrete Laplacian following ideas in [13]. Otherwise the absence of such terms leads to boundary layer effects of the discrete pressure.

For solving the nonlinear discrete problem (11),(12) with $a = u_h$, we use a successive approximation technique. For given $\hat{u}_h^{(m)} = (u_h^{(m)}, p_h^{(m)}) \in X_h, m \in N_0$ find $\hat{u}_h^{(m+1)} = (u_h^{(m+1)}, p_h^{(m+1)}) \in X_h$ from

$$B_G(\hat{u}_h^{(m+1)}, \hat{v}_h) + \sum_{j=1}^{N} \delta_j \Big(N(a, \hat{u}_h^{(m+1)}) - f, N(a, \hat{v}_h)\Big)_{K_j} \qquad (24)$$

$$+ \alpha(p_h^{(m+1)}, r)_\Omega \qquad = (f, v_h)_\Omega$$

$$(\nabla \circ u_h^{(m+1)}, q_h)_\Omega + \sum_{j=1}^{N} \tau_j (\nabla \circ u_h^{(m+1)}, \nabla \circ v_h)_{K_j} \qquad = 0 \qquad (25)$$

a) pressure

b) velocity

c) horizontal velocity at $x = 0.5$

d) vertical velocity at $y = 0.5$

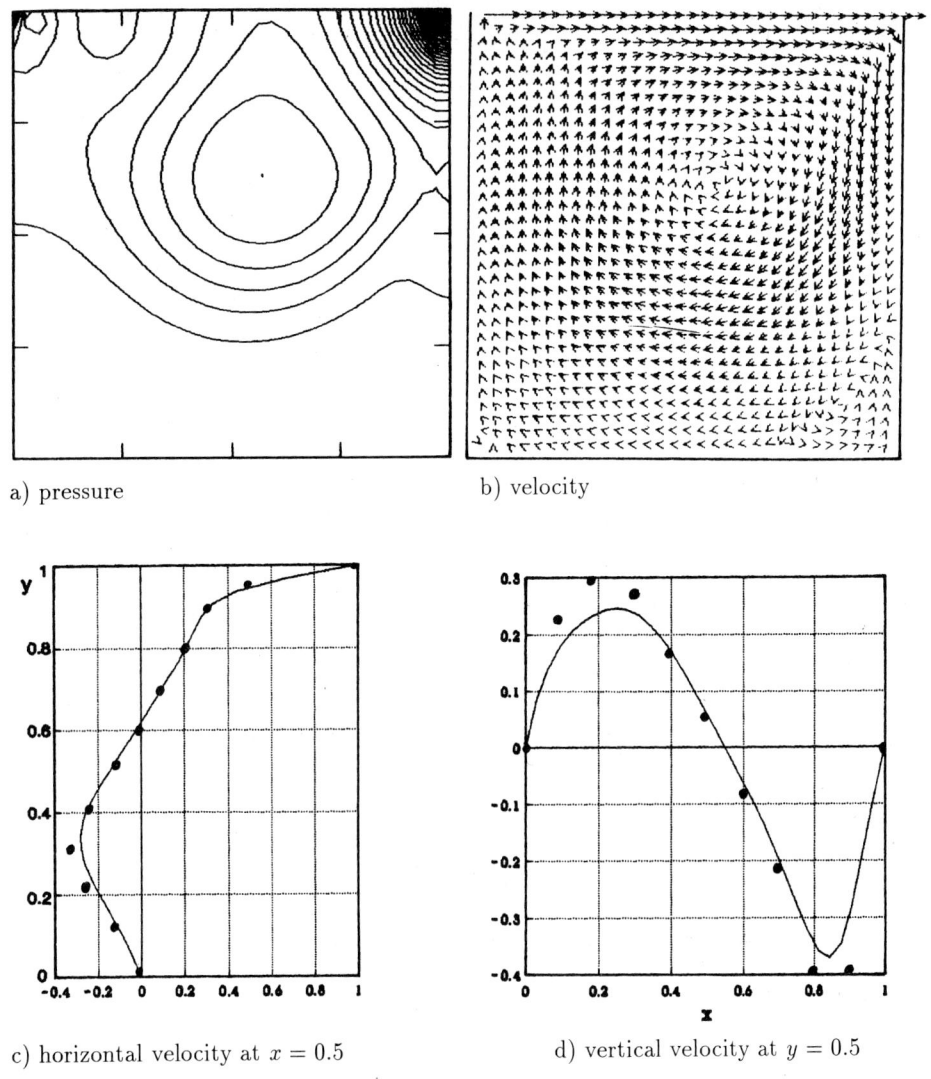

Figure 2: Driven cavity flow at $Re = 400$
— own results (32×32 mesh)
• cf. [14] (256×256 mesh)

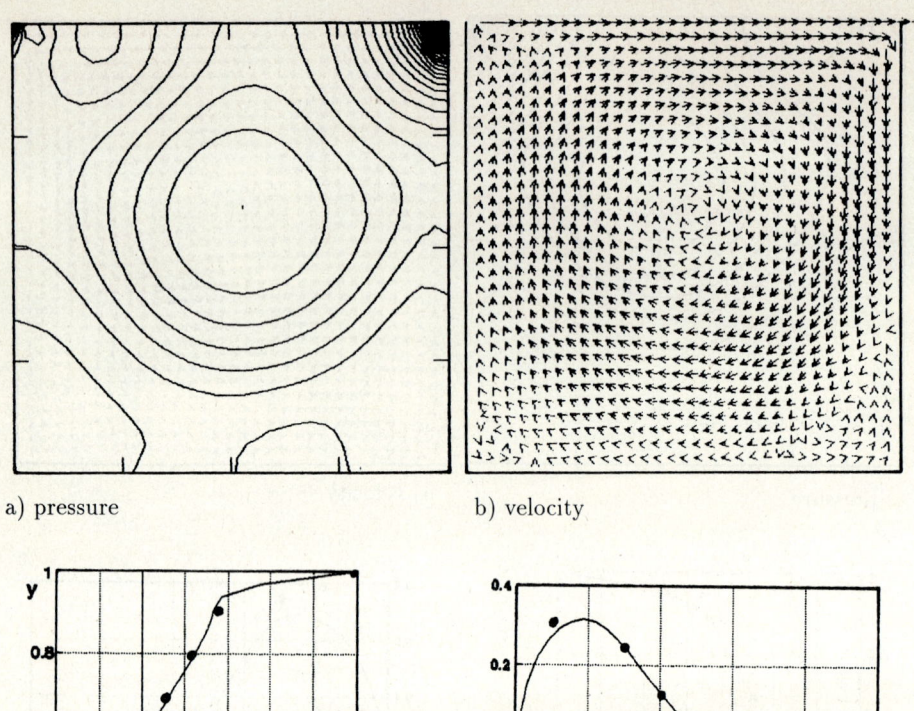

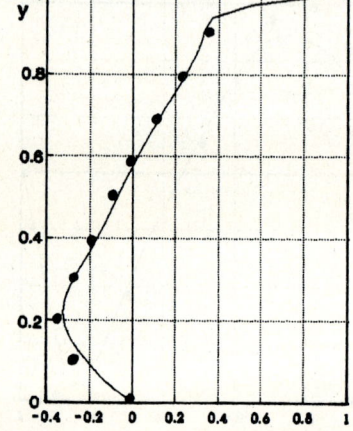

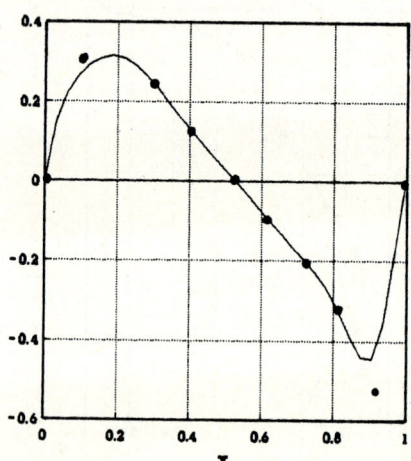

a) pressure

b) velocity

c) horizontal velocity at $x = 0.5$

d) vertical velocity at $y = 0.5$

Figure 3: Driven cavity flow at $Re = 1000$
– own results (32×32 mesh)
• cf. [14] (256×256 mesh)

Table 2: Numerical convergence rates (averaged) for Stokes flow

Example norm	(a1)	(a2)	(b)	(c)
$\|u - u_h\|_{0,2}$	1.88	1.82	1.41	1.77
$\|u - u_h\|_{0,\infty}$	1.60	1.85	1.57	1.94
$\|p - p_h\|_{0,2}$	1.89	1.64	1.42	1.75
$\|p - p_h\|_{0,\infty}$	1.11	1.09	1.08	1.21
$\|\nabla \circ (u - u_h)\|_{0,\infty}$	1.00	0.98	0.99	0.85

with $a = u_h^{(m)}$. Note that (24),(25) correspond to a linearized Navier–Stokes problem of type (11),(12). As an example we considered the standard driven cavity flow problem in the unit square. In fig. 2 to 4 we present the results for Reynolds numbers 400, 1000 and 3000 on an equidistant 32×32 mesh. These results which are comparable with those given in [12], [4] and further results in [12], [4] motivate us to proceed our investigations with low and equal-order elements. Theoretical considerations and computational implementation of coupling of the transport equation of momentum with that of energy are in preparation.

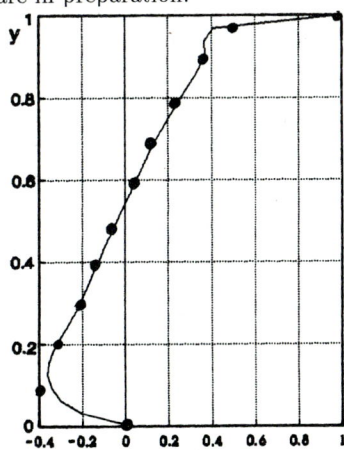

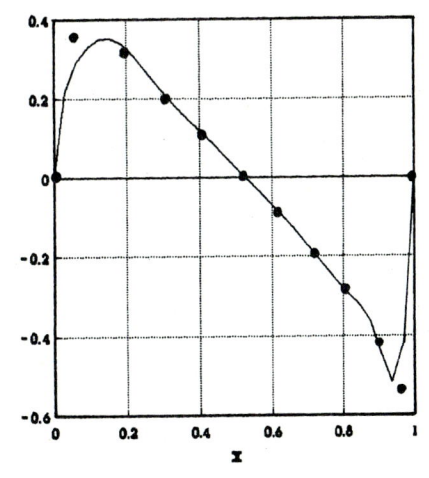

a) horizontal velocity at $x = 0.5$

b) vertical velocity at $y = 0.5$

Figure 4: Driven cavity flow at $Re = 3000$
– own results (32×32 mesh)
• cf. [14] (256×256 mesh)

References

[1] T.J.R. Hughes, L.P. Franca, and M. Balestra. A new finite element formulation for computational fluid dynamics: V. A stable Petrov–Galerkin formulation of the Stokes problem accomodating equal–order interpolations. *Computer Methods in Applied Mechanics and Engineering*, 59:85–99, (1986).

[2] T.J.R. Hughes. Recent progress in the development and understanding of SUPG methods with special reference to the compressible Euler and Navier–Stokes equations. *Journal of Numerical Methods in Fluids*, 7:1261–1275, (1987).

[3] C. Johnson. Numerical solution of partial differential equations by the finite element method. Studentliteratur, Sweden, 1987.

[4] T.E. Tezduyar. Stabilized finite element formulations for incompressible flow computations. Lecture Series 1991-01, 1991. von Karman Institute for Fluid Dynamics.

[5] R. Temam. Navier–Stokes equations and nonlinear functional analysis. In *CBMS-NSF Regional Conference Series in Applied Mechanics*. SIAM, 1983.

[6] F. Brezzi and M. Fortin. *Mixed and Hybrid Finite Element Methods*. Springer Verlag, 1991.

[7] L.P. Franca and R. Stenberg. Error analysis of some Galerkin–least–squares methods for the elasticity equations. To appear in *SIAM Journal of Numerical Analysis*.

[8] G. Lube and A. Auge. Regularized mixed finite element approximations of incompressible flow problems, II. Navier–Stokes flow. Preprint 15/91, University of Technology Magdeburg, PF 4120, O-3010 Magdeburg, June 1991.

[9] L.P. Franca, S.L. Frey and T.J.R. Hughes. Stabilized finite element methods: I. Application to the advective–diffusive model. To appear in *Computer Methods in Applied Mechanics and Engineering*.

[10] J. Douglas and J. Wang. An absolutely stabilized finite element for the Stokes problem. *Math. Comp.*, 52:495–508, (1989).

[11] C. Johnson. Adaptive finite element methods for diffusive and convective problems. *Computer Methods in Applied Mechanics and Engineering*, 82:301–322, (1990).

[12] T.E. Tezduyar, R. Shih, S. Mittal, and S.E. Ray. Incompressible flow computations with stabilized bilinear and linear equal–order interpolation velocity–pressure elements. Preprint UMSI 90/165, University of Minesota, 1990.

[13] R. Pierre. Simple C^0 approximations for the computation of incompressible flows. *Computer Methods in Applied Mechanics and Engineering*, 68:205–227, (1988).

[14] L.B. Zhang. A second–order upwinding finite difference scheme for the steady navier–stokes equations in primitive variables in a driven cavity with a multigrid solver. *Math. Modell. Numer. Anal.*, 24:133–150, (1990).

METHODS USING UNSTRUCTURED GRIDS

A RUNGE-KUTTA TVD FINITE VOLUME METHOD FOR STEADY EULER EQUATIONS ON ADAPTIVE UNSTRUCTURED GRIDS

K. Riemslagh and E. Dick
Department of machinery, University of Ghent
Sint Pietersnieuwstraat 41, B-9000 Gent, Belgium

SUMMARY

A TVD-time dependent discretization of the Euler equations is formulated for unstructured grids. The grid is generated by Delaunay triangulation. The spatial discretization is based on the vertex-centred finite volume method with an upwind definition of the fluxes based on polynomial flux-difference splitting. The time dependent system is integrated with the standard Runge-Kutta type multistage time stepping method. Four stages are used with standard time step lengths. Adaptive refinement of the grid is done based on a pressure difference criterion. The method is illustrated on a supersonic wedge flow.

INTRODUCTION

The polynomial flux-difference splitting introduced by the second author [1], is used here to construct a TVD-time dependent discretization of the Euler equations on unstructured grids. This flux-difference splitting technique is of Roe-type, i.e. it satisfies the requirements formulated by Roe [2]. It is however simpler. Its simplicity follows from dropping the secondary requirement of having a unique definition of average flow variables. This secondary requirement defines the original Roe-splitting within the class of methods allowed by the primary requirements. The secondary requirement is however not necessary. The grids considered are composed of triangles. The vertex-centred finite volume method is employed. On each face of a control volume, the first order flux is defined taking into account the two vertices on both sides of the face. The second order correction is constructed based on the flux-extrapolation concept introduced by Chakravarthy and Osher [3], using flux limiting.

Unstructured triangular grids are very attractive for two-dimensional flow calculations. One reason is that geometries of arbitrary complexity can be meshed. The other is that mesh adaption by addition of points is very simple. The generation of unstructured triangular grids can be done in several ways. One of the techniques is based on the well known Delaunay triangulation algorithm formulated by Bowyer [4]. This algorithm, which is very popular nowadays [5,6,7], is used here. The algorithm iteratively generates a grid, by bringing point after point into the grid. Every time, a small portion of the grid is deleted and reconnected to include the new point. This feature of the algorithm allows for easy grid adaption through refinement.

THE TVD-SCHEME

Vertex-centred finite volume discretization.

Figure 1 shows part of an unstructured grid. In the vertex-centred finite volume method nodes are located at the vertices of the grid. The control-volumes are formed by connecting the centres of the cells in the interior of the domain. At the boundaries, centres of cells and centres of boundary sides are used. The fluxes at a side (ab) of a control volume are defined based on function values and gradients of function values in the adjacent nodes i and j.

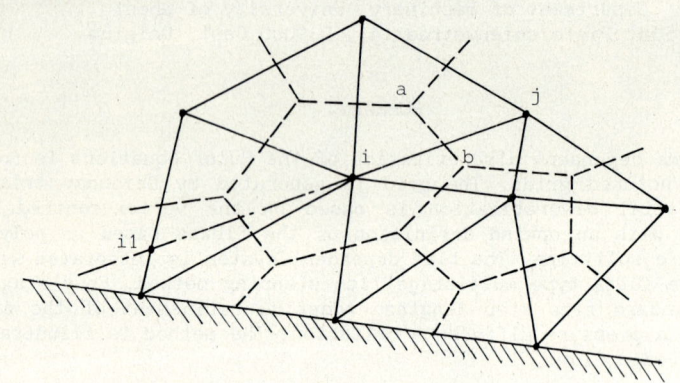

Fig. 1 : Vertex-centred discretization.

Upwind flux-definition.

To define the flux through a side of a control volume, use is made of the flux-difference splitting principle.

Euler equations in two dimensions take the form

$$\frac{\partial U}{\partial t} + \frac{\partial f}{\partial x} + \frac{\partial g}{\partial y} = 0 , \qquad (1)$$

where U is the vector of conserved variables

$$U^T = \{ \rho, \rho u, \rho v, \rho E \} , \qquad (2)$$

and where the flux-vectors are

$$f^T = \{ \rho u, \rho u u + p, \rho u v, \rho H u \} , \quad g^T = \{ \rho v, \rho u v, \rho v v + p, \rho H v \} . \qquad (3)$$

ρ is density, u and v are Cartesian velocity components, p is pressure, $E = p/(\gamma-1)\rho + u^2/2 + v^2/2$ is total energy, $H = \gamma p/(\gamma-1)\rho + u^2/2 + v^2/2$ is total enthalpy and γ is the adiabatic constant.

A flux through a side of a control volume can be written as a combination of Cartesian flux-vectors by

$$F = (n_x f + n_y g)\Delta s , \qquad (4)$$

where n_x and n_y are the components of the unit normal to the side and Δs is the length of the side.

For the side (ab) in figure 1, using the normal in the sense i to j, a flux-difference can be written as

$$\Delta F_{i,j} = F_j - F_i = A_{i,j}(U_j - U_i)\Delta s_{i,j} . \qquad (5)$$

To define the discrete Jacobian A, we use here the polynomial flux-difference splitting. This splitting technique is based on the polynomial character of the vector of the conserved variables (2) and the Cartesian flux-vectors (3), with respect to the primitive variables ρ, u, v and p. Full details of this splitting are given in [8,9].

The Jacobian matrix A can be written as $A = R \Lambda L$, where R and L are right and left eigenvector matrices in orthonormal form and where Λ is the eigenvector matrix. The eigenvalues are discrete equivalents to the normal velocity on the side (twice) and to the normal velocity plus and minus velocity of sound.

The splitting of the Jacobian is defined by the splitting of the eigenvalue matrix through

$$A^+ = R \Lambda^+ L, \qquad A^- = R \Lambda^- L,$$

where Λ^+ and Λ^- contain positive and negative parts of the eigenvalues.

A first order upwind flux is defined by

$$F^1_{i,j} = \frac{1}{2}(F_i + F_j) - \frac{1}{2}(A^+_{i,j} - A^-_{i,j})(U_j - U_i)\Delta s_{i,j}. \qquad (6)$$

In order to define a second order flux, the second part in the right hand side of (6), which contains the positive and negative parts of the flux-difference, is decomposed into components along the eigenvectors of the Jacobian, according to

$$\Delta F^{\pm}_{i,j} = A^{\pm}_{i,j}(U_j - U_i)\Delta s_{i,j} = \sum_n r^n_{i,j} \lambda^{n\pm}_{i,j} \ell^n_{i,j}(U_j - U_i)\Delta s_{i,j}$$

$$= \sum_n \Delta F^{n\pm}_{i,j}, \qquad (7)$$

where r^n and ℓ^n are right and left eigenvectors associated to the eigenvalue λ^n. The second order flux is then defined by

$$F^2_{i,j} = \frac{1}{2}(F_i + F_j) - \frac{1}{2}\sum_n \Delta F^{n+}_{i,j} + \frac{1}{2}\sum_n \Delta F^{n-}_{i,j} + \frac{1}{2}\sum_n \Delta \widetilde{F}^{n+}_{i,j} - \frac{1}{2}\sum_n \Delta \widetilde{F}^{n-}_{i,j} \qquad (8)$$

where $\Delta \widetilde{F}^{n\pm}$ are limited combinations of flux-difference components $\Delta F^{n\pm}$ and shifted differences $\Delta \widetilde{F}^{n\pm}$, in the sense of i for positive components and in the sense of j for negative components.

The limiter used here is the minmod-limiter. The shifted flux-difference components are constructed based on shifted differences of conserved variables. These are obtained by elongating the line segment (ij) into the adjacent triangles and constructing the intersections with the opposing sides (point i1 for the shifting in i sense in figure 1). The shifted flux-differences are defined by

$$\Delta \widetilde{F}^{n\pm}_{i,j} = r_{i,j} \lambda^{n\pm}_{i,j} \ell^{n\pm}_{i,j} \Delta \widetilde{U}^{n\pm}_{i,j} \Delta s_{i,j}.$$

The foregoing technique to define the second order flux commonly is called the flux-extrapolation technique.

Time-stepping.

The fourth-order modified Runge-Kutta time stepping with step parameters 1/4, 1/3, 1/2 and 1 is used here. Local time stepping is employed. The CFL-criterion is based on the monotonicity condition of a single step first order time stepping.

The single step first order method, for interior nodes, is

$$\frac{Vol_i}{\Delta t}(U^{n+1}_i - U^n_i) + \sum_j A^-_{i,j}(U^n_j - U^n_i)\Delta s_{i,j} = 0,$$

where Vol_i is the surface of the control volume around node i.

The monotonicity condition reads (where ρ means spectral radius)

$$\rho\left[\frac{Vol_i}{\Delta t}I + \sum_j A^-_{i,j}\Delta s_{i,j}\right] \geq 0 . \tag{9}$$

A sufficient condition to satisfy (9) is given by

$$\frac{Vol_i}{\Delta t} \geq \sum_j \rho(-A^-_{i,j})\Delta s_{i,j} = \sum_j \max(0,c-v_n)\Delta s_{i,j} , \tag{10}$$

where c is the velocity of sound and v_n is the normal outgoing velocity component on a side.

Boundary conditions

At solid boundaries, impermiability is imposed by setting the convective part of the flux equal to zero. At inflow and outflow, first, fluxes are defined based on nodal values. Second, after each stage of the time stepping, the boundary conditions are imposed. At subsonic inlet, Mach number is retained, while stagnation conditions and flow direction are imposed. At subsonic outflow the reverse is done. In the example discussed below, the flow is supersonic. So, at inflow all values are prescribed while at outflow no boundary conditions are to be imposed.

It is easy to see that at solid boundaries, the CFL-criterion is to be modified in the sense that in the sum (10) no contributions from the boundary-sides enter.

THE MESH GENERATION

Delaunay triangulation.

The automatic triangulation of an arbitrary set of points can be achieved using the Delaunay triangulation. Robust algorithms to construct this triangulation are nowadays fully developed. The triangulation is based upon the concept of the in-circle criterion. The triangulation is the geometrical dual of the Voronoi diagram. This diagram is the construction of tiles in which a region is associated with every point, in such a way that each region is closer to a point than to any other point in the field. The bounding line segments form the Voronoi diagram. If points with common line segments are connected then the Delaunay triangulation is formed. The vertices of the Voronoi diagram are at the centres of the circles, each passing through three points which form an individual triangle in the triangulation. One of the properties of such a construction is that no point lies inside the circumcircle of a triangle. It is this feature that is used for the construction of the triangulation. The Delaunay triangulation of an arbitrary set of points is the most equiangular, or smoothest triangulation that can be formed with that set.

The basic building block of the grid generator is a routine which allows to add a point to an existing Delaunay triangulation. The result of this operation is again a Delaunay triangulation. This routine can be split into several simple steps [4]. The first step is a search over the triangles to find all of the triangles inside whose circumcircle the new point lies. These triangles are deleted from the triangulation leaving a hole or insertion polygon. The new point is then added to the mesh by connecting every point of the insertion polygon with the new point. One can prove that the resulting triangulation is again a Delaunay triangulation [5].

Initial grid

The construction of the initial grid can be decomposed into several steps. Each step has its own specific algorithm. The steps are : definition of boundary points; construction of a starting triangulation; adding interior points with the area criterion; adding interior points with the aspect ratio criterion; swapping to optimal connections, followed by smoothing based on a min-max criterion followed by swapping to Delaunay triangulation; smoothing based on an averaging criterion followed by swapping to Delaunay triangulation.

The definition of the boundary points for the test case described later on, is trivial because the boundary is a polygon. If the boundary would be curved, the boundary point density would be made proportional to the local curvature. For a polygon, boundary points are added so that local spacing is less than a certain value.

The second step is to create a Delaunay triangulation of this initial set of boundary points. The starting grid that results typically is atrocious, but this is of no concern as the subsequent connection of interior points into the mesh dramatically improves the mesh quality. To construct this starting grid a small procedure is followed. Four points defining a rectangle in which all the boundary points lie, are used to construct two triangles. Now, all the boundary points are added to this triangulation using the earlier described algorithm. At this stage the whole rectangle is triangulated, the inside of the domain to triangulate, as well as the outside. A visual check is done to make sure all the boundary edges appear in the triangulation. Now, the triangles that lie outside the domain are removed from the grid. To detect these triangles extra points are added outside the domain, but inside the rectangle. Triangles that are formed with at least one of these points are removed. The extra points give the notion of an outer normal to the domain.

What we have now is a Delaunay triangulation of the domain defined by boundary points only. In the third step the grid is improved by adding points into the interior. Indeed, triangles of which the area is too big might appear in the grid. For this, points are added, and placed in the centre of the circumcircle of the triangles that are selected by the area-criterion. Again the previous described algorithm is used to include these new points. Points are added until all triangles have an area that is less than a certain value.

During the fourth step, points are added, again in the centre of the circumcircle of a triangle. These triangles are now selected by the aspect-ratio criterion. As a measure of the aspect-ratio, the ratio of the circumcircle radius over twice the incircle radius is used. This gives a measure of the skinniness of the triangle. Triangles are refined until every triangle has an aspect-ratio less than 1.7 or has an area less than a minimum value, or until the triangle cannot be further refined.

In the fifth step of the generation of the initial grid, no more points are added. But through the use of three subprograms the number of surrounding triangles a point has, is optimized. The goal is a grid where this number is six for internal points, and the internal angle divided by 60 degrees for boundary points.

The first subprogram that is used, reconnects some points in the grid. The diagonal of the quadrilateral formed by two neighbouring triangles is swapped if the following expression (A) will decrease with

$$A = \sum_{i=1}^{4} (N_{opt} - N_i)^2,$$

where N_{opt} is 6 for an internal node, and N_i is the number of surrounding triangles of node i. The sum is taken over the four points of the quadrilateral.
The second subprogram repositions the points towards

$$x_o = (\max x_i + \min x_i)/2 \ , \ y_o = (\max y_i + \min y_i)/2,$$

where maximum and minimum are taken over all the surrounding nodes the point has.
The third subprogram makes the triangulation again a Delaunay triangulation. This is done by swapping the diagonal of a quadrilateral formed by two neighbouring triangles, if the minimum of the six internal angles increases. This criterion is equivalent to the Delaunay criterion.

Since only local reconnections and repositionings of points were done during the first two subprograms, usually two loops over all the edges reconstruct the Delaunay triangulation. This group of three subprograms is executed 10 times. How the number of edge swappings evolves during execution, can be seen in table 1 for the initial grid of the wedge test case, described in the next section.

Table 1 : History of fifth step in construction of the initial grid of the wedge test case.

execution number	1	2	3	4
# swap connections	19	3	2	3
max Δs after relaxation	2.38 E-1	8.39 E-2	5.92 E-2	1.04 E-1
# swaps to Delaunay	11	2	3	1
5	6	7	8	9	10
---	---	---	---	---	---
1	2	2	2	2	1
3.76 E-2	5.52 E-2	6.02 E-2	8.43 E-1	5.45 E-2	3.08 E-2
2	2	3	2	1	0

The sixth and final step of the initial grid generation, is very similar to the previous step. Two subprograms are alternatively executed.
The first subprogram consist of looping over the points, shifting every point towards

$$x_o = \frac{1}{n} \sum_{i=1}^{n} x_i, \qquad y_o = \frac{1}{n} \sum_{i=1}^{n} y_i.$$

In other words, the potential energy of the set of springs, every edge being a spring, is minimized.
The next subprogram is again the edge-swapping routine to make the triangulation Delaunay.
These two routines are executed 10 times. This cycle converges very rapidly as can be seen in table 2 for the wedge test case. Typically only during the first two executions edges are swapped.

Table 2 : History of sixth step in construction of the initial grid of the wedge test case.

execution number	1	2	3	4
max Δs after relaxation # swaps to Delaunay	2.06 E-1 2	4.86 E-1 1	1.15 E-2 0	2.78 E-3 0
5	6	7	8	9	10
---	---	---	---	---	---
6.92 E-4 0	1.75 E-4 0	7.43 E-5 0	7.83 E-5 0	7.82 E-5 0	7.82 E-5 0

Adaptivity - mesh refinement

One of the most important advantages of unstructured grids is the possibility to refine locally the mesh during the computation. Succesive mesh refinements in critical zones may be performed. The optimal way is to start the computation on a coarse and smooth grid which is not at all specific to a hypothetical solution, and refine during the run, according to criteria adapted to the particular physical solution.

In the physical problem treated here, supersonic flow, the most interesting regions are of course the shock waves, which have to be precisely located. Criteria can be easily defined characterizing such flow features. The criterion used here is the pressure difference over an edge :

$$\text{abs}(\Delta p) > \frac{P_{max} - P_{min}}{C}.$$

The constant C is a sensitivity factor.

Edges are refined by placing a new point into the centre of the edge. To include the new point in the mesh, the earlier described procedure is used.

NUMERICAL RESULTS

Figure 2 shows the grid obtained after the fourth stage of the initial grid generator for a channel with a wedge type constriction. The top grid shown in figure 3 is the grid after the sixth stage. The improvement of the quality of the grid due to the two last stages in the grid generator is clear. The number of surrounding triangles for an interior node is almost everywhere equal to 6 and the smoothness of the grid is obvious.

The wedge height is 8,816 % of the wedge basis. The channel height is 60 % of the wedge basis. The incoming flow is uniform with Mach number 2.

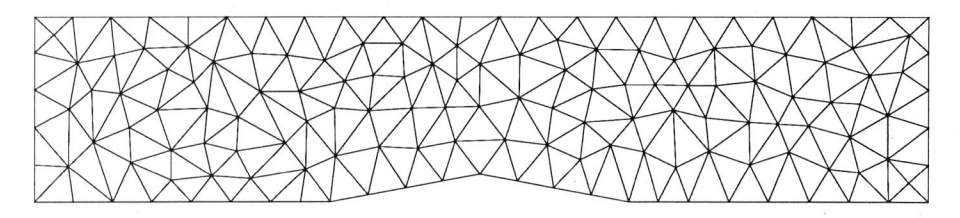

Figure 2. Initial grid after four stages of the grid generator.

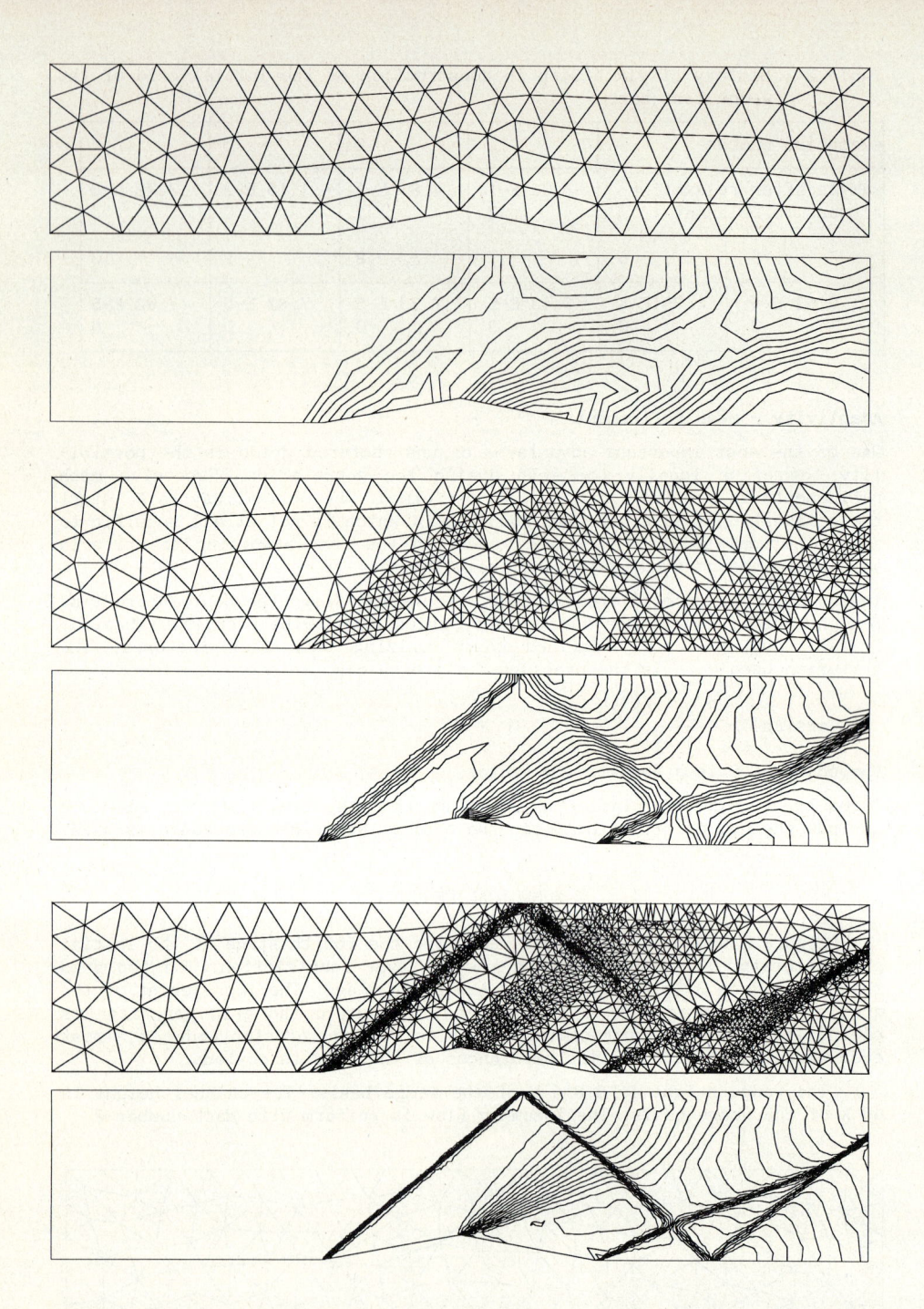

Figure 3. Initial grid (completed-154 points), an intermediate grid (769 points) and final grid (2746 points) together with iso-Machline results (increment 0.05).

Figure 3 shows the initial grid, an intermediate grid and the final grid, together with the iso-Machline results. Table 3 gives some detail on the evolution of the computation. Four refinement stages are performed. Local time stepping is used with CFL = 1.3. As is well known, the allowable time step for the standard multistage time stepping is much lower when used with upwind discretizations than when used with central discretizations. Restrictions come from stability requirements and from requirements of preservation of the TVD properties of the basic single step method. We do not enter here the discussion of the optimization of the time stepping for upwind discretizations. It is however clear that by a suitable choice of the time stepping parameters, a large gain in efficiency can be obtained.

In every stage of the refinement, the solution is computed up to a convergence of the increment of ρE to 10^{-12}. This value of ρE is scaled by the value at inflow.

Table 3. Evolution of the refinement for the wedge test-case.

Phase	# points	# triangles	# iterations
0	154	249	254
1	345	605	584
2	769	1420	1189
3	1534	2924	1940
4	2746	5326	3174

Figure 4 shows the corresponding convergence behaviour.

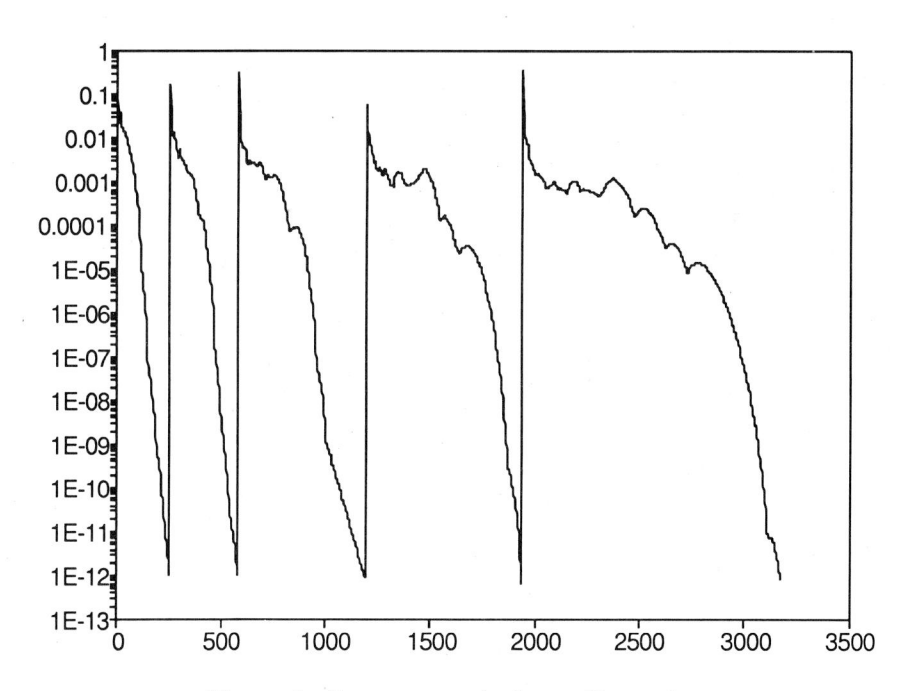

Figure 4. Convergence during refinement.

Figure 5 shows an enlarged view of the final grid in the vicinity of the leading edge.

Figure 5. Enlarged part of the final grid.

CONCLUSION

It has been shown that by the combination of an adaptive unstructured grid generation technique and a high resolution flow solver, high quality solutions for Euler equations can be obtained.

ACKNOWLEDGEMENT

The research reported here was granted by the Belgian Science Policy Office (Diensten voor Programmatie van het Wetenschapsbeleid) under contract IT/SC/13, as part of the national program for large scale scientific computing (informatietechnologie, supercomputing) and under contract IUAP/17 as part of the national program for interuniversity fundamental research (Interuniversitaire Attractiepolen van Fundamenteel Onderzoek).

REFERENCES

1. Dick E., J. Comp. Phys. 76 (1988), 19-32.
2. Roe P.L., J. Comp. Phys. 43 (1981), 357-372.
3. Chakravarthy S.R. and Osher S., AIAA paper 85-0363, 1985.
4. Bowyer A., Computer Journal, 24, (1981), 162-166.
5. Mitty T.J., Baker T.J., Jameson A., Proc. 3rd. Intern. Conf. on Numerical Grid Generation in CFD and related Fields, Barcelona, 1991.
6. Mavriplis D.J., idem as 5.
7. Weatherill N.P., idem as 5.
8. Dick E., J. Comp. Phys. 91 (1990), 161-173.
9. Dick E., Multigrid Methods III, Birkhäuser Verlag Basel, 1991, 1-20.

COMPUTATION OF THE VISCOUS FLOW AROUND MULTI-ELEMENT AEROFOILS USING UNSTRUCTURED GRIDS

Luca Stolcis and Leslie J. Johnston

Department of Mechanical Engineering, UMIST
PO Box 88, Manchester M60 1QD, England

SUMMARY

A solution method is presented for the compressible Reynolds-averaged Navier-Stokes equations on unstructured grids incorporating a high-Reynolds number $k - \epsilon$ turbulence model. A cell-centred finite-volume spatial discretisation and an explicit time-stepping scheme are utilised. The influence of added numerical dissipation is discussed and an empirical remedy for the reduction of its influence in viscous layers is proposed. The method is applied to predict the turbulent flow around multi-element aerofoil sections. Results for both the low-speed NLR 7301 and transonic SKF 1.1 wing-flap configurations are presented, showing an overall good agreement with experiment.

INTRODUCTION

High-lift devices such as trailing-edge flaps and leading-edge slats provide the additional lift necessary to fulfil the low-speed performance requirements associated with take off and landing. In addition, these devices can also be used to enhance the manoeuvrability of combat aircraft at transonic conditions. This increase in lifting capability is achieved by changing the aerofoil camber, thereby increasing the overall circulation, and by splitting the pressure rise over several elements to suppress the tendency for flow separation. The result is a multi-element aerofoil section, the flowfield around which possesses a rather high degree of complexity. There can be strong interactions between wakes from upstream elements and the upper surface boundary layers developing on downstream elements. Thick viscous layers are present over the upper surface of a trailing-edge flap, and their development may be significantly affected by flow curvature. Furthermore, large regions of separated flow can also be present, even for conditions well below maximum lift. The flow around a leading-edge slat can become locally supersonic, even for low freestream Mach numbers, due to the large suction levels induced in this region. Such complex flow features, together with the geometric complexity arising from the closely-coupled multi-element configuration, make the development of a suitable computational method a non-trivial task.

In the past, the prediction of multi-element aerofoil flows involved the use of viscous-inviscid interaction methods, which are computationally very efficient, but are unable to describe many of the complex viscous flow features present. Furthermore, the coupling between the inviscid and viscous flow solvers tends to break down in the presence of anything other than limited separated flow regions. A more complete description of the physical phenomena can be achieved only by methods based on a solution of the full Reynolds-averaged Navier-Stokes equations, in conjunction with a suitable turbulence model [1]. The generation of suitable computational grids around multi-element configurations is not an easy task because of the multiple connectivity of the flow domain. By employing unstructured computational grids, as has been done previously in several mainly inviscid flow methods, most of these problems can be overcome.

The aim of the present work is the development of a numerical method able to predict the complex turbulent flow around multi-element, high-lift configurations, both at low-speed and transonic conditions. The approach adopted consists of solving the compressible Reynolds-averaged Navier-Stokes equations using an unstructured computational grid and a two-equation turbulence model. The paper describes the governing mean-flow and turbulence-transport equations, their spatial discretisation and numerical solution using an explicit time-stepping scheme. The influence of numerical dissipation on the resulting viscous flow solutions is considered, together with the steps taken to minimise its influence in the present method. Results are presented for the low-speed flow around a high-lift aerofoil-flap configuration, and the transonic flow around an aerofoil equipped with a manoeuvre flap.

GOVERNING FLOW EQUATIONS

The Reynolds-averaged Navier-Stokes equations in mass-averaged form are employed to enable the computation of compressible turbulent flows. The additional Reynolds stress terms, resulting from the time-averaging procedure, are modelled according to the Boussinesq hypothesis by introducing a turbulent eddy viscosity μ_t. The flow around multi-element aerofoils is so complex that algebraic turbulence models for μ_t cannot be applied in a satisfactory manner, such models indeed being restricted to simple wall-bounded or wake flows. Therefore, the present method employs a two-equation $k - \epsilon$ turbulence model, wherein the eddy viscosity is expressed in terms of the turbulent kinetic energy k and its dissipation rate ϵ:

$$\mu_t = c_\mu \rho k^2 / \epsilon \qquad (1)$$

c_μ being a model constant. Two modelled transport equations for k and ϵ are added to the four mean-flow equations, so that the system of governing equations becomes:

$$\frac{\partial}{\partial t} \int_\Omega \vec{W} d\Omega + \oint_{\partial \Omega} (\vec{F} dy - \vec{G} dx) = \int_\Omega \vec{S} d\Omega \qquad (2)$$

where $\vec{W}$ is the vector of time-averaged conserved variables:

$$\vec{W} = \begin{pmatrix} \rho \\ \rho u \\ \rho v \\ \rho E \\ \rho k \\ \rho \epsilon \end{pmatrix} \qquad (3)$$

and $\partial \Omega$ is the boundary of the computational domain Ω. u and v are the mean-velocity components in the x,y cartesian coordinate directions, ρ is the density and ρE the total energy. The flux vectors are composed of two contributions:

$$\vec{F} = \vec{F^I} + \vec{F^V} \; , \; \vec{G} = \vec{G^I} + \vec{G^V} \qquad (4)$$

the convective fluxes (superscript I) and diffusive fluxes (superscript V) being given by:

$$\vec{F^I} = \begin{pmatrix} \rho u \\ \rho u^2 + p \\ \rho u v \\ \rho u H \\ \rho u k \\ \rho u \epsilon \end{pmatrix} , \; \vec{F^V} = \begin{pmatrix} 0 \\ \tau_{xx} \\ \tau_{xy} \\ u\tau_{xx} + v\tau_{xy} + q_x \\ \beta_{kx} \\ \beta_{\epsilon x} \end{pmatrix} ,$$

$$\vec{G^I} = \begin{pmatrix} \rho v \\ \rho u v \\ \rho v^2 + p \\ \rho v H \\ \rho v k \\ \rho v \epsilon \end{pmatrix} , \; \vec{G^V} = \begin{pmatrix} 0 \\ \tau_{xy} \\ \tau_{yy} \\ u\tau_{xy} + v\tau_{yy} + q_y \\ \beta_{ky} \\ \beta_{\epsilon y} \end{pmatrix} \qquad (5)$$

where p is the static pressure and ρH the total enthalpy. The viscous stresses can be written as:

$$\tau_{xx} = -(\mu + \mu_t)s_{xx} + \frac{2}{3}\rho k \ , \quad \tau_{yy} = -(\mu + \mu_t)s_{yy} + \frac{2}{3}\rho k \ , \quad \tau_{xy} = -(\mu + \mu_t)s_{xy} \qquad (6)$$

evaluating the molecular viscosity μ using Sutherland's law. The components of the mean-strain tensor are:

$$s_{xx} = 2\frac{\partial u}{\partial x} - \frac{2}{3}\left(\frac{\partial u}{\partial x} + \frac{\partial v}{\partial y}\right) \ , \quad s_{yy} = 2\frac{\partial v}{\partial y} - \frac{2}{3}\left(\frac{\partial u}{\partial x} + \frac{\partial v}{\partial y}\right) \ , \quad s_{xy} = \frac{\partial u}{\partial y} + \frac{\partial v}{\partial x} \ . \qquad (7)$$

The heat-flux terms are given by:

$$q_x = -\gamma\left(\frac{\mu}{Pr} + \frac{\mu_t}{Pr_t}\right)\frac{\partial T}{\partial x} \ , \quad q_y = -\gamma\left(\frac{\mu}{Pr} + \frac{\mu_t}{Pr_t}\right)\frac{\partial T}{\partial y} \ . \qquad (8)$$

Pr and Pr_t are the laminar and turbulent Prandtl numbers respectively, γ is the ratio of specific heats and T is the static temperature. The diffusive fluxes for the k and ϵ equations are:

$$\beta_{kx} = -\left(\mu + \frac{\mu_t}{\sigma_k}\right)\frac{\partial k}{\partial x} \ , \quad \beta_{ky} = -\left(\mu + \frac{\mu_t}{\sigma_k}\right)\frac{\partial k}{\partial y} \ ,$$

$$\beta_{\epsilon x} = -\left(\mu + \frac{\mu_t}{\sigma_\epsilon}\right)\frac{\partial \epsilon}{\partial x} \ , \quad \beta_{\epsilon y} = -\left(\mu + \frac{\mu_t}{\sigma_\epsilon}\right)\frac{\partial \epsilon}{\partial y} \qquad (9)$$

σ_k and σ_ϵ being further model constants. The vector $\vec{S}$ in equation (2) contains the source terms deriving from the two modelled turbulence-transport equations for k and ϵ:

$$\vec{S} = \begin{pmatrix} 0 \\ 0 \\ 0 \\ 0 \\ P_k - \rho\epsilon \\ c_{\epsilon 1}P_k\epsilon/k - c_{\epsilon 2}\rho\epsilon^2/k \end{pmatrix} \ . \qquad (10)$$

P_k represents the production of turbulent kinetic energy:

$$P_k = \mu_t\left[s_{xx}\frac{\partial u}{\partial x} + s_{xy}^2 + s_{yy}\frac{\partial v}{\partial y}\right] - \frac{2}{3}\rho k\left(\frac{\partial u}{\partial x} + \frac{\partial v}{\partial y}\right) \ . \qquad (11)$$

Finally, the turbulence model contains the following five constants:

$$c_\mu = 0.09 \ , \quad c_{\epsilon 1} = 1.44 \ , \quad c_{\epsilon 2} = 1.92 \ , \quad \sigma_k = 1.0 \ , \quad \sigma_\epsilon = 1.3 \ . \qquad (12)$$

The high-Reynolds number form of the $k - \epsilon$ turbulence model is employed in the present work, and wall functions based on the semi-logarithmic law-of-the-wall are used to provide the near-wall boundary conditions for k and ϵ [2],[3]. No-slip and adiabatic wall boundary conditions are applied at solid walls for the mean-flow equations.

NUMERICAL SCHEME

The computational domain Ω is subdivided into a set of non-overlapping polygonal cells, and the governing flow equations, represented by equation (2), are applied to each cell in turn. A cell-centred, finite-volume spatial discretisation is adopted and subsequent to this the system becomes a large set of ordinary differential equations in time:

$$\frac{d\vec{W}_K}{dt} + \frac{\vec{Q}_K^I + \vec{Q}_K^V}{\Omega_K} = \frac{\vec{S}_K}{\Omega_K} \qquad (13)$$

where $\vec{W}_K$ contains the cell-centre values of the conserved variables for the K-th cell, whilst $\vec{Q}_K^I$ and $\vec{Q}_K^V$ are respectively the discretised forms of the convective and diffusive fluxes:

$$\vec{Q}_K^I = \sum_{i=1}^{nedge} \left[(\vec{F}^I)_i \Delta y_i - (\vec{G}^I)_i \Delta x_i \right], \quad \vec{Q}_K^V = \sum_{i=1}^{nedge} \left[(\vec{F}^V)_i \Delta y_i - (\vec{G}^V)_i \Delta x_i \right] \qquad (14)$$

and $\vec{S}_K$ is the discretised source vector. The summations in equation (14) are over the edges (nedge) forming the K-th computational cell. The convective fluxes at the i-th edge are evaluated using a simple average of the conserved variables for the cell-centres (subscripts K and P) adjacent to the edge:

$$(\vec{F}^I)_i = \vec{F}^I(\frac{\vec{W}_K + \vec{W}_P}{2}) \quad , \quad (\vec{G}^I)_i = \vec{G}^I(\frac{\vec{W}_K + \vec{W}_P}{2}) . \qquad (15)$$

The discretised diffusive fluxes $(\vec{F}^V)_i$ and $(\vec{G}^V)_i$ involve the first derivatives of variables at cell edges, and these are computed using the auxiliary control volume APBK of Fig. 1, as outlined in [3], A and B being the grid vertices forming the i-th edge:

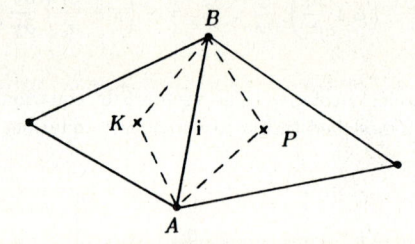

Fig. 1 Auxiliary control volume (− − −−) for the viscous fluxes

The above spatial discretisation scheme leads naturally to a solution algorithm written purely in terms of edges. This enables the use of polygonal cells containing an arbitrary number of sides. The turbulence-transport equations are discretised in space following a similar procedure to the mean-flow equations. However, in order to enhance stability and to ensure convergence to a steady-state solution, the centred scheme for the mean-flow convective fluxes is replaced by a hybrid central/upwind scheme, similar to that widely-used for the discretisation of the incompressible Navier-Stokes equations in primitive variables. The hybrid scheme uses a combination of streamwise-upwind (first order) and centred (second order) schemes, where the switch from one scheme to the other is dependent upon the local cell Reynolds or Peclet number. The convective flux evaluated by the streamwise-upwind scheme can be written as:

$$(Q_\Phi^I)_i = max[0, A_i \Phi_K] + min[0, A_i \Phi_P] \qquad (16)$$

where Φ is the scalar quantity convected (ρk or $\rho \epsilon$), and $A_i = u_i \Delta y_i - v_i \Delta x_i$. The flux computed by the centred scheme has a form equivalent to that for the mean-flow equations.

The above hybrid scheme has been adopted for its robustness and because it allows both the mean-flow and turbulence-transport equations to be marched in time with the same local time step. In contrast, the centred scheme with added numerical dissipation (see next section) requires a globally-constant minimum time step to be used for the k and ϵ equations [3]. Experience indicates that the use of the hybrid scheme (for the k and ϵ equations only) does not have a significant effect on the structure or quality of the computed flow-field, which is much more significantly affected by the order of the convection scheme employed for the mean-flow equations.

The steady-state solution of equation (13) is obtained by marching in time using an explicit four-stage scheme [4]. Local time step and implicit residual smoothing techniques are used to increase the convergence rate.

NUMERICAL DISSIPATION

The above discretisation scheme for the mean-flow equations is equivalent, at least for uniform cartesian grids, to a central-difference scheme. As such, and because it is applied to flows characterised by high Reynolds numbers, the damping provided by the physical viscous stresses is insufficient. For this reason additional numerical dissipation is required to prevent odd-even decoupling, to damp oscillations around shock waves and to enhance convergence rate. The present method employs a numerical dissipation, $\vec{D}_K$, of a form associated with Jameson [4], consisting of a blending of second- and fourth-difference operators. This is modified for use on unstructured grids [5],[6], and added explicitly to equation (13) which becomes:

$$\frac{d\vec{W}_K}{dt} + \frac{\vec{Q}_K^I + \vec{Q}_K^V - \beta_K \vec{D}_K}{\Omega_K} = \frac{\vec{S}_K}{\Omega_K} \tag{17}$$

β being a scaling parameter to be defined below.

The influence of the numerical dissipation terms on the computed solution is a matter of considerable importance, especially when dealing with complex viscous flows. Indeed, in the boundary layer and wake regions there can be significant flow curvature, so that even in the absence of numerical oscillations, the numerical dissipation can be 'switched on', since it is based upon even derivatives of the conserved variables. A higher than natural total level of dissipation can have dramatic effects on the computed flowfield, delaying or preventing the onset of separation, reducing the size of recirculation regions, reducing the velocity defect in near-wake regions, etc.

An advantage of the above explicitly-added numerical dissipation is that it is possible to monitor its magnitude in each cell and at any time, by comparing $\vec{D}_K$ with the value of the viscous fluxes $\vec{Q}_K^V$. Common sense and some numerical investigations suggest that in viscous layers the ratio $\vec{D}_K/\vec{Q}_K^V$ must be at least less than 0.1. Computational experiments indicate that, for laminar flows in the presence of mild pressure gradients, this ratio is always less than 0.1-0.01. However, for complex turbulent flows, where the gradients are more severe, this is not necessarily the case, and an appropriate scaling of the numerical dissipation needs to be introduced. The technique adopted in the present method consists of using the ratio of eddy viscosity to laminar viscosity as a scaling parameter, to reduce the level of $\vec{D}_K$ in high physical-viscosity flow regions.

This approach is found to be superior to other scalings based solely on geometrical considerations (i.e. distance from the wall, etc.) because it will also reduce numerical dissipation levels in far-wake and recirculating flow regions remote from the aerofoil surfaces. The scaling parameter in equation (17) takes the form:

$$\beta = 1 - e^{-(\alpha \mu / \mu_t)} \tag{18}$$

α being a constant, which takes a value of around 50 for the calculations presented below.

RESULTS

In this present section, results are presented for the computation of the viscous flow development around two multi-element aerofoil sections, covering both low-speed and transonic flow conditions.

a) NLR 7301 wing-flap

van den Berg [7] reports detailed mean-flow measurements for a high-lift wing-flap configuration at low speed. The geometry of the flap-cove region has been designed explicitly to eliminate complex flow features such as separated regions and mixing between wakes and boundary layers. For this reason, this configuration has been widely-used in the past for the validation of viscous-inviscid interaction methods for multi-element aerofoils. Preliminary results using the present method, for $M_\infty = 0.185, \alpha = 6.0^o, Re = 2.51x10^6$, have been presented previously in [3] where, even with a relatively coarse grid, the surface pressure distribution and wing mean-velocity profiles are in excellent agreement with experiment. However, the mean-velocity profiles downstream of the wing trailing-edge show a substantial overprediction of the near-wake velocity deficit. By using the adaptively-generated finer grid of Fig.2a (6267 nodes, 18567 edges and 12299 cells) a much better agreement is obtained in the near-wake region, as shown in Fig.2c. The excellent agreement with the experimental pressure distribution is retained, Fig.2b.

b) SKF 1.1 wing with manoeuvre flap

The SKF 1.1 multi-element aerofoil, derived from a basic aerofoil section of supercritical design, has been tested for a range of transonic speeds and flap geometries in the DFVLR 1m x 1m wind tunnel [8]. Computational results are presented for configuration 5, with a flap gap of 1.55 % chord and flap angle of 10°, at three different Mach numbers and for a fixed geometrical angle of incidence of 3°. The experimental data have been corrected for wind tunnel wall interference effects following the guidelines in the original report [8]. These corrections amount to a smaller incidence angle and, due to the high-lift coefficients ($C_L > 1.0$), these corrections are very large.

The surface pressure distribution for $M_\infty = 0.6$ is shown in Fig.3a. The leading-edge suction peak and the small supersonic region of the wing are well captured, whilst the downstream pressure levels are somewhat higher than experiment. The wing and flap lower surface pressures are well-predicted in the separated flow region near the flap cove, but the flap leading-edge suction peak is slightly underpredicted, probably due to the relatively coarse grid employed. Fig.3b presents the results for $M_\infty = 0.7$. Overall, good agreement with the measured surface pressure distribution is achieved, apart from the shock wave position on the upper surface of the wing, which is predicted 15 % chord too far downstream. Note, however, that the discrepancies are associated only with the position of the shock wave and not with its strength. Grid refinement in the streamwise as well as in the normal direction were tested, but no significant changes in shock wave position were observed. For $M_\infty = 0.76$, Fig.3c, a much stronger shock wave is present on the wing and there may be also a weak shock wave on the flap. The computed shock position is still closer to the wing trailing edge than experiment, and this presumably affects the pressure distribution on the flap further downstream. Fig.4 shows velocity vectors in the large recirculation region near the flap cove, for the $M_\infty = 0.7$ case. The computational grid employed for all the calculations is depicted in Fig.5a. It contains 5031 nodes, 9712 cells, 14742 edges, and 130 surface nodes on each element, with near-wall cell-centre values of y^+ less than 100. Iso-Mach contours for the $M_\infty = 0.7$ case are shown in Fig.5b, illustrating the complexity of viscous transonic multi-element flowfields.

For the three cases considered in these preliminary calculations, the overall agreement with experiment is very good, apart from the shock-wave position for the transonic cases and the wing upper surface pressures levels for the subcritical case. The reason for these discrepancies is under investigation, but the rather large tunnel interference corrections that are required may be a contributing factor.

CONCLUSIONS

A method for the computation of compressible turbulent flow over multi-element aerofoils has been presented. The Reynolds-averaged Navier-Stokes equations are solved in conjunction with a two-equation turbulence model, to allow the computation of the complex flow features associated with such configurations. The implementation of the $k - \epsilon$ turbulence model within the unstructured solver framework was quite straightforward. The results obtained confirm the ability of the method to predict the complex flow around high-lift configurations even in the presence of shock waves. Anticipated future work will involve the inclusion of a low-Reynolds number turbulence model to remove the uncertainities related to the use of wall function boundary conditions, a study of the effects of grid refinement, and the possible use of Reynold-stress based turbulence models for improved prediction of shock-wave/boundary-layer interactions.

ACKNOWLEDGEMENT

The work of the first author (L. Stolcis) was supported by the Commission of the European Communities, Directorate General for Science, Research and Development (Science Plan), under grant SC1/900369.

REFERENCES

[1] KING, D.A. and WILLIAMS, B.R. "Developments in computational methods for high-lift aerodynamics", Aeronautical Journal, 92 (1988), pp. 265-288.
[2] LAUNDER, B.E. and SPALDING, D.B. " The numerical computation of turbulent flows", Comput. Meth. Appl. Mech. Eng., 3 (1974), pp. 269-289.
[3] STOLCIS, L., and JOHNSTON, L.J. "Compressible flow calculations using a two-equation turbulence model and unstructured grids", 7th Int. Conf. on Num. Meth. in Laminar and Turbulent Flow, Stanford, California, USA, 15-19 July 1991.
[4] JAMESON, A. "Numerical solution of the Euler equations for compressible inviscid fluids", Numerical Methods for the Euler equations of fluid dynamics, Ed. Angrand, F. et al., SIAM, (1985), pp. 199-245.
[5] MAVRIPLIS, D.J. "Multigrid solution of the two-dimensional Euler equations on unstructured triangular meshes", AIAA Journal, 26 (1988), pp. 824-831.
[6] STOLCIS, L., JOHNSTON, L.J. "Solution of the Euler equations on unstructured grids for two-dimensional compressible flow", Aeronautical Journal, 94 (1990), pp. 181-195.
[7] VAN DEN BERG, B. "Boundary layer measurements on a two-dimensional wing with flap", NLR TR 79009 U (1979).
[8] STANEWSKY, E. and THIBERT, J.J. 'Airfoil SKF 1.1 with maneuver flap', AGARD AR 138 (1979), pp. A5-1 to A5-29.

a) Computational grid

b) Surface pressure distribution

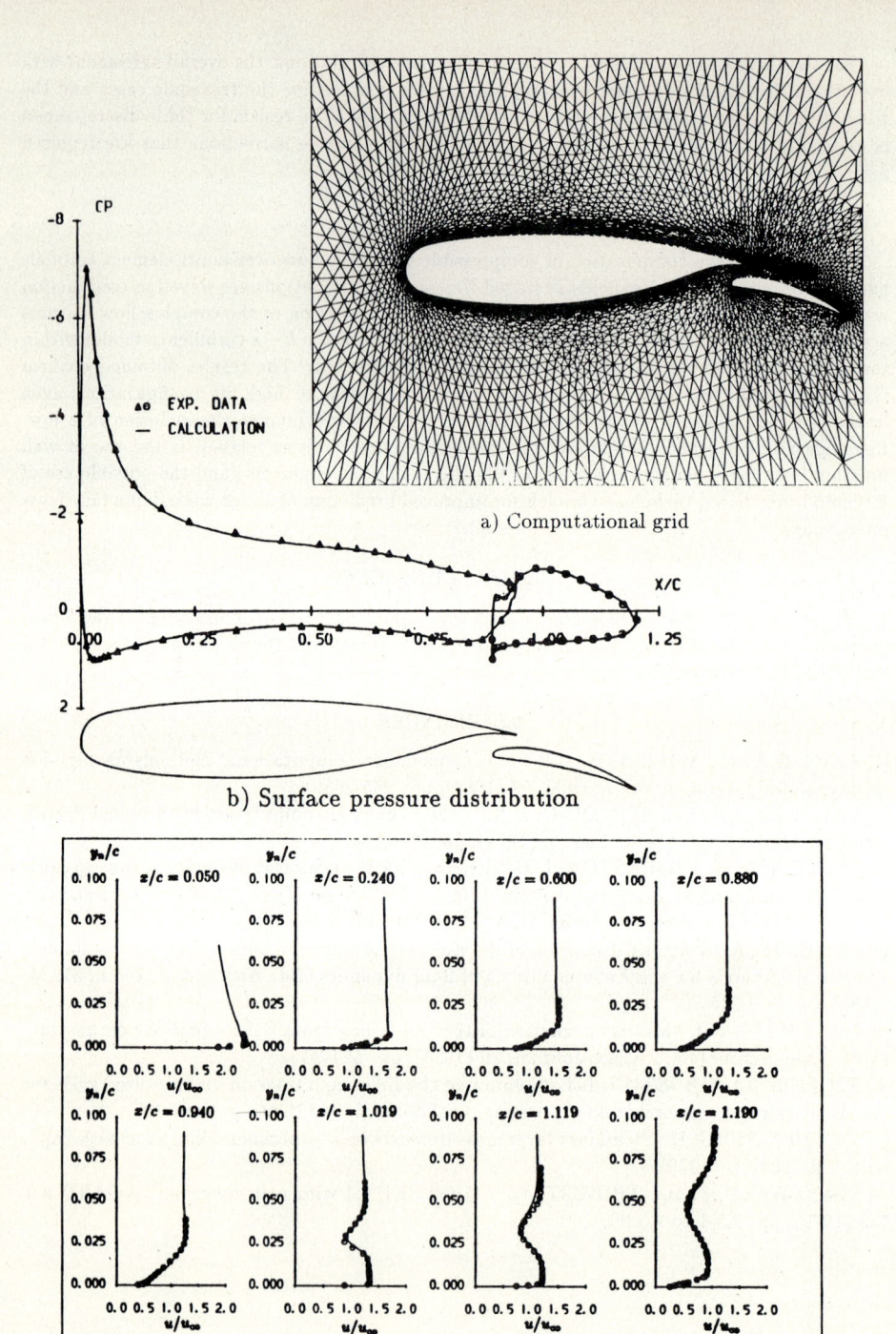

c) Upper surface velocity profiles

Fig. 2 NLR 7301 wing-flap configuration ($M_\infty = 0.185, \alpha = 6.0°, Re = 2.51 x 10^6$)

a) ($M_\infty = 0.60, \alpha = -0.05°, Re = 2.00x10^6$)

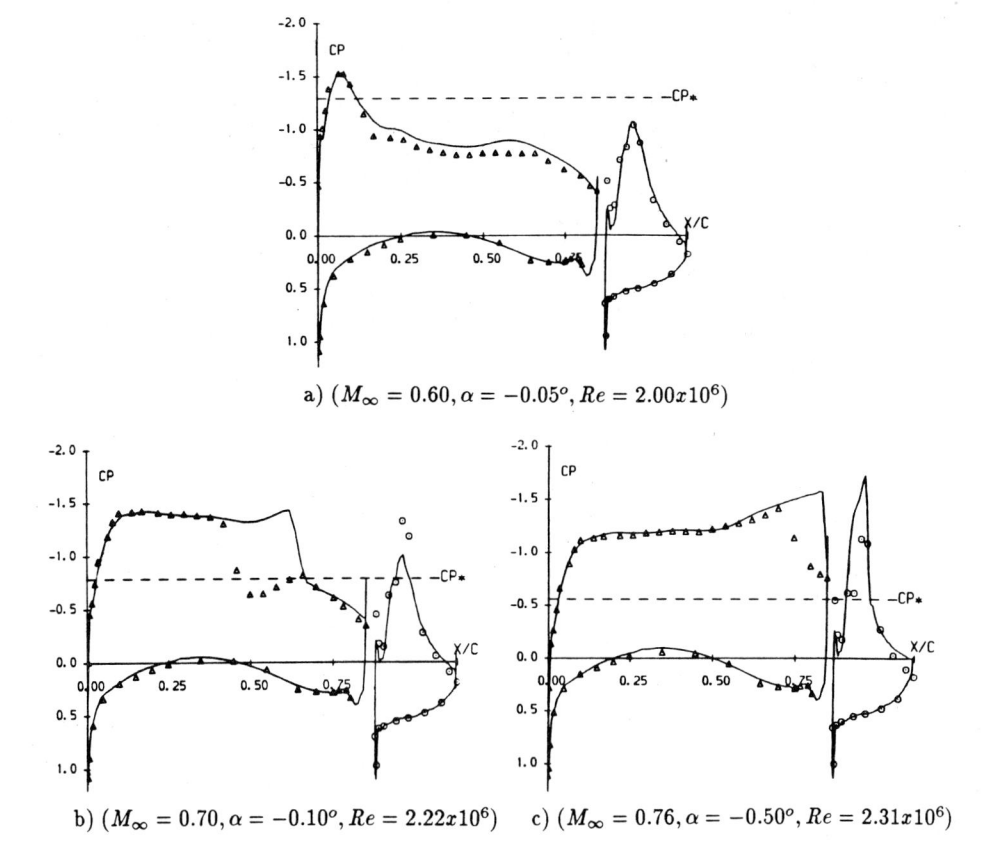

b) ($M_\infty = 0.70, \alpha = -0.10°, Re = 2.22x10^6$)   c) ($M_\infty = 0.76, \alpha = -0.50°, Re = 2.31x10^6$)

Fig. 3   SKF 1.1 wing-flap configuration (Surface pressure distribution)

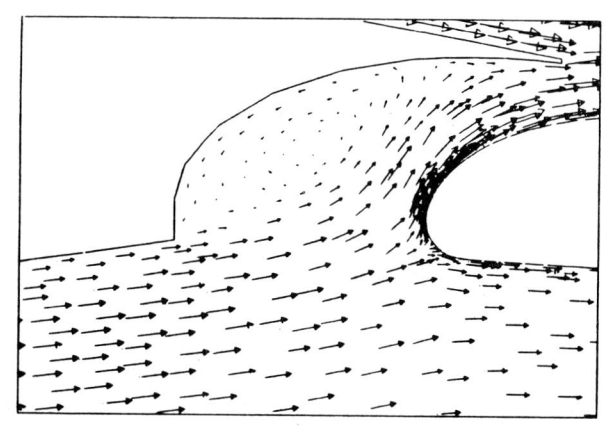

Fig. 4   SKF 1.1 wing-flap configuration ($M_\infty = 0.7, \alpha = -0.1°, Re = 2.22x10^6$)
Velocity vectors (partial view)

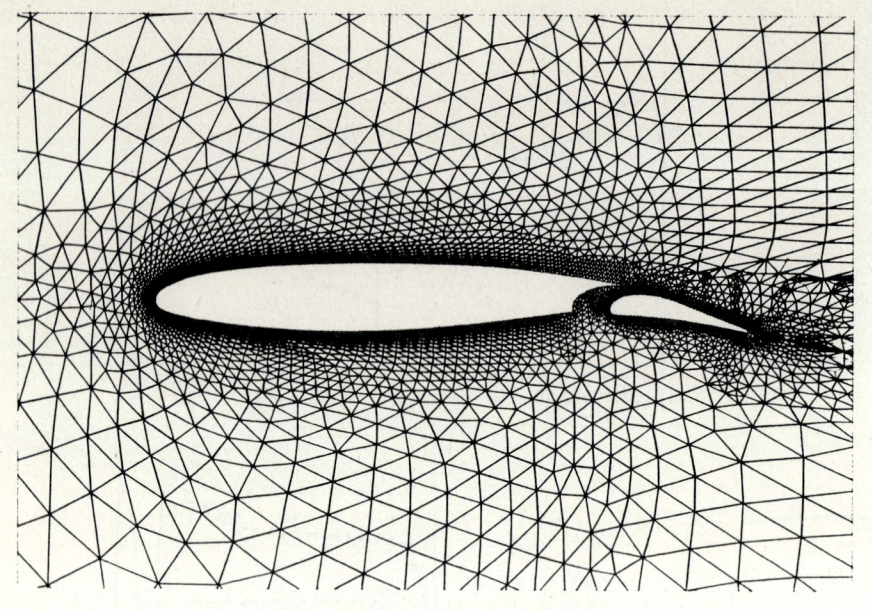

a) Computational grid (partial view)

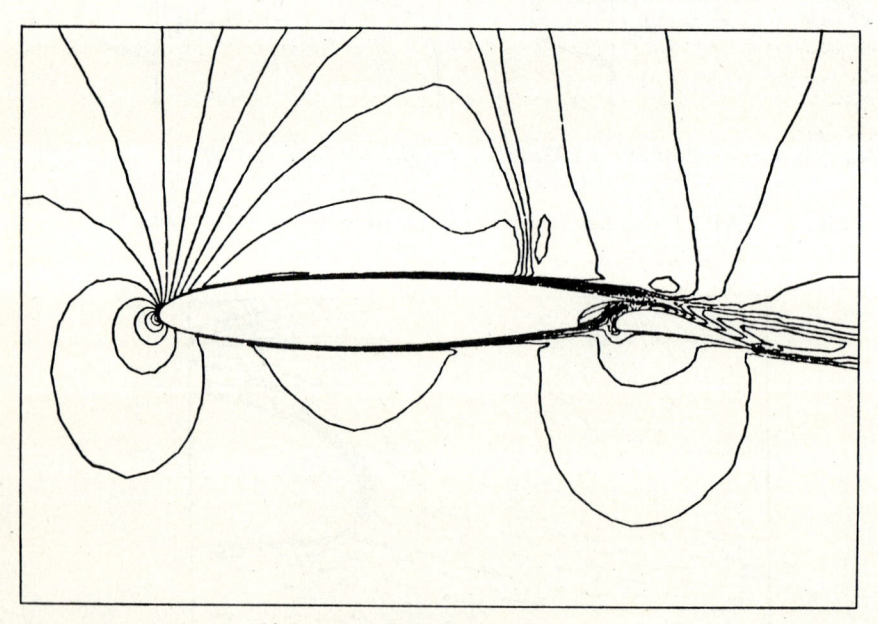

b) Iso-Mach contours ($\Delta M = 0.05$)

Fig. 5 SKF 1.1 wing-flap configuration ($M_\infty = 0.7, \alpha = -0.1^\circ, Re = 2.22 x 10^6$)

Adaptive Solutions of the Conservation Equations on Unstructured Grids

R. Vilsmeier, D. Hänel

Aerodynamisches Institut, RWTH Aachen,
Templergraben 55, D 51 Aachen, Germany

Abstract

The Euler and Navier-Stokes equations are solved on unstructured triangular grids. The Finite-volume method is used for the discretization on cell-vertex and node-centered arrangements of control volumes. For comparisons the flux-vector splitting concept as well as central formulations of the fluxes are employed. Viscous fluxes are computed in a central way. Mesh generation techniques are briefly discussed. Furthermore three mesh adaptation methods are presented. Simple test cases were chosen to compare the solutions of different methods.

Introduction

Solutions of the conservation equation on unstructured triangulated grids enable geometrical flexibility and adaptive meshing for a high degree. Such methods have proved valuable for solutions of the Euler equations in complex 2-D and 3-D flows, e.g. [1], [2], [3], [4]. The same concept applied to the compressible Navier-Stokes equations for high Reynolds numbers has been not so far developed as for inviscid flows. A major reason is the presence of very different scalings in viscous flows with the consequences of higher grid resolution, the directionality of gradients in viscous layers and the corresponding deformation of grid cells. Despite of the difficulties mentioned above (and because of the advantages against structured meshes) the number of computations for the Navier-Stokes equations on fully unstructured meshes shows an increasing tendency, computations on unstructured grids are shown e.g. in [5], [6], [7], [8], [9] and many other more.

1 Method of solution

The numerical solution is based on the 2-D, time-dependent Euler or Navier-Stokes equations written in conservative integral form.

$$\int_\tau Q_t d\tau + \oint_A (F - S)dy - \oint_A (G - R)dx = 0 \ . \tag{1}$$

Herein is

$$Q = \begin{pmatrix} \rho \\ \rho u \\ \rho v \\ \rho E \end{pmatrix} \quad F = \begin{pmatrix} \rho u \\ \rho u^2 + p \\ \rho uv \\ \rho u H_t \end{pmatrix} \quad G = \begin{pmatrix} \rho v \\ \rho vu \\ \rho v^2 + p \\ \rho v H_t \end{pmatrix} \quad S = \begin{pmatrix} 0 \\ \tau_{xx} \\ \tau_{xy} \\ s_4 \end{pmatrix} \quad R = \begin{pmatrix} 0 \\ \tau_{xy} \\ \tau_{yy} \\ r_4 \end{pmatrix}$$

where Q is the vector of the conservative variables, F and G describe the inviscid flux contributions (Euler terms), and S and R are the viscous terms in a Cartesian frame (x,y,t). The gas is assumed to be perfect. More details of the formulations can be found elsewhere.

The equations are discretized on triangulated, unstructured grids using the Finite-Volume approach. The control volumes are arranged for a cell-vertex discretization, as well as for a node-centered one. The inviscid (Euler) terms of the equations are approximated by flux-vector splitting, and by central formulations with artificial damping, respectively. The discrete equations for the volume-averaged conservative variables Q on a node P yield:

$$\tau \left.\frac{\Delta Q}{\Delta t}\right|_P + Res_\Delta(P) = 0 \qquad (2)$$

where τ is the area of the control volume and Δt is the time step for a point P.
The solution is achieved with an explicit Runge-Kutta time stepping scheme.

1.1 Mesh generation techniques

The generation of a basic mesh, representing the geometry and covering the computational domain with a reasonable distribution of points is the necessary basis for any solution method. In the present study two different grid generation techniques were developed. These methods were presented in [10] with more detail, therefore only a short overview is given.

The first method is based on a given set of points which are connected by Delaunay triangulation. This method becomes efficient, when special techniques for reducing the searching time are used, such as the domain decomposition [11] or space marching. The essential drawback is, that a reasonable point distribution has to be provided in advance. The flexibility is therefore low.

The other mesh generation concept uses only a prescribed distribution of boundary points. In a first step these points are connected to triangles. Thereafter new points are added in the center of local too coarse triangles and connected to new cells. The mesh is rearranged by diagonal swapping with some restrictions for the number of triangles surrounding a common node and final smoothing is applied. Repeated application of this procedure results in an efficient and flexible grid generator. Fig. 1 shows an example with the first triangulation and the final, smoothed mesh.

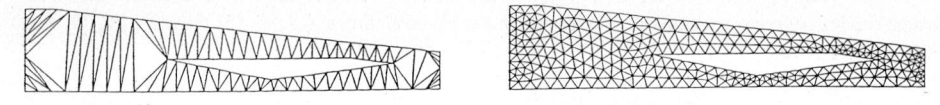

Fig. 1 Grid generation for a scram jet inlet. First triangulation (left) and final mesh (right).

The quality of a first mesh for Euler computations may be judged by evaluating the quotient of the covered area divided by the sum total of the circumcircle areas of the triangles. Although the meshes generated with the second method show good values, the critical importance of the quality of the basic mesh can be reduced, if adaptation and techniques (see chapter 2) are available.

1.2 Finite-Volume discretization

1.2.1 Arrangement of control volumes

The grid generation defines a distribution of discrete points and the connection to their neighbours. To preserve the conservative properties of Eq. (1), finite control volumes have to be defined within the triangulation. Here, two different formulations, the cell-vertex and the node-centered formulations are used in computations. The boundaries of the control volumes for the cell-vertex formulation consist of natural edges of the mesh, as sketched in Fig. 2 (outer contour). The control volumes in node-centered formulations are surrounded by a set of lines from the centers of the triangles to the centers of the edges, Fig. 2 (dashed line).

For each natural non-boundary edge the connectivity is stored as a basic cell consisting of the two points forming the edge, as well as the opposite points for the two neighboring triangles,

Fig. 3. In the cell-vertex case the fluxes will be computed over the edge $P1$-$P2$, and considered for updating the points $P3$ and $P4$. In the case of a node-centered control volume, each edge $P1$-$P2$ supports two boundary segments of two neighboring control volumes. These are the lines from the middle of the edge $P1$-$P2$ to the middle of both neighboring triangles, Fig. 3, dashed lines. In the node-centered case the fluxes influence the points $P1$ and $P2$ when performing a time step.

Boundary points require a special control volume. The natural boundary edges of the domain are treated in a similar way. As they have only one neighboring triangle, the basic cell for these edges consists only of three points, forming the boundary triangles.

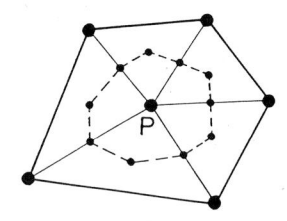

Fig. 2 Control volumes for a point P. 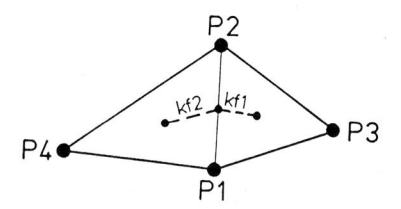   Fig. 3 Basic cell for the edge $P1$-$P2$.

1.2.2 Conservative discretization of the fluxes

The discretization of the residual $Res_\Delta(P)$ for a non boundary point on a cell-vertex control volume reads:

$$Res_\Delta(P) = -\sum_{k=1}^{km} jc_k(F_k \Delta Y_k - G_k \Delta X_k)$$

and corresponding for the node-centered scheme:

$$Res_\Delta(P) = -\sum_{k=1}^{km} jn_k(F_{kf1(k)} \Delta Y_{kf1(k)} - G_{kf1(k)} \Delta X_{kf1(k)} + F_{kf2(k)} \Delta Y_{kf2(k)} - G_{kf2(k)} \Delta X_{kf2(k)}).$$

Herein the index k refers to the first edge ($P1$ - $P2$) and km to the maximum number of edges of the control volume around point P. The indices $kf1(k), kf2(k)$ refer to the two boundary segments of the control volume supported by the edge k.

The switchers jc_k and jn_k have the discrete values -1, 0, 1, depending on the location of the point in the basic cell. The Cartesian differences ΔX read:

$\Delta X_k = X_{P2} - X_{P1}$ (cell-vertex)

$\Delta X_{kf1(k)} = \frac{1}{3}X_{P3} - \frac{1}{6}(X_{P1} + X_{P2})$ $\Delta X_{kf2(k)} = \frac{1}{6}(X_{P1} + X_{P2}) - \frac{1}{3}X_{P4}$ (node-centered).

The differences ΔY are formulated in the same way.

1.2.3 Numerical formulation of the inviscid fluxes

The Euler equations form a hyperbolic system of equations. Due to the nonlinearity of the equations, the numerical approach allows several different formulations of the inviscid flux functions. In the present study a central scheme with artificial damping, and a shock capturing upwind scheme (flux-vector splitting) are studied.

Central discretization

The central discretization is based on the unsplit inviscid fluxes F and G, as defined in Eq. (1). The numerical flux on the edge of a control volume is defined as an arithmetical average of neighbouring data without considering the characteristic wave spreading. The fluxes may first be computed at the points and later interpolated on the center of the corresponding boundary section of the control volume (flux-averaging), or the variables may be first interpolated, the fluxes later computed (variable-averaging). In the present computations the variable-averaging is preferred. For the cell-vertex scheme the flux formulation reads: (G in analogy)

$$F_k = F(\frac{1}{2}(Q_{P1} + Q_{P2})).$$

For the node-centered scheme the fluxes are calculated as e.g.:

$$F_{kf1(k)} = F(\frac{5}{12}(Q_{P1} + Q_{P2}) + \frac{1}{6}Q_{P3}) \qquad F_{kf2(k)} = F(\frac{5}{12}(Q_{P1} + Q_{P2}) + \frac{1}{6}Q_{P4}).$$

Algorithms with central discretizations require additional artificial damping terms to avoid undesired numerical oscillations. The damping terms are computed as shock capturing terms with second differences, and as high frequency filters using fourth differences of the conservative variables. The present formulation corresponds in essential to that in [12].

Flux-vector splitting

In the flux–vector splitting concept the flux is split in two parts with corresponding positive and negative eigenvalues of their Jacobian.

$$\tilde{F}_k = F^+(Q_k^+) + F^-(Q_k^-). \tag{3}$$

The variables $Q^\pm$ are extrapolated from the left (+) and the right (−) to the cell interface. In the present study the flux–vector splitting concept as proposed by van Leer [13] and modified by the author [14] is employed.

For the higher order extension of upwind schemes, van Leer has suggested the so called MUSCL approach [15], where the variables $Q^\pm$ are projected with a higher order polynomial from both sides to the cell interface.

Flux-vector splitting was formulated for the cell-vertex concept. Several possibilities of projecting the variables and the computing the split fluxes were tested.

For the computation of the flux components the conservative variables are projected on the edges of the mesh. The projection may be performed achieving first or quasi second order accuracy.

<u>Methods for first order accuracy:</u> The basic cell stored for each natural edge of the mesh, Fig. 3, provides the required information, as only the points $P3$ and $P4$ are additionally involved in the projection.

<u>Methods of quasi second order accuracy:</u> These methods require additional information, to be taken from neighboring cells. For this purpose, additional supporting points have to be determined. Several possibilities exist. The points $P3S$ and $P4S$, obtained by crossing the lines from the middle of the edge (point D) and $P3$, respectively $P4$ with the next natural edge of the mesh behind $P3$ or $P4$, showed to be suitable, Fig. 4.

Furthermore several possibilities for projecting the conservative variables along the predefined points exist. The linear extrapolation of the conservative variables from $P3S$ and $P3$ to the point D as well as from $P4S$ and $P4$ to the point D is found to be a good approach, Fig. 5. To reduce the CPU-time, the addresses of the crossed edges as well as the inter/ extrapolation coefficients are stored.

The upwind influence may be reduced by later computing the split fluxes taking a weighted average of the projected variables and the ones interpolated for a central discretization. This allows to use the upwinding as a damping term for central computations.

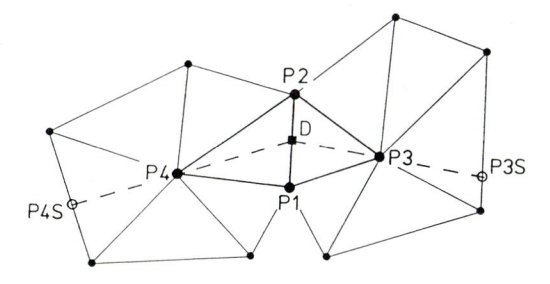

Fig. 4 Additional supporting points $P3S$ and $P4S$.

Fig. 5 Projection of variables.

1.2.4 Discretization of the viscous terms

The computation of viscous fluxes requires first derivatives in space. The node centered control volume is suitable since their interfaces are located within the triangles formed by the nodes. The derivatives can be computed directly on the triangles and used without averaging. The points involved in the computation of viscous residuals are restricted to the direct neighbors of the discretized point. Therefore a good accuracy is expected, specially for shear and boundary layers. The discretization follows chapter 1.2.2 by analogy.

The computation of inviscid residuals showed good results and good efficiency using the cell-vertex control volume. The idea is therefore to combine the inviscid residual computation on cell-vertex control volumes with the viscous residuals computed as described above. In fact the results, comparing this mixed approach with a full node-centered formulation, did not show any disadvantage. Therefore all the results presented here were obtained using the mixed approach. As the viscous residuals may not be computed at each step of the Runge-Kutta algorithm, the higher CPU-time spent to compute them on the node centered control volume is easily afforded.

1.3 Results and Comparisons for inviscid flow

Node-centered versus Cell-vertex

For the central schemes, node-centered and cell-vertex control volumes were used. The results are almost identical. Since the interfaces of the node-centered control volumes consist of twice as much segments, the CPU-time for the calculation of residuals is higher, without remarkable advantage.

Variable-averaging versus Flux-averaging

The comparisons were made on the node-centered, central scheme. The results obtained are very similar for both methods. Flux-averaging is less CPU-time consuming, as a lower amount of flux computations have to be performed each time step. On the other hand, its storage request is higher, as the fluxes are stored per point. Furthermore the method using variable-averaging showed to be more robust. As therefore only a lower amount of time steps is required to obtain a solution, both methods have similar efficiency.

Flux-vector splitting and central formulations

Based on the cell-vertex control volume, comparisons between flux-vector splitting and central formulations are made. First order Flux-vector splitting has shown to be very viscous, and therefore only suitable as damping terms for central discretizations.

Flux-vector splitting of quasi second order showed better results. On the other hand these methods are very expensive in storage and CPU-time. The CPU-time consumption for explicit algorithms is almost proportional to the time spent to compute the fluxes and to perform the interpolations. This effort is about four times higher than for the central methods. Additional CPU-time is required to compute the additional supporting points and the interpolation coefficients. On the other hand the number of time steps to obtain a solution is usually lower as in the central case, and no additional damping terms are required.

For subsonic applications no substantial advantage of the Flux-vector splitting compared to the centered scheme was found. Remarkable advantages may be found for supersonic flows in some cases. Results obtained with central differences show a good shock resolution, but gave high errors for total enthalpy, depending on the formulation of the damping terms. Also unphysical solutions may be produced. As a simple example the supersonic, inviscid flow past a circular cylinder is studied. The free stream Mach number is $Ma_\infty = 2.0$. The behaviour behind the rear shocks on the surface of the cylinder is very sensitive. The separation of the flow could not be avoided, when using the central method with artificial damping, that means the results conflict with the regarded Euler equations. Fig. 6 compares the results of a central scheme with those obtained, using flux-vector splitting of quasi second order.

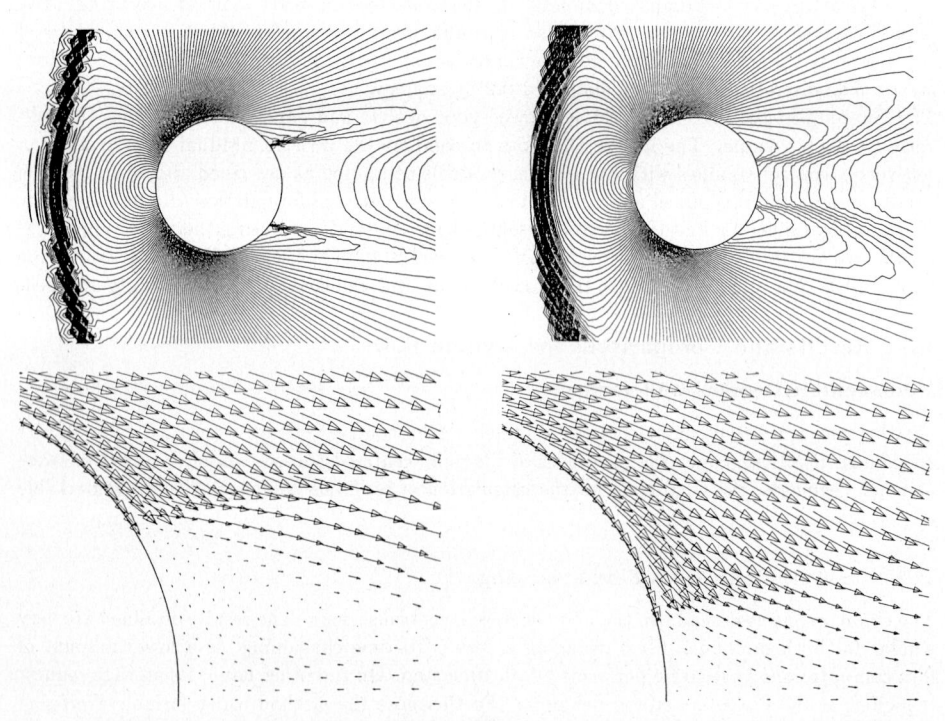

Fig. 6 Inviscid flow over a circular cylinder at $Ma_\infty = 2.0$. Isobars (top) and speed vectors on the rear part (below) obtained with a central formulation (left) and quasi second order flux-vector splitting (right).

1.4 Results for viscous flow

As a simple test case the flow past a flat plate is studied, Fig 8. The results were obtained, using the mixed cell-vertex node-centered approach and low artificial quasi-fourth order damping. The most important feature on this computations is the mesh adaption, therefore the results are discussed in chapter 2.

2 Mesh adaption techniques

The geometric flexibility using unstructured grids allows a very sensitive adaption of meshes to the local requirements. Three methods will be compared here. Two of those were already presented [10], therefore only a brief overview is given.

<u>Classical refinement and Delaunay reconnection</u> is a very simple method to adjust a mesh. All edges of triangles, exceeding a maximum difference of a key variable are refined. The mesh is reconnected by diagonal swapping to fit Delaunay's criterion. However the produced cells do not show any directionality in the plane. Therefore a high number of additional cells may be created to obtain a good accuracy in space.

<u>Central refinement and cell-orientation</u> combines a sparse insertion of additional grid points in the center of critical triangles, and a solution influenced reconnection with the aim of reducing the maximum difference of the chosen key variables. The method is very suitable for Euler computations as the reconnection process reacts on discontinuities and steady changes in the flow field according to the gradients. As the location of all points is fixed since their insertion, the local density of the mesh may never be reduced.

Both methods showed to be less suitable to adapt meshes for Navier-Stokes computations, where extreme flat triangles are required, without the presence very high gradients as they may appear on discontinuities. Therefore a third method is in development. The idea is based on [16] and extended to full dependency of a previous solution.

2.1 Flow-dependent local virtual stretching

The adaption is to be considered as a continuation of the mesh generation process where only boundary points are required, briefly described in chapter 1. It outlines as follows:

A) For each point of the previous mesh the length of the longest surrounding edge is computed, multiplied with a statistic factor and stored. This quantity L_k provides the information about the local mesh density.

B) Based on the solution obtained on the previous mesh, one or more stretching vectors may be defined. In the present version the vector of the pressure gradient and the vector of maximum shear are computed per triangle and stored per point after averaging. These two vectors are used for virtual stretching. The virtual stretching is performed in both directions.

C) The old mesh remains as a background mesh, providing the local quantities computed in A) and B), whenever required.

D) Additional points are inserted in the middle of all boundary edges, which in stretched space are longer than the minimum value L_k for the two points forming the edge.

E) Additional points are inserted in the center of all triangles with at least one edge, which, in stretched space, is longer than the minimum value L_k of the two points forming the edge.

F) Several reconnection tools are employed based on diagonal swap. The first and simplest is to fit Delaunay's criterion in the stretched space. A second method modifies the mesh in order to obtain a triangulation with as far as possible six triangles surrounding one point, while, in stretched space no extreme angles appear. Other tools are concerned with the preparation of the mesh, to avoid invalid triangulations on later smoothing.

G) Smoothing is applied in the stretched space, to increase the grid quality.

H) Correction of possible errors, which may appear while smoothing.

The tools D, E and F as well as F, G and H are applied in recurrence. Therefore the computational time is much higher than for the other two methods described. Specially the application of the tools F, G and H, representing a mixed discrete homogenous optimization, is expensive. Fortunately, it is not necessary to obtain full convergence at these steps.

2.2 Results and Comparisons

The different adaption strategies are studied on two simple cases, where analytical results are available.

2.2.1 Supersonic flow over a ramp

The results were obtained using a central cell-vertex scheme with second order damping (this damping theoretically only reacts near the shocks, as the flow apart the shocks is piecewise uniform). The first solution is obtained on a coarse, Delaunay-triangulated mesh, Fig. 7a. Solutions obtained, using the described adaption methods, are shown in Figs. 7b to 7d.

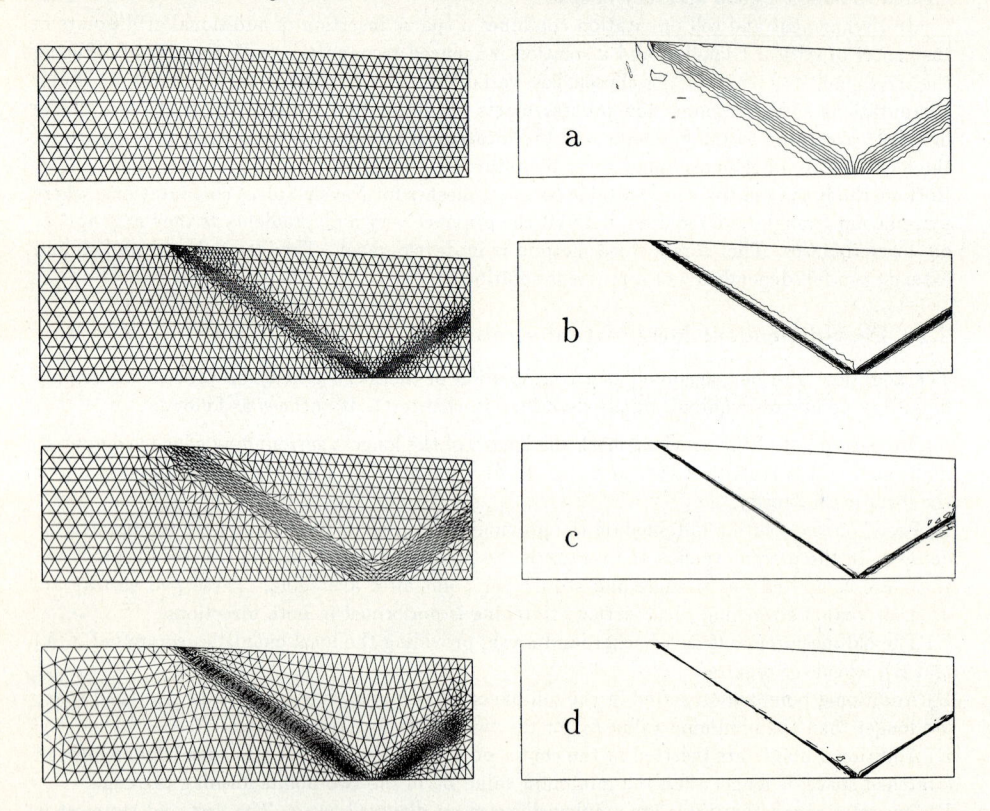

Fig. 7a/b/c/d Shock reflection ($Ma_\infty = 2.0$). Meshes and isobars:
    a) First, unadapted mesh, 657 points.
    b) After two classical refinement cycles with Delaunay reconnection, 1632 points.
    c) After one central refinement cycle with cell-orientation, 888 points.
    d) After one adaption cycle with virtual stretching, 1291 points.

The quality of the solutions as well as the final number of points and the number of refinement cycles have to be taken into account. The classical refinement using Delaunay's criterion is the less suitable, as it requires most points, most number of refinement cycles and produces the worst result. Both other methods show good results. As the method using cell-orientation is much simpler and faster, it is the best choice.

2.2.2 Flow over a flat plate

The subsonic viscous flow, $Ma_\infty = 0.3$ and Reynolds number $Re = 1.0 \cdot 10^5$, is investigated.

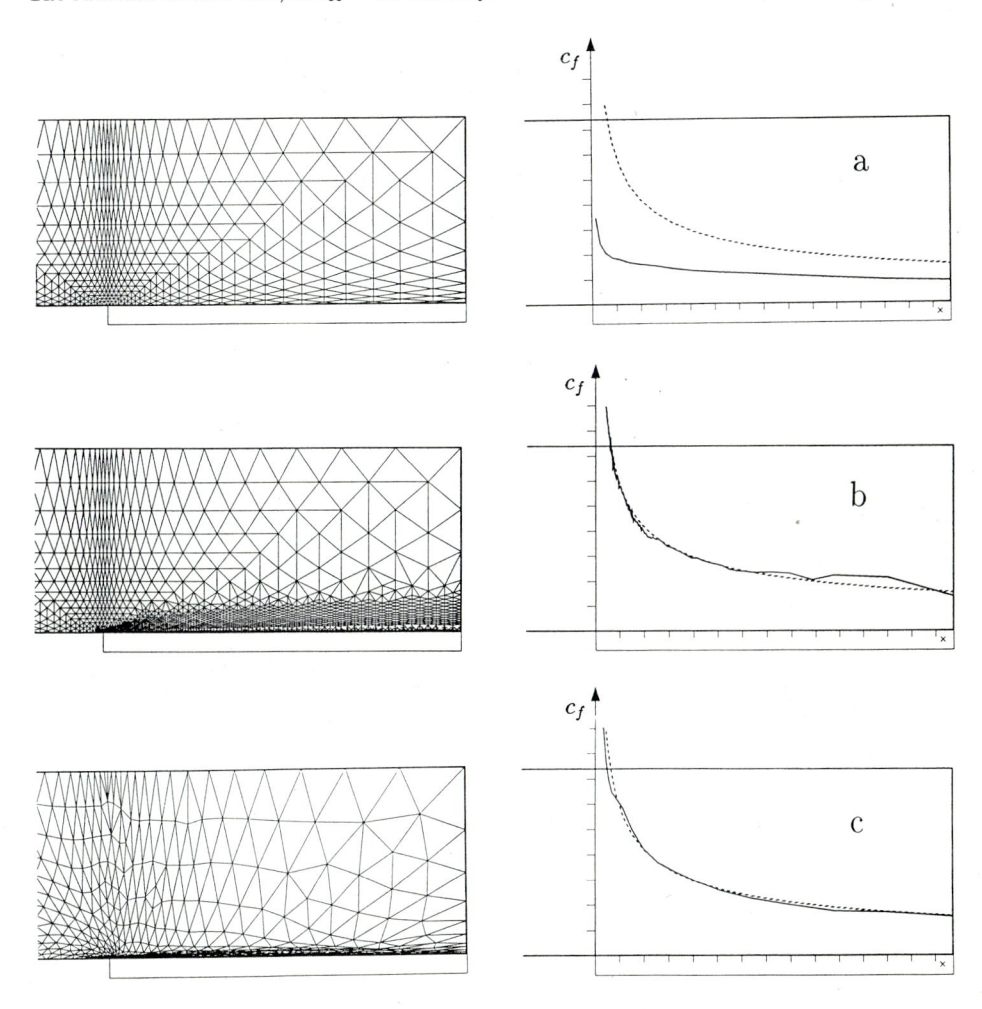

Fig. 8a/b/c Subsonic flow over a flat plate, $Ma_\infty = 0.3$, $Re = 1.0 \cdot 10^5$. Meshes and c_f distributions. Solutions (solid lines) and Blasius (dotted lines).
    a) Unadapted coarse mesh, triangulated by Delaunay's criterion, 553 points.
    b) Mesh after four classical refinement cycles, Delaunay's criterion, 4422 points.
    c) Mesh after one adaption cycle, solution dependent virtual stretching, 1272 points.

The computations were started on a mesh, far to coarse for an accurate resolution of the boundary layer. Fig. 8a shows the initial mesh, and the c_f distribution in comparison to the Blasius solution. Solutions on adapted meshes are shown in Fig. 8b/c.

The results show clearly, that triangulations with Delaunay's criterion are not suitable for Navier-Stokes computations. Results obtained by virtual stretching may be compared to those obtained on structured grids. For similar results on structured grids, the number of points is in the same range as used here. Due to indirect addressing and the fact, that on unstructured control volumes more fluxes have to be computed per time step, the efficiency is still higher on the structured mesh.

References

[1] Löhner, R., Baum, J.D.: *Numerical Simulation of Shock Interaction Using a new Adaptive H-Refinement Scheme on Unstructured Grids.* AIAA paper No. 90-0700, (1990).

[2] Periaux, J.: *Finite Element Simulations of Three-Dimensional Hypersonic Reacting Flows around Hermes.* Contrib. on Second Joint Europe-US Short Course on Hypersonics. Org. by GAMNI-SMAI, Univers. of Texas at Austin, and U.S. Air Force Academy, Colorado Springs, Jan. 1989,1989).

[3] Baker, T.J., Jameson, A.: *Improvements to the Aircraft Euler Method,* AIAA paper No. 87-0452, (1987).

[4] Peraire J., Formaggio L., Morgan K., Zienkiewicz: *Finite Element Euler Computations in Three Dimensions.* AIAA paper No. 88-0032, (1988).

[5] Mavriplis D. J., Jameson A.: *Multigrid Solution of the Navier-Stokes Equation on Triangular Meshes.* AIAA Journal, vol. 28, pp. 1415-1425, (1990).

[6] Prabhu, P.K., Stewart, J.R., Thareja, R.R.: *A Navier-Stokes Solver for High Speed Equilibrium Flows and Application to Blunt Bodies.* AIAA paper 89-0668, (1989).

[7] Koschel, W., Vornberger, A.: *Turbomachinery Flow Calculation on Unstructured Grids Using Finite Element Methods.* Notes on Numerical Fluid Mechanics, vol. 25, Vieweg Verlag, Braunschweig/Wiesbaden, (1989).

[8] Lohner R., Morgan K., Peraire j., Vahdati M.: *Finite Element Flux-Corrected Transport for the Euler and Navier-Stokes Equations.* Int. J. for Numer. Meth. in Fluid Mech., vol. 7, pp. 1093-1109, (1987).

[9] Argyris, J., Doltsinis, I., Friz, H.: *Hermes Space Shuttle: Exploration of Reentry Aerodynamics.* Comput. Meth. Appl. Mech. Eng., vol 73, (1989).

[10] Vilsmeier R., Hänel D.: *Generation and Adaptation of 2-D Unstructured Meshes.* In: A.S. Arcila, J. H user, P.R. Eiseman, J.F. Thompson (Ed.): Numerical Grid Generation in Computational Fluid Dynamics and Related Fields, pp. 55-66, North-Holland, Amsterdam, 1991.

[11] Barth T. J.: *On unstructured grids and solvers,* VKI-Lecture Series 1990-03, 1990.

[12] Mavriplis D. J.: *Solution of the Two-Dimensional Euler Equations on Unstructured Triangular Meshes,* Thesis, Princeton University, 1987.

[13] van Leer, B.: *Flux-Vector Splitting for the Euler Equations.* Lecture Notes in Physics vol. 170, pp. 507-512, (1982).

[14] Schwane, R., Hänel, D.: *An Implicit Flux-Vector Splitting Scheme for the Computation of Viscous Hypersonic Flow.* AIAA-paper No. 89-0274, (1989).

[15] van Leer, B.: *Towards the Ultimate Conservative Difference Scheme. A second-order sequel to Godunov's method.* J. Comp. Phys. vol.32, pp.101-136, (1979).

[16] Mavriplis D. J.: *Unstructured and Adaptive Mesh Generation for High Reynolds Number Viscous Flows.* In: A.S. Arcila, J. H user, P.R. Eiseman, J.F. Thompson (Ed.): Numerical Grid Generation in Computational Fluid Dynamics and Related Fields, pp. 79-91, North-Holland, Amsterdam, 1991.

CONVECTIVE-ADVECTIVE FLOWS

ANALYSIS OF NATURAL CONVECTION WITH LARGE TEMPERATURE DIFFERENCE USING A FAST MATRIX SOLVER

Yukinori AKIBA
C&C Information Technology Research Laboratories, NEC Corporation
1-1 Miyazaki-4chome, Miyamae-ku, Kawasaki, 216 Japan

Shun DOI
Second C&C System Development Division, NEC Corporation
34-2 Shiba-5chome, Minato-ku, Tokyo, 108 Japan

Kunio KUWAHARA
The Institute of Space and Aeronautical Science
3-1-1 Yoshinodai, Sagamihara, Kanagawa, 229 Japan

SUMMARY

A scheme for computing 3D unsteady flow with large temperature difference is presented. The compressible Navier-Stokes equations are approximated, assuming that the density variation due to pressure difference is small, so that the effect of the acoustic wave propagation, which is not of primary importance in low Mach-number flows, is removed. In this scheme, linear systems which result from the discretization of the pressure equation are large, sparse, nonsymmetric and, in addition, not diagonally dominant. In this study, a preconditioned conjugate gradient (PCG) method suited to vectorsupercomputers is applied for solving these stiff linear systems. Numerical experiments are done to investigate the thermal convection in the chemical vapor deposition reactor used for the manufacture of electronic devices. Results show the effect of unsteady buoyant flow caused by a large temperature difference. The efficiency of the PCG linear solver on vectorsupercomputers is also demonstrated.

INTRODUCTION

Natural convection due to large temperature differences plays an important role in the manufacturing process of semiconductor devices. For example, the chemical vapor deposition (CVD) process is generally used for the crystal growth of gallium arsenide. The substrate is heated to about 900K, so that the carrier gas near the substrate is also heated by thermal conduction. Thus, unsteady thermal convection occurs caused by a temperature difference of about 600K. The thermal boundary layer near the substrate has a great influence on the uniformity of the crystal. Although flow in CVD have been observed experimentally, numerical simulation has begun only in recent years[1].

The Boussinesq approximation is widely applied in natural convection ploblems. However, this approximation cannot be applied in the flow with large temperature differences, since the condition of incompressibility does not hold. In this paper, we present a new scheme in order to analyze flow with relatively large temperature difference. Kuwahara proposed a method[4] by which the effect of acoustic wave propagation due to the pressure is efficiently removed from the compressible Navier-Stokes equations, and they are discretized based on the MAC formulation[2][3]. In this method, the pressure equation is obtained by taking the divergence of the momentum equation. Since it includes the gradient of temperature, the coefficient matrices which result from the finite difference discretization of the pressure equation are large, sparse, nonsymmetric and, in addition, not diagonally dominant. Hence, the solution of the system of linear equations is the most time consuming part of the calculation. We also present the bi-conjugate gradient (BCG) method[5] combined with the tridiagonal factorization (TF) preconditioner to efficiently solve the stiff linear system on vectorsupercomputers[6].

These methods are applied to investigate the unsteady thermal convection in a CVD reactor. Numerical experiment results on the NEC SX-2 supercomputer are presented. We also discuss the comparison of efficiency between the tridiagonal factorization and incomplete LU (ILU) decomposition preconditioners[8].

GOVERNING EQUATIONS

We shall consider gas flow in a three-dimensional region. Although the flow has a low Mach-number, the present scheme is based on the compressible equations. The compressible Navier-Stokes equations are written as follows:

The continuity equation:

$$\frac{D\rho}{Dt} = -\rho \frac{\partial u_i}{\partial x_i}, \tag{1}$$

where ρ is the density, u_i the velocity component in the x_i direction and,

$$\frac{D}{Dt} = \frac{\partial}{\partial t} + u_j \frac{\partial}{\partial x_j} \tag{2}$$

is the material derivative.

The momentum equation:

$$\frac{Du_i}{Dt} = -\frac{1}{\rho}\frac{\partial p}{\partial x_i} + \frac{1}{\rho}\frac{\partial}{\partial x_j}\{\mu(\frac{\partial u_i}{\partial x_j} + \frac{\partial u_j}{\partial x_i})\} + f_i, \tag{3}$$

where p is the pressure, μ the viscosity, and f_i the external body force.

The energy equation:

$$\frac{DT}{Dt} = \frac{1}{\rho c_p}\frac{\partial}{\partial x_j}(k\frac{\partial T}{\partial x_j}), \tag{4}$$

where T is the temperature, c_p the specific heat, and k the thermal conductivity.

Here, the gas is considered to be perfect, so that equation of state is given by

$$p = \rho R T ,\qquad(5)$$

where R is the gas constant.

The pressure equation is obtained by taking the divergence of the momentum equation (3) as

$$\frac{1}{\rho}\frac{\partial}{\partial x_j}\frac{\partial p}{\partial x_j} + \frac{\partial p}{\partial x_j}\frac{\partial}{\partial x_j}(\frac{1}{\rho}) = \frac{\partial}{\partial x_i}\left[-\frac{Du_i}{Dt} + \frac{1}{\rho}\frac{\partial}{\partial x_j}\{\mu(\frac{\partial u_i}{\partial x_j} + \frac{\partial u_j}{\partial x_i})\} + f_i\right].\qquad(6)$$

For simplicity, the viscosity μ and the thermal conductivity k are assumed to be constant. The dissipation function term is neglected because it is negligible so long as the flow is not supersonic.

A SCHEME FOR NATURAL CONVECTION WITH LARGE TEMPERATURE DIFFERENCE

Boussinesq approximation In order to analyze natural convection, the Boussinesq approximation is generally used in incompressible flow. In this method, the buoyancy caused by thermal expansion is taken into account only by the term of external gravity. Thus the momentum equation (3) becomes

$$\frac{Du_i}{Dt} = -\frac{1}{\rho_0}\frac{\partial p}{\partial x_i} + \frac{\mu}{\rho_o}\frac{\partial}{\partial x_j}\frac{\partial u_i}{\partial x_j} + \frac{\rho - \rho_0}{\rho_0} f_i ,\qquad(7)$$

where ρ_0 is the density at the reference temperature T_0. Using the coefficient β of thermal expansion, ρ is given by

$$\rho = \rho_0[1 - \beta(T - T_0)].\qquad(8)$$

In this approximation, the condition of incompressibility must be satisfied, that is

$$\beta \Delta T \frac{\kappa}{U_0 L_0} \ll 1 ,\qquad(9)$$

where L_0 is the reference length, U_0 the reference velocity, $\kappa = k/\rho c_p$, and $\Delta T = T - T_0$. When the gas is air, $L_0 = 0.01(m)$, $U_0 = 0.1(m/s)$ and $T_0 = 300K$, the order of magnitude of the left hand side of (9) becomes $O(10^{-2})$ when $\Delta T = 100K$, and $O(10^{-1})$ when $\Delta T = 1000K$. Hence, the Boussinesq approximation can not be applied when the temperature difference in the gas becomes large.

New formulation When the compressible Navier-Stokes equation is employed to compute low Mach-number flow, numerical instability arises because the velocity of sound is typically hundred times faster compared to the flow speed. We now introduce a new scheme to analyze low Mach-number flow with large temperature differences which avoid the numerical instability. We can divide the pressure p into two parts

$$p = p_\infty + p_d ,\qquad(10)$$

where p_∞ is the constant atmospheric pressure and p_d the pressure variation due to the fluid motion. Since p_d is sufficiently small compared to p_∞, p may be assumed to be constant. Hence, the perfect gas equation (5) is approximated as

$$p = \rho R T \approx p_\infty = const. \tag{11}$$

By taking the material derivative, this equation gives

$$\frac{1}{\rho}\frac{D\rho}{Dt} = -\frac{1}{T}\frac{DT}{Dt}. \tag{12}$$

Associating with the continuity equation (1), we obtain

$$\frac{\partial u_j}{\partial x_j} = \frac{1}{T}\frac{DT}{Dt}. \tag{13}$$

From equation (13) and the energy equation (4), the continuity equation may then be represented as

$$\frac{\partial u_j}{\partial x_j} = \frac{k}{\rho c_p}\frac{1}{T}\frac{\partial}{\partial x_j}\frac{\partial T}{\partial x_j}. \tag{14}$$

Using the above approximation, we can deal with the effect of compressibility caused by thermal expansion and efficiently eliminate the effect of acoustic wave propagation from the compressible Navier-Stokes equations. No numerical instability due to the effect of acoustic wave propagation occurs any longer.

Regarding the calculation of the density ρ, there is no need to solve the continuity equation (1). Let ρ_0 be the density at T_0. Equation of state may then be written as

$$p = \rho_0 R T_0. \tag{15}$$

From equations (15) and (5), the relation between ρ and T is obtained as

$$\frac{\rho}{\rho_0} = \frac{T_0}{T}. \tag{16}$$

Nondimensionalizing the density and the temperature as

$$\rho^* = \frac{\rho}{\rho_0}, \quad T^* = \frac{T}{T_0}, \tag{17}$$

ρ^* is given by

$$\rho^* = \frac{1}{T^*}, \tag{18}$$

instead of the continuity equation.

Discretization The momentum equation (3), the energy equation (4) and the pressure equation (6) are solved using an implicit scheme based on the MAC method. A first order difference is used for the time derivatives. The diffusive terms are discretized by second order differences. In the convective terms, the third order upwind scheme proposed by Kawamura and Kuwahara[7] is employed to cope with high Reynolds-number flow, i.e.

$$u\frac{\partial u}{\partial x} \approx u_i \frac{-u_{i+2} + 8u_{i+1} - 8u_{i-1} + u_{i-2}}{12\Delta x} + |u_i|\frac{u_{i+2} - 4u_{i+1} + 6u_i - 4u_{i-1} + u_{i-2}}{4\Delta x}, \tag{19}$$

where u_i is the velocity at the i-th grid node and, Δx the mesh size. This scheme yields small numerical diffusion compared to the first order upwind method. The velocity, the pressure and the temperature are defined on the grid node. Fluid motion is computed by the following algorithm. Note that the discretization in time is considered, and n stands for the n-th time step.

Step1: Give the initial conditions for velocity, pressure and temperature.

Step2: Calculate the temperature T^{n+1} by the energy equation i.e.

$$\frac{T^{n+1} - T^n}{\Delta t} + u_j^n \frac{\partial T^{n+1}}{\partial x_j} = \frac{k}{\rho c_p} \frac{\partial}{\partial x_j} \frac{\partial T^{n+1}}{\partial x_j}, \qquad (20)$$

where Δt is the time interval.

Step3: The pressure equation is solved as

$$\frac{1}{\rho} \frac{\partial}{\partial x_j} \frac{\partial p^{n+1}}{\partial x_j} + \frac{\partial p^{n+1}}{\partial x_j} \frac{\partial}{\partial x_j}(\frac{1}{\rho}) = -(\frac{\partial u_j^{n+1}}{\partial x_j} - \frac{\partial u_j^n}{\partial x_j})/\Delta t$$
$$+ \frac{\partial}{\partial x_i}(u_j \frac{\partial u_i^n}{\partial x_j}) + \frac{\partial}{\partial x_i}(\frac{\mu}{\rho} \frac{\partial}{\partial x_j} \frac{\partial u_i^n}{\partial x_j}), \qquad (21)$$

where $\rho = 1/T^{n+1}$. $\partial u_j^{n+1}/\partial x_j$ is calculated by equation (14), which is usually assumed to be zero in the original MAC method.

Step4: Compute the velocity by the momentum equation, that is

$$\frac{u_i^{n+1} - u_i^n}{\Delta t} + u_j^{n+1} \frac{\partial u_i^n}{\partial x_j} = -\frac{1}{\rho} \frac{\partial p^{n+1}}{\partial x_i} + \frac{\mu}{\rho} \frac{\partial}{\partial x_j} \frac{\partial u_i^{n+1}}{\partial x_j} + f_i. \qquad (22)$$

Step5: Repeat Step2-Step4.

The flow in an arbitrary-shaped three-dimensional region is calculated using a system of curvilinear coordinates.

PRECONDITIONED ITERATIVE METHOD

By the finite difference discretization in 3D general curvilinear systems, the pressure equation gives a difference equation where a node is connected to 18 adjacent nodes as shown in Fig. 1. Similarly, the momentum and energy equations in which the third order upwind technique is applied to the convective terms give 25-point difference equations.

The coefficient matrices of these difference equations are sparse and nonsymmetric, and the solution of these equations consume about 99% of the CPU time. Since the matrices of the momentum and energy equations, which include the time derivative, are diagonally dominant, they give rise to good convergence and cost only about 10%

19-point defference   7-point difference

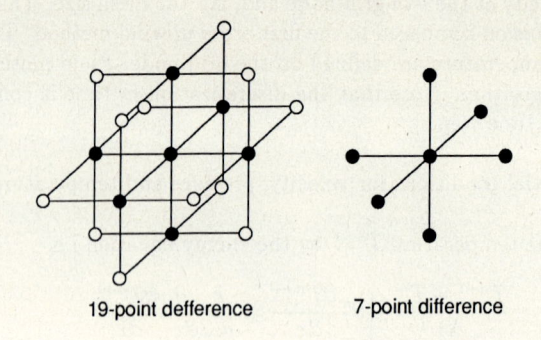

Fig. 1: Grid structure of finite difference equations

of the CPU time. On the other hand, in the pressure equation, Neumann condition is imposed on most of the boundaries. Further, since the gradient of density is included in the presented scheme, the coefficient matrix of the pressure equation is not diagonally dominant in regions where large density variation due to thermal expansion is present.

Usually successive over-relaxation (SOR) methods are used for solving these linear systems. However, SOR methods do not give good convergence for stiff matrices. For solving the stiff linear systems arising from the pressure equation on vectorsupercomputers, we now propose a conjugate gradient method associated with a preconditioner, which uses the matrix components corresponding to the 7-point finite difference matrix in the 3D rectangular grid system.

Let the linear system of the 19-point matrix of the pressure equation be

$$Au = b \ . \tag{23}$$

The approximating matrix M, which can be easily inverted by forward-backward substitution, is written as

$$M = A + R \ , \tag{24}$$

where R is the remainder matrix. An equation equivalent to equation (23) is given by

$$M^{-1}Au = M^{-1}b \ . \tag{25}$$

Then, we apply the bi-conjugate gradient (BCG) method to equation (25).

It is important how to vectorize the forward and backward substitution of M. The use of all the 19-point matrix elements of A as the preconditioner results in poor vectorization and excessive calculation. Hence, we shall use only the 7-point matrix elements as the preconditioner and adopt the tridiagonal factorization (TF) algorithm developed by Doi and Harada[7].

Let A consist of the diagonal D, A_{x_i} ($x = 1, 2, 3$) related to the x_i direction, and the remaining A_{others},

$$A = D + A_{x_1} + A_{x_2} + A_{x_3} + A_{others} \ . \tag{26}$$

In the TF method, the preconditioner M is defined by

$$M = (D + A_{x_1})D^{-1}(D + A_{x_2})D^{-1}(D + A_{x_3}) \ . \tag{27}$$

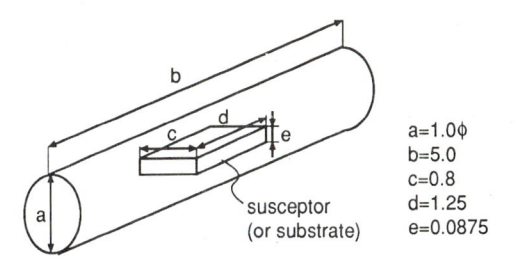

Fig. 2: CVD reactor model

a=1.0φ
b=5.0
c=0.8
d=1.25
e=0.0875

Each direction component is factorized into lower and upper matrices to give

$$M = L_{x_1} U_{x_1} L_{x_2} U_{x_2} L_{x_3} U_{x_3} \ . \tag{28}$$

The inversion of $L_{x_1} U_{x_1}$ is independent of the x_2 and x_3 directions, and similarly for the x_2 and x_3 directions. Therefore the inversion of the TF preconditioner can be easily vectorized.

NUMERICAL EXPERIMENTS

Numerical experiments were carried out on the NEC SX-2 supercomputer to investigate the natural convection in the cylindrical CVD reactor model shown in Fig. 2. At the inlet boundary, the velocity according to a given Reynolds-number (Re) was imposed and the temperature was set to 300K, while Neumann conditions were imposed at the outlet. The temperature of the reactor surface was set to 300K. The substrate was heated to 900K. The carrier gas is nitrogen whose Prandtl-number is 0.716. The diameter of the cylinder and the flow speed at the inlet are chosen as the reference length and velocity, respectively. The region was discretized by $81 \times 51 \times 51 = 210681$ grid points.

Fig. 3-6 show the temperature and velocity distribution of the unsteady thermal convection at $Re = 1000$ and 10. When the inlet flow is slow, the natural convection governs the flow in the reactor. In Fig. 3 and 5, backward flow is seen in the front end of the substrate. Plumes rise periodically from the substrate as shown in Fig. 4.

Table 1 demonstrates the efficiency of our linear solver, which uses the TF preconditioner. It is 1.86 times as fast as the method using the incomplete LU (ILU) decomposition preconditioner, which is vectorized using the hyper-plane method.

CONCLUSION

In this paper, we proposed a method for analyzing natural convection with large temperature differences. In this method, the effect of sound wave propagation is efficiently removed from the compressible Navier-Stokes equations. Hence, we can calculate the fluid motion using an extension of the MAC method. Regarding the solution of

Table 1: Iterations and computation time

solver	iteration counts	iteration ratio	cpu time sec.	cpu time ratio
ILUBCG	83	1.00	10.58	1.00
TFBCG	98	1.18	5.68	0.54

the linear systems arising from 19-point finite difference scheme, the proposed 7-point TFBCG method is almost twice as fast as the conventional ILUBCG method.

We plan to compare our simulation results with experimental data in order to verify our method.

REFERENCES

[1] Jansen, N., Orazem, M. E., Fox, B. A. and Jesser, W. A.:"Numerical study of the influence of reactor design on MOCVD with a comparison to experimental data", J. Cryst. Growth, Vol. 112 (1991), pp.316-336.

[2] Harlow, F. H. and Welch, J. E.:"Numerical calculation of time-dependent viscous incompressible flow of fluid with free surface", Phys. Fluids, Vol. 8 (1965), pp. 2182-2189.

[3] Chorin, A. J.:"Numerical Solution of the Navier-Stokes Equations", Math. Comput., Vol. 22 (1968), pp. 745-762.

[4] Kuwahara, K.:"Computation of Thermal Convection with a Large Temperature Difference", Proc. of 4th Int. Conf. on Appl. Num. Modeling (1984).

[5] Fletcher, R.:"Conjugate Gradient Method for Indefinite Systems", Lecture Notes in Mathematics 506, pp. 73-89, Springer-Verlag (1976).

[6] Doi, S. and Harada, N.:"Tridiagonal Factorization Algorithm: A Preconditioner for Nonsymmetric System Solving on Vectorcomputers", J. Info. Process., Vol. 11 (1988), pp. 138-146.

[7] Kawamura, T. and Kuwahara, K.:"Computation of High Reynolds Number Flow around a Circular Cylinder with Surface Roughness", AIAA paper 84-0340 (1984).

[8] Van der Vorst, H. A.:"Iterative Solution Methods for Certain Sparse Linear Systems with a Non-Symmetric Matrix Arising from PDE-Problems", J. Compt. Phys., Vol. 44 (1981), pp. 1-19.

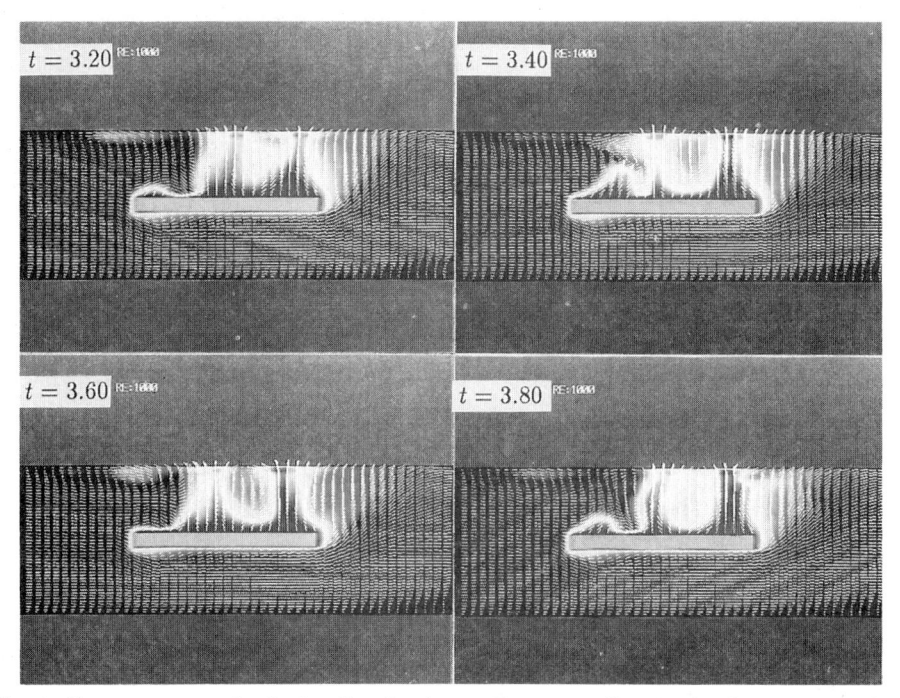

Fig. 3: Temperature and velocity distribution at $Re = 1000$ (longitudinal cross section)

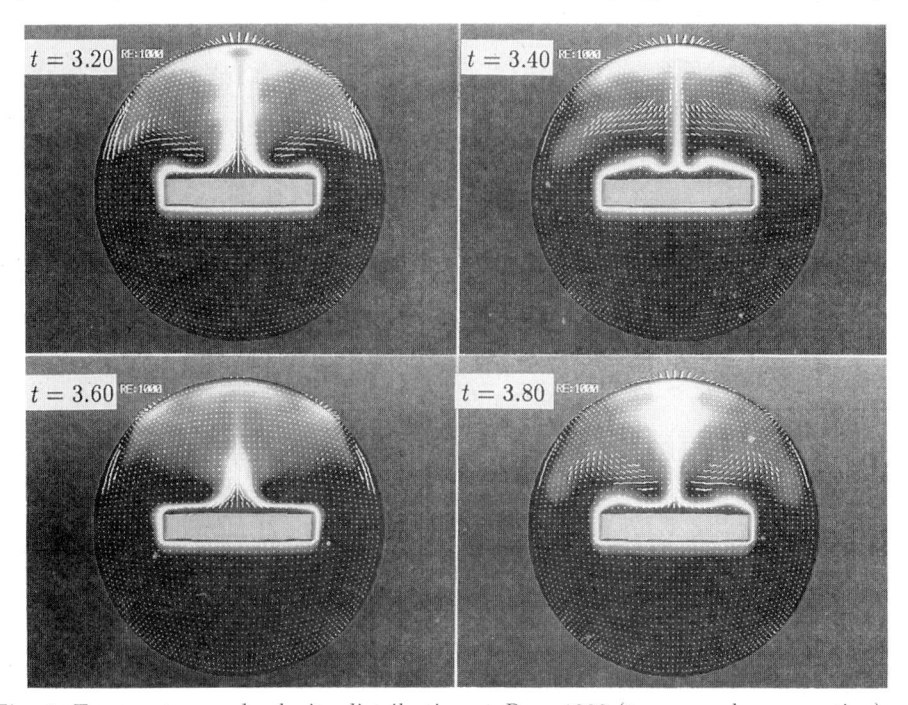

Fig. 4: Temperature and velocity distribution at $Re = 1000$ (transversal cross section)

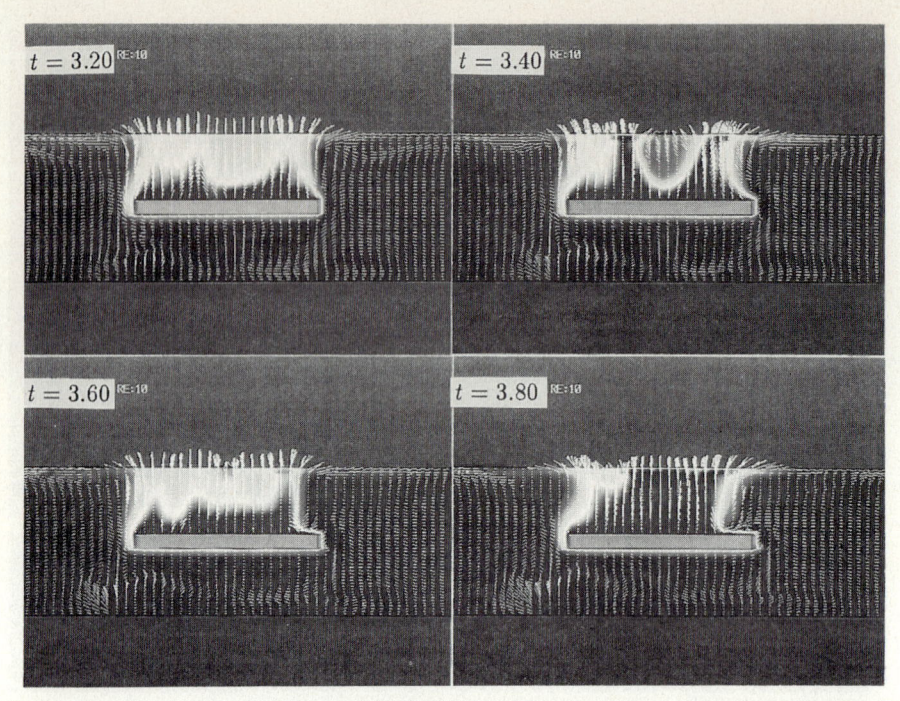

Fig. 5: Temperature and velocity distribution at $Re = 10$ (longitudinal cross section)

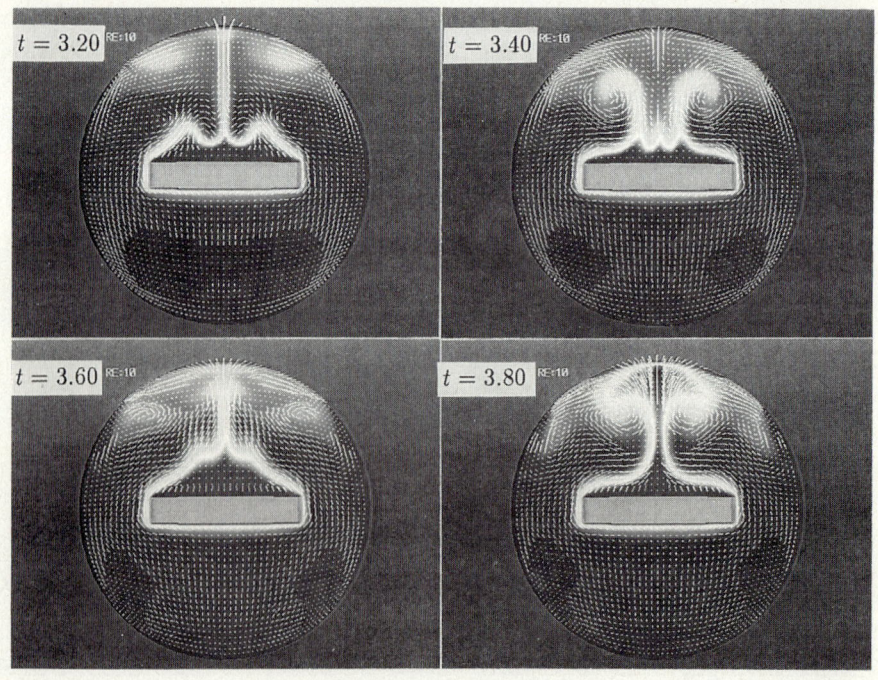

Fig. 6: Temperature and velocity distribution at $Re = 10$ (transversal cross section)

COMPARISON OF SOME NUMERICAL SCHEMES FOR THE ADVECTION OF A PASSIVE POSITIVE SCALAR ON A COARSE GRID.

M. Pourquié and F.T.M. Nieuwstadt
Delft University of Technology
Laboratory for Hydro- and Aerodynnamics
Rotterdamseweg 145
2628 AL Delft
The Netherlands

Summary

We seek a numerical scheme to describe the transport of a passive scalar. Our scheme should satisfy the requirement that sharp gradients in the concentration field are not smoothed out. Furthermore the scalar should always be positive. The field of application is the calculation of turbulent transport in large-eddy simulation of a turbulent flow, but the results apply also to other fields of fluid dynamics. Several candidates for numerical schemes are described and examined with the aid of test problems.

Statement of the problem

We consider the solution of the advection-diffusion equation:

$$\frac{\partial c}{\partial t} + \frac{\partial uc}{\partial x} + \frac{\partial vc}{\partial y} + \frac{\partial wc}{\partial z} = \frac{\partial}{\partial x}\left(K\frac{\partial c}{\partial x}\right) + \frac{\partial}{\partial y}\left(K\frac{\partial c}{\partial y}\right) + \frac{\partial}{\partial z}\left(K\frac{\partial c}{\partial z}\right).$$

This equation describes the transport and diffusion of a passive scalar c in a velocity field where u, v, w are the given velocity-components in the x-, y- and z-direction and K is the diffusion coefficient. This is a familiar problem in fluid mechanics. Although we have only one equation which is moreover linear, the numerical solution presents nevertheless problems. This is due to the character of the equation and the conditions we impose on the solution. These two points are discussed in more detail in the next paragraph.

Difficulties in the solution of the advection-diffusion equation

First something about the character of the equation. Difficulties in solving the advection-diffusion equation arise when the Péclet number, Pé, is large (say ≥ 2). Here Pé $= UL/K$, with U and L characteristic velocity and length scales of the velocity field and K the diffusion coefficient. Pé is a measure of the importance of advection relative to diffusion. For large Pé the equation is dominated by the advection part of the equation. This may be represented by the advection-diffusion equation with K=0. This advection part is, mathematically speaking, of a hyperbolic nature. Physically this means that steep gradients in a solution can travel over large distances before being damped out. This brings us to the conditions under which our numerical solver has to work.

For our application in large-eddy simulation we ordinarily have Pé numbers ≥ 20, which is very large. Also, we expect that steep gradients are present in our problem, e.g. due to initial conditions. Then the numerical solution should be able to represent them and should not unnecessarily smooth out these gradients. Further, we require that the solution remains always positive. This is important, for instance, if we describe the transport of some chemical species, because negative concentrations do not exist in reality.

Other conditions come from our special application, i.e. large-eddy simulation (LES) of turbulent flow. LES is 3-D and time-dependent and gives a complicated velocity field. Furthermore LES typically works on a coarse grid.

Presentation of possible solutions

For simplicity of presentation the rest of this paper will concentrate on the case of $K = 0$, or Pé $= \infty$, which is in principle the most difficult case.

First we show two straightforward solution methods and where they go wrong. Our first attempt is, to use central differences for space and time derivatives. This will give second-order accuracy in space and time. However, a steep gradient in c will lead to wiggles (due to dispersion) and therefore to undesired negative values.

On the other hand, a first order scheme such as upwind can be considered. The solution is always positive now, but we see extreme smearing (due to numerical diffusion).

Either the effect of dispersion (wiggles) or of diffusion (smearing) or both are present more or less in all finite difference methods. (See figure 1, which shows the advection of a rectangle and a sine wave)

However, we try to keep both effects to a minimum. We will now briefly present three advanced finite difference methods which are designed to do this, and in addition a method which cannot be classified as a finite difference method.

First something general about finite difference methods. We see that simple linear approaches do not give satisfactory results. In fact, Godunov [5] showed that there is no linear finite difference method, of order higher than 1, which is free of dispersion. Therefore all methods we examine are non-linear. Another thing which all the difference methods we treat have in common is the fact that they work with the fluxes of the concentration at the cell boundaries. They try to adapt the fluxes to remove undesired phenomena (diffusion, wiggles).

For the sake of presentation we show a 1-D picture of the (staggered) grid we use, indicating the places where we put the concentration grid values and the velocities/fluxes (see figure 2).

The method of Smolarkiewicz.

Introduction.

Smolarkiewicz [1] uses the first order upwind scheme as basis. As we have seen, this scheme is positive but very diffusive. The error of this scheme is estimated with Taylor series. The lowest order term in the error is a diffusion term. This term is subsequently removed with a second, corrective upwind step. The use of the upwind method to remove the lowest order term of the error guarantees that the scheme remains positive.

Correction step for the upwind method.

Using a Taylor series expansion we find, that to higher order the standard first order upwind solution satisfies not the pure advection equation, but another equation, namely:

$$\frac{\partial c}{\partial t} + \frac{\partial uc}{\partial x} = \frac{\partial}{\partial x}\left(0.5(u\Delta x - \Delta t u^2)\frac{\partial c}{\partial x}\right) + H.O.$$

The term H.O. stands for terms of $O(\Delta^2 x, \Delta x \Delta t, \Delta^2 t)$. The lowest order part of the error term has the form of a diffusion term. It is responsible for the large diffusivity of the upwind method, so we aim to remove it.

Following Smolarkiewicz [1] we now take another point of view and regard the error term as an advection term:

$$\frac{\partial u_* c}{\partial x}$$

with advection velocity defined as:

$$u_* = -\frac{0.5(u\Delta x - \Delta t u^2)\frac{\partial c}{\partial x}}{c} .$$

To remove this undesired term, we add another upwind step, using the additional velocity u_* with a minus sign. This removes the lowest order part of the error term. Of course, the extra upwind step will give an error term too. However, because the velocity u_* we use is already small $O(\Delta)$ the error will be $O(\Delta^2)$. So we have in fact a second order method. The anti-diffusive velocity $-u_*$ is approximated numerically by:

$$-u_*(i) = [u(i)\Delta x - u(i)^2\Delta t]\frac{c(i) - c(i-1)}{c(i) + c(i-1) + \epsilon}\frac{1}{\Delta x} .$$

ϵ is a small number (for instance 10^{-15}) which is added for the case that $c(i)$ and $c(i-1)$ are both 0.

Further remarks on the method.

We will now give some possible improvements of this method. Firstly, the correction is an upwind step, so it will give an error too. This can be corrected with another upwind step etc. However, such procedure did not improve the solutions very much. Secondly, we can also multiply the anti-diffusive velocity u_* by a number, greater than 1 but near 1 to enhance the anti-diffusive effect. This improves the solution somewhat. Thirdly, because the method is non-linear, adding a constant before advection and subtracting it after advection will change the solution. In fact, it is less diffusive now, but leads to wiggles (dispersion!) and more importantly to unwanted negative values!

Van Leer's Second Scheme

Introduction

The method of Van Leer [2] works with two schemes. One is the familiar (first-order accurate) upwind scheme, the other is the second order accurate upwind scheme of Fromm. The method works with a monitor which inspects the known current solution. This monitor looks for the onset of wiggles. If this onset is observed, the upwind scheme is invoked to temper them. If not, then we use the (tempered) fluxes of the scheme of Fromm. The method tries to get a monotonic solution rather than a positive solution: in other words, local maxima and minima are flattened.

Explanation of the method.

The method of Van Leer calculates the fluxes at every cell boundary in the following way. First an effective value for the concentration (C) is calculated at the cell boundary. This is multiplied by the velocity at the cell boundary to get the flux. The calculation of the effective concentration value is as follows; we limit ourselves here to a positive velocity at the cell boundary. First a smoothness monitor inspects the old solution for wiggles. See figure 3. This is done by comparing

$$I1 = |c_u - 2c_c + c_d|$$

and

$$I2 = |c_u - c_d| .$$

The first quantity can also be written as

$$I1 = |(c_u - c_c) - (c_c - c_d)|.$$

Both quantities are compared in figure 3 for a monotone, smooth function and for a function with wiggles (maxima/minima) present. Apparently, if

$$|c_u - 2c_c + c_d| \geq |c_u - c_d|$$

then there is a maximum/minimum present in the interval containing c_u, c_c, c_d. If this is the case, then C is set equal to c_c. This corresponds to the first order upwind flux.
If

$$|c_u - 2c_c + c_d| < |c_u - c_d|$$

then there is no wiggle present. Then we use the tempered Fromm fluxes (the original Fromm fluxes were adapted (tempered) so that a monotonic concentration profile remains monotonic for a constant velocity):

$$C = c_c + (1 - Cou)\frac{(c_c - c_u)(c_d - c_c)}{(c_d - c_u)}.$$

Here Cou is the Courant number $u\dfrac{\Delta t}{\Delta x}$ calculated at the cell boundary. Now we can calculate the concentration at the new timelevel with the fluxes using the C just calculated.

Further remarks on the method.

We can expect with this method that physical maxima/minima are damped, if they have the form of a sharp peak. (Sharp means: extending over a few cells only). For our problems in LES, which use initial concentration distributions extending over a few gridcells only, this seems a disadvantage.

The multi-dimensional flux corrected transport (MFCT) method.

The MFCT method [3] works with two schemes, a lower order scheme which is positive but diffusive and a higher order difference scheme which is less diffusive but not necassarily positive. The lower order fluxes and solution are calculated. The higher order fluxes are calculated too. The difference between the higher order and lower order fluxes is then used to improve the lower order solution, i.e. to make it less diffusive. To prevent wiggles/negative values only a limited portion of the flux difference is used.

The lower order and higher order fluxes.

For the lower order fluxes and solution we take the upwind scheme. For the higher order solution we use central differences in space and time. We use the lower order fluxes to calculate the lower order solution, c_{lower}.

The limited higher order fluxes.

Call the higher order fluxes $fhigh$ and the lower order fluxes $flow$. The difference between $fhigh$ and $flow$ is called A.

$$A(i) = fhigh(i) - flow(i).$$

To get an improved upwind solution without all the troubles of central differences we use a limited part of A

$$c_{t+\Delta t}(i) = c_{lower}(i) - \frac{\Delta t}{\Delta x}(L(i+1)A(i+1) - L(i)A(i))$$

with

$$0 \leq L(i) \leq 1 \text{ for all i .}$$

L is the (flux-)limiter. L may vary from point to point. It depends on local circumstances. The right way to calculate L is the key to the success of the method.

Briefly, the limiter L is found as follows. The current, known solution and the newly calculated lower order solution are used to construct an upper bound and a lower bound. The flux limiter is then found by requiring that the new solution must remain between these two bounds.

Further remarks on the method.

In the actual implementation we used a higher order scheme which is second order in time and fourth order in space in the interior and second order in space at the boundary. This way the treatment of boundary conditions did not differ from the treatment for the other, second order difference schemes.

The Second Moment method.

Introduction.

All the methods described so far are difference methods. The only values used in the formulas are the values at gridpoints. The idea of the Second Moment Method [4] is, that for really coarse grids this may not be enough. We need to use subgrid information, i.e. we take into account how the concentration varies within a cell. Let us assume that within a cell the concentration has the form of a rectangle. See figure 4. The rectangle need not fill a cell completely. Every timestep the rectangles are moved with the average velocity in the cell. After this advection step some cells will have parts in more than 1 cell. The next thing we do is form one rectangle of all parts present a single cell. After this has been completed, one timestep has been finished (every cell contains one concentration rectangle) and we are ready for the next timestep. The formation of new rectangles out of the old parts is treated in the next section.

Formation of one rectangle out of parts of several rectangles.

Now we explain how we make a new rectangle out of parts of old rectangles. See figures 4 and 5. We use for each cell a local, normalized coordinate system. The midpoint of the cell is at x=0. The boundaries are at x=$-\frac{1}{2}$ and at x=$+\frac{1}{2}$. In every cell the concentration has the form of a rectangle. The rectangle is known, if we know the total amount of concentration, the position of the center of gravity and the width. So in every cell we have three parameters which characterize the concentration distribution within the cell. All three parameters must be given in the initial condition. Some rectangles have moved just within a cell. Others have crossed the cell boundary. See figure 4. Out of the parts within one cell we will form new rectangles. To define a rectangle, we need total concentration, center of gravity and width. We use for the total concentration C: (for a cell with length 1 and origin at the midpoint)

$$C = \int_{-\frac{1}{2}}^{\frac{1}{2}} c\,dx .$$

The center of gravity F is calculated with:

$$F = \int_{-\frac{1}{2}}^{\frac{1}{2}} xc\,dx$$

and the width R is calculated using the variance:

$$R^2 = \frac{1}{12} \int_{-\frac{1}{2}}^{\frac{1}{2}} x^2 c\,dx \ .$$

The factor $\frac{1}{12}$ is introduced so that the width is 1 for a rectangle which fills a cell completely. If all calculations have been performed, we have new rectangles and we can take the next timestep. The formulas given above will be used for distributions of c consisting of rectangles. For this case we can express the quantities for the entire distribution in a cell in terms of the quantities for each rectangle part separately. For the moment give all rectangle parts in a cell a number k. If we denote the amount of concentration c in rectangle part k by C_k, the center of gravity by F_k and the width by R_k we can derive the following formulas:

$$C_{new} = \sum_k C_k \tag{1}$$

$$F_{new} = \frac{\sum_k C_k F_k}{C_{new}} \tag{2}$$

$$R_{new}^2 = \frac{C_k R_k^2}{C_{new}} + 12\frac{(C_k F_k^2 - C_{new} F_{new}^2)}{C_{new}} \ . \tag{3}$$

Here the index *new* is short for: new out of old parts.

Further remarks on the method.

Due to computer problems a vectorized version of the Second Moment Method could not be included in the test problems at this moment.

Presentations of some test results

In the literature, in which the preceding methods are given, test results are described. However, different tests are shown for the different methods so that no direct comparison is possible. Here we apply the same tests to all methods so that we can make a good comparison. The tests are designed in such a way that exact solutions can be calculated and compared to the numerical results. The test are 2-D, permitting easy visualization while still being multi-dimensional.

We present graphical results for two test cases. They refer to pure advection (K=0). The first test shows a rectangular shaped concentration block advected by a velocity field with vector (1,1) (this means in the figure that the advection velocity points from the upper left to the lower right corner; see figure 6). The other test uses a cosine hill shaped concentration profile in a rotating velocity field (the so-called Molenkamp test). This test shows the capability of Second Moment to advect non-rectangular concentration distributions in non-uniform velocity fields. (See figure 7)

Discussion and conclusion

The results of the tests shown in figure 6 and 7 and other tests are summarized in four tables. For the details of the calculation of these tables see Pourquié [6].

There are tables for accuracy, positivity, CPU and memory requirements. No method is the best in all four tables. (See tables 1-4) However, if CPU and memory requirements are considered of lesser importance, then we may conclude, that for advection of nonsmooth data on a coarse, rectangular grid the Second Moment Method is the most accurate and at the same time positive. The difference methods fail to represent gradients extending over less than, say, two gridcells. Second in accuracy is MFCT, which gives negative values of machine accuracy magnitude. As second moment is not easily adaptable to curvilinear grids, we consider MFCT for our purposes as the best candidate for curvilinear grids.

References

[1] Smolarkiewicz P.K. (1984) A fully multi-dimensional positive definite advection transport algorithm with small implicit diffusion.
J.Comp.Phys. 54, 325-326.

[2] Leer, B. Van (1974) Towards the Ultimate Conservative Difference Scheme. II. Monotonicity and Conservation Combined in a Second-Order Scheme.
J.Comp.Physics 14,361-370.

[3] Zalesak, S.T. (1979) A fully multidimensional flux-corrected transport algorithm for fluids.
J.Comp.Phys. 31, 335-362.

[4] Egan, B.A. and Mahoney J.R. (1972) Numerical modelling of advection and diffusion of urban area source pollutants.
J.Appl.Meteor. 11, 312-322.

[5] Godunov, S.K. (1959) Finite difference method for numerical computation of discontinuous solutions of the equations of fluid dynamics.
Mat.Sb. 47, 271.

[6] Pourquié, M.J.B.M. (1990) Numerical comparison of advection schemes to be used on a coarse grid. Internal report MEAH-94, Laboratory for Aero- and Hydrodynamics, Delft University of Technology.

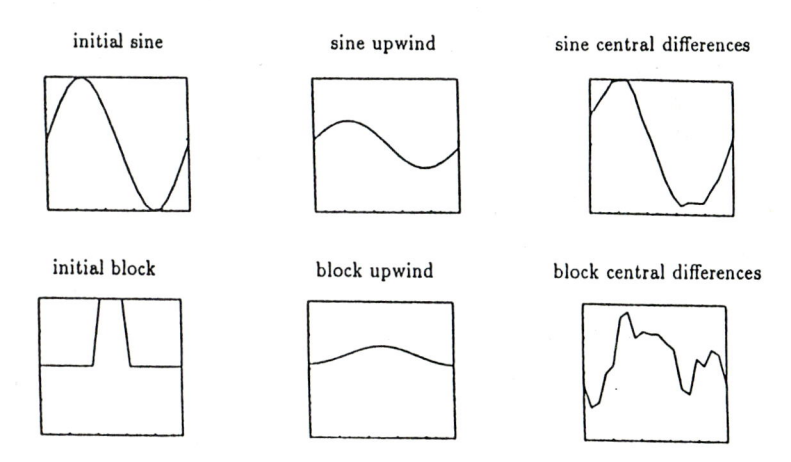

Figure 1: Solution of the 1-D advection equation ($K = 0$) for a rectangle and a sine by upwind and central differences, showing wiggles (dispersion) and smearing (diffusion). The velocity is 1, the gridspacing 1, the timestep 0.5. Advection took place over 40 timesteps. At the boundary periodic conditions are applied.

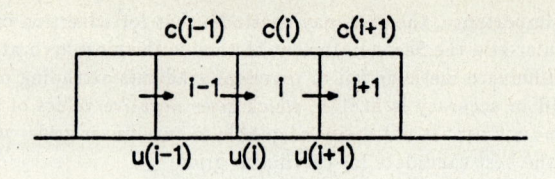

Figure 2: The staggered grid used for the schemes. Arrows indicate fluxes.

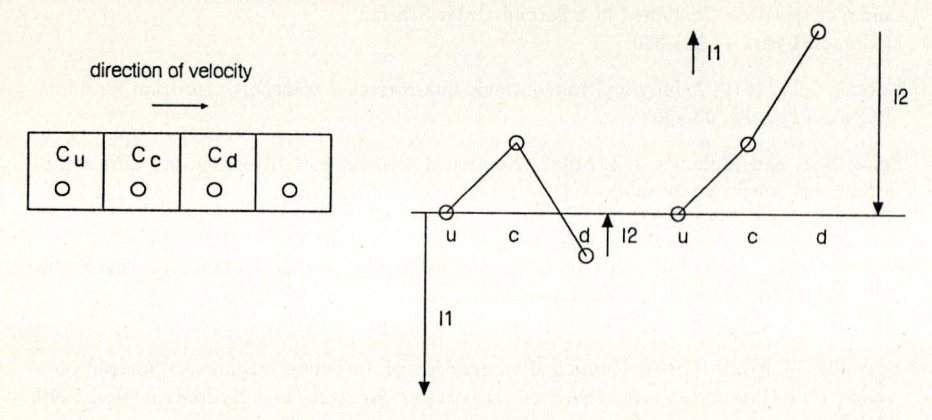

Figure 3: Smoothness monitor for Van Leer's scheme

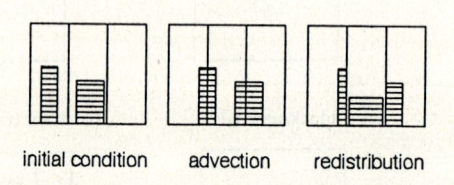

Figure 4: The Second Moment Scheme

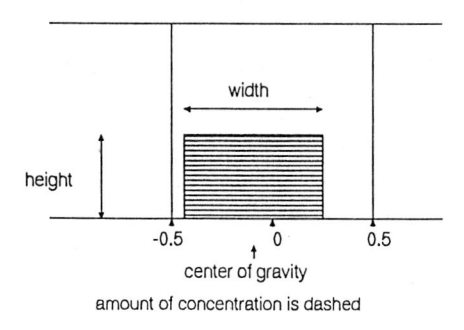

Figure 5: Parameters of the Second Moment Scheme

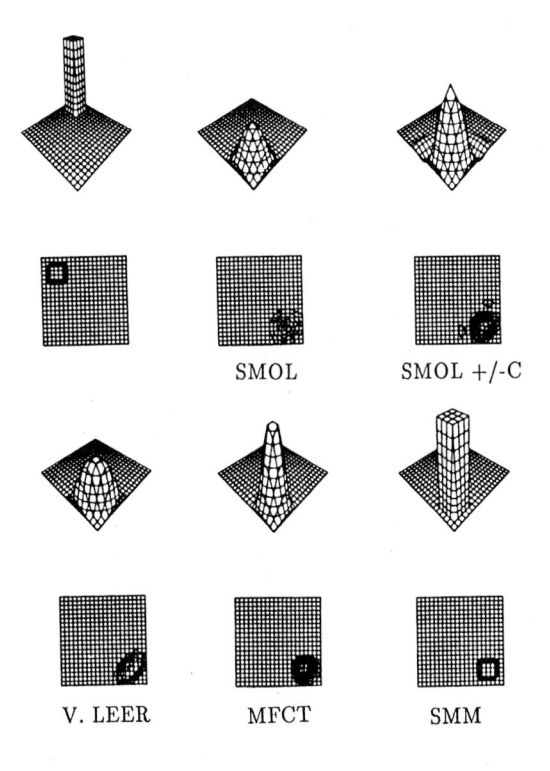

Figure 6: 2-D test: transport of a rectangle by the numerical methods treated for a velocity (1,1). The gridspacing is 1, the maximum timestep 0.3. Numerical solutions are shown after 15 time-units (= 50 time-steps). Shown are, from left to right and top to bottom, 3-D and 2-D contour plots of the initial conditions and the solution by Smolarkiewicz, Smolarkiewicz+,-constant, Van Leer, MFCT and SMM. SMM is identical to the exact solution and thus gives the best solution, second comes MFCT. Of the other (difference) methods Smolarkiewicz+,-constant shows dispersion (wiggles); all difference methods show diffusion (smearing).

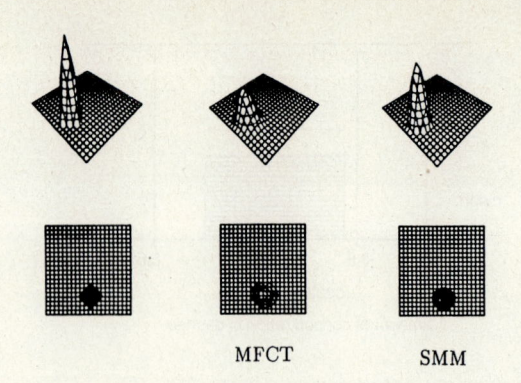

Figure 7: 2-D test: transport of a cosine hill in a rotating velocity field. Shown are, from left to right, 3-D and 2-D contourplots of: initial and (exact) end condition, the solution by MFCT (the best of the difference methods) and the solution by SMM. The results pertain to one complete rotation (340 time-steps, maximum Courant number is 0.43). This picture shows the ability of SMM to advect in a non-uniform velocity field.

Table 1: Accuracy comparison (coarse data). Methods in order of accuracy. The accuracy comparison was made on basis of calculated errors (maximum error, mean error and mean square error) and consideration of the peak height.

SMM	1
MFCT	2
Van Leer	3
smolarkiewicz,+/-constant	4
smolarkiewicz,sec	8

Table 2: Positivity comparison (coarse data). Largest negative value compared to the peak value of the exact solution.

SMM	0
smolarkiewicz,sec	0
MFCT	-10^{-19}
Van Leer	10^{-3}
smolarkiewicz,+/-constant	-0.7

Table 3: CPU time comparison; CPU measured relative to Van Leer's method

smolarkiewicz,sec	1
smolarkiewicz,+/-constant	1
Van Leer	1
MFCT	1.5
SMM	10

Table 4: Memory comparison, number of arrays; the first number between brackets is the minimum number of arrays needed, the second number is the number of help arrays used.

smolarkiewicz,sec	3 (1+2)
smolarkiewicz+,- constant	3 (1+2)
Van Leer	3 (1+2)
MFCT	7 (2+5)
SMM	10 (5+5)

NUMERICAL METHOD FOR CONDUCTION-CONVECTION-RADIATION HEAT TRANSFER IN TWO-DIMENSIONAL IRREGULAR GEOMETRIES

D. R. Rousse and B. R. Baliga

McGill University, 817 Sherbrooke W., Montréal, Canada, H3A 2K6

SUMMARY

The research presented in this paper is the first step of a program aimed at broadening the range of possible applications of Control-Volume Finite Element Methods (CVFEMs) for fluid flow and heat transfer. In this work, numerical solutions for two-dimensional combined modes of heat transfer are obtained by combining two solution procedures developed earlier: one for the solution of radiative transfer in participating media, and the other for the prediction of convection-diffusion phenomema in irregular geometries. The proposed method accounts for absorbing, emitting, and isotropically scattering media. In the formulation, the Radiative Transfer Equation (RTE) is approximated by a set of 12 partial differential equations. Simple problems have been investigated to assess the accuracy of the method. Solutions are presented for: (1) convection-diffusion; (2) radiation only; (3) conduction-radiation; and (4) convection-diffusion-radiation. Results indicate that the proposed CVFEM provides approximate solutions that are accurate and economical enough to pursue further investigations using extensions of this method.

NOMENCLATURE

ρ	Medium density, $[Kg/m^3]$	$\vec{\Omega}$	Direction of propagation
c_p	Medium specific heat, $[J/Kg\,K]$	τ	Optical depth
T	Temperature, $[K]$	ω	Solid angle, $[sr]$
$\vec{u}$	Velocity, $[m/s]$	$\vec{n}$	Unit vector normal to a surface
k	Medium conductivity, $[W/m\,K]$	N	Stark number or # of discrete directions
$\vec{q}$	Heat flux, $[W/m^2]$	Pe	Peclet number
s	Internal heat generation, $[W/m^3]$	τ_0	Optical thickness
g	Incident radiant energy, $[W/m^2]$	ω_0	Albedo for single scattering
e	Emissive power, $[W/m^2]$	$\vec{J}$	Dimensionless flux
i	Radiative intensity, $[W/m^2\,sr]$	ϕ	Dependent variable
α	Absorption coefficient, $[m^{-1}]$		
σ	Scattering coefficient, $[m^{-1}]$		
κ	Extinction coefficient, $[m^{-1}]$		
ϵ	Emissivity of a surface		

	Subscripts		Superscripts
r	Radiation	$\rightarrow$	Vectorial quantity
b	Black body	$'$	Incoming direction
m	Discrete direction	$-$	Bulk value
$*$	Reference quantity		
B	Boundary		

INTRODUCTION

The propagation of radiation through participating media has been the focus of a large number of diverse studies, particularly in the field of engineering heat transfer [1,2]. This arises from the relative importance of radiative heat transfer in industrial processes. Efficient designs of furnaces, ovens and combustion chambers call for numerical methods capable of solving propblems that involve conduction-convection-radiation, or combined modes of heat transfer.

Several general methods have been put forward to investigate multidimensional radiative transport in participating media. Among them, the zonal method and the Monte-Carlo method are two of the most widely and successfully used. However, as reported in [3], both these methods have proved difficult to incorporate in other discrete numerical methods for conduction and convection heat transfer. Consequently, alternate ways have been considered for the calculation of radiative heat transfer in the context of practical applications. Accurate solutions have been obtained using approximate methods such as discrete-ordinates methods [4,5] and differential approximations [6]. The finite element method [7] and other weighted residual based methods, such as CVFEMs, also overcome the aforementioned coupling difficulty, and they permit accurate and yet affordable numerical solutions of radiative exchange in participating media.

The objective of this work is to establish the capabilities and limitations of a CVFEM formulation for the solution of the equations that govern combined modes of heat transfer, including radiative heat transfer, in a fluid medium that emits, absorbs and scatters thermal radiation. Guidance for the formulation of a solution procedure for radiation heat transfer has been obtained from discrete ordinates formulations proposed by Chandrasekhar [8], Carlson and Lathrop [9], and Fiveland [4].

The work involves two principal coupled tasks: (1) solution of the radiative transfer equation (RTE) in order to obtain the distribution of intensity of radiation in the domain of interest; and (2) solution of the energy equation that governs the temperature field in the presence of conduction, convection and radiation. In the present formulation, the radiative heat transfer contribution to the heat balance is incorporated as a constant volumetric heat generation term in the discretized energy equation. The suggested CVFEM is tested by applying it to several simple problems involving the rectangular geometry depicted in Fig. 1(a).

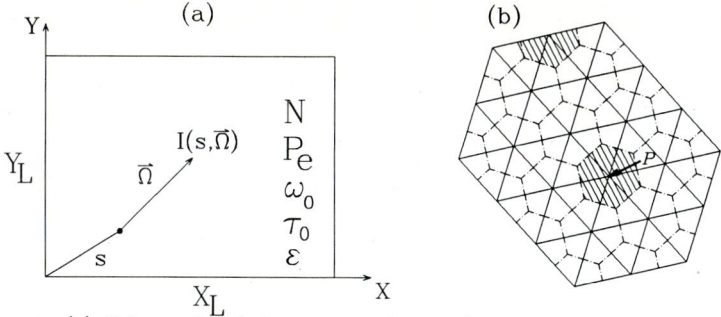

Figure 1: (a) Schematic of the rectangular enclosure surrounding a radiating fluid; (b) general discretized irregularly shaped domain.

ANALYSIS

Governing equations: Consider steady conduction, convection, and radiation in a fluid bounded by the above mentioned gray-diffuse enclosure. The fluid participates in the radiatiation process, and it is assumed to be gray and isotropic. In addition, its thermophysical properties are assumed to remain constant. For this problem, the steady-state heat balance can be written as follows:

$$\rho c_p \vec{\nabla} \cdot (\vec{u}T) = k\vec{\nabla}^2 T - \vec{\nabla} \cdot \vec{q_r} + s \tag{1}$$

where the first term accounts for advection, the second term represents the diffusion contribution, the third term involves radiation heat transfer, and the last term describes the internal heat generation within the medium. All variables are described in the nomenclature.

The divergence of the radiative flux vector, $\vec{q_r}$, is a linear function of the excess of the local black-body emissive power, e_b, over the incident radiant energy from the medium and the walls, g, which in turn is obtained by integrating the intensity of radiation, i, over all directions. The required set of governing equations for radiative transfer calculation is then:

$$\vec{\nabla} \cdot \vec{q_r} = \alpha [4 e_b - g] \tag{2}$$

$$g = \int_{4\pi} i \, d\omega \tag{3}$$

$$\vec{\nabla} \cdot (\vec{\Omega} i) = -\kappa i + \alpha i_b + \frac{\sigma}{4\pi} \int_{4\pi} i \, d\omega' . \tag{4}$$

Employing the following dimensionless variables and parameters

$$Y = \kappa y; \quad X = \kappa x; \quad \Theta = \frac{T}{T_*}; \quad U = \frac{u}{\bar{u}}; \quad Q_r = \frac{q_r}{4\sigma T_*^4};$$

$$I = \frac{i}{4\sigma T_*^4}; \quad E_b = \frac{e_b}{4\sigma T_*^4}; \quad G = \frac{g}{\sigma T_*^4}; \quad S = \frac{s}{4\sigma T_*^4 \kappa} \tag{5}$$

$$N = \frac{k\kappa}{4\sigma T_*^3}; \quad Pe = \frac{\rho c_p \bar{u} L_*}{k}; \quad \tau_0 = \kappa L_*; \quad \omega_0 = \frac{\sigma}{\kappa} \tag{6}$$

the governing equations, Eq. 1–4, become

$$\vec{\nabla}\cdot(\vec{U}\Theta) = \frac{\tau_0}{Pe}\vec{\nabla}^2\Theta - \frac{\tau_0}{Pe\,N}\vec{\nabla}\cdot\vec{Q}_r + \frac{\tau_0}{Pe\,N}S \qquad (7)$$

$$\vec{\nabla}\cdot\vec{Q}_r = (1-\omega_0)\left[E_b - G\right] \qquad (8)$$

$$G = \frac{1}{4}\int_{4\pi} I\,d\omega \qquad (9)$$

$$\vec{\nabla}\cdot(\vec{\Omega}I) = -I + (1-\omega_0)I_b + \frac{\omega_0}{4\pi}\int_{4\pi} I\,d\omega'. \qquad (10)$$

The dimensionless radiative boundary conditions for all test problems involving radiation transport is

$$I_B = \epsilon I_b + \frac{1-\epsilon}{\pi}\int_{2\pi} |\vec{\Omega}'\cdot\vec{n}_B|\,I'_B\,d\omega'. \qquad (11)$$

Domain discretization: For two-dimensional geometries, the calculation domain is spatially divided into three-node triangular elements. Then the centroids of the elements are joined to the mid-points of the corresponding sides. This creates polygonal control-volumes, around each node, that collectively fill the calculation domain entirely and exactly. A sample domain spatially discretized is shown in Fig. 1(b); the solid lines denote the domain and element boundaries, the dashed lines represent control-volume faces, and the shaded areas show the control volumes associated with one internal node. In this discretization scheme, curved boundaries are approximated by piecewise linear curves [10]. In discretizing the integral conservation equation of radiative transport, Eq. 10, it is found that the directional dependence of intensity, I, requires angular discretization. Therefore, the calculation domain is also divided into N discrete directions, $\vec{\Omega}_m$, each associated with solid angle, ω_m [3]. This gives a set of N partial differential equations approximating the original integro-differential RTE and in which integrals are replaced by a quadrature summed over each direction [3].

Conservation equations: Consider a typical node in the calculation domain. It could be a boundary node, as well as node P depicted in Fig. 1(b). Integral conservation equations corresponding to Eq. 7 and Eq. 10 for each of the N directions can be obtained by applying the conservation principle to spatial control-volumes. At each node, the weighting function is unity (one) over its associated control-volume and zero elsewhere, leading to a sub-domain type of Method of Weighted Residuals (MRW). Hence, the method produces discretization equations that satisfy a global conservation requirement.

The general form of the integral equations around node P is the following

$$\int_{A_P} \vec{J}_\phi\cdot\vec{n}\,dA = \int_{V_P} S_\phi\,dV \qquad (12)$$

where $\vec{J}_\phi$ and S_ϕ for the dependent variables relevant to this study are given in the following table

	$\vec{J}_\phi$	S_ϕ
Eq.7	$\vec{U}\Theta - \frac{\tau_0}{Pe}\vec{\nabla}\Theta$	$\frac{\tau_0}{Pe\,N}(S - \vec{\nabla}\cdot\vec{Q}_r)$
Eq.10	$\vec{\Omega}_m I_m$	$-I_m + (1-\omega_0)I_b + \frac{\omega_0}{4\pi}\sum \omega'_m I'_m$

Discretization equations: The derivation of the algebraic approximation to the above mentioned conservation equation calls for the specification of interpolation functions, within the elements, for the dependent variables. In the particular case of this study, no interpolation function was needed for properties because they are assumed constant. However, the formulation generally permits variation of thermophysical properties. The element-based interpolation functions employed in this formulation for diffusion, convection, and radiation are: (1) linear; (2) MAss-Weighted average (MAW) or FLow Oriented (FLO) [11]; and (3) one-dimensional integrated RTE [3], respectively. The intensity of radiation being assumed constant over a sufficiently small discrete solid angle, ω_m, algebraic equations are obtained for each discrete direction when intensity is considered. Algebraic representations of the governing energy equation and of the N approximate RTE are obtained in a manner that avoids negative coefficients [11].

Solution procedure: Solutions of relevant algebraic equations are obtained iteratively because of nonlinearity. The investigated problem is highly nonlinear since the in-scattering source terms and the boundary conditions in Eq. 10 depend on intensity, the emission term depends on the temperature field provided by Eq. 7, and the temperature field calculation requires an evaluation of the intensity of radiation. Relatively weak nonlinearities also arise due to temperature dependent radiative and thermophysical properties.

The algebraic representation of the energy equation is solved first incorporating the divergence of the radiative flux vector as a source term based on a guessed intensity field. The intensity of radiation, for all discrete directions, is computed next based on the newly calculated temperature field and the source of radiative intensity. Coefficients and source related terms in the algebraic system of equations are then updated for each dependent variable, based on the latest available temperature and intensity fields. The procedure is repeated until the variations in the discrete dependent variables, Θ and I_m, between successive iterations, are within prescribed tolerances.

RESULTS

Convection-diffusion: After preliminary tests, a simple problem was chosen to validate the formulation for convection-diffusion. The domain depicted in Fig.1(a) was confined between two rotating concentric cylinders on which prevailed different given temperatures. The fluid in the domain was considered to undergo solid-body rotation, along with the bounding cylinders. The resulting one-dimensional problem in the cylindrical polar coordinate system was solved analytically. This problem becomes two-dimensional in a Cartesian formulation. The appropriate boundary conditions in the cartesian formulation can be imposed using the analytical solution. The two-dimensional cartesian formulation was solved numerically using the proposed CVFEM. Table 1 presents the maximum relative error, $\Delta = \frac{|\theta - \theta_a|}{\theta_a}$, for different Peclet numbers, Pe, and internal heat generation, S, for a 21x21 grid and two types of interpolation functions. The dimensionless temperature, Θ, of the outer cylinder was two and the convergence criteria, based on the summation of all errors, was 10^{-9}. The MAW scheme introduces errors of about one percent when the source term is ten times the order of magnitude of the dependent variable and the Peclet number is relatively high. Nevertheless, this very stable scheme of interpolation is employed to begin the computation from scratch for all test problems involving convection-diffusion. Then the more accurate FLO scheme is used until convergence is arrived at.

Table 1: Maximum computational relative error in computing temperature field for various Pe and S, and two types of interpolation functions.

	Error = $\Delta * 100$					
	FLO scheme			MAW scheme		
Pe	S=0	S=1	S=10	S=0	S=1	S=10
0.01	9.83E-4	1.59E-3	4.59E-3	9.87E-4	1.43E-3	3.65E-3
1.00	9.94E-4	1.61E-3	4.66E-3	1.43E-3	1.78E-2	1.13E-1
100.00	8.07E-4	1.26E-3	3.70E-3	2.43E-2	4.23E-1	2.66E00
10000.00	6.27E-4	1.01E-3	3.26E-3	3.69E-2	5.15E-1	3.36E00

Radiation: The geometry shown in Fig.1(a) was subsequently assumed to be a gray rectangular enclosure filled with a radiatively absorbing and scattering medium. Two cases were examined: (1) scattering only; and (2) absorption only. Results for several sets of parameters are given in previous work [3,12]. A sample of these results is given in Fig. 2 to show the relative accuracy of the proposed CVFEM. In this example, the dimensionless emissive power of the bottom wall was set to one while those of other surfaces were at zero. The enclosure was considered to be square and filled with scattering medium, $\omega_0 = 1$. The suggested formulation slightly underpredicts the radiative heat flux when compared to the discrete ordinate method of Fiveland [4]. However, the latter method [4], does not comply with the half-range first moment restriction that has to be imposed on the quadrature set.

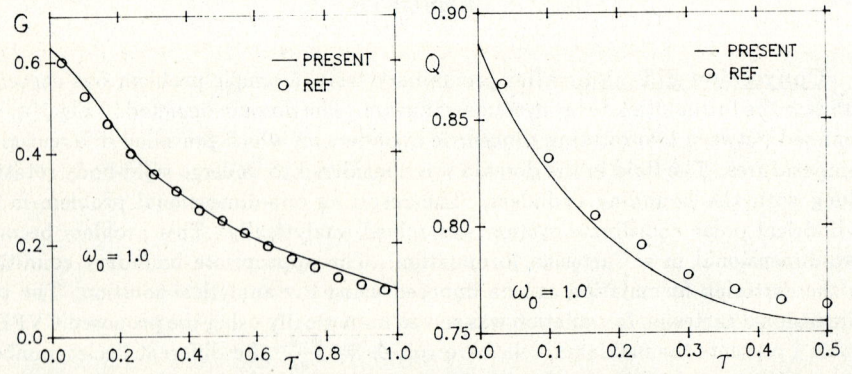

Figure 2: Radiation in a square black enclosure with scattering media: (a) Centerline incident radiant energy; (b) Radiant heat flux on the hot wall.

Conduction-radiation: Coupled radiative and conductive heat transfer was then considered in a rectangular grey enclosure. The nondimensional temperature was set to unity on the bottom surface, $Y = 0$, the other surfaces were at $\theta = 0.5$, and the lower corners had a temperature of 0.75. The medium was assumed to absorb and emit, but it did not scatter radiation, $\omega_0 = 0$. The nondimensional vertical centerline temperature and the hot wall heat flux distributions for a square black enclosure, with different values of N, the Stark number, are presented in Fig. 3. Solutions provided by the suggested CVFEM compare favorably with those obtained by investigators employing a FEM formulation [7].

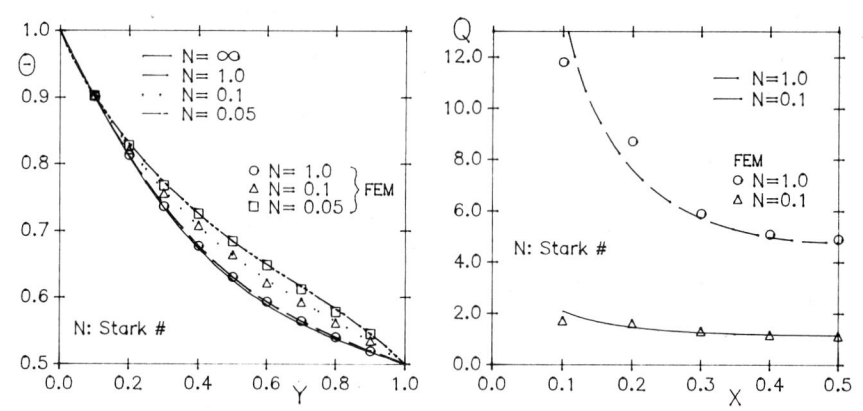

Figure 3: Conduction-radiation in a square black enclosure with absorbing media and different N parameters: (a) Centerline temperature profiles; (b) Total heat flux on the hot wall.

Convection-diffusion-radiation: Combined convection-diffusion-radiation was considered in a fully developed laminar flow bounded by the geometry depicted in Fig.1(a), which was considered to be a straight gray channel in which the fluid medium participated in the radiative transfer. The nondimensional temperature was set to unity on the top and bottom surfaces. The inlet and outlet sections of the channel were assumed to be imaginary porous black surfaces, through which the medium flowed without restrictions. For purpose of comparison, attention was focused on a location of the fully developed heat transfer region where the centerline dimensionless temperature is 0.5. The dimensionless vertical temperature is presented for various Stark number in Fig. 4. The effect of the optical thickness variations on the total heat flux on the bottom wall is also presented in the same figure. The results obtained using the method put forward in this paper are compared with those produced by considering fully developed heat transfer [13]. In that reference [13], the axial temperature gradient was replaced by $\left(\frac{T_w-T}{T_w-\bar{T}}\right)\frac{dT}{dx}$ and conduction as well as radiation in the axial direction were neglected. An examination of the results suggests that the CVFEM proposed in this paper adequately predicts temperature and heat fluxes in simple convection-diffusion-radiation problems. The convergence was found to be slow for small Stark numbers. This situation arises from the relatively high coefficient that multiplies the source term in the energy equation whenever the problems are radiatively dominating. Slow convergence was also obtained when reflecting boundaries were considered.

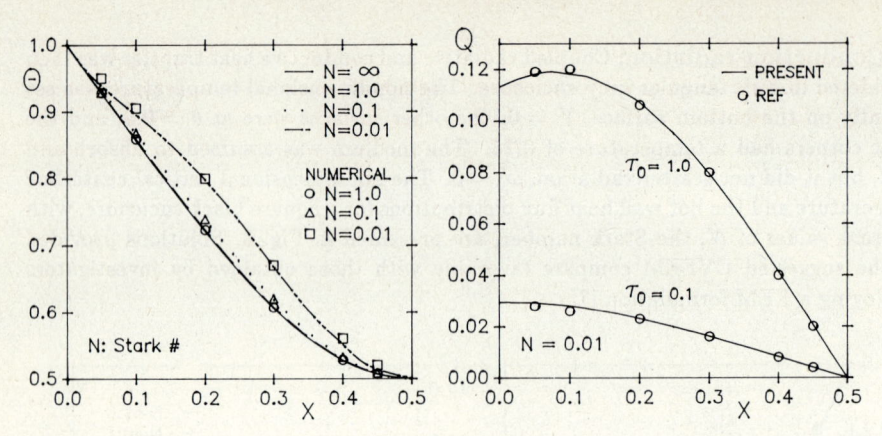

Figure 4: Convection-diffusion-radiation in a gray channel with absorbing media: (a) Stark number effect on temperature; (b) Optical thickness effect on the lower wall radiant heat flux.

CONCLUSION

A novel CVFEM for combined conduction-convection-radiation heat transfer in two-dimensional plane geometries, filled with emitting, absorbing, and scattering media, has been presented in the current study. It has been shown that the method can provide accurate temperature and heat flux distributions in regular two-dimensional geometries. The proposed method, like most others, requires a substantial amount of CPU time to achieve convergence when either low emissivity or low Stark number are considered. Further investigations concerning the selection of appropriate interpolation functions for intensity within elements still have to be carried out, and more test cases have to be benchmarked to corroborate the proposed method thoroughly. Extensions of angular discretization and boundary condition treatment are being undertaken to allow applications to problems with irregular domains.

Acknowledgment: The authors greatfully acknowledge the Natural Sciences and Engineering Research Council and the "Fonds pour la formation de chercheurs et l'aide à la recherche": the first author for graduate studies fellowships, and the second author for operating research grants.

REFERENCES

[1] Siegel, R., and Howell, J.R., – Thermal Radiation Heat Transfer, McGraw-Hill, New York, 1972.

[2] Viskanta, R., Radiation heat transfer: Interaction with conduction and convection and approximate methods in radiation, *Proc. 7th Int. Heat Transfer Conf.*, 1, Toronto, p.103-121, (1982).

[3] Rousse, D.R. and Baliga, B.R., – Formulation of a Control-Volume Finite Element Method for Radiative Transfer in Participating Media, *Proc. 7th Int. Conf. Num. Meth. Thermal Problems*, p.786, Stanford, 1991.

[4] Fiveland, W.A., – Discrete-ordinates solutions of the radiative transport equation for rectangular enclosures, *ASME J. Heat Transfer*, 106, no.2, 699-706, (1984).

[5] Truelove, J.S., – Discrete-Ordinate Solutions of the Radiation Transport Equation, *ASME J. Heat Transfer*, 109, no.4, 1048-52, (1987).

[6] Modest, M. F., – Radiative Equilibrium in a Rectangular Enclosure Bounded by Gray Walls, *J. Quant. Spectros. Radiat. Transfer*, 15, 445-61, (1975).

[7] Razzaque, M.M., and al., – Coupled radiative and conductive heat transfer in two-dimensional rectangular enclosure with gray participating media using finite element, *J. Heat Transfer*, vol.106, no.3, p.613, 1984.

[8] Chandrasekhar, S., *Radiative Transfer*, Clarendon Press, Oxford, 1950.

[9] Carlson, B.G., and Lathrop, K.D., Transport Theory – The method of Discrete-Ordinates in:*Computing Methods in Reactor Physics*, Gordon and Breach, NY, 1968

[10] Baliga, B.R. and Patankar, S.V., "Elliptic systems: Finite-Element Method II", *Handbook of Numerical Heat Transfer*, Wiley, New-York, Chap. 11, 1988.

[11] Saabas, H.J., – *A Control Volume Finite Element Method for Three-Dimensional, Incompressible, Viscous Fluid Flow*, Ph.D. Thesis, Dept. of Mechanical Engineering, McGill University, 1991.

[12] Rousse, D.R. and Baliga, B.R., – Radiation Heat Transfer in an Absorbing, Emitting, and Scattering media, *Proc. 4th Int. Conf. Nonlinear Eng. Comp.*, Swansea, 1991.

[13] Viskanta, R., – Interaction of Heat Transfer by Conduction, Convection, and Radiation in a Radiating Fluid, *ASME J. Heat Transfer*, vol.85, pp.318-328, 1963.

MESH GENERATION AND ADAPTATION

SENSORS FOR SELF-ADAPTING GRID GENERATION IN VISCOUS FLOW COMPUTATIONS

J. Fischer
Lehrstuhl für Fluidmechanik der TU München
Arcisstr.21, 8000 München 2

SUMMARY

When an adaptive method in computational fluid dynamics is used, it is necessary to formulate appropriate refinement criteria. Usually the discretization is adapted using the magnitude of the gradient of a "physically relevant" variable. In this paper a different approach is presented. Looking closely at the features of the different flow phenomena, sensors can be derived, that are able to detect and distinguish the relevant flow phenomena. Although problems arise with the detection of weak phenomena on coarse grids, remarkable improvements can be observed compared to the standard indicators.

INTRODUCTION

The goal of an adaptive method is the improvement of the numerical solution of a given physical problem. This improvement is normally considered to be equivalent to the reduction of the integral approximation error. Using the actual high-resolution finite-volume-schemes [1], [2] with interpolation functions of second or third order, the first term of the truncation error is represented by third or fourth order derivatives. The numerical evaluation of these approximation errors will not give any numerically or physically reasonable result. Therefore the discretization is normally adapted using physical criteria. The gradient of the density [3] or of the pressure [4] is often used for this purpose. It is not difficult to criticize this approach: Firstly, by the confinement to one physical variable those flow regions are neglected in the adaption process, in which other variables vary strongly, while the chosen one is nearly constant. Secondly, fluid dynamical phenomena normally are characterized by the fact, that the <u>absolute</u> rates of change of a certain variable are not independent of the state in front of the phenomenon. Taking the local gradients magnitude as a measure of the error, two shocks of equal normal Mach number but different pre-shock-states may have completely different weight regarding the adaption of the discretization. These observations show, that greater attention has to be devoted to the formulation of appropriate refinement criteria and to the correct identification of flow phenomena.

REDISTRIBUTION METHODS AND METHODS OF LOCAL REFINEMENT

When structured grids are used, there exist two main alternatives regarding grid adaption. Firstly, the redistribution of a given set of points is possible by means of some kind of optimization method [5], secondly, the strategy of locally refining the mesh can be employed. Every redistribution method is based on the demand for the equidistribution of the approximation error in the field. In onedimensional form this can be formulated as

$$\Delta x \cdot w = \text{const.} \qquad (1)$$

or in continuous form as

$$x_\xi \cdot w = \text{const.} \qquad (2)$$

Hereby x represents the physical coordinate, with ξ being the coordinate in computational space. w represents the weight function and should be an appropriate measure of the local approximation error. Using the magnitude of the gradient of a scalar function Φ as weighting function, in one dimension we trivially get

$$w = \|\nabla\Phi\| = |\Phi_x| \stackrel{\text{discr.}}{=} \frac{|\Delta\Phi|}{\Delta x}, \qquad (3)$$

Δx being always positive. Combining equations (1) and (3), it follows, that in the onedimensional case the demand of the equidistribution of error is nothing than the claim of constant differences in computational space:

$$|\Delta\Phi| = \text{const.} \qquad (4)$$

Employing methods of local refinement, in addition to that, criteria can be used, that exploit physical properties of the single phenomena in order to detect them. These criteria will be called sensors in the following. For this purpose the phenomena are sorted out by the use of a preprocessing criterion and then are classified. After the classification different refinement criteria can be employed for the different flow phenomena.

FLOW PHENOMENA AND THEIR DETECTION BY SENSORS

An adaptive method to be used in computational fluid dynamics has to be able to improve the solution by means of grid refinement in those subregions of the domain, where the solution is strongly depending on the discretization. It is known in the field of computational fluid dynamics, that the solution at gasdynamical discontinuities and at poorly resolved viscous layers varies substantially with increasing grid refinement. Therefore the assumption can be made, that the global solution can be improved by improving the representation of these phenomena. The flow phenomena included in the following discussion are:

- Shock waves,
- Contact discontinuities,
- Shear layers (including the special case of velocity boundary layers),
- Vortex layers,

and opposed to these phenomena:

- Isentropic flow.

Keeping apart the domain of isentropic flow, which is included in the above list only in opposition to the other phenomena, all remaining phenomena have the character of layers, obviously depending on the Reynolds-number concerning the viscous ones. Therefore the normal direction on these layers represents a preferential direction, in which the greatest changes in the characteristic variables occur.

The boundary layers represent a special case of the above list, because their position, and therefore the normal direction on them, is known a priori. Since the changes across the boundary layer are also known normally, care for an appropriate discretization in normal direction can be taken in advance. If the changes cannot be determined a priori, as may be the case in hypersonic calculations using a radiation adiabatic boundary condition, where strong temperature changes arise across boundary layers, it should be no problem to formulate appropriate refinement criteria along the normal direction on the solid body surface. In the main flow direction the sensors, that are described in the following, can be used.

The reliable detection of shock discontinuities has an application in redistribution methods, too. Since the shocks have no macroscopic characteristic length scale, the value of the weight function will be increasingly higher with increasing refinement. This is the reason why the weight function has to be bounded at the shocks in order to avoid an overrefinement at the shock waves to the debit of the viscous layers in the field. Since the value of the weight function in the viscous layers can reach very high values in case of high-Reynolds-number calculations, the weight function cannot be bounded globally. Therefore the shock waves have to be detected in order to limit the weight function values only in these regions.

Previous to the employment of the sensors to identify the different phenomena the solution has to be sorted. Hereby it would be advantageous to separate the regions of continuous flow from the discontinuities as far as possible. A possible and obvious choice would be the use of the entropy as characteristic variable for this subdivision. But the entropy increase is very small at weak shocks. An at least equally suited variable is therefore represented by the density. Using the densities gradient [6], those cells, in which

$$\|\nabla \rho\|_{i,k} \geq C_1 \overline{w} \tag{5}$$

is not valid, with

$$\overline{w} = \frac{1}{(i \cdot k)} \sum_{i,k} \|\nabla \rho\|_{i,k} \tag{6}$$

being the average of the densities gradients magnitude, are sorted out. The constant C_1 is added by the author in order to achieve generality. Using instead the differences in computational space [7], only those cells are considered, where

$$(\Delta \rho)_{i,k} = \max\left(|\rho_{i+1,k} - \rho_{i,k}|, |\rho_{i-1,k} - \rho_{i,k}|, |\rho_{i,k+1} - \rho_{i,k}|, |\rho_{i,k-1} - \rho_{i,k}|\right) \tag{7}$$

is greater than the threshold

$$(\Delta \rho)_{\lim} = C_2 \max_{i,k}(\Delta \rho)_{i,k} \tag{8}$$

where on the right-hand side the maximum over all cells is searched. It is evident, that both approaches are equally suited for the separation of discontinuities from regions of isentropic flow, if the cell size is constant. Then the success of the subdivision depends only on the strength and the resolution of the single phenomena. In case, that the resolution near the discontinuities differs from that in the regions of isentropic flow, the following is valid: If the discretization at the discontinuity is fine and the discretization in the regions of isentropic flow is coarse, the gradients magnitude, viceversa the differences in computational space are the better measure. Since the latter are consistent with the principles of equidistribution, it is the opinion of the authors, that this kind of measure should normally be preferred in an adaptive method.

Vorozhtsov and Yanenko [6] propose to determine the normal direction on the layers by means of the density gradient:

$$\underline{n} = (\rho_x, \rho_z)^T (||\nabla \rho||)^{-1} \qquad (9)$$

The left and right states are defined to be in those cells (i*, k*), whose direction vector

$$\underline{r} = \begin{pmatrix} x_{i^*,k^*} - x_{i,k} \\ z_{i^*,k^*} - z_{i,k} \end{pmatrix} \qquad (10)$$

has the smallest angle to the normal vector $\underline{n}$, the left state being determined with the positive $\underline{n}$-direction and the right state with the negative one. The search is conducted in all eight cells surrounding the cell (i,k). Sensors based partially on the ideas of [6] can then be formulated as follows. Taking

$$S = \frac{p}{\rho^\gamma} \qquad (11)$$

as a measure of the entropy, the Sensor 1

$$w_1 = |S_l - S_r| > (\Delta S)_{lim} \qquad (12)$$

with

$$(\Delta S)_{lim} = \delta_1 \max_{i,k} (\Delta S)_{i,k} \qquad (13)$$

distinguishes between regions of isentropic flow and regions of discontinuities. $(\Delta S)_{i,k}$ is determined as in equation (7). In order to distinguish shocks from contact discontinuities the antiparallelity of the gradients of the density and of the temperature through the contact discontinuity is used in the Sensor 2

$$w_2 = \text{sign}(\rho_l - \rho_r) + \text{sign}(T_l - T_r) \begin{cases} = 2 \text{ at shocks} \\ = 0 \text{ at contact disc.} \end{cases} \qquad (14)$$

Other sensors, with whose help the contact discontinuities can be distinguished from the shocks and the isentropic changes of state, may be formulated with the help of pressure and normal velocity differences, both being of smaller order at contact discontinuities.

Reliable sensors for the detection of shocks can be constructed by means of the relation of pressure to density or temperature changes, this relation showing higher values at shock discontinuities compared to regions of isentropic flow and viscous layers. At shocks the dimensionless relation of pressure to density changes is a function of the normal mach number only:

$$\overline{\left(\frac{\Delta p}{\Delta \rho}\right)} = \left(\frac{\Delta p}{\Delta \rho}\right) \frac{\rho_1}{p_1} = \frac{\gamma(2 + (\gamma - 1) M_1^2)}{(\gamma + 1)} = f(M_1) \qquad (15)$$

the index 1 indicating the pre-shock-state. Similarly the relation of pressure to temperature changes is a function of the normal Mach number only:

$$\overline{\left(\frac{\Delta p}{\Delta T}\right)} = \left(\frac{\Delta p}{\Delta T}\right)\frac{T_1}{p_1} = \frac{\gamma(\gamma+1)M_1^2}{(\gamma-1)(\gamma M_1^2 + 1)} = f(M_1) \qquad (16)$$

In isentropic flow the above mentioned relations are wellknown:

$$\left(\frac{\partial p}{\partial \rho}\right)\frac{\rho}{p} = \gamma \, , \, \left(\frac{\partial p}{\partial T}\right)\frac{T}{p} = \frac{\gamma}{\gamma - 1} \qquad (17)$$

The dependency of the relation of the pressure to the density and the temperature rise on the Mach number (equation (15)) is shown in figure 1 (for $\gamma = 1.4$). Since over contact discontinuities the pressure rise is of smaller order, the above relations can be utilized for the detection of shocks in the solution. The above mentioned criteria are weakened by the smearing of the shocks over a certain number of cells. For a first estimation a linear increase of pressure and density over the cells representing the shock can be assumed. If the shock is considered to be represented in four cells (see figure 2 for the qualitative distribution of pressure and density), the relation between pressure and density changes for the single cells is that shown in figure 3a). It can be seen easily, that in case of linear increase this criterion would be able to identify a shock only between the cells n=0 and n=1. Therefore, in order to achieve reliability, the determination of the relation between pressure and density changes is changed as follows:

Let i* and k* represent the coordinates of the neighbour, in whose direction the greatest change in the density occurs. Hereby the search is conducted only in the four neighbours at $(i^*,k^*) = (i+1,k), (i-1,k), (i,k+1), (i,k-1)$. Writing i* and k* as

$$i^* = i + \Delta i \, , \, k^* = k + \Delta k \qquad (18)$$

obviously with

$$\Delta i, \Delta k \in \{-1, 0, 1\} \wedge |\Delta i + \Delta k| = 1 \qquad (19)$$

and defining

$$i' = i + 2\Delta i \, , \, k' = k + 2\Delta k \, , \, i'' = i - \Delta i \, , \, k'' = k - \Delta k, \qquad (20)$$

the relation of the pressure to the density changes

$$\overline{\left(\frac{\Delta p}{\Delta \rho}\right)}_{i,k} = \left(\frac{|p_{i^*,k^*} - p_{i,k}|}{|\rho_{i^*,k^*} - \rho_{i,k}|}\right)\left(\frac{\rho_{min}}{p_{min}}\right) \qquad (21)$$

is corrected with the help of the following two relations computed over a distance of two cells:

$$\overline{\left(\frac{\Delta p}{\Delta \rho}\right)}_{i,k}^{'} = \left(\frac{|p_{i',k'} - p_{i,k}|}{|\rho_{i',k'} - \rho_{i,k}|}\right)\left(\frac{\rho_{min}^{'}}{p_{min}^{'}}\right), \, \overline{\left(\frac{\Delta p}{\Delta \rho}\right)}_{i,k}^{''} = \left(\frac{|p_{i^*,k^*} - p_{i'',k''}|}{|\rho_{i^*,k^*} - \rho_{i'',k''}|}\right)\left(\frac{\rho_{min}^{''}}{p_{min}^{''}}\right) \qquad (22)$$

Φ_{min} represents the minimum of the two values, with whom the difference is calculated. Finally the Sensor 3 is determined as the maximum of the three relations:

$$w_3 = \max\left(\overline{\left(\frac{\Delta p}{\Delta \rho}\right)}_{i,k}, \overline{\left(\frac{\Delta p}{\Delta \rho}\right)}_{i,k}^{'}, \overline{\left(\frac{\Delta p}{\Delta \rho}\right)}_{i,k}^{''}\right) \qquad (23)$$

If the value of w_3 is higher than a threshold δ_3 a shock is identified.

The sensor based on the relation of pressure to temperature changes can be implemented similarly. It is well-known that pressure, temperature and density are coupled by the equation of state so that the changes in two of these variables determine the change in the third one completely. In figure 3b) the dependency of the relation of pressure to temperature changes is plotted under the asssumption of linearly increasing pressure and density for the shock smeared over four cells nominated before. Surprisingly the increase in pressure in relation to the increase in temperature is higher in the cells with higher values of n. Therefore the authors think, that by the combination of both criteria a reliable sensor for detecting shocks in a hydrodynamical solution can be constructed.

Previously it was mentioned, that the absolute changes in the values of a variable cannot be a correct measure for the identification of flow phenomena. In this sense the dimensionless change in density

$$\frac{\Delta\rho}{\rho_1} = \frac{2(M_1^2 - 1)}{2 + (\gamma - 1)M_1^2} \tag{24}$$

is a better measure for the preliminary detection of shock discontinuities. In figure 4 the dependency of the above mentioned criterion on the parameter n is shown (shock as in figure 2). It is easy to see, that using Sensor 4 with the threshold

$$w_4 = \frac{(\Delta\rho)_{i,k}}{\rho_{min}} > \delta_4 = 0.1 \tag{25}$$

shocks with a normal Mach number of approximately 1.2 can be detected in all four cells. Hereby $(\Delta\rho)_{i,k}$ is determined with equation (7) and ρ_{min} is the smaller of the two values with whom the difference is calculated. Obviously using this sensor not only shocks, but also regions of strong isentropic expansion or compression and compressible shearlayers are detected.

Confronting the geometrical shape and the behaviour of the key variables across contact discontinuities and viscous layers the familiarities between these flow phenomena are evident. The authors therefore think, that for the detection of viscous layers the same sensors may be employed as for the detection of contact discontinuities. The only exception may be represented by the center of the vortex layer, where the key variables achieve a local extremum. This can be circumvented by the evaluation of the vorticity.

RESULTS

For the validation of the previously proposed sensors two test cases are chosen: First the problem of shock-shock-interaction at hypersonic speed is examined with an inviscid calculation. Hereby the method of Eberle [1] is used. The free stream Mach number is 4.6, the free stream temperature is equal to 120 K. The impinging shock has an angle of 20.9° to the horizontal plane. In figure 5 the isolines of density, entropy and temperature determined by a calculation on a 50x50 grid are shown. Because of the coarse discretization the contact discontinuities arising in the solution are badly resolved, especially concerning their representation in the density field. Looking at figure 5b) it is evident, that the entropy used in the sensor 1 is much more suited for the detection of contact discontinuities than the density. Comparing figure 5 with the isolines depicted in figure 6, where the results of a calculation on a 200x200 grid are shown, it is confirmed, that the solution changes strongly only near the discontinuities. Therefore, when the whole grid is refined, large regions are uselessly refined. Applying the sensor based on the dimensionless density rise (sensor 4) together with the sensor of the relation of pressure to density changes (sensor 3), the shocks in the solution can be detected as shown in figure 7. Hereby all those cells are plotted, in which the criterion is fulfilled. It must be emphasized, that the solution on the coarse grid with 50x50 cells is used. In figure 8 the detecting of contact discontinuities in the field is demonstrated. Here 0.5 times the mean density gradient ($C_1 = 0.5$ in relation (5)) is used as threshold for the preprocessing of the solution (see figure 8a)). In the figures 8b) and c) is shown, that even on the coarse 50x50 grid the shear layers emanating from the contact discontinuities can be detected with the sensor of the entropy differences. Furthermore the detection seems to be quite independent of the choice of δ_1. The distinguishing of shocks from contact discontinuities by means of the antiparallelity of density and temperature gradient is demonstrated in figure 9. Although several cells, that surely do not represent shocks can be seen in figure 9.a), the separation is quite satisfactory. Nevertheless the determination of the shock positions by means of the sensors 3 and 4 (see figure 7) seems to work better.

The second test case is represented by the transonic viscous flow around an airfoil (CAST 7). This calculation has been performed with the Navier-Stokes-Method NSFLEX [8] on a C-type grid with originally 257x66 cells. For the following calculations the grid has been split in three blocks, whose boundaries are shown in the figures as dotted lines. The free stream Mach number is 0.7, the free stream temperature 300 K, the Reynolds-number is 4,000,000, the angle of attack is 2°. The calculation has been performed with the algebraic turbulence model of Baldwin and Lomax, the transition is fixed at 7% of the chord length. The density isolines are shown in figure 7. Using as criterion the differences in the density in the computational domain with the threshold C_2 equal to 0.05 (see relation (8)), 17.9% of the cells are detected (for the distribution of these cells see figures 11a) and b)). Since large regions are characterized by strong isentropic changes of state the additional application of sensor 1 diminishes the number of detected cells to 8.2%. Nevertheless, as can be seen from figure 12b), the vortex layer behind the air foil is detected as well as in the previous case. The detection of the vortex layer almost in the whole domain behind the airfoil is surprising, because the grid boundary on the right side is more than 17 chord lengths away from the trailing edge and the discretization therefore very coarse. In order to detect shocks in this solution it is sufficient to use sensor 4 with $\delta_1=0.1$. In the figures 13a) and b) the shock position is shown in the physical and the computational domain. Only 0.44% of the total number of cells are detected.

At last the uselessness of refining isentropic regions in the present case is demonstrated. In figure 14a) the density isoline distribution of an inviscid calculation is shown. Applying the criterion of the density differences in computational space (equations (7) and (8) with $C_2=0.1$), the cell distribution shown in figure 14b) is detected. On a locally refined grid (13 blocks) a solution as depicted in figure 15a) can be obtained. Reapplying the criterion used before only the shock points are detected in the refined region (see figure 15b)). Comparing the figures 14a) and 15a) it is evident, that the solution shows remarkable differences only in the region of the shock discontinuity. Thus the isentropic regions have been refined uselessly.

CONCLUSIONS

Several possibilities of formulating sensors for the detection of flow phenomena have been presented. These sensors can for example be used in an adaptive method for the determination of those regions in the flow field, where a local refinement may be useful for the efficient improvement of the solution. Compared to the standard indicators the proposed sensors show remarkable improvements in the detection of weak phenomena on coarse grids. The separation of isentropic flow from the discontinuities in order to avoid useless refinement in the former regions is possible, too. The authors think, that in combination with an adaptive method basing on local refinement as the one proposed in [7] a reliable and efficient tool for the treatment of fluid dynamic problems can be obtained.

ACKNOWLEDGEMENTS

The authors wish to thank the German Research Association for supporting these studies in its Priority Research Programme 'Strömungssimulation mit Hochleistungsrechnern'.

REFERENCES

[1]   A. Eberle: "MBB EUFLEX, a new Flux Extrapolation Scheme solving the Euler Equations for arbitrary Geometry and Speed", MBB/LKE122/S/PUB140, 1984.

[2]   P.L. Roe: "Characteristic-Based-Schemes for the Euler Equations", Ann. Rev. Fluid Mechanics, 18 (1986), pp. 337-365.

[3]   K. Nakahashi, G.S.Deiwert: "A Three-Dimensional Adaptive Grid Method", AIAA-85-0486, 1985.

[4]   H.J.Kim, J.F. Thompson: "Three-Dimensional Adaptive Grid Generation on a Composite-Block-Grid", AIAA-88-0311, 1988.

[5]   J.U. Brackbill, J.S. Saltzmann: "Adaptive Zoning for Singular Problems in Two Dimensions", Journal of Computational Physics, 46, 1982.

[6]   E.V. Vorozhtsov, N.N. Yanenko: "Methods for the Localization of Singularities in Numerical Solutions of Gas Dynamic Problems", Springer Verlag, New York-Berlin-Heidelberg, 406 S., 1990.

[7]   J.J. Quirk: "An Adaptive Grid Algorithm for Computational Shock Hydrodynamics", Ph. D. Thesis, College of Aeronautics, Cranfield Institute of Technology (1991).

[8]   M.A. Schmatz: "NSFLEX-an implicit Relaxation Method for the Navier-Stokes Equations for a wide range of Mach numbers", Proc. 5th GAMM-Seminar on the "Numerical treatment of the Navier-Stokes-Equations", NNFM, Vol. 30, Vieweg, Braunschweig-Wiesbaden.

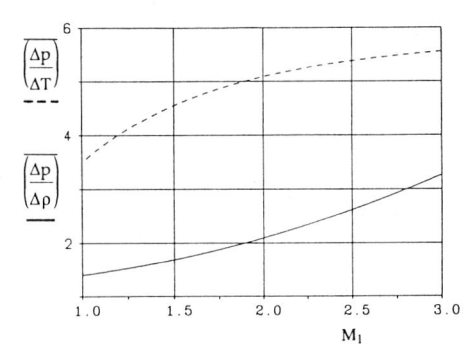

Fig.1: Dependency of the dimensionless relation of pressure to temperature and pressure to density differences on the Mach number.

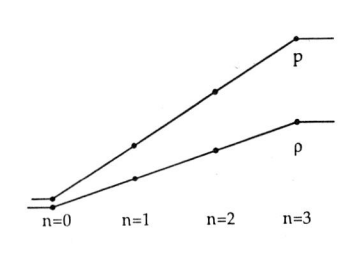

Fig.2: Assumed linear increase of density and pressure.

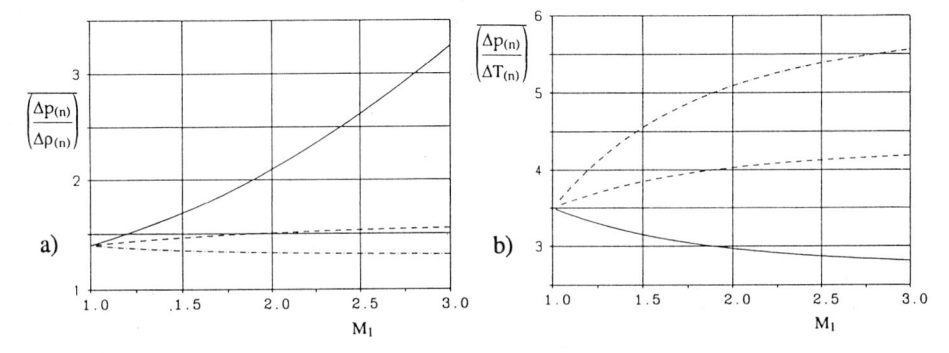

Fig.3: Weakening of the criteria for a shock smeared over four cells ($\Delta\Phi(n) = \Phi(n+1)-\Phi(n)$):
a) Pressure to density change, b) Pressure to temperature change,
—— : $n = 0$, ---- : $n = 1$, —·—·— : $n = 2$.

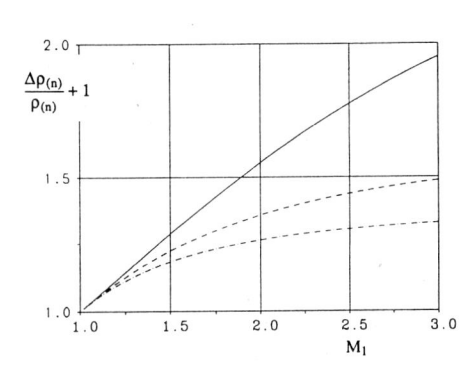

Fig.4: Dimensionless density rise for a shock smeared over four cells
—— : $n = 0$, --- : $n = 1$, —·— : $n = 2$.

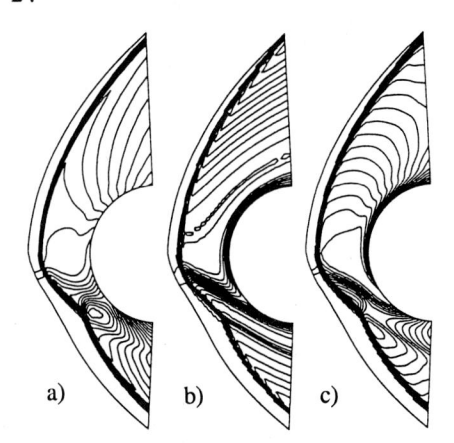

Fig.5: Isolines of shock-shock-interaction on a 50x50-grid: a) Density, b) Entropy, c) Temperature (Values as in figure 6).

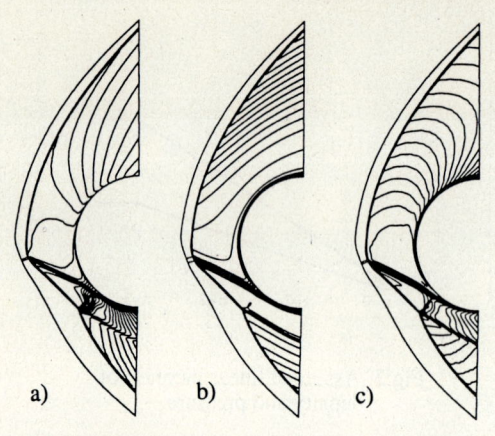

Fig.6: Isolines of shock-shock-interaction on a 200x200-grid:
a) Density, b) Entropy, c) Temperature.
$\rho_{min}/\rho_\infty = 0.8$, $\rho_{max}/\rho_\infty = 12.5$, $\Delta\rho/\rho_\infty = 0.418$,
$S_{min}/S_\infty = 1.04$, $S_{max}/S_\infty = 2.94$, $\Delta S/S_\infty = 0.068$,
$T_{min} = 130$ K , $T_{max} = 630$ K , $\Delta T = 17.9$ K .

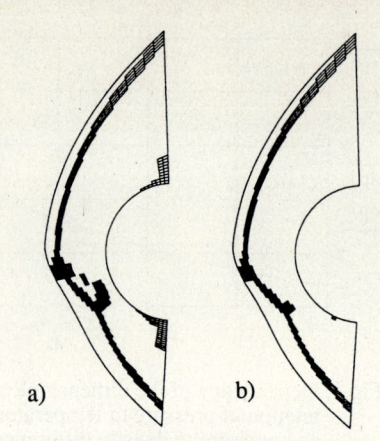

Fig.7: Detecting shocks in the solution on the 50x50-grid:
a) Sensor 4 with $\delta_4 = 0.1$,
b) a) + sensor 3 with $\delta_3 = 1.5$.

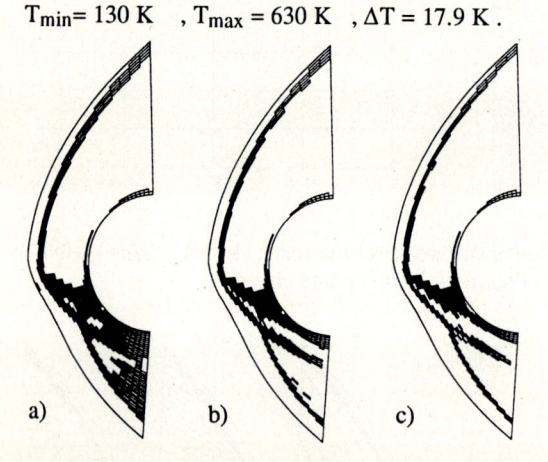

Fig.8: Detecting contact discontinuities and shearlayers:
a) $|\nabla\rho| > 0.5\,|\overline{\nabla\rho}|$, b) a) + sensor 1 with $\delta_1 = 0.03$,
c) a) + sensor 1 with $\delta_1 = 0.05$.

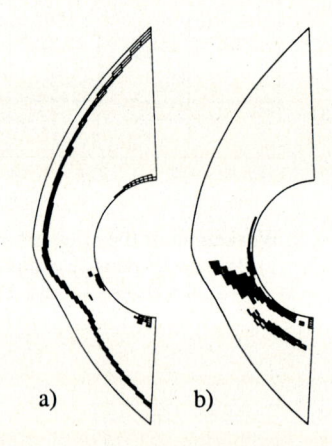

Fig.9: Dividing the phenomena by the sensor 2:
a) Shocks, b) Shear layers.

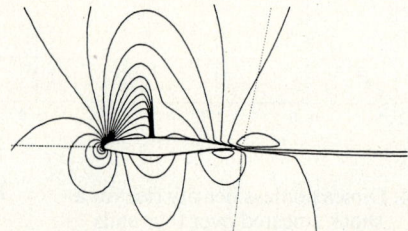

Fig.10: Density isolines of transsonic viscous flow around an airfoil: $\rho_{min}/\rho_\infty = 0.5$, $\rho_{max}/\rho_\infty = 1.3$, $\Delta\rho/\rho_\infty = 0.0286$.

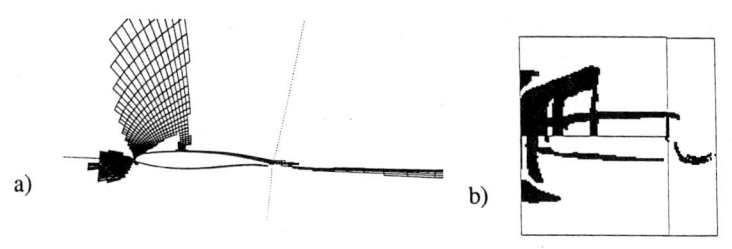

Fig.11: Cells to refine using $\Delta\rho > 0.05$ $(\Delta\rho)_{max}$: a) Phys. domain, b) Comput. domain.

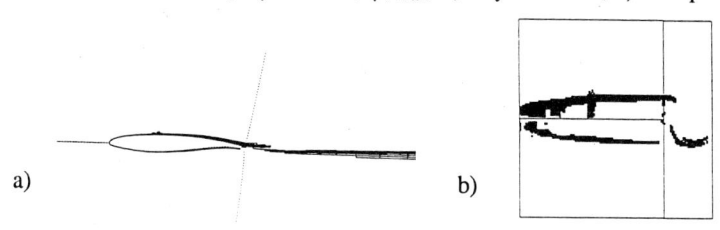

Fig.12: Cells to refine using criterion of fig.11 + sensor 1 with $\delta_1 = 0.05$:
a) Physical domain, b) Computational domain.

Fig.13: Cells to refine using sensor 4 with $\delta_4 = 0.1$: a) Phys. domain, b) Comput. domain.

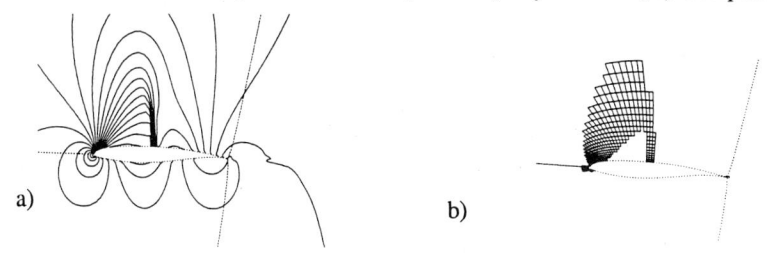

Fig.14: Inviscid calculation (values as in Figure 10): a) Density isolines, b) Cells to refine.

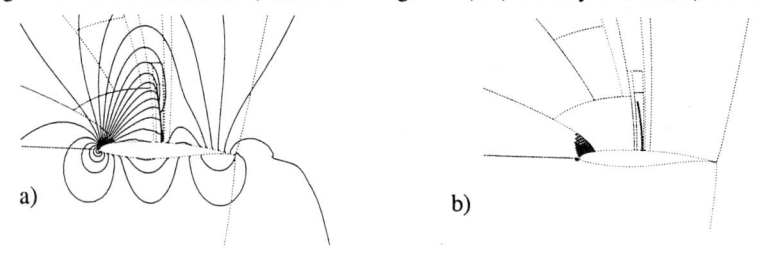

Fig.15: Locally refined inviscid calculation: a) Density isolines, b) Cells to refine.

Grid Alignment Effects and Rotated Methods for Computing Complex Flows in Astrophysics

R. J. LeVeque, Department of Applied Mathematics,
University of Washington, Seattle WA 98195, USA

R. Walder, Seminar für Angewandte Mathematik,
ETH Zürich, 8092 Zürich, Switzerland

Summary

We discuss some advantages and difficulties of Cartesian grid calculations on high Mach number flows. We show and explain strong grid alignment effects even in smooth flow regions and in steady state solutions. We present first order rotated schemes for the Euler equations which are able to overcome some of the difficulties, but still display strong dependence of shock positions and speeds on the underlying scheme.

Introduction

This work is inspired by astrophysical calculations showing severe grid alignment problems with high Mach number flows. In particular, we have studied colliding stellar winds in a double star system and also point blasts into different surrounding density distributions. These are important astrophysical problems since over 50 percent of all stars exist in double star systems and are typically losing mass due to stellar winds. Supernovae- and novae-explosions are the most important examples of astrophysical point blasts.

Below we will present calculations from symbiotic double star systems in the latest stage of stellar evolution, where stellar winds are an important determining factor in the interaction and the evolution of the double star system. Figure 2 illustrates the situation. The density contours in a logarithmic scale shows the two bow shocks and the slip line in between, where the two flows come together. The star on the right side of the figure is a Red Giant that produces a strong stellar wind, losing about 10^{-5} - 10^{-7} solar masses per year. This mass flow is slightly supersonic (around Mach 2 or 3) and is the source of a dense nebula which surrounds both stars. Suddenly, by processes which are still not well understood, the other (left) star also begins to lose mass. This new flow drives a shock into the surrounding nebula. The characteristics of this wind are very different from those of the other: it is much less dense (by a factor of perhaps 100), but highly supersonic with Mach numbers in the region of 100. The momenta of the two winds may be of similar magnitude. Heat transfer via radiation is an important factor in these flows, but one which has not yet been included in our model.

The numerical calculations involve a number of difficulties. One wind is highly supersonic, producing very strong shocks and strong convection. The gradients of density in the winds are very steep: the density of one wind is about two orders of magnitude higher than the other and densities over the entire computational domain differ by 6 or 7 orders of magnitude. There are several different time and length scales involved. Moreover, we would like to do simulations advancing to times at which the interesting shock structure is far from the stars.

We have chosen to use a Cartesian grid refinement method based on a code developed by Berger and LeVeque[1]. Grid refinement is necessary for this problem in order to efficiently capture the interesting features. Although the flow in the region of each star is radially symmetric, the interesting shock features are not aligned with either of these directions. The Cartesian grid approach has the advantage that automatic grid refinement is easily implemented and that very efficient vectorized integrators can be used on each grid. The Cartesian grid is cut by two circles representing boundaries of each star along which the flow is assumed to be known. This results in some rectangular cells being replaced by irregular polygons that may be arbitrarily small relative to the regular cells. We need to specify boundary conditions in these cells that are accurate and stable with a time step chosen relative to the regular cells.

For solid wall boundaries (e.g. an airfoil embedded in a Cartesian grid), Berger and LeVeque[2] have developed such boundary conditions. In this work we have successfully modified these boundary conditions to simulate the stellar wind from the stars. The fluxes at the solid wall segments of Cartesian cells used in [2] have been replaced by the given supersonic outflow fluxes.

This approach has been tested and works well with moderate Mach numbers. Unfortunately, difficulties appear with the very high Mach number flows that are physically relevant for this problem, as seen in Figures 2 for example. Results show a large grid alignment effect due to the use of a Cartesian grid in a region where the flow is radially symmetric with very steep density gradients. The method used in the original code is a typical high resolution method, based on work of Colella [3]. This is a very good method for moderate problems, but breaks down for the problems considered here.

To avoid this difficulty, we have implemented a rotated method following ideas of [5], [8] and [10]. In this method, Riemann problems are solved in directions relevant to the flow direction, rather than in the coordinate directions. The work reported here is for first order accurate methods. We are currently working on extending these ideas to high resolution second order accurate methods. In the next section, we will give a brief review of the Godunov scheme and describe the basic ideas of rotated difference schemes.

In Section 3 we will show that a first order rotated difference scheme can give a significant improvement over the standard Godunov method. However, the use of rotated schemes gives rise to some interesting new difficulties. In particular, we see that the computed solution (on underresolved grids) can be highly dependent on the particular rotation method used. For example, in stationary solutions the locations of the shock and contact discontinuities can depend on the particular method used and the choice of rotation angle.

In Section 4 we describe and explain one grid alignment effect seen in the figures, the appearance of "bumps" along the coordinate axes that result from strong supersonic outflow that is interpreted as transonic expansions in the one dimensional problems solved to compute the numerical fluxes.

First order Godunov and rotated schemes

We consider the Euler equations in two space dimensions,

$$\vec{u}_t + f(\vec{u})_x + g(\vec{u})_y = 0, \tag{1}$$

where $\vec{u} = (\rho, \rho u, \rho v, E)$ is the vector of conserved quantities and f and g are the flux functions. As a simplified model we consider an ideal gamma-law gas.

We consider finite volume methods on Cartesian grids, taking the form

$$U_{ij}^{n+1} = U_{ij}^n - \frac{\Delta t}{A}\left[F_{i+1/2,j} - F_{i-1/2,j} + G_{i,j+1/2} - G_{i,j-1/2}\right]. \tag{2}$$

Here U_{ij}^n represents the cell average over the (i,j) cell $[x_{i-1/2}, x_{i+1/2}] \times [y_{j-1/2}, y_{j+1/2}]$ on a uniform grid with spacing $\Delta x = \Delta y = h$. The numerical flux $F_{i+1/2,j}$ is an approximation to

$$\frac{1}{\Delta t}\int_{t_n}^{t_{n+1}} \int_{y_{j-1/2}}^{y_{j+1/2}} f(u(x_{i+1/2}, y, t)) \, dy \, dt \approx h f(u(x_{i+1/2}, y_j, t_{n+1/2})) \tag{3}$$

at the right side of the (i,j) cell. Similarly, $G_{i,j+1/2}$ is an approximation to the flux at the upper boundary.

Godunov's method is a standard first order numerical method in which $F_{i+1/2,j}$ is computed by solving the one-dimensional Riemann problem $u_t + f(u)_x = 0$ with data $u_l = U_{ij}^n$ and $u_r = U_{i+1,j}^n$. See, e.g., [6] for a description of such methods.

In the one-dimensional analog of Godunov's method, the wave structure in the solution of the Riemann problem is very revealing and gives a method that is strongly based on physics and behaves quite well, although solutions are strongly smeared due to the inherent dissipation of the method.

In two space dimensions, however, the structure of the solution to a one-dimensional Riemann problem obtained by taking data from adjacent cells in the x- or y-direction can be misleading in terms of understanding the two-dimensional structure of the flow. This leads to the grid alignment effects seen in Figure 2 in a way that will be explained.

In two dimensional flow there is often a single direction which is the "dominant" direction in some sense. Examples include the radial direction away from a star or the direction normal to a shock. An attractive idea is to use a method that makes use of this dominant direction.

Rotated schemes take advantage of the fact that the form of the Euler equations remains invariant under rotation by an arbitrary angle. We can define a new coordinate system with axes in some ξ-direction (at angle Θ to the x-axis) and an orthogonal η-direction. The equation (1) then becomes

$$u_t + f^\xi(u)_\xi + g^\eta(u)_\eta = 0$$

where

$$f^\xi(u) = f(u)\cos\Theta + g(u)\sin\Theta$$
$$g^\eta(u) = -f(u)\sin\Theta + g(u)\cos\Theta.$$

If we can compute numerical fluxes F^ξ and G^η in the ξ- and η-directions, then we can recover fluxes F and G (see Figure 1a for an example) in the x- and y-directions using

$$F = F^\xi \cos\Theta - G^\eta \sin\Theta \tag{4}$$
$$G = F^\xi \sin\Theta + G^\eta \cos\Theta. \tag{5}$$

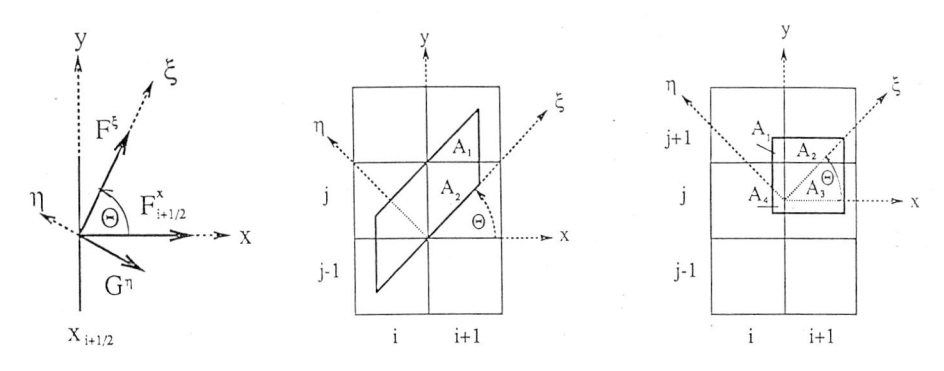

Figure 1: a) Transformation of fluxes, b) and c) Interpolation methods A, B

In a first order rotated generalization of Godunov's method, we continue to use (2), but now determine F or G by performing the following steps at each interface:

1. Choose appropriate directions ξ and η, based on the flow.

2. Compute data u_l^ξ, u_r^ξ in the ξ-direction and solve the Riemann problem $u_t + f^\xi(u)_\xi = 0$ with this data to find F^ξ. Repeat in the η-direction to find G^η.

3. Use (4) or (5) to compute the flux normal to the interface.

Clearly there are several choices to be made. The physical properties of a complex flow may suggest several different rotation angles and it is not clear which should be used. As we will see, the choice of angle can make a dramatic difference in the computed solution.

In Step 2 of the algorithm, we must interpolate the data from the underlying Cartesian grid to obtain appropriate data in the ξ- and η-directions. There are several ways that we might do this. Here we mention two possibilities. Method A is illustrated in Figure 1b. We construct a box extending distance h in the ξ-direction and use the areas of overlap with each cell to determine interpolation weights. This gives

$$u_r^\xi = (A_2 \times U_{i+1,j} + A_1 \times U_{i+1,j+1})/(A_1 + A_2) = (1 - \sin\Theta)U_{i+1,j} + \sin\Theta U_{i+1,j+1}. \quad (6)$$

In the same way one gets u_l^ξ, and u_l^η, u_r^η in the η-direction.

Method B is illustrated in Figure 1c. A box of size h^2 is centered distance $h/2$ from the interface in the ξ-direction, and the overlap with each of four cells are used to define weights,

$$u_r^\xi = (A_1 \times U_{i,j+1} + A_2 \times U_{i+1,j+1} + A_3 \times U_{i+1,j} + A_4 \times U_{i,j})/h^2. \quad (7)$$

This is equivalent to bilinear interpolation between the four neighboring cell centers to the point lying distance $h/2$ from the center of the interface in the ξ-direction.

In smooth regions the results with Method B are much better. On problems with strong shocks, however, we see oscillations near the shocks. This is presumably due to the fact that (7) takes data from both sides of the shock and does not fully preserve the upwind nature of Godunov's method.

Figure 2: a) Solution of colliding winds using standard Godunov method compared with b) solution using a rotated scheme with rotation in radial direction with respect to the left star and c) solution using a rotated scheme, with rotation in velocity direction.

Numerical results

Figure 2 shows the steady state solutions for a double star calculation with different methods on a 200×200 grid. On the boundary of the two stars we set the densities $\rho_{left} = 0.0067$ and $\rho_{right} = 1$, with velocities $v_{left} = 39.1$ and $v_{right} = 2.17$ and pressures $p_{left} = 0.04$ and $p_{right} = 0.6$. The polytropic index γ is 1.6. The flow from the left star has a Mach number of 12.65, from the right one of 2.14. The separation of the stars is 0.2 and each star has a radius of 0.038 in a domain which is the unit square.

With Godunov's method 2a we see strong bumps aligned with the x- and y-axis, which are the directions in which the Riemann problems are solved.

We also see grid alignment effects in the plots of the solution, calculated with the rotated scheme (method A), but they look different. Instead of bumps we have depressions. Figure 2b using the rotated scheme with Θ chosen based on the radial direction relative to the star on the left. In Figure 2c we have instead chosen the rotation angle at each interface based on the average flow direction from the cells on either side.

The difference between these solutions is quite striking. In particular, notice that the shock location is different in each case. Also significant is the different resolution of the slip line where the two flows come together. This phenomenon has not yet been studied in sufficient depth. One interesting question is whether the choice of rotation angle based on the characteristics of the flow might cause some form of feedback that in turn affects the flow.

The use of an appropriate rotated scheme seems to alleviate some of the difficulties seen with bumps in the above calculations. To see that it can also improve the location of shocks, we show another set of calculations.

In Figure 3 we present solutions of a point blast calculated with different methods. The initial data of the point blast are : density $\rho = 0.1$, velocity $v \equiv 0$ in the whole domain and pressure $p_{in} = 600$ inside of a circle with radius $r_0 = 0.05$ and $p_{out} = 0.06$ outside of this circle. The polytropic index γ is equal to 1.6. The figure shows plots from calculations with different methods on a 100×100 grid, all taken at the same computational time. The first plot shows calculatons with Godunov's method, the second one with a rotated scheme, where the rotation angle was chosen to be 45 degrees

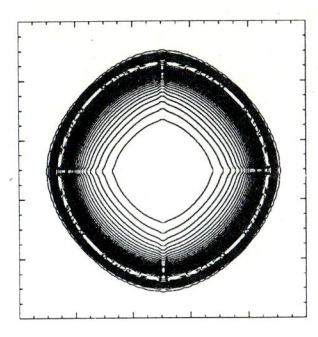

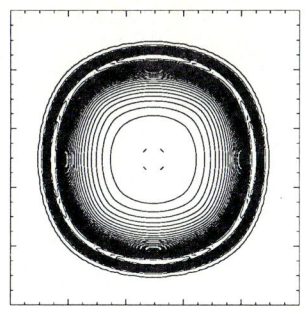

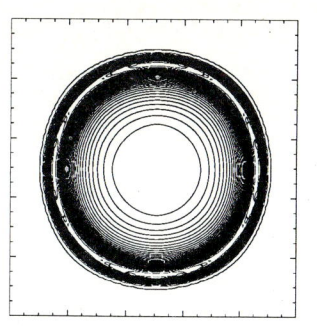

Figure 3: Point blast calculated with Godunov method, scheme with 45 degree rotation and scheme with rotation in flow direction.

everywhere. The third plot shows results obtained with the rotated scheme in which the ξ-direction at each interface is chosen in the direction of velocity. One can see clearly the dependence of the shock position of the rotation angle. The shock should be circular. With Godunov's method it tends to become diamond shaped, indicating that propagation is faster along the coordinate axes than at 45°. With rotation at 45° the propagation is now fastest in this direction, and so the shock becomes square. Rotating in the velocity direction gives relatively nice results, with a roughly circular shock.

Description of the grid alignment effects

One of the obvious grid alignment effects seen in the above experiments is the appearance of "bumps" along the coordinate axes. This can be explained by the fact that one-dimensional cross-sections of the flow can exhibit quite different physical flow characteristics than the full two dimensional flow. In particular, a flow that is everywhere highly supersonic in two dimensions can yield transonic behavior in computing the one-dimensional fluxes. This type of difficulty is well-known in other contexts, [4], [9], [11].

This effect can be nicely illustrated with a simple model problem, a single time step on a flow with energy and pressure that are initially uniform and should remain roughly so over short times, but for which steep gradients in density and velocity give rise to the bump phenomenon in a striking way. Here we can explicitly calculate the energy flux and see that bumps arise from one-dimensional transonic rarefaction waves.

To isolate the grid alignment effects from other possible sources of difficulty, such as boundary effects from the star boundaries, we use a radially symmetric outflow problem with the star center off the grid, at $(x_0, y_0) = (0.5, -0.2)$, while the computational grid is the unit square. We use the following initial conditions:

$$\rho(\vec{x}, 0) = \rho_0/r^2 \qquad u(\vec{x}, 0) = \tilde{x} q_0$$
$$p(\vec{x}, 0) = p_0 \qquad v(\vec{x}, 0) = \tilde{y} q_0$$

with $r^2 = (x - x_0)^2 + (y - y_0)^2$ and $\tilde{x} = x - x_0$, $\tilde{y} = y - y_0$. We use $\rho_0 = 1.4$, $p_0 = 1$, and various values for q_0. For a gamma-law gas with $\gamma = 1.4$, the Mach number is then q_0 everywhere. Energy is inintially also constant in x and y with the value

$$E(\vec{x}, 0) = E_0 = \frac{p_0}{\gamma - 1} + \frac{1}{2} \rho_0 q_0^2. \qquad (8)$$

With these initial conditions, we can easily compute that at time $t = 0$,

$$\begin{aligned}
\rho_t &= 0 & (\rho u)_t &= -\rho_0 q_0^2 \tilde{x}/r^2 \\
p_t &= (\gamma - 1)(-2K + \rho_0 q_0^3) & (\rho v)_t &= -\rho_0 q_0^2 \tilde{y}/r^2 \\
E_t &= -2K
\end{aligned} \qquad (9)$$

where $K = q_0(E_0 + p_0)$. In particular, the derivatives of pressure and energy are spatially uniform, so that these quantities should remain nearly spatially uniform over small times. With an explicit numerical method we might expect these quantities to remain uniform at the end of one time step, particularly the energy which is itself one of the conserved variables. However, the energy flux depends on the velocity, which has steep gradients. We can at least hope for smooth solutions after one time step.

Figures 4a and 4b show results with the standard Godunov method for $q_0 = 4$ and $q_0 = 10$, respectively. We see that E remains constant only outside of the wedge

$$|x - x_0| < \frac{|y - y_0|}{\sqrt{q_0^2 - 1}}. \qquad (10)$$

A discontinuity arises along the boundary of this wedge. The pressure remains only approximately constant outside the wedge and again displays discontinuties at the wedge boundary. For large values of the Mach number q_0, this wedge shrinks to a thin strip along the coordinate axis $x = x_0$, giving bumps of the form seen in the earlier results.

The appearance of these discontinuities is easy to predict if we compute the numerical fluxes being used. The true energy flux in the x- and y-directions is given by

$$\begin{aligned}
u(E + p) &= (x - x_0)K \\
v(E + p) &= (y - y_0)K.
\end{aligned} \qquad (11)$$

With Godunov's method, we solve 1D Riemann problems in the x- and y-directions separately. For q_0 large enough, the flow is everywhere supersonic in the y-direction. So the computed flux at the top of each cell is simply the flux function evaluated at the cell value. In particular, the energy flux is

$$G_{i,j+1/2} = hv(E + p)\Big|_{(x_i,y_j)} = hK(y_j - y_0). \qquad (12)$$

In solving the x-Riemann problems, the flow is only supersonic for $|\tilde{x}|$ large enough, since $u = 0$ at $\tilde{x} = 0$. For supersonic flow we require (c is the adiabatic sound speed)

$$u = \tilde{x} q_0 > c = \sqrt{\frac{\gamma p}{\rho}} = \sqrt{\frac{\gamma p_0}{\rho_0}(\tilde{x}^2 + \tilde{y}^2)} = \sqrt{(\tilde{x}^2 + \tilde{y}^2)}. \qquad (13)$$

From this, we see that the region where the flow appears to be *subsonic* in the x-direction is precisely the wedge (10).

Outside of this wedge, where the flux is supersonic in both x and y, solving the Riemann problem in x will return the flux at the upwind cell center as the interface flux $F_{i+1/2,j}$. For example, for $\tilde{x} > |\tilde{y}|/\sqrt{q_0^2 - 1}$ we will obtain

$$F_{i+1/2,j} = hu(E + p)\Big|_{(x_i,y_j)} = hK(x_i - x_0). \qquad (14)$$

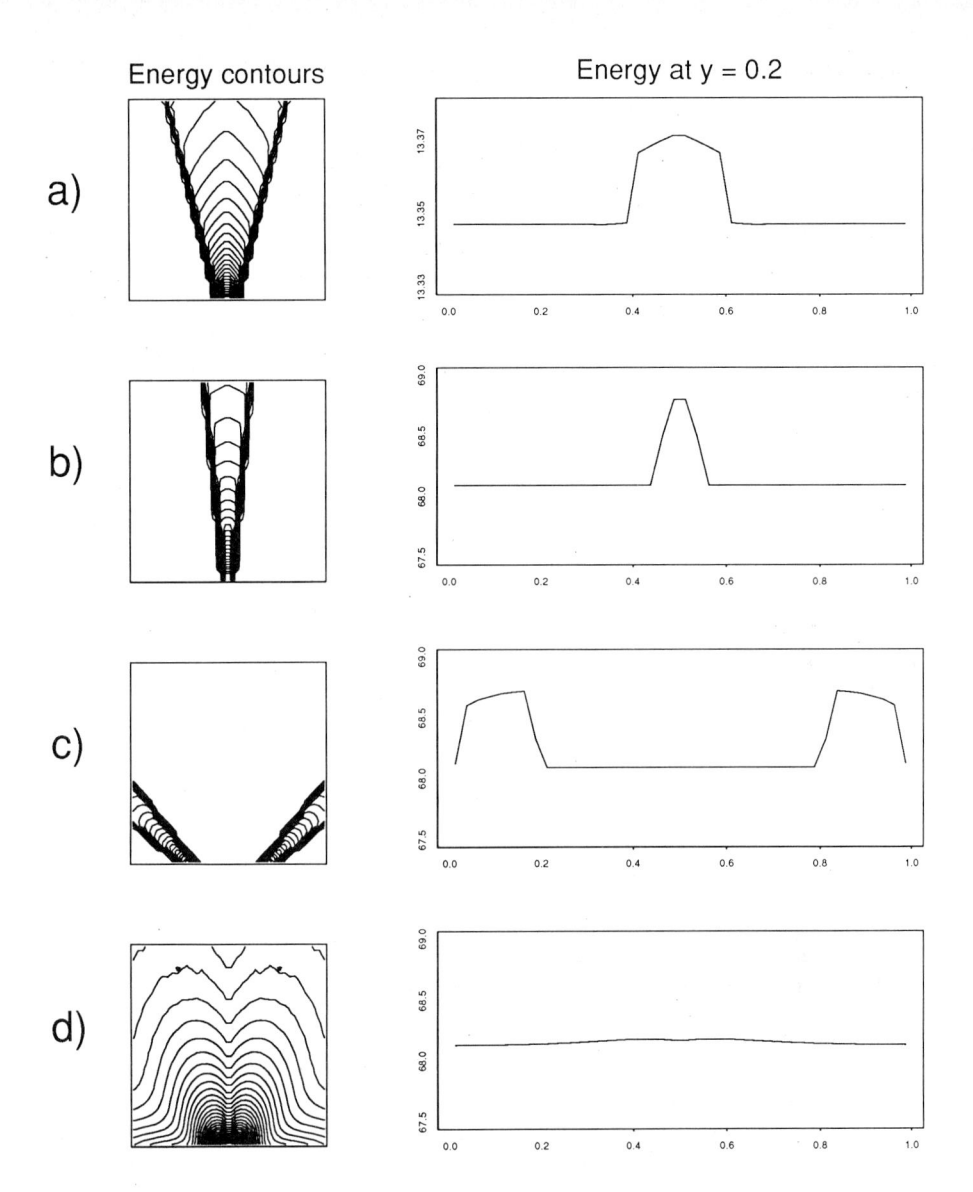

Figure 4: Results for the model problem. Energy contour plots and a slice at $y = 0.2$ are shown after one time step with various methods. a) Godunov method for Mach number $q_0 = 4$. b) Godunov method for $q_0 = 10$. c) Rotated method with $\Theta = 45°$ everywhere ($q_0 = 10$). d) Rotated method with rotation in the radial direction ($q_0 = 10$).

Combining this with (12) and computing the updated cell value via (2), we obtain

$$E_{ij}^1 = E_{ij}^0 - \frac{\Delta t}{h^2}(2h^2 K) = E_0 - 2\Delta t K.$$

This is constant in x and y and is consistent with the value of E_t from (9).

Inside the wedge, where the flow is subsonic in x, the flux is not simply the flux function evaluated at the upwind cell center. Instead, solving the Riemann problem at the interface will give an intermediate value somewhere between the values at the neighboring cell centers.

As an extreme example, consider the case where q_0 is very large, large enough that the only subsonic cell interfaces are along the coordinate line $\tilde{x} = 0$. Let I be the index for which $\tilde{x}_{I-1/2} = 0$. Then we have

$$F_{i+1/2,j} = h(i+1/2)K \quad \text{for } i \geq I,$$
$$F_{i+1/2,j} = h(i+3/2)K \quad \text{for } i \leq I-2$$

from (14), while

$$F_{I-1/2,j} = 0.$$

As a result, flux differencing gives

$$F_{i+1/2,j} - F_{i-1/2,j} = h^2 K \quad \text{for } i \neq I-1, I$$

as desired away from the central cells, but

$$F_{i+1/2,j} - F_{i-1/2,j} = \frac{1}{2}h^2 K \quad \text{for } i = I-1 \text{ and } I.$$

The flux difference in these cells is only half what it should be, leading to

$$E_{ij}^1 = E_0 - \frac{3}{2}\Delta t K$$

for $i = I-1$ and I. This leads to the bump along the coordinate line $x = x_0$.

The fact that Godunov's method gives nonsmooth fluxes in transonic rarefaction waves is also responsible for the appearance of so-called "entropy glitches" or "dog-legs" in certain calculations. See [7] for a description of this problem and some analysis similar to what is presented here.

Figure 4c shows results computed with the rotated scheme presented above, where we have first chosen a rotation angle $\Theta = 45°$ everywhere. We now see bumps appearing along the edges of wedges $|\tilde{x}-\tilde{y}| < |\tilde{y}|/\sqrt{q_0^2 - 1}$ at 45° to the grid. In the rotated Riemann problems, the flow is supersonic in both the ξ and η directions outside these wedges, while inside the wedges the flow in the η direction appears as a transonic rarefaction. Rotating at other fixed angles gives bumps in the corresponding direction. This is also true for the steady state double star calculations.

For the model problem, the most natural choice of rotation direction would be the radial direction. In this case the Riemann problem in the η-direction gives a transonic rarefaction everywhere. This gives results that are no longer exactly constant anywhere, but are nearly constant everywhere, as seen in Figure 4d. This is a substantial improvement over the other calculations.

Clearly much work remains to be done in developing high resolution methods to solve these difficult problems. It seems that rotated difference methods have some advantages, but there are still many intriguing difficulties to be understood and overcome.

Acknowledgements

R. LeVeque was supported in part by NSF Grant DMS-8657319 and was in residence at ETH during the course of much of this work. R. Walder thanks the Institute of Astronomy of ETH, where he has a half time position. He is supported by a research grant of ETH.

References

[1] M. J. Berger and R. J. LeVeque. An adaptive cartesian mesh algorithm for the Euler equations in arbitrary geometries. In *AIAA 9th Computional Fluid Dynamics Conference, Buffulo, NY*, June 1989.

[2] M. J. Berger and R. J. LeVeque. Stable boundary conditions for cartesian grid calculations. *Computing Systems in Engineering*, 1(Nos-4):305-311, 1990.

[3] P. Colella. Multidimensional upwind methods for hyperbolic conservation laws. 1984. Preprint.

[4] M. G. Crandall and A. Majda. The method of fractional steps for conservation laws. *Math. Comp.*, 34:285-314, 1980.

[5] S. F. Davis. A rotationally biased upwind difference scheme for the Euler equations. *Journal of Computational Physics*, 56:65-92, 1984.

[6] R. J. LeVeque. *Numerical methods for conservation laws. Lectures in Mathematics, ETH Zürich*, Birkhäuser Verlag, Basel, 1990.

[7] R. J. LeVeque and J. B. Goodman. TVD schemes in one and two space dimensions. *Lectures Appl. Math.*, 22:51-62, 1985.

[8] D. W. Levy, K. G. Powell, and B. van Leer. An implementation of a grid-independent upwind scheme for the Euler equations. In *AIAA 9th Computional Fluid Dynamics Conference, Buffolo, NY*, June 1989.

[9] S. W. C. Noelle. On the limits of operator splitting: numerical experiments for the complex Burgers' equation. Preprint, Konrad-Zuse-Zentrum, Berlin, 1990.

[10] K. G. Powell and B. van Leer. *A genuinely Multi-Dimensional Upwind Cell-Vertex Scheme for the Euler Equations*. Technical Report Technical Memorandum 102029, NASA, 1989.

[11] P. L. Roe. Discontinuous solutions to hyperbolic systems under operator splitting. ICASE Report No. 87-64, NASA Langley Research Center, 1987.

ADAPTIVE MESH COUPLING OF EULER EQUATION AND BOUNDARY LAYER SOLUTIONS FOR TRANSONIC AIRFOILS

M. Mokry
High Speed Aerodynamics Laboratory
Institute for Aerospace Research
National Research Council Canada
Ottawa, Ontario, Canada, K1A 0R6

SUMMARY

A mesh displacement procedure for coupling the Euler equations solution and viscous boundary layer calculation for a transonic, high Reynolds number flow past an airfoil is described. Comparisons with experiment are presented for the RAE 2822 airfoil.

INTRODUCTION

The concept of transpiration velocity or equivalent sources [1], which is the most common approach to coupling of boundary layer and potential flow solutions, becomes somewhat ambiguous for the Euler equation solutions, as the total pressure and total enthalpy of the transpiring fluid need to be specified on the inflow portions of the boundary. Selection of appropriate boundary conditions for these quantities can be made on the basis of asymptotic approximations [2]-[5], but there are other weaknesses inherent in the concept. Numerical scatter of the transpiration velocity, arising from the streamwise differentiation of the displacement thickness, can be a source of inaccuracy particularly if the evaluation of the surface pressure requires the use of the normal momentum relation [6],[7]. Another disadvantage of this concept is that there is no direct 'visual' control of the displacement boundaries and that modeling of deflected wakes is not feasible.

Similar limitations are not encountered in the displacement surface coupling. The matching principle [8] implies that the boundary-layer displacement surface can be treated as a stream surface in the outer inviscid flow, provided that the gradients in the equivalent inviscid flow are negligible within the boundary layer region. In the strong-interaction regions, where this is no longer the case, the additional transpiration term can be eliminated by suitably modifying the slope of the displacement surface.

The most frequent criticism of the displacement concept is that a new mesh has to be generated after each boundary layer solution. However, since there is no matrix inversion involved in the explicit time stepping solution of the Euler equations, the mesh can conveniently be updated together with local flow quantities, as the solution advances towards the steady flow case. Coupling of the Euler and boundary layer equations using boundary-layer displaced meshes has earlier been reported in Refs.[3] and [9].

EULER CODE

The numerical solution of the Euler equations for transonic flow past an airfoil is obtained by an explicit finite volume Runge-Kutta time stepping according to Jameson et al., [10]-[11]. The finite volume scheme is cell-centered. Convergence to steady state is accelerated by the local time stepping, implicit residual averaging, enthalpy damping, and multigrid strategy.

In the interactive procedure developed here, the boundary layer calculations are invoked on the most refined mesh, after all preliminary multigrid cycles on the coarser meshes have been completed. No changes to inviscid flow boundary conditions are necessary: the solid wall boundary condition is retained over the displaced airfoil surface and the boundary cells on the upper and lower sides of the wake transfer mass, momentum and energy, as if they continued to have common faces. The sequence of operations in a typical 'sawtooth' cycle [10] with viscous correction is: Euler time step on the fine mesh, boundary layer calculation, mesh modification, recalculation of cell volumes, collection of residuals from the fine mesh into the updated coarser mesh, Euler time step on the coarser mesh, collection of residuals to the next coarser mesh, and so on. After reaching the coarsest mesh, the residual corrections are successively interpolated back from each mesh without any intermediate Euler calculation. The possibility of improving convergence by using a W-cycle [12], which postpones return to the modified fine mesh after a viscous correction has been made, is being explored.

BOUNDARY LAYER CODE

The boundary layer calculation is carried out in a direct mode and precludes, at this stage, modeling of separated flows. A two-equation integral formulation of Drela and Giles [13] is used for laminar flow (integral momentum and kinetic energy shape parameter equations), and a three-equation formulation of Green et al. [14] is used for turbulent flow (integral momentum, entrainment, and lag equations). The wake is treated as two separate boundary layers with zero skin friction [14]. An explicit Runge-Kutta code [15] with step size control, based on the formulas of Dormand and Prince, is used to integrate the corresponding systems of the first order differential equations. Work [16] is also underway to replace the integral method by the finite difference method, using the mixing length and $k - \epsilon$ models for turbulent flow.

RELAXATION

The empirical parameters, obtained here by numerical experimentation, confirm the earlier findings [17] that in transonic flow the viscous corrections to the Euler boundary input need to be implemented with the relaxation factor $\omega \leq 0.1$ to keep the interactive process convergent.

The first calculation of the displacement thickness δ^* from the computed surface pressure distribution is performed after about 30 Euler multigrid cycles, before a fully converged inviscid solution has been reached. In order to establish the background level of mesh displacement as quickly as possible, the first application of the displacement thickness is direct,

$$\delta^{(1)} = \delta^{*(1)},$$

subject to the constraint $\delta^{(1)} \leq K$, where K is the average value of $\delta^{*(1)}$ in the wake.

The subsequent boundary layer calculations are invoked no more often than every 5th cycle and not before the residual has dropped below the level reached in the initial 30 Euler cycles. The mesh displacement in the n-th viscous iteration is evaluated from its previous value and the new displacement thickness as

$$\delta^{(n)} = [1 - \omega^{(n)}]\delta^{(n-1)} + \omega^{(n)}\delta^{*(n)}, \quad n = 2, 3, \ldots$$

using a linearly decreasing relaxation factor

$$\omega^{(n)} = C(N - I)/N,$$

where I is the current multigrid cycle number, $N (= 1200)$ is the total of multigrid cycles, and $C = 0.05$. Based on the theoretical [18] and experimental [19] observations that the outer (inviscid) flow in the vicinity of the trailing edge is aligned either with the lower or upper side of the airfoil, the growth constraint adopted at the trailing edge is

$$d\delta^{(n)}/ds \leq \theta,$$

where s is the streamwise distance from the leading edge and θ is the trailing edge angle.

The slope of the wake trace is relaxed in a similar fashion, permitting $C = 0.15$ or more.

MESH ADAPTION

The points of a structured O-mesh are indexed by $i = 1,\ldots,m+1$ in the circumferential direction and by and $j = 1, 2,\ldots$ in the radial direction, as indicated in Fig.1. In the method described here two meshes are used: the inviscid-flow mesh x_0, y_0, which is aligned with the wake trace (trailing edge streamline), and the viscous-flow mesh x, y, which is both wake aligned and boundary-layer displaced. The mesh lines $i = 1$ and $i = m + 1$ in the inviscid-flow mesh are identical and represent the wake trace; in the viscous-flow mesh they differ and represent the lower and upper displacement surfaces of the wake respectively. All Euler equations calculations are performed on the viscous-flow mesh; the inviscid-flow mesh serves as relaxation reference.

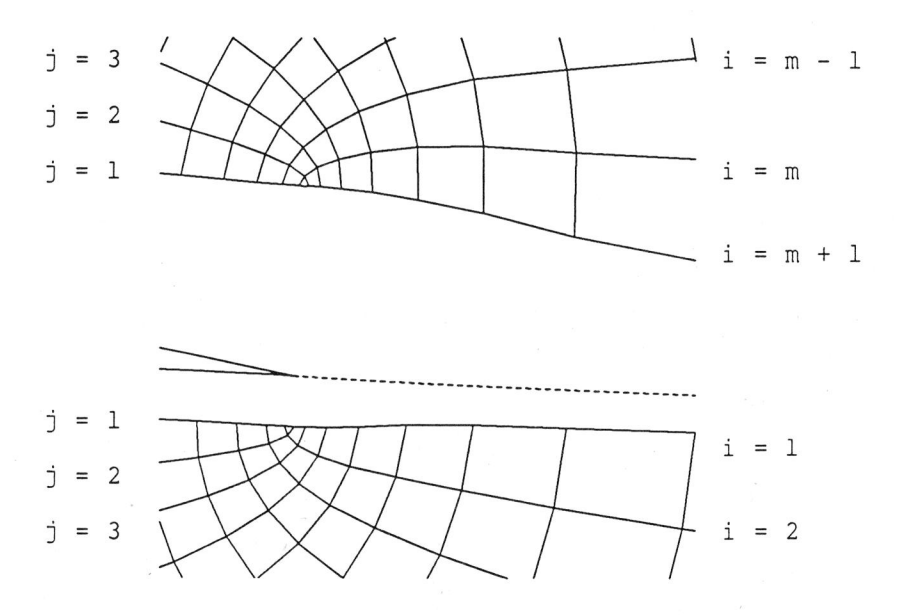

Fig.1   Trailing edge detail of viscous-flow mesh.

Flow direction, used to construct the wake trace, is obtained by averaging the velocity vectors in upper and lower wake-adjacent cells. The interactive mechanism for the reduction of the lift force due to viscosity is provided by directing the first wake trace segment along the bisector of the angle formed by the upper and lower displacement surfaces at the trailing edge (Kutta condition). The mesh points are redistributed in the circumferential direction using the local wake deflection and keeping their distances constant.

Adaption of an O-mesh for the boundary-layer and wake displacement effects is slightly more complex since it involves both the radial and circumferential displacements. An appropriate shearing transformation providing a spatially smooth superposition of such displacements is

$$x^{(n)}(i,j) = x_0^{(n)}(i,j) + \frac{(i-1)\delta_x^A(i) + (j-1)\delta_x^L(j)}{m(i+j-2)}(m-i+1)$$
$$+ \frac{(m-i+1)\delta_x^A(i) + (j-1)\delta_x^U(j)}{m(m-i+j)}(i-1),$$

where δ_x^A, δ_x^L, and δ_x^U are the x-components of the vector $\mathbf{n}\delta^{(n)}$ on the airfoil and lower and upper surfaces of the wake respectively. The unit vector $\mathbf{n}$, pointing into the flowfield, is taken normal to the wake and most of the airfoil surface. At the trailing edge, where the slope of the inviscid-flow streamline changes discontinuously, the airfoil normals are blended with wake normals to avoid overlapping of the displaced cells.

On the airfoil surface the transformation formula gives

$$x^{(n)}(i,1) = x_0^{(n)}(i,1) + \delta_x^A(i)$$

and on the lower and upper surfaces of the wake

$$x^{(n)}(1,j) = x_0^{(n)}(1,j) + \delta_x^L(j)$$
$$x^{(n)}(m+1,j) = x_0^{(n)}(m+1,j) + \delta_x^U(j)$$

as required. Analogous formulae apply to the y-displacements of the mesh.

The mesh adaption residual in the n-th viscous iteration is defined here as the RMS of $(\delta^{*(n)} - \delta^{(n)})/\delta^{*(n)}$ for the airfoil surface points where $\delta^{(n)}$ is not subjected to growth constraints.

RESULTS

A comparison is made with the experimental data [20] for the supercritical 12.1% thick RAE 2822 airfoil, tested in the RAE 2.44m×1.83m wind tunnel with 1.6% slotted sidewalls. The data, which is one of the best documented and frequently used in airfoil code validations [21], include surface pressure measurements and boundary layer and wake parameters deduced from traverses of pitot and static pressure measuring probes. The 0.61m chord airfoil, mounted vertically between the solid top and bottom walls, gave the width to chord ratio of 4 and span to chord (aspect) ratio of 3. The unseparated-flow 'Case 6' data, which are analyzed here, were obtained in a wind tunnel test at Mach number $M = 0.725$, angle of attack $\alpha = 2.92°$, Reynolds number based on chord $Rc = 6.5 \times 10^6$, and transition fixed at 3% chord length on both surfaces. The experimental values of lift, drag, and pitching moment coefficients are $C_L = 0.740$, $C_D = 0.0126$, and $C_M = -0.095$ respectively.

The computational and experimental data presented in Figs.2-7 are to a large extent self-explanatory so that only a few comments will be made here.

The inner 18 layers of the fully adapted 160×32 viscous-flow mesh, obtained after 1200 multigrid cycles with 180 viscous corrections, are shown in Fig.2. A trailing edge detail of the same mesh is shown in Fig.3.

The computed isomach field is plotted in Fig.4, where the sonic line and wake trace are indicated by broken lines. Because of alignment of the mesh with the wake, the discontinuity of the isomachs across the wake is very 'clean', i.e. there are no visible secondary effects caused by artificial dissipation in the Euler code.

Comparison of computed (lines) and experimental (symbols) surface pressure distributions is shown in Fig.5. The value $M = 0.732$, used in the computation, was found to provide a better overall agreement with experimental pressure distributions than $M = 0.729$, corrected for wall interference ($\Delta M = 0.004$). The free-air value $\alpha = 2.48°$ was found indirectly, by matching the computed C_L with the experimental one; the corresponding $C_D = 0.0116$ and $C_M = -0.0867$ underestimate the experimental values.

The failure to better predict the experimentally observed compression behind the leading edge suction peak is also noticeable on plotted results for the same case in Refs.[4],[9],[17],[21]. The observed discrepancy is most likely due to a three-dimensional contamination of the experiment by a system of lateral Mach waves originating in the boundary layers at the junction of the airfoil and the solid walls [22], an effect that cannot be accounted for by a two-dimensional computation.

The reason for overprediction of the trailing edge pressure is not clear: several modification of trailing edge conditions have been tried with little difference. A part of the problem may be a streamwise pressure gradient in the wind tunnel, but no data is available to substantiate this claim.

Comparison of the computed boundary layer parameters (lines) with the sparse experimental points (symbols) in Fig.6 confirms that the integral boundary layer method works well for unseparated flows [4].

Figure 7 shows the convergence history for the most refined mesh level, using the nomenclature of Ref.[10]. The solid curve is the RMS residual in density. The spikes, triggered by viscous corrections, are seen to decay in the logarithmic scale. No viscous corrections were appled in the last 60 cycles, allowing the density residual to reach machine zero.

The broken curve is the number of supersonic cells, normalized by their count in the final, 1200th cycle. This practical measure of convergence for transonic flows indicates that for technical applications a reasonably converged interactive solution is obtained in 600-800 cycles.

The mesh adaption residual (not plotted) was found to decrease monotonically from 38% to 0.5% between the second and last viscous iteration. The computation required about 30 CPU minutes on a RS/6000-320 workstation.

CONCLUSIONS

The procedure described here combines the efficiency of the finite volume solution of the Euler equations, integral boundary layer methods, and simple mesh transformations. For engineering predictions, a sufficiently converged solution is obtained in 600-800 multigrid cycles with approximately 100 boundary layer updates.

ACKNOWLEDGEMENTS

This work is based on Jameson's code FLO52. The computations were performed on a RS/6000-320 workstation, available to the Institute for Aerospace Research through a NRC-IBM CFD Joint Study Agreement.

REFERENCES

[1] Lighthill, M.J., "On Displacement Thickness," Journal of Fluid Mechanics, Vol.4, 1958, pp.383-392.

[2] Johnson, W. and Sockol, P., " Matching Procedure for Viscous-Inviscid Interactive Calculations," AIAA Journal, Vol.17, June 1979, pp.661-663.

[3] Whitfield, D.L., Thomas, J.L., Jameson, A., and Schmidt, W., "Computation of Transonic Viscous-Inviscid Interacting Flow," *Numerical and Physical Aspects of Aerodynamic Flows II*, ed. Cebeci, T., Springer-Verlag, 1984, pp.285-295.

[4] Whitfield, D.L. and Thomas, J.L., "Transonic Viscous-Inviscid Interaction Using Euler and Inverse Boundary-Layer Equations," *Computational Methods in Viscous Flows*, ed. Habashi, W.G., Pineridge Press, 1984, pp.451-474.

[5] Le Balleur, J.C., "Numerical Flow Calculation and Viscous-Inviscid Interaction Techniques," *Computational Methods in Viscous Flows*, ed. Habashi, W.G., Pineridge Press, 1984, pp.419-450.

[6] Rizzi, A., "Numerical Implementation of Solid Body Boundary Conditions for the Euler Equations," ZAMM, Vol. 58, 1978, pp.301-304.

[7] Thomas, J.L., "Transonic Viscous-Inviscid Interaction Using Euler and Inverse Boundary-Layer Equations," Mississippi State University, Ph.D. Thesis, 1983.

[8] Murman, E.M and Bussing, R.A., "On the Coupling of Boundary-Layer and Euler Equation Solutions," *Numerical and Physical Aspects of Aerodynamic Flows II*, ed. Cebeci, T., Springer-Verlag, 1984, pp.313-325.

[9] Drela, M., "Two-Dimensional Transonic Aerodynamic Design and Analysis Using the Euler Equations," Massachusetts Institute of Technology, Gas Turbine Laboratory Rept. 187, Feb.1986.

[10] Jameson, A., "Solution of the Euler Equations for Two Dimensional Transonic Flow by a Multigrid Method," Applied Mathematics and Computation, Vol.13, 1983, pp.327-355.

[11] Jameson, A. and Schmidt, W., "Some Recent Developments in Numerical Methods for Transonic Flows," Computer Methods in Applied Mechanics and Engineering No.51, 1985, pp.467-493.

[12] Jameson, A., "Computational Transonics," Communications on Pure and Applied Mathematics, Vol.16, 1988, pp.507-549.

[13] Drela, M. and Giles, M.B., "Viscous-Inviscid Analysis of Transonic and Low Reynolds Number Airfoils," AIAA Journal, Vol.25, Oct.1987, pp.1347-1335.

[14] Green, J.E., Weeks, D.J., and Brooman, J.W.F., "Prediction of Turbulent Boundary Layers and Wakes in Compressible Flow by a Lag-Entrainment Method", R.& M. No.3791, 1977.

[15] Hairer, E., Nørsett, S.P., and Wanner, G., *Solving Ordinary Differential Equations I, Nonstiff Problems*, Springer-Verlag, 1987, pp.433-435.

[16] Khalid, M., "Turbulent Boundary Layer Solution for Two-Dimensional Compressible Flow Using Mixing Length and $K - \epsilon$ Turbulence Models," IAR-AN-69, National Research Council Canada, Nov.1990.

[17] Conway, J.T., "Modelling of Boundary Layer and Trailing Edge Thickness Effects For the Euler Equations Using Using Surface Transpiration," The CASI First Canadian Symposium on Aerodynamics, Ottawa, Dec.1989, pp.14.1-14.11.

[18] Schmidt, W., Jameson, A., and Whitfield, D., "Finite-Volume Solutions to the Euler Equations in Transonic Flow," Journal of Aircraft, Vol.20, Feb.1983, pp.127-133.

[19] Poling, D.R. and Telionis, D.P., "The Traling Edge of a Pitching Airfoil at High Reduced Frequencies," Transactions of the ASME, Vol.109, Dec.1987, pp.410-414.

[20] Cook, P.H., McDonald, M.A., and Firmin, M.C.P., "Aerofoil RAE 2822 - Pressure Distributions, and Boundary Layer and Wake Measurements," *Experimental Data Base for Computer Program Assessment*, AGARD-AR-138, May 1979, pp. A6.1 - A6.76.

[21] Jones, D.J., "The CAARC Cooperative Project on on Aerofoil Code Validation," CC.No.AE.1000, Commonwealth Advisory Aeronautical Research Council, Aug.1990.

[22] "Two-Dimensional Transonic Testing Methods," NLR TR 83086 U, July 1981, pp.64-80.

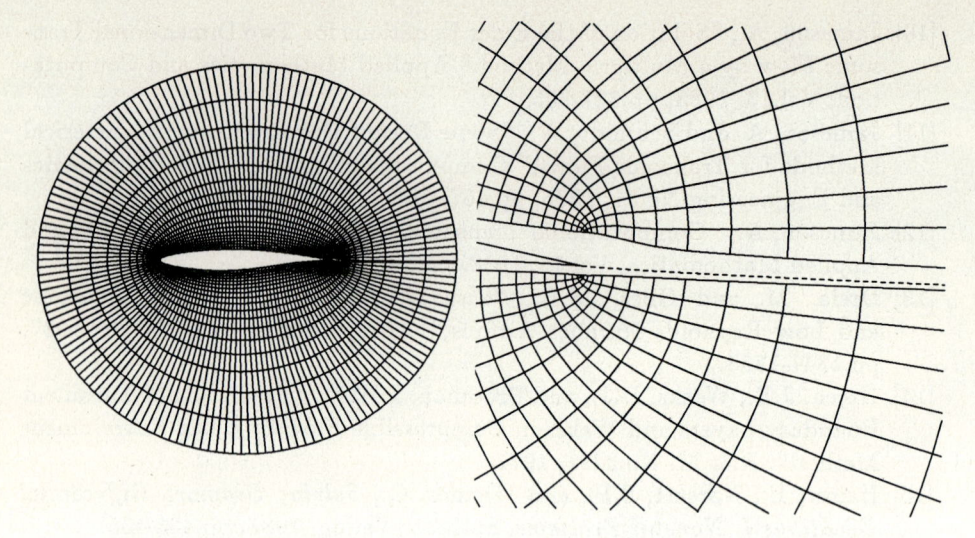

Fig.2 Inner part of grid for RAE 2822
Mach 0.732   α 2.48°   Rc 6.500*10⁶
160x32 grid   1200 cycles

Fig.3 Trailing edge detail of grid for RAE 2822
Mach 0.732   α 2.48°   Rc 6.500*10⁶
160x32 grid   1200 cycles

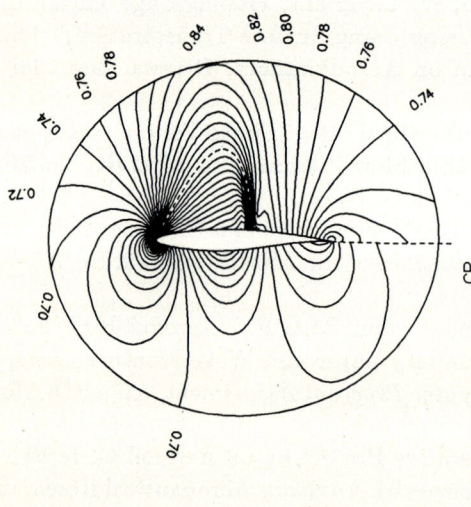

Fig.4 Isomachs for RAE 2822
Mach 0.732   α 2.48°   Rc 6.500*10⁶
CL 0.7413   CD 0.0116   CM −0.0867
160x32 grid   1200 cycles

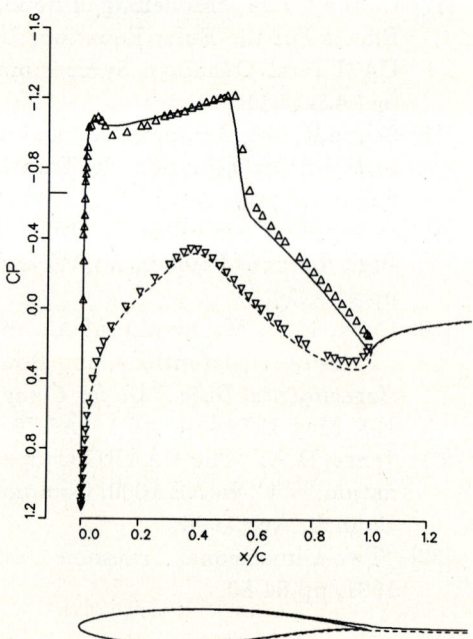

Fig.5 Pressure distribution for RAE 2822
Mach 0.732   α 2.48°   Rc 6.500*10⁶
CL 0.7413   CD 0.0116   CM −0.0867
160x32 grid   1200 cycles

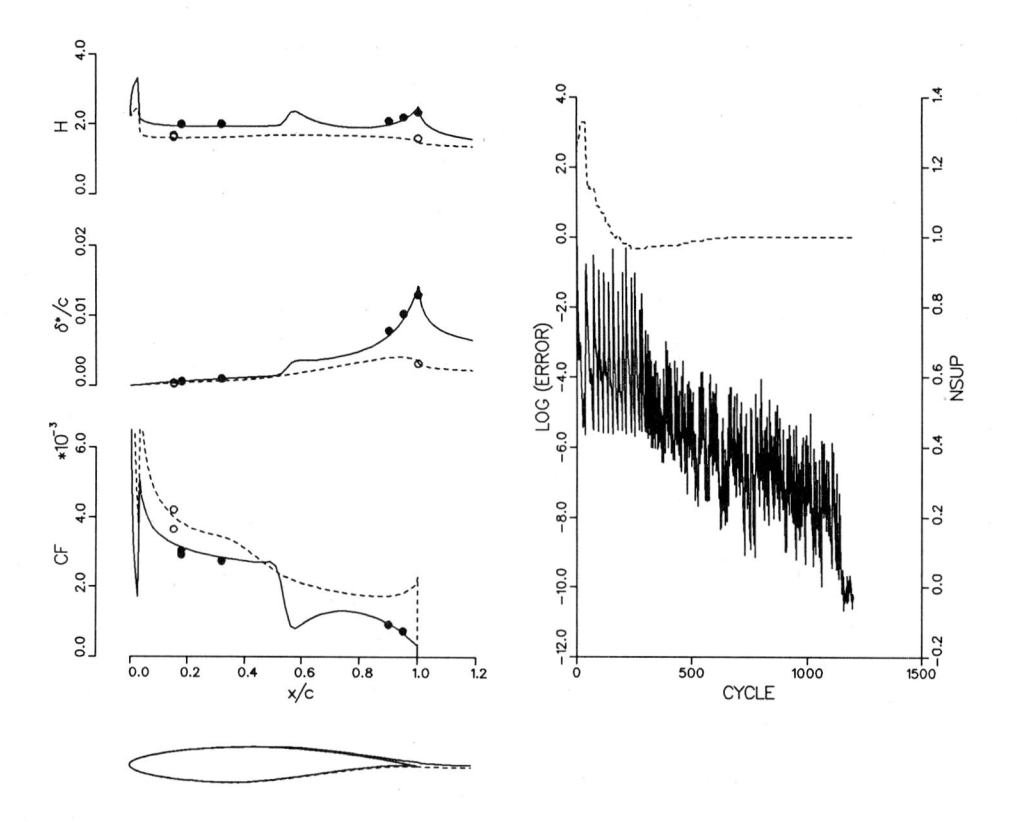

Fig.6 Boundary layer parameters for RAE 2822
Mach 0.732   α 2.48°   Rc 6.500∗10⁶
CL 0.7413   CD 0.0116   CM −0.0867
160x32 grid   1141 cycles

Fig.7 Convergence history for RAE 2822
Mach 0.732   α 2.48°   Rc 6.500∗10⁶
160x32 grid   1200 cycles
Residual 0.073x10⁰ →   0.024x10⁻⁴

An Attempt to the Surface Grid Generation Based on Unstructured Grid

Masahiro Suzuki
Institute of Computational Fluid Dynamics
1-16-5 Haramachi,Meguro-ku,Tokyo 152,JAPAN

Summary

The surface grid generation method, which can utilized the surface definition of Computer Aided Design (CAD) system directly, is presented. First, the triangulation is performed on the each CAD defined patches. Then linear partial differential equations are solved on the linear triangle elements. Finally, the surface grids are constructed by searching for the contours inside the solution domain. The coordinate values of grid points are obtained by converting the parametric coordinate into the Cartesian coordinate through the parametric forms of the CAD defined patches.

Introduction

There have been noticeable advances in supercomputer technology and concomitant progress in Computational Fluid Dynamics (CFD) in recent years. An increasing number of numerical investigations have been conducted to tackle complex, three-dimensional flow configurations. However, one formidable stumbling block in these endeavors has been associated with grid generation. It has been well recognized that setting grids on the surface is the most time-consuming task in grid generation process and it is a bottle-neck of the CFD analysis. This is attributed to the gap between geometric generation and surface grid generation.

The geometric data for complex configurations, which are often utilized in engineering applications, are customarily produced by use of a CAD system. The representation of three-dimensional geometries in CAD system uses concepts from two popular forms of solid representations: (1) Constructive Solid Geometry (CSG) and (2) Boundary representation (B-rep). CSG represents the geometry by combining (via unions, differences and intersections) many copies of a few basic primitives solids (blocks, cylinders, cones and spheres). In B-rep systems, the solids are described by a set of patches. Each patches can be defined by each parametric forms and, thus, arbitrary curved surfaces can be designed. Aerodynamical designs, e.g., airplanes and automobiles, are usually design by B-rep environment. The transformation from CSG form to B-rep form is possible, but in general the inverse is not. Taking account of the above situations, surface grid generation on the set of patches, which is designed by B-rep environment, is considered as a basic research here. It is not purpose of this study to link a specific CAD system to the surface grid generation.

At this time, the existing techniques generally treat generating the surface grid as a two-dimensional boundary value problem on a curved surface, which is specified by a quadrangular patch [1]. The generation of the surface grid is accomplished over several stages. First, the Cartesian coordinate values $(x_{i,j}, y_{i,j}, z_{i,j})$ of the boundary points on the four edges of the surface grid are specified, converting these values to the surface parametric coordinate values $(u_{i,j}, v_{i,j})$ on the edges. Then, the interior values in the array $u_{i,j}$ and $v_{i,j}$ from the edge values are determined by interpolation or partial differential equations' solution. Finally, these parametric coordinate values are converted to the Cartesian coordinates $(x_{i,j}, y_{i,j}, z_{i,j})$. This procedure requires a well-structured surface, i.e., a quadrangular patch, for the input data. This surface usually coincides with none of the original patches, which are made within the CAD system. Therefore, some data processors, which use some forms of interpolations, are required to produce a new quadrangular patch for the input data from the original surface. This is a cumbersome task and, in addition, there are no guarantees that the new patch fits with the original surface. Though it is possible to conform the produced surface grid to the original surface by projections, it requires iterative calculations for every surface grid points and thus it costs much. In view of the stringent requirement for the input data, the conventional surface grid generation is inefficient and laborious.

It would be highly advantageous if a way can be found in which the surface definition of the CAD system could be directly utilized for the surface grid generation. To achieve that, the requirement of conventional surface grid generators for the input data must be taken away. From this point of view, investigations have been pursued to devise a scheme, which is dealt with plural polygonal surface patches (referred to as the unstructured grid) by using Finite Element Method (FEM) [2]. In [2], it is possible to treat several kinds of surface patches, e.g., triangle and quadrangular, it is, however, noted that the triangle element is most flexible to treat complex geometries. In addition, though the accuracy of triangle element is, by and large, less than that of quadrangle's in FEM, the grid generator does not require high accuracy. Therefore, triangulation by the advancing front method [3] is adopted as a unstructured grid generator in this paper. Introducing the concept of surface parametric coordinate into the method of [2] makes the linkage of geometric generation and surface grid generation feasible. The proposed method is consist of three procedures. First, triangulations are performed within whole/part of each CAD defined patches, on which a surface grid is wanted to generate. Subsequently, the partial differential equations are mapped on these triangular patches, and numerical solutions are acquired by the finite element procedure. Finally, the surface grid is constructed by searching for the contours inside the solution domain. Each step is conducted through converting values between the parametric coordinate and Cartesian coordinate.

The method

2.1 Unstructured grid generation

The triangulation is performed as the following steps:

a. The Cartesian coordinate values of the boundary points on the four edges of the surface grid are specified. The number and location of these points are not necessarily coincident with that of the surface grid points. (Figure 1.a)

b. Points are distributed on the boundaries of the original patches (the CAD defined patches), which are contained within the surface grid. The coordinate values are specified in the Cartesian coordinate. (Figure 1.b)

c. On each original patches, these Cartesian coordinate values are converted into parametric coordinate values. (Figure 1.c)

d. The interior nodes are distributed within part of each patches, which is bounded by the edge boundary of the surface grid and boundaries of the original patches, in the parametric plane. (Figure 1.d) The distributing points is done in an automatic manner. First, the points are set equidistantly within the whole area of the original patch. Then points outside of the surface grid are excluded by the shortest distance searching method [3].

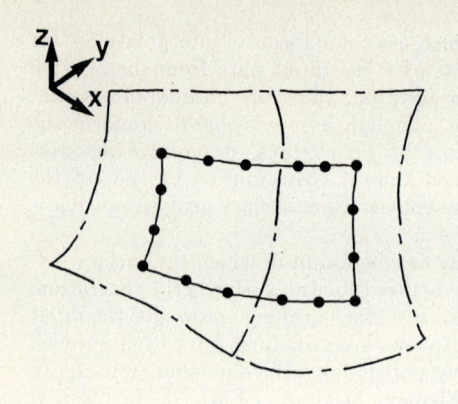

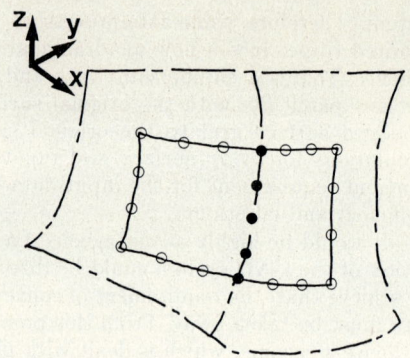

(a) Boundary points on the four edges of the surface grid are specified.

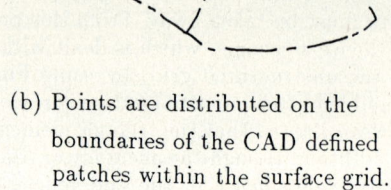

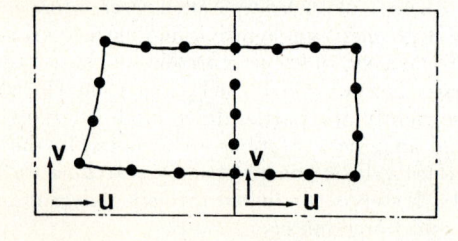

(b) Points are distributed on the boundaries of the CAD defined patches within the surface grid.

(c) The Cartesian coordinate values are converted into the parametric values.

(d) Interior nodes are set.

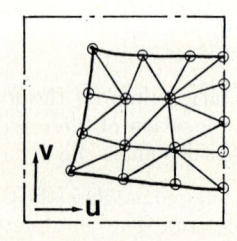

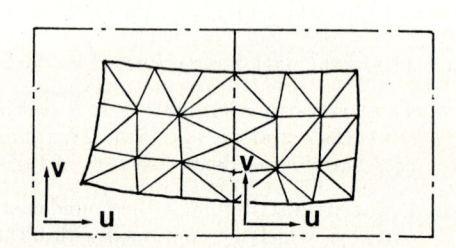

(e) The triangulation is performed by the advancing front method.

(f) Each triangulations are combined.

Figure 1. Construction of the unstructured grid. Dashed lines show the CAD defined patches.

398

e. The triangulation is performed on each patches by two-dimensional version of the advancing front method [4] in the parametric plane. (Figure 1.e)

f. Each triangulations are combined into a global network. (Figure 1.f) Each node number is renumbered in the global network to perform the finite element procedure. To search for the contour lines quickly and to transform the parametric coordinate values into the Cartesian coordinate values, the following data structure is employed: Each elements has the informations of the element number, the original patch number in which the element is contained, node numbers, the parametric coordinate values of each nodes and neighboring element numbers. To start searching for the contour lines, the numbers of the element, which are located on the $\eta = \eta_o$ edge boundary of the surface grid, are also stored.

(a) The parametric coordinate (u, v)

(b) The Cartesian coordinate (x, y, z)

(c) The two-dimensional plane (X, Y)

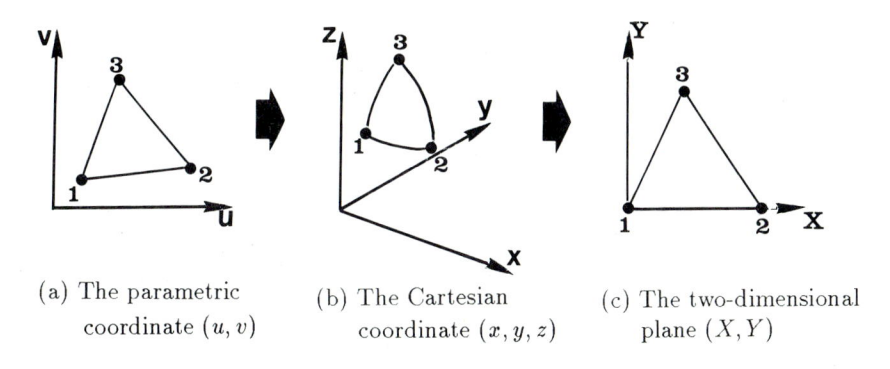

Figure 2. Mapping of the triangular element

2.2 Finite element solution procedure

The partial differential equations shown below are mapped on the unstructured grid and solved by FEM [2]:

$$\nabla(\lambda_1 \nabla \xi) = 0, \qquad \nabla(\lambda_2 \nabla \eta) = 0. \tag{1}$$

where λ is weight function. The Galerkin's weighted residual approach is employed. The linear interpolation function of three-node triangular elements is adopted. The coordinate values of nodes are converted from the parametric coordinate (u, v), which is set in §2.1, into the Cartesian coordinate (x, y, z) by using the parametric form of each original patches. Then these values are transformed onto a two-dimensional plane (X, Y). (Figure 2.) The integrations are done by using these values. In contrast to the well-known elliptic grid generation equations solved in the computational domain, these equations are solved in the physical domain. This avoids the need to solve non-linear equations. The Incomplete Cholesky decomposition Conjugate Gradient (ICCG) method is used to solve the linear system of equations. In conjugate with solving the linear equations, the ICCG method drastically reduces the computational times in comparison with the common direct method. The solution contours of each equation define the curvilinear coordinate lines of ξ and η, respectively. Therefore, the boundary conditions specify the topology of the surface grid. The surface grid itself forms the boundary in the three-dimensional flow field; therefore, it is imperative to achieve the attraction of coordinate lines to the interior fixed areas rather than on the edge boundaries. The attraction areas, e.g., big curved segments of the surface, are specified in the physical domain in the case of the initial grid generation, not in the case of the flow solution adaptation. Since the equations are solved in the physical domain, the control of the grid space can be directly implemented by use of the weight functions λ. As λ increases, the gradient of the solution decreases, i.e., the grid spacing becomes large. In practice, the weight function can be related to the curvature of surface or the distance from the boundaries.

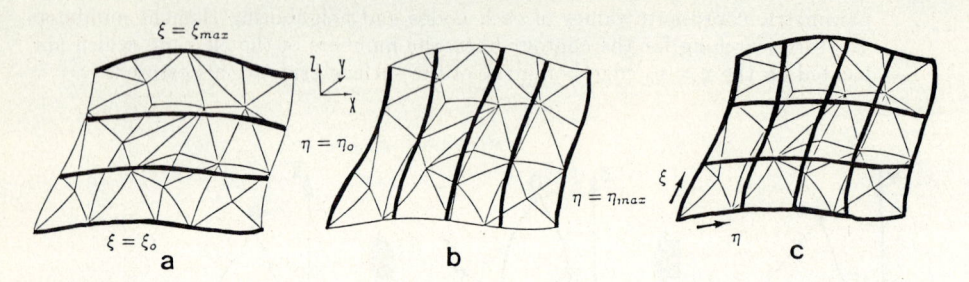

Figure 3. Construction of the surface grid. (a) Construction of $\xi - constant$ lines. (b) Construction of $\eta - constant$ lines. (c) The surface grid. (c) is obtained by combining (a) and (b).

2.3 Surface grid construction

In order to obtain the surface grid, the curvilinear coordinate lines are constructed by searching for the contours inside the solution domain (Figure 3). It does not take much computational time to search for the contours; the network information of the unstructured grid and a hierarchical data structure are executed in advance (§2.1(f)). The searching for the contours is proceeded as follows:

a. Find an element, in which a line of $\xi = \xi_i$ crosses, among the elements, which are located on the $\eta = \eta_o$ edge boundary and are listed up in advance (§2.1(f)).

b. Check whether a line of $\eta = \eta_j$ crosses the element. If not, go to (g).

c. Obtain the coordinate value of a intersection of $\xi = \xi_i$ and $\eta = \eta_j$. As the linear interpolation function of three-node triangular elements is adopted, the values of ξ and η within the element are as follows:

$$\xi = \alpha_1 + \alpha_2 u + \alpha_3 v, \qquad \eta = \beta_1 + \beta_2 u + \beta_3 v. \tag{2}$$

where, α and β are constant and specified by

$$\begin{pmatrix} \alpha_1 \\ \alpha_2 \\ \alpha_3 \end{pmatrix} = \begin{pmatrix} 1 & u_1 & v_1 \\ 1 & u_2 & v_2 \\ 1 & u_3 & v_3 \end{pmatrix}^{-1} \begin{pmatrix} \xi_1 \\ \xi_2 \\ \xi_3 \end{pmatrix}, \tag{3}$$

$$\begin{pmatrix} \beta_1 \\ \beta_2 \\ \beta_3 \end{pmatrix} = \begin{pmatrix} 1 & u_1 & v_1 \\ 1 & u_2 & v_2 \\ 1 & u_3 & v_3 \end{pmatrix}^{-1} \begin{pmatrix} \eta_1 \\ \eta_2 \\ \eta_3 \end{pmatrix}. \tag{4}$$

Here, subscripts of u, v, ξ and η indicate the node number of an element. Therefore, the coordinate value of a intersection of $\xi = \xi_i$ and $\eta = \eta_j$ are obtained by solving the linear system of equations (2).

$$\begin{pmatrix} u_{i,j} \\ v_{i,j} \end{pmatrix} = \begin{pmatrix} \alpha_2 & \alpha_3 \\ \beta_2 & \beta_3 \end{pmatrix}^{-1} \begin{pmatrix} \xi_i - \alpha_1 \\ \eta_j - \beta_1 \end{pmatrix}. \tag{5}$$

d. Check whether the point $(u_{i,j}, v_{i,j})$ is located inside the element. If not, go to (g).

e. Converting $(u_{i,j}, v_{i,j})$ into the Cartesian coordinate values $(x_{i,j}, y_{i,j}, z_{i,j})$ though using the parametric form of the original patch, which contains the element. Then register the coordinate values as that of the surface grid points. This procedure guarantees the surface grid points always fit with the original surface.

f. $j = j + 1$. Then go to (b).

g. Find the next element, in which the line of $\xi = \xi_i$ crosses, through utilizing the information of the neighboring element number (§2.1(f)). Then go to (b).

The above procedures are performed recursively for searching all i, j points.

Examples

Though the presented method can deal with any kinds of parametric surfaces, the bilinearly blended Coons patches are treated here as an example. The surface grid generation on two bilinearly blended Coons patches are presented in Figure 4. The parametric form of the Coons patch is expressed by,

$$\mathbf{x}(u, v) = \begin{pmatrix} 1 - u & u \end{pmatrix} \begin{pmatrix} \mathbf{x}(0, v) \\ \mathbf{x}(1, v) \end{pmatrix} + \begin{pmatrix} \mathbf{x}(u, 0) & \mathbf{x}(u, 1) \end{pmatrix} \begin{pmatrix} 1 - v \\ v \end{pmatrix}$$
$$- \begin{pmatrix} 1 - u & u \end{pmatrix} \begin{pmatrix} \mathbf{x}(0, 0) & \mathbf{x}(0, 1) \\ \mathbf{x}(1, 0) & \mathbf{x}(1, 1) \end{pmatrix} \begin{pmatrix} 1 - v \\ v \end{pmatrix}. \tag{6}$$

The following boundary curves are adopted here.

For Surface No.1:

$$\mathbf{x}(u, 0) = \begin{pmatrix} x(u, 0) \\ y(u, 0) \\ z(u, 0) \end{pmatrix} = \begin{pmatrix} u - 1 \\ -1 \\ (\cos -1 - 1)(1 - u) \end{pmatrix},$$

$$\mathbf{x}(u, 1) = \begin{pmatrix} x(u, 1) \\ y(u, 1) \\ z(u, 1) \end{pmatrix} = \begin{pmatrix} u - 1 \\ 1 \\ (\cos 1 - 1)(1 - u) \end{pmatrix},$$

$$\mathbf{x}(0, v) = \begin{pmatrix} x(0, v) \\ y(0, v) \\ z(0, v) \end{pmatrix} = \begin{pmatrix} -1 \\ 2v - 1 \\ \cos(2v - 1) - 1 \end{pmatrix},$$

$$\mathbf{x}(1, v) = \begin{pmatrix} x(1, v) \\ y(1, v) \\ z(1, v) \end{pmatrix} = \begin{pmatrix} 0 \\ 2v - 1 \\ 0 \end{pmatrix}.$$

$$(7)$$

where, $0 \leq u \leq 1$, $0 \leq v \leq 1$.

For Surface No.2:

$$\mathbf{x}(u,0) = \begin{pmatrix} x(u,0) \\ y(u,0) \\ z(u,0) \end{pmatrix} = \begin{pmatrix} u-1 \\ -1 \\ 0 \end{pmatrix},$$

$$\mathbf{x}(u,1) = \begin{pmatrix} x(u,1) \\ y(u,1) \\ z(u,1) \end{pmatrix} = \begin{pmatrix} u-1 \\ 1 \\ \sin 1(1-u) \end{pmatrix},$$

$$\mathbf{x}(0,v) = \begin{pmatrix} x(0,v) \\ y(0,v) \\ z(0,v) \end{pmatrix} = \begin{pmatrix} 0 \\ 2v-1 \\ 0 \end{pmatrix},$$

$$\mathbf{x}(1,v) = \begin{pmatrix} x(1,v) \\ y(1,v) \\ z(1,v) \end{pmatrix} = \begin{pmatrix} 1 \\ 2v-1 \\ \sin v \end{pmatrix}.$$

(8)

where, $0 \leq u \leq 1$, $0 \leq v \leq 1$.

The surface grid is generated on an area on both Coons patches. First, a part of both Coons patches are triangulated in Figure 4.a,b. The equations (1) are solved on the unstructured grid and then surface grid are constructed (Figure 4.c). The points of the surface grid are precisely fit with the Coons patches (Figure 4.b). Once the points on the edge boundary lines of the surface grid are specified, the whole computations are done within a few seconds on the IRIS 4D/220GTX workstation.

Concluding remarks

The method, which link the geometric generator and the surface grid generation, is proposed. The surface definition of the geometric generator is directly utilized. There are no needs of converting the surface definition of the geometric generator and it results in not only eliminating the error of the geometry representation but also reducing the human labor. Adopted linear equations and the network information of the unstructured grid helps to cut down the CPU time of generating the surface grid. The presented method may be suitable for the interactive environment.

References

[1] Thompson, J. F., "Some Current Trends in Numerical Grid Generation," *Numerical Methods for Fluid Dynamics III*, K. W. Morton and M. J. Baines (eds.), Clarendon Press, Oxford, (1988).
[2] Suzuki, M., "Surface Grid Generation Based on Unstructured Grid," to appear in AIAA Journal.
[3] Nagashima, S., "Drawing for CG", Pixcel, 71, 173-177, (1988).
[4] Löhner, R. and Parikh, P., "Generation of Three-dimensional Unstructured Grids by the Advancing Front Method", AIAA-88-0515.

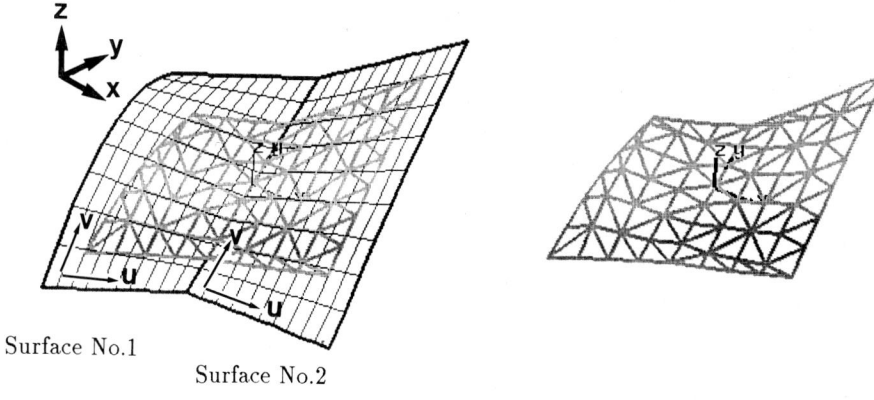

(a) The unstructured grid on the Coons patches.

(b) The unstructured grid.

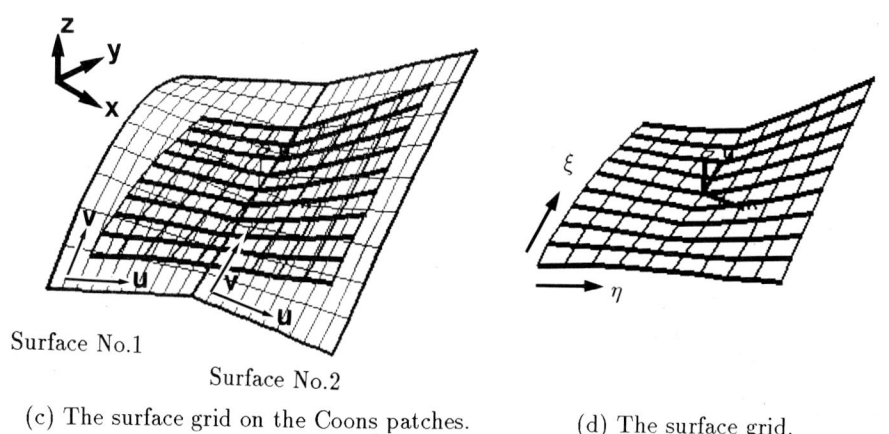

(c) The surface grid on the Coons patches.

(d) The surface grid.

Figure 4. The surface grid generation on the bilinearly blended Coons patches.

403

TURBULENT COMPRESSIBLE FLOWS

LAMINAR AND TURBULENT VISCOUS COMPRESSIBLE FLOWS USING IMPROVED FLUX VECTOR SPLITTINGS

D. Drikakis and S. Tsangaris

National Technical University of Athens
Lab. Aerodynamics, P.O Box 640 70 ,157 10, Athens, Greece

SUMMARY

Improvements of Flux Vector Splitting (FVS) methods for the Navier-Stokes equations are presented. FVS methods are developed in combination with the well known MUSCL upwind scheme as well as with a hybrid upwind scheme. The hybrid upwind can be used with third or fourth order of accuracy. Flow test cases show that improvement of a modified Steger-Warming FVS as well as of the van Leer FVS can be obtained in viscous flow computations using the combination of the FVS methods with the third order hybrid upwind. The Navier-Stokes equations are solved by an implicit unfactored method using Newton iterations and Gauss-Seidel relaxation. Results are presented both for laminar and turbulent flows.

INTRODUCTION

The last years, inaccuracies in viscous flow computations using Flux Vector Splitting (FVS) methods have been observed [1],[2],[3]. FVS methods present accurate results for inviscid flows but cause the adverse effect in viscous flow calculations. MacCormack et al. [1] has reported that the inaccuracies of Steger-Warming FVS [4] are caused by the large numerical mixing of the fluid in the boundary layer and the development of a fictitious pressure gradient. Hänel et al. [2] has observed inaccuracies in the boundary layers using van Leer's FVS [5] in combination with the MUSCL upwind scheme, while van Leer et al. [3] has found that Roe's and Osher's flux formulas are more accurate than van Leer's FVS method [4].
The inaccuracies of the FVS methods due to the construction of the flux functions which ignore the linear waves and badly diffuse the boundary layers. Inspite of these inaccuracies attempts to improve the FVS methods are continued because the flux formulas are relatively easy to linearize in implicit time marching schemes as well as are simpler in the extension of real gas effects. In the present paper modifications of the FVS methods for viscous flows are examined. The van Leer FVS method [3] as well as a modified Steger-Warming FVS [6] are developed in combination with a hybrid upwind extrapolation scheme. FVS methods have also used with the well known MUSCL upwind scheme [7]. Comparative study between the upwind schemes in laminar and turbulent flows shows that the third order hybrid upwind improves the FVS methods in viscous flows.The unfactored Navier-Stokes equations are solved by an implicit first order accurate in time scheme, using Gauss-Seidel relaxation technique.

The changes in the implicit part of the equations are minimum using the present improved FVS methods. The present extensions do not affect the stability and the convergence of the Navier-Stokes solver.

GOVERNING EQUATIONS

The governing equations are the time dependent Navier-Stokes equations for a compressible fluid. These equations can be written in conservation dimensionless form and for a generalized coordinate system as:

$$JU_t + (E_{inv})_\xi + (G_{inv})_\zeta = \frac{1}{Re}[(E_{vis})_\xi + (G_{vis})_\zeta] \tag{1}$$

where R_e is the Reynolds number and $U = (\rho, \rho u, \rho w, e)^T$ is the conservative solution unknown vector. E_{inv}, G_{inv} are the inviscid flux vectors while E_{vis}, G_{vis} are the viscous flux vectors. $J = x_\xi z_\zeta - z_\xi x_\zeta$ is the Jacobian of the transformation $\xi = \xi(x,z)$ and $\zeta = \zeta(x,z)$ from Cartesian coordinates x,z to generalized coordinates ξ, ζ.

FVS METHODS

1. Modified Steger-Warming Flux Vector Splitting (MSW-FVS)

The Steger-Warming flux vector splitting decomposes the inviscid flux into two parts, positive and negative, in accordance with the sign of the eigenvalues. In the original Steger-Warming FVS [4] the positive E^+ and the negative E^- flux are calculated on the nodes $i, i+1$ respectively. MacCormack et al. [1] has proved that this formulation produces large errors in viscous flow calculations while improvement in the boundary layer results is obtained defining the Jacobians and the eigenvector matrices on the cell face $i+1/2$:

$$E_{i+\frac{1}{2}} = (T\Lambda^+ T^{-1})_{i+1/2}(U^-_{i+1/2}) + (T\Lambda^- T^{-1})_{i+1/2}(U^+_{i+1/2}) \tag{2}$$

where T, T^{-1} are the left and right eigenvector matrices respectively while Λ^+, Λ^- are the diagonal positive and negative eigenvalue matrices. Better results in the conservation of the total temperature, in inviscid hypersonic flows are obtained using the above formulation (eq.2), [6]. The above formulation is considered in the present viscous flow calculations. U^-, U^+ are the conservative variables of the left and right states on the cell face $i+1/2$.

Because at vanishing zeroth eigenvalue mass flux is not differentiable and flux vector splitting can not be applied, the fluxes are modified as a continuous function of the first and the second eigenvalue $\hat{\lambda}_1, \hat{\lambda}_2$ by splitting the zeroth eigenvalue as:

$$\hat{\lambda}_0^\pm = \frac{\hat{\lambda}_1^\pm + \hat{\lambda}_2^\pm}{2}. \tag{3}$$

Negative influence of the FVS method on the conservation of the total temperature, emerges by the formulation of the energy flux. Improvement of the total temperature conservation is obtained by the discretization of the energy flux in terms of the total enthalpy . The same conclusion has also been presented by other authors [2] for the van Leer's FVS method. Finally the convective flux can be defined as :

$$E^*_{i+\frac{1}{2}}(U^*) = J\rho|\nabla\xi| \begin{pmatrix} \frac{1}{2}(\hat{\lambda}^+_1 + \hat{\lambda}^+_2) \\ \left(u + \frac{s\tilde{\xi}_x}{\gamma}\right)\frac{\hat{\lambda}^+_1}{2} + \left(u - \frac{s\tilde{\xi}_x}{\gamma}\right)\frac{\hat{\lambda}^+_2}{2} \\ \left(w + \frac{s\tilde{\xi}_z}{\gamma}\right)\frac{\hat{\lambda}^+_1}{2} + \left(w - \frac{s\tilde{\xi}_z}{\gamma}\right)\frac{\hat{\lambda}^+_2}{2} \\ \frac{1}{2}H^*(\hat{\lambda}^+_1 + \hat{\lambda}^+_2) \end{pmatrix} \quad (4)$$

where s is the speed of sound and H the total enthalpy. The terms $\lambda_{0,1,2}$ are defined as:

$$\hat{\lambda}_0 = u\tilde{\xi}_x + w\tilde{\xi}_z \quad , \quad \hat{\lambda}_1 = \hat{\lambda}_0 + s \quad , \quad \hat{\lambda}_2 = \hat{\lambda}_0 - s \quad (5)$$

$$\tilde{\xi}_x = \frac{\xi_x}{\sqrt{\xi_x^2 + \xi_z^2}} \quad , \quad \tilde{\xi}_z = \frac{\xi_z}{\sqrt{\xi_x^2 + \xi_z^2}} \; .$$

2. Van Leer Flux Vector Splitting method (VL-FVS)

Van Leer's [5] FVS method constructs the fluxes as a function of the local Mach number. The inviscid flux E is splitted according to the contravariant Mach number M_ξ in ξ-direction. For supersonic flows:

$$E^+ = E, E^- = 0 \quad for \quad M_\xi \geq 1 \quad and \quad E^- = E, E^+ = 0 \quad for \quad M_\xi \leq -1 \quad (6)$$

For subsonic flows ($|M_\xi| \leq 1$):

$$E^*_{i+\frac{1}{2}} = \begin{pmatrix} f^*_{mass} \\ f^*_{mass}\left(u - \lambda_0\frac{\tilde{\xi}_x}{\gamma} \pm 2s\frac{\tilde{\xi}_x}{\gamma}\right) \\ f^*_{mass}\left(w - \lambda_0\frac{\tilde{\xi}_z}{\gamma} \pm 2s\frac{\tilde{\xi}_z}{\gamma}\right) \\ f^*_{mass}H \end{pmatrix} \quad (7)$$

where

$$f^{\pm}_{mass} = \pm|\nabla\xi|\frac{\rho s}{4}(M_\xi \pm 1)^2 \quad , \quad M_\xi = \frac{u\xi_x + w\xi_z}{s|\nabla\xi|} \; .$$

3. Generalized FVS methods for hypersonic flows

Generalized FVS methods for hypersonic flows have been constructed by the superposition of these methods with a second order artificial model in the region of the strong shock wave. The generalized fluxes are defined as:

$$E_{i+1/2} = [E_{FVS}(U)]_{i+1/2} + [D^2_\xi U]_{i+1/2} \quad (8)$$

where $(D^2_\xi U)$ is an artificial dissipation model defined by the eigenvalues. The above hybrid flux has also been used in the past for inviscid hypersonic flows [8].

HIGH ORDER EXTRAPOLATION SCHEMES

For the caclulation of the conservative variables U^-, U^+ from the left and right state of the cell face, two families of high order extrapolation schemes have been considered. One of the most widely used extrapolation schemes is the Monotone Upstream centered Scheme for Conservation Law (MUSCL) [7]:

$$U^-_{i+1/2} = U_{i,j} + \frac{S_{i,j}}{4}[(1-kS_{i,j})\nabla + (1+kS_{i,j})\Delta]U_{i,j} \qquad (9)$$

$$U^+_{i+1/2} = U_{i+1,j} - \frac{S_{i+1,j}}{4}[(1+kS_{i+1,j})\nabla + (1-kS_{i+1,j})\Delta]U_{i+1,j} \qquad (10)$$

with $\Delta U_{i,j} = U_{i+1,j} - U_{i,j}$, $\nabla U_{i,j} = U_{i,j} - U_{i-1,j}$. S is the van Albada limiter function [9]. The spatial accuracy depends on the parameter k. For instance k=-1 produces fully upwinded, k=0 symmetric, k=1/3 third order biased and k=1 centered scheme. MUSCL scheme is second order accurate for two dimensional flows while the third order upwind-biased formulation of this scheme is strictly third order for one dimensional calculations. This scheme has been used in the present paper in combination with the MSW-FVS and VL-FVS methods for viscous flow fields.

The second upwind extrapolation scheme which has been developed in the present paper in order to improve the FVS methods is a hybrid upwind extrapolation scheme. This scheme has been used in the past by Eberle [10] for the solution of the Euler equations in combination with a Riemann solver. The hybrid scheme is constructed by the superposition of the first, second, third and fourth order extrapolation schemes:

$$U^* = AU^{1,*} + (1-A)[BU^{2,*} + (1-B)[CU^{3,*} + (1-C)U^{4,*}]]. \qquad (11a)$$

The superscripts 1,2,3,4 denote the several order of the extrapolation. For instance the third and fourth order extrapolation are defined as:

Third order: $\left(U^3_{i+\frac{1}{2}}\right)^- = \frac{1}{6}(5U_i - U_{i-1} + 2U_{i+1})$, $\left(U^3_{i+\frac{1}{2}}\right)^+ = \frac{1}{6}(5U_{i+1} - U_{i+2} + 2U_i)$

Fourth order: $\left(U^4_{i+\frac{1}{2}}\right)^- = \left(U^4_{i+\frac{1}{2}}\right)^+ = \frac{1}{12}(7U_i + 7U_{i+1} - U_{i-1} - U_{i+2})$.

The terms A, B are limiter functions defined by the second order derivatives of the pressure:

$$A = \min(1, d^*|p^2_{\xi\xi,i+1} - p^2_{\xi\xi,i}|) \quad , \quad B = \min(1, b^*|p^2_{\xi\xi,i+1} + p^2_{\xi\xi,i}|). \qquad (11b)$$

The values of the constants d,b, C are d=4.5,b=2.5,C=2.25. If the parameter b has large values the scheme swithces on second order of accuracy.

UNFACTORED IMPLICIT RELAXATION SOLUTION
OF THE NAVIER - STOKES EQUATIONS

An unfactored implicit method has been developed in combination with the FVS methods for the solution of the system of equations. The implicit method is first order accurate in time. The unfactored equations are solved by a Newton method constructing a sequence of approximations q^ν such that $\lim_{\nu>1} q^\nu \to U^{n+1}$ where ν is the subiteration state. A Newton form is obtained by the linearization of the equation (1), around the known subiteration state ν, as follows:

$$J\frac{\Delta q^{\nu+1}}{\Delta t} + (A_{inv}^\nu \Delta q^{\nu+1})_\xi + (C_{inv}^\nu \Delta q^{\nu+1})_\zeta + (C_{vis}^{th} \Delta q^{\nu+1})_\zeta = J\frac{U^n - q^\nu}{\Delta t} - RHS \qquad (12)$$

$$RHS = (E_{inv})_\xi + (G_{inv})_\zeta - \frac{1}{Re}(E_{vis})_\xi - \frac{1}{Re}(G_{vis})_\zeta. \qquad (13)$$

$q^\nu, q^{\nu+1}$ are the solution vectors at the subiteration states $\nu, \nu+1$ respectively. Gauss-Seidel relaxation using 4 subiteration states is applied on the left hand side (LHS) of equation (12) while the RHS is held constant. On the LHS the thin layer viscous Jacobian C_{vis}^{th} is used for steady state calculations, saving computational time. The inviscid fluxes on the LHS of equation (13) are splitted. In order to retain the stability of the implicit solution using the FVS methods, the splitted eigenvalue matrices are defined as:

$$\Lambda^+ = \max(F, \Lambda), \Lambda^- = \min(-F, \Lambda) \quad with \quad F = Bh\max[|\lambda_1|, |\lambda_2|]. \qquad (14)$$

The sensor B is defined from the values of the RHS. If the FVS methods are used with the MUSCL upwind the sensor B is defined by the squares of the Mach number on the left and right states of the cell face. If the FVS methods are used in combination with the hybrid upwind the sensor B is the limiter function (eq. 11b). The constant h is 0.5.

RESULTS

a. Laminar flows

Analytical validation of the FVS methods in combination with the high order extrapolation schemes has been obtained for the laminar flow ($M_\infty = 0.5, Re = 10^4$) over a flat plate. The computational mesh is 44x44. In Fig. 1a the predictions for the velocity profile using the MUSCL upwind, the third order and the fourth order hybrid upwind in combination with the MSW-FVS, are presented. Comparisons with Blasius analytical solution show that the third order hybrid upwind is more accurate than MUSCL upwind as well as than fourth order scheme, improving the accuracy of the MSW-FVS method. The results for the velocity profile are the same using the VL-FVS method (Fig. 1b). The contribution of the third order hybrid scheme is more significant in the skin friction predition. In Fig. 2a comparisons of high order schemes (using the MSW-FVS method) show that third order hybrid scheme predicts accurate skin friction distribution while the fourth order version of the scheme as well as the MUSCL upwind present inaccuracies. MUSCL upwind has been used in the third order biased formulation. Third order hybrid upwind improves also the VL-FVS method predicting with accuracy the skin friction distribution (Fig. 2b). The generalized formulation (eq.8) of the FVS methods can also be used in hypersonic flows. Iso-Mach lines for the laminar hypersonic flow $M_\infty = 10, Re = 1.2 \times 10^4$ over a hyperbola are plotted in Fig. 3a. Comparison of the present results with the corresponding results by Mundt et al. [11] for the skin friction distribution is shown in Fig. 3b. The results between the MSW-FVS and VL-FVS are similar. The prediction of the skin friction is found in good agreement with the corresponding results by Mundt et al. [11]. The computed stagnation temperature is 4650 °K using the VL-FVS and 4695 °K using the MSW-FVS. The analytic solution is 4620 °K. The convergence history for the hypersonic test case is shown in Fig. 3c.

b. Turbulent flows

Validation of the FVS methods has also been obtained in transonic turbulent flow fields over 18% circular arc-airfoil. An algebraic eddy-viscosity model [12] is used. Experimental results for the circular arc-airfoil can be found in Ref. [13]. The computational boundaries are located 6 chords upstream of the leading edge and 9 chords downstream of the trailing edge so that all gradients in the flow direction may be assumed negligible ($\frac{\partial \phi}{\partial \xi} = 0$). On the surface of the airfoil no-slip boundary conditions and adiabatic wall conditions are considered. On the upstream boundary and along the far transverse boundary (6 chords lengths away) the flow is assumed uniform. The computational mesh is 60x38 with 10 points upstream of the leading edge and 20 points behind of the airfoil. Ahead and behind the airfoil the flow is considered symmetric.

Experimental results [13] have indicated that the flow is steady with trailing edge separation for the turbulent flow $M_\infty < 0.755$ while for $0.755 < M_\infty < 0.782$ may be periodically unsteady. Numerical results for the the flow $M_\infty = 0.775, Re_c = 2 \times 10^6$ using the MSW-FVS method with the MUSCL scheme in two different formulations (third order biased, fully upwind) are shown in Fig. 4a. The two formulations of the MUSCL scheme give the same results presenting inaccuracies especially in the shock wave region. In Fig.4b comprarisons with the experimental results for the pressure coefficient distribution using the MSW-FVS method with the third and fourth order hybrid scheme are presented. The MSW-FVS method with the third odrer scheme predicts with accuracy the pressure distribution over most of the airfoil but inaccuracies occur in the separation region near the trailing edge. MSW-FVS method with the fourth order hybrid scheme presents accurate results in the separation region (although the inadequate turbulence model) and ahead of the shock wave but overpredicts the pressure distribution in the peak of the shock wave transposing the location of the latter. In the same figure comparison is also presented with the numerical prediction by Deiwert [12]. Deiwert uses MacCormack method for the solution of the Navier-Stokes equations. From the above results is shown that the third order hybrid scheme improves the MSW-FVS method in comparison with the MUSCL scheme over the most of the airfoil. The results using either the VL-FVS or the MSW-FVS method are the same (Fig 4c). The inaccuracies of the fouth order hybrid scheme in the shock wave region may result from the limiter functions of (eq.11b). Further investigation for a better formulation of these limiters in the turbulent boundary layers is needed. The shock turbulent boundary layer interaction plotting the iso-Mach lines is shown in Fig. 5. The location of the shock - turbulent boundary layer interaction is located about 0.740 chord on the airfoil surface, induced a separation region with reattachment nearly 0.1 chord downstream of the trailing edge.

Results for higher Reynolds numbers are presented in Figures 6a,b using the VL-FVS with the third order hybrid scheme. Pressure coefficient distribution for the transonic turbulent flows $M_\infty = 0.72, Re_c = 11 \times 10^6$ [14] and $M_\infty = 0.751, Re_c = 17 \times 10^6$ [13], over the 18% circular arc-airfoil are shown in Figs 5a,b respectively. In the lower Mach number cases a weaker shock is formed while trailing edge separation occurs. The differences with the experimental results for the case $M_\infty = 0.72, Re_c = 11 \times 10^6$ are not large. Differences in the prediction of the shock location exist for the second flow case $M_\infty = 0.751, Re_c = 17 \times 10^6$ while the pressure in the separation region and ahead of the shock wave are predicted with accuracy. Numerical experiments showed that there are not differences in the results usin the MSW-FVS method. This flow case, in actuality, is pseudo-steady presenting small variations on the shock location and the pressure distribution [14].

CONCLUSIONS

In the present paper a MSW-FVS method as well as the VL-FVS method are developed using a hybrid extrapolation scheme and the MUSCL scheme. Validation of the implicit FVS methods with the extrapolation schemes show that the third order hybrid scheme improves the FVS methods, in laminar and turbulent flows, in comparison with the MUSCL scheme. The present MSW-FVS has similar behaviour with the VL-FVS method. Further investigation is needed especially in the turbulent flow fields using more adequate turbulence models as well as better limiter functions, especially, in the case of the fourth order hybrid scheme, both for laminar and turbulent flows.

REFERENCES

[1] MACCORMACK, R.W., CANDLER, G. : "The solution of the Navier-Stokes equations using Gauss-Seidel line relaxation", Comp. and Fluids, Vol. 17, No. 1 (1989), pp. 135-150.

[2] HÄNEL, D., SCHWÄNE, R., SEIDER, G. : "On the accuracy of upwind schemes for the solution of the Navier-Stokes equations", AIAA-Paper 87-1105, (1987).

[3] VAN LEER, B., THOMAS, J.L., ROE, P.L., NEWSOME, R.W. : "A Comparison of Numerical Flux Formulas for the Euler and Navier-Stokes equations", AIAA-Paper 87-1184, (1987).

[4] STEGER, J.L., WARMING, R.F. : "Flux Vector Splitting of the inviscid Gasdynamic Equations with Application to Finite-Difference Methods", J. Comp. Phys., vol. 40, (1981), pp. 263-293.

[5] VAN LEER, B. : "Flux-Vector Splitting for the Euler equations", Proc. 8th Int. Conf. on Numerical Methods in Fluid Dynamics, Aachen, (1982), Lecture Notes in Physics, vol. 170, Springer, Berlin (1982), pp. 507-512.

[6] Drikakis, D., submitted Ph.D thesis NTUA, (1991)

[7] VAN LEER, B. : "Towards the Ultimate Conservative Difference Scheme V.: A Second - Order Sequel to Godunov's Method", J. of Comp. Phys., vol. 32 (1979), pp. 101-136.

[8] Drikakis, D., Tsangaris, S. : "Shock Capturing Method for Hypersonic flows and Real Gas Effects", Int. Conf. on Computational Engineering Science, Melbourne, Australia, 12-16 August, (1991).

[9] van Albada,G. D., van Leer, B., Roberts, W.W., "A Comparative stydy of Computational methods in cosmic gasdynamics", Astron. Astrophys. , 108, 76-84, (1982)

[10] EBERLE, A. : "Characteristic Flux Averaging Approach to the Solution of Euler's Equations", VKI Lecture Series, Comp. Fluid Dynamics, 1987-04 (1987).

[11] Mundt, Ch., Pfintzer, M., Schmatz, M.A. : "Calculation of viscous hypersonic flows using a coupled Euler / second order boundary layer method" 8th GAMM Conference on Numerical methods in Fluid Mechanics, (1989).

[12] Deiwert, G.S. : "Numerical simulation of high Reynolds Number Transonic Flows", AIAA Journal, vol. 13, (1975), pp. 1354-159.

[13] McDevitt, J.B., Levy Jr., L.L., Deiwert, G.S. : "Transonic flow about thick circular-arc airfoil", AIAA Journal, vol. 14, (1976), pp. 606-613.

[14] Levy Jr., L.L. : "Experimental and computational steady and unsteady transonic flows about a thick airfoil", AIAA Journal, vol. 16, (1978), pp. 564-572.

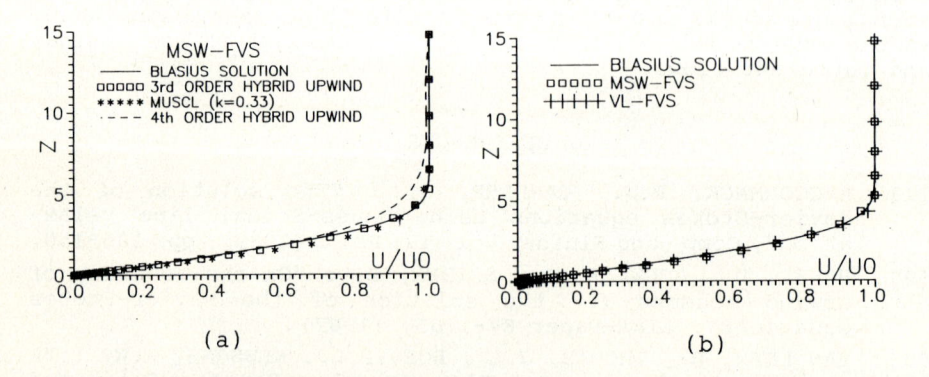

Fig. 1a,b: Velocity profiles over a flat plate
(a) Comparisons between the upwind schemes using MSW-FVS
(b) Comparison between VL-FVS and MSW-FVS using the third order hybrid upwind

Fig. 2a,b: Skin friction distributions over a flat plate

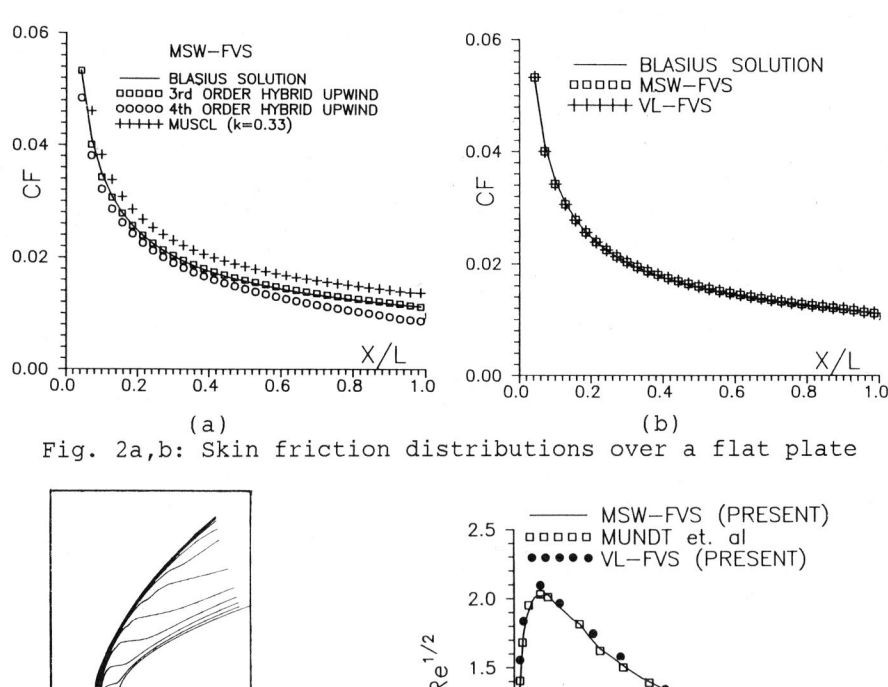

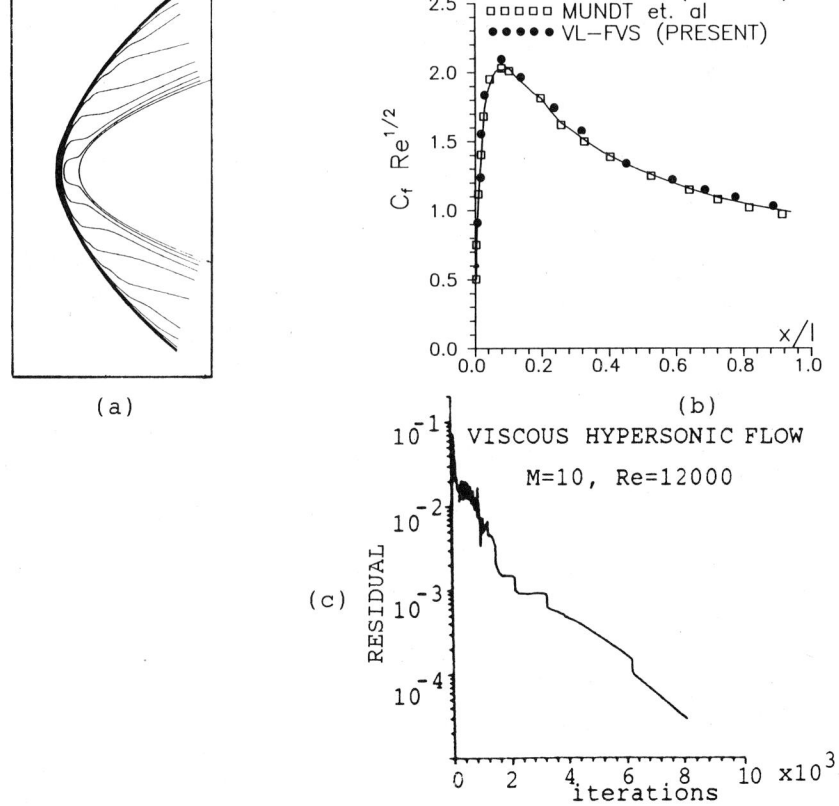

Fig. 3a,b,c: Hypersonic flow (M=10, Re=12000) over a hyperbola. (a) Iso-Mach lines (MSW-FVS) (b) Skin friction distribution (VL-FVS, MSW-FVS) (c) Convergence (MSW-FVS)

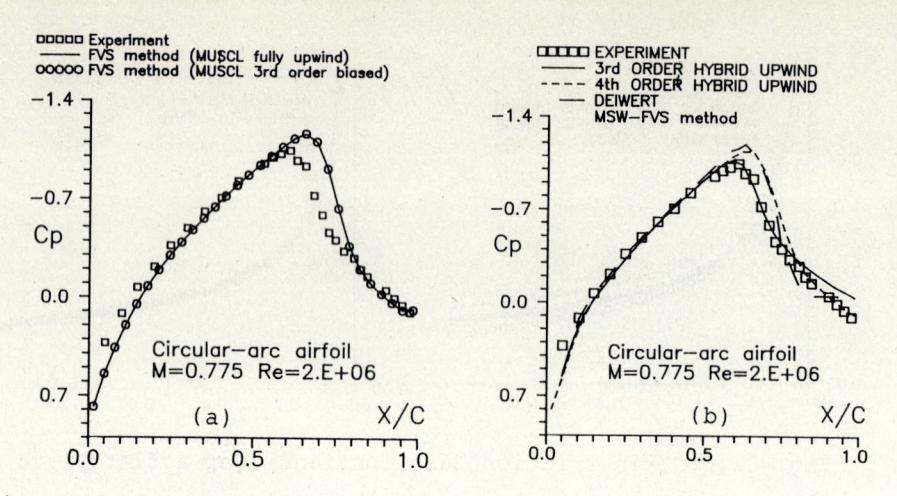

Fig. 4a,b: Turbulent flow. Comparison between the upwind schemes.

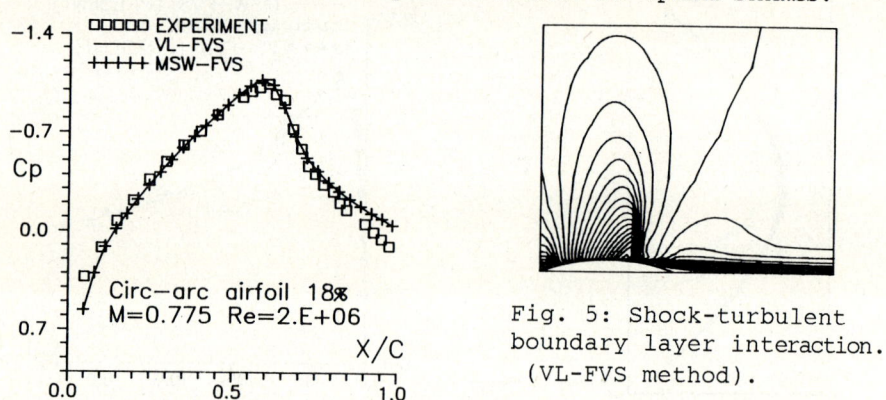

Fig. 4c: Comparison between VL-FVS and MSW-FVS. (3rd order upwind)

Fig. 5: Shock-turbulent boundary layer interaction. (VL-FVS method).

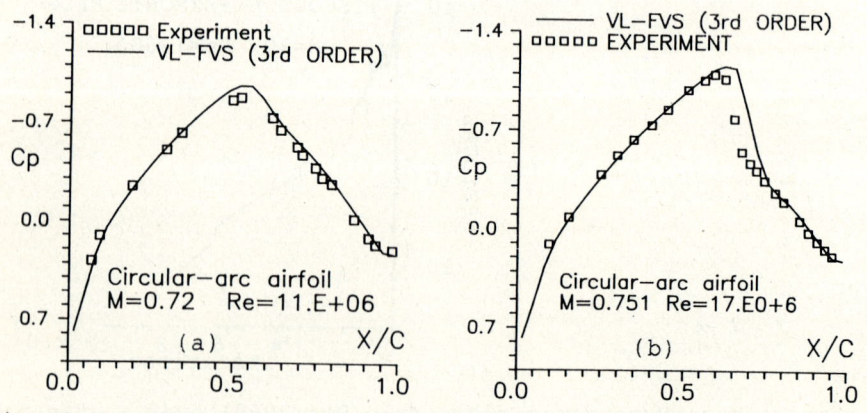

Fig. 6a,b: High Reynolds number flows using the VL-FVS and the 3rd order hybrid upwind.

416

SUPERSONIC TURBULENT BOUNDARY LAYER
WITH PRESSURE GRADIENTS

F. Hanine, A. Kourta, H. Haminh
Institut de Mécanique des Fluides de Toulouse (CNRS U.A. 0005)
Avenue du Professeur Camille Soula, 31400 Toulouse Cedex, France

SUMMARY

The current work presents computational studies of a compressible boundary layer in the presence of non zero pressure gradients. The aim of this investigation is to study and analyse the behaviour of near-wall compressible flows under the effects of both favourable and adverse pressure gradients. Results are reported using two alternative two-equation turbulence closures and a second order closure in order to provide an evaluation of the various models. These calculations, conducted through the boundary layer, have allowed a better understanding of the mechanisms controlling this kind of flow. The effects of streamwise pressure gradients on the mean and turbulent quantities and on the turbulence anisotropy, are studied, particularly in the viscous sublayer. It is concluded that there are a specific pressure gradient effects present in the bondary layer. The results of the analysis show that the effects found in incompressible flows are also present in this case, and have been qualitatively reproduced. There are, in addition, other influences due to the compressibility. From a comparative analysis with the various models, we find that the results with the Chien model containing the Nichols correction gave a better prediction than Chien model under a favourable pressure gradient.

INTRODUCTION

A very accurate calculation of the turbulent boundary layer characteristics in supersonic flows under pressure gradients effects is necessary for designing modern flight vehicles with high performances. A large literature now exists for subsonic flows along flat surfaces in both adverse and favourable pressure gradients. However, there is still a need for more work in understanding the mechanisms which appear when the velocity is supersonic. In these flows, the variation in both the density and the temperature lead to a more complex physical behaviour. It is generally accepted that the influence of pressure gradients on the turbulence is in the region close to the wall called the viscous wall region. The structural aspects of this region play an important role in understanding how turbulence is generated at the wall. So, in order to improve the understanding of the near wall boundary layer flows including compressibility and pressure gradients effects, considerable attention must be given to the viscous wall region. The existence of organised eddies, attached to the wall, bringing high momentum fluid to the wall and ejecting low momentum fluid from the wall is the most important feature since it controls the turbulent exchanges in the boundary layer.

The motivation for this study is, then, the computation and the analysis of the compressible boundary layer under pressure gradient effects. Particular attention will be paid to the near wall region. From the analysis we want to identify some features relating to the effects of the pressure gradients in the viscous sublayer.This should enable improvements to be made in modelling compressibility and pressure gradient effects on the near wall boundary layer. In order to provide a better understanding of the physics of these phenomena, a number of turbulence closure models are applied close to the wall. A further aim of the study is to evaluate the performance of these models and possibly to propose some improvements. The first turbulence model used is the k-ε Chien model (CH) [1], the second one is the Chien model including Nichols compressibility correction (CH(corr)) [2]. The last model used is a full second order closure Reynolds stress model (RSE) [3] to allow the study of the anisotropic effects of the pressure gradient on the Reynolds stress components and the evaluation of the other models. The three models contain "low Reynolds number" corrections and have been tested before to enable their use at low turbulent Reynolds numbers and to predict a turbulent compressible flow close to a solid wall with zero pressure gradient [4]. In this study the pressure gradient effects are introduced and analysed.

TURBULENCE MODELS

The governing equations solved are the mass-weighted averaged Navier-Stokes equations with a thin layer approximation. To close this set of equations different turbulence closure models are used. The compressibility conditions are introduced in the various equations by use of Favre averaging and accounting for the non-zero mass weighted fluctuating velocities. In all models viscosity and thermal conductivity fluctuations are neglected.

First order turbulence models

In these models, the turbulent velocity correlations are given by the following Boussinesq approximation :

$$-\overline{\rho\, u_i u_j} = \mu_t \left(\frac{\partial U_i}{\partial x_j} + \frac{\partial U_j}{\partial x_i} \right) - \frac{2}{3} \mu_t \delta_{ij} \frac{\partial U_l}{\partial x_l} - \frac{2}{3} \delta_{ij} \overline{\rho}\, k. \qquad (1)$$

For the first order k-ε model, the turbulent viscosity μ_t is expressed as :

$$\mu_t = c_\mu f_\mu \overline{\rho} \frac{k^2}{\varepsilon} ; \qquad (2)$$

The turbulent kinetic energy k and its dissipation rate, ε, are given by two transport equations:

$$\frac{\partial}{\partial x_k}(\overline{\rho}\, U_k k) = \frac{\partial}{\partial x_k}\left((\mu + \frac{\mu_t}{\sigma_k}) \frac{\partial k}{\partial x_k} \right) - \overline{\rho\, u_i u_k} \frac{\partial U_i}{\partial x_k} - \overline{u_i}\frac{\partial \overline{P}}{\partial x_i} - \overline{\rho}\, \varepsilon^*, \qquad (3)$$

$$\frac{\partial}{\partial x_k}(\overline{\rho}\, U_k \varepsilon^*) = \frac{\partial}{\partial x_k}\left((\mu + \frac{\mu_t}{\sigma_\varepsilon}) \frac{\partial \varepsilon^*}{\partial x_k} \right) + c_{\varepsilon_1} f_1 \frac{\varepsilon^*}{k} \overline{\rho\, u_i u_k} \frac{\partial U_i}{\partial x_k} - c_{\varepsilon_2} f_2 \overline{\rho} \frac{\varepsilon^{*2}}{k}$$

$$- c_{\varepsilon_3} \frac{\varepsilon^*}{k} \overline{u_i}\frac{\partial \overline{P}}{\partial x_i} + \overline{\rho}\, E . \qquad (4)$$

where the use of $\varepsilon^* = \varepsilon - D$ as the "dissipation variable" leads to a numerically convenient boundary condition since D is chosen such that $\varepsilon^* = 0$ at the wall.
The various constants and functions used in the Chien model are given by the following :

$c_\mu = 0{,}09$; $c_{\varepsilon_1} = 1{,}35$; $c_{\varepsilon_2} = 1{,}8$; $c_{\varepsilon_3} = 2{,}0$; $\sigma_k = 1{,}0$; $\sigma_\varepsilon = 1{,}3$;

$f_\mu = 1 - \exp(-0{,}0115\, y^+)$; $f_1 = 1$; $f_2 = 1 - 0{,}22\exp\left(-\left(\frac{R_y}{6}\right)^2\right)$;

$E = -2\nu \dfrac{\varepsilon}{y^2} \exp(-y^+/2)$; $D = 2\nu \dfrac{k}{y^2}$.

In these functions, R_t is the turbulent Reynolds number, y^+ is the wall coordinate and u_τ is the shear velocity. All these are given by :

$$R_t = \frac{k^2}{\nu \varepsilon} \ ; \ y^+ = \frac{y \, u_\tau}{\nu} \ ; \ u_\tau = \sqrt{\frac{\tau_w}{\rho_w}} \ .$$

For the Chien model including Nichols compressibility correction, we introduce in the turbulent kinetic energy production the following correction :

$$P_{k(\text{corr})} = P_k - 4\,(\gamma-1)\,M\frac{k}{a^2}\,P_k \ .$$

This correction acts to improve the prediction of the turbulent production level by directly including the effect of Mach number (M).

Second order model

In the second order model, the turbulence stress tensor is obtained by solving a modelled form of the transport equation:

$$\frac{\partial}{\partial x_k}(\overline{\rho\, u_i u_j}\, U_k) = T_{ij} + P_{ij} + \varepsilon_{ij} + \phi_{ij1} + \phi_{ij2} + \phi_{ijw} - \overline{u_i}\frac{\partial \overline{P}}{\partial x_j} - \overline{u_j}\frac{\partial \overline{P}}{\partial x_i} \ . \tag{5}$$

A version of the Launder-Reece and Rodi model is used, where the diffusion (T_{ij}), the production (P_{ij}) and the dissipation (ε_{ij}) terms are given by the relationships (6).

$$T_{ij} = \frac{\partial}{\partial x_k}\left(c_{s'}\overline{\rho}\frac{k}{\varepsilon}\overline{u_k u_l}\frac{\partial}{\partial x_l}(\overline{u_i u_j}) + \mu\frac{\partial}{\partial x_k}(\overline{u_i u_j})\right) ,$$

$$P_{ij} = -\overline{\rho}\,(\overline{u_i u_k}\frac{\partial U_j}{\partial x_k} + \overline{u_j u_k}\frac{\partial U_i}{\partial x_k}) ,$$

$$\varepsilon_{ij} = -\overline{\rho}\frac{\varepsilon}{k}(\overline{u_i u_j}\, f_s + \frac{2}{3}(1-f_s)\delta_{ij}\,k) \ . \tag{6}$$

The pressure strain term $\Phi_{ij} = \Phi_{ij1} + \Phi_{ij2} + \Phi_{ijw}$, which represents an interaction between the fluctuating pressure field and the fluctuating velocity field, serves to redistribute the Reynolds stresses. This term contains three contributions : a term Φ_{ij1} due to the distorsion of the Reynolds stress field; Φ_{ij2} due to the distorsion of the mean velocity field, and a term Φ_{ijw} arising because of the proximity of the wall. These models are written in the relationships (7).

$$\phi_{ij1} = -c_1 f_1 \overline{\rho}\frac{\varepsilon}{k}(\overline{u_i u_j} - \frac{2}{3}\delta_{ij}\,k) ,$$

$$\phi_{ij2} = -\frac{c_2+8}{11}(P_{ij} - \frac{2}{3}\delta_{ij}P) - \frac{8c_2-2}{11}(D_{ij} - \frac{2}{3}\delta_{ij}D)$$

$$- \frac{30\,c_2-2}{55}\overline{\rho}\,k\,(\frac{\partial U_i}{\partial x_j} + \frac{\partial U_j}{\partial x_i} - \frac{2}{3}\delta_{ij}\frac{\partial U_i}{\partial x_j}) ,$$

$$\phi_{ijw} = (c_3 \overline{\rho}\frac{\varepsilon}{k}(\overline{u_i u_j} - \frac{2}{3}\delta_{ij}\,k) + c_4(P_{ij} - D_{ij}) + c_5 \overline{\rho}\,k\,(\frac{\partial U_i}{\partial x_j} + \frac{\partial U_j}{\partial x_i} - \frac{2}{3}\delta_{ij}\frac{\partial U_i}{\partial x_j}))\frac{k^{3/2}}{\varepsilon\, x_n} \ , \tag{7}$$

with

$$D_{ij} = -\overline{\rho}\,(\overline{u_i u_k}\frac{\partial U_k}{\partial x_j} + \overline{u_j u_k}\frac{\partial U_k}{\partial x_i}) \ ; \ P = D = -\overline{\rho}\,\overline{u_i u_j}\frac{\partial U_i}{\partial x_j} \ .$$

In equation (5), the two last terms are related to the effects of the pressure gradient.

The velocity density correlation is given by : $\overline{u_i} = \dfrac{1}{(n-1)c_p T} U_j \overline{u_i u_j}$, where n represents the polytropic coefficient, equal to 1.2, and c_p the specific heat transfer constant.

A function f_1 is introduced in the Φ_{ij1} term in order to reduce the coefficient c_1 and to improve the anisotropy in the vicinity of the wall. The model also takes into account the effect of anisotropies in ε_{ij}.

In this model, the form of the dissipation of kinetic energy, ε, is :

$$\dfrac{\partial}{\partial x_k}(\overline{\rho U_k}\varepsilon) = \dfrac{\partial}{\partial x_k}((\mu + c_\varepsilon \dfrac{k}{\varepsilon}\overline{\rho u_k u_l})\dfrac{\partial \varepsilon}{\partial x_l}) - c_{\varepsilon_1}\dfrac{\varepsilon}{k}\overline{\rho u_i u_k}\dfrac{\partial U_i}{\partial x_k} - c_{\varepsilon_2} f_\varepsilon \overline{\rho}\dfrac{\varepsilon}{k}(\varepsilon - 2\nu(\dfrac{\partial k^{1/2}}{\partial x_n})^2)$$

$$+ c_{\varepsilon_3}\mu\dfrac{k}{\varepsilon}\overline{u_j u_k}(\dfrac{\partial^2 U_i}{\partial x_j \partial x_l})(\dfrac{\partial^2 U_i}{\partial x_k \partial x_l}) - c_{\varepsilon_4}\dfrac{\varepsilon}{k}\overline{u_i}\dfrac{\partial \overline{P}}{\partial x_i}.\qquad(8)$$

The modelling coefficients and functions used in these equations are :
$c_1 = 1,5;\quad c_2 = 0,4;\quad c_3 = 0,125;\quad c_4 = 0,015\ ;\quad c_5 = 0;\quad c_\varepsilon = 0,375;$
$c_{\varepsilon_1} = 1,28;\quad c_{\varepsilon_2} = 1,8;\quad c_{\varepsilon_3} = 1\ ;\quad c_{\varepsilon_4} = 2;\quad c_s' = 0,25;$
$f_1 = \exp(-2/(1+R_t/30))\quad f_\varepsilon = 1\ ;\quad f_s = 1/(1+R_t/10)$ where $R_t = k^2/(\nu \varepsilon)$.

In all the equations listed earlier the mean quantities, except the mean pressure and density, are represented by capital characters.

RESULTS AND DISCUSSION

The turbulence models listed earlier were solved together with the momentum and energy equations for a two dimensional boundary layer. The numerical method used is an adapted version of the implicit marching procedure of Patankar and Spalding [5]. The calculations were performed on a smooth adiabatic wall. Table 1 gives the variation of the edge Mach number and of the dimensionless pressure gradient parameter $p^+ = \nu(dp/dx)/(\rho_w u_\tau^3)$ from the initial section to the end of the plate.

Table 1 : flow conditions for favourable and adverse pressure gradients.

	dp/dx<0	dp/dx>0
M	2,79 → 3,5	2,79 → ≈ 1,4 (RSE model)
		2,79 → ≈ 1,1 (k-ε models)
p⁺	-1.235 10⁻⁴ → -3.5 10⁻⁴	4.55 10⁻² → 3.606 10⁻⁴

Before describing the results of the calculation, we will discuss some aspects that may be helpful later in the interpretation of the pressure gradient effects on the structure of the viscous sublayer. In the boundary layer flows, it is assumed that the outer edge of the viscous wall region is defined as the location where viscous momentum transfer becomes negligible compared to turbulent transfer : it resides in the logarithmic law region. Consequently, the logarithmic layer is located at the distance from the surface where the viscous shear stress becomes small compared to the total shear stress. Hence a rapid thickening of the viscous layer suggests that the logarithmic layer disappears and that the flow will go to a laminarized state. Conversly, an increase of the turbulence leads to a slimming of the viscous region and the logarithmic region becoming more developed.

As observed by Finnicum [6], a physical analysis of the flow shows that the turbulence generation results from the interaction between the wall eddies and the mean flow kinetic energy. The viscous interaction with the wall is controlled by disturbances in the spanwise plane which are elongated in the flow direction. These disturbances bring high momentum fluid toward the wall and create velocity deficient fluid away from the wall. The wall eddies, by this process, convert mean flow energy transported to the viscous wall region into large streamwise velocity fluctuations u. This is manifested by the observed near wall maximum in the u profile. The streamwise velocity fluctuations created in the viscous wall layer are then converted into normal and spanwise fluctuations by means of the pressure strain correlation which represents a redistribution between the various fluctuation velocities. These, in turn, interact with the wall to repeat the cycle.

When we analyse the effects of the pressure gradient on the transport equations of the flow, we find that these features are confirmed. A negative pressure gradient in the momentum equation causes a decrease of the turbulent momentum diffusion in the logarithmic region. In the viscous region the molecular viscosity increases due to the increase of the temperature at the wall. This behaviour explains the disappearance of the logarithmic region, the increase of the viscous sublayer and the tendency of the velocity profile to go to a laminar state. When the pressure gradient is positive, the opposite behaviour appears: the turbulent momentum diffusion near the wall increases and the velocity becomes rapidly uniform, leading to a slimming of the boundary layer. In the presence of a favourable pressure gradient, the decrease of the turbulence production, expressed as a function of turbulent shear stress uv and mean velocity gradient, is related to the decrease of the uv component and mean velocity gradient. This induces a decrease of the normal Reynolds stresses u^2, v^2 and w^2. The transport equation of ε, the dissipation rate of the turbulent kinetic energy, shows that the favourable pressure gradient gives an increase of the generation term of this quantity. This is in accordance with the decrease of k. On the other hand, in the presence of an adverse pressure gradient the opposite effect is noted.

The computed results for the test cases are represented by the figures (1) to (16). Figures (1) and (2) respectively show the distribution of the density in favourable and adverse pressure gradients. We note a decrease of the magnitude of this parameter in the accelerated flow and an important increase in the presence of deceleration. For the first case the variation is approximately between 0,0125 and 0,0625. For the second, the variation is between 2,1 and 4. This is in agreement with the behaviour of the temperature in the boundary layer which increases and decreases respectively for accelerated and decelerated flows (fig.3,4). The skin friction distribution for both cases is given in figures (5) and (6). The position of the beginning of the laminarization ($x \approx 0,95m$) is well predicted by all models: at this station the turbulence level decrease strongly. The Chien model predicts a higher level of the skin friction distribution. For the adverse pressure gradient case, the decrease of this parameter with distance is well predicted by all models, although the first order models predict a higher level. The main influence of the pressure gradient appears in the distribution of the total shear stress which is plotted in figures (7) and (8). Although the models show the same qualitative behaviour, the k-ε models, in line with the skin friction distribution, predict higher shear stress levels. The decrease in the presence of acceleration and increase in the adverse case are well predicted by all models. In this last case, the maximum of the total shear stress is not at the wall. Figures (9) and (10) show the velocity profiles in wall coordinates. The mean velocity is transformed to account for compressibility by the Van Driest transformation given

by : $$u^+ = \frac{U}{u_\tau} = \frac{\int_0^u \sqrt{\frac{\bar{\rho}}{\rho_\omega}}\, du}{u_\tau}.$$ (9)

These figures also include the analytical curve with the generally accepted semi-logarithmic law represented by the straight line. In the wall region all models show an increase in the sublayer thickness under the favourable pressure gradient. A disappearance of the logarithmic region is predicted by all the models. Conversly, in the presence of the adverse pressure gradient, the size of the viscous sublayer decreases, the logarithmic region is still present and appears more developed. These features are in good agreement with the prediction of the production to dissipation ratio for the turbulent kinetic energy (Fig. 11,12). Figures (13) and (14) show that the influence that the pressure gradient has on the turbulence level which decreases under the acceleration and increases for the decelerated case. The position of the maximum is also further removed from the wall in the second case. The generation of the longitudinal fluctuating velocity is based upon the interaction between the wall eddies and the mean kinetic energy. When the flow is accelerated these eddies become less important and the level of the u^2 component decreases. Since the production of the normal and spanwise fluctuating velocities are related to the longitudinal component via the pressure strain correlation, the level of these quantities will also decrease. In the presence of an adverse pressure gradient we note an opposite effect on these quantities. Figures (15) and (16) show the distribution of the normal Reynolds stresses. The level of the various quantities decreases under an acceleration while it increases under deceleration. The position of the maximum is located far from the wall in this later case. Near the wall, until $y^+ \approx 100$, the anisotropy appears to decrease in the presence of a favourable pressure gradient, and to increase in the presence of adverse pressure gradients. For clear representation we do not plot on the figures the results of a zero pressure gradient: the reader can refer to [4] for comparison.

CONCLUDING REMARKS

A numerical solution of the turbulent boundary layer equations based on turbulence modelling is presented for compressible flows under pressure gradients. We can say that some features found in incompressible flows under favourable and adverse pressure gradients are also present in the turbulent compressible boundary layer. The effects of a favourable pressure gradient induce a decrease of the turbulence levels in the boundary layer. The heat transfer is increased and induces an increase in the dissipation level. This causes the disappearance of the logarithmic region and an increase of the near wall viscous region. Conversely, when the flow is decelerated, the opposite behaviour is observed. Analysis of the compressibility effects shows that the acceleration of the flow causes an increase of the far field Mach number along the flat plate, while the deceleration causes a decrease of this parameter. In the presence of higher Mach numbers, the compressibility effects are more pronounced. Hence, the decrease of the turbulence levels which characterizes the compressibility effects will increase under the effects of the acceleration of the flow and be damped by the deceleration. Another new feature appearing in the compressible flow concerns the thickening of the boundary layer under acceleration and its slimming in the presence of deceleration. This is related to the decrease of the density in the first case and its increase in the second. All the tested models gave reasonably good results. The correction introduced to the Chien model seems to be less efficient when the flow is decelerated but gives better results in the presence of an acceleration.

REFERENCES

[1] Chien K.Y., "Predictions of channel and boundary layer flows with a low Reynolds number turbulence model", AIAA Journal, VOL.20 Jan. 1982, PP. 33-38.

[2] Nichols R.H.,"A Two-Equation Model for Compressible Flows", AIAA Paper 90-0494.

[3] Launder B.E., Reece G.J. and Rodi W., "Progress in the development of a Reynolds stress turbulence closure".J.F.M.68-3,PP 537-566, 1975.

[4] Hanine F., Kourta A., "Performance of turbulence models to predict supersonic boundary layer flows", CMA Journal , 1991.

[5] Patankar S.V. and Spalding D.B., "Heat and Mass Transfer in Boundary Layers", 1970 London Intertext.

[6] Finnicum D. S., "Pressure gradient effects in the viscous wall region of a turbulent flow", Ph. D. thesis, Univ. of Illinois, Urbana, 1986.

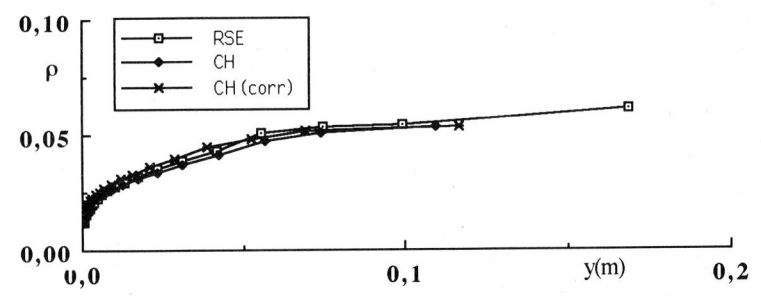

Figure 1: Density profiles, favourable pressure gradient

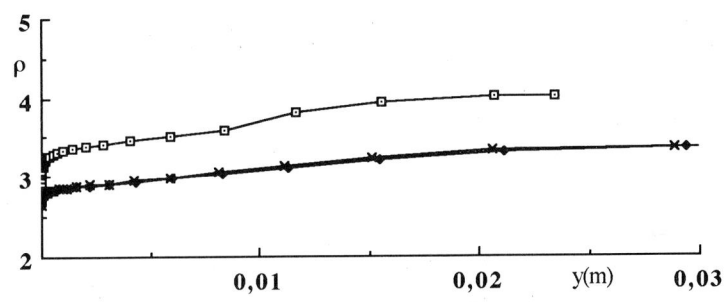

Figure 2: Density profiles, adverse pressure gradient

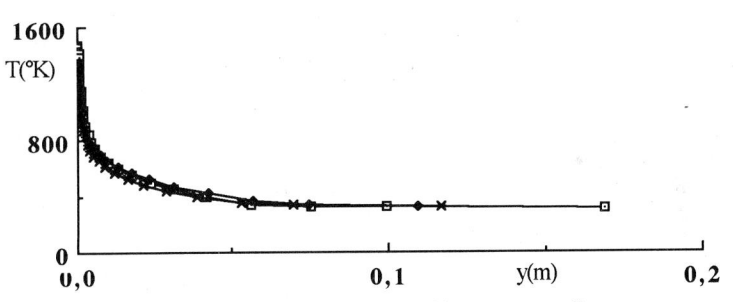

Figure 3: Tempetature profiles, favourable pressure gradient

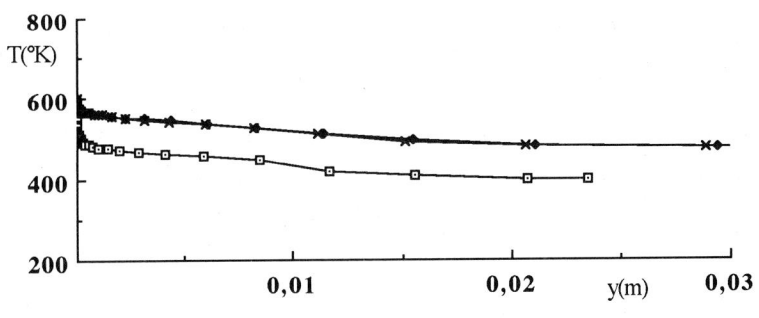

Figure 4: Temperature profiles, adverse pressure gradient

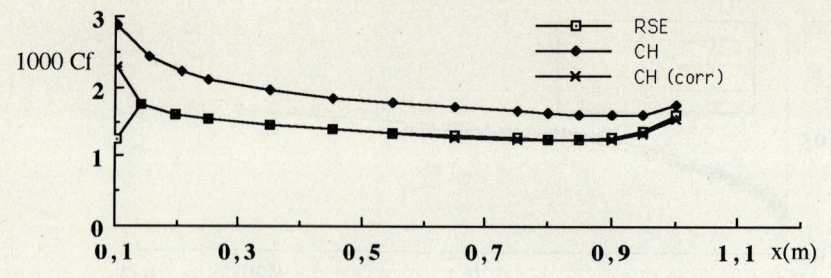

Figure 5: Skin friction distribution, favourable pressure gradient

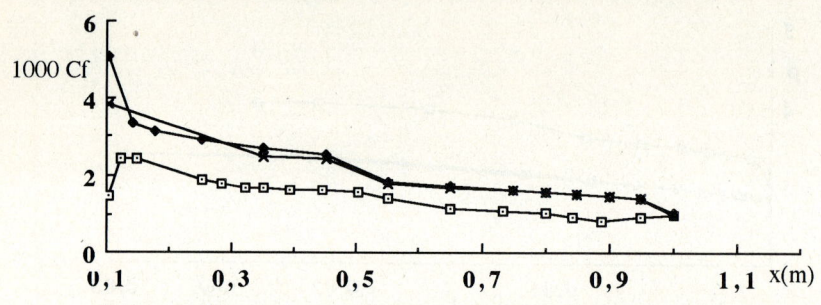

Figure 6: Skin friction coefficient distribution, adverse pressure gradient

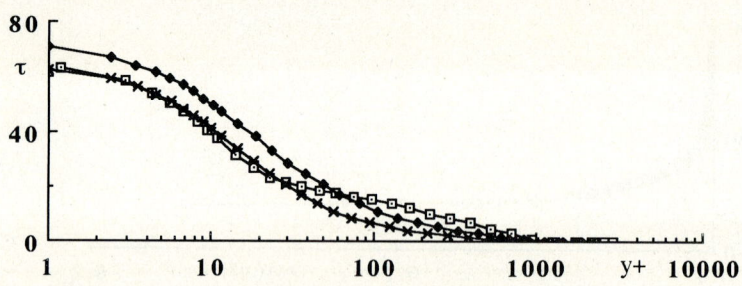

Figure 7: Total shear stress profiles, favourable pressure gradient

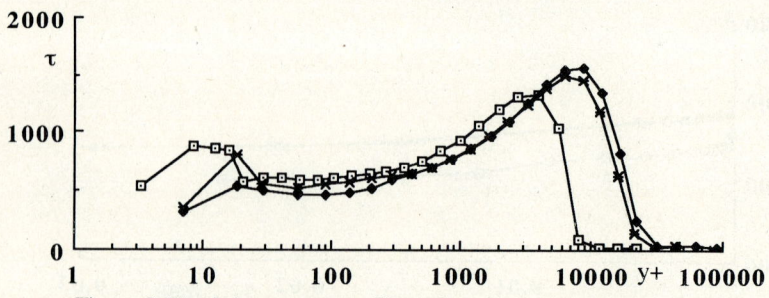

Figure 8: Total shear stress profiles, adverse pressure gradient

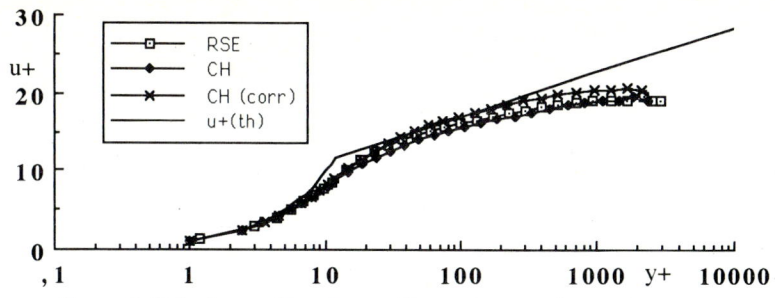

Figure 9: Velocity profiles, favourable pressure gradient

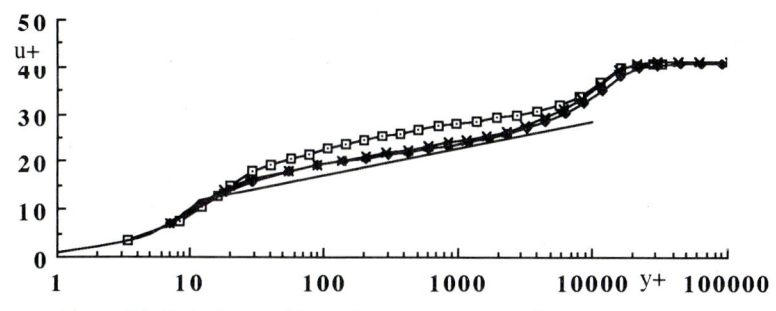

Figure 10: Velocity profiles, adverse pressure gradient

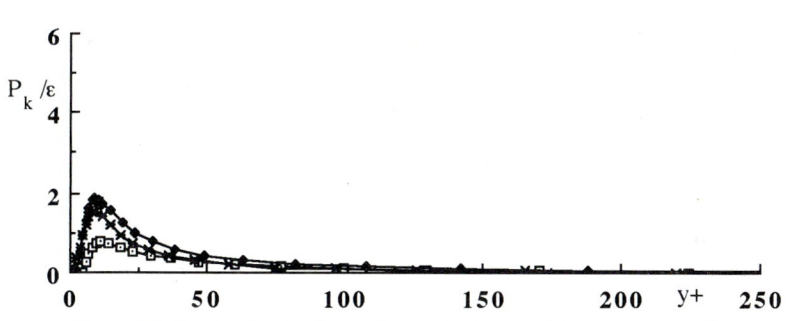

Figure 11: Production to dissipation ratio, favourable pressure gradient

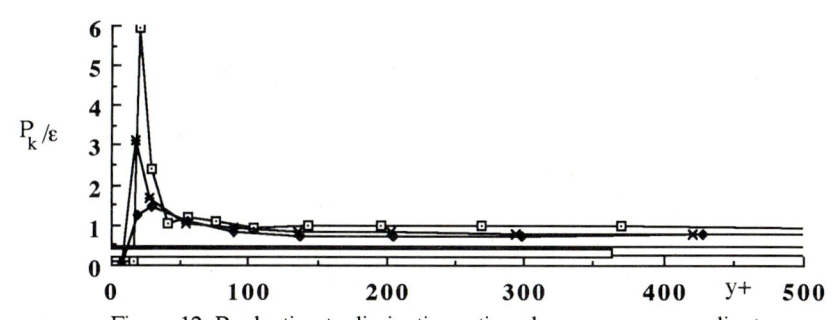

Figure 12: Production to dissipation ratio, adverse pressure gradient

Figure 13: Turbulent kinetic energy, favourable pressure gradient

Figure 14: Turbulent kinetic energy profiles, adverse pressure gradient

Figure 15: Reynolds stresses profiles, favourable pressure gradient

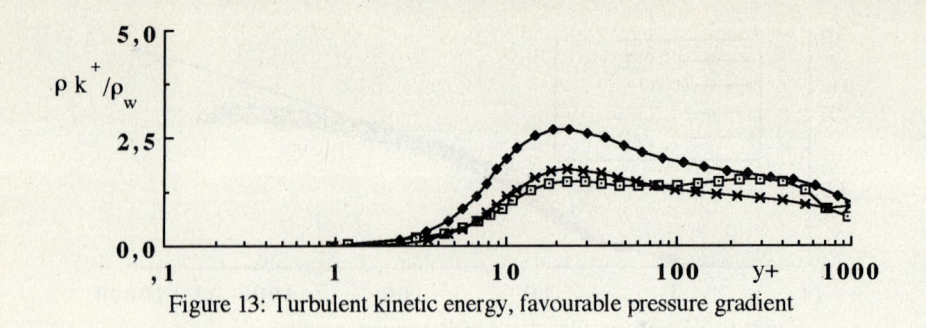

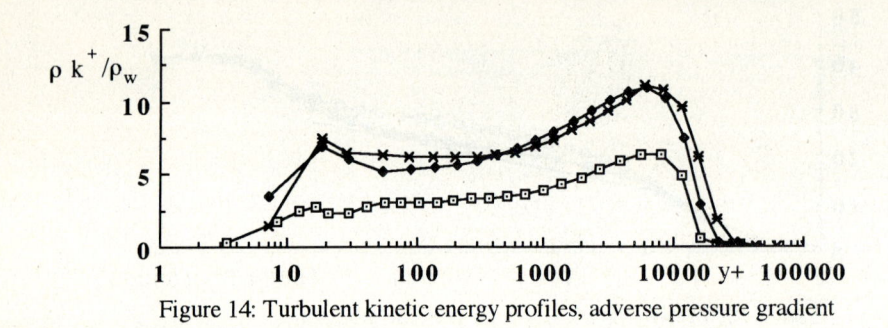

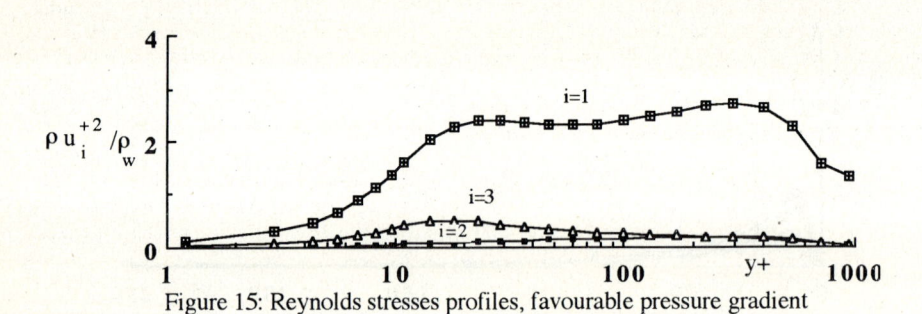

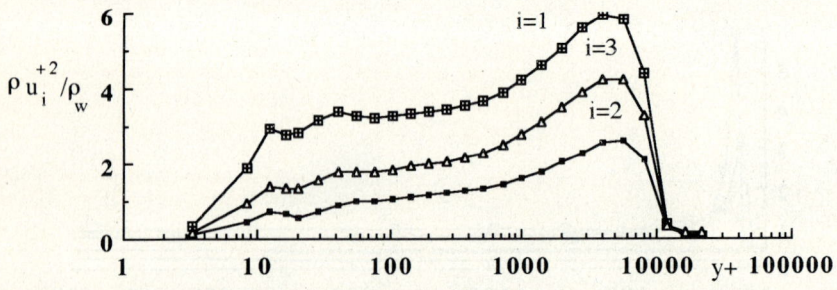

Figure 16: Reynolds stresses profiles, adverse presssure gradient

TRANSONIC AEROFOIL PERFORMANCE BY SOLUTION OF THE
REYNOLDS-AVERAGED NAVIER-STOKES EQUATIONS WITH A
ONE-EQUATION TURBULENCE MODEL

L. J. Johnston

Department of Mechanical Engineering, UMIST
P O Box 88, Manchester M60 1QD, England

SUMMARY

A computational method for the viscous flow around transonic aerofoil sections is described. The Reynolds-averaged Navier-Stokes equations are discretised in space using a cell-centred finite-volume procedure, steady-state solutions being obtained by marching the unsteady flow equations in time using an explicit multi-stage scheme. Turbulence modelling is currently at the one-equation level, solving an additional transport equation for the turbulent kinetic energy. Results are presented for the transonic flow development around the MBB-A3 supercritical aerofoil section, covering a range of conditions from fully-subcritical to shock-induced separation. Surface pressure distributions are in reasonable agreement with experiment, but drag levels are overpredicted at higher Mach numbers.

1. INTRODUCTION

The development of methods to compute accurately the viscous transonic flow around aerofoil sections continues to receive much attention in the Computational Fluid Dynamics (CFD) community. This is particularly true for modern supercritical sections, the significant rear loading of which leads to an increase in the importance of viscous effects relative to more traditional sections. Supercritical aerofoils are designed to feature weak upper surface shock waves at the conditions of practical interest, generally positioned downstream of mid-chord to increase the lifting capability of the section. The pressure rise to the trailing-edge is thus confined to a short chordwise region, resulting in strong adverse pressure gradients. These, in turn, give rise to the development of thick turbulent boundary layers which are close to separation in the trailing-edge region.

The practical application of supercritical aerofoil technology has been particularly successful for civil transport aircraft wings. Such applications are dominated by the requirement to minimise drag at the design cruise condition, by ensuring a separation-free flow with only weak shock waves. Under these conditions, design and analysis methods combining solutions of the potential equation for the inviscid flow and the integral boundary-layer equations for the viscous flow have been highly successful. However, the coupling procedures between the inviscid and viscous flow algorithms tend to break down in the presence of significant regions of shock-induced and/or trailing-edge separation. The trend in the CFD community is now towards the solution of the Reynolds-averaged Navier-Stokes equations, to enable the routine calculation of flows near the buffet boundary, where strong shock waves and extensive separated-flow regions are likely to be present. These methods require the explicit use of a turbulence model, in order to close the set of mean-flow equations.

The aim of the present paper is to describe a recently-developed solution procedure for the Reynolds-averaged Navier-Stokes equations. A simple, but relatively efficient, numerical algorithm is adopted for the governing flow equations, enabling effort to be concentrated on turbulence model implementation and evaluation. Results are presented of computations for the MBB-A3 supercritical aerofoil section over a wide range of flow conditions, a one-equation turbulence model being employed. A more complete description of the present calculation method, together with an extensive validation for a range of transonic aerofoils, can be found in Johnston [1], [2].

2. GOVERNING FLOW EQUATIONS

The two-dimensional Reynolds-averaged Navier-Stokes equations, in terms of mass-weighted average variables for application to compressible turbulent flows, are to be solved. The set of governing flow equations, augmented by an additional modelled transport equation for the turbulent kinetic energy, can be written in time-dependent integral form as follows

$$\frac{\partial}{\partial t} \int_{\Omega} \underline{W} \, d\Omega + \int_{\Omega_s} \underline{H} \cdot \underline{n} \, d\Omega_s + \int_{\Omega} \underline{S}^v \, d\Omega = 0. \tag{1}$$

Ω is the two-dimensional flow domain, Ω_s is the boundary to the domain and $\underline{n}$ is the unit outward normal to this boundary. $\underline{W}$ is the vector of conserved variables

$$\underline{W} = \begin{bmatrix} \rho \\ \rho U \\ \rho V \\ \rho E \\ \rho k \end{bmatrix} \tag{2}$$

where ρ, E and k are the density, total energy per unit mass and turbulent kinetic energy respectively; U and V are the cartesian mean-velocity components. $\underline{H}$ contains the flux vectors

$$\underline{H} = \underline{F} \cdot \underline{i} + \underline{G} \cdot \underline{j} = (\underline{F}^i + \underline{F}^v) \cdot \underline{i} + (\underline{G}^i + \underline{G}^v) \cdot \underline{j}. \tag{3}$$

$\underline{i}$ and $\underline{j}$ being unit vectors in the X- and Y-directions of the cartesian coordinate system. The flux vectors $\underline{F}$ and $\underline{G}$ include both convective transport (superscript i) and viscous diffusive transport (superscript v) terms

$$\underline{F}^i = \begin{bmatrix} \rho U \\ \rho U^2 + P \\ \rho UV \\ \rho UH \\ \rho Uk \end{bmatrix}, \quad \underline{F}^v = \begin{bmatrix} 0 \\ \sigma_{xx} \\ \sigma_{xy} \\ U\sigma_{xx} + V\sigma_{xy} + q_x \\ -(\mu + \mu_t/\sigma_k) \, \partial k/\partial X \end{bmatrix},$$

$$\underline{G}^i = \begin{bmatrix} \rho V \\ \rho UV \\ \rho V^2 + P \\ \rho VH \\ \rho Vk \end{bmatrix}, \quad \underline{G}^v = \begin{bmatrix} 0 \\ \sigma_{xy} \\ \sigma_{yy} \\ U\sigma_{xy} + V\sigma_{yy} + q_y \\ -(\mu + \mu_t/\sigma_k) \, \partial k/\partial Y \end{bmatrix}. \tag{4}$$

P is the static pressure and H the total enthalpy per unit mass. σ_{xx}, σ_{yy} and σ_{xy} are components of the stress tensor, whilst q_x and q_y are components of the heat-flux vector. Diffusive transport of the turbulent kinetic energy has been modelled by assuming an isotropic gradient diffusion process, σ_k being a model constant. Reynolds stresses and turbulent heat fluxes in the mean-flow equations are modelled by introducing an isotropic eddy viscosity μ_t and a turbulent Prandtl number Pr_t. Thus, the viscous stress terms in equation (4) become

$$\sigma_{xx} = -(\mu + \mu_t) s_{xx} + \frac{2}{3} \rho k,$$

$$\sigma_{yy} = -(\mu + \mu_t) s_{yy} + \frac{2}{3} \rho k, \quad \sigma_{xy} = -(\mu + \mu_t) s_{xy} \tag{5}$$

where the components of the mean-strain tensor are

$$s_{xx} = 2\frac{\partial U}{\partial X} - \frac{2}{3}\left[\frac{\partial U}{\partial X} + \frac{\partial V}{\partial Y}\right] \;,\; s_{yy} = 2\frac{\partial V}{\partial Y} - \frac{2}{3}\left[\frac{\partial U}{\partial X} + \frac{\partial V}{\partial Y}\right] \;,\; s_{xy} = \left[\frac{\partial U}{\partial Y} + \frac{\partial V}{\partial X}\right]. \quad (6)$$

Similarly, the components of the heat-flux vector in equation (4) become

$$q_x = -\gamma\,(\mu/Pr + \mu_t/Pr_t)\,\partial T/\partial X \;,\; q_y = -\gamma\,(\mu/Pr + \mu_t/Pr_t)\,\partial T/\partial Y. \quad (7)$$

Pr and Pr_t are the laminar and turbulent Prandtl numbers respectively, γ being the ratio of specific heats. The static temperature T in equation (7) is evaluated using the perfect gas relation, with the pressure P determined using the definition of the total energy. Sutherland's law is used to evaluate the laminar viscosity μ.

$\underline{S}^V$ in equation (1) contains the source terms in the modelled transport equation for k

$$\underline{S}^V = \begin{bmatrix} 0 \\ 0 \\ 0 \\ 0 \\ S_k \end{bmatrix}. \quad (8)$$

The surface boundary conditions for equation (1) are the no-slip conditions, together with the assumption of an adiabatic wall

$$U_w = V_w = k_w = (\partial T/\partial y_n)_w = (\partial P/\partial y_n)_w = 0. \quad (9)$$

Subscript w denotes conditions at the wall and y_n is the surface normal distance. Non-reflecting farfield boundary conditions are applied at the outer boundary to the computational domain. These are constructed using the Riemann invariants for a one-dimensional flow normal to the outer boundary.

3. SPATIAL DISCRETISATION

The computational domain Ω is divided into a finite number of non-overlapping quadrilateral cells. Fig 1 shows a typical cell, which has four edges (1,2,3,4) and four vertices (a,b,c,d). The conserved variables, equation (2), within a cell are represented by their average cell-centre values, such quantities being denoted by suffices (i,j) in the local curvilinear coordinate directions (ξ,η). The governing flow equations are now applied to each computational cell in turn. Performing the finite-volume spatial discretisation of equation (1) before the time discretisation leads to a large set of ordinary differential equations with respect to time

$$\frac{d\underline{W}_{i,j}}{dt} = -\frac{1}{h_{i,j}} \sum_{k=1}^{4} (\underline{F}_k \Delta Y_k - \underline{G}_k \Delta X_k) - \underline{S}^V_{i,j} \quad (10)$$

where $h_{i,j}$ is the area of the cell, and the summation (k) over the four edges is a discrete approximation to the contour integral in equation (1). Cell-edge values of the flux terms are approximated by averaging the two adjacent cell-centre values

$$\underline{F}_1 = (\underline{F}_{i,j} + \underline{F}_{i,j-1})/2 \qquad \text{etc.} \quad (11)$$

The viscous flux terms in equation (10) require cell-edge values of first derivatives of U, V and T with respect to X and Y. Derivatives at the cell vertices (a,b,c,d) are calculated using auxiliary cells surrounding each vertex, Fig 2. For example, a discrete application of the divergence theorem to the auxiliary cell surrounding vertex b results in

$$(\partial U/\partial X)_b = \frac{1}{h_b}\sum_{k=5}^{8} U_k\,\Delta Y_k \;,\; (\partial U/\partial Y)_b = -\frac{1}{h_b}\sum_{k=5}^{8} U_k\,\Delta X_k \quad (12)$$

where h_b is the area of the auxiliary cell. Cell-edge mean velocities are approximated by the relevant cell-centre values

$$U_5 = U_{i+1,j} \quad , \quad U_6 = U_{i+1,j+1} \quad \text{etc.} \quad (13)$$

Cell-edge values of the derivatives are then approximated by

$$(\partial U/\partial X)_1 = [(\partial U/\partial X)_a + (\partial U/\partial X)_d]/2 \quad \text{etc.} \quad (14)$$

4. TURBULENCE MODELLING

The present one-equation turbulence model is similar to that of Mitcheltree et al [3]. The near-wall formulation is a compressible flow adaptation of an incompressible flow model due to Wolfshtein [4]. A modelled transport equation for the turbulent kinetic energy k is solved in conjunction with the mean-flow equations. Turbulent length scales, required to close the turbulence model, are defined in algebraic form. The source term S_k in equation (8) contains production terms and the rate of turbulent kinetic energy dissipation ϵ

$$S_k = -(\mu_t s_{xx} - \frac{2}{3}\rho k)\frac{\partial U}{\partial X} - (\mu_t s_{yy} - \frac{2}{3}\rho k)\frac{\partial V}{\partial Y} - \mu_t s_{xy}^2 + \rho\epsilon \quad (15)$$

the mean strains s_{xx}, s_{yy} and s_{xy} being defined in equation (6). The eddy viscosity and dissipation rate of k are modelled as follows

$$\mu_t = c_\mu \rho k^{1/2} L_\mu \quad , \quad \rho\epsilon = \rho k^{3/2}/L_\epsilon . \quad (16)$$

The two length scales L_μ and L_ϵ are defined by algebraic relations, their formulation being in two parts for wall-bounded flows. In the inner region

$$L_{\mu i} = c_1 y_n [1 - \exp(-R_k/A_\mu)] \quad ,$$

$$L_{\epsilon i} = c_1 y_n [1 - \exp(-R_k/2c_1)] \quad , \quad R_k = \rho k^{1/2} y_n/\mu . \quad (17)$$

Both length scales take constant values in the outer region, $L_{\mu o}$ and $L_{\epsilon o}$, scaling with the local boundary layer thickness; see Johnston [1] for further details. Blending functions are used to ensure a smooth matching of the inner and outer formulations

$$L_\mu = L_{\mu o} \tanh(L_{\mu i}/L_{\mu o}) \quad , \quad L_\epsilon = L_{\epsilon o} \tanh(L_{\epsilon i}/L_{\epsilon o}), \quad (18)$$

Finally, the one-equation model involves five constants

$$\sigma_k = 1.0 \quad , \quad c_\mu = 0.09 \quad , \quad \kappa = 0.41 \quad , \quad A_\mu = 76 \quad , \quad c_1 = \kappa/c_\mu^{3/4} \quad (19)$$

The practical implementation of the turbulence model in wake regions is a problem. For simplicity, the trailing-edge eddy viscosity and dissipation rate distributions are imposed at all stations downstream in the wake. This is a reasonable approximation for the immediate near-wake region, and the far wake is expected to have little influence on the flow development around the aerofoil.

5. SOLUTION PROCEDURE

Following the spatial discretisation described in Section 3, the resulting large set of ODEs, equation (10), is integrated in time to a steady-state solution, using an explicit four-stage scheme. Additional numerical dissipation terms are used to suppress odd-even point decoupling and prevent oscillations around shock waves. These terms are constructed following the approach of Jameson et al [5] for the Euler equations, but with explicit control of their magnitude in viscous flow regions; see Johnston [1] for details.

6. RESULTS

Bucciantini et al [6] present surface pressure measurements and integrated loads for the MBB-A3 supercritical aerofoil section, covering a wide range of transonic flow conditions.

Test results from the ARA Bedford 8in x 18in transonic wind tunnel are chosen for comparison, since the tunnel corrections to incidence angle, lift and pitching moment coefficients are well-defined for these cases. Fig 3 shows the aerofoil geometry, together with the inner region of the computational grid. This consists of 272 x 64 cells in the surface tangential and near-normal directions respectively, 224 of the cells being around the aerofoil itself. Initial cell spacings of 0.002 and 0.00005 chord, in the tangential and near-normal directions, are used in the leading- and trailing-edge regions, the latter spacing corresponding to a maximum y^+ of about 5 at the cell-centres adjacent to the surface. The outer boundary to the computational domain is 15 chords away from the aerofoil and 10 chords downstream of the trailing edge. Transition is fixed at 0.03 and 0.4 chord on the upper and lower surfaces respectively for the computations, whilst the experiments involve free transition. The freestream Reynolds number is 6×10^6.

Fig 4 compares predicted surface pressures with experiment for a range of incidence angles at a Mach number of 0.70. Differences are essentially confined to shock-wave position, which is predicted too far downstream at the lower incidences. The reason for this discrepancy is currently under investigation, but may be associated with the low aspect ratio of the wind tunnel model. Shock-induced separation is predicted for the two highest incidence cases, with chordwise extents of $0.4 < X/c < 0.44$ and $0.42 < X/c < 0.49$ respectively. Fig 5 shows that the lift-drag and lift-pitching moment polars are well-predicted for this freestream Mach number. The second set of results, Figs 6 and 7, involve a range of Mach numbers at a fixed incidence angle $\alpha \approx 1.7°$. Again, the computed shock-wave position is too far downstream, with the shock-wave strength being significantly over-predicted. The flow is fully-attached in all cases apart from at $M = 0.798$, for which a large shock-induced separated-flow region is predicted for $0.68 < X/c < 0.85$. The over-prediction of shock-wave strength results in premature drag divergence, as shown in Fig 7. Note, however, that the variation of pitching moment with Mach number is well-predicted.

7. CONCLUSIONS

The development of a method to solve the compressible Reynolds-averaged Navier-Stokes equations has been described. The method involves a finite-volume spatial discretisation, together with an explicit multi-stage time-stepping procedure. A modelled transport equation for the turbulent kinetic energy is integrated directly into the set of mean-flow equations. No changes to the basic flow algorithm are found to be necessary, to accommodate this additional equation. Computations for transonic aerofoil flows indicate a need for improved turbulence modelling of shock wave/boundary layer interactions.

REFERENCES

[1] JOHNSTON L J, "Solution of the Reynolds-averaged Navier-Stokes equations for transonic aerofoil flows", The Aeronautical Journal, October 1991.
[2] JOHNSTON L J "Computation of viscous transonic aerofoil flows using eddy-viscosity based turbulence models", 7th International Conference on Numerical Methods in Laminar and Turbulent Flow, Stanford, California, USA, 15-19 July 1991.
[3] MITCHELTREE R, SALAS M and HASSAN H "A one equation turbulence model for transonic airfoil flows", AIAA Paper 89-0557, 1989.
[4] WOLFSHTEIN M "The velocity and temperature distribution in one-dimensional flow with turbulence augmentation and pressure gradient", Int J Heat and Mass Transfer, Vol 12, 1969, pp 301-318.
[5] JAMESON A, SCHMIDT W and TURKEL E "Numerical solutions of the Euler equations by finite volume methods using Runge-Kutta time-stepping schemes", AIAA Paper 81-1259, 1981.
[6] BUCCIANTINI G, OGGIANO M S and ONORATO M "Supercritical airfoil MBB-A3 surface pressure distributions, wake and boundary condition measurements", AGARD AR 138, May 1979, A8-1 to A8-25.

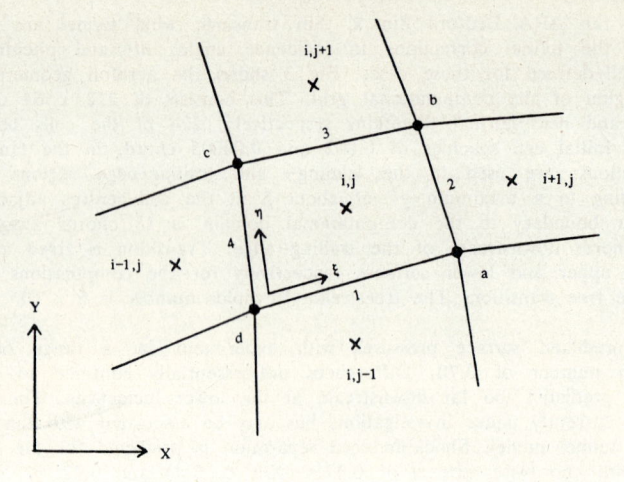

Fig 1   Notation for a typical computational cell
x = cell centres; ● = cell vertices

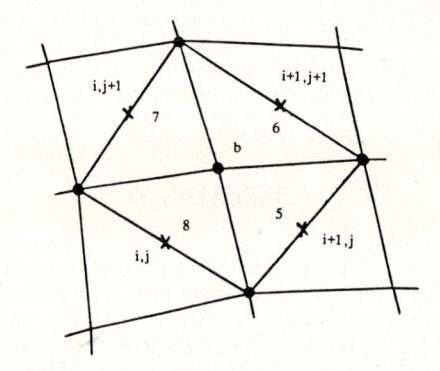

Fig 2   Auxiliary cell, for viscous flux computation

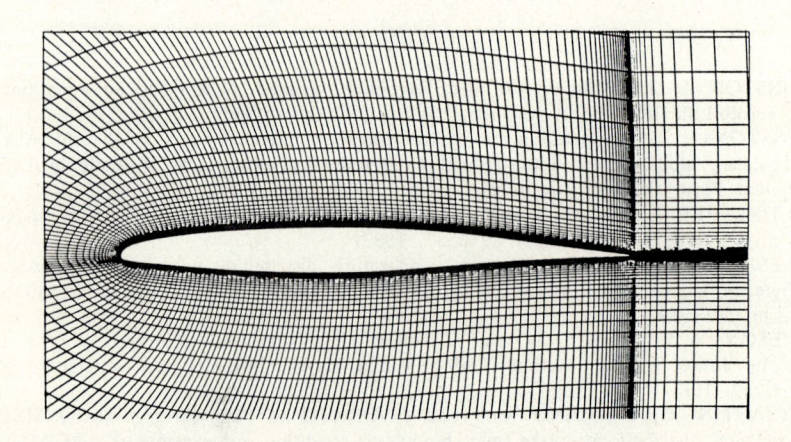

Fig 3   Computational grid for MBB-A3 aerofoil

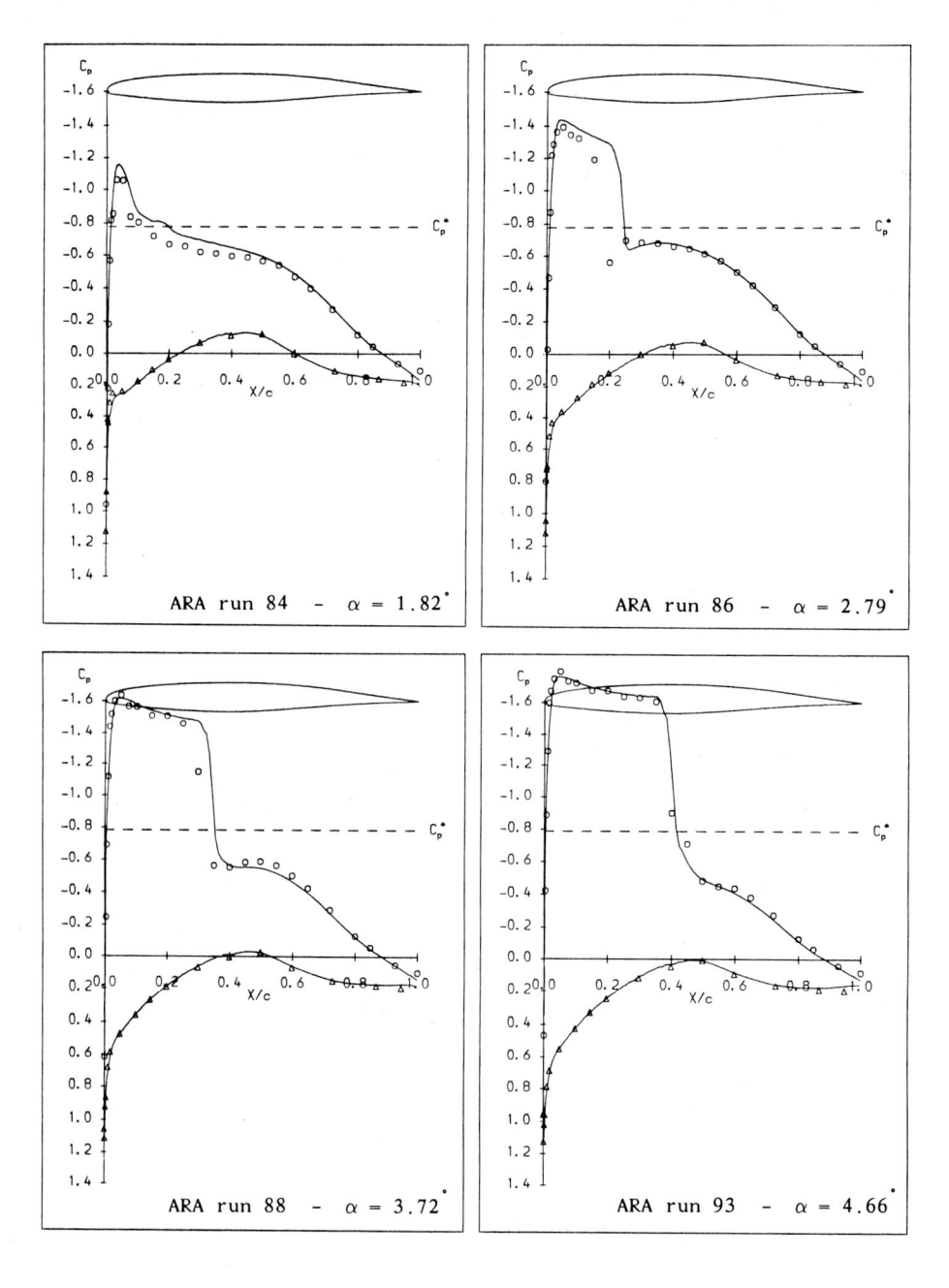

Fig 4 Surface pressure distributions for $M = 0.70$, $R = 6 \times 10^6$
o, △ experiment; ——— present calculations

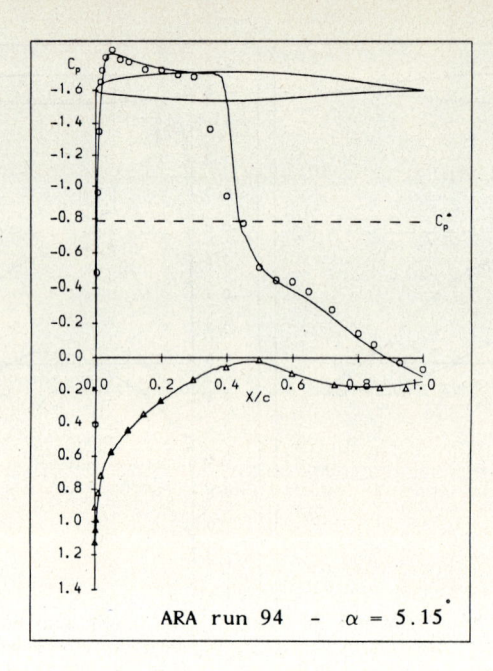

Fig 4   concluded

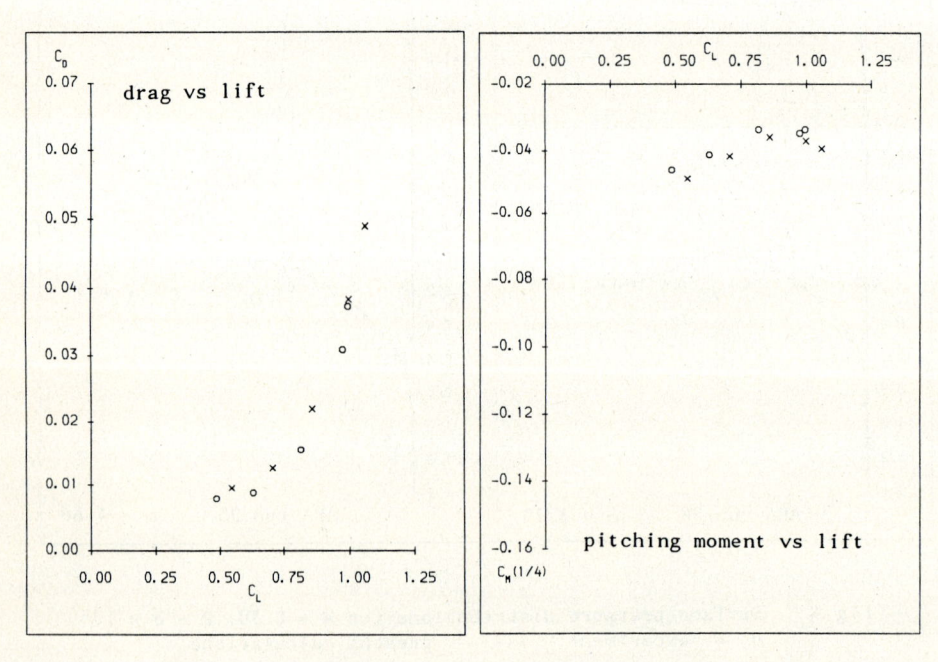

Fig 5   Integrated loads for $M = 0.70$, $R = 6 \times 10^6$
o experiment; x present calculations

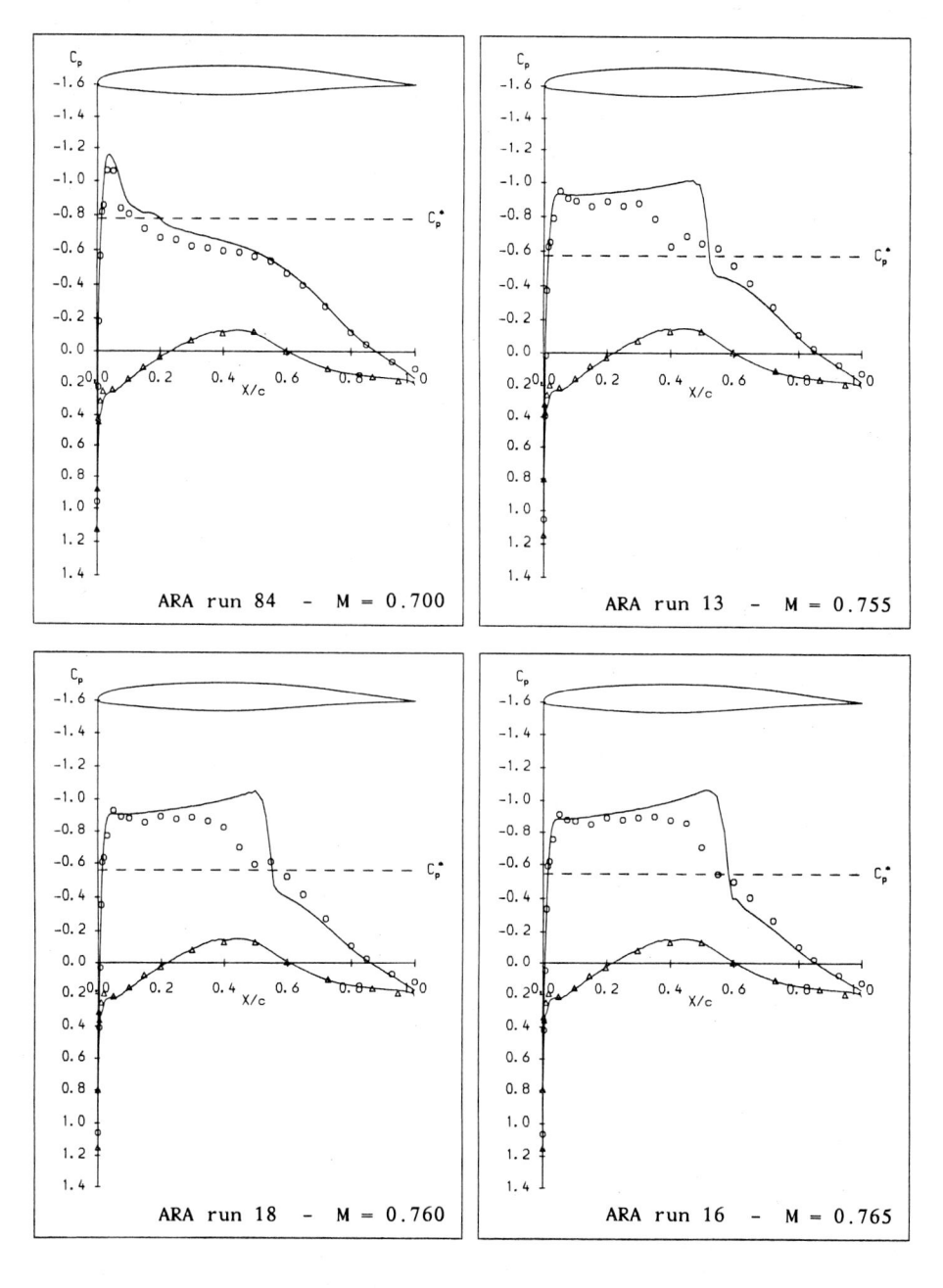

Fig 6 Surface pressure distributions for $\alpha \approx 1.7°$, $R = 6 \times 10^6$
o, △ experiment; ——— present calculations

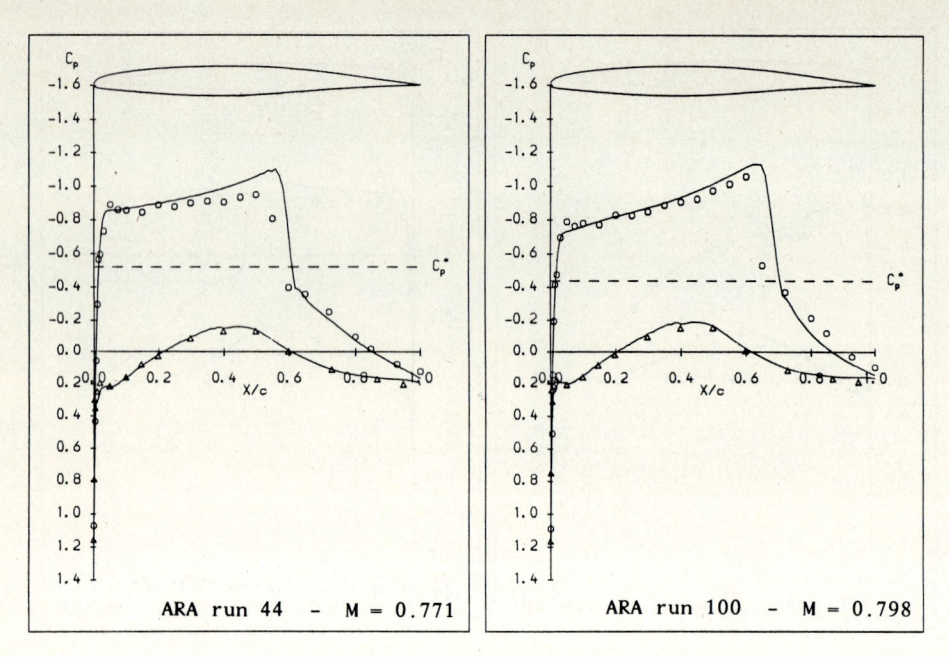

Fig 6   concluded

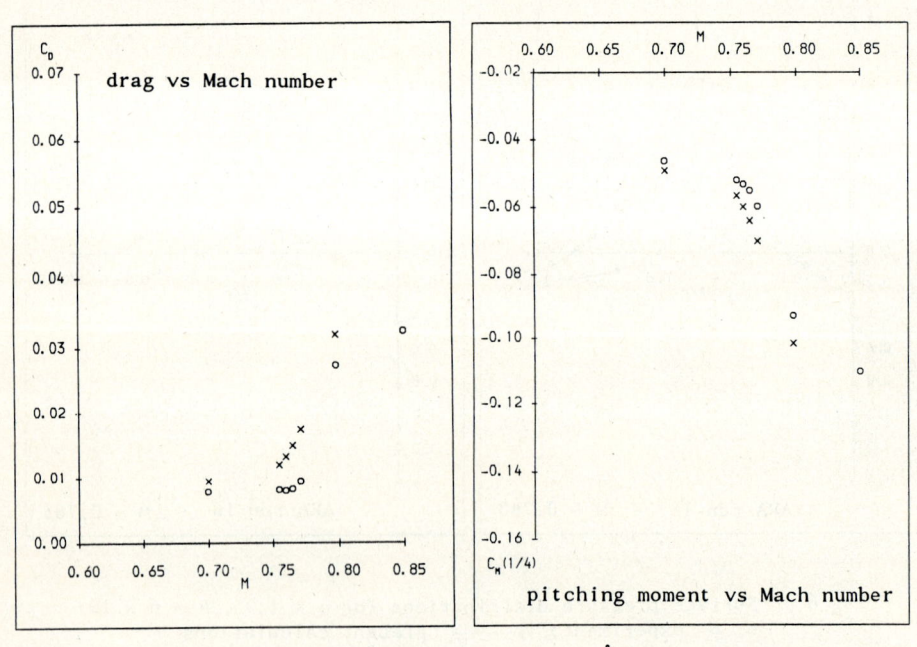

Fig 7   Integrated loads for $\alpha \approx 1.7°$, $R = 6 \times 10^6$
o experiment;   x present calculations

COMPLEX FLOWS

Numerical Simulation of Wind Tunnel Flows Past a Flame Holder Model

X.S. Bai and L. Fuchs
Dept. of Gasdynamics, The Royal Institute of Technology S–100 44
Stockholm, Sweden

Summary

In this paper the numerical simulation of cold (isothermal) flows past a flame holder in a wind tunnel is presented. In the calculations the turbulent flow field is treated by means of approximated modelling methods, namely the k–ε equations and Large Eddy Simulations (LES). In order to get high resolution, very fine grids are required. To handle such large scale problems we use both Multigrid methods and Local grid refinements technique. The results show that k–ε model gives a steady flow field when the inlet flow is steady and boundaries are fixed. However the results from LES show unsteady flows at the same inlet and boundary situations. When compared with experiments, it is found that the k–ε model is in agreement with the (averaged) velocity distributions, while LES can capture the same fluctuation frequencies flows as measurements do. It is also shown that Multigrid method and Local grid refinements improve the computational efficiency.

1. Introduction

The recirculation region and intense turbulent mixing behind a bluff body have positive effects for flame stabilization in combustion chambers. The flame stability characteristics of the bluff body are largely affected by the aerodynamics of the flow around the body. So it is of great interest to better understand such flow fields quantitatively for engineering purposes. Some relevant experimental research works have been carried out [1–2]. Currently, "cold" and "hot" measurements are being carried out at Volvo Flygmotor AB [3] on flame holder models placed in a wind tunnel ("Validation Rig" (VR)). Here, we consider the numerical simulations of the cold flow on the VR model.

The flow in the VR model is inevitably turbulent since the Reynolds number is very high. It is impossible for current supercomputers to simulate such turbulence flow by directly solving the full Navier–Stokes equations because of the huge size of required storage. It is also not always necessary to do so since in many engineering cases only the averaged characteristics of the flow field are of interest. Therefore one may use modelling methods for the small scale variations. In the current calculations we use both the (more or less standard) k–ε model [4–6] and Large Eddy Simulations (LES).

The basic problem in LES as in the k–ε model is modelling the behavior of the unresolved scales, in terms of an effective viscosity. An often used model in LES is the Smagorinsky model, which is simple, but has certain limitations.

Our main goal here is to calibrate and assess the sensitivity of these models to parameter variations, for the family of flows under consideration.

2. Mathematical Models

2.1 Turbulence Modeling Equations

The Reynolds averaged Navier–Stokes equations are used to describe the turbulent flow. In cartesian coordinates the equations for steady incompressible flow can be written as:

$$\frac{\partial u_j}{\partial x_j} = 0 \tag{1}$$

$$u_j \frac{\partial u_i}{\partial x_j} = -\frac{\partial p}{\partial x_i} + \frac{\partial}{\partial x_j}(v_{eff} \frac{\partial u_i}{\partial x_j}) . \tag{2}$$

The turbulence kinetic energy k and turbulence kinetic energy dissipation rate ε are given by the two–equation k–ε model:

$$u_j \frac{\partial k}{\partial x_j} = \frac{\partial}{\partial x_j}(\frac{v_{eff}}{c_k}\frac{\partial k}{\partial x_j}) + v_t(\frac{\partial u_i}{\partial x_j} + \frac{\partial u_j}{\partial x_i})\frac{\partial u_i}{\partial x_j} - \epsilon \tag{3}$$

$$u_j \frac{\partial \epsilon}{\partial x_j} = \frac{\partial}{\partial x_j}(\frac{v_{eff}}{c_\epsilon}\frac{\partial \epsilon}{\partial x_j}) + C_1 v_t \frac{\epsilon}{k}(\frac{\partial u_i}{\partial x_j} + \frac{\partial u_j}{\partial x_i})\frac{\partial u_i}{\partial x_j} - C_2 \frac{\epsilon^2}{k} \tag{4}$$

where v_{eff} is effective viscosity $v_{eff} = v_L + v_t$. v_L is laminar viscosity, v_t is turbulent eddy viscosity:

$$v_t = C_\mu \frac{k^2}{\epsilon} . \tag{5}$$

$C_1, C_2, C_\mu, C_\epsilon, C_k$ are constants.

Equations (1) – (5) form a nonlinear system, which requires boundary conditions to be specified along all boundaries. The boundary conditions at inlet and outlet have given profiles, and no slip velocity is applied at solid walls. The boundary conditions for k and ε at solid walls are generally hard to give, mostly they are calculated at the first grid point (p) near the wall (wall functions). The following wall function model is used.

$u, v, w :$ $\quad (\tau_w/\varrho)^{1/2} \equiv u_* = V_p \kappa / \ln(\frac{Eu_* y_p}{v_L})$ $\quad$ for $y^+ > 11.63$

$\quad\quad\quad\quad \tau_w/\varrho = V_p v_L / y_p$ $\quad\quad\quad\quad\quad\quad\quad$ for $y^+ \leq 11.63$

$k :$ $\quad\quad k_p = C_\mu^{-1/2} u_*^2$ $\tag{6}$

ϵ:
$$\epsilon_p = u_*^3/(\kappa y_p)$$
$$y^+ = \frac{u_* y_p}{\nu_L}.$$

y_p is distance between the first grid and wall, E and κ are constants. The density ρ is constant for incompressible flows, τ_w is wall shear–stress. V_p is the velocity component at point p parallel to the wall, u_* is the wall shear velocity. y^+ is the dimensionless distance from the solid wall.

2.2 Large Eddy Simulation

The equations for LES are the unsteady Navier–Stokes equations with a sub–grid scale model, i.e. Smagorinsky's model. The Smagorinsky constant is taken to be 0.2, and the filter bandwidth has been varied to study the effect of this parameter. In the calculations reported here, the filter bandwidth is taken to be equal to one mesh spacing. As part of the calculation, we have computed the mean values of the dependent variables, and the correlations of the fluctuations. Details of these results are to be reported elsewhere. In LES calculations, non–slip boundary conditions are applied on solid boundaries.

3. Solution Procedure

The system of equations (1) – (6) is approximated using a finite difference method. Second order central differences are used to approximate the diffusive terms. Hybrid (central/upwind) differences are used to approximate convective terms, depending on the local Peclet number. The discrete system of equation is solved by a Multi–Grid (MG) method [7]. The smoother consists of a sequence of sweeps (pointwise relaxation) on each equation. When the continuity equation is updated at a given cell, the velocity and pressure are updated simultaneously. The transfer to coarse grids is done by volume averaging and corrections are interpolated to fine grids by trilinear interpolations.

The LES calculations are done using different discrete approximations to the convective terms: The non–conservative formulation and the formulation of Arakawa [8]. The latter formulation is the discrete approximation of the convective terms written as:

$$0.5(u_j \frac{\partial u_i}{\partial x_j} + \frac{\partial u_j u_i}{\partial x_j}).$$

Both forms are approximated with central differences applied in a defect correction mode. The basic time–dependent flow solver uses hybrid discretization, but higher order schemes can be introduced by correcting the lowest order truncation errors in each time step. The time–integration is done by solving implicitly (the Stokes part or the full Navier – Stokes equations), using a Fixed MG method in each time step. The time steps that we use are such that we can detect fluctuations of flow variables up to several hundreds Hz. If smaller time steps are of interest, an explicit scheme can be used to integrate the momentum equations, while the MG solver can be used to solve continuity equation (for updating the pressure). The amount of computational work for each time step can be restricted to some 3 to 6 times the work required to compute the residuals.

In order to get much better resolution at near wall regions, a local grid refinement technique is used [7,9]. The boundary conditions on locally refined grids are either given (at physical boundary) or derived from the values on the next coarser grid (at internal boundary).

4. Results and Discussion

The calculations that we present here have been done on the geometry of the VR [3]. Figure 1 shows the VR model. A triangular prism (with a base height of 40 mm) has been placed in a wind tunnel of height of 120 mm (in z- direction), and width of 240 mm (in y- direction). The inlet velocity profile was uniform with an air speed of about 17 m/s. The experimental data was taken at mid–plane in y- direction (test plane).

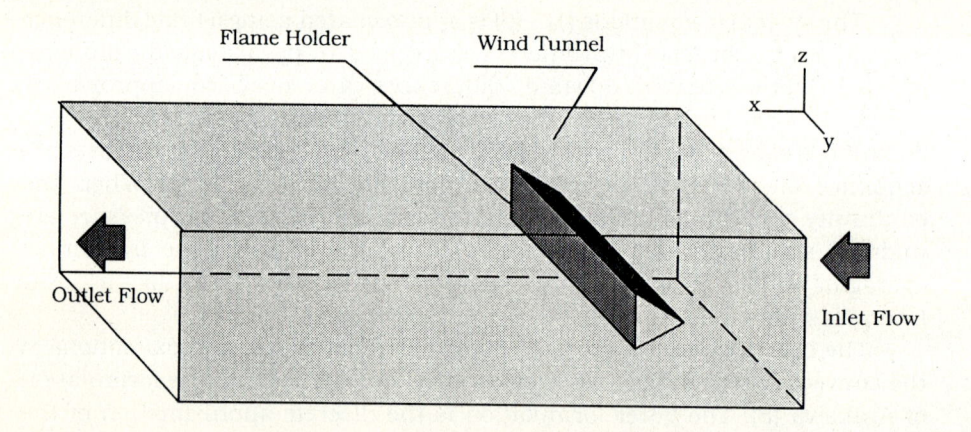

Figure 1 An illustration plot of the validation rig (VR)

Initial calculations were made by 2-D codes: using either primitive variables and k–ε model on a cartesian grid, or a LES using streamfunction and vorticity formulation on an overlapping grid system. In the later case, an O–mesh (body fitted) was used around the flame holder, and a cartesian mesh was used inside the wind–tunnel. The LES calculations showed that the flow was unsteady. Because of this unsteadyness, there was no reason to believe that the 2-D assumption would be adequate for LES calculations and therefore later calculations were all done on the 3-D geometry.

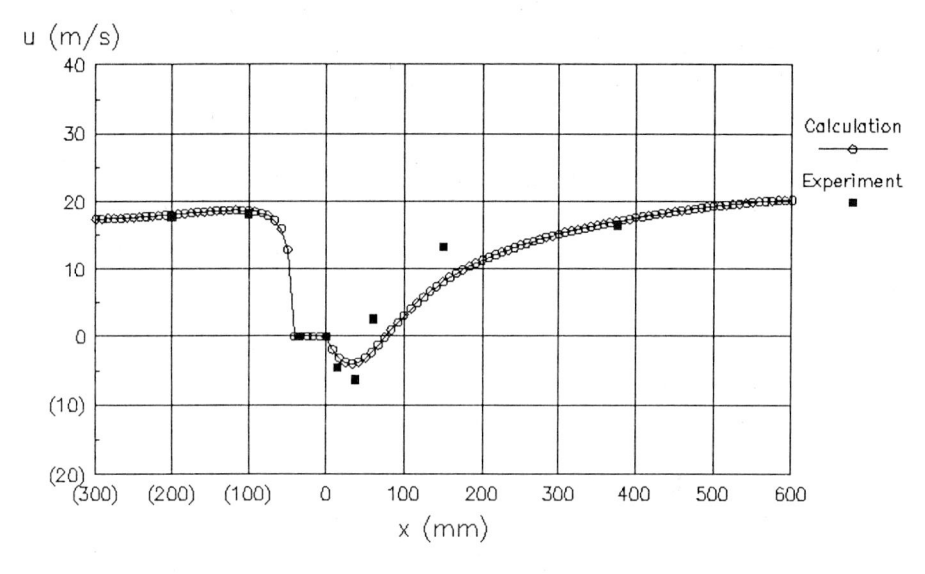

Figure 2 u- velocity component distribution along streamwise x- direction

The numerical results calculated by using the k-ε model are carried out on a grid 122x26x30 with 3 Multigrid levels. A locally refined grid 72x50x58 was added around the flame holder. As reported in [7], Multigrid methods generally accelerate the convergence rate, and locally refined grids improve the numerical accuracy with minimal cost of computer resource.

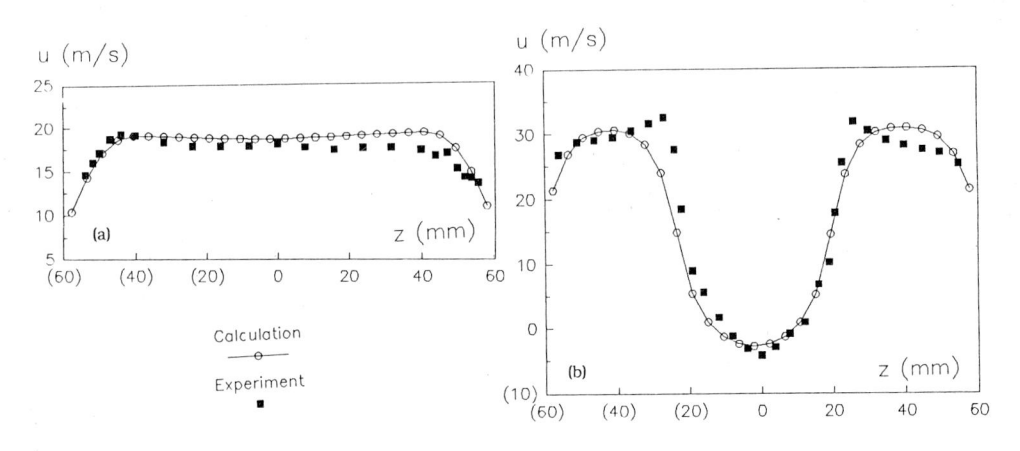

Figure 3 u- z distributions in various x sections

443

A comparison between computed results by using the k–ε model and experimental data is shown in figures 2. This figure depicts the stream-wise velocity component along the center of the tunnel. Figure 3 shows the u-velocity component vs. z- axis distribution in various sections along x-direction. The relevant experimental data is also plotted in this figure. As seen the computational results are comparable with the experimental ones.

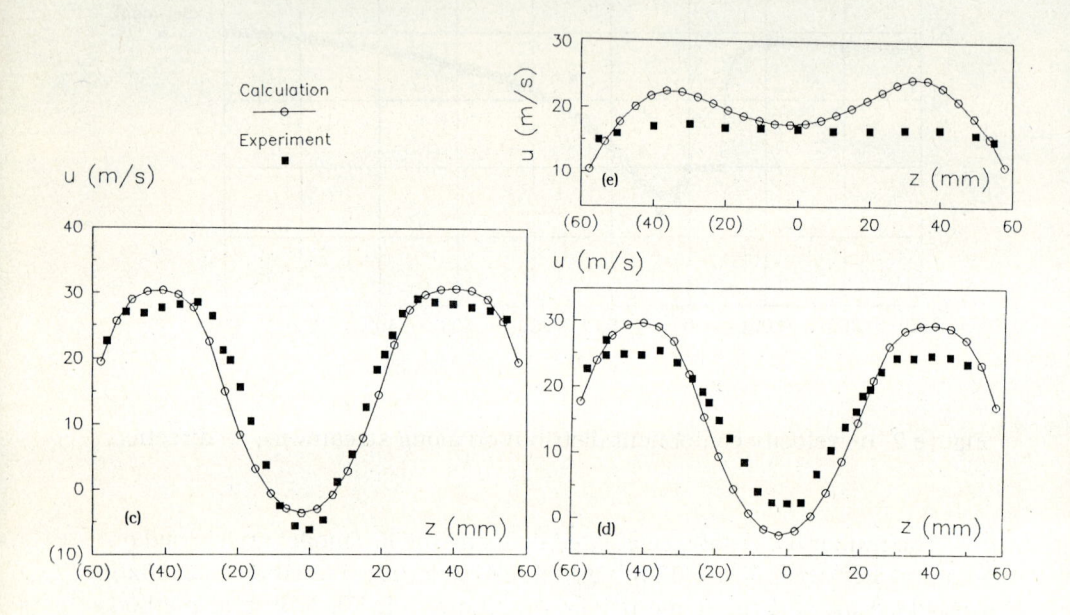

Figure 3 (continued)
(a) x=–100 mm, (b) x=15 mm, (c) x=38 mm, (d) x=61 mm, (e) x=376 mm

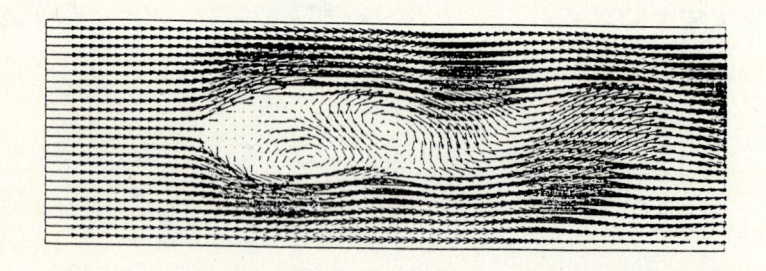

Figure 4 Instantaneous velocity vector at the test plane (LES)

LES calculations (using Smagorinsky sub-grid scale model, with small filtering length scales leads to an unsteady flow, and a symmetric mean velocity distribution. The calculations have been performed on a rather fine grids, with local grid refinements. In most calculations the total number of computational cells varies between one half and one million. Figure 4 shows the instantaneous velocity vector field at the mid-plane. The averaged field is shown in figure 5.

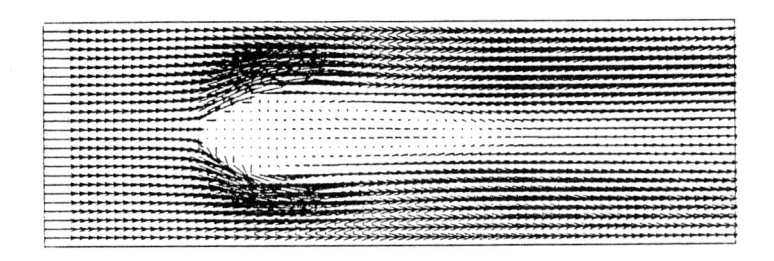

Figure 5 The (time) averaged velocity field at the test plane (LES)

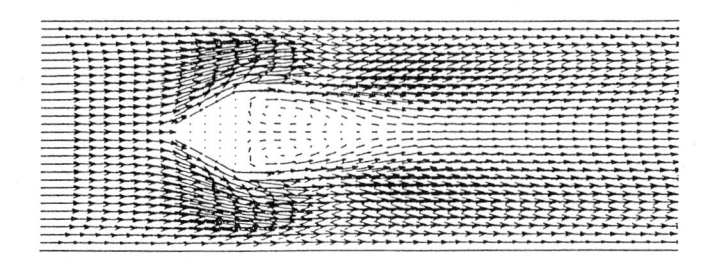

Figure 6 Steady state velocity field at the test plane (k–ε model)

The velocity vector field calculated by the k–ε model, is shown in figure 6. As seen from these figures, these LES calculations overpredict the length of the separation distance, compared to the k–ε model. The predicted spectrum of the fluctuation in the transverse velocity components at a distance of 120 and 180 mm behind the flame holder is shown in figures 7 and 8, respectively. The clear peaks at frequencies lower than 30 Hz are found to be in agreement with reported experimental results [3]. In the LES calculation it is found that the choice of parameters have large influence on the results. Figure 9 shows the results with Arakawa terms, which gives the length of wake region behind the holder shorter and in better agreement with the experimental data. Figure 10 shows the corresponding results but using the non-conservative discretisation. A Longer wake region is obtained, compared with figure 9 and experimental data. This is shown also in the previous figure 5. In figure 9-10 the results are shown in non- dimensional form.

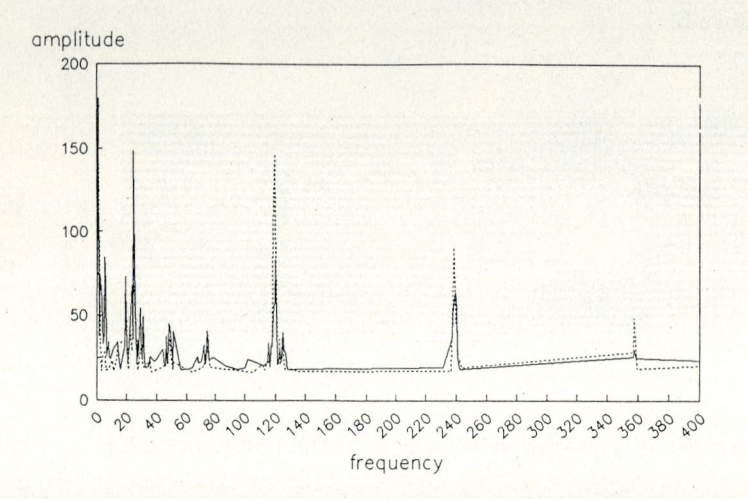

Figure 7 Spectrum of the fluctuation in the transverse components of the velocity vector (120 mm downstream of flame holder)
_____ u- component --- v- component

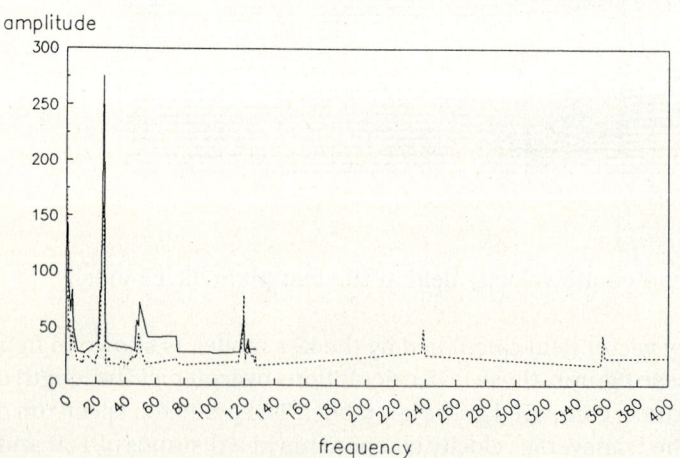

Figure 8 Spectrum of the fluctuation in the transverse components of the velocity vector (180 mm downstream of flame holder)
_____ u- component --- v- component

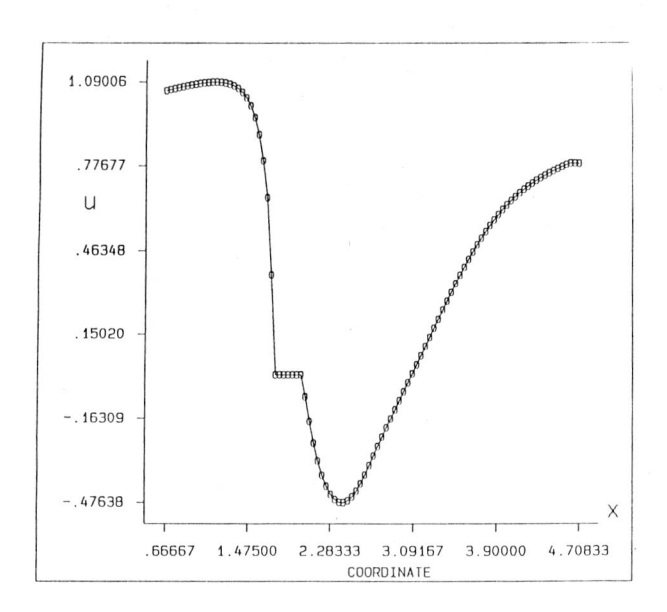

Figure 9. u- velocity component distribution along streamwise x- direction. (Non-dimensional axis; With Arakawa terms)

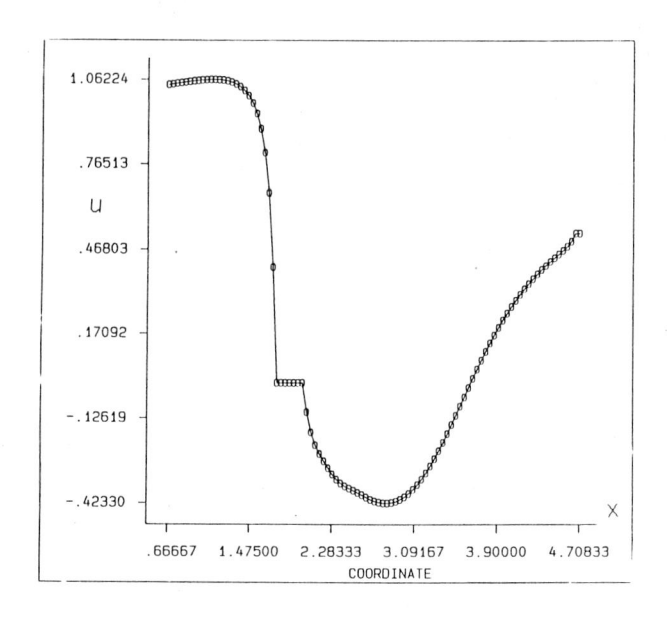

Figure 10. u- velocity component distribution along streamwise x- direction. (Non-dimensional axis; Non-conservative discretisation)

5. Concluding Remarks

This paper presents the numerical calculations of turbulent cold flows (isothermal) in a wind–tunnel with a flame holder model placed in the tunnel. Both the k–ε model method and Larger Eddy Simulations equipped with the Smagorinsky sub–grid scale model have been used. The results of k–ε model are in agreement with the experimental data in terms of the wake length behind the flame holder. The vortex shedding computed by LES calculations using Arakawa's formulation is found to be in good agreement with the measurement in the experimental setup.

Acknowledgement

The work is supported by Swedish National Board for Technical Development (STU) under the grant No. 89-2269.

Reference

1. Sullerey R.K, Gupta A.K. and Moorthy C.S.:"Similarity in The Turbulent Near Wake of Bluff Bodies", AIAA J. Vol.13, No.11 1975.
2. Fujii S., Geomi M. and Eguchi K.:"Cold Flow Tests of a Bluff Body Flame Stabilizer", Transactions of the ASME J. of Fluid Engineering, Vol.100, 1978, pp. 323-332.
3. Olovsson S. Private Communication 1990.
4. Speziale C.G.:"Discussion of Turbulence Modeling: Past and Future", NASA CR 181884, 1989.
5. Launder B.E. and Spalding D.B.:" The Numerical Computation of Turbulent Flows", Computer Methods in Applied Mechanics and Engineering, Vol.3, 1974, pp.269-289.
6. Rodi, W.:"Turbulence Models and Their Application in Hydaulics – A State of The Art Review", Institut for Hydromechanik, 1980.
7. Bai X.S. and Fuchs L.:"A Fast Multi–Grid Method for 3–D Turbulent Incompressible Flows", Proc. of 7th Int. Conf. on Numer. Method in Laminar and Turbulence Flows", Stanford, USA, July, 1991.
8. Arakawa A.: "Computational Design for Long–Term Numerical Integration of the Equations of Fluid Motion: Two–Dimensional Incompressible Flow. Part I", J. of Computational Physics 1, 1966, pp.119-143.
9. Fuchs L.:"A Local Mesh Refinement Technique For Incompressible Flows", Computers & Fluids, Vol. 14, No.1, 1986, pp.69-81.

NUMERICAL SIMULATION OF H2-O2 DIFFUSION FLAMES IN A ROCKET ENGINE

D. CHARGY[(1)], B. LARROUTUROU[(1)], M. LORIOT[(2)]

(1) CERMICS, INRIA, SOPHIA-ANTIPOLIS, 06560 VALBONNE, FRANCE
(2) SIMULOG, SOPHIA-ANTIPOLIS, 06560 VALBONNE, FRANCE

ABSTRACT

We describe an upwind finite-element method for the numerical simulation of reactive flows. The method uses in particular a triangular finite-element mesh and an upwind non oscillatory scheme based on an approximate Riemann solver for the evaluation of the convective terms for all species. Results concerning the reactive flows which occur in the combustion chamber of a rocket engine are presented.

1 INTRODUCTION

Many industrial flows are so complex that the detailed understanding of the flow features is a very difficult task. In many of these situations, the numerical simulation of these flows can play an essential role in this understanding and enables a better description and explanation of the phenomena involved. This is especially true for engineering reactive flows. As a consequence, there is an increasing need for computer codes modelling the reactive flow of a gas mixture at various Mach numbers.

The ignition process in a rocket engine is certainly among the most complex engineering flow systems: it involves indeed three-dimensional effects, multi-phase flows, compressible effects (with in particular strong pressure ratios, supersonic regions, shock waves), turbulence, unsteadiness, a complet set of chemical reactions, involving a large number of chemical species, high temperature and pressure ratios, etc... (see [10]).

Because of the problem complexity, we will limit ourselves in this study to a simplified model problem. We consider here the ignition of the subsonic flow generated by two H_2-O_2 injectors in a two-dimensional combustion chamber.

The flow is assumed to be gaseous and inviscid; all species behave as perfect gases but have different specific heats and molecular weights. In the experiments, we first compute the inert steady flow, and then initiate the ignition process (with a local artificial energy deposit) and observe the flame propagation and the convergence to a steady reactive flow with a diffusion flame around each injector.

The system of governing equations therefore includes the Euler equations, with additional continuity equations for each of the gaseous species and with additional diffusive and reactive terms in the energy and in the species equations. For the source term, we use the Eddy Break Up model, with an ignition temperature.

The numerical method used to solve this model problem uses an upwind non oscillatory scheme constructed using a multi-component extension of Roe's approximate Riemann solver for the evaluation of the convective terms of the Euler equations and a scheme based on the exact solution of Riemann problem for the convection of the species [8]. This scheme preserves the maximum principle (and in particular the positivity) for the mass fractions of all species. Lastly, the diffusive terms are solved by a standard finite-element formulation. The second-order accuracy is obtained by using piecewise-linear variables with limited slopes instead of piecewise-constant variables, following the "MUSCL" approach of Van Leer [14]. This scheme operates on an unstructured finite-element triangulation of the computational domain, which makes it possible to use whenever necessary an adaptive mesh refinement procedure which locally divides the elements in order to improve the spatial resolution of the method in the thin reaction zone [3].

This numerical method is presented in the next section, while several numerical results are presented and discussed in Section 3.

2 NUMERICAL METHOD

2.1 The equations

As was said earlier, we will concentrate on the simulation of multi-component inviscid gaseous flows. The system of governing equations will therefore include the Euler equations, with additional continuity equations for each of the gaseous species, and with additional diffusive and reactive terms in the energy and in the species equations. In fact, the viscous effects could be included without particular difficulty in our method (see [6]), which would essentially affect the numerical solution in the neighbourhood of the chamber walls.

In order to present the numerical method, we consider a mixture made of three species Σ_1 (H_2), Σ_2 (O_2) and Σ_3 (the products of reaction), whose mass fractions will be denoted Y_1, Y_2 and Y_3 (that is, ρY_1, ρY_2 and ρY_3 are the separate densities of the three species, ρ being the mixture density). Since $Y_3 = 1 - Y_1 - Y_2$, we will only consider in the sequel Y_1 and Y_2.

The two-dimensional reactive flow of this three-component mixture is described by the following system of governing equations (see [2], [15]):

$$\begin{cases} \rho_t + (\rho u)_x + (\rho v)_y = 0 \ , \\ (\rho u)_t + (\rho u^2 + p)_x + (\rho u v)_y = 0 \ , \\ (\rho v)_t + (\rho u v)_x + (\rho v^2 + p)_y = 0 \ . \\ E_t + [u(E+p)]_x + [v(E+p)]_y = \overrightarrow{\nabla}.(\lambda \overrightarrow{\nabla} T) + \Omega_T + \sum_{k=1}^{3} \overrightarrow{\nabla}.(\rho D C_k^p T \overrightarrow{\nabla} Y_k) \ , \\ (\rho Y_1)_t + (\rho u Y_1)_x + (\rho v Y_1)_y = \overrightarrow{\nabla}.(\rho D \overrightarrow{\nabla} Y_1) + \Omega_{Y_1} \ , \\ (\rho Y_2)_t + (\rho u Y_2)_x + (\rho v Y_2)_y = \overrightarrow{\nabla}.(\rho D \overrightarrow{\nabla} Y_2) + \Omega_{Y_2} \ , \end{cases} \quad (1)$$

$$\begin{cases} E = \sum_{k=1}^{3} \rho Y_k C_k^v T + \tfrac{1}{2}\rho(u^2 + v^2) \ , \\ p = \sum_{k=1}^{3} \dfrac{\rho Y_k R T}{M_k} \ . \end{cases} \quad (2)$$

Our notations are classical : u and v are the components of the mixture velocity $\overrightarrow{U}$, p is the pressure, E is the sum of the internal and kinetic energies per unit of volume, λ is the mixture thermal conductivity, T the temperature, D is the molecular diffusion coefficient of species. We have assumed that all species behave as perfect gases, but that

Σ_1, Σ_2 and Σ_3 may have different specific heats C_1^v, C_2^v, C_3^v, C_1^p, C_2^p and C_3^p and different molecular weights M_1, M_2 and M_3. Lastly, the source terms Ω_T and Ω_{Y_k} represent the contribution of the chemical reactions to the energy and mass fraction equations (these terms will be made precise later for the specific problem considered in our calculations). We will assume below that the quantities λ, ρD, C_k^p, and C_k^v are constant.

2.2 Spatial approximation

We now present the numerical scheme used to solve equations (1)-(2); this scheme is a mixed finite-element / finite-volume scheme. More precisely, it uses for the hyperbolic terms appearing in the left-hand side of system (1) an upwind formulation, derived from the extension to mixtures of some of the usual "approximate Riemann solvers" used to solve the Euler equations of a single gas (see [7]). Moreover, it operates on an unstructured finite-element triangulation, which makes it possible to employ a computational mesh fitted to complex geometries and adapted to complex solutions (involving for instance strong shocks or thin flames). This scheme can be extended to three-dimensional calculations in a straightforward way (see for instance [13]).

To specify the scheme, we first write system (1)-(2) under the following form:

$$W_t + F(W)_x + G(W)_y = P(W)_x + Q(W)_y + S(W) , \tag{3}$$

with: $W = (\rho, \rho u, \rho v, E, \rho Y_1, \rho Y_2)^T$,

$$F(W) = \left(\rho u, \rho u^2 + p, \rho u v, u(E+p), \rho u Y_1, \rho u Y_2\right)^T ,$$

$$G(W) = \left(\rho v, \rho u v, \rho v^2 + p, v(E+p), \rho v Y_1, \rho v Y_2\right)^T ,$$

$$P(W) = \left(0, 0, 0, \lambda T_x + \sum_{k=1}^{3} \rho D C_k^p T(Y_k)_x, \rho D(Y_1)_x, \rho D(Y_2)_x\right)^T ,$$

$$Q(W) = \left(0, 0, 0, \lambda T_y + \sum_{k=1}^{3} \rho D C_k^p T(Y_k)_y, \rho D(Y_1)_y, \rho D(Y_2)_y\right)^T ,$$

$$S(W) = (0, 0, 0, \Omega_T, \Omega_{Y_1}, \Omega_{Y_2})^T .$$

Then we introduce a finite-element triangulation of the computational domain. In order to derive a finite-volume formulation, we consider a dual partition of the domain in control volumes or cells : a cell C_i is constructed around each vertex S_i by means of the medians of the neighbouring triangles, as shown on Figure 1. Integrating system (3) on the control volume C_i, we get:

$$\iint_{C_i} W_t + \int_{\partial C_i} (F\nu_i^x + G\nu_i^y) = \int_{\partial C_i} (P\nu_i^x + Q\nu_i^y) + \iint_{C_i} S(W) , \tag{4}$$

where $\vec{\nu_i} = (\nu_i^x, \nu_i^y)$ is the outward unit normal on ∂C_i. It now remains to specify how the four integrals in (4) are evaluated.

The time derivative and source terms integrals are approximated using a mass-lumped approximation:

$$\iint_{C_i} W_t = \frac{dW_i}{dt} area(C_i) , \quad \iint_{C_i} S(W) = S(W_i) area(C_i) . \tag{5}$$

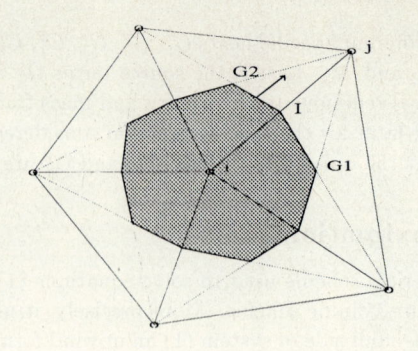

Figure 1: Control volume C_i around vertex S_i

In addition to its simplicity, the mass-lumped approximation has two advantages: first, it allows us to employ an explicit time integration scheme, which is no longer possible when the consistent non diagonal mass matrix is used; moreover, the mass-lumped approximation of the heat equation keeps the positiveness of the unknowns, while a consistent finite-element formulation does not.

To approximate the second integral in (4), we begin by noticing that (2) yields (using Mayer's relation $M_k(C_{pk} - C_{vk}) = R$):

$$p = (\gamma - 1)[E - \frac{1}{2}\rho(u^2 + v^2)], \qquad (6)$$

where γ is the local specific heat ratio of the mixture:

$$\gamma = \frac{\sum\limits_{k=1}^{3} Y_k C_{pk}}{\sum\limits_{k=1}^{3} Y_k C_{vk}}. \qquad (7)$$

It is now a well-known fact that the system:

$$W_t + F(W)_x + G(W)_y = 0, \qquad (8)$$

with the pressure p given by (6)-(7), is a nonlinear *hyperbolic* system of conservation laws (see [1], [4], [9]). Using this fact, we can define an "approximate Riemann solver" for system (6)-(8). We use here an extension to mixture of Roe's scheme [12] (we refer to [1], [4] and [9] for more details on the extension of the classical Godunov-type schemes and flux vector splittings to system (6)-(8)). Given two values W_L and W_R of W, and a vector $\vec{\eta} = (\eta^x, \eta^y)$, we define a numerical flux function Φ by:

$$\Phi(W_L, W_R, \vec{\eta}) = \frac{1}{2}[\mathcal{F}_\eta(W_L) + \mathcal{F}_\eta(W_R)] + \frac{1}{2}|\tilde{H}_\eta|(W_R - W_L). \qquad (9)$$

In this expression, we have set $\mathcal{F}_\eta(W) = \eta^x F(W) + \eta^y G(W)$, and $\tilde{H}_\eta = \tilde{H}_\eta(W_L, W_R)$ is a diagonalisable matrix satisfying Roe's property:

$$\mathcal{F}_\eta(W_L) - \mathcal{F}_\eta(W_R) = \tilde{H}_\eta(W_L, W_R)(W_R - W_L); \qquad (10)$$

the matrix $|\tilde{H}_\eta|$ is defined using the diagonalization of $\tilde{H}_\eta$: if $\tilde{H} = T\tilde{\Lambda}T^{-1}$ with $\tilde{\Lambda}$ diagonal, $\tilde{\Lambda} = (\tilde{\lambda}_k)_{k=1,5}$, then we set $|\tilde{H}| = T|\tilde{\Lambda}|T^{-1}$, with $|\tilde{\Lambda}| = (|\tilde{\lambda}_k|)_{k=1,5}$.

In fact, as analysed in [8], the scheme based on numerical flux function (9) does not preserve the positivity of the mass fraction, and this may induce problems, especially in the case of combustion. We thus use instead of (9) a modified flux function; we evaluate the total mass, momentum and energy fluxes as in (9), and modify the separate mass fluxes as proposed in [8]. The resulting scheme has then the property of preserving the maximum principle (and in particular the positivity) for all mass fractions.

To evaluate the second integral in (4), we first write it in the form:

$$\int_{\partial C_i} (F\nu_i^x + G\nu_i^y) = \sum_{j \in \mathcal{K}(i)} \int_{\partial C_{ij}} (F\nu_i^x + G\nu_i^y), \quad (11)$$

where $\mathcal{K}(i)$ is the set of neighbouring nodes of S_i, and where $\partial C_{ij} = \partial C_i \cap \partial C_j$. Then, defining the vector $\overrightarrow{\nu_{ij}} = (\nu_{ij}^x, \nu_{ij}^y)$ by:

$$\nu_{ij}^x = \int_{\partial C_{ij}} \nu_i^x, \quad \nu_{ij}^y = \int_{\partial C_{ij}} \nu_i^y, \quad (12)$$

we obtain a first-order accurate upwind approximation of the convective flux (13) by:

$$\int_{\partial C_i} (F\nu_i^x + G\nu_i^y) = \sum_{j \in \mathcal{K}(i)} \Phi(W_i, W_j, \overrightarrow{\nu_{ij}}). \quad (13)$$

A second-order accurate upwind extension can be derived by using a MUSCL-type approximation instead of a constant by cell approximation. In this case, (13) becomes:

$$\int_{\partial C_i} (F\nu_i^x + G\nu_i^y) = \sum_{j \in \mathcal{K}(i)} \Phi(W_{ij}, W_{ji}, \overrightarrow{\nu_{ij}}), \quad (14)$$

where the same numerical flux function Φ is used, but where W_{ij} (resp.: W_{ji}) is a second-order accurate approximation of W at the cell interface ∂C_{ij} inside the cell C_i (resp.: inside the cell C_j), written as:

$$\begin{cases} W_{ij} = W_i + \dfrac{1}{2} \overrightarrow{\nabla} W(i).\overrightarrow{S_i S_j}, \\ W_{ji} = W_j + \dfrac{1}{2} \overrightarrow{\nabla} W(j).\overrightarrow{S_j S_i}, \end{cases} \quad (15)$$

Here, $\overrightarrow{\nabla} W(i)$ and $\overrightarrow{\nabla} W(j)$ are some limited approximations of the gradient of the dependant variables W in the cells C_i and C_j respectively. We refer to [5] for the details.

Finally, we consider the integral of the diffusive fluxes $\int_{\partial C_i} (P\nu_i^x + Q\nu_i^y)$. In view of the definitions (3) of P and Q, this integral reduces to expressions like $\int_{\partial C_i} \overrightarrow{\nabla} T.\overrightarrow{\nu_i}$, $\int_{\partial C_i} \overrightarrow{\nabla} Y.\overrightarrow{\nu_i}$ and $\int_{\partial C_i} T\overrightarrow{\nabla} Y.\overrightarrow{\nu_i}$. To evaluate these terms, we consider here that the gradients are constant in each triangle τ of the triangulation. More precisely, we consider that, in a triangle τ with vertices S_j ($1 \leq j \leq 3$), we have:

$$\nabla T|_\tau = \sum_{j=1}^{3} T_j \nabla \phi_j, \quad \nabla Y|_\tau = \sum_{j=1}^{3} Y_j \nabla \phi_j, \quad (16)$$

where ϕ_j is the P1 finite-element basis function associated to vertex S_j. It is easy to check that this approximation is equivalent to the classical P1 finite-element discretization of the diffusive terms.

2.3 Boundary conditions

Let Γ be the boundary of the computational domain and $\vec{n}$ be the outward unit normal on Γ. We assume that Γ is divided into two parts $\Gamma = \Gamma_0 \cup \Gamma_\infty$, on which different boundary conditions will be used. In this partition, Γ_0 represents a solid wall, while Γ_∞ represents the far field (inflow or outflow) boundaries.

In our scheme, we do not treat a boundary condition by forcing the value of a variable to a prescribed boundary value, but consider instead the integral formulation (4) and apply the boundary conditions by modifying the flux integrals on ∂C_i for those cells C_i such that $\partial C_i \cap \Gamma \neq \emptyset$.

For instance, for a vertex S_i located on Γ_0, we do not impose the slip condition $\vec{U} \cdot \vec{n} = 0$, but take this condition into account in the evaluation of the convective fluxes, getting:

$$\int_{\partial C_i \cap \Gamma_0} Fn^x + Gn^y = \left(0, \int_{\partial C_i \cap \Gamma_0} pn^x, \int_{\partial C_i \cap \Gamma_0} pn^y, 0, 0, 0\right)^T ;$$

the pressure integrals are computed as:

$$\int_{\partial C_i \cap \Gamma_0} pn^x = p_i \int_{\partial C_i \cap \Gamma_0} n^x , \quad \int_{\partial C_i \cap \Gamma_0} pn^y = p_i \int_{\partial C_i \cap \Gamma_0} n^y . \tag{17}$$

Moreover, assuming that the wall is adiabatic and non catalytic ($\vec{\nabla} T = 0$ and $\vec{\nabla} Y = 0$ on Γ_0), we set:

$$\int_{\partial C_i \cap \Gamma_0} Pn^x + Qn^y = 0 . \tag{18}$$

For a vertex S_i located on Γ_∞, we again set:

$$\int_{\partial C_i \cap \Gamma_\infty} Pn^x + Qn^y = 0 , \tag{19}$$

which leads to assuming that the temperature and mass fraction gradients vanish at the far field boundary Γ_∞.

For the convective terms, the outflow boundary does not need a particular treatment in our problem since the flow is supersonic there; for the subsonic inflow boundary, we wish to impose the inflowing mass flux. Calling Γ_e the subsonic inflow boundary, we simply impose the boundary flux on Γ_e as follows:

$$\int_{\partial C_i \cap \Gamma_e} (F\nu_i^x + G\nu_i^y) = F(\tilde{W}_i)\eta^x + G(\tilde{W}_i)\eta^y ; \tag{20}$$

our first way of defining $\tilde{W}_i$ was as follows (notice that, according to the theory of characteristics, we use exactly one value from W_i, namely the pressure p_i for the definition of $\tilde{W}_i$):

$$\tilde{W}_i = \left(\rho_e, (\rho u)_e, (\rho v)_e, \frac{p_i}{\gamma_e - 1} + \frac{1}{2}\rho_e(u_e^2 + v_e^2), (\rho Y_1)_e, (\rho Y_2)_e\right)^T . \tag{21}$$

In fact, we slightly modified this definition: it appeared from the numerical calculations that a more stable approximation was obtained if we force the pressure of state $\tilde{W}_i$ to be the same for all vertices of a given inflow boundary (for the value of this pressure, we take the average of the inner pressure p_i at this boundary).

2.4 Time integration

Since the objective of our study is to simulate the unsteady ignition process, we do not attempt to use large time steps, and simply use the classical first-order accurate forward Euler scheme to advance in time the numerical solution. We then have a stability restriction on the time step. The analysis of the stability of the whole nonlinear scheme being too complex, we derive three approximate stability conditions by considering separately the hyperbolic terms, the diffusive terms and the reactive terms. The time step is then chosen as the smallest of the hyperbolic, diffusive and reactive time steps.

3 NUMERICAL RESULTS

We first present the steady-state solution on a fine mesh of 5281 nodes (Figure 2-a) obtained by solving the Euler equations (with no reaction term). On the inflow boundary, we impose a mass flux of $250kg/m^2/s$ for the oxygen (in the center of each injector) and $156kg/m^3$ for the hydrogen, the density being respectively of $12.5kg/m^3$ and $0.78kg/m^3$; the pressure of the two injected species is $9.74\ 10^5 Pa$. At t=0, the chamber is full of gaseous water at rest under a pressure of $10^3 Pa$. We show on Figure 2 the mass fraction contours of oxygen, the density field and the Mach numbre contours. A sonic line is visible at the throat. The Mach number in the chamber is around 0.15, and the temperature varies between 15 and 48 ^oK.

Figure 3 shows the solution obtained with a diffusion and combustion model. As said earlier, we use for this simulation the Eddy break up model. With the notations used before, this model writes as:

$$\Omega_T = Q\omega, \quad \Omega_{Y_1} = -m_1/m_3\omega, \quad \Omega_{Y_2} = -m_1/m_2\omega$$

with $\omega = K\rho Min(Y_1, \frac{Y_2}{8})$, and $K = 4\ 10^6$ (resp. $K = 0$) if the temperature is above (resp. lower than) the ignition temperature $T_i = 900^oK$. The heat released by the chemical reaction is $Q = 13.9\ 10^3 kJ/kg$; for the diffusion coefficients, we take $\lambda = 1.2 J/m/s/^oK$ and $(\rho D) = 1.3\ 10^{-3}\ m^2/s$. The ignition is started by means of an energy deposit near the throat of the nozzle. A premixed flame then climbs up towards the injectors (Figure 3-a). The unburnt gases are compressed and reach the ignition temperature. On Figure 3-b, the ignition can be observed near the injectors, followed by a strong unsteady evolution (Figure 3-c). At convergence (Figure 3-d), we obtain a diffusion flame in front of eaqch injector. The maximum ejection speed, which was of 430 m/s in the pure Euler computation, is now of $3800m/s$. Again, a sonic line may be observed at the throat. The Mach number in the chamber is also close to 0.15.

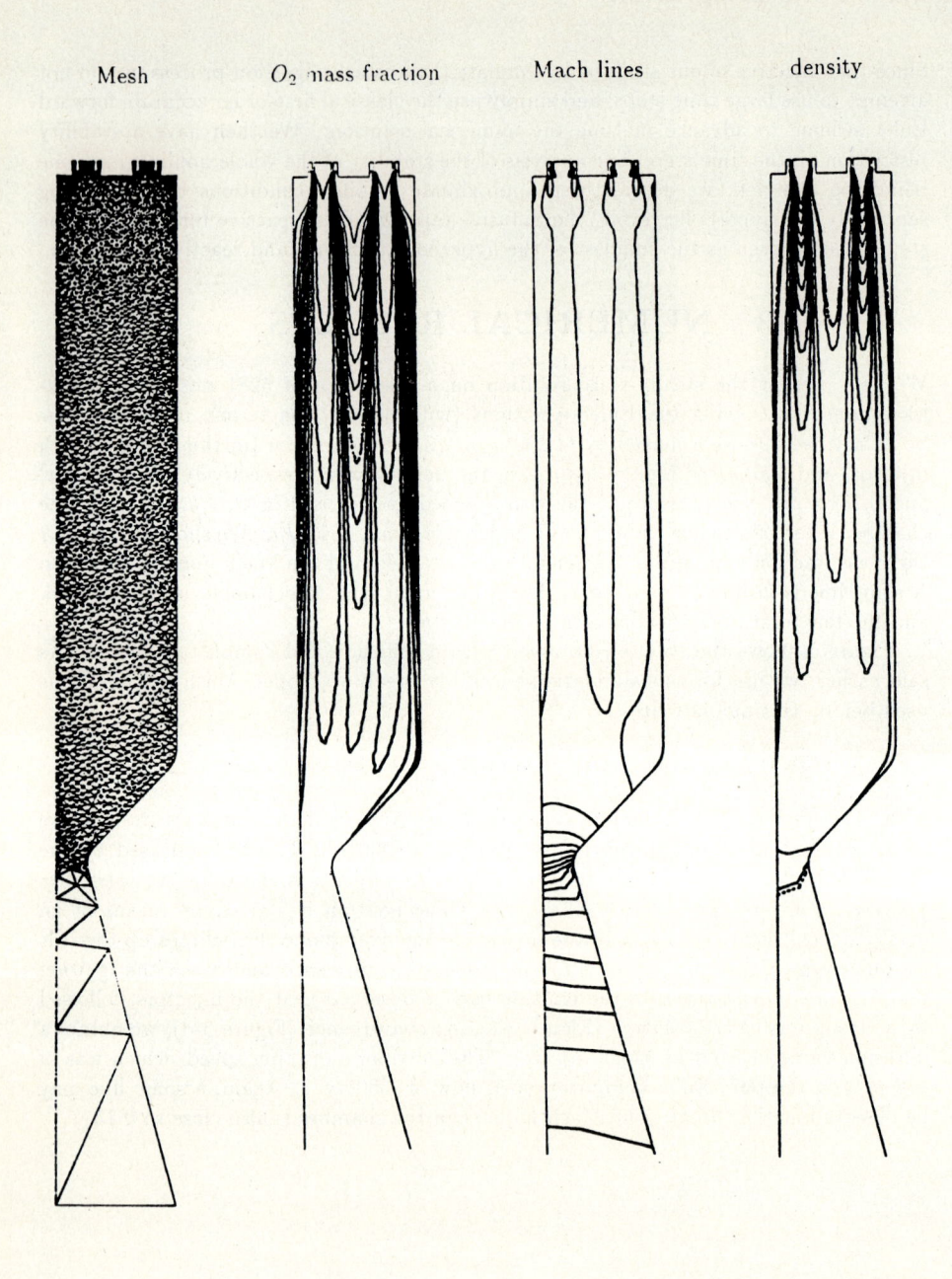

Figure 2: mesh and Euler Solution

(a) (b) (c) (d)

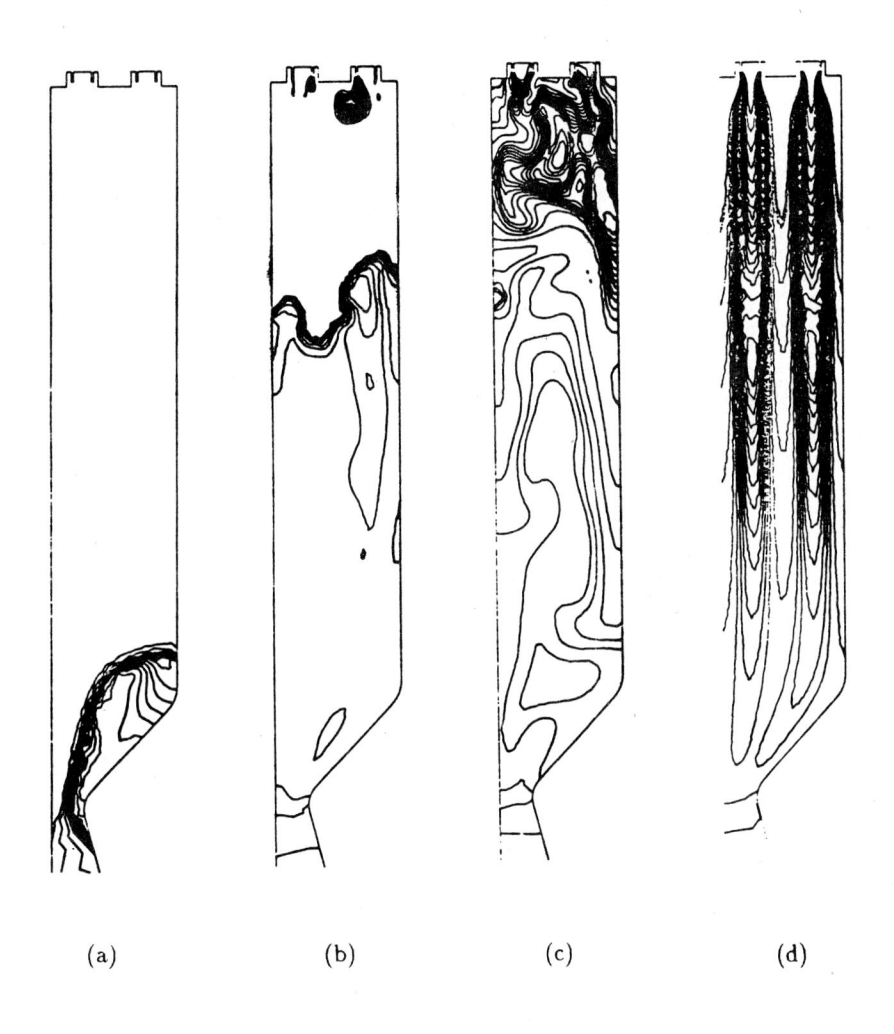

Figure 3: Temperature field at various times for the reactive calculation

REFERENCES

[1] R. ABGRALL, "Généralisation du schéma de Roe pour le calcul d'écoulements de mélanges de gaz à concentrations variables", La Recherche Aérospatiale, pp. 31-43, **6**, (1988).

[2] J. D. BUCKMASTER & G. S. S. LUDFORD, "Theory of laminar flames", Cambridge Univ. Press, Cambridge, (1982).

[3] D. CHARGY, A. DERVIEUX & B. LARROUTUROU, "Upwind adaptive finite element investigations of the two-dimensional reactive interaction of supersonic gaseous jets", Int. J. Num. Meth. Fluids. **11**, pp. 751-767, (1990).

[4] G. FERNANDEZ & B. LARROUTUROU, "Hyperbolic schemes for multi-component Euler equations", Nonlinear hyperbolic equations - Theory, numerical methods and applications, Ballmann & Jeltsch eds., pp. 128-138, Notes on Numerical Fluid Mechanics, **24**, Vieweg, Braunschweig, (1989).

[5] F. FEZOUI, "Résolution des équations d'Euler par un schéma de Van Leer en éléments finis", INRIA Report 358, (1985).

[6] F. FEZOUI, S. LANTERI, B. LARROUTUROU & C. OLIVIER, "Résolution numérique des équations de Navier-Stokes pour un fluide compressible en maillage triangulaire", INRIA Report 1033, (1989).

[7] A. HARTEN, P. D. LAX & B. VAN LEER, "On upstream differencing and Godunov type schemes for hyperbolic conservation laws", SIAM Review, **25**, pp. 35-61, (1983).

[8] B. LARROUTUROU, "How to preserve the mass fraction positivity when computing compressible multi-component flows", J. Comp. Phys., to appear.

[9] B. LARROUTUROU & L. FEZOUI, "On the equations of multi-component perfect or real gas inviscid flow", "Non linear hyperbolic problems", Carasso Charrier Hanouzet Joly eds., Lecture Notes in Mathematics, Springer Verlag, Heidelberg, (1989).

[10] P. LIANG, S. FISHER & Y. M. CHANG, "Comprehensive modeling of a liquid rocket combustion chamber", AIAA paper 85-0232, (1985).

[11] B. PALMERIO, "Self adaptive F. E. M. algorithms for the Euler equations", INRIA Report 338, (1984).

[12] P. L. ROE, "Approximate Riemann solvers, Parameter vectors and difference schemes", J. Comp. Phys, **43**, p. 357, (1981).

[13] B. STOUFFLET, J. PERIAUX, F. FEZOUI & A. DERVIEUX, "Numerical simulation of 3-D hypersonic Euler flow around a space vehicle using adapted finite elements", AIAA paper 87-0560, (1987).

[14] B. VAN LEER, "Flux vector splitting for the Euler equations", Eight international conference on numerical methods in fluid dynamics, Krause ed., pp 507-512, Lecture notes in physics, **170**, Springer-Verlag, (1982).

[15] F. A. WILLIAMS, "Combustion theory", second edition, Benjamin Cummings, Menlo Park, (1985).

A Modified Semi-implicit Method for Two-phase flow Problems

Xin Kai Li[1]

Oxford University Computing Laboratory

11 Kebel Road, Oxford

England

SUMMARY

A modified semi-implicit finite difference method for the solution of two-phase flow problem is presented. Example calculations for the two-phase flow model problems are also presented.

1. Introduction

The numerical calculations of two-phase flow problems play an essential role in various areas of modern science and engineering. The major numerical schemes used in this field have been based on semi–implicit finite difference schemes. The basic guidelines employed in the development of the semi–implicit numerical method have been discussed in [1, 2, 3, 4]. However, all the applications of semi–implicit numerical methods to two–phase flow problems mentioned in the literature are limited to solve continuous problems. For resolving discontinuous problems, they exhibit spurious oscillations near discontinuities. In the present work, we attempt to modify a semi–implicit scheme and overcome this difficulty.

2. Mathematical two-phase flow model

In this paper, the following gas-liquid two-phase flow model is considered.
The continuity conservation equations are

$$\frac{\partial}{\partial t}\alpha_g + \frac{\partial}{\partial x}(\alpha_g u_g) = M_g, \qquad (2.1)$$

$$\frac{\partial}{\partial t}\alpha_l + \frac{\partial}{\partial x}(\alpha_l u_l) = M_l. \qquad (2.2)$$

The momentum equations are

$$\frac{\partial}{\partial t}(\alpha_g \rho_g u_g) + \frac{\partial}{\partial x}(\alpha_g \rho_g u_g^2) = -\alpha_g \frac{\partial p_g}{\partial x} + (p_i - p_g)\frac{\partial \alpha_g}{\partial x} + F_{gl} + F_{gw}, \qquad (2.3)$$

---

[1]The current address is Department of Engineering, University of Leicester, Leicester, LE1 7RH, England.

$$\frac{\partial}{\partial t}(\alpha_l \rho_l u_l) + \frac{\partial}{\partial x}(\alpha_l \rho_l u_l^2) = -\alpha_l \frac{\partial p_l}{\partial x} + (p_i - p_l)\frac{\partial \alpha_l}{\partial x}$$

$$+ F_{lg} + F_{lw} - \alpha_l(\rho_l - \rho_g)g\sin\theta. \tag{2.4}$$

The gas and liquid volume fractions satisfy the geometrical condition, i.e.

$$\alpha_g + \alpha_l = 1; \quad h = \alpha_l H. \tag{2.5}$$

In equations (2.1) through (2.5) the nomenclature is as follows: the suffix l denotes the liquid phase, and the suffix g the gas phase; α is the volume void fraction, ρ is the phase density, u is the phase velocity, p is the pressure, g is the gravity; F_{lg}, F_{gw} and F_{lw} represent the friction factors for the liquid–gas, gas–wall and liquid–wall interfaces respectively; θ denotes the angle between the pipe axis and the horizontal; H represents the height of the channel and h the height of the liquid, the distance from the bottom of the channel to the liquid–gas surface, and satisfying $0 \leq h \leq H$, and M_g and M_l are the mass sources of gas and liquid, and will depend on the physical situations of interest. It can be derived from (2.1), (2.2) and (2.5) and, if $M_g = M_l = 0$, takes the form

$$\frac{\partial}{\partial x}(\alpha_g u_g + \alpha_l u_l) = 0. \tag{2.6}$$

3. Modified semi-implicit method

The modified semi–implicit method consists of the following five steps for the two-phase flow equations given in Section 2.:

- Step 1. Predictor momentum equations with forcing terms:

$$\frac{\hat{u}_g^{n+1} - u_g^n}{\Delta t} + u_g^n \nabla_{j+1/2} u_g^n + \beta(\hat{u}_g^{n+1} - u_g^n)\nabla_{j+1/2} u_g^n \tag{3.7}$$

$$+ \frac{1}{\rho_g}\frac{p_{j+1}^n - p_j^n}{\Delta x} = \frac{1}{(\alpha_g \rho_g)_{j+1/2}^n}(\hat{F}_{gl}^{n+1} + \hat{F}_{gw}^{n+1}) + R_g(\alpha_g^n),$$

$$\frac{\hat{u}_l^{n+1} - u_l^n}{\Delta t} + u_l^n \nabla_{j+1/2} u_l^n + \beta(\hat{u}_l^{n+1} - u_l^n)\nabla_{j+1/2} u_l^n \tag{3.8}$$

$$+ \frac{1}{\rho_l}\frac{p_{j+1}^n - p_j^n}{\Delta x} = \frac{1}{(\alpha_l \rho_l)_{j+1/2}^n}(\hat{F}_{lg}^{n+1} + \hat{F}_{lw}^{n+1}) + R_l(\alpha_l^n),$$

where

$$\beta = \begin{cases} 0, & \text{if } \nabla_{j+1/2} u_{j+1/2}^n < 0 \\ 1, & \text{if } \nabla_{j+1/2} u_{j+1/2}^n \geq 0 \end{cases}$$

$$\hat{F}_{gl}^{n+1} = \hat{F}_{gl}^{n+1}(\hat{u}_g^{n+1}, \hat{u}_l^{n+1}),$$

$$\hat{F}_{lg}^{n+1} = \hat{F}_{lg}^{n+1}(\hat{u}_g^{n+1}, \hat{u}_l^{n+1}),$$

$$R_g(\alpha_g) = \frac{1}{\rho_g}\xi_g \nabla_j \alpha_g,$$

$$R_l(\alpha_l) = \frac{1}{\rho_l}(\nabla_j \xi_l - \nabla_j \xi_g) + \frac{\xi_l}{\alpha_l \rho_l}\nabla_j \alpha_l + g\sin\theta\frac{(\rho_l - \rho_g)}{\rho_l}.$$

In the above difference equations, the variables $\hat{F}_{gl}^{n+1}$, $\hat{F}_{lg}^{n+1}$, $\hat{F}_{gw}^{n+1}$, and $\hat{F}_{lw}^{n+1}$ are obtained from the linearized equations about F_{gl}, F_{lg}, F_{gw} and F_{lw} as functions of $\hat{u}_g^{n+1}$ and $\hat{u}_l^{n+1}$, respectively. This predictor procedure is employed in an implicit fashion for the forcing terms in order to provide better interfacial drag force terms for the next step.

- Step 2. Corrector mass and momentum equations.

$$\frac{\tilde{\alpha}_g^{n+1} - \alpha_g^n}{\Delta t} + \frac{(\alpha_g u_g)_{j+1/2}^n - (\alpha_g u_g)_{j-1/2}^n}{\Delta x} = 0, \quad (3.9)$$

$$\frac{\tilde{\alpha}_l^{n+1} - \alpha_l^n}{\Delta t} + \frac{(\alpha_l u_l)_{j+1/2}^n - (\alpha_l u_l)_{j-1/2}^n}{\Delta x} = 0, \quad (3.10)$$

$$\frac{\tilde{u}_g^{n+1} - u_g^n}{\Delta t} + u_g^n \nabla_{j+1/2}\tilde{u}_g^{n+1} + \beta(\tilde{u}_g^{n+1} - u_g^n)\nabla_{j+1/2}u_g^n \quad (3.11)$$
$$+\frac{1}{\rho_g}\frac{p_{j+1}^n - p_j^n}{\Delta x} = \frac{1}{(\tilde{\alpha}_g \rho_g)_{j+1/2}^{n+1}}(\hat{F}_{gl}^{n+1} + \hat{F}_{gw}^{n+1}) + R_g(\tilde{\alpha}_g^{n+1}),$$

$$\frac{\tilde{u}_l^{n+1} - u_l^n}{\Delta t} + u_l^n \nabla_{j+1/2}\tilde{u}_l^{n+1} + \beta(\tilde{u}_l^{n+1} - u_l^n)\nabla_{j+1/2}u_l^n \quad (3.12)$$
$$+\frac{1}{\rho_l}\frac{p_{j+1}^n - p_j^n}{\Delta x} = \frac{1}{(\tilde{\alpha}_l \rho_l)_{j+1/2}^{n+1}}(\hat{F}_{lg}^{n+1} + \hat{F}_{lw}^{n+1}) + R_l(\tilde{\alpha}_l^{n+1}).$$

These finite difference equations are the same as those solved in the semi–implicit scheme [1], with two major changes based on physical phenomena. First, the mass conservation equations are added and evaluated explicitly by donor cell differencing. Although this change increases computation costs, it provides new values $\tilde{\alpha}_g^{n+1}$ and $\tilde{\alpha}_l^{n+1}$ which will be used in the momentum equations to determine values of $R_g(\tilde{\alpha}_g^{n+1})$ and $R_l(\tilde{\alpha}_l^{n+1})$. Secondly, the convective terms in the momentum equations are calculated implicitly in a linearized form instead of in an explicit form as is done in the semi–implicit scheme. Clearly, if the volume fractions $\tilde{\alpha}_g$ and $\tilde{\alpha}_l$ are given from the conservation mass equations (3.9) and (3.10), it remains to solve a couple of tridiagonal matrix systems with the normal pressure p^n as the known variables for all the velocities $\tilde{u}_g^{n+1}$ and $\tilde{u}_l^{n+1}$ from the momentum equations (3.11) and (3.12).

- Step 3. Stabilize momentum equations.

$$\frac{u_g^{n+1} - u_g^n}{\Delta t} + u_g^n \nabla_{j+1/2}\tilde{u}_g^{n+1} + \beta(u_g^{n+1} - u_g^n)\nabla_{j+1/2}u_g^n \quad (3.13)$$
$$+\frac{1}{\rho_g}\frac{p_{j+1}^{n+1} - p_j^{n+1}}{\Delta x} = \frac{1}{(\tilde{\alpha}_g \rho_g)_{j+1/2}^{n+1}}(F_{gl}^{n+1} + F_{gw}^{n+1}) + R_g(\tilde{\alpha}_g^{n+1}),$$

$$\frac{u_l^{n+1} - u_l^n}{\Delta t} + u_l^n \nabla_{j+1/2}\tilde{u}_l^{n+1} + \beta(u_l^{n+1} - u_l^n)\nabla_{j+1/2}u_l^n \quad (3.14)$$
$$+\frac{1}{\rho_l}\frac{p_{j+1}^{n+1} - p_j^{n+1}}{\Delta x} = \frac{1}{(\tilde{\alpha}_l \rho_l)_{j+1/2}^{n+1}}(F_{lg}^{n+1} + F_{lw}^{n+1}) + R_l(\tilde{\alpha}_l^{n+1}).$$

The main difficulty in solving the above equations comes from the spatial coupling of each node to its neighbours for the pressures at new time level p^{n+1}. The easiest way to compute these equations is to eliminate velocities by combining the mass equations with the momentum equations, then to solve for the pressures. This procedure is accomplished in the following way. First, by means of equations (2.1) and (2.2), an implicit finite–difference form for the constraint equation (2.6) can be taken to obtain

$$\nabla_j(\tilde{\alpha}_g^{n+1} u_g^{n+1}) + \nabla_j(\tilde{\alpha}_l^{n+1} u_l^{n+1}) = 0 \qquad (3.15)$$

which ensures that the evolved volume fractions are all positive and sum to unity. Secondly, since the pressure terms in this step are implicit but the convective terms are explicit in the momentum equations, it is easy to verify that, from the momentum differencing equations (3.13) and (3.14), the velocities u_g and u_l can also be written as functions of the pressures at the advanced time level $n+1$:

$$u_g^{n+1} = T_g + S_g(p_{j+1}^{n+1} - p_j^{n+1}), \qquad (3.16)$$

and

$$u_l^{n+1} = T_l + S_l(p_{j+1}^{n+1} - p_j^{n+1}) \qquad (3.17)$$

where T_g, T_l, S_g, and S_l are obtained from Step 2 as functions of $\tilde{\alpha}^{n+1}$, $\tilde{u}_g^{n+1}$ and $\tilde{u}_l^{n+1}$. Equations (3.16) and (3.17) are then used to eliminate the advanced time level $(n+1)$ velocities u_g and u_l from equation (3.15). Then a tridiagonal matrix system involving only the advanced time level $(n+1)$ pressure is obtained. Once this system is solved, p^{n+1} is known, and hence both u_g^{n+1} and u_l^{n+1} may be obtained by substitution back into equations (3.16) and (3.17).

This step is efficient because the reduced pressure problem contains all the coupled effects of interfacial exchanges; even when the coupling is very strong, the pressure problem can be solved relatively cheaply.

- Step 4. Stabilize mass equations.

$$\frac{\alpha_g^{n+1} - \alpha_g^n}{\Delta t} + \frac{(\alpha_g u_g)_{j+1/2}^{n+1} - (\alpha_g u_g)_{j-1/2}^{n+1}}{\Delta x} = 0, \qquad (3.18)$$

$$\frac{\alpha_l^{n+1} - \alpha_l^n}{\Delta t} + \frac{(\alpha_l u_l)_{j+1/2}^{n+1} - (\alpha_l u_l)_{j-1/2}^{n+1}}{\Delta x} = 0. \qquad (3.19)$$

This step uses the mass equations only. It is clear that each equation only involves one unknown variable α_g^{n+1} or α_l^{n+1}. This is because the new time $(n+1)$ level velocities, u_g^{n+1} and u_l^{n+1}, are known from the third step. Hence both equations (3.18) and (3.19) are simple tridiagonal linear systems, with unknowns α_g^{n+1} and α_l^{n+1}, respectively.

- Step 5. Limited correction.

We construct a simple limited correction method which is based on the theory of Total Variation Diminishing upwind finite difference schemes.
1. For each grid interval $j + 1/2$ compute

$$L_{j+1/2} = \{K_{j+1/2}^+(r_j^+) + K_{j+1/2}^-(r_{j+1}^-)\}(\psi_{j+1}^n - \psi_j^n).$$

2. Add the dissipative terms to the old solution ψ_j^{old}

$$\psi_j^n = \psi_j^{old} + [L_{j+1/2} - L_{j-1/2}],$$

where ψ represents the functions α_g, α_l, u_g, u_l and p; K is defined by

$$K^{\pm}(r^{\pm}) = \frac{1}{2}\nu(\lambda)[1 - \phi(r^{\pm})],$$

where

$$\lambda = \max\ \{|u_g|\frac{\Delta t}{\Delta x},\ |u_l|\frac{\Delta t}{\Delta x}\},$$

$$\nu(\lambda) = \begin{cases} \lambda(1-\lambda), & \text{if } \lambda \leq 0.5, \\ 0.25, & \text{if } \lambda > 0.5, \end{cases} \quad \phi(r) = \begin{cases} \min\ (2r,\ 1) & \text{if } r > 0, \\ 0 & \text{if } r \leq 0. \end{cases}$$

$$r^+ = \frac{\psi_j^n - \psi_{j-1}^n}{\psi_{j+1}^n - \psi_j^n}, \quad r^- = \frac{\psi_{j+1}^n - \psi_j^n}{\psi_j^n - \psi_{j-1}^n}.$$

We note that, since the terms involving the K's are second order difference terms with positive coefficients, they will clearly be dissipative.

4. Concluding remarks

A new modified semi-implicit method for two-phase flow problems has been presented. The following remarks should be made:

First, since the corrector mass equations in the second step are explicit, when donor cell differencing is used, it is easy to show that the solutions $\tilde{\alpha}_g$ and $\tilde{\alpha}_l$ satisfy a discrete admissible condition, i.e. there exists a constant M such that

$$\sum_j \tilde{\alpha}_g^n \leq M \quad \text{and} \quad \sum_j \tilde{\alpha}_l^n \leq M \quad \forall n, \tag{4.20}$$

provided

$$\frac{\Delta t}{\Delta x} \max\{|u_g|,\ |u_l|\} \leq 1.$$

This implies the existence of invariant regions, and that the finite–difference approximations converge to the solution of the differential equations. And a related work can be found in [5].

Secondly, we wish the volume fractions α_g and α_l to be less than or equal to unity. This can only be achieved if the velocities satisfy equation (2.6), which in turn implies the constraint equation (2.5). Therefore, provided the velocities u_g and u_l used in equations (3.9), (3.10) and (3.15) are consistent in this sense, the volume fractions α_g and α_l will always be less than or equal to unity.

Finally, we note that a limited correction has been used in the last step for the purpose of resolving discontinuous solutions, because the semi–implicit method is only appropriate to continuous problems. When solving discontinuous problems, such a method exhibits spurious oscillations near discontinuities. The new procedure, referred to as the limited correction, has advantages over both the conventional artificial viscosity scheme and the TVD upwind scheme, and can eliminate the spurious oscillations near discontinuities. This shows that the modified semi–implicit method can be used to efficiently simulate problems where shocks must be considered, as well as to simulate problems in which disturbances propagate at very high speeds without forming shocks. In particular, the method does not contain the problem dependent parameters of conventional artificial viscosity schemes and it does not require the complex logic of upwind schemes.

5. Numerical examples

We conclude this paper by giving two examples.

- Test problem 1.

This problem is designed to examine a wave propagation in a confined channel by using the system of the two–phase flow equations given in Section 2.. A typical configuration of the gas and the liquid two–phase flow region in a horizontal pipe is drawn in Figure 1. In this case, at the initial instant, $t = 0$, both the gas and the liquid are stationary, and the interface has a discontinuity at the mid section of the channel. The expected behaviour of the system is that two disturbance waves propagate in opposite directions from mid section.

From theoretical considerations, the analytical speed of the interfacial wave is given [5]. The computed speed of this wave in the two–phase flow motion can be deduced from the following relation:

$$\text{Computed wave speed} = \frac{\text{distance of the wave propagation}}{\text{time required for moving such a distance} \times 2}. \quad (5.21)$$

All the calculations for Test problem 1 are performed using the mesh ratio $\Delta t/\Delta x = 0.005$ and $\Delta x = \frac{1}{5}$. The computed wave speed is $v_{\text{appro}} = 2.24719$; while the theoretical value, calculated in [5], is $v = 2.2658$. Thus the numerical results agree with the analytical solution.

Figure 3 shows the predicted positions of the air–water interface with a limited correction $\lambda = 0.100$. Numerical solutions obtained without this limited correction are shown in Figure 4. We see that the semi–implicit scheme is only appropriate to the continuous problems, because numerical solutions obtained by using the semi–implicit scheme exhibit severe oscillations near the discontinuities as demonstrated by Figure 4. However, the new procedure, referred to as the limited correction, can overcome this difficulty, and Figure 3 shows that it tends to remove such spurious oscillations. Thus our scheme is the more efficient one for the simulation of a wave or shock in two–phase flow.

Figures 5(a) and 5(b) show diagrammatic representations of the propagation of the wave by illustrating the calculated liquid heights as functions of time and distance. Figure 5(a) shows that there are no oscillations in the numerical solutions when the limited correction is added. In contrast to Figure 5(a), Figure 5(b) shows that there exist spurious oscillations near discontinuities for the numerical solutions without adding the limited correction. Thus this modified semi–implicit scheme is more robust than the unmodified one.

- Test problem 2.

As a final test problem, we examine the behaviour of a wave formation and propagation in a plane channel with one end closed and the other open which is displayed in Figure 2. In this case, at the initial instant $t = 0$, we assume that the gas–liquid interface is horizontal and the two fluids move with equal but opposite velocities at each point. At the open end of the channel, a constant influx of the liquid is maintained, from $t = 0$ onwards. It is expected that the liquid level will start rising at the closed end and, that a wave front of the liquid will move towards the inlet.

All the calculations for Test problem 2 are performed using the mesh ratio $\Delta t/\Delta x = 0.005$ and $\Delta x = \frac{1}{3}$. The liquid levels at different time intervals are displayed in Figure 6. The analytical velocity of the wave is $v = -1.21239$ which is calculated from [5]. If we assume that the wave front is located at a point where the level has risen by a value equal to one half of the maximum level risen, then we find that the computed velocity of the wave is $v_{appro} = -1.21457$. Thus the numerical solution is quite accurate, being in error by approximately 0.2 per cent.

It should be noted that, although these two test problems represent idealized two–phase flow models, the numerical solutions have demonstrated the potentiality of the modified semi–implicit method for calculating the propagation of the waves in more general the two–phase flow motions. In particular, the numerical approximations have indicated that the modified semi–implicit method very effectively eliminate the spurious oscillations in the solution near discontinuities.

Acknowledgement: The author is greatly indebted to Prof. K. W. Morton and Dr. J. D. Donnelly for their kind and helpful advice. Work partially supported by the Chinese Academy of Sciences and British Council.

References

[1] John H. Mahaffy, *A stability-enhancing two-step method for fluid flow calculations*, J. Comput. Phys., 1982, 46, p329-341.

[2] H. B. Stewart and B. Wendroff, *Two-phase flow: models and methods*, J. Comput. Phys., 1984, 56, 3, p363-409.

[3] J. A. Trapp and R. A. Riemke, *A nearly–implicit hydrodynamic numerical scheme for two–phase flows*, J. Comput. Phys., 1986, 66, 1, p62–82.

[4] N. N. Yanenko, *The Method of Fractional Steps*, Springer–Verlag, 1971, New York Heidelberg Berlin.

[5] X. K. Li, *Mathematical modelling and numerical approximations for two-phase flow problems*, D.Phil. thesis, University of Oxford, 1990.

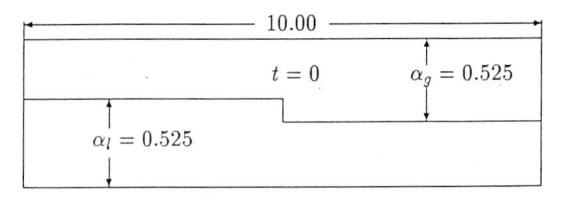

Figure 1: Test problem 1.

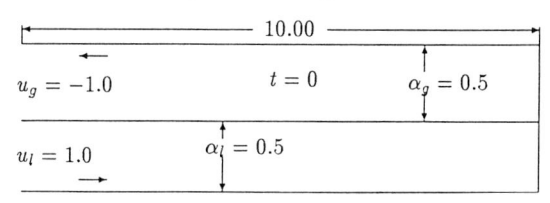

Figure 2: Test problem 2.

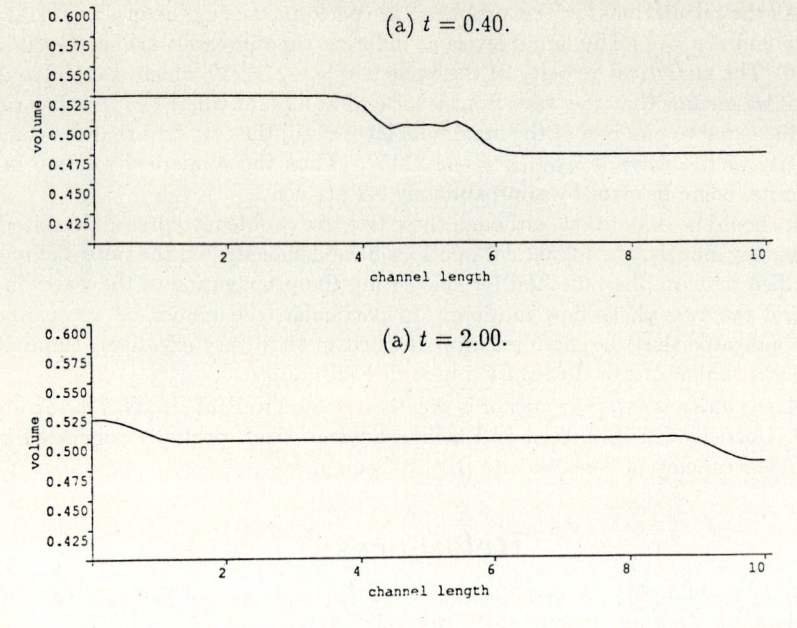

Figure 3: Liquid levels predicted for Test problem 1. The limited correction coefficient is $\lambda = 0.1$.

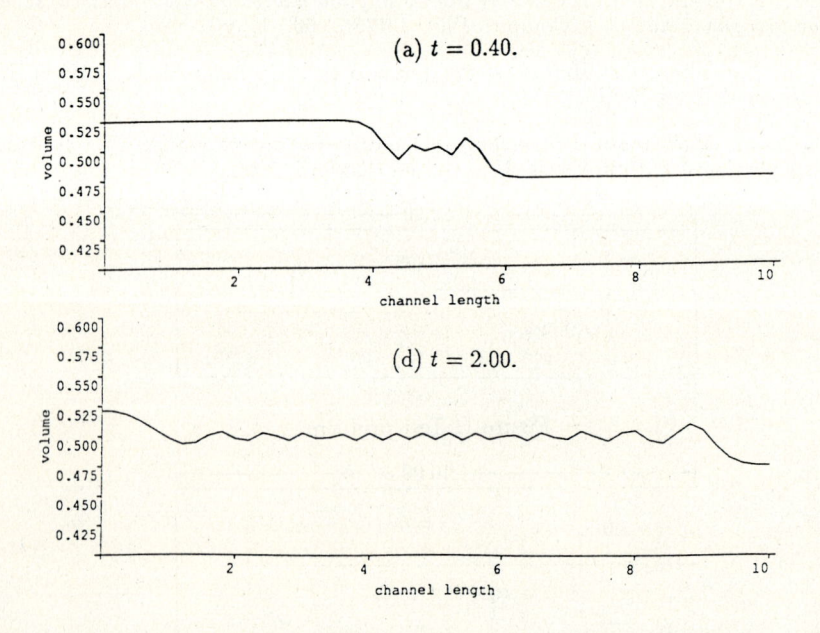

Figure 4: Liquid levels predicted for Test problem 1 without the limited correction.

(a) $\lambda = 0.1$

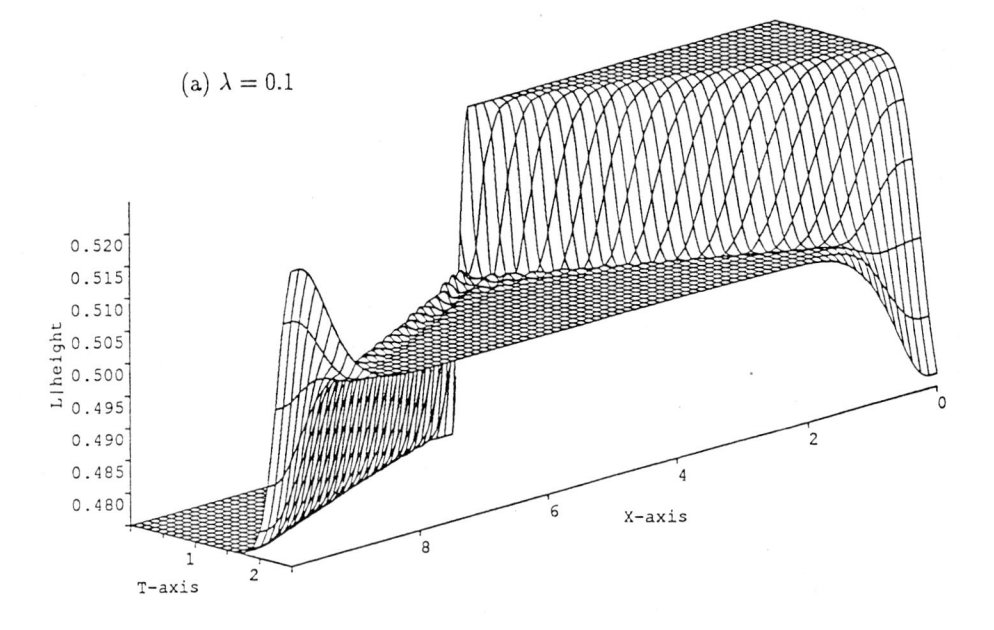

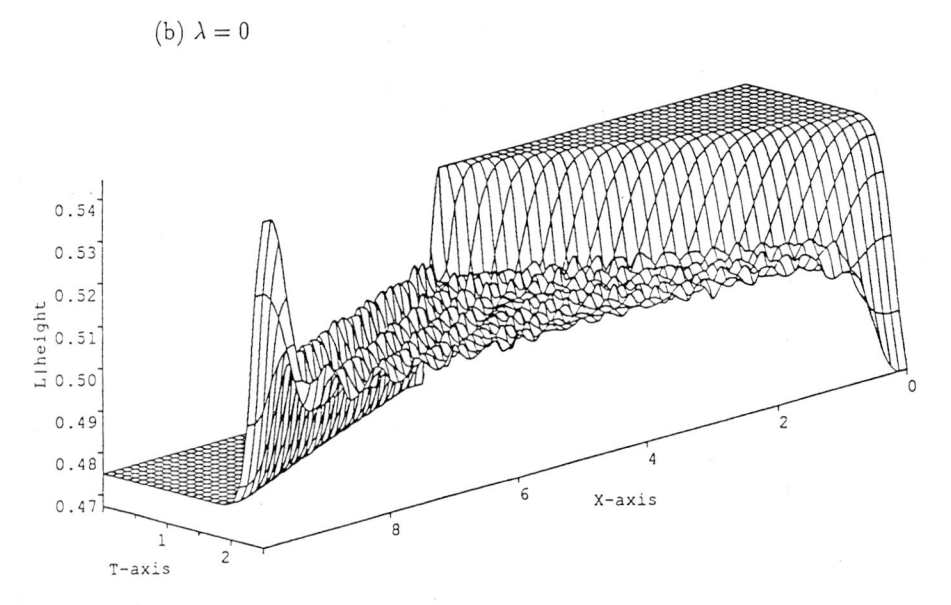

(b) $\lambda = 0$

Figure 5: Liquid heights as functions of time and distance for Test problem 1.

Figure 6: Liquid levels predicted for Test problem 2. The limited correction coefficient is $\lambda = 0.3$.

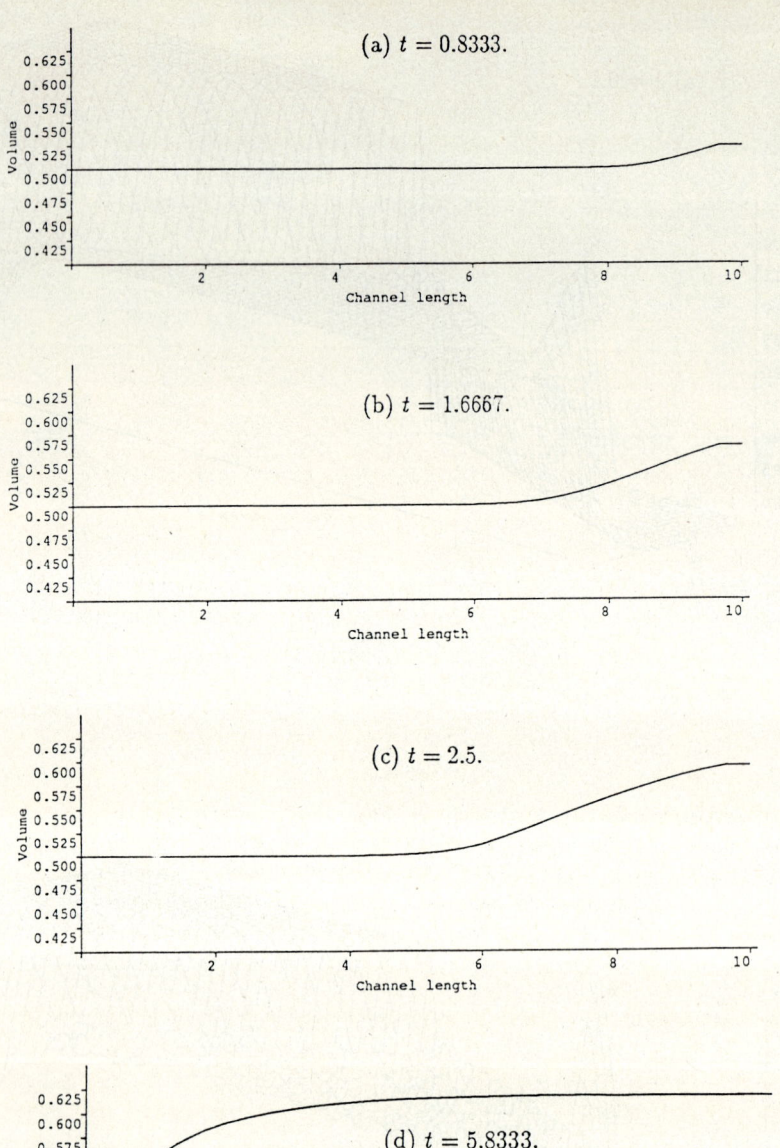

NUMERICAL CALCULATIONS OF FLOWS IN CURVED-DUCT INTAKE PORTS AND COMBUSTION CHAMBERS

J.Y. Tu and L. Fuchs

Department of Gasdynamics, The Royal Institute of Technology
S-100 44, Stockholm, Sweden

SUMMARY

A numerical scheme is presented which uses a multigrid method and overlapping grids. The current scheme has been applied for computations of three-dimensional incompressible flows in multicomponent configurations of internal combustion engines. A time-independent grid system is constructed for the moving boundary, i.e., the moving piston in the engine. This grid system is entirely different from others used to solve similar problem. The performance of the present scheme has been validated by comparing results with those from an equivalent, single grid method and those from experiments. In addition, the flexibility and the potential of the method has been demonstrated by calculating several cases which would be very difficult to handle by other approaches.

INTRODUCTION

The performance and the efficiency of an internal combustion (IC) engine depends strongly upon the flow pattern within the cylinder [1]. It has also been shown experimentally that the in-cylinder flow structure of IC engines depends on the geometry of the intake port and the location of the intake axis with respect to the chamber centreline. One of the major difficulties in calculating the engine flow arises due to a complex time-dependent geometry. Most of the numerical methods for simulating flows in IC engines have been based on the approach that a single global time-dependent grid system is generated for the whole computational domain [2,3]. Such an approach requires geometrical simplifications of the computational model. The computational domain has to be reduced to the extent of what numerical grid generation techniques can handle due to such a simplification.

These geometrical difficulties can be overcome by resorting to an overlapping grid system by which the region of interest is divided into several geometrically simpler subregions (or zones). Overlapping grid techniques have been used to calculate in-cylinder flows in simplified engine models with hole ports during the intake and the exhaust processes [4]. In order to extend the developed methodology to flow simulations in more realistic configurations of IC engines, both the curved-duct intake port flow and the in-cylinder flow are simultaneously considered in this paper. The multigrid (MG) method is also incorporated into the overlapping grid technique to accelerate the convergence of the solution. Thus, an efficient solver for calculating three-dimensional flow fields in IC engines is obtained.

In the IC engine problem, there is a moving piston which is needed to handle the moving boundary problem. In most of the previous work [2,3], a time-dependent grid that expands and compresses with the piston motion was employed. This approach needs more computational time for regenerating new grids at each time step during the solution. Moreover, the accumulation of error due to the interpolation from the solution at the last time step to the new grids would impair the accuracy of the unsteady solution. In the present case, however, a time-independent grid system in which completely new grids are not necessarily generated at each time step, is constructed by using a local computational region being attached to the moving piston.

A case for which the flow can be calculated equally well using an equivalent, single grid method is chosen to validate and assess the performance of the MG method on overlapping grids. The flow in the model geometry of an engine with a central curved-duct intake port is calculated to validate the computer code by comparison with existing experimental results. The numerical results for more geometrically complex cases are presented to demonstrate the flexibility and potential of the present scheme.

NUMERICAL METHODS

Governing equations

At this stage the flow within the curved-duct intake port and the cylinder is assumed to be incompressible and laminar. We consider the three-dimensional unsteady Navier-Stokes equations:

$$\nabla \cdot \mathbf{U} = 0 \qquad (1)$$

$$\frac{\partial \mathbf{U}}{\partial t} + \mathbf{U} \cdot \nabla \mathbf{U} = -\nabla P + \frac{1}{Re}\nabla^2 \mathbf{U} \qquad (2)$$

where $\mathbf{U}=(u,v,w)$ is the Cartesian velocity in the x-, y- and z-directions, respectively; t indicates the dimensionless time; P represents the dimensionless pressure; Re is the Reynolds number based on a reference velocity U_o, a reference length D and the kinematic viscosity v.

The boundary conditions are: the velocity vector is zero relative to solid boundaries, and is given at inlet. From the overall mass balance, the velocity at the inlet section is adjusted to the motion of the piston:

$$V_{in} = S_{piston}(\pi D^2/4)/A_e \qquad (3)$$

where S_{piston} is the piston speed, D the cylinder diameter and A_e the effective intake area. The initial conditions are as follows: At t=0, the piston is stationary at the top dead center (TDC) of the cylinder and the flow everywhere is at rest. The flow inside the cylinder is driven by the motion of the piston away from TDC, according to a simple harmonic motion, i.e. the motion of the piston follows a cosine wave while its velocity follows a sine wave.

Overlapping grid system

The grid topology for engine configurations with a curved-duct intake port is illustrated in Fig.1. The overlapping grid system consists of a body-fitted grid for the chamber of the engine and a local grid fitted to the curved-duct intake port. As seen, the two grid systems overlap each other without any restriction of grid point co-location. Information exchange among different grids is done by interpolation.

In order to construct a time-independent grid system for the moving boundary problem, the moving piston is treated artificially as a variable solid body in the computational domain. An overlapping grid system is first generated for the piston lying in the bottom dead center (BDC) where the size of the body disappears. When the piston moves towards the top dead center (TDC) or from the TDC to the BDC, the size of the body will vary and those grid points lying in the body will be flagged as the unused points which are excluded from the calculation. Care should be taken for the case in which the piston face does not exactly lie on the grid plane. A local computational region, (i.e. a thinner cell layer as illustrated in Fig. 1), is allowed to attach to the piston. The main advantage of this time-independent grid system is that no mesh regeneration is required as the piston moves.

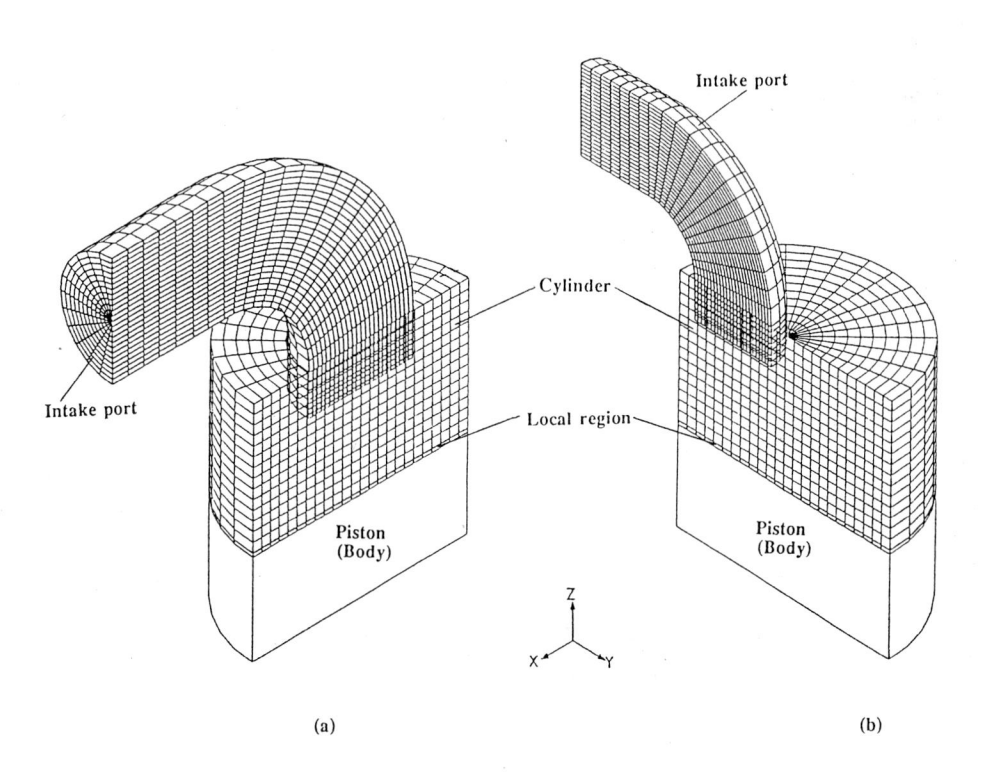

Figure 1 Illustration of overlapping grids for engine configurations with a curved-duct intake port

Numerical procedure

The finite volume method is used to discretize the Navier–Stokes equations using cartesian velocity components. By this approach one avoids the need for transformation of coordinates and simplifies the information exchange procedure among different grids. We adopt a semi–staggered grid system in which all velocity components are defined at the cell vertex while pressure is defined at the cell center. The control volume for the continuity equation is the cell element itself. For the momentum equations, the control volume is formed by joining the cell centers surrounding the calculation point. Wedged–shaped control volumes are used in case of degeneration (as near the axis of cylindrical coordinates). The main advantage of using a semi–staggered grid system is that on the one hand, the treatment of cells of general shape (co–location of velocity components) is simplified. On the other hand one does not have to specify boundary conditions of the pressure. The co–location of velocity components requires an additional "dissipation" term into the continuity equation, to eliminate oscillation ("odd–even decoupling") in pressure. This additional term is formally of fourth order accuracy and therefore does not alter the total accuracy of the results.

An implicit time discretization is applied to eq. (2). The resulting equations are solved by a MG solver at each time step [5]. The MG scheme allows for smoothing, symmetric line relaxation on the moment equations and pointwise relaxation on the continuity equation resulting in an updating of all dependent variables at that step as well. Fine to coarse grid transfer is done by volume averaging whereas the transfer of the corrections from coarse to fine grids is done by trilinear interpolations. Information exchange among the different grids is carried out by multi–dimensional Lagrange interpolation. A flux correction to the appropriate velocity components is added to guarantee global continuity.

COMPUTED EXAMPLES

A curved–duct intake port

A case for which the flow can be calculated by an equivalent, single grid method is chosen to validate and assess the performance of the method presented above. The geometry consists of a 90 degree bend with a mean radius of curvature of 3.2 times the diameter of the pipe, and inlet and outlet extensions of 2.0 and 3.2 times the diameter, respectively. The fully developed flow profile is given at the inlet and a zero–gradient condition is applied to the outlet. The Reynolds number, based on the maximum inlet velocity and the diameter of the inlet, is 150.

For reference purposes, the grid is artificially divided into three sections, each of which consists of a sequence of three level grids. The finest grids are 9 x 26 x 17 for the inlet segment, 9 x 26 x 29 for the bend and 9 x 26 x 21 for the outlet segment. Figure 2 depicts the overlapping grid system and some of the computed results. The scale factor of velocity vectors in Figs. 2 (a)–(c) is two hundred times larger than in Fig. 2 (d). These results show good agreement with the previous numerical [6] and experimental data. [7]

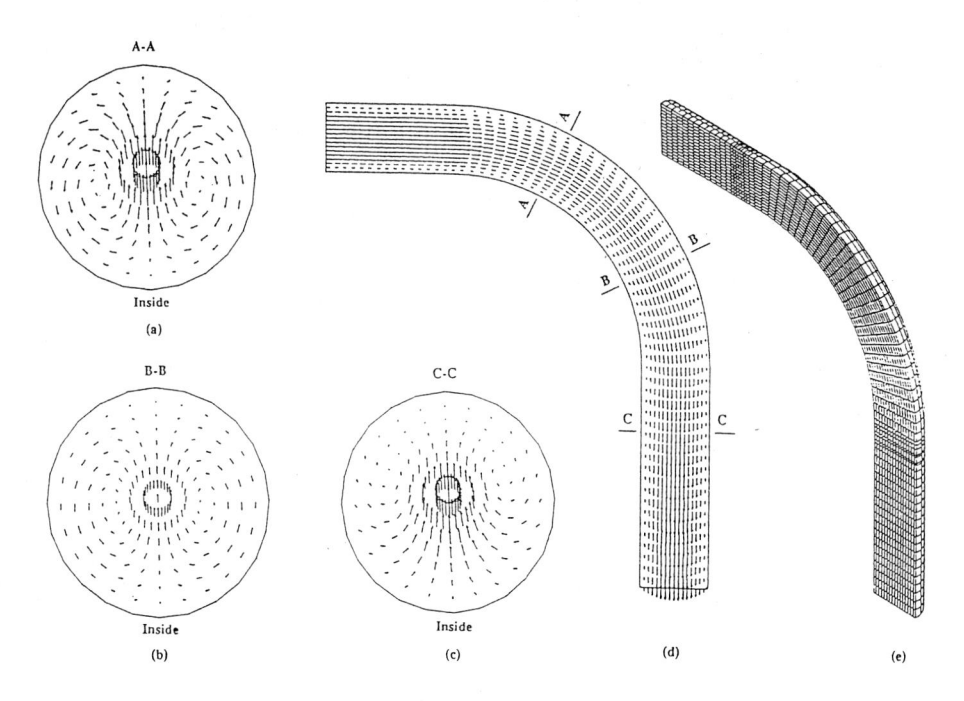

Fig. 2   Velocity vector plots and 3-D view of the overlapping grid system

In addition, a calculation is done using a single grid system that is, in all other respects, equivalent to the method described in this report. The number of computational cells used in this calculation is 9 x 26 x 65, which is close to the total number of all the three finest grids used in the overlapping grid system. The MG procedure is also employed to this single grid system by constructing a sequence of coarser grids. In this case, the performance of the overlapping grid scheme can be assessed to comparing single grid calculations.

For the purpose of comparison, we freeze the pressure field and relax the momentum equations. Figure 3 shows the convergence histories of the different calculations as the function of work units (WU). One WU is the amount of work to perform one sweep of symmetrical line relaxation on the finest grids. The work units required by the implementation of the interfaces on the overlapping grid system are estimated according to the required CPU time. It can be concluded that the current scheme is more efficient than the single grid scheme. The information exchange among the zones leads to some deteriorated convergence rate. The overlapping grid system is, however, far more flexible to deal with the complex geometry than a global single grid system.

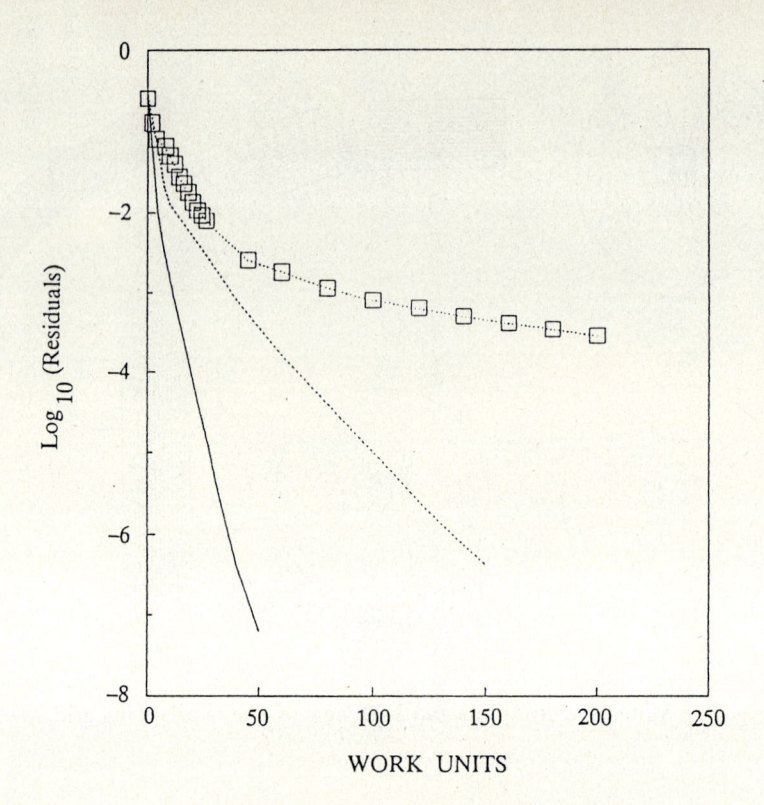

Fig.3   Convergence histories of the different schemes
 ——— : MG in the single grid system
 .... : MG in the overlapping grid system
 ...□... : Single grid scheme in the single grid system

A chamber with a central curved-duct intake port

The flow field in a model engine whose configuration has been used in the experimental investigation of Sanatian and Moss [8] is calculated here. The Reynolds number, based on the maximum piston velocity and the chamber diameter, is taken as Re = 100. The overlapping grid system for this case is illustrated in Fig.1 (a). The finest grids used are 13 x 26 x 29 both for the intake port and for the chamber. A fixed valve, 0.45 times the diameter of the cylinder, is located at the distance 0.12 times the stroke away from the head of the chamber. In the numerical calculation, it is assumed that the valve is fully opened during the entire intake process.

Figure 4 shows a three-dimensional velocity vector plot of the intake stroke flow at an equivalent crank angle (CA) $\theta = 90°$ The interaction of the intake jet with the wall produces large scale rotating flow patterns within the cylinder volume. The upper corner of the chamber contains a second smaller vortex rotating in the opposite direction. The flow around the valve periphery is found to be no-uniform. The computed flow structure is graphically similar to the experimental pictures [8].

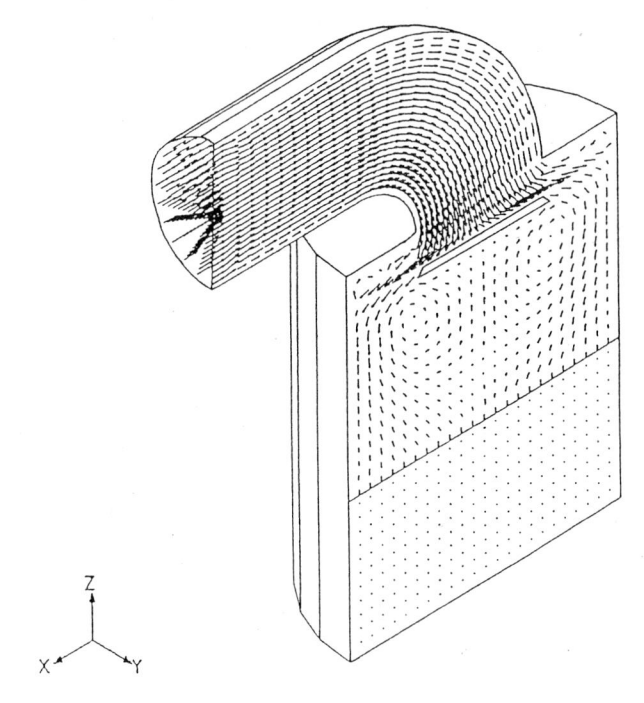

Fig. 4 Velocity vector plots at an equivalent crank angle $\theta = 90°$ for the case with a central curved–duct intake port

A chamber with an off–centre curved–duct intake port

A more practical configuration which is a chamber fitted with an off–centre curved–duct intake port has been investigated to demonstrate the flexibility of the present method. The Reynolds number, based on the maximum piston velocity and the chamber diameter, is taken as Re = 150. In order to increase the effect of swirl within the combustion chamber, a arrangement between the intake port and the combustion chamber is made similarly to that in the experimental investigation of Hirotomi et al. [9]. The overlapping grid system is illustrated in Fig.1 (b), where the only difference from the present case is due to the different location of the curved–duct port with respect to the cylinder axis.

The grids used for this case are composed of three level grids for each component grid in which the finest grids are 13 x 26 x 29 for the chamber and 9 x 26 x 29 for the curved–duct intake ports. Figures 5 and 6 depict the flow fields computed for this case which are exhibited in terms of the instantaneous velocity vectors at different equivalent crank angles $\theta = 90°$ and $180°$.

At the crank angle $\theta = 90°$ (Fig.5), a tangential flow with respect to the cylinder axis from the outlet of the curved–duct intake port, as shown in Fig. 5 (c), is created so that a strong swirling flow near the head of the cylinder is formed as revealed in Fig. 5 (d). A complex flow showing a small vortex near the inlet of the cylinder is also found. Figure 5 (e) depicts that a nearly plane–symmetrical swirl flow close to the piston is produced by the interaction of the intake jet flow with the piston face.

(a) Y=0.25

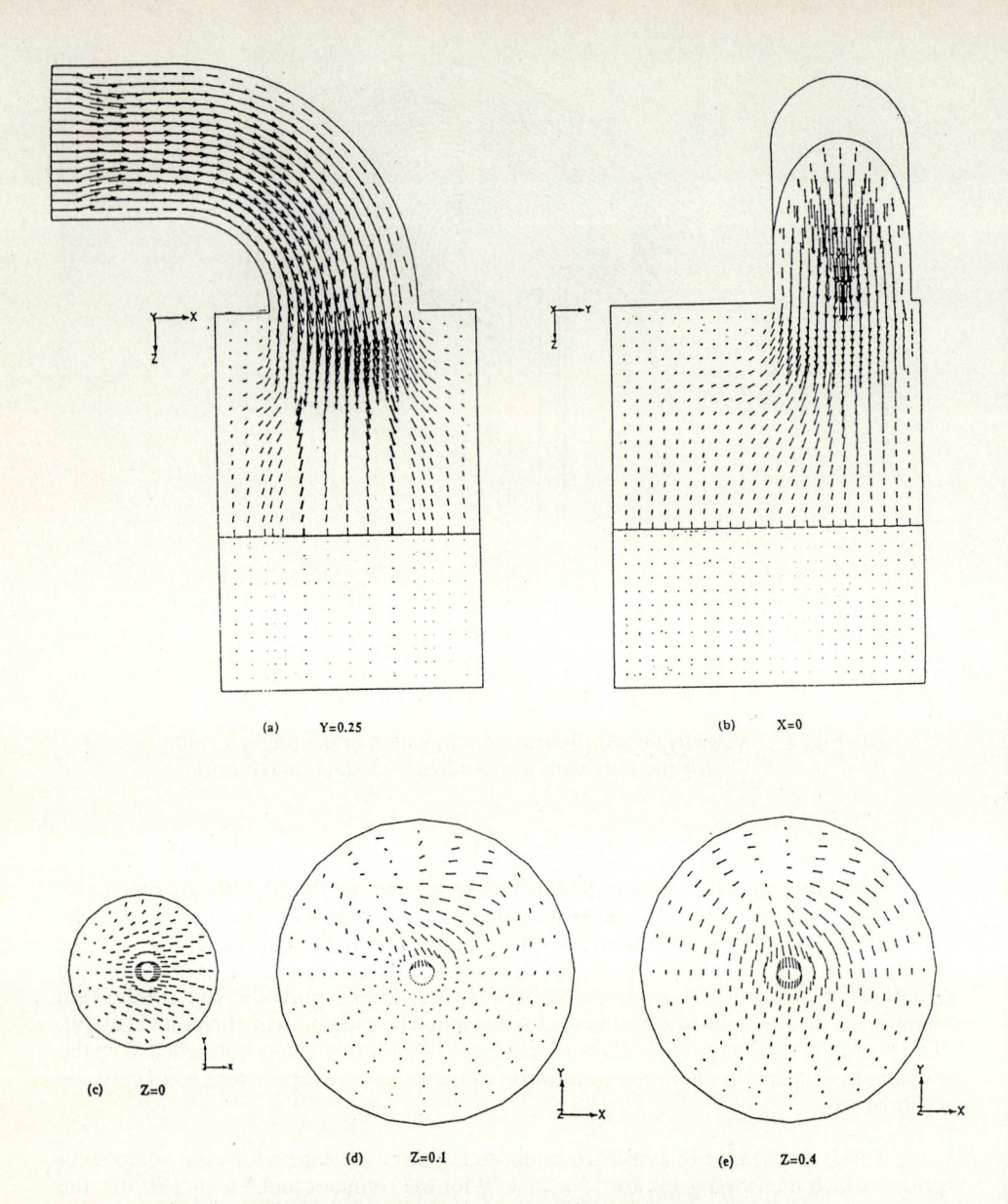

(b) X=0

(c) Z=0

(d) Z=0.1

(e) Z=0.4

Fig. 5   Velocity vector plots at an equivalent crank angle $\theta = 90°$ for the case with a off-center curved-duct intake port

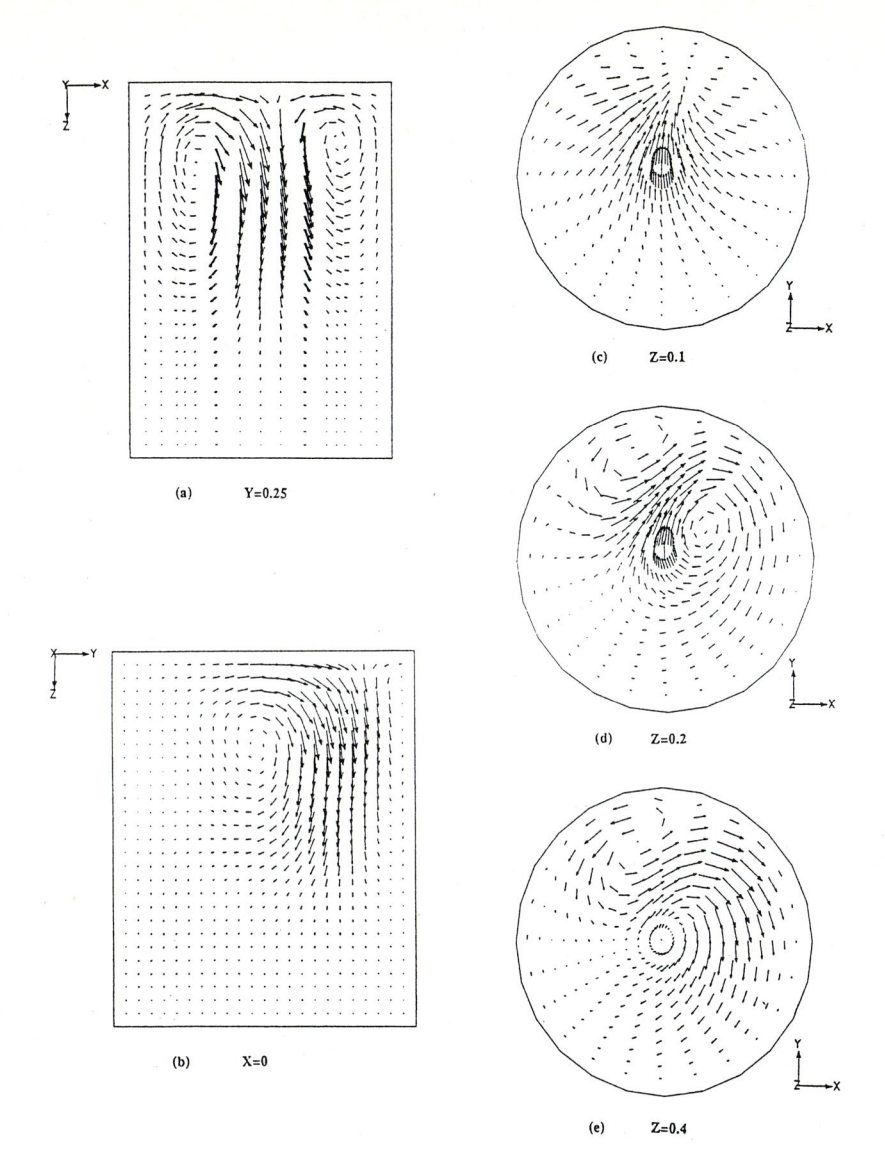

Fig. 6   Velocity vector plots at an equivalent crank angle $\theta = 180°$ for the case with a off-center curved-duct intake port

Until the full induction, the large scale rotating flow patterns within the cylinder volume are clearly evident, as exhibited in Figs. 6 (a) and (b). Several small vortices revealing a complex flow in the diameter plane are illustrated in Figs. 6 (c)–(e). The pair vortex flow pattern, shown in Fig. 6 (d), strongly resembles that observed experimentally in water analog of intake process by Hirotomi et al. The scalar factor in Figs. 6 (d) and (e) is four times larger than those in Figs. 6 (a)–(c).

CONCLUDING REMARKS

The flow in an engine cylinder equipped with a curved-duct intake port system has been investigated by using an overlapping grid technique and a multigrid method. The current overlapping grid technique has enhanced flexibility and capability in dealing with three-dimensional, multicomponent, time-dependent configurations. The MG scheme is mandatory for the overlapping grid system to compensate for the additional reduction of computational speed due to information exchange. The numerical results demonstrate that the current scheme is more appropriate and efficient in handling complex problems with multicomponent geometries compared with conventional single grid schemes. Currently, a ($k-\epsilon$) turbulence model is being incorporated into the computer code.

ACKNOWLEDGMENT

This work is financially supported by the Swedish National Board for Technical Development (STU grant no.8902268).

REFERENCES

1. J. B. Heywood, 'Fluid Motion Within the Cylinder of Internal Combustion Engines', *Journal of Fluids Engineering*, **109**, 3-64 (1987).
2. A.D. Gosman, 'Computer modeling of flow and heat transfer in engines, progress and prospects', *JSME, Int. Symp. on Diagnostics and Modeling of Combustion in Reciprocating Engines*, Tokyo, 1985, pp. 15-26.
3. T. Wakisaka, Y. Shimamoto and Y. Isshiki, 'Three-dimensional numerical analysis of in-cylinder flows in reciprocating engines', *SAE paper 860464*, (1986).
4. J.Y. Tu and L. Fuchs, 'Application of Overlapping Grids to 3-D Flow Calculations in a Model Engine', *Numerical Methods in Laminar and Turbulent Flow*, (Eds., C. Taylor, et al.), Pineridge Press, 1991, pp. 1198-1208.
5. J.Y. Tu, L. Fuchs and X.S. Bai, 'Finite Volume and Multigrid Methods on 3-D Zonal Overlapping Grids', *Multigrid Methods: Special Topics and Applications II*, GMD-Study, No.189, (Eds. W. Hackbusch, U. Trottenberg), 1991, pp. 337-348.
6. M. Reggio and R. Camarero, 'A calculation scheme for three-dimensional viscous incompressible flows', *ASME J. Fluids Eng.*, **109**, 345-352 (1987).
7. M. Enayet, M. Gibson, A. Taylor and M. Yianneskis, 'Laser-doppler measurements of laminar and turbulent flow in a pipe bend', *Int. J. Heat Fluid Flow*, **3**, 213-219 (1982).
8. R. Sanatian and J.B. Moss, 'The experimental and computer simulation of induction flows in spark ignition engines', *Heat and Mass Transfer in Gasoline and Diesel Engines*, (Eds. D.B. Spalding and N.H. Afgan), 1989, 275-289.
9. T. Hirotomi, I. Nagayama, S. Kobayashi and Yamamasu, 'Study of induction swirl in a spark ignition engine', *SAE paper 810496, SAE Trans.* **90**, 1851-1867 (1981).

HYPERSONIC FLOWS

Hypersonic Flow Simulation by Marching on an Adapting Grid

by

B. Müller*, P. Niederdrenk″, H. Sobieczky″

* Institute of Fluid Dynamics, ETH Zürich, Switzerland
″ DLR, Institute for Theoretical Fluid Mechanics, Göttingen, Germany

SUMMARY

An accurate and robust solver of the three-dimensional Euler equations has beeen developed based on upwind relaxation strategies. Coupling the flow solver to a fast adaptive grid generator yields an efficient numerical tool. Results for inviscid hypersonic flow past the forebody of a generic aerospace plane demonstrate, how the solution marches downstream on its self constructed and adapted grid.

INTRODUCTION

The aero-space transport system 'Sänger' implies a considerable test bed for hypersonic research in Germany. From the first stage of this system a generic configuration has been derived and modelled with the new geometry generator [1]. While in the present paper the forebody of this geometry merely serves as a realistic test case for the new CFD-tool, following studies will be intended to compare CFD results for a number of leading edge shapes, lower surface compression ramps and the resulting three-dimensional shock system. One goal will be a suitably taylored quality of the flow at the location of the inlet for ramjet propulsion.

The present paper describes a fast, adaptive grid generation and an accurate marching flow solution procedure. These components have been coupled in such a way that the solution marches downstream on automatically created and adapted cross section grids.

GRID GENERATION

The three dimensional grid is constructed from a sequence of planar cross-sectional grids. Since the generation of these planar grids is described in detail in reference [2], we will only strive for the main ideas here. We start from a second order hyperbolic equation for the position vector $\underline{r}$ as a function of the computational variables η, ζ:

$$r_\zeta^2 (\underline{r}_{\eta\eta} - \phi \underline{r}_\eta) - r_\eta^2 (\underline{r}_{\zeta\zeta} - \psi \underline{r}_\zeta) = 0 \ . \tag{1}$$

Defining the angle between the grid lines by θ, the curvature of a line $\eta = $ const by $\kappa^{(\eta)}$ and the arc length increments along a line $\zeta = $ const by s_η and along a line $\eta = $ const by ℓ_ζ, respectively, the appertaining control functions may be expressed as

$$\phi = \frac{s_{\eta\eta}}{s_\eta} + \frac{s_\eta}{\sin\theta} \left(\kappa^{(\eta)} - \cos\theta \, \kappa^{(\zeta)} \right) \ , \quad \psi = \frac{\ell_{\zeta\zeta}}{\ell_\zeta} - \frac{\ell_\zeta}{\sin\theta} \left(\kappa^{(\zeta)} - \cos\theta \, \kappa^{(\eta)} \right) \ . \tag{2}$$

As long as curvature terms are small, the grid will mainly be influenced by the specifications of the relative changes of the arc length increments along the grid lines. Choosing ζ as the time-like marching direction starting at the body surface, we split the control function ψ into two terms

$$\psi = \psi_e + \psi_c \quad \text{with} \quad \psi_e = \frac{\ell_{\zeta\zeta}}{\ell_\zeta} \;,\; \psi_c = \frac{-\ell_\zeta}{\sin\theta}\left(\kappa^{(\zeta)} - \kappa^{(\eta)}\cos\theta\right).$$

Applying an upwind difference approximation in the ζ-k-direction and a central difference approximation in the circumferential η-j-direction on the unknown level $k+1$ results in an implicit, first order accurate scheme for equation (1)

$$\Delta_\zeta \underline{r}_{j,k} = \frac{AR^{-2}}{1 - \psi_e/2}\left[(\delta_{\eta\eta} - \phi\,\delta_\eta)\underline{r}_{j,k+1} + AR^2\,\psi_c\,\delta_\zeta\underline{r}_{j,k}\right] + \frac{1 + \psi_e/2}{1 - \psi_e/2}\,\nabla_\zeta \underline{r}_{j,k}\,, \qquad (3)$$

where the symbols ∇, δ, Δ denote upwind, central and downwind differences, respectively, and $AR = s_\eta/\ell_\zeta$ is the cell aspect ratio. The restriction to first order accuracy corresponds to an introduction of first order artificial viscosity suppressing the propagation of slope discontinuities into the field. For not too large deviations from orthogonality the most important term in ψ_c is that due to the curvature of a line $\zeta =$ const, which is usually largest at the body surface. Assume we know an approximation for the term containing ψ_c in equation (3), the solution could easily be marched in ζ-direction.

Instead of solving equation (3) in a hyperbolic sense we manipulate its last term in order to make the grid feel a prescribed outer boundary. To this end we expand ψ_e in terms of differences of prescribed arc length ℓ, write the source term as

$$S_{j,k} = \frac{1 + \psi_e/2}{1 - \psi_e/2}\,\nabla_\zeta \underline{r}_{j,k} = \frac{\Delta_\zeta \tilde{\ell}_{j,k}}{\nabla_\zeta \tilde{\ell}_{j,k}}\,(\nabla_\zeta \ell_{j,k})\,\underline{e}_{j,k}\,,$$

where $\underline{e}$ is a unit vector in ζ-direction, and approximate the ratio of upwind differences of prescribed over calculated arc lengths by corresponding distances from the just calculated level to the outer boundary

$$\frac{\nabla_\zeta \ell_{j,k}}{\nabla_\zeta \tilde{\ell}_{j,k}} \approx \frac{\ell_{j,k\max} - \ell_{j,k}}{\tilde{\ell}_{j,k\max} - \tilde{\ell}_{j,k}}\,.$$

Neglecting $\kappa^{(\eta)}$ simple estimations of $\kappa^{(\zeta)}_{j,k}$, $\ell_{j,k\max} - \ell_{j,k}$ and $\underline{e}_{j,k}$ do not only allow to generate grids in a single sweep but simultaneously force them to meet - at least approximately - a prescribed outer boundary.

GRID ADAPTATION

Once a solution is known on a starting grid, new values for the control functions are obtained from one-dimensional relocations of points according to equidistributions of a solution dependent weight along the lines of the starting grid.

Consider, for example, a line $\eta =$ const and let f be a quantity of the physical solution, which is known at arc length increments ℓ_ν of the starting grid. The new spacing ℓ_ζ follows from a solution to the non-linear equation

$\ell_\zeta w(\ell) = \text{const}$ with $w = 1 + c |\nabla f|$.

Expanding $\ell_\zeta = v_\zeta \ell_v$, we determine $\zeta(v)$ from the linear equation

$$\zeta_v = \frac{w[\ell(v)]l_v}{\text{const}}, \quad \text{where} \quad \text{const} = \int_1^{\bar{v}} w[\ell(v)] \ell_v dv/(\zeta_{max} - 1)$$

and $\bar{v}$ defines the value $\ell(\bar{v})$, up to which the new point distribution is required along a line of the last grid. While keeping the total number of points fixed, $\zeta_{max} = v_{max}$, such a situation, i.e. $\bar{v} < v_{max}$, may occur along lines radiating away from the body, when in connection with the usual adaptation the extension of the flow field in physical space is simultaneously reduced.

Subsequent interpolation of $\ell(v)$, $\zeta(v)$ to unit increments in ζ yields the desired new arc length distribution $\ell(\zeta)$. Evaluation of $s(\eta)$ follows from an analogous procedure applied to the other family of lines and serves to determine the adaptive part of the control function ϕ.

For pointed bodies only an initial guess of the flow field's outer boundary is needed in the plane i = 2 to create a starting grid.

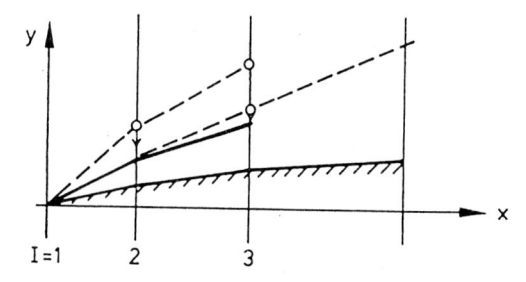

Figure 1. Grid boundary relocation Outer boundaries of starting (---) and final (—) grid

From the physical solution on this cross section a new outer grid boundary is defined, which will generally fall closer to the body surface. Having adapted the grid on the reduced cross-sectional field at i = 2 this new grid is projected downstream onto the cross sections at i = 3 and i = 4, from whereon the procedure is repeated starting with the redefinition of the outer grid boundary at i = 3.

Since the projected grid is an already pre-adapted one, only one adaptation step is needed to obtain an appreciable flow field resolution.

SPACE MARCHING METHOD

In steady inviscid hypersonic flow simulations we can exploit the fact that the Euler equations are hyperbolic in the streamwise direction. Thus, space marching can be employed, which is much more economical in terms of CPU-time and memory then time stepping.

Instead of solving the steady Euler equations we follow here a time-iterative approach [3]. The time-dependent Euler equations are solved for the steady state in each crossflow plane starting from a conical solution near the apex. Thus, linearization and factorization errors inherent in conventional implicit marching codes can be reduced to arbitrary low levels at the cost of more iterations. Another advantage of this procedure is that it can be easily relaxed in subsonic pockets to allow upstream and downstream sweeps.

3D Euler Equations

The 3D Euler equations for perfect gas ($\gamma = 1.4$) read in dimensionless conservation law form and in general coordinates:

$$\frac{\partial J^{-1} q}{\partial t} + \frac{\partial E^{(\xi)}}{\partial \xi} + \frac{\partial E^{(\eta)}}{\partial \eta} + \frac{\partial E^{(\zeta)}}{\partial \zeta} = 0 \tag{4}$$

where $q = (\rho, \rho u, \rho v, \rho w, e)^T$ is the vector of the conservation variables, and J is the Jacobian of the transformation from Cartesian to general coordinates (Fig. 2).

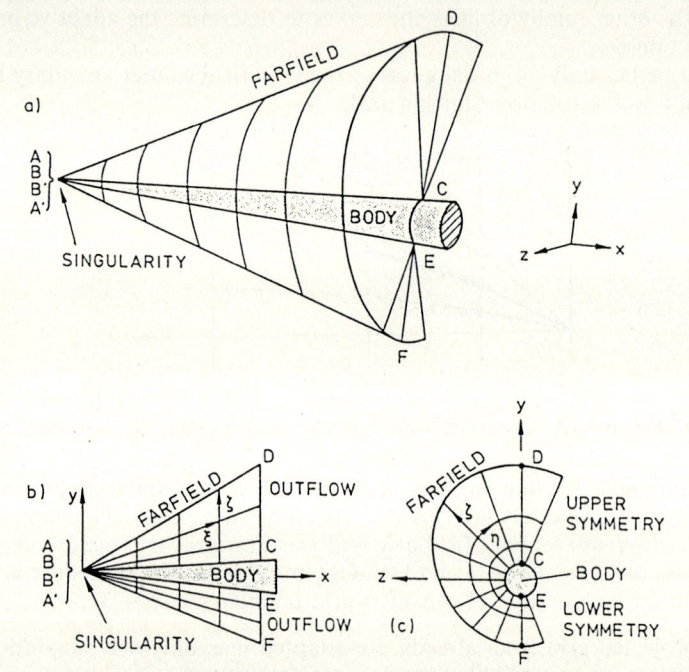

Figure 2. Boundaries of physical domain and coordinate system. (a) Global view, (b) Plane of symmetry, (c) Outflow surface

$E^{(\xi)}$, $E^{(\eta)}$, and $E^{(\zeta)}$ are the fluxes in the streamwise ξ-, circumferential η-, and near normal ζ-directions, respectively. They can be expressed by means of the Cartesian flux tensor $\underline{F}$ and the side normals, e.g.

$$E^{(\xi)} = \underline{F} \cdot J^{-1} \underline{\nabla} \xi \ . \tag{5}$$

Flux Discretization

The fluxes are discretized by a cell-vertex finite volume method. The control volume of a grid point is defined by the midpoints of the adjacent 8 hexahedra (cf. Fig. 3 for the illustration in 2D). Once the grid has been adapted at crossflow plane i, the interface i + 1/2 is kept fixed. The fluxes are determined by the second-order accurate upwind total variation diminishing (TVD) method of Harten and Yee [4],[5]. For hypersonic flow special care in enforcing the entropy condition is necessary for Roe's approximate Riemann solver, on which the upwind TVD method of Harten and Yee is based [6].

For a space marching method, the discretization of the streamwise flux $E^{(\xi)}_{i+\frac{1}{2}}$ (suppressing the j and k indices) needs special consideration. Assuming that the steady state solution of the Euler equations (4) is known at the crossflow planes 1 to i-1, the Euler equations (4) at crossflow plane i are intialized by q_{i-1} and iterated to the steady state. In order to avoid upstream influence with the five point stencil q_{i+1} and q_{i+2} at the two downstream crossflow planes are set equal to the actual value of q_i. Thus, the upwind TVD method for the streamwise flux reduces to the first-order upwind discretization

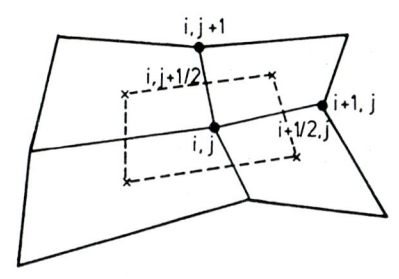

Figure 3. Control volume (dashed line) of grid point i,j in 2D

$$E^{(\xi)}_{i+\frac{1}{2}} = \underline{F}_i \cdot \left(J^{-1} \underline{\nabla} \xi \right)_{i+\frac{1}{2}}. \tag{6}$$

DD-ADI Method and Characteristic Underrelaxation

Employing the first-order Euler implicit formula for the time derivative in (4), the fluxes are approximately linearized with first-order in space. Thus, we arrive at a block-pentadiagonal linear system in the crossflow plane i. We solve the system by the vectorized diagonally dominant alternating direction implicit (DD-ADI) method [7].

A linear stability analysis of the DD-ADI method (5) applied to the 2D linear wave equation suggests that the relaxation factor should be $\omega = 1$ or $0 < \omega \leq 1/2$, if the fluxes in the η- and ζ-directions are first- or second-order accurate, respectively. The Harten-Yee upwind TVD scheme reduces to first-order in a certain component and direction, if that component of the characteristic variables in the respective direction encounters a local extremum. Thus, characteristic underrelaxation is proposed: in $B^{(\eta)\pm} = R^{(\eta)} \Lambda^{(\eta)\pm} R^{(\eta)-1}$, the split Jacobian matrices of $E^{(\eta)}_j$, the split eigenvalues

$$\lambda^{(\eta)\pm}_\ell, \quad \ell = 1, ..., 5, \quad \text{are replaced by} \quad \lambda^{(\eta)\pm}_\ell / \omega^{(\eta)}_\ell, \quad \ell = 1, ..., 5, \tag{7}$$

where $\omega_\ell^{(n)}$ is sensed by the ℓ-th component of $\alpha_{j+\frac{1}{2}}^{(n)} = R_{j+\frac{1}{2}}^{(n)}{}^{-1}(q_{j+1} - q_j)$:

$$\omega_{\ell j+\frac{1}{2}}^{(n)} = 1 - \frac{\frac{1}{4}\left(|g_{\ell j}^{(n)}| + |g_{\ell j+1}^{(n)}|\right) + \varepsilon}{|\alpha_{\ell j+\frac{1}{2}}^{(n)}| + 2\varepsilon} \tag{8}$$

with $g_{\ell j}^{(n)} = \text{min mod}\{\alpha_{\ell j-\frac{1}{2}}^{(n)}, \alpha_{\ell j+\frac{1}{2}}^{(n)}\}$, $\varepsilon = 10^{-12}$ to avoid division by zero.

Characteristic underrelaxation is used analogously in the ζ-direction.

RESULTS

The space marching code was validated for laminar hypersonic flow past a circular cone by comparison with Tracy's experimental investigation [6]. The few results presented in this paper are mainly intended to illustrate the code's coupling with the adaptive grid generator. They refer to inviscid flow and one simple choice of the adaptation weight only. Viscous flow applications are on the way.

Fig. 4 shows the generic forebody and three coarse cross section grids (16 x 41 points), which are adapted with respect to the absolute value of the local density gradient ($M_\infty = 4.5$, $\alpha = 6°$). For the middle cross section of fig. 4 the fine grid (31 x 81) solution (Fig. 5) may be compared to the coarse adapted grid solution (Fig. 6). Though the fine grid's point number is eight times that of the coarse grid, its bow shock resolution (Fig. 5b) is obviously poorer than that of the coarse, projected, preadapted grid (Fig. 6b). The following adaptation step (Fig. 6c,d) adjusts the outer boundary on the upper side and slightly sharpens the gradient across the bow shock. It has only a relatively small corrective effect with respect to the gain already obtained by downstream projection of a preadapted grid. In comparison to an adaptation procedure starting from an unadapted grid [5], projection of preadapted grids saves at least one adaptation step or in other words one third of the computational time. Here as well as in [5] the constant c appearing in the adaptation weight w was determined from a prescribed grid line spacing ratio $\Delta\ell_{max}/\Delta\ell_{min} = 15$ on all cross sections.

The coarse adapted grid (Fig. 7a) clearly reflects the position of the bow shock in the symmetry plane. As the radius of the leading edge, non-dimensionalized by the span, decreases downstream, strong gradients develop near the wing's tip, while the bow shock looses in strength. Thus, the bow shock location can no longer be identified in the grid near the outflow section. Here, due to peaky gradients at the tip, the adapted grid is clustered only very locally towards the leading edge. Wiggles in the outer grid bounding surface are due to the solution adaptive cross-sectional specification of the calculation field. They do not influence the solution. Isobars obtained from the adapted coarse grid (Fig. 7b) and the unadapted fine grid solution (Fig. 7c) show compression waves emanating from the lower body surface downstream of x/L = 0.5. They are due to a small compression ramp.

CONCLUSIONS

1. Adaptive grids are created in crossflow planes by approximate, non-iterative solutions to hyperbolic grid generation equations.
2. Convergence of the space marching flow solver to the steady state in each crossflow plane is enhanced by characteristic underrelaxation.

Grid generation and adaptation have been combined with the space marching flow solver such that the solution marches downstream on automatically created and adapted cross section grids.

The extension to viscous flow simulation is straight forward, except for the parabolized streamwise flux in the subsonic part of the boundary layer and the simultaneous adaptation of a number of layers, e.g. shock waves, boundary layers, vortex layers etc..

Reference

[1] Sobieczky, H., Stroeve, J.C.: "Generic Supersonic and Hypersonic Configurations", AIAA 91-3301, Baltimore, 1991

[2] Niederdrenk, P.: "Solution - Adaptive Grid Generation by Hyperbolic / Parabolized Hyperbolic P.D.E.s.". In: Numerical Grid Generation in Computational Fluid Dynamics and Related Fields, Eds.: A.S. Arcilla, J. Häuser, P. Eisemann, J. Thompson; North-Holland, 1991

[3] Chakravarthy, S.R., Szema, K.Y.: "An Euler Solver for Three-Dimensional Supersonic Flows with Subsonic Pockets", AIAA Paper 85-1703, 1985

[4] Yee, H.C., Harten, A.: "Implicit TVD Schemes for Hyperbolic Conservation Laws in Curvi-linear Coordinates", AIAA J., Vol. 25, pp. 266-274, 1987

[5] Müller, B., Niederdrenk, P., Sobieczky, H.: "Simulation of Hypersonic Waverider Flow", First International Hypersonic Waverider Symposium, University of Maryland, 1990

[6] Müller, B.: "Upwind Relaxation Method for Hypersonic Flow Simulation", DLR Report, to be published 1991

[7] Lombard, C.K., Bardina, J., Venkatapathy, E., Oliger, J.: "Multi-Dimensional Formulation of CSCM-An Upwind Flux Difference Eigenvector Split Method for the Compressible Navier-Stokes Equations", AIAA Paper 83-1895, 1983

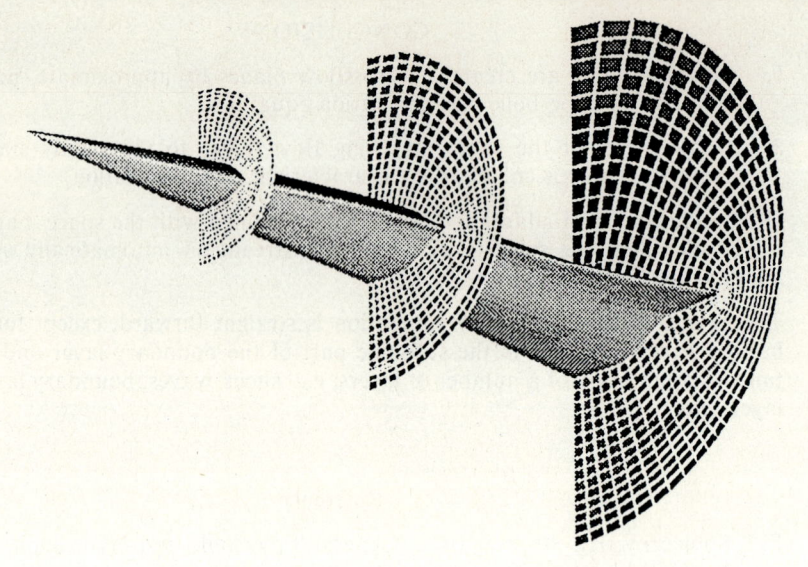

Figure 4. generic forebody; adapted coarse cross section grids (16 x 41 x 16)

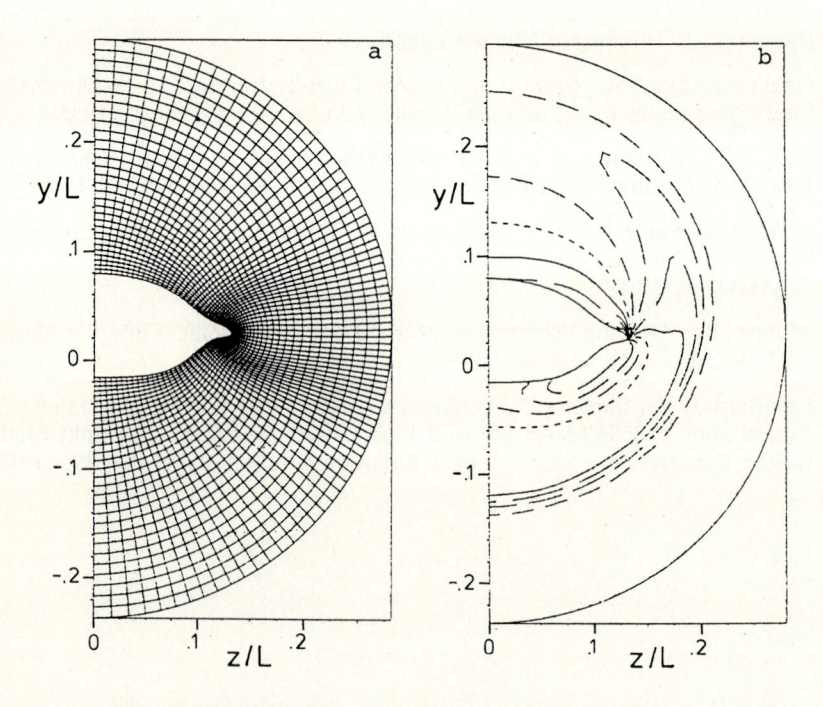

Figure 5. mid-plane at x = 0.63 a) fine unadapted grid (31 x 81)
b) isochors

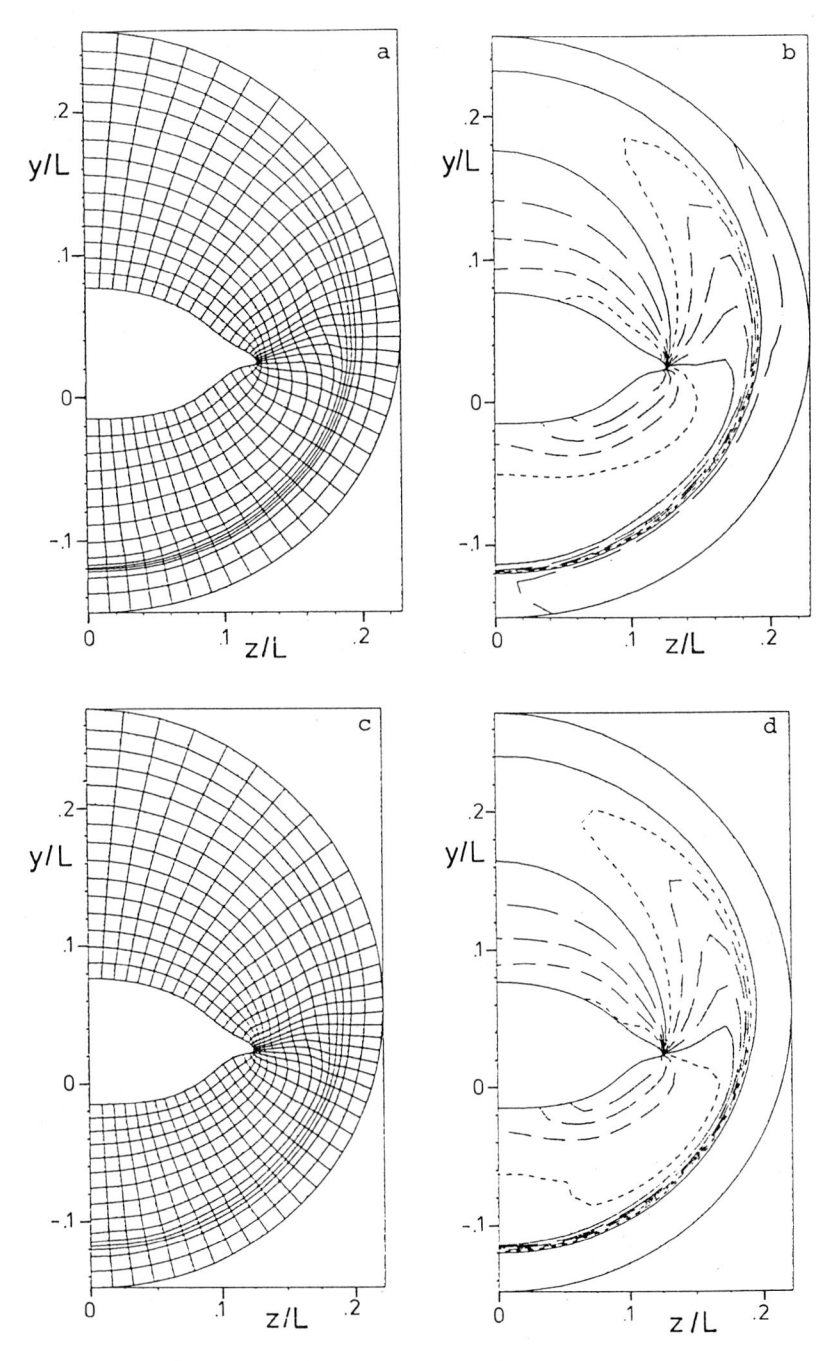

Figure 6. mid-plane at x = 0.63:
projected coarse grid (a) and isochors (b);
adapted coarse grid (c) and isochors (d)

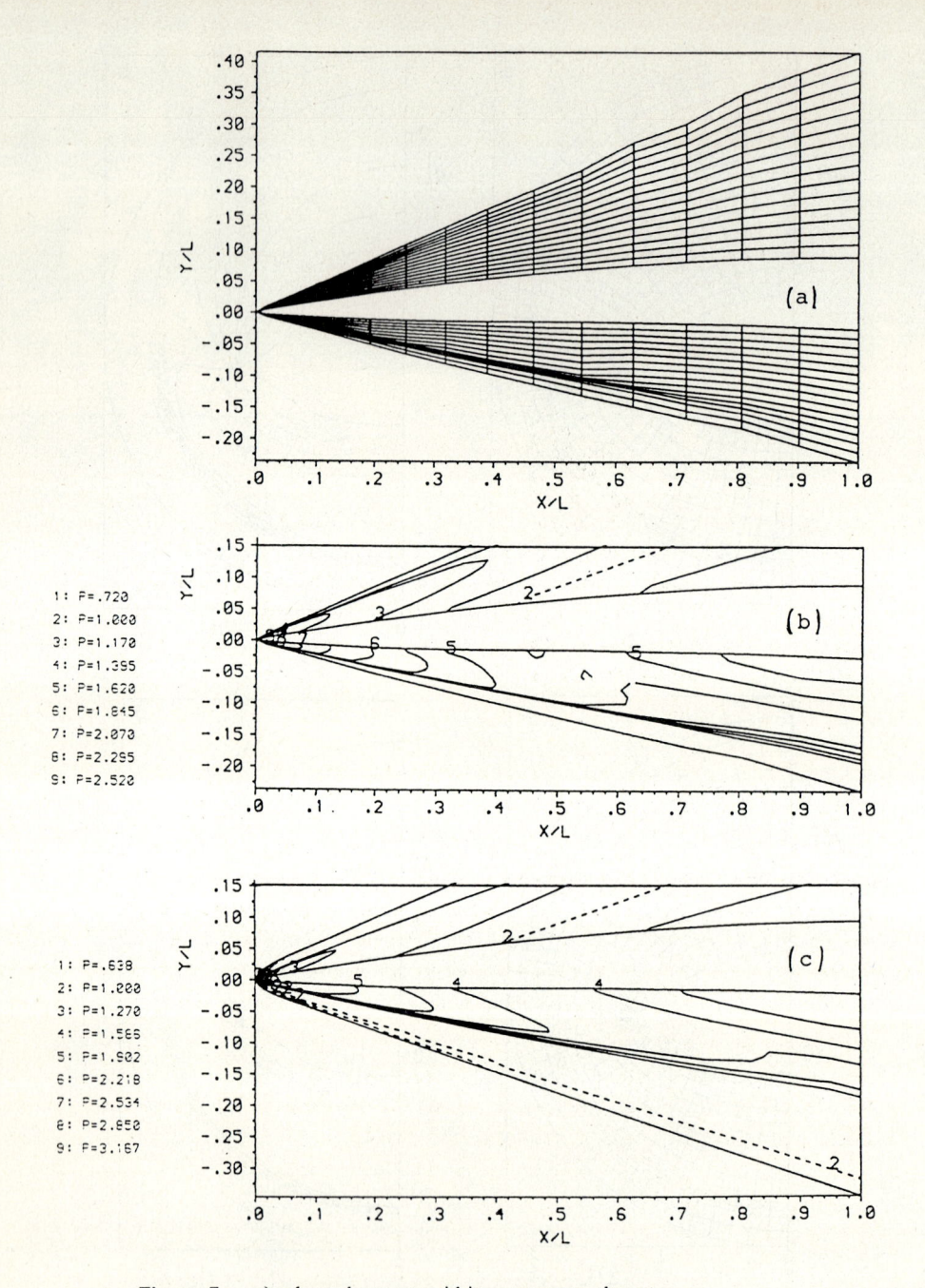

1: P=.720
2: P=1.000
3: P=1.170
4: P=1.395
5: P=1.620
6: P=1.845
7: P=2.070
8: P=2.295
9: P=2.520

1: P=.638
2: P=1.000
3: P=1.270
4: P=1.565
5: P=1.902
6: P=2.218
7: P=2.534
8: P=2.850
9: P=3.167

Figure 7. a) adapted coarse grid in symmetry plane
b) isobars from adapted coarse grid solution
c) isobars from unadapted fine grid solution

Calculation of viscous hypersonic flows in chemical non-equilibrium

Ch. Mundt

Messerschmitt-Bölkow-Blohm GmbH
Postfach 80 11 60
D - 8000 München 80

Summary

An efficient method enabling the calculation of viscous flows at high Mach- and moderate to high Reynolds numbers in chemical non-equilibrium by a solution of the Euler- and second order boundary layer equations is presented. This approach is valid for flows showing no strong viscous/inviscid interaction and is, for example suited for determining the flow at the windward side of reentry vehicles. Applications are discussed together with the modelling of effects of practical interest, such as catalytic walls and wall radiative equilibrium.

1 Introduction

The development of hypersonic aircraft and reentry vehicles depends to a high degree on the availability of numerical methods predicting the flow field as well as the chemical and thermodynamic properties of the fluid. Due to the advent of powerfull computers, improved numerical methods and the progresses in the modelling of the important high temperature effects it is nowadays possible to use numerical methods to determine aerodynamic and thermal loads at a vehicle without the drawbacks of experimental facilities or free flight experiments.

Depending on the application and the accuracy needed or required, the calculation of viscous flows can be performed using different methods. Most often, the Navier Stokes (NS) equations are solved for this purpose. For general, three-dimensional configurations a huge CPU-time consumption and large memory requirements are standard, even if the spatial resolution leaves much to be desired. In situations showing only weak viscous/inviscid interaction, such as the flow on the windward side of reentry vehicles, the solution of the Euler and second order boundary layer equations is advantageous, and yields similar results compared to NS calculations [1].

Thus, the method [1,2] developed and used successfully for ideal and equilibrium flows was extended to treat effects of chemical non-equilibrium [3,4]. The present paper presents the further development to three-dimensional flows. In the following, the numerical method and the modelling of high temperature effects are discussed and results are shown.

2 Numerical procedure

2.1 Inviscid part

The method [4] is used, which is sketched in the following. The Euler equations in quasi-linear form written for a curvilinear coordinate system state

$$\vec{U}_{,\tau} + \mathbf{A}\vec{U}_{,\xi} + \mathbf{B}\vec{U}_{,\eta} + \mathbf{C}\vec{U}_{,\zeta} = \vec{S}. \tag{1}$$

The vector U contains the conservative variables

$$\vec{U} = (\rho, \rho m_1, .., \rho m_s, \rho v^{1'}, \rho v^{2'}, \rho v^{3'}, \rho(e + \Delta e^0 + 0.5\,\vec{v}^2))^T, \tag{2}$$

and S is the chemical source vector

$$\vec{S} = (0, \dot{\Omega}_1, .., \dot{\Omega}_s, 0, 0, 0, 0)^T. \tag{3}$$

The symbols used above ($\rho, m_s, v^{1'}, v^{2'}, v^{3'}, e, \Delta e^0$ and $\dot{\Omega}_s$) denote density, species mass fractions, the three cartesian velocity coordinates, the internal energy, the energy of formation and chemical source terms for each species s. The Jacobians $\mathbf{A}, \mathbf{B}, \mathbf{C}$, generalized as $\mathbf{K}$ in the following, are split according to the sign of their eigenvalues

$$\mathbf{\Lambda}_K = \mathbf{T}_K^{-1}\,\mathbf{K}\,\mathbf{T}_K = \mathbf{\Lambda}_K^+ + \mathbf{\Lambda}_K^-, \tag{4}$$

with the matrices of the eigenvectors $\mathbf{T}$. The spatial derivatives, evaluated with the third order upwind biased formula

$$\vec{U}_{,\xi\ i}^{\pm} = (\vec{U}_{i\mp 2} - 6\vec{U}_{i\mp 1} + 3\vec{U}_i + 2\vec{U}_{i\pm 1})\,/\,6\Delta\xi\;, \tag{5}$$

are split accordingly. The resulting system of equations

$$\begin{aligned}\vec{U}_{,\tau} &+ \mathbf{T}_A\mathbf{\Lambda}_A^+\mathbf{T}_A^{-1}\vec{U}_{,\xi}^+ + \mathbf{T}_A\mathbf{\Lambda}_A^-\mathbf{T}_A^{-1}\vec{U}_{,\xi}^- \\ &+ \mathbf{T}_B\mathbf{\Lambda}_B^+\mathbf{T}_B^{-1}\vec{U}_{,\eta}^+ + \mathbf{T}_B\mathbf{\Lambda}_B^-\mathbf{T}_B^{-1}\vec{U}_{,\eta}^- \\ &+ \mathbf{T}_C\mathbf{\Lambda}_C^+\mathbf{T}_C^{-1}\vec{U}_{,\zeta}^+ + \mathbf{T}_C\mathbf{\Lambda}_C^-\mathbf{T}_C^{-1}\vec{U}_{,\zeta}^- = \vec{S}\end{aligned} \tag{6}$$

is then integrated in time using a second order accurate three step Runge-Kutta scheme:

$$\begin{aligned}\vec{U}^{(1)} &= \vec{U}^n - a_1\Delta t P(\vec{U}^n) \\ \vec{U}^{(2)} &= \vec{U}^n - a_2\Delta t P(\vec{U}^1) \\ \vec{U}^{n+1} &= \vec{U}^n - a_3\Delta t P(\vec{U}^2).\end{aligned} \tag{7}$$

The operator P contains the space derivatives and the chemical sources.

The bow shock, identical to the outer boundary of the computational region, is fitted using the Rankine-Hugoniot equations. Since these equations do not include conditions for the species mass fractions, these have to be found by physical arguments. For general non-equilibrium flows, where the chemical relaxation length is comparable to the size of the vehicle, the composition is considered as frozen. For flows approaching equilibrium, or for other cases where this assumption is unsuitable, for example due to largely differing relaxation lenghts of different processes, an one-dimensional integration is performed for the species concentrations in the first grid cell. At the body, a normal mass flux is prescribed as boundary condition.

Inside the computational domain, either an explicit or an point implicit representation of the chemical source terms is possible. Total enthalpy (including the enthalpy of formation) correction and pseudo space-marching in the supersonic part of the flow are used to accelerate the convergence.

2.2 Viscous part

The viscous part of the flow is calculated with a method solving the second order boundary layer equations in the formulation for chemical non-equilibrium. They are written in a locally monoclinic coordinate system using tensor notation and write [3]:

Global continuity equation

$$(\sqrt{g}\rho v^\alpha)_{,\alpha} + (\sqrt{g}\rho v^3)_{,3} = 0, \tag{8}$$

Species continuity equations

$$\rho v^i m_{s,i} + \frac{1}{Sc\, Re} \sum_l \frac{\hat{m}_s \hat{m}_l}{\hat{m}^2} \left((\rho D_{sl}\, \hat{n}_{l,3})_{,3} + \Gamma^\beta_{\beta 3}\, \rho D_{sl}\, \hat{n}_{l,3} \right) = \dot{\Omega}_s, \tag{9}$$

x^α - momentum equations

$$\rho \left(v^\alpha_{,\beta} v^\beta + v^\alpha_{,3} v^3 + \Gamma^\alpha_{\beta\gamma} v^\gamma v^\beta + 2\Gamma^\alpha_{\beta 3} v^\beta v^3 \right) =$$

$$= -g^{\alpha\beta} p_{,\beta} + \frac{1}{Re} \left(\left(\mu v^\alpha_{,3} \right)_{,3} + 2\Gamma^\alpha_{\beta 3}\, \mu v^\beta_{,3} + \Gamma^\beta_{\beta 3}\, \mu v^\alpha_{,3} \right), \qquad \alpha = 1,2 \tag{10}$$

x^3 - momentum equation

$$\Gamma^3_{\beta\gamma}\, \rho v^\beta v^\gamma = -p_{,3}, \tag{11}$$

Energy equation

$$\rho c_p \left(v^\alpha T_{,\alpha} + v^3 T_{,3} \right) + \rho (v^\alpha \sum_s h_s m_{s,\alpha} + v^3 \sum_s h_s m_{s,3}) =$$

$$= Ec \left(v^\alpha p_{,\alpha} + v^3 p_{,3} \right) + \frac{Ec}{Re} g_{\alpha\beta}\, \mu v^\alpha_{,3} v^\beta_{,3} + \frac{1}{Pr\, Re} \left((kT_{,3})_{,3} + \Gamma^\beta_{\beta 3}\, kT_{,3} \right)$$

$$- \frac{1}{Sc\, Re} \sum_s (h_s \sum_l \frac{\hat{m}_s \hat{m}_l}{\hat{m}^2} ((\rho D_{sl}\, \hat{n}_{l,3})_{,3} + \Gamma^\beta_{\beta 3}\, \rho D_{sl}\, \hat{n}_{l,3}))$$

$$- \sum_s \dot{\Omega}_s\, \Delta h^0_s. \tag{12}$$

In addition to the variables explained before, $T, h, c_p, v^\alpha, v^3, \mu, k$, and D_{sl} denote the temperature, enthalpy, specific heat at constant pressure, surface tangential contravariant velocity coordinates and surface normal velocity coordinate, viscosity, heat conductivity and diffusion coefficient, respectively. $\Delta h_s^0, \hat{m}_s$ and $\hat{n}_s$ are the species' formation enthalpy, mole weight and mole fraction and $\hat{m}$ is the mole weight of the mixture. The Eckert, Prandtl, Schmidt and Reynolds numbers Ec, Pr, Sc and Re are determined from the free stream values. Quantities unexplained up to now describe the geometry. Terms of second order are underlined. However, they are not the only effects of second order, because the metrics and the boundary conditions change, too, when switching from second to first order.

These equations are of parabolic type in space and can therefore be solved efficiently by space marching. Depending on the crossflow in the boundary layer, a selection between five difference schemes is made to achieve a stable integration. However, the detailed description of the space marching technique is beyond the scope of this paper and can be found in [5]. The equations are discretized using the second order accurate Crank-Nicolson scheme and those containing second derivatives (species continuity equations, x^α-momentum equations and energy equation) are solved simultaneously using the implicit Thomas algorithm. The chemical source terms are treated explicitly. The continuity and x^3-momentum equations are solved decoupled. This solution process is iterated along one normal together with the determination of the chemical, thermodynamic and transport properties until convergence is reached. The computational domain extents from the boundary layer thickness, where the outer boundary conditions are interpolated from the inviscid solution, to the body surface, where several boundary conditions, such as catalytic or non-catalytic, cooled, adiabatic or radiating walls can be prescribed.

3 Modelling of chemical, thermodynamic and transport properties

3.1 Modelling of chemical properties

A mixture of five species (N_2, N, O_2, O and NO), each modelled as a thermally perfect gas, is generally considered appropriate to describe the behaviour of high temperature, chemically reacting air. In the present method, 17 reactions r are considered in the gas phase (homogeneous chemistry):

$$N_2 + M \leftrightarrow 2N + M, \qquad N_2 + O \leftrightarrow NO + N,$$
$$O_2 + M \leftrightarrow 2O + M, \qquad NO + O \leftrightarrow N + O_2,$$
$$NO + M \leftrightarrow N + O + M, \qquad M = N_2, N, O_2, O, NO.$$

The magnitude of the mass production or destruction is obtained from

$$\dot{\Omega}_s = \hat{m}_s \sum_r \nu (k_{fr} \prod_s \hat{v}_s^{-\nu'_{sr}} - k_{br} \prod_s \hat{v}_s^{-\nu''_{sr}}), \quad \text{with } \nu = \nu'' - \nu', \tag{13}$$

the stochiometric coefficients and the mole specific volume $\hat{v}$. The forward reaction rate is described by an extended Arrhenius law

$$k_f = c_f \, T^{\eta_f} \, e^{-T_a/T}, \tag{14}$$

and the backward reation rate is determined from the principle of detailed balance

$$k_b = \frac{k_{eq}}{k_f}. \tag{15}$$

The constants c_f, η_f and T_a of [6] are used.

In many cases of practical interest, the effects of catalytical walls are important. The thermal loads are increased considerably with increasing recombination probability γ, because the heat of reaction, which is consumed and stored during the reactions behind the bow shock is now released at the interface between wall and fluid. The catalycity is determined by both the gas species and the wall material and surface properties.

The discontinuity between gas and solid leads to considerable changes in gas and chemical kinetics. Important points are the availability of a virtually infinite number of thermodynamic degrees of freedom of the solid, which is suitable to dissipate energy (heat of reaction), and the existence of potential wells, which emerge as the result of pertubations of the cristal lattice and may change activation energies. Also, the absorbed species are much closer at the surface than in the gas, in general, and stay at this small distance for a longer time.

Considering the modelling of the heterogenous chemical reactions, there have been attempts, [7,8] for example, to obtain a detailed description of each or at least some of the basic steps, which are important: absorption of the reactants, reaction at the surface, and finally desorption of the products. Sometimes also the diffusion of the gas particles on the surface is modeled. However, the results are not very encouraging, since too many parameters are unknown or are known with great uncertainity.

Therefore, the modelling of the comprehensive reaction probability, which is defined as the ratio of reacting mass flux to the total mass flux impinging on the wall, is preferred:

$$\gamma = \frac{j_{reac}}{j} = \frac{k_w \, \rho \, m_s}{j}. \tag{16}$$

The probability is modelled using an Arrhenius-equation:

$$\gamma = c e^{-\frac{T_a}{T_W}}, \tag{17}$$

and together with the Herz-Knudsen equation

$$k_w = \gamma \sqrt{\frac{\hat{R} \, T}{2 \, \pi \, \hat{m}_s}}, \tag{18}$$

a similar structure as used for the homogeneous chemical reactions is obtained:

$$j_{reac} = \gamma \sqrt{\frac{\hat{R} \, T}{2 \, \pi \, \hat{m}_s}} \, \rho \, m_s = c_f \, T^\eta \, e^{-\frac{T_a}{T_w}} \, \rho \, m_s. \tag{19}$$

3.2 Modelling of thermodynamic properties

The thermodynamic properties of the species are accounted for by the usual polynomial for the mole specific heat at constant pressure

$$\hat{c}_p(T) = \sum_i a_i T^{i-1} . \tag{20}$$

By integration and application of basic thermodynamic laws one obtains:

$$\hat{h}_s = \int \hat{c}_{ps}(T) dT, \tag{21}$$

$$\hat{s}_s = \int \frac{\hat{c}_{ps}(T)}{T} dT, \tag{22}$$

$$\hat{g}_s = \hat{h}_s - T\hat{s}_s . \tag{23}$$

From these state variables, the equilibrium constant is calculated by the van't Hoff equation:

$$\frac{d(\ln k_{eq})}{dT} = \frac{\Delta \hat{g}_r}{\hat{R} T^2} . \tag{24}$$

Another possibility is to use fits to the results of this equation.

3.3 Modelling of transport properties

The transport properties are obtained from the results of collision theory, see [9]. Assuming the Lennard-Jones potential as governing the collision process, the following expressions are obtained for the species' viscosity, heat conductivity and diffusion coefficients:

$$\mu_s = 2.669 \cdot 10^{-6} \frac{\sqrt{\hat{m}_s \cdot 1000\, T}}{\sigma_s^2\, \Omega^{*(2,2)}(T^*)}, \tag{25}$$

$$k_s = 8.334 \cdot 10^{-2} \frac{\sqrt{T/(\hat{m}_s \cdot 1000)}}{\sigma_s^2\, \Omega^{*(2,2)}(T^*)}, \tag{26}$$

$$D_{sl}^{bin} = 2.264 \cdot 10^{-3} \frac{\sqrt{T\, 10^{-3} \cdot (1/\hat{m}_s + 1/\hat{m}_l)}\, \hat{m}}{\sigma_s^2\, \Omega^{*(1,1)}(T^*)\, \rho} . \tag{27}$$

The numerical values of the potential parameter ϵ needed to calculate the reduced temperature T^*, and the collision crosssection σ found in literature, [9] for example, apply only for low temperatures (some hundred degrees). Therefore, new values (see table1) have been determined from the high temperature viscosities' of N_2, N, O_2 and O [10,11], which lead to a considerable improvement for describing high temperature transport. So far, no corresponding data has been found for NO. Thus, there is still a need for improvement of the data basis. However, due to the small content of NO, the influence is supposed to be small. For the fit of the collision integrals $\Omega^{*(1,1)}, \Omega^{*(2,2)}$, see [12].

Since these formulas apply only for Maxwellian gases, the transport of energy by internal degrees of freedom has to be accounted for by the semiempirical Eucken correction:

$$k_s = \mu_s \frac{\hat{c}_{ps} + \frac{5}{4}}{\hat{R}} . \tag{28}$$

The mixture's viscosity and heat conductivity are obtained from the Wilke law, and if a computation considering polynary diffusion is intended, the polynary diffusion coefficients are given by [13]:

$$D_{sl} = \frac{\hat{n}_s}{\frac{\hat{n}_l}{D_{sl}^{bin}\sqrt{D_{ss}^{bin}/D_{ll}^{bin}}} + \sum_{k \neq l} \frac{\hat{n}_k}{D_{kl}^{bin}}} + \frac{\frac{\hat{m} - \hat{m}_s \hat{n}_s}{\hat{m}_l}}{\frac{\hat{n}_l}{D_{sl}^{bin}\sqrt{D_{ll}^{bin}/D_{ss}^{bin}}} + \sum_{k \neq s} \frac{\hat{n}_k}{D_{sk}^{bin}}}. \qquad (29)$$

4   Results

In this section results of the method described are discussed. In its three-dimensional version the method has been applied to the front part of the reentry vehicle Hermes, see fig.1. For a discussion of two-dimensional results, see [14]. The free stream parameters are corresponding to an altitude of 75 km and are given by

$$p_{ref} = 2.49 \ Pa \quad , \quad \rho_{ref} = 4.33 \cdot 10^{-5} \ \frac{kg}{m^3},$$

$$T_{ref} = 200.2 \ K \quad , \quad m_{N_2} = 0.767,$$

$$m_{O_2} = 0.233 \quad , \quad v_{ref} = 7020 \ m/s,$$

$$Ma = 25 \quad , \quad Re/m = 2. \cdot 10^4,$$

$$Pr = 0.72 \quad , \quad Sc = 0.88,$$

$$L = 6.5 \ m \quad , \quad \alpha = 30^0.$$

Laminar flow is considered for the complete configuration because of the high altitude, low Re-number flow. The complete mesh consists of 110x33x16 points in the i,j,k direction for the Euler calculation, and in principle of 110x33x101 points for the solution of the boundary layer equations. However, the solution is only obtained in regions where no strong viscous/inviscid interaction is occuring, that is in the region of the nose and at the windward side of the body. The region accessible by this method can be seen in fig.2, where the isobars are plotted in the zone of the boundary layer solution.

Fig.3 depicts the result of a calculation for a cooled ($T_w/T_{ref} = 6$) non-catalytic wall. The lines of constant N mass fractions show a slowly recombining flow at the windward side of the body. This changes drastically, if a recombination process at finite rate is assumed to take place at the wall. The parameters of [15], which apply for the Space shuttle in principle, have been used in the modelling described before. From fig.4 it may be concluded, that the strongest recombination occurs in the stagnation region, but also along the body considerable recombination is found. Note that the oscillations in the isolines are caused by the coarse grid, which has only 11 points in the crossflow direction at the windward side of the aft body.

The assumption of a radiation equilibrium wall, where the heat transported to the wall is radiated to the surroundings without any heat flux through the wall, again changes the picture. The N-isolines in the front part are similar to the cooled wall case, but in the aft part the is higher (fig.5). Of interest is the resulting temperature for this case, see fig.6. It decreases rapidly from the stagnation region.

5 Conclusion

For regions of the flowfield where weak viscous/inviscid interaction is found, the solution of the Euler/second order boundary layer equations provides an efficient tool for studying hypersonic flows. Homogeneous and heterogeneous chemical reactions are accounted for. The results show the influence of the different boundary conditions, as radiating walls, for example.

In the future, further validation will be done and the introduction of vibrational non-equilibrium is planned.

Acknowledgement

The author wishes to thank Dr. F. Monnoyer for providing the method SOBOL and Dr. M. Pfitzner for providing the non-equilibrium Euler method.

References

[1] Ch. Mundt, M. Pfitzner, M.A. Schmatz: *Calculation of viscous hypersonic flows using a coupled Euler / second order boundary layer method*, P. Wesseling (ed.): Proc. 8th GAMM-Conference on Num. Meth. in Fluid Mechanics, NNFM Vol. 29, Vieweg, 1990, pp. 419-429.

[2] F. Monnoyer, Ch. Mundt, M. Pfitzner: *Calculation of the hypersonic viscous flow past reentry vehicles with an Euler-boundary layer coupling method*, AIAA-paper 90-417, 1990.

[3] Ch. Mundt, F. Monnoyer: *Calculation of hypersonic non-equilibrium viscous flow using second order boundary layer theory*, International Aerospace Congress, Melbourne, 1991.

[4] M. Pfitzner: *A 3-d non-equilibrium shock-fitting algorithm using effective Rankine-Hugoniot equations*, AIAA-paper 91-1467, 1991.

[5] F. Monnoyer: *Calculation of three-dimensional attached viscous flow on general configurations using second-order boundary layer theory*, Z. Flugwiss. Weltraumforsch. 14, pp. 95-108, 1990.

[6] C. Park: *On the convergence of computation of chemically reacting flows*, AIAA-paper 85-247, 1985.

[7] W.A. Seward, E.J. Jumper: *Oxygen recombination on Space Shuttle thermal-protection-tile like surfaces*, AIAA-paper 90-54, 1990.

[8] H.J. Kreuzer: *Diffusion, adsorption and desorption at surfaces*, J. Chem. Soc. Faraday Trans., vol.86 no.8, pp.1299-1305, 1990.

[9] Hirschfelder, J.O.; Curtiss, Ch.F.; Bird, R.B.: *Molecular theory of gases and liquids*, J. Wiley, New York, 1964.

[10] L. Biolsi, D. Biolsi: *Transport properties for the nitrogen system: N_2, N, N^+, and e*, AIAA-paper 83-1474, 1983.

[11] L. Biolsi: *Transport properties for the oxygen system: O_2, O, O^+, and e*, AIAA-paper 88-2657, 1988.

[12] Ch. Mundt, G. Hein: *Aerodynamic Numerical Analysis - C1 Non-equilibrium and turbulent transport properties I*, MBB-FE122-HERMES-R-12, 1990.

[13] E. Obermeier, A. Schaber: *A simple formula for multicomponent gaseous diffusion coefficients derived from mean free path theory*, Int.J.Heat Mass Transfer, vol. 20, 1977, pp.1301-1306.

[14] Ch. Mundt, E.H. Hirschel: *Modelling of chemical and physical effects with respect to flows around reentry bodies*, Proc. 3rd Aerosp. Symp., Braunschweig, 1991.

[15] C.D. Scott: *Catalytic recombination of nitrogen and oxygen on high-temperature reusable surface insulation*, AIAA-paper 80-1477, 1980.

Table

Table 1: Parameters for the Lennard-Jones 6-12 potential

species	temperature interval (K)	ε/k (K)	σ (Å)
N_2	200 - 300	91.5	3.681
	2000 - 3000	659.2	2.925
N	200 - 300	66.5	2.940
	2000 - 3000	819.3	2.327
O_2	200 - 300	113.0	3.430
	2000 - 3000	528.0	2.970
O	200 - 300	210.0	2.330
	2000 - 3000	792.7	2.270
NO	200 - 300	119.0	3.470

Figures

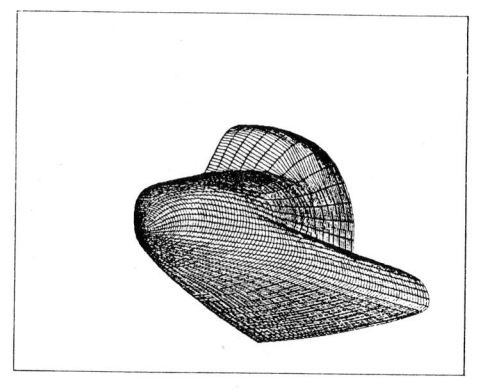

Figure 1: Grid at the front part of Hermes

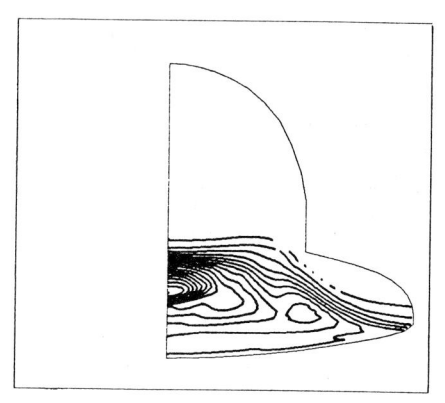

Figure 2: Frontal view of the isobars at Hermes

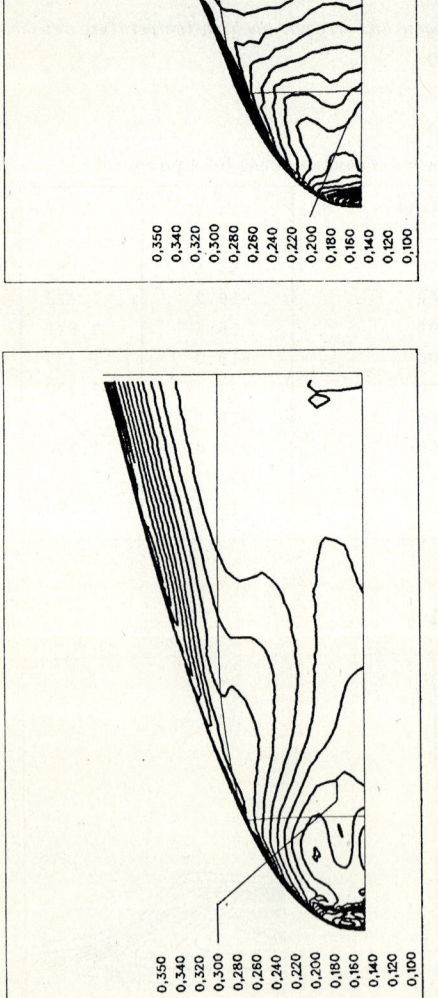

Figure 3: Iso N-lines, cooled, non-catalytic wall

Figure 4: Iso N-lines, cooled, catalytic wall

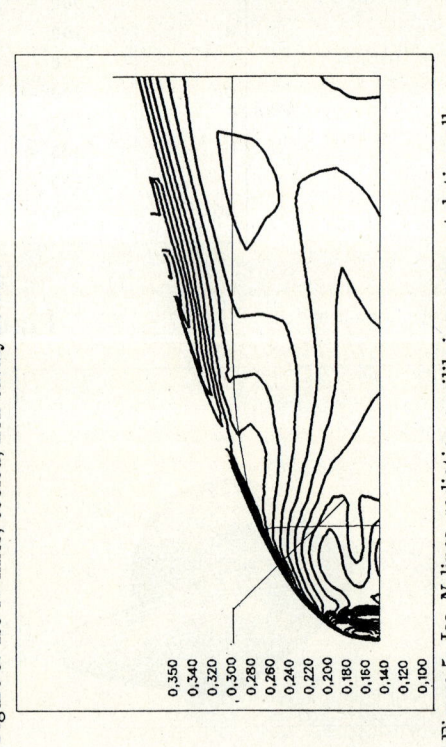

Figure 5: Iso N-lines, radiation equilibrium, non-catalytic wall

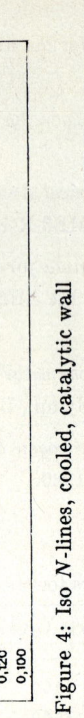

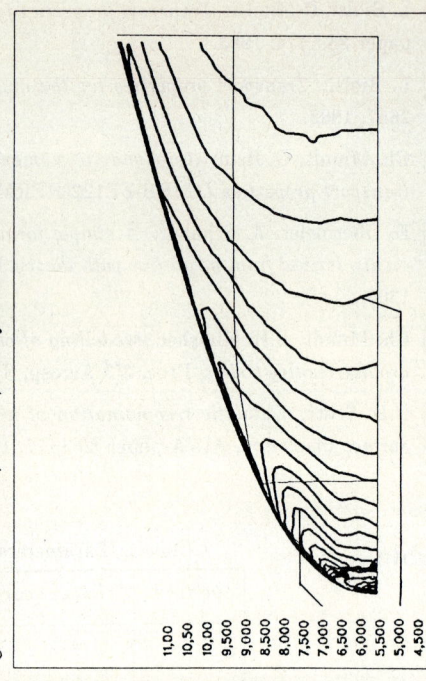

Figure 6: Temperature lines, radiation equilibrium, non-catalytic wall

Adaptive Mesh Refinement for Unsteady, Nonequilibrium, High Speed Flows

M. Valorani [1] and M. Di Giacinto [2]

Dipartimento di Meccanica ed Aeronautica, Università di Roma "La Sapienza"
Via Eudossiana 18, 00184 Roma

Summary

An adaptive mesh refinement technique using embedded patches is presented. The technique is especially oriented to the numerical simulation of high speed reactive flows in unsteady regimes. Structured, uniform, orthogonal subgrids of increasing resolution are obtained by means of a recursive, modular procedure. The main merits of the technique are (i) the low computational overhead for subgrid generation and management, (ii) the general formulation of the space-time conditions at the subgrid interfaces, (iii) a suitable formulation of the refinement criterium for unsteady problems, (iv) the straightforward applicability to multidimensional problems. The performances of the method are analyzed in onedimensional test cases with special attention to the accuracy obtainable in the description of propagation phenomena.

1 Introduction

Numerical simulations of hypersonic flows and high speed combustion must face the challenges prompted by the wide range of characteristic scales brought in by relaxation processes and chemical reactions. Implicit, stiffly stable solvers may contribute to stabilize the calculations. Nevertheless, in reactive flows there always occur chemical or relaxation layers, possibly of unsteady character, which for both accuracy and stability reasons impose severe resolution requirements [1,2]. When these flow features are not scattered throughout the entire domain, these requirements can be efficiently fullfilled by means of adaptive mesh refinements.

Adaptive refinements can be achieved by techniques based on *structured* or *unstructured*, *moving* or *embedded* grids. The need for solutions in increasingly complex geometries have directed the research towards the use of solvers based on *unstructured* grids. Although this approach looks promising, it seems at the moment more suited for steady solutions than for time accurate calculations. *Moving* grids increase locally the accuracy of the calculation by letting a fixed number of grid points free to move and cluster around the main features of the flow. *Embedded* grids of increasing resolution pursue the same goal by placing extra computational points there where significant flow variations occur.

The embedding of new computational points can be obtained on a *cell by cell* basis when a cell needing an higher resolution is furtherly subdivided in two or more parts. This approach is typical of finite element techniques [3,8,9], and is also widely used in conjunction with finite volumes methods [7]. The grids which are then obtained are usually of the unstructured type. The numerical solution on such grids requires linked lists with an indirect addressing of the nodes which might hinder the vectorization of the codes. An alternative route of embedding makes use of *patches*. This term designates a subset of cells singled out by a refinement criterium, which *as a whole* is subjected to subdivision. The rationale along with these patches are defined characterizes the different techniques presented in the literature.

In Berger [4] the patches are built around the main features of the flow *as the solution evolves*, following a certain set of rules which optimize the number, the sizes and the topological properties of these patches. The optimization is based on elements of graph theory, pattern recognition, and artificial intelligence. This process demands a surplus of computational work, generally performed in scalar mode, and usually implies a certain number of heuristic criteria whose general validity is not always ascertainable. Patches allow the use of off-the-shelf vectorizable integrators and provide uniform grids which avoids the deterioration of the signals. In order to better describe the features of the flow, they

---
[1] Research Scientist
[2] Associate Professor

can be either aligned or rotated with respect to the coordinate system of the coarse grid according to the kind of problem to be solved.

A different approach is followed by Kallinderis [5] with the goal of developing implicit schemes of integration on the patches and of reducing the number of "randomly" embedded cells generated by a cell by cell technique. Their *embedded patches* are specific regions of the domain (e.g. boundary layers) enclosing a defined, fixed number of cells of the initial mesh selected *at the beginning* of the calculation. The remaining of the domain is still discretized with an unstructured grid obtained by cell by cell embedding. Quoting Kallinderis [6] "if the majority of the cells belonging to a given patch (say 90%) are flagged for division then all cells currently belonging to the patch are also divided; conversely, if very few cells (say less than 10%) are flagged, none of those within the patch are divided".

Since our interest is mainly directed towards unsteady flows, we propose a technique of adaptive mesh refinement based on *structured, embedded* grids. This technique has been developed within the context of hyperbolic problems, because of the intended application in reactive high speed flows [10]. Although the method shares features of both the above mentioned approaches, namely it uses patches *throughout* the whole domain as in [4] and defines the topology and the sizes of the patches *at the beginning* of the calculation as in [5], it has been derived on a rather different conceptual basis. In fact, in order to devise a general purpose technique of adaptive mesh refinement, computationally efficient and suited to tridimensional problems, we have resorted to a modular, recursive concept to define the embedded grids. Similar procedures in different contexts have been presented in [8,9].

The modular, recursive approach reduces the cost of performing frequent adaptations of the mesh, which can therefore follow more tightly the temporal evolution of the flow features. Thus, regions requiring higher resolution become narrower. Moreover, when overlapping between grids of different resolution occurs, the numerical integration is performed on the finer grids only. These properties reduce the number of computational points on which the integration is actually performed. Conversely, the *a-priori* definition of the subgrid topology may increase the computational points because of a rougher spatial tailoring of the embedded grids around the flow features.

The propagation of signals is taken into account in the refinement criterium which drives the adaptive embedding. An accurate and efficient formulation of the space-time conditions at the subgrids interfaces is also proposed.

The performances of the technique will be analyzed with a special attention to the generation of spurious waves at the subgrid interfaces and as regards the accuracy of unsteady solutions. These kind of analyses are more simple and clear if carried out in a onedimensional context. Therefore, we studied the unsteady case on a 1D scalar model equations (inviscid and viscous Burgers equations), and the steady case on the 1D system of equations of the isentropic, inviscid, nozzle flow.

2 Spatial Adaptation

Let we consider the problem of the numerical integration of an (hyperbolic) system of PDE on a rectangular, possibly multidimensional, computational domain $D \subseteq R^n$, which needs to be discretized with different resolution levels. The rationale of the generation of the patches, which we name *subgrids*, is based on a *modular, recursive* partition of a relatively coarse root grid. This grid is structured, uniform, orthogonal and contains a number of cells $N_0 = \prod_{i=1}^{n} n_{0,i}$, where the $n_{0,i}$ are constrained to be integer powers of two. The recursive partition defines a tree of subgrids of different resolution levels, according to the following procedure. At first, the root grid is subdivided in each space dimension into an even number (N_{part}) of equally sized parts. The value N_{part}, preset by the user as an integer power of two, is the module of the partitioning. Then, by halving all cells of these parts, N_{part}^n new subgrids with twice the resolution of the root grid are defined. The recursive application of this same rule defines subgrids of always increasing resolution. The final result of the recursion is a system of subgrids hierarchically organized in a tree, as sketched in fig. 1 for the onedimensional case. For the case $N_{part} = 2$, this procedure is equivalent to the one proposed by Schmidt [8] for a cell by cell embedding in a finite element context.

Let N_{cell} be the minimum value among the $n_{0,i}, i = 1, n$. The pair (N_{cell}, N_{part}) implicitly defines the maximum number of halvings (l_{max}) of the root grid that pertains to the subgrids of highest resolution. In fact, the recursivity requirement constrains the number of cells of a subgrid to be not smaller than the partition module N_{part} in all dimensions. Hence, the maximum level of achievable resolution (l_{max}) is given by the relation $l_{max} = (\log_2 N_{cell} - \log_2 N_{part}) / (\log_2 N_{part} - 1)$. Once the pair (N_{cell}, N_{part}) is given, the total number of subgrids of the tree, their geometry, size and resolution are uniquely determined *once and for all at the beginning of the calculation, independently of any refinement criterium and of the solution evolution.* This modular, a-priori approach *eliminates almost completely the need of any post-processing* of the information provided by the refinement criterium. The refinement criterium "simply" acts as a mean to activate or deactivate the integration over the already defined subgrids.

The activated subgrids are denoted with the symbol $G_{(m)}^l$, where the subscript m is a progressive number which identifyes the subgrid and the superscript l is an integer that indicates the resolution level. The root grid has the identification number $m = 1$ and its resolution level is zero. It is sufficient that just one cell belonging to a part[3] $g_{(m)j^n}^l$ of a subgrid $G_{(m)}^l$ needs an higher resolution to activate the integration over a new subgrid $G_{(q)}^{l+1}$ stemming from it (fig. 2). The integration of the PDE is then performed on the activated subgrid $G_{(q)}^{l+1}$, while it is halted on the part $g_{(m)j^n}^l$ of $G_{(m)}^l$. In this way the spoiling of computational work due to the overlapping of patches is avoided. Clearly, only the active subgrids, togheter with their pointers, need to be actually allocated in the computer memory. The modular approach makes rather straithforward the algorithmic implementation of these two features, which save both CPU and computer memory.

Each selection of the partition module provides trees of subgrids having different performances. For a given level of resolution a small value of N_{part} generates a tree with fewer subgrids working at a relatively low efficiency (measured for example by the ratio of flagged over total cells in the subgrid). An higher value of N_{part} produces more subgrids but of higher efficiency. The first choice realizes a rougher tailoring of the embedding on the features of the flow than the second one, requiring then a number of (embedded) grid points higher than strictly necessary, but a lower overhead cost for the subgrid management. The second choice may achieve performances close to those obtainable with more sophisticated codes, but uses an higher number of subgrids (patches). However, the overall overhead due to the subgrids management is contained in this case as well, since the modular, recursive approach defines the subgrids already at the beginning of the calculation and the activation of a new subgrid requires a very limited amount of work.

Each subgrid, once activated, is allocated in sectors of the computer memory which are not necessarily contiguous. Therefore, the topological properties of the physical domain have to be transferred to the system of active subgris by means of pointers. The modular approach allows the definition of a small number of pointers, even for the multidimensional case. Additional pointers handle the dynamic storage allocation required by the subgrid activation and deactivation process. More details can be found in [10].

3 Temporal adaptation

In unsteady hyperbolic problems the time dimension plays a role different from that of the spatial dimensions and therefore has to be discretized in observance of the following issues: (i) the upper bound of the time step of integration is defined by the CFL stability conditions; (ii) time steps much smaller than those allowed by the local CFL increase the numerical diffusion and dispersion and waste computational work; (iii) the interface conditions between two adjacent subgrids are more accurate

---

[3]$j^n \stackrel{\text{def}}{=} (j_1, \ldots, j_n)$ is defined as a multi-index which has the same dimensions of $D \subset R^n$; each component of j^n takes values ranging from 1 to N_{part} ($j_i^n = 1, N_{part}; i = 1, n$).

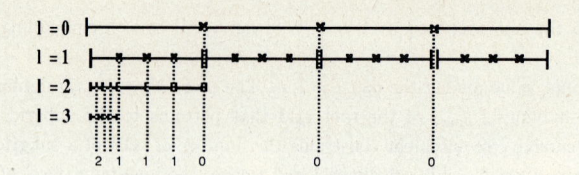

Figure 1: Structure of the subgrid tree ($N_{part} = 4$).

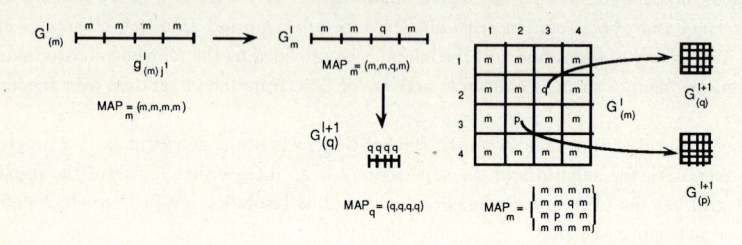

Figure 2: The modular, recursive definition of subgrids, with the relevant pointers.

when enforced at the same time level; (iv) the regridding can be more efficiently performed when the numerical solution throughout the flowfield is available at the same time level, otherwise numerical complications arise.

The time discretization, the order of integration and the enforcement of the interface conditions are defined in the present technique on the basis of the previous considerations and by taking advantage of the modular discretization of space. By combining the cell size definition with the global CFL limit, the time steps of integration of each subgrid become also defined as multiples of integers powers of two. The time step of the coarser grid defines the basic unit of integration or, following the nomenclature proposed by Pervaiz et al [7], the *time stride* ΔT. The order of integration here adopted is illustrated for a two step algorithm in fig. 3, and correspond to the asymptotic concept of keeping the solution frozen on the coarser subgrids while letting it evolve on the finer ones. The results obtained at the end of the predictor of the coarser grids are used to assign the boundary conditions at the adjacent finer subgrids. This in turn reduces the size of the stencil required to impose the boundary conditions among subgrids.

The adaptation of the subgrid tree to the flow is made in the present work only at the beginning of a time stride, in accordance to the issue (iv). The time adaptation becomes less and less efficient as the highest level of resolution needed increases. In fact, in order to follow propagating signals requiring the finest resolution, it is necessary to widen the region of fine grids with respect to that defined on a simple local instantaneous analysis of the flow. The larger is the widening, the poorer is the time adaptation and less efficient is the overall calculation. For many other techniques which performs the remeshing only every many time strides the problem is even more serious. For very high resolution level differences, it can make almost worthless to try to obtain an accurate *istantaneous* tailoring of the subgrids on the signals.

4 Conditions at interfaces

A domain discretized by means of embedded subgrids is affected by the presence of artificial cuts which pose numerical problems quite similar to those created by true boundaries of the domain. In particular, the interface conditions must be enforced by respecting the direction of propagation of the signals and

Figure 3: Time integration; p and c denote the predictor and corrector steps respectively; numbers between brackets specify the order of integration among subgrids in a *time stride* ΔT ($l = 2$).

by preserving the conservation of integral quantities when the embedding procedure is applied to shock capturing schemes. The first problem can be easily solved with upwind schemes formulated in term of characteristic variables, while the second problem has not been faced yet, since so far the adaptive technique has been implemented to schemes using a quasilinear formulation only. A further problem specifically caused by the presence of internal boundaries is the production of spurious wave numerically generated. The complete elimination of these waves is impossible, since their generation is structurally associated to the loss of information determined by joining grids of different resolution. The coarser to finer transition acts as a filter in the space-time domain producing reflected and refracted waves when a signals crosses through it. The theoretical analysis of these phenomena is a rather specific topic which exceeds the limits of the present work. A deeper analysis can be found in Vichnevetsky [11]. However, in our work we have experimentally observed that these spurious waves generally have a small amplitude and a frequency close to the highest frequency resolvable by the grid. These small oscillations become evident only for very smooth solutions, when their intensity attains the same magnitude of the flow signals.

For an hyperbolic problem the solution in a point uniquely depends on the data inside its domain of dependence. This property can be exploited to integrate each grid "independently" from the others, providing that the mixed initial-boundary value problem defined on the subgrid and by the space-time conditions at the subgrid interfaces is converted in a purely initial value problem. This is accomplished by simply adding at the sides of the grid the computational points defining the numerical domain of dependence. These buffer zones are named *domains of connection* (fig. 4). The possibility of advancing the solution on each grid separately could be exploited in the context of parallel computing. A further advantage of this approach is that the integration scheme does not need to be modified at the subgrid interfaces. This allows in turn to obtain conservative interface conditions in a rather straightforward way.

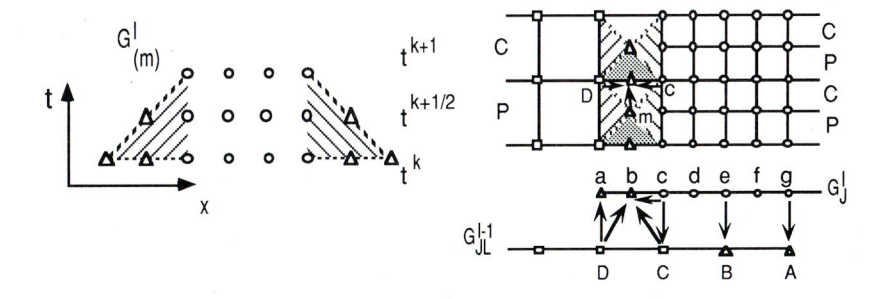

Figure 4: Shadowed zones mark the *domains of connection* pertaining to the grid G_m^l. Tri- angles realize the space-time link to others grids.

Figure 5: Interconnecton between grids of different resolution. Data transfer follows the arrows. Dark zone marks overlapping of adjacent domains of connection.

505

The additional points inside the domain of connection of two adjacent grids with the same resolution correspond exactly one another, at the same time level. Therefore a simple injection of data from one grid to the neighbouring gives precisely the same results of those obtainable on a uniform grid formed by joining the two subgrids together. This circumstance eliminates, as far as the accuracy is concerned, all problems of piecing together many different subgrids of equal resolution.

The problem is more delicate when the resolution of the two subgrids is different. Note in fig. 5 that the coarse grid nodes can receive data from the finer grid without problems since the nodes coincide. Instead, the point b belonging to the finer grid must be initialized by interpolating values of the coarse grid somehow. Point b can be averaged between the two different pairs (D, c) or (D, C), since the value of the variables in the point astride the adjacent subgrids can be defined on the basis of the calculation performed on either grids. It can also be left unchanged, but this corresponds to assign the value m of the previous time step to it (fig. 5). None of the above options is without faults. In fact, depending on the direction of propagation of the signal, all interpolations introduce a perturbation travelling behind or in front of the signal itself. It has been observed that, at least for steady solution, averages lead to more accurate solutions.

The options of using structured grids, together with that of increasing the grid resolution by consecutive halvings only, reduce to the minimum the number of interpolations required to let the grids follow the propagating signals and allow an easier coordination of the truncation errors among grids of different resolutions. Both effects improve the space-time accuracy of the calculation.

The use of the outcome of a consistent, first order predictor step to define the initial conditions of the finer grids has the effect of halving the spatial extent of the domain of connection of a second order scheme. When the solver does not use a predictor-corrector scheme, it is necessary to perform some kind of time interpolations to obtain an analogous reduction of extra points.

5 Refinement criteria

The refinement criterium here adopted is based on a simple local analysis of the normalized gradients, but some provisions to account for the propagation of signals are also discussed.

Local criteria. An evaluation of the resolution level locally required to properly solve numerically the equations in a certain region of the domain can be obtained either by performing a Richardson extrapolation to estimate the truncation error [4], or by sizing the flow features in terms of variable differences or gradients [5,6]. The first approach is based on the extrapolation of the results obtained by integrating first the equations over a time step and then over the half of the same time step. It is therefore an *a-posteriori* criterium and as such it wastes the computational work for obtaining the intermediate values at the half time step. Conversely, the analysis on the instantaneous flow features allows the definition of *a-priori* criteria. In the present work we used the criterium proposed by Pervaiz et al. [7] which is based on a statistical analysis of the flow gradients. When the criterium is applied on a simple scalar equation in, say, the variable u, it reduces to a gradient normalization by means of their average $(\overline{\nabla u})$ and root mean square $(\sigma_{\nabla u})$ values computed on the whole domain of integration D:

$$\mu(\overline{x}) = \left|\nabla u(\overline{x}) - \overline{\nabla u}\right|/\sigma_{\nabla u} \qquad \overline{x} \in D. \tag{1}$$

When $\mu(\overline{x}) \ll 1$ the flow is locally characterized by gradients of intensity close to the average. For $\mu(\overline{x}) \geq 1$ the gradients are one or many times the rms value higher than the average. Note also that when the number of cells is sufficiently high then the average value $\overline{\nabla u}$ tends to zero. In this case the condition $\mu(\overline{x}) \ll 1$ defines regions of very weak gradients, while $\mu(\overline{x}) \geq 1$ those of strong gradients which presumably may feature stronger nonlinearities needing finer resolutions. The definition of a suitable resolution level based on the information carried by $\mu(\overline{x})$ can be very diversely approached. So far to assign the resolution level to each subgrid, we take the integer part of the continuous function $\mu(x)$

$$lev(x) = int(\mu(x)) \tag{2}$$

to obtain an istogram of integers lev. This estimate is carried out at the beginning of each time stride. A new subgrid will therefore inherits the values of $lev(x)$ already estimated on its parent subgrid. The process of activation of subgrids of higher and higher level stops when the resolution level l of the finest subgrid is equal to the estimated lev.

Upwind criteria In section 3 we have observed that regions of fine grids wider than those predicted by local criteria are needed to anticipate the signal propagation. One way to cope with this problem is by simply increasing the size of the patches of a certain percentage, which can be proportional, for example, to the number of time strides between two consecutive mesh adaptations [4]. In the present technique this is accomplished by activating in advance a suitable number of subgrids in the region where the signal is going to enter. To accomplish this, it is required a refinement criterium which can account for both the intensity and the direction of propagation of the signals. In order to obtain a guideline to include these features in a criterium, let we consider a simple scalar equation having only one characteristic field, defined by the eigenvalue u. Then, since the mesh adaptation is performed only at the beginning of a time stride ΔT, the maximum distance that the fastest signal can travel in such interval of time is $L = |u_{max}|\Delta T$. Therefore, a conservative criterium can be obtained by modifing the function $\mu(x)$ in a given point $\overline{x}$ as follows

$$\hat{\mu}(\overline{x}) = max \left[\mu(x) \quad \text{for} \quad \left\{ \begin{array}{ll} \overline{x} - L \leq x \leq \overline{x} & \text{if} \quad u > 0 \\ \overline{x} \leq x \leq \overline{x} + L & \text{if} \quad u < 0 \end{array} \right. \right]. \tag{3}$$

This change does not account neither for a possible change of sign of u, nor for nonlinearities which could either weaken or strengthen the signals during a time stride. Moreover, since system of equations have many characteristic fields the criterium may be furtherly changed in the following

$$\tilde{\mu}(\overline{x}) = max\{\mu(x) \quad \text{for} \quad \overline{x} - L \leq x \leq \overline{x} + L \quad ; \forall u\} \tag{4}$$

where now L is defined on the basis of the fastest of the system eigenvalues. The effects of using a local or an upwind criterium will be evidentiated in a test case presented in the next section.

6 Results

The validation of the adaptive mesh refinement technique has been carried out on test cases based on 1D model equations possessing exact solutions characterized by the presence of strong gradients. The numerical solvers used with the adaptive mesh refinement technique are all based on the quasilinear formulation of the equations and are numerically solved with explicit, predictor-corrector schemes. The convective terms are discretized with upwind differences, and the viscous terms with centered differences. The space-time accuracy is verified on the 1D, inviscid and viscous Burgers equation with different initial condition (expansions, a travelling shock and the N-wave solution [12]). The convergence properties and the accuracy at steady state for a system of equations are tested on an isentropic, inviscid nozzle flow.

Expansion wave: smooth initial conditions (Inviscid Burgers Equation). An initial condition defined by an hyperbolic tangent may generate a smooth expansion wave. This test case provides information on the generation of spurious waves at the subgrid interfaces. At later times the gradients of the solution become increasingly weak. It is under these circumstances that the generation of spurious waves become more pronounced. Fig. 6(I)a shows the overall error distribution between exact and numerical solution with an enlargement in correspondence of a subgrid interface. The presence of oscillations of small amplitude and frequency defined by the grid spacing is evident. It has been observed that the adaptive grid refinement was activating too many grids of a resolution exceeding that really

Table 1: Accuracy for the simple expansion wave.

	N_{cell}	N_{part}	N_{tot}	N_{iter}	$\|u\|_1 \cdot 10^3$	$\|u\|_2 \cdot 10^3$	l_{max}	l
Uniform grid	64	1	65	24	1.6	2.30	-	-
Uniform grid	128	1	129	48	0.56	1.00	-	-
Local	64	4	97	24	0.39	0.88	3	1
Upwind	64	4	97	24	0.41	0.92	3	1
Local	64	8	89	24	0.77	1.70	1	1
Upwind	64	8	97	24	0.57	1.00	1	1

needed for describing this smooth solution. This is a fault of the refinement criterium as formulated in equation (1). Because of the gradient normalization, the criterium looses the information on the actual amplitude of the gradients and retains only how they mutually compare. In the case of very smooth solutions, it detects only the noise associated to the weak variations of the gradients. From this result is clear that the criterium has to be modified in order to account for the actual amplitude of the gradients also.

Expansion wave: non smooth initial conditions (Inviscid Burgers Equation). A simple expansion wave, characterized by two gradient discontinuities, defines the initial conditions for this test. At a later time, Fig. 6(I)b shows the spatial distribution of the scalar function u, propagating according to the Burgers equation, and the error distribution measured as $(u_{num}(x) - u_{exact}(x))$. Note that no significant perturbations arise when the gradient discontinuities cross the subgrids interfaces. Table 1 reports the errors obtained after a certain number of iterations (N_{iter}) with different initial grid resolution N_{cell}, different partition modules N_{part}, and using either the local or the upwind refinement criterium. For $N_{part} = 1$ the adaptive mesh refinement is deactivated and the solution is found on a standard uniform grid. The error obtained with the adapted mesh using a root grid with 64 cells, both with N_{part} equal to 4 and 8 is comparable with that of a uniform grid with 128 cells. Observe that $N_{part} = 4$ gives errors smaller than $N_{part} = 8$, and that in this test case the use of the upwind criterium not always improves the overall performances. However, the maximum resolution level l actually used by the adaptive technique is smaller than the maximum[4] available $l_{max} = 3$ and this precludes an otherwise possible further reduction of the error. This is again caused by a not yet optimized refinement criterium.

Travelling shock (Viscous Burgers Equation). The asymptotic stage of a compression wave defined initially by an hyperbolic tangent is a travelling wave modelling a viscous shock transition. The speed of propagation of the shock is defined as the average between the values of the states immediately in front and behind the shock itself. In the viscous case the solution across the shock remains continuous, and therefore it is in principle always possible to obtain correct solutions by integrating the Burgers equation in the quasilinear form. However, the gradients in the shock grow steeper and steeper as the Reynolds number increases and soon the resolution required to properly describe their variations become hardly obtainable even by using adapted meshes. In the example shown in fig. 6(II), the upstream state is defined by $u = 0.5$ and the downstream state by $u = 1.5$. Therefore, the shock speed is $w = 1$. The Reynolds number is 400, $N_{cell} = 256$ and $N_{part} = 4$. The finest grids used have a resolution level $l = 5$, and this locally corresponds to a uniform mesh of $256 \cdot 2^5 = 8192$ intervals. The overall distribution and the strong enlargements immediately upstream and downstream of the transition confirm the satisfactory performances delivered by the adaptive technique. The strong curvature is properly described and so are the contributions of the viscous terms. The storage memory actually allocated for this test has been of 582 positions, and the integration has been performed on 425 nodes only.

---

[4]N.B.: in some calculations using $N_{part} = 4$, l_{max} has been lowered of one unit with respect to the value provided by the relation of section 2.

The accuracy of the solution is checked by monitoring the constancy of the shock speed evaluated numerically as the temporal variation of the area underneath the solution $u(x,t)$. The calculation has been performed both with the local and the upwind refinement criterium. The results demonstrate that, within a certain tolerance, the shock speed remains sensibly constant around the exact value of $w = 1$. The solution fluctuates around the correct value probably because of the inaccuracies due to spurious waves generated by changes in the mesh resolution. The use of an upwinded criterium, which reduces the chances that the shock could leave the finer grids, smooths somewhat the solution. Nevertheless, a periodic disturbance still persists. This is associated to a periodic reduction of the highest resolution level assigned by the refinement criterium. It is likely that a better calibration of the treshold values used in the criterium would eliminate this problem.

N-Wave (Viscous Burgers Equation). This case is also characterized by the presence of a steep shock transition. However, the state downstream the shock is not (re)set by the (left) boundary condition and therefore numerical inaccuracies which lower this value are immediately translated in a smaller shock speed. Thus, the numerical solution lags more and more behind the exact one as time advances. The case in fig. 6(III) has a Reynolds number of 6000, $N_{cell} = 512$ and $N_{part} = 8$, $l_{max} = 3$. A uniform grid having the same resolution should have $512 \cdot 2^3 = 4096$ cells. Clearly, the spatial lag, shown by the enlargements upstream and downstream the shock, is bound to increase as time progresses. Nevertheless, the comparison with the exact solution shows an excellent overall agreement while the enlargements display the accuracy level obtained even under these severe conditions.

Nozzle flow (Inviscid, Isentropic, Quasi -1D Euler Equation). The geometry of the nozzle is defined by the relation $A(x) = (x + 2/x)/2$ which is singular for $x = 0$. By placing the nozzle inlet x_0 close and close to $x = 0$, it is possible to grade the intensity of the source area term in the Quasi-1D Euler equations. In this way the nozzle flow may provide a model to test the behaviour of the adaptive refinement technique in the presence of strong nonlinear source terms, such as those characterizing reacting flows. In fig. 6(IV) $x_0 = 0.005$, $N_{cell} = 128$ and $N_{part} = 4$, $l_{max} = 5$. All available resolution levels have been used, and therefore the resolution just around x_0 is the same of a uniform grid with 4096 cells. At steady state, the storage memory allotted has been of 280 positions and the integration has been actually carried out on 209 nodes. The error distribution $(u_{num}(x) - u_{exact}(x))$ show jumps at the changes of resolution level due to the different truncation errors which cannot be eliminated. The residual drops properly, showing that the adaptive mesh refinement does not disturb the convergence of the solution. The same calculation has been repeated with a uniform grid with the same number of cells actually used by the adapdated mesh (209). No solution has been obtained for $x_0 < 0.04$, a value one order of magnitude larger than the one used for the test case.

7 Conclusions

The test cases have demonstrated the feasibility of the technique and the aspects which still deserve further analysis. Future work must be devoted to a deeper study of the influence of the space-time discretization on the propagation phenomena, and to the optimization of the interaction between the subgrid activation process and the refinement criterium. The actual implementation of the technique to multidimensional problems will follow.

References

[1] Kassoy, A., Roytburd, V., *Numerical Combustion*, p.374, Dervieux, A., Larrouturou, B. (eds.), (Springer 1989).

[2] Valorani, M., IIIrd Int. Conf. on Hyperbolic Problems, Engquist, B., Gustaffson, B. (eds.), pp. 899-912 (Chartwell-Bratt 1990).

[3] Lohner, R. et al., *AIAA Paper*, 85-1531 (1985).

[4] Berger, M.J., Colella, P., *J. Comput. Phys.*, vol. 82 (1989) pp. 64-84.

[5] Kallinderis, J.G., Baron, J.R., *Proc. 11th Int. Conf. Num. Meth. in Fluid Dynamics*, Williamsborough (Virginia 1988).

[6] Kallinderis, J.G., Baron, J.R., *Computational Methods in Viscous Aerodynamics*, Murthy, T.K.S. and Brebbia, C.A. (eds.) (Elsevier 1990), pp.163-196.

[7] Pervaiz, M.M., Baron, J.R., *AIAA J.*, vol. 27, n. 10 (1989).

[8] Schmidt, G.H., *Proc. VIII GAMM Conf.*, Wesseling, P. (ed.), NNFM, Vol.29 (Vieweg 1990), pp.493-502.

[9] Bieterman, M.B., et al., *Proc. VIII GAMM Conf.*, Wesseling, P. (ed.), NNFM, Vol.29 (Vieweg 1990), pp.22-30.

[10] Valorani M., *Ph.D. Thesis*, Dipartimento di Meccanica ed Aeronautica, Università di Roma "La Sapienza" (1991).

[11] Vichnevetsky, R., *Int. J. for Numer. Meth. in Fluids*, vol. 7 (1987) pp. 409-452.

[12] Whitham, G.B., *Linear and Nonlinear Waves*, (Wiley 1974), pp.96-109.

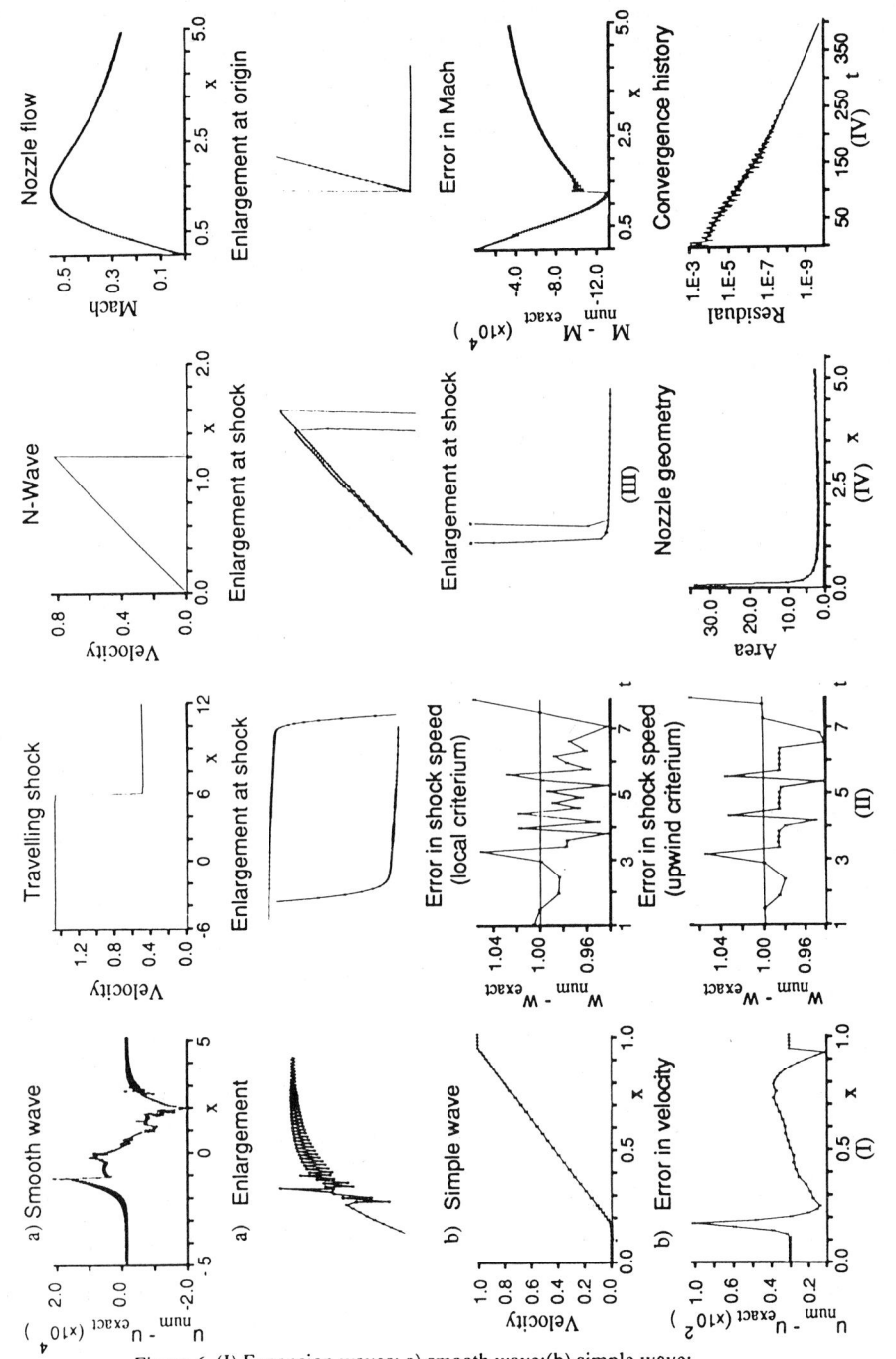

Figure 6. (I) Expansion waves: a) smooth wave; (b) simple wave;
(II) travelling shock; (III) N-Wave; (IV) nozzle flow.

ON RELIABILITY OF THE TIME-DEPENDENT APPROACH TO OBTAINING STEADY-STATE NUMERICAL SOLUTIONS

H.C. Yee[1]

NASA Ames Research Center, Moffett Field, CA 94035, USA

and

P.K. Sweby[2]

University of Reading, Whiteknights, Reading RG6 2AX, England

Summary

When a time-dependent approach is used to obtain steady-state numerical solutions of a fluid flow or a steady nonlinear partial differential equation (PDE), a boundary value problem is transformed into an initial-boundary value problem with unknown initial data. If the steady PDE is strongly nonlinear and/or contains stiff nonlinear source terms, then slow convergence, nonconvergence or spurious steady state numerical solutions can occur even though the time step is well below the CFL limit and the initial data are physically relevant. Here the term "time-dependent approach" is used loosely to include some of the iteration procedures (due to implicit time discretizations), relaxation procedures and preconditioners for convergence acceleration strategies in numerically solved steady PDEs. This is due to the fact that most of these procedures can be viewed as approximations of time-dependent PDEs. Using nonlinear dynamics and local and global bifurcation theory, much of the asymptotic behavior of the resulting nonlinear discretized counterparts from finite discretizations of model PDEs can be analyzed and explained [1-9].

The goal of this paper is to utilize these tools to reveal some of the spurious behavior of asymptotic numerical solutions due to time discretizations for a model viscous Burgers' equation. Here we use the term "spurious asymptotic numerical solutions" to mean asymptotic solutions that satisfy the discretized counterparts but do not satisfy the underlying ODEs or PDEs. Asymptotic solutions here include steady-state solutions, periodic solutions, limit cycles, chaos and strange attractors. Although the study is restricted to an artificially small number of spatial grid points, it is only by understanding the mechanisms in such simple situation that one can gain insights and appreciate the potential of erroneous results in practical computations. It can be shown that with the same initial data, time step, grid spacing and spatial discretizations, but different time discretizations, the resulting schemes (associated with a nonlinear PDE) can converge to different stable asymptotic numerical solutions. Moreover, it is possible that none of these numerical solutions is a close approximation to the true solution of the PDE. The unique property of the separate dependence of the solutions on initial data for the individual PDE and its discretized counterparts is essential for employing a time-dependent approach to the steady state in fluid dynamics problems and, in particular, for hypersonic reacting flows and combustion related computations, since these problems are strongly nonlinear and stiff.

Other spurious asymptotic behavior of numerical solutions due to time/spatial discretizations is reported in our companion papers [1-9]. These works in conjunction with the present work imply that operating with a time step below the linearized stability limit does not necessarily result in a true approximation to the exact solution even though the initial data might be physically relevant. In addition, the conventional concept of a divergent solution when operating with a time step above

---

[1] Staff Scientist, Fluid Dynamics Division.
[2] Lecturer, Department of Mathematics.

the linearized stability limit is not always correct since spurious asymptotes can occur. See the next paragraph and tables 1 and 2 for a comparison between linear stability and nonlinear stability analysis. Consequently when linearized stability limits are used as a guide for a time step constraint for highly coupled nonlinear system problems, this time step might exceed the actual linearized stability limit of the coupled equations. In particular, when one tries to stretch the maximum limit of the linearized allowable time step for highly coupled nonlinear systems, it is possible that all of the different types of spurious asymptotes can be achieved in practice depending on the initial conditions, the form of the nonlinear PDE and the numerical scheme. This is compounded in practical situations when the exact linearized stability limit is usually not computed, but rather a frozen coefficient procedure at each time step with a fixed grid spacing is used to estimate the true stability of the algorithm.

Depending on the initial data, the possible asymptotic solution behavior of numerical schemes operating at time steps below or above the CFL limit or linearized stability limit for nonlinear ordinary differential equations (ODEs) and/or time-dependent nonlinear PDEs is summarized in Table 1. Each of the phenomena listed in Table 1 can be supported by simple model differential equations (DEs) with commonly used finite discretization in computational fluid dynamics (CFD) (cf. [1-4] and the next section). Table 2 compares the various guidelines, assumptions, usage and applicability of four different methods in obtaining stability criterion for time step constraints for the time-dependent approach to the steady-state numerical solutions. Note that the third row under the assumption and usage heading (concerning inside and outside the stability interval) reflects the conventional thinking and practice in CFD rather than the statement of truth. A high percentage of the current computer codes in CFD are operated under the guidelines of the first stability criterion. Without the source term, the first three stability procedures are equivalent for 1-D linear initial value problems. In the absence of source terms in nonlinear problems, the first three procedures are equivalent when the iterated solutions are very near the exact steady-state solution, since perturbing around the iterated solution at the "frozen coefficient or time level n" is approximately equal to perturbing around the exact steady state. When multiple steady states of the governing DE and/or spurious asymptotes exist and/or in the presence of source terms, these procedures are different. Although the last three stability procedures cannot be realized in practice, they are in the order of increasing importance in providing closer stability behavior for strongly nonlinear problems. As evidenced from the present results and the results of [1-9] where the exact steady states of the model DE are known, nonlinear analysis uncovers much of the nonlinear phenomena which linearized analysis is not capable of predicting.

Although more theoretical development and extensive numerical experimentation are needed, we believe that these findings could have important implications in the interpretation of numerical results from existing computer codes and widely used CFD algorithms in combustion, reacting flows and certain turbulence models in compressible Navier-Stokes computations.

Spurious Asymptotic Numerical Solutions of a Model PDE

Quite a few 2×2 systems of first-order autonomous nonlinear coupled ODEs were examined. Detailed nonlinear analysis of four of these systems is reported in the expanded version of this paper [2]. Part of the analysis was published in an internal report [10]. The aims of the present research and [2,10] are to investigate the dynamical behavior of time discretizations by considering various common ODE solvers applied to simple 2×2 systems of nonlinear coupled ODEs, since it is only by understanding the mechanism in such simple situations that we can hope to fully appreciate the potential of erroneous results in the solution of the large systems of nonlinear coupled equations encountered in full discretizations of PDEs. The underlying goal is to investigate the effect of time discretizations on the existence and stability of spurious asymptotic numerical solutions of PDEs when the time splitting or method of lines approach is used to numerically approximate the PDE solutions.

In addition, a basic understanding of the difference in the dynamical behavior of ODE solvers between scalars and systems of strongly nonlinear first-order autonomous ODEs is acquired via this study. Due to the fact that there is no limit cycle or higher dimensional tori counterparts for the scalar first-order autonomous ODEs, spurious limit cycles and higher dimensional tori can only be introduced by the numerics when solving nonlinear ODEs other than scalar first-order autonomous type, systems of nonlinear coupled ODEs, and/or by using a scheme with higher than two time levels for the scalar first-order autonomous ODEs. We conjecture that the existence of stable spurious limit cycles and/or other forms of spurious periodic solutions including numerical chaos might be one of the contributing factors in nonconvergence of the time-dependent approach to the steady state. See the following numerical example for some explanations.

Due to space limitation, only the viscous Burgers' equation

$$\frac{\partial u}{\partial t} + \frac{1}{2}\frac{\partial (u^2)}{\partial x} = \epsilon \frac{\partial^2 u}{\partial x^2}, \quad \epsilon > 0 \qquad (1)$$

is considered. Let $u_j(t)$ represent an approximation to $u(x_j, t)$ where $x_j = j\Delta x, j = 1, ..., J$ with Δx the uniform grid spacing. Consider the three-point central difference in space with periodic boundary condition $u_{J+j} = u_j$ and assume $\sum_{j=1}^{J} u_j =$ constant which implies that $\sum_{j=1}^{J} \frac{du_j}{dt} = 0$. For simplicity, take $J = 3$ and $\Delta x = 1/3$ [11]. Then with $\beta = 9\epsilon$

$$\frac{du_1}{dt} + \frac{3}{4}\left(u_2^2 - u_3^2\right) = \beta(u_2 - 2u_1 + u_3) \qquad (2a)$$

$$\frac{du_2}{dt} + \frac{3}{4}\left(u_3^2 - u_1^2\right) = \beta(u_3 - 2u_2 + u_1) \qquad (2b)$$

$$\frac{du_3}{dt} + \frac{3}{4}\left(u_1^2 - u_2^2\right) = \beta(u_1 - 2u_3 + u_2) \qquad (2c)$$

$$\sum_{j=1}^{3} \frac{du_j}{dt} = 0. \qquad (2d)$$

This system can be reduced to a 2×2 system of first-order nonlinear autonomous ODEs. In this case, the nonlinear convection term is contributing to the nonlinearity of the ODE system (2). Equation (2) has four steady-state solutions (fixed points) of which three are saddles and one is a stable spiral at (1/3, 1/3) for $\beta \neq 0$. For $\beta = 0$ the stable spiral becomes a center. Figure 1 shows the phase portrait ($u_1 - u_2$ plane) of these exact asymptotic solutions.

A specialized computer code was written for the Connection Machine to study the bifurcation phenomena of the corresponding steady-state numerical solutions and the strong dependence of the numerical solution on initial data when widely used time discretizations in CFD are applied to equation (2) as the time step is varied. To check against our computation from the Connection Machine, some mathematical analysis has been performed using the algebraic manipulation package in Mathematica [12]. Also two existing interactive computer programs were modified to validate qualitatively computations for situations which were too complex for Mathematica to analyze. One of the interactive computer programs, written by Creon Levit of NASA Ames, was originally designed to aid the study of flow visualization in CFD [13]. The other computer program "AUTO" [14], written by Eusebius Doedel, is a software package for continuation and bifurcation problems in ODEs. These two computer programs only can provide spot checks on the wealth of dynamical behavior that the Connection Machine provides. Also, the study cannot be realized in a reasonable time frame without the highly parallel architecture of the Connection Machine or similar parallel machine.

The time discretizations that we are considering are similar to our scalar study in references [1,5]. These schemes include the explicit Euler, modified Euler (a second-order Runge-Kutta method), improved Euler (another second-order Runge-Kutta method), Heun (a third-order Runge-Kutta method),

Kutta (another third-order Runge-Kutta method) and a fourth-order Runge-Kutta method. Our investigation indicates that all of the studied Runge-Kutta methods exhibit spurious limit cycles and other spurious periodic solutions. For the Kutta method, spurious asymptotes can occur below the linearized stability of the scheme. The bifurcation diagrams and basins of attraction of the various ODE solvers are reported in detail in [2]. Here the basin of attraction is a domain of a set of initial conditions whose solution curves (trajectories) all approach to the same asymptotic state. The wealth of dynamical behavior is beyond the page limit of these proceedings. However, one important aspect on the interplay between initial data, spurious asymptotes, basins of attraction and the time-dependent approach to the asymptotic numerical solutions is discussed here, since only with a solid understanding of this concept can one determine the reliability of the time-dependent approach to obtaining steady-state numerical solutions.

The discussion are restricted to two Runge-Kutta methods, namely the second-order improved Euler and the third-order Kutta method with initial data confined to $-1 \leq u_1 \leq 1.5$ and $-1 \leq u_2 \leq 1.5$. Here the improved Euler method applied to $\frac{du}{dt} = \alpha S(u)$, where α is a parameter, is

$$u^{n+1} = u^n + \frac{r}{2}\left[S^n + S(u^n + rS^n)\right] \quad (3)$$

with $r = \alpha \Delta t$ and $S^n = S(u^n)$. The Kutta method is

$$u^{n+1} = u^n + \frac{r}{6}\left(4k_2 + k_1 + k_3\right)$$
$$k_1 = S^n$$
$$k_2 = S\left(u^n + \frac{r}{2}k_1\right)$$
$$k_3 = S\left(u^n + \frac{r}{2}k_2\right). \quad (4)$$

The phase plane is divided into 512×512 points of initial data. Each of the basins of attraction are obtained by integrating the resulting discretized counterpart 10,000 times, keeping track of where each initial data asymptotically approaches, and color coding according to the individual asymptotes.

Case I. The Improved Euler

To give an example of the existence of spurious limit cycles and its effect on the numerical basins of attraction for the exact steady state (stable spiral fixed point at (1/3, 1/3)), figure 2 shows the basins of attraction of the improved Euler method for 4 different fixed $\Delta t = 0.1, 1, 2.25, 2.35$ with $\beta = .1$. By a bifurcation comptation, we found that the first two time steps are below the linearized stability around the exact steady state (1/3, 1/3) and the last two time steps are above the limit. Unlike the semi-discrete system (2) shown in figure 1, only the stable solutions are plotted here. Also, the three unstable saddle points are located outside of the plotting area. On the plots $(u_1^n, u_2^n) = (u_n, v_n)$. Through this numerical study, the following phenomena were observed.

(a) Below the linearized stability limit of the scheme, no spurious stable steady states were observed. However, this does not preclude the existence of unstable spurious asymptotes that can influence the numerical basins of attraction of the true steady states. As a matter of fact, this is precisely the case. Figure 2a shows the approximate exact basin of attraction for the stable spiral (the white dot at (1/3, 1/3)), since this basin is the same when one integrates equation (2) by the explicit Euler (which does not produce spurious steady states [15,1]) with a very small time step. One can obtain the exact basin for (1/3,1/3) by integrating backward and forward in time along the unstable and stable maniforlds (separatrices) of the three saddle points. The red regions (or grey region for the black and white plot) are the numerical basin of attraction for (1/3, 1/3) for $\Delta t = 0.1$ and 1 respectively. The black regions are the numerical basins of divergent solutions. Increasing the time step to $\Delta t = 1$ (below the linearized stability limit) resulted in an enlargement of the numerical basin of attraction (1/3, 1/3). In other words, what was expected to be nonphysical initial data can now actually be in the numerical basin for (1/3, 1/3).

(b) Above the linearized stability limit, spurious limit cycles (cf. figure 2c) and higher dimensional periodic solutions (cf. figure 2d) were observed. A further increase in Δt resulted in numerical chaos type phenomena and eventually divergence (with additional increase in Δt). What is expected to be physical initial data now can either converge to a spurious limit cycle or other periodic solution or diverge. Now the red (grey) or multicolor (different shade of grey) are the basins of the spurious limit cycle (the irregular white closed curve for figure 2c) and the spurious periodic solution (white dots for figure 2d). For these latter two time steps, the numerical basins for (1/3, 1/3) by the improved Euler disappear. However, if the initial data are in the red or multicolor region, one get nonconvergence of the numerical steady state instead of what the linearized stability predicts. The phenomena observed above might well be an explanation on the nonrobustness or nonconvergence of numerical methods encountered in practice.

(c) Although no spurious steady-state numerical solutions exist for this case, the existence of unstable asymptotes below the linearized stability limit and/or the existence of stable and unstable asymptotes above the linearized stability limit is just as detrimental to the numerical basins of attraction for the true steady state as if a spurious steady state occured below the linearized stability limit. In the latter case, however, a spurious steady-state can be mistaken for the true steady state in practical computations. See [1,2,10] for details

Case II. The Kutta Method

To give an example of the existence of spurious asymptotes below the linearized stability limit of the scheme as well as existence of spurious limit cyles above the linearized stability limit, figure 3 shows the basins of attraction of the Kutta method for 4 different fixed time steps $\Delta t = 0.1, 1, 1.826, 1.85$ (all below the linearized stability of the scheme) with $\beta = 0.1$. The red regions are the numerical basins of attraction of the asymptotic state (1/3,1/3) for the various Δt. The same domain, and same number of initial data and iterations as case I were used. All of the phenomena observed for case I hold true for the current case. In addition, new phenomena arise that complicate the numerical basin of attraction tremendously. This occurs in the form of stable and unstable spurious asymptotes below the linearized stablility of the scheme (cf. Figures 3c and 3d). The numerical basin for (1/3, 1/3) has become fractal like with the birth of fragmented, isolated new basins of attraction due to the presence of spurious periodic solutions (the three white complicated closed curves with the associated purple, green and blue basins shown in figure 3c). For the case of the existence of <u>unstable</u> spurious asymptotes, the numerical basins for (1/3, 1/3) is fractal like (cf. figure 3d). Not shown are the basins for the spurious limit cycles above the linearized stability limit. In general these basins have similar structure and features as case I except no higher order periodic spurious solution exists. Note also, with the same time step $\Delta t = 1$, both ODE solvers behave the same.

From the above study one can conclude that for a given Δt below the linearized stability limit, the numerical solution can converge to either (a) an exact steady state, (b) a spurious periodic solution, or (c) diverge even though the initial data are physically relevant. In general for different nonlinear DE and numerical method combinations, in addition to the above phenomena, the numerical solution can have all the phenomena discussed in table 1 (cf. [1-4]).

Concluding Remarks

One might argue that for the model problem that was considered, it may be trivial to check whether an asymptote is spurious or not. However, the main purpose of this study is to set the baseline dynamical behavior of the scheme so that one can use it wisely in other more complicated settings such as when nonlinear systems of PDEs are encountered for which the exact solutions are not known. Under this situation, spurious asymptotes could be computed and mistaken for the correct asymptotic solutions. Hopefully, with this simple example, we can demonstrate the importance of the subject and, most of all, the importance of knowing the general dynamical behavior of asymptotes of the schemes for genuinely nonlinear model DEs before applying these schemes in practical calculations.

References

[1] Yee, H.C., Sweby, P.K., Griffiths, D.F.: "Dynamical Approach Study of Spurious Steady-State Numerical Solutions for Nonlinear Differential Equations, Part I, The Dynamics of Time Discretizations and its Implications for Algorithm Development in Computational Fluid Dynamics," NASA TM-102820, April 1990, also J. Comput. Phys., Vol. 97 (1991).

[2] Yee, H.C., Sweby, P.K.: "Dynamical Approach Study of Spurious Steady-State Numerical Solutions for Nonlinear Differential Equations, Part II, The Dynamics of Numerics of Systems of 2×2 ODEs and its Connections to Finite Discretizations of PDEs," NASA Technical Memorandum, 1991.

[3] Lafon, A., Yee, H.C.: "Dynamical Approach Study of Spurious Steady-State Numerical Solutions for Nonlinear Differential Equations, Part III, The Effects of Nonlinear Source Terms and Boundary Conditions in Reaction-Convection Equations," NASA TM-103877, July 1991.

[4] Lafon, A., Yee, H.C.: "Dynamical Approach Study of Spurious Steady-State Numerical Solutions of Nonlinear Differential Equations, Part IV, Stability vs. Numerical Treatment of Nonlinear Source Terms," NASA Technical Memorandum, in preparation.

[5] Sweby, P.K., Yee, H.C., Griffiths, D.F.: "On Spurious Steady-State Solutions of Explicit Runge-Kutta Schemes," Numerical Analysis Report 3/90, March 1990, Univ. of Reading, also NASA TM-102819, April 1990.

[6] Yee, H.C.: "A Nonlinear Dynamical Approach to Algorithm Development in Hypersonic CFD," Proceedings of the 4th International Symposium on Computational Fluid Dynamics, Davis, Calif., Sept. 9-12, 1991.

[7] Griffiths, D.F, Stuart, A.M., Yee, H.C.: "Numerical Wave Propagation in Hyperbolic Problems with Nonlinear Source Terms," University of Bath Report, March 1991.

[8] Lafon, A., Yee, H.C.: "Nonlinear Reaction-Convection BVPs vs. Their Finite Difference Approximations: Spurious Steady-State Numerical Solutions," Proceedings of the 4th International Symposium on Computational Fluid Dynamics, Davis, Calif., Sept. 9-12, 1991.

[9] Griffiths, D.F, Stuart, A.M., Sweby, P.K., Yee, H.C.: "Stability of Spurious Steady-State Solutions of Runge-Kutta and Related Methods for PDEs," in preparation.

[10] Sweby, P.K., Yee, H.C.: "On Spurious Asymptotic Numerical Solutions of 2×2 Systems of ODEs," Numerical analysis Report 7/91, August, 1991, University of Reading, England.

[11] Griffiths, D.F., Stuart, A.M.: private communications.

[12] Wolfram, S.: *Mathematica, A System Doing Mathematics by Computer*, Addison Wesley, California, 1988.

[13] Globus, A., Levit, C., Lasinski, T.: "A Tool for Visualizing the Topology of Three-Dimensional Vector Fields," to appear in the proceedings of Visualization '91, October 22-25, 1991, San Diego, CA.

[14] Doedel, E.: "AUTO: Software for Continuation and Bifurcation Problems in Ordinary Differential Equations," Cal. Tech. Report, Pasadena, Calif., May, 1986.

[15] Iserles, A.: "Nonlinear Stability and Asymptotics of O.D.E. Solvers," International Conference on Numerical Mathematics, Singapore, R.P. Agarwal, ed., Birkhauser, Basel, 1989.

Table 1. Genuinely nonlinear behavior of asymptotic numerical solutions of nonlinear PDEs vs. time steps (ss: steady state)

BELOW CFL or Linearized Stability Limit	ABOVE CFL or Linearized Stability Limit
Can converge to (a) correct exact ss (b) incorrect exact ss (c) incorrect exact asymptotes (d) spurious asymptotes (e) divergent solution	Same as **BELOW** **Except** (a)

Table 2. Stability guidelines for time step constraints for time-dependent approach to steady-state numerical solutions (ss: steady state)

Type	CFL Limit (von Neumann analysis)	Linearized Stability I	Linearized Stability II	Nonlinear Stability
Assumption & Usage	perturbed around u^n	perturbed around exact SS		
	ignore the source term if exists		include the source term if exists	
	inside the stability interval $\Rightarrow$ converges to the correct SS outside the stability interval $\Rightarrow$ diverges			spurious asymptotes exist
	local behavior weakly nonlinear			global behavior strongly nonlinear
Stability Region	consists of a single continuous interval		can consist of disjoint intervals	
Initial Data (IC)	no concept of strong dependence on IC			strong dependence on IC
Applicability	insufficient for strongly nonlinear ODEs & PDEs		closer to nonlinear analysis	for strongly nonlinear ODEs & PDEs

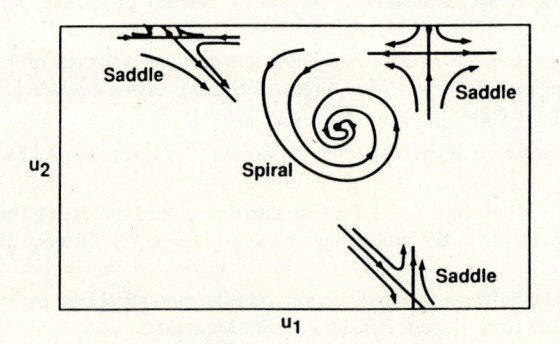

Figure 1. Phase portrait of system (2)

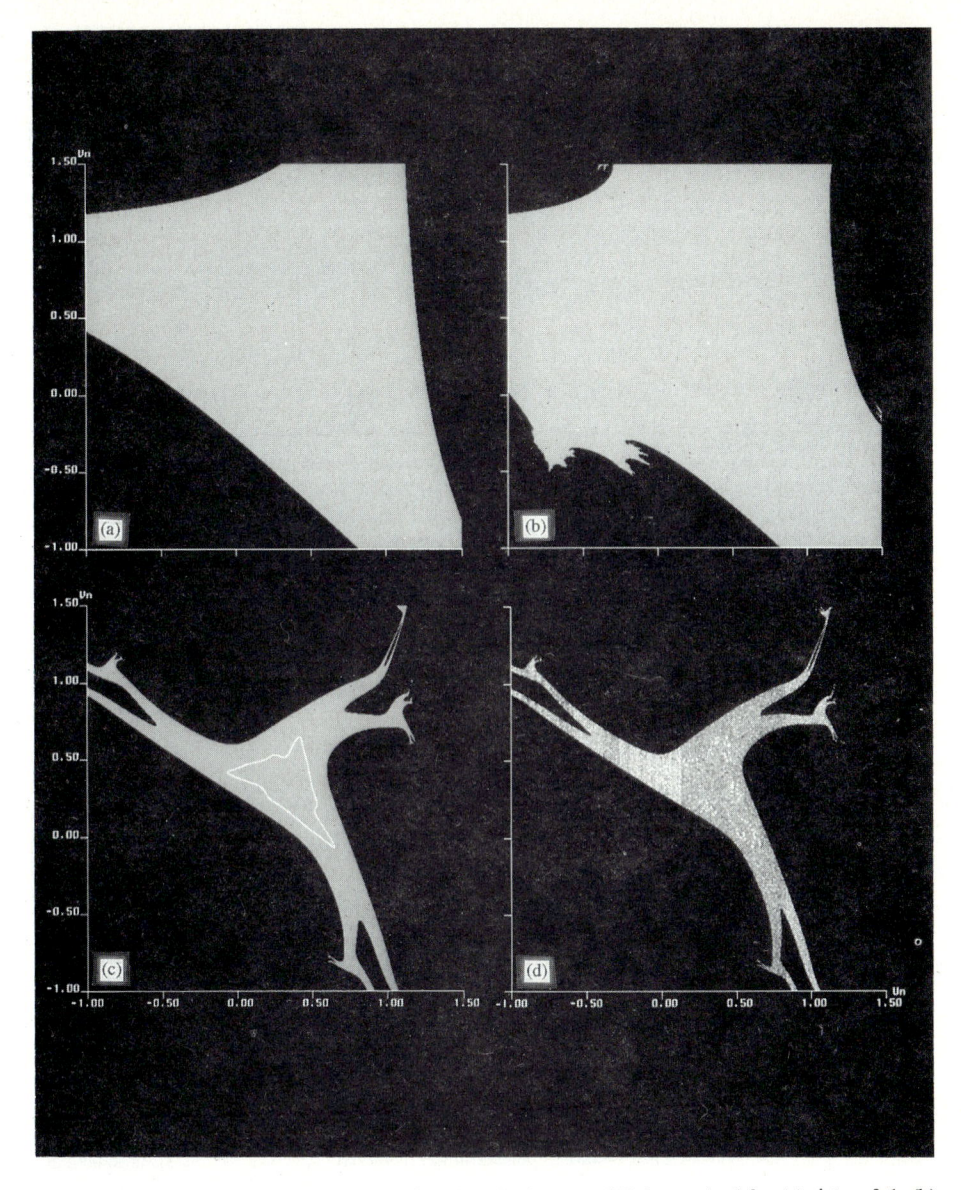

Figure 2. Basins of attraction for system (2) using the improved Euler method for (a) $\Delta t = 0.1$, (b) $\Delta t = 1.0$, (c) $\Delta t = 2.25$, and (d) $\Delta t = 2.35$

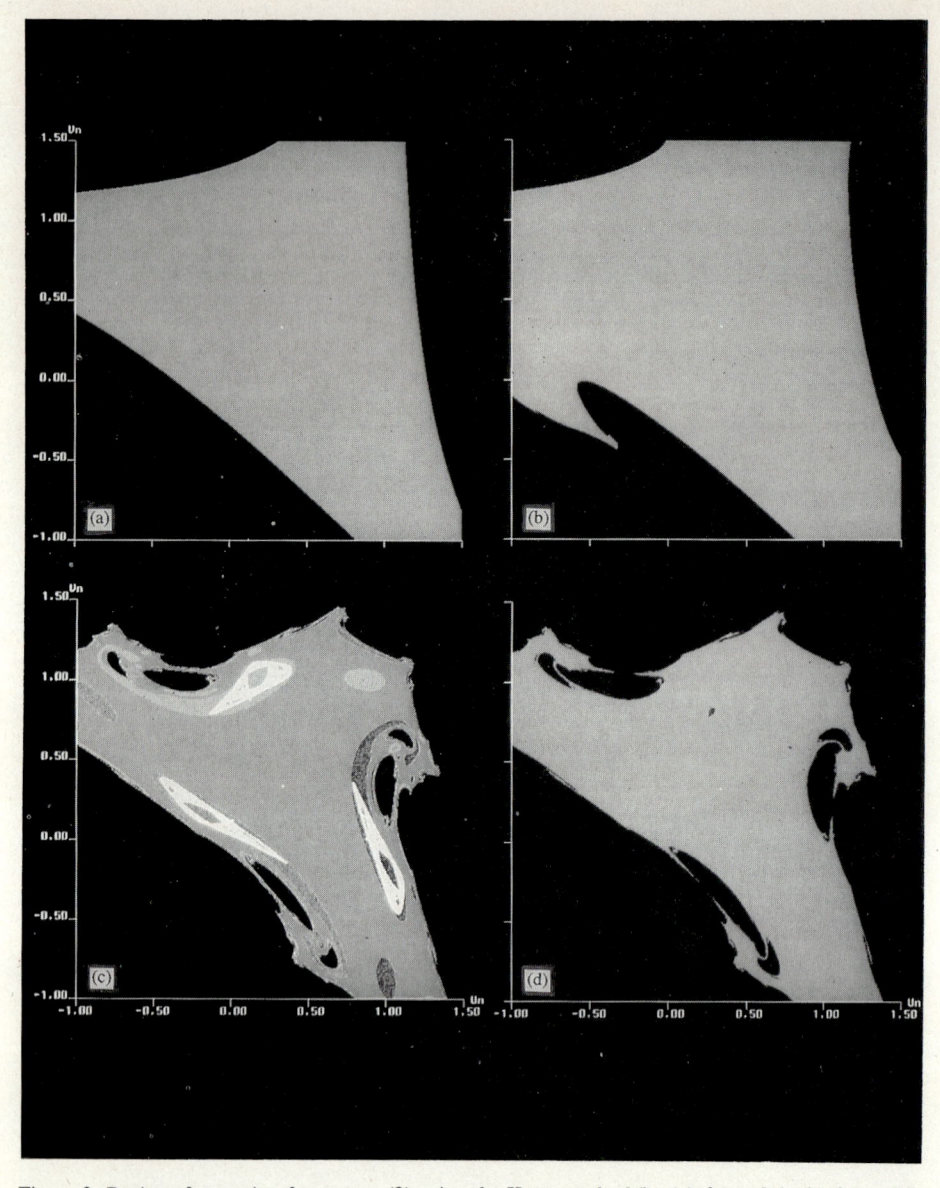

Figure 3. Basins of attraction for system (2) using the Kutta method for (a) $\Delta t = 0.1$, (b) $\Delta t = 1.0$, (c) $\Delta t = 1.826$, and (d) $\Delta t = 1.85$

DIRECT SIMULATION OF TURBULENT FLOWS

A METHOD FOR DIRECT NUMERICAL SIMULATION OF COMPRESSIBLE BOUNDARY-LAYER TRANSITION

N. A. Adams, N. D. Sandham and L. Kleiser
DLR, Institute for Theoretical Fluid Mechanics, Göttingen, Germany

SUMMARY

A numerical method for computing three-dimensional time-dependent compressible transitional flow over a flat plate is presented. A direct comparison between Chebyshev and Padé schemes for the spatial derivative is made. The method is validated by comparison with linear instability theory. Some nonlinear results for a second mode in high supersonic flow are presented.

1. INTRODUCTION AND PHYSICAL MODEL

The objective of our work is to extend the method of direct numerical simulation of transition to turbulence from incompressible to compressible flow. To achieve this goal a code has been developed to solve the compressible, three-dimensional, unsteady Navier-Stokes equations (NSE) for a wall-bounded flow. We hope to establish a tool to get new insights into the essentially unknown nonlinear stages of supersonic transition and of fully-developed turbulence.

At the time t_0 the simulation starts in a reference frame moving with a certain velocity c_g downstream. A cartesian coordinate system is used with the streamwise direction denoted with x, the spanwise with y and the wall-normal direction with z. The flow over an infinite flat plate develops into a two dimensional laminar boundary layer which is used as the base flow for the simulations. At present a Chapman-Rubesin [1] base flow, assuming a linear viscosity law, is being used because of its semi-analytic formulation. The base flow is made an exact solution by adding a forcing term to the NSE. This is equivalent to solving the full nonlinear equations for disturbances from the given base flow. The computational box is assumed to be small in its extents so that the base flow may be treated as locally parallel. The streamwise and spanwise flow extents are chosen to meet the largest wavelengths appearing in the linear stability analysis. The flow is assumed to be periodic in the x- and y-directions. The advantage of this approach over the spatial description used for example in [2] is the considerably smaller computational domain leading to a significantly better flow resolution. Also, as remarked in [3], it approximates well the physical reality.

2. MATHEMATICAL MODEL

For the conservative variables $U = \{\rho, \rho u, \rho v, \rho w, e_T\}$, where ρ is the density, e_T is the total internal energy and u, v, w are the velocity components, the conservation equations can be written in dimensionless form by

$$\frac{\partial U}{\partial t} = \frac{\partial F}{\partial x} + \frac{\partial G}{\partial y} + \frac{\partial H}{\partial z} + Z, \qquad (1)$$

augmented by the equation of state for ideal gases. The definitions of e_T, the flux variables F, G and H and the corresponding normalization quantities can be found in [4]. The ratio of specific heats is $\kappa = 1.4$. The freestream Mach number is denoted with M_∞ and the local values of pressure and temperature with p and T respectively. Sutherland's law of viscosity is assumed to be valid. The Prandtl number is set constant as $Pr = 0.72$. For the forcing term Z one can derive the relation

$$Z = \begin{bmatrix} c_g \frac{\partial \bar{p}}{\partial \xi} \\ c_g \frac{\partial (\bar{p}\bar{u})}{\partial \xi} - \frac{1}{Re}\frac{\partial}{\partial z}\left(\bar{\mu}\frac{\partial \bar{u}}{\partial z}\right) \\ 0 \\ 0 \\ c_g \frac{1}{2}\frac{\partial (\bar{p}\bar{u}^2)}{\partial \xi} - \frac{1}{(\kappa-1)M_\infty^2 Pr Re}\frac{\partial}{\partial z}\left(\bar{\mu}\frac{\partial \bar{T}}{\partial z}\right) - \frac{1}{Re}\frac{\partial}{\partial z}\left(\bar{\mu}\bar{u}\frac{\partial \bar{u}}{\partial z}\right) \end{bmatrix} \quad (2)$$

by substituting the laminar boundary-layer solution into the NSE. The local viscosity $\bar{\mu}$ is calculated from $\bar{T}$ by Sutherland's law, Re is the Reynolds number referred to the displacement thickness, $\bar{\rho}(z)$ is the base flow density profile and $\bar{u}(z)$ is the base velocity profile. The downstream variable before the Galileian transformation is denoted by $\xi = x + c_g t$. In the calculations presented in this paper the forcing term is time independent, since c_g is set to zero.

The boundary conditions are:
(a) No-slip condition at the impermeable wall, $u(x,y,0) = v(x,y,0) = w(x,y,0) = 0$.
(b) Either (i) isothermal condition at the wall $T_W = const$ or (ii) zero heat transfer condition at the wall $\frac{\partial T}{\partial z}\big|_W = 0$.
(c) Non-reflecting boundary conditions at the truncation plane $z = z_{max}$.
(d) Periodicity in x and y.

As an initial condition the flowfield is imposed with the base flow to be investigated at a certain downstream position ξ_0 of the plate. This solution is then superimposed with eigenfunctions from the linear stability analysis and/or with random noise. In the former case the initial condition takes the form

$$U(x,y,z;0) = \bar{U}(\xi_0, z) + A \cdot \mathbf{Re}(\hat{U}(z) \cdot e^{i\alpha x + i\beta y}), \quad (3)$$

where $\hat{U}(z)$ are the complex eigenfunctions of linear stability analysis, normalized with the maximum of $\hat{u}(z)$, and α and β are the wavenumbers in x and y respectively. **Re** stands for the real part, A is a certain initial amplitude of the disturbance and $\bar{U}$ is the base flow. For the extents of the computational box the relations $L_x = 2\pi/\alpha$ and $L_y = 2\pi/\beta$ are valid. In the latter case

$$U(x,y,z;0) = \bar{U}(\xi_0, z) + A \cdot z^2 \cdot e^{-4z} \cdot \Omega(x,y,z) \quad (4)$$

is used. Ω stands for a uniform distribution of random numbers $-1 < \Omega < 1$. The amplitude is modulated with an analytical function of z to fulfill the boundary conditions.

3. NUMERICAL METHOD

The numerical solution procedure uses a method of lines formulation. For time integration a third-order explicit Runge-Kutta scheme with low storage requirement is used. The whole scheme, including scratch storage for FFT's needs 19 storage units for every grid point. Details concerning the Runge-Kutta scheme can be found in [5].

In the periodic directions a Fourier collocation discretization is used. For the wall-normal direction three different discrete spatial derivative operators have been tested:
(A) Chebyshev collocation (as used in [7] and [8]),
(B) tridiagonal compact Padé scheme of 6th order at the inner points, 3rd order at the boundary points and 4th order at the adjacent points (applied to compressible mixing layer simulation in [5], [6]),
(C) uniformly 6th order Padé scheme.

Scheme (A) could be classified as a pseudospectral approach for the determination of the temporal derivative in the NSE. It uses FFT's together with a recursion formula for the derivatives (instead of a matrix operator [9]). The Padé schemes use a hermitian-like interpolation of the function to be derived. They are linear schemes of the form $\mathbf{A}\mathbf{f}' = \mathbf{B}\mathbf{f}$, where $\mathbf{f}$ is the vector of gridpoint values of a function $f(z)$ and $\mathbf{f}'$ the corresponding derivative vector. $\mathbf{A}$ and $\mathbf{B}$ are constant matrices. For scheme (B) $\mathbf{A}$ is tridiagonal and $\mathbf{B}$ is pentadiagonal. The higher order at the boundaries for scheme (C) leads to 5×5 blocks in $\mathbf{A}$ in the upper left and lower right corners and also a 5-point stencil in $\mathbf{B}$ at the boundaries. This requires a slightly modified solution algorithm. Fundamentals about the Padé schemes in general can be found in [10] and about the schemes used in the present method in [11].

At the wall the boundary conditions used are in agreement with those found in [12] to be sufficient for well-posedness of the NSE. The Dirichlet boundary conditions (a) and (b)(i) are imposed after every time step (including Runge-Kutta substeps). The density is calculated from the interior. It was confirmed that an additional (characteristic) correction for the density is a stationary source of spurious waves emanating from the boundary, since the boundary conditions are overspecified [12]. The isothermal condition (b)(i) is enforced at every time step (including substeps) by overwriting the total internal energy with the boundary value, using the equation of state. In case of the Neumann condition (b)(ii), the spatial derivative operator is directly imposed with the known boundary values. For the (implicit) Padé schemes this leads to a cancellation of the boundary rows and usage of the known derivative at the adjacent point. For scheme (A) the boundary values are overwritten. The adjacent points are not affected since the Chebyshev-derivative operator has an explicit form. The boundary conditions (c) are a straightforward adaptation of the formulation for hyperbolic systems from [13]. The Eulerian terms of the NSE are transformed into characteristic form at $z = z_{max}$ in the (z,t)-plane. The outgoing waves are calculated from interior points, while the amplitudes of incoming waves are kept constant in t. Afterwards the boundary terms are transformed back into conservation form. For this to be correct it is necessary that the effects of viscosity are negligible at the upper boundary, which can be obtained by choosing z_{max} sufficiently large. These boundary conditions do not correspond exactly to the condition of exponential decay usually enforced in linear stability analysis. If z_{max} is large enough so that the flow is essentially governed by the inviscid equations, the time-dependent non-reflecting boundary conditions at the outer edge are consistent with the sufficient conditions for stability derived in [12], because they automatically switch from inflow to outflow type and vice-versa. The periodic conditions (d) are naturally fulfilled by the Fourier ansatz in x and y.

For flexibility in arranging the grid points in physical space a mapping of the physical space onto the computational space is desirable. In the present method this concerns only the z-direction. The benefits are the possibility of relaxing the stringent time-step limitation of method (A), and of improving the resolution of the z-derivative scheme in some region of the domain. Let z be the independent variable in the physical domain and ζ the corresponding transformed variable in the computational domain. As in [8] an algebraic mapping is defined by

$$z = \frac{z_{max} z_{0.5}(\zeta + 1)}{z_{max} - \zeta(z_{max} - 2z_{0.5})} \tag{5}$$

for method (A), and

$$z = \frac{z_{max} z_{0.5} \zeta}{z_{max} - z_{0.5} - \zeta(z_{max} - 2z_{0.5})} \tag{6}$$

for methods (B) and (C). (Recall that for method (A) $-1 \leq \zeta \leq 1$ is a mapping of the equidistantly spaced variable $0 \leq \phi \leq \pi$ by $\zeta = -\cos(\phi)$, while for methods (B) and (C) $0 \leq \zeta \leq 1$ is equidistantly spaced). In addition a transcendental mapping can be defined by

$$\eta = a \cdot \zeta + b + c \cdot \sinh\left(\frac{\zeta - d}{e}\right). \tag{7}$$

The intermediate variable η is mapped by $z = z_{max} \frac{\eta+1}{2}$ for method (A) and $z = z_{max}\eta$ for methods (B) and (C) onto the physical domain. The parameters a and b depend on the method used. They are functions of the parameters c, d and e and are determined by the values of z at the endpoints of the physical interval. The constant d is determined so that the grid point condensation is most effective at a defined location z_{mv} (this requires the solution of a transcendental equation). Parameters c and e are free and can be used for tuning. Actually these mappings can be used in combination by applying the algebraic after the transcendental mapping. In the current work the constants have been chosen so that a nearly equidistant grid is used within the boundary layer, with not too many points wasted outside.

A CFL-criterion is used at every time step to calculate an admissible step size, $\Delta t = CFL/R$, where

$$R = \pi \cdot \max_{x,y,z} \left| \frac{|u|+a}{\Delta x} + \frac{|v|+a}{\Delta y} + \frac{|w|+a}{\Delta z} + \pi \psi \left(\frac{1}{\Delta x^2} + \frac{1}{\Delta y^2} + \frac{1}{\Delta z^2}\right) \right|. \tag{8}$$

Here $\max_{x,y,z}$ stands for the maximum over the computational domain, a is the local speed of sound and for the NSE of the form (1) one gets [5]

$$\psi = \max\left(\frac{\mu}{\rho \text{Re}}, \frac{\mu}{(\kappa-1)\rho M_\infty^2 \text{PrRe}}\right). \tag{9}$$

4. VALIDATION AND RESULTS

4.1 Spatial derivative operators

To test the accuracy of differentiation, and particularly the influence of numerical boundary conditions, a sample function which resembles a disturbed boundary layer profile with a high gradient at the wall boundary was used (constructed by superimposing a hyperbolic tangent with damped Fourier modes) [11]. To allow comparison with the Chebyshev scheme, the grid points of the Padé schemes were mapped with a cosine function, clipped to $[0.2, \pi - 0.2]$ in order to be invertible, onto the domain of the sample function $[0,1]$. The root-mean-square (rms) errors of the different schemes are shown in fig.1a. Scheme (B) has an overall convergence rate of $1/N^5$, where N is the number of grid points. The lower order of approximation at the boundary reduces the global convergence rate. Scheme (C) possesses the theoretical convergence rate of $1/N^6$. For coarse grids ($N \leq 40$) the rms error is larger for the 6th order scheme (C) as well as for the Chebyshev scheme (A). Scheme (A) convergences spectrally as $3 \cdot e^\sigma$, where $\sigma = -6 \cdot 10^{-7} N^4$. The effect of roundoff error becomes noticeable for $N \geq 90$ (machine accuracy is 14 decimals). The convergence behaviour of the different schemes at the boundary is shown in fig.1b, where the rms error of the two boundary points is plotted against N. The advantages of the spectral scheme (A) at the boundary can be seen clearly. The Padé schemes show the theoretically expected convergence rates. Scheme (B) has the lowest absolute error for small N.

From the point of view of highest possible accuracy scheme (A) is the best, but it has the significant drawback of an operations count of $(5\, log_2 N + 8 + 2\nu)N$, where ν is the order of the derivative, and also of a serious time-step limitation due to the grid point concentration at the boundary. The operations count for scheme (B) is $11 \cdot N$ and for scheme (C) approximately $1.2 \cdot (11 \cdot N)$. *Fig.1c* shows the CPU time versus rms error for the different schemes. The CPU time refers to 10 differentiation-operations in z on a CRAY-YMP (single processor) with a discretization of 129^3, vectorized in x and y. As expected the Padé schemes show an algebraic dependence between CPU time and error. The spectral scheme requires only a small amount of additional time to improve the answer by eight orders of magnitude, though at a relatively high cost.

The numerical stability of the fully discretized scheme was tested by monitoring the energy growth of the (1,0) Fourier mode in 2D calculations (with a fully 3D code) at $M_\infty = 5.0$ and $Re = 10000$. A discretization of $N_x = 4$ and $N_z = 129$ points was used with scheme (A) and of $N_z = 193$ with schemes (B) and (C). The time-step size was calculated by the CFL condition of equation (8). A perturbation of the (1,0) mode with a linear eigensolution of small amplitude was used as the initial condition. In practice scheme (A) was found to be stable in the linear sense at least up to $CFL = 10$, schemes (B) and (C) at least up to $CFL = 4$, since the growth of first mode-energy was constant over about 700 time steps. The latter result corresponds to the observations reported in [5]. With scheme (A) condition (8) seems to be too conservative. The calculations of the following sections are performed with $CFL = 1.5$, below the theoretical linear stability limit for the third order Runge-Kutta scheme presently used.

4.2 Comparison with linear stability theory

The NS code should be able to reproduce the linear stages of instability. Comparison with temporal linear stability theory is applicable, since we have periodic boundary conditions in x and y. The accuracy is determined by comparing the frequency and the growth rate of the kinetic energy of the Fourier modes with reference solutions from linear stability analysis (computed with a spectral boundary value method). The comparison is made between t_0 and $t_0 + \delta t$, where δt is of order 10^{-4}. For schemes (B) and (C) this occurs after the first time step, for scheme (A) after the first 100 time steps. The set of mapping parameters was adapted for every scheme to maximize the resolution in the boundary layer.

The first test case is at $M_\infty = 0.3$, $Re = 700$ ($Re_{Blasius} = 398$, based on the Blasius length scale). The wavenumbers in x and y were $\alpha = 0.27$ and $\beta = 0.4$ respectively. The algebraic mapping (5), (6) was used. The results for this Mach number are condensed in *table 1*, including the discretization and the mapping parameters used. Here the growth rate ω_i is calculated as the growth rate of the square root of the kinetic energy of the disturbance integrated over z. Obviously less than $N_z = 65$ points are required to resolve this first mode eigensolution. For $N_z = 129$ scheme (A) shows a roundoff-error effect as mentioned in the previous section. The relatively strongly damped 3D-mode is also accurately calculated.

The second test case is at $M_\infty = 5.0$, $Re = 10000$ ($Re_{Blasius} = 1042$). The wavenumber is $\alpha = 2.217$ (2D only). In this case we employ a combination of the algebraic (5), (6) and the transcendental mapping (7), which leads to a nearly equidistant gridpoint distribution in the boundary layer. *Table 2* contains the results for this case. Scheme (B) with $N_z = 33$ behaves accidentally well at this first time step but the error increases in time due to insufficient resolution in z. Also scheme (C) with $N_z = 33$ is in error. Naturally, the highest accuracy can be achieved with scheme (A), but schemes (B) and (C) are acceptable with $N_z = 129$.

In agreement with [7] we observed that the local (in z) growth rate is very sensitive to the interpolation scheme that is used as interface between the linear solver and the NS code. Spectral interpolation was found to work well. When starting with linear eigensolutions a slight transient behaviour with temperature and pressure was observed. The local linear growth rate for the velocity components remains for a sufficiently well resolved computation very close to the value calculated with the linear stability code ($\sim 0.005\%$). However, for the temperature and the pressure the deviation from this value rises locally to $\sim 2.5\%$. These differences appear for all spatial schemes, and may be a result of the form of NSE used. The present code uses an internal energy formulation, while in the linear stability code the energy equation is in terms of temperature. They do not affect the above results concerning the primary-mode kinetic energy, since only velocity components over a short time interval are considered. These differences do not originate from the boundary treatment, since the wall-boundary is treated stably and z_{max} is taken large enough so that there are no disturbances emanating from the upper boundary. For z_{max} chosen too low, a mild growth of the primary instability at the boundary indicates that the boundary condition does not match the exponential decay imposed in the linear stability theory as mentioned in section 3.

A more general test was made with a random-noise initial condition for the $M_\infty = 5.0$ case mentioned before. Scheme (B) was used with a discretization of 8×161 points and a combination of algebraic and transcendental mapping. The initial amplitude was $A = 10^{-8}$. The most amplified disturbance has a period of $2\pi/\omega_r = 3.11$ time units. After about 10 periods the exact linear growth rate has developed. About 75 periods later the amplitude has increased to 0.01%, which is big enough to accurately extract the linear eigenfunctions. The local difference between the linear eigensolution and the NS solution is less than 0.15% and therefore invisible in *fig.2*. For all flow quantities the local growth rate is less than 0.27%, the frequency less than 0.007% larger than the values from the linear stability theory. Also they are constant in $z \leq 2$ (boundary-layer thickness is $z_\delta = 1.34$) within 0.04% and 0.05% respectively. At about 110 periods the amplitude has increased to 2.8% and weakly nonlinear effects become visible, for example in the local eigenvalue for the u-disturbance shown on *fig. 3*. Growth rate and frequency develop a kink near the critical layer (at $z = 1.07$). The frequency decreases slightly over the whole boundary layer with increasing time. However the growth rate decreases only above the kink and its overshoot near the critical layer becomes more intense.

4.3 A Mach 5 nonlinear test case

As a 2D nonlinear problem the $M_\infty = 5.0$ case from the previous section was recalculated with the linear eigensolution of $\alpha = 2.217$ as the initial condition. This is the most amplified 2D mode at this Mach number and is known as the second (or Mack) mode [14]. The initial amplitude was $A = 0.01$ and the resolution was $N_z = 193$ in z, while in x it was increased during the simulation up to $N_x = 128$. *Fig.4a* shows that the primary mode achieves a saturated state after approximately 23 periods. At $t \simeq 75$ the (0,0)-mode has decreased by nonlinear mean flow distortion, while (with $c_g = 0$) the base flow is temporally constant. The repeated grid refinements in x have only a very local effect. *Fig.4b* shows the mode energy versus wavenumber. The resolution in x is sufficient, since the mode energy decays rapidly with k_x and there is no aliasing effect visible. Similar behaviour was observed when the resolution in x was limited to $N_x = 64$. *Fig.5* shows a time series of isoline plots of pressure $\Delta p = p - p_\infty$, density and vorticity $\omega_y = (\partial u/\partial z - \partial w/\partial x)$. The pressure distribution at $t = 0$ has the second-mode appearance with one sign change of the eigenfunction in z (negative contours are dashed). The vorticity is dominated by the mean flow near the critical layer and by the perturbation near the wall. The appearance of Δp does not essentially change until $t \simeq 65$, when a strong pressure gradient begins to build up at $z = 0.3$. Near the wall a strong vorticity layer forms, and close to the critical layer near $z = 1.2$ a vortex develops. This vortex lies in the supersonic region relative to the wall (the sonic line is near $z = 0.6$). At $t \simeq 75$ the pressure gradient at $z = 0.3$ strengthens into what is believed to be a small oblique shock, and the vorticity layer near the wall becomes more intense. The wiggles in the pressure plot at $t \simeq 75$ show that resolution in z is marginal. The problems first appear near the region of high strain in the critical layer, and the simulation had to be stopped at $t \simeq 100$. It appears that the resolution in z has to be improved to accurately calculate nonlinear saturation at $M_\infty = 5.0$ and the relatively large Reynolds number $Re = 10000$.

5. CONCLUSIONS

In this paper we have compared several schemes accurate enough for the direct simulation of transition and turbulence in compressible boundary-layer flows. Chebyshev methods proved to be the most accurate, but have serious time-step limitations when used together with explicit time advancement. Padé schemes were found to be an efficient and flexible alternative, capable of computing the linear instabilities accurately. One of the Padé schemes (scheme (B)) has been used to compute the nonlinear growth and saturation of a 2D second-mode instability in a boundary-layer at $M_\infty = 5.0$.

REFERENCES

[1] *Chapman, D.R., Rubesin, M.W.*, Temperature and Velocity Profiles in the Compressible Laminar Boundary Layer with Arbitrary Distribution of Surface Temperature, J. Aer. Scien. 16, 1949, 547-565.
[2] *Thumm, A., Wolz, W., Fasel, H.*, Numerical Simulation of Spatially Growing Three-Dimensional Disturbance Waves in Compressible Boundary Layers, Laminar-Turbulent Transition, eds. D. Arnal & R. Michel, Springer, New York, 1990, 303-308.
[3] *Kleiser, L., Zang, T.A.*, Numerical Simulation of Transition in Wall-Bounded Shear Flows, Ann. Rev. Fluid Mech. 23, 1991, 495-537.
[4] *Anderson, D.A., Tannehill, J.C., Pletcher, R.H.*, Computational Fluid Mechanics and Heat Transfer, Hemisphere, New York, 1984.
[5] *Sandham, N.D., Reynolds, W.C.*, A Numerical Investigation of the Compressible Mixing Layer, Report No. TF-45, Mechanical Engineering Dept., Stanford University, 1989.
[6] *Sandham, N.D., Reynolds, W.C.*, Three Dimensional Simulations of Large Eddies in the Compressible Mixing Layer, J. Fluid Mech. 244, 1991, 133-158.
[7] *Erlebacher, G., Hussaini, M.Y.*, Non-linear Evolution of a Second Mode Wave in Supersonic Boundary Layers, ICASE Rep. No. 89-15, NASA Langley, 1989.
[8] *Erlebacher, G., Hussaini, M.Y.*, Numerical Experiments in Supersonic Boundary-Layer Stability, Phys. Fluids **A2**, 1990, 94-104.
[9] *Canuto, C., Hussaini, M.Y., Quarteroni A., Zang, T.A.*, Spectral Methods in Fluid Dynamics, Springer, New York, 1988.
[10] *Lele, S.K.*, Compact Finite Difference Schemes with Spectral-like Resolution, CTR Manuscript 107, Stanford University, 1990 (submitted to J. Comp. Phys.).
[11] *Adams, N.A.*, Vergleich von Padé- und Tschebyscheff- Approximation als diskrete Ableitungsoperatoren, DLR Internal Report, IB 221-91 A 16, Göttingen, 1991
[12] *Oliger, J., Sundström, A.*, Theoretical and Practical Aspects of Some Initial Boundary Value Problems in Fluid Dynamics, SIAM J. Appl. Math., 35, 1978, 419-446.
[13] *Thompson K.W.*, Time Dependent Boundary Conditions for Hyperbolic Systems, J. Comp. Phys. 68, 1987, 1-24.
[14] *Mack L.M.*, Boundary-Layer Linear Stability Theory, Special Course on Stability and Transition of Laminar Flow, AGARD Report 709, 1984, 3-1 - 3-81.

Table 1 Comparison of growth rate ω_i with the result of linear stability theory for $M_\infty = 0.3$, $\alpha = 0.27$, $\beta = 0.4$ and $z_{max} = 25$. Initial amplitude is $A = 10^{-4}$ and $z_{0.5}$ is a mapping parameter. Δ is the percentage error from linear results, which are $\omega_{i,2D} = 1.668639 \times 10^{-3}$ and $\omega_{i,3D} = -0.6329422 \times 10^{-3}$.

Scheme	N_z	$z_{0.5}$	$\omega_{i,2d}[10^{-3}]$	$\Delta_{i,2d}[\%]$	$\omega_{i,3d}[10^{-3}]$	$\Delta_{i,3d}[\%]$
A	065	8	1.668583	0.033	−6.329452	0.00047
A	129	10	1.668074	0.034	−6.329399	0.00035
B	065	2	1.669618	0.059	−6.329299	0.00193
B	129	2	1.668789	0.009	−6.329417	0.00007
C	065	2	1.668998	0.022	−6.329349	0.00115
C	129	2	1.668749	0.007	−6.329421	0.00001

Table 2 Comparison of growth rate ω_i with the result of linear stability theory for $M_\infty = 5.0$, $\alpha = 2.217$. Initial amplitude is $A = 10^{-4}$ and $z_{0.5}$, c, d and z_{mv} are mapping parameters. Δ is the percentage error from the linear result, which is $\omega_i = 4.310313 \times 10^{-2}$.

Scheme	N_z	$z_{0.5}$	c	d	z_{mv}	$\omega_i[10^{-2}]$	$\Delta[\%]$
A	033	1.5	0.7	0.85	1.1	4.302534	0.1805
A	065	1.5	0.7	0.85	1.1	4.310338	0.0006
A	129	1.5	0.7	0.85	1.1	4.310307	0.0002
B	033	1.5	1.0	0.80	1.1	4.332859	0.5231
B	065	1.5	1.0	0.80	1.1	4.351698	0.9601
B	129	1.5	1.0	0.80	1.1	4.314990	0.1085
C	033	1.5	1.0	0.80	1.1	4.572299	6.0781
C	065	1.5	1.0	0.80	1.1	4.327678	0.4029
C	129	1.5	1.0	0.80	1.1	4.310434	0.0028

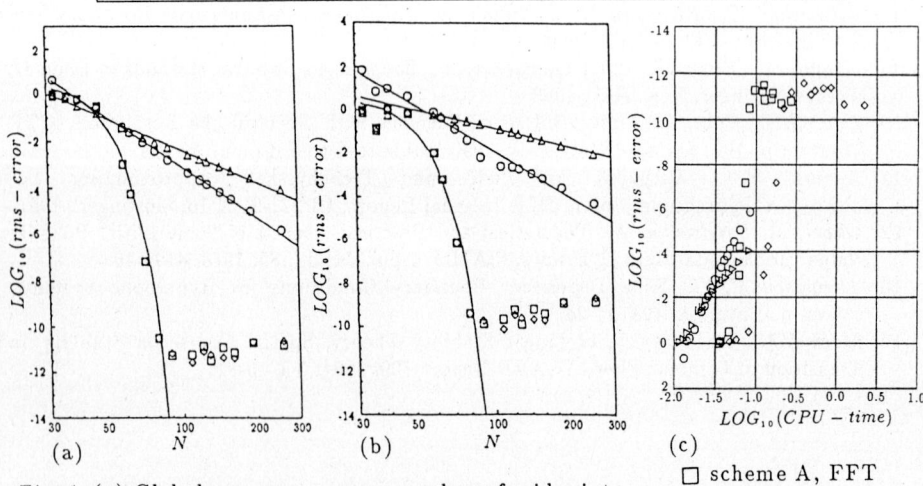

Fig. 1 (a) Global rms-error versus number of gridpoints.
(b) Boundary rms-error versus number of gridpoints.
(c) CPU-time versus global rms-error.

□ scheme A, FFT
◊ scheme A, matrix
△ scheme B
o scheme C

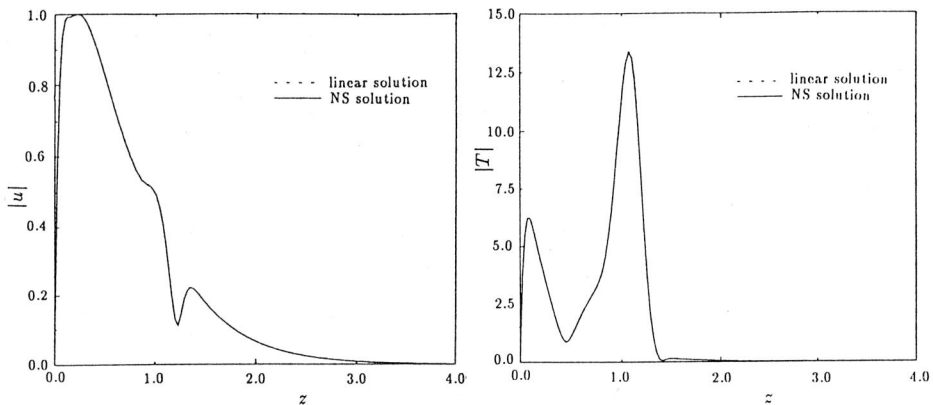

Fig. 2 Absolute value of u(1,0)-mode and T(1,0)-mode compared with linear eigenfunction. Navier-Stokes solution starting from random noise.

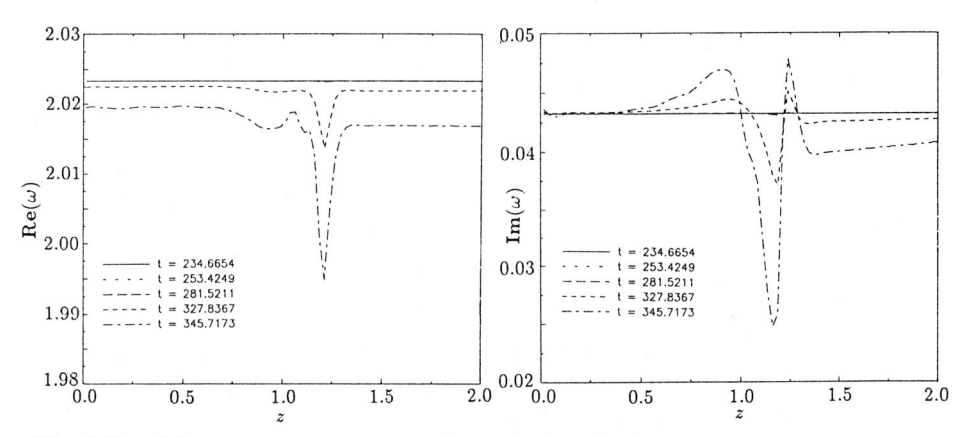

Fig. 3 Local frequency and growth rate of u(1,0)-mode.

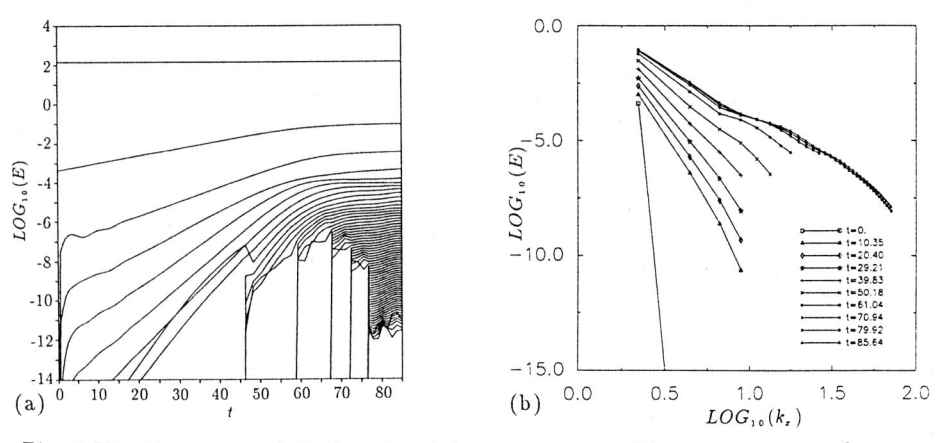

Fig. 4 Kinetic energy of (1,0)-mode: (a) versus time t, (b) versus wavenumber $k_x = k_1 \cdot \alpha$ (k_1 are the Fourier modes in x).

(a) Pressure.   (b) Density.   (c) Vorticity.

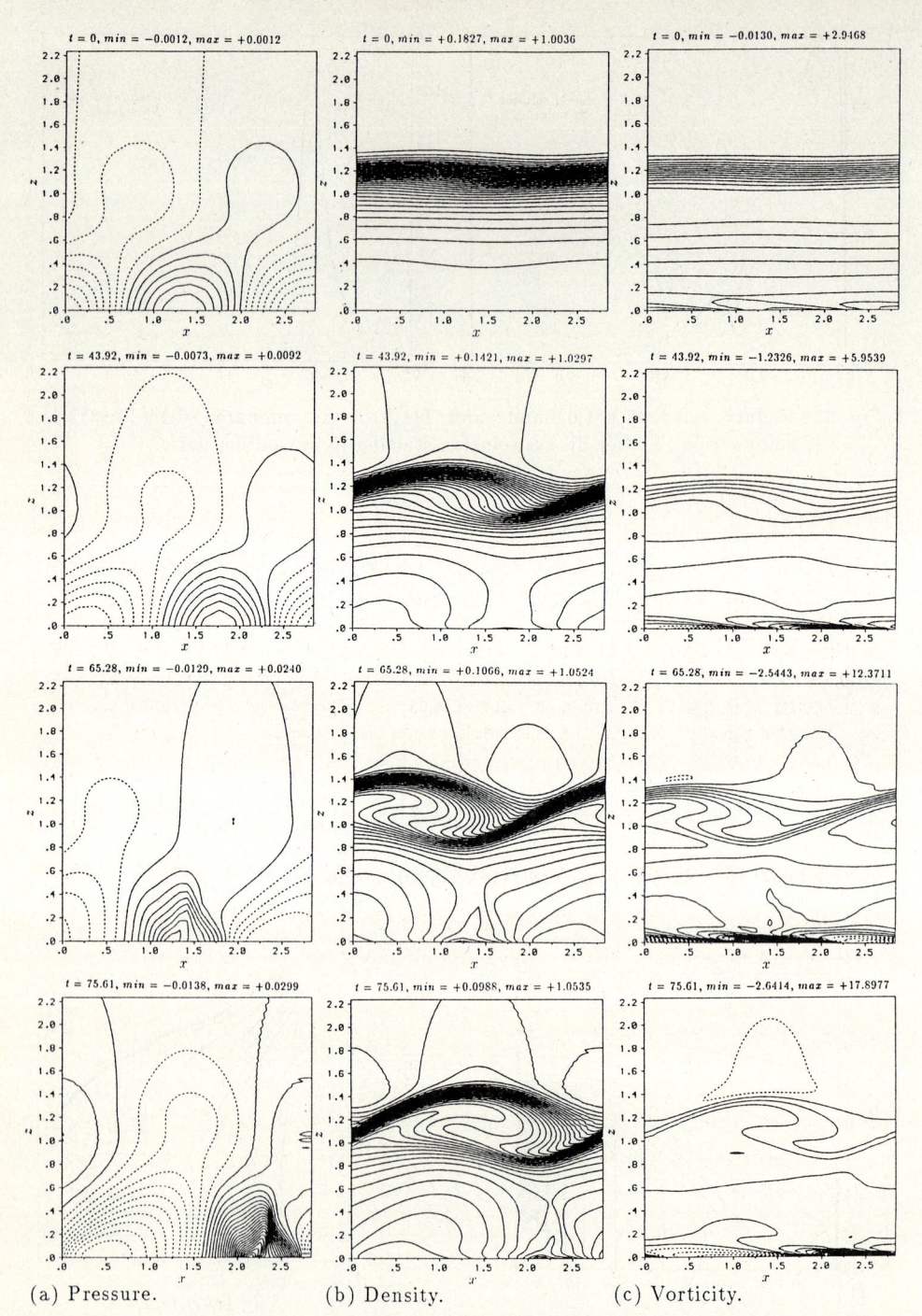

Fig. 5 Lines of (a) constant pressure $\Delta p = p - p_\infty$, (b) constant density ρ and (c) constant spanwise vorticity $\omega_y = \left(\frac{\partial u}{\partial z} - \frac{\partial w}{\partial x}\right)$.

DIRECT SIMULATION OF INCOMPRESSIBLE, VISCOUS FLOW THROUGH A ROTATING SQUARE CHANNEL

T.H. Lê, J. Ryan, K. Dang-Tran

ONERA, B.P. 72, 92322 Châtillon Cedex, France.

SUMMARY

This paper deals with the direct simulation of the three-dimensional unsteady flow of an incompressible viscous fluid driven by a pressure gradient through a square channel that rotates about an axis perpendicular to the channel roof. The dimensionless, three-dimensional, time-dependent Navier-Stokes equations, written in a frame of reference rotating with the channel, are solved by a finite difference method with non-staggered variable arrangement. Central difference schemes are used for temporal and spatial discretization and are both second-order accurate. Pressure is obtained by solving a Poisson-like equation with a discrete divergence velocity minimization algorithm. An implicit residual smoothing technique is used to avoid restrictive stability conditions. The incompressibility constraint is solved, preserving numerically the mass conservation. Results have been obtained for an extended range of dynamical parameters. In a known typical steady state, the two and four vortex patterns were found. As the Rossby number is increased to 3 and the Ekmann number decreased to .001 a new state, at the limit of the onset of turbulence, is obtained at Reynolds number Re = 3000.

INTRODUCTION

Secondary flows frequently occur in rotating machinery : examples are coolant flow within turbine blades, flow within pipes used in components such as heat exchangers and flow inside impellers of centrifugal pumps. The pressure-driven viscous flow through a square duct, with the pipe rotating about an axis perpendicular to its own, can serve as an ideal representation of such flows because of its simple geometry. The secondary flow pattern is generated and sustained by the Coriolis force introduced by the duct rotation. Two dynamical parameters govern the structure of the flow in the rotating duct : the Ekman number, defined as $Ek = \dfrac{\nu}{b^2 \Omega}$, which represents the ratio of viscous to Coriolis force and the Rossby number, defined as $Ro = \dfrac{W}{b\Omega}$, which represents the ratio of the convective acceleration to the Coriolis force. Here Ω is the magnitude of the angular velocity, b is the length scale representing the channel width, ν is the kinematic viscosity, and W is the axial velocity scale. The Reynolds number Re is then equal to the ratio of the Rossby number to the Ekman number. Fig. 1 shows the geometry and the flow definitions, where L is the channel length.

Comprehensive studies of secondary flows in a rotating square duct may be found in [1, 2]. In these papers the bifurcation structure of these flows are studied by numerically solving the two-dimensional equations of motion obtained in the approximation of axial (translation) invariance of the velocity field for $0 \leq Ro \leq 15$, $0.001 \leq Ek \leq 0.1$, $0 \leq Re \leq 1500$ and $0 \leq Ro \leq 5$, $Ek = 0.01$, $0 \leq Re \leq 500$ respectively.

In the present study, three-dimensional calculations of the unsteady flow field were performed using the *PEGASE* code, developed at ONERA for the direct simulation of separated flows [3] and turbulent flows [4], for an extended range of dynamical parameters up to a Reynolds number Re = 3000. . This code has been modified to take into account the Coriolis terms for the present application. A new numerical scheme has been introduced for the calculation of the continuity equation.

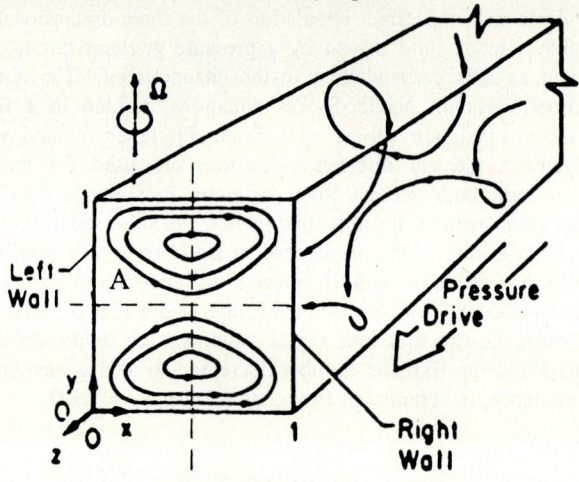

Fig. 1 : Geometry and qualitative character of the rotating square channel flow

NUMERICAL METHOD

Governing equations

The equations governing the flow of an incompressible viscous fluid are the Navier-Stokes equations in a primitive variable formulation, written here in rotational form. The dimensionless equations written in a frame of reference rotating with the channel are :

$$\frac{\partial U}{\partial t} + (Ro\ \nabla \times U + 2\omega) \times U + \nabla Q = Ek\ \nabla^2 U + F$$

$$\nabla \cdot U = 0$$

where $U = U(x, y, z, t) = (u, v, w)^T$ is the three-dimensional velocity field, $\omega = (0, 1, 0)^T$ is the angular velocity, and $F = (0, 0, 1)^T$ is an external force related to an imposed mean pressure gradient $\frac{\partial \bar{P}}{\partial z}$. Through its potential the centrifugal term has been combined with the total pressure in the variable Q ($Q = P + \frac{1}{2} U^2 - \frac{1}{2}(x^2+y^2+z^2)$) . The characteristic scales are $t_c = \frac{1}{\Omega}$ for time, $L_c = L = b$ for length, $P_c = -L \frac{\partial \bar{P}}{\partial z}$ for pressure, $W = \frac{P_c}{\Omega L} = -\frac{1}{\Omega} \frac{\partial \bar{P}}{\partial z}$ for velocity.

In order to study the flow field with grid refinement, generalized coordinates (ξ, η, ζ) are introduced but, as in the case without refinement, the dependent variables remain the cartesian velocity components and the scalar field Q . The geometry of a rectangular duct allows us to use a simple orthogonal grid, thus x (resp. y and z) depends only on ξ (resp. η and ζ).

Grid generation

Two-dimensional grid clustering is employed in order to resolve adequately the velocity gradients in the boundary layers. Grid points are also concentrated in regions where the vorticity changes rapidly, then the following transformation are introduced :

$$x = 0.5 \left[1 + \frac{\tanh(\alpha_1 \xi)}{\alpha_0} \right] ; \quad \xi = -1 + \frac{2(i-1)}{i_{max}-1} ; \quad 1 \leq i \leq i_{max}$$

$$y = 0.5 \left[1 + \frac{\tanh(\alpha_1 \eta)}{\alpha_0} \right] ; \quad \eta = -1 + \frac{2(j-1)}{j_{max}-1} ; \quad 1 \leq j \leq j_{max}$$

$$z = \zeta ; \quad \zeta = \frac{k-1}{k_{max}} ; \quad 1 \leq k \leq k_{max}$$

$\alpha_0 = 0.9 \qquad \alpha_1 = \operatorname{atanh} \alpha_0$.

Transformed equations

This partial transformation of equations yields for the x -momentum equation :

$$u_t + w \left[Ro \left(\zeta_z u_\zeta - \xi_x w_\xi \right) + 2 \right] - v \, Ro \left(\xi_x v_\xi - \eta_y u_\eta \right) + \xi_x Q_\xi =$$
$$Ek \left(\xi_x^2 u_{\xi\xi} + \eta_y^2 u_{\eta\eta} + \zeta_z^2 u_{\zeta\zeta} + \xi_{xx} u_\xi + \eta_{yy} u_\eta \right)$$

and analogous expressions for y and z -momentum equations. In the z-direction a constant grid spacing is used.

In contrast to the momentum equations the continuity equation is written in strong conservation form. It is because of this formulation that numerical mass conservation is achieved.

$$(J \, \xi_x \, u \,)_\xi + (J \, \eta_y \, v \,)_\eta + (J \, \zeta_z \, w \,)_\zeta = 0$$

where J is the Jacobian of the transformation from the stretched coordinates to the cartesian coordinates, given by the simple expression $J = \xi_x \, \eta_y \, \zeta_z$.

Discretization procedure

Transformed equations, including the continuity equation, are discretized on a non-staggered grid : velocity components and Q are located at the same node which greatly alleviates the coding burden and enhances the applicability of the method. The numerical scheme is based on a method proposed in [5] with some improvements concerning the numerical scheme used for the discrete continuity equation. The convective and diffusive terms are both discretized in time by an explicit second order accurate Adams-Bashforth scheme. Space derivatives are approximated by second order accurate finite difference schemes, centered at the inner nodes and non-centered at boundary nodes. An implicit residual technique is used to reduce stability constraints on the time step when small grid spacing is employed. The discretized equations are written as follows :

(1) $\quad \mathbf{A} \left(\dfrac{U^{n+1} - U^n}{\Delta t} \right) + \mathbf{C} (U^*) + \mathbf{G} \, Q^{n+\frac{1}{2}} = Ek \, \mathbf{L} (U^*)$

(2) $\quad \mathbf{D} (T \, U^{n+1}) = 0.$

- $Q^{n+\frac{1}{2}}$ is defined at time $(n+\frac{1}{2}) \, \Delta t$,
- $U^* = 1.5 \, U^n - 0.5 \, U^{n-1}$,
- $\mathbf{G} = (\xi_x \delta_\xi , \eta_y \delta_\eta , \zeta_z \delta_\zeta)$, where δ is the classical second order central-difference operator ,
- $\mathbf{C} = (C_u , C_v , C_w)$ are the "convective" terms ,
- $\mathbf{L}$ is a seven point discrete operator for diffusive terms ,
- $\mathbf{D} = (\delta_\xi , \delta_\eta , \delta_\zeta)$ the discrete divergence operator ,
- $T = (J \, \xi_x , J \, \eta_y , J \, \zeta_z)$,
- $\mathbf{A}$ is an implicit smoothing operator .

Boundary conditions

The Navier-Stokes equations, for an incompressible flow, require boundary conditions on velocity, which are sufficient to allow the determination of both velocity and pressure. No-slip boundary conditions are assigned at the solid walls. In the driven pressure gradient direction a periodic condition is used, i.e. the value at the point just outside the boundary is equal to the value which is on the opposite boundary, derivatives at these boundaries are approximated with a centered formula.

The problem specification is completed by prescribing initial conditions for the velocity field. Two kinds of initial conditions are used. The first one considers the flow is at rest initially and the second one consists in imposing the semi-analytical solution of [6] for the stationary laminar flow in a square duct. Computational results show no-dependency in regards to these initial conditions.

Pressure calculation

Computing the pressure is usually the most time consuming part of the method [7]. In the present work, a new scheme has been developed to solve the elliptic pressure equation, which yields an accurate solution of the mass conservation equation.

Applying operator ($\mathbf{D} T$) to equation (1), the following equation is obtained :

$$(3) \quad \mathbf{D} (T\ U^{n+1}) = \mathbf{D} (T\ U^n) + \Delta t\ \mathbf{D}\ T\ \mathbf{A}^{-1} [\ Ek\ \mathbf{L} (U^*) - \mathbf{C} (U^*) - \mathbf{G}\ Q^{n+\frac{1}{2}}].$$

Should the left hand side of (3) be null which is equivalent to the following Poisson like equation in $Q^{n+\frac{1}{2}}$:

$$(4) \quad - \mathbf{D}\ T\ \mathbf{A}^{-1}\ \mathbf{G}\ Q^{n+\frac{1}{2}} = - \frac{\mathbf{D} (T\ U^n)}{\Delta t} - \mathbf{D}\ T \mathbf{A}^{-1} [\ Ek\ \mathbf{L} (U^*) - \mathbf{C} (U^*)],$$

then the mass conservation equation (2) will be satisfied.

The solution of (4) is obtained by means of eliminating the spurious modes and minimizing iteratively a discrete norm of the velocity divergence.

These spurious modes consist of the discretization modes due to the choice of a non-staggered grid (i.e : velocity and pressure are computed at the same nodes) and of the constant physical mode. The former (discretization) modes may be suppressed by lowering the number of degrees of freedom in pressure : pressure is computed only on inner nodes. Because of the later mode, equation (4) must be minimized in a space modulo constants, this implies that the residue $\mathbf{D} (T\ U^{n+1})$ may be a constant. This constant is annulled because the central difference scheme applied to the conservative divergence formulation is used with an odd number of points in the x-y Dirichlet directions. This scheme satisfies a discrete compatibility condition , for more details see [8, 9]. A first application was done in [10] for a plane channel flow. As a result , because an iterative residual solver [11] is used , the constant pressure mode is automatically suppressed.

RESULTS AND DISCUSSION

The secondary flow structure of the square rotating duct results from a complex interaction between convective inertial forces, viscous and Coriolis forces. A bifurcation structure at low Reynolds number has been numerically evidenced for this flow in [1] by exploring a wide range of Rossby numbers ($0 \leq Ro \leq 15$) . A transition from a two vortex flow to a four vortex flow pattern takes place for $1.24 \leq Ro \leq 1.76$ and $Ek = 0.01$. This solution branch has been confirmed in [2] at the same Ekman number, for $0 \leq Ro \leq 5$ and completed : up to five solution branches have been put into evidence for $3 \leq Ro \leq 5$.

Our purpose in the present study is to confirm the existence of the two flow patterns considered as test cases. A third simulation at higher Reynolds number is also carried out in order to assess the feasability of the three-dimensional simulation technique for future engineering computations.

Calculation details

All calculations were performed on a Cray XMP 416 computer. Three meshes have been used, depending on flow conditions : a 31x31x11 grid for $Ro = 2$, $Ek = 10^{-2}$ ($Re = 200$) , a 61x61x5 grid for $Ro = 3$, $Ek = 10^{-2}$ ($Re = 300$) , and a 91x91x5 grid for $Ro = 3$, $Ek = 10^{-3}$ ($Re = 3000$). As will be seen later, computed flows are quasi two-dimensional up to $Re = 3000$. As a consequence a minimum of 5 grid points has been used in the z-direction. In the x-y grid plane, a minimum spacing of 0.00356 is achieved near the solids walls and a maximum spacing of 0.01817 near the center of the cross section. For this case the time step size is 0.003 and the total number of time steps used is 1400. The CPU time varies from 90 seconds for the first case to 3000 seconds for the third case .

Results

The first case is related to the simulation of a flow at moderate Rossby number $Ro = 2$ and at Ekmann number equal to 0.01 corresponding to a Reynolds number equal to 200 . The flow field in a cross-section (not shown here) exhibits a two-vortex structure as that can be observed in [1]. This structure is stable in time and the two vortices remain nearly centered on the midplane parallel to the rotation axis. At higher Rossby number, the two-vortex flow is unstable.

The second case concerns the temporal transition from a two-vortex to a four-vortex structure. With the previous Ekman number, solutions were computed at a Rossby number equal to 3 which corresponds to the upper branch bifurcation of the diagram depicted in [1].

Figure 2-a shows the time evolution of the u velocity component at the point A (0.03,0.5,0.0) of fig. 1. The flow structure evolves from a two vortex pattern ($u \leq 0$) to a four vortex pattern ($u \geq 0$) in a way similar to those found in [1] , although more oscillatory. Figure 2-b shows the mass flow rate time evolution. It evolves from zero (initial flow at rest) to a constant value corresponding to the steady state four vortex pattern.

Figure 3 shows the velocity field in the (Ox, Oy) plane at times t_0 and t_1 (see figure 2) corresponding to a transient two vortex pattern and to a steady four vortex pattern.

Figure 4 shows the axial velocity distributions in the (Ox, Oy) plane, along the vertical centerplane and along the horizontal centerplane at the same times as mentioned above. Results obtained are very similar to those in [1]. The axial velocity shows a rectilinear distribution across the core when the two-vortex structure is present, it increases in the center of the duct with a local maximum as the four-vortex structure appears.

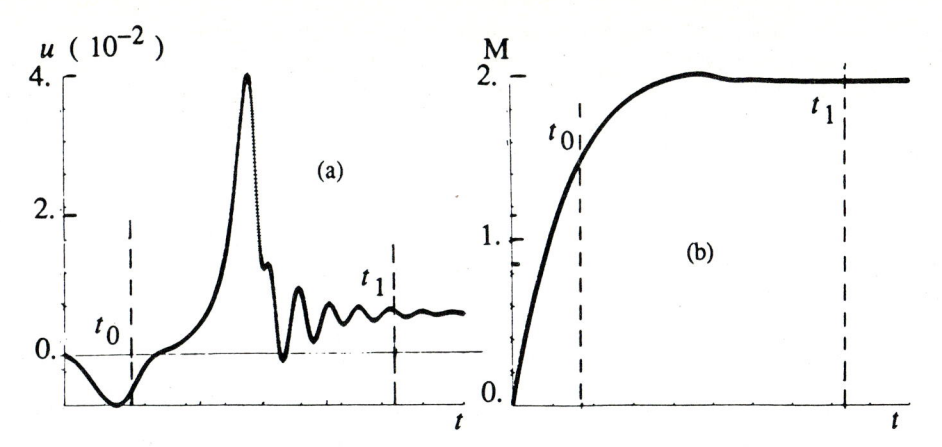

Fig. 2 : Transition from two-vortex to four-vortex structure flow

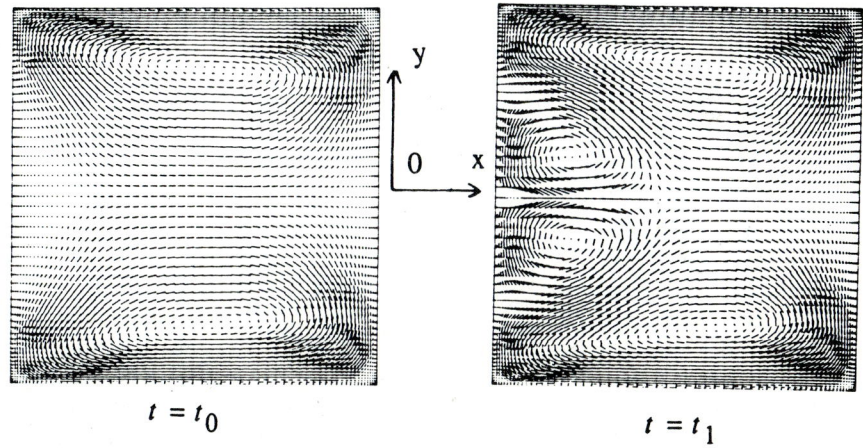

$t = t_0$ $t = t_1$

Fig. 3 : Visualization of velocity fields

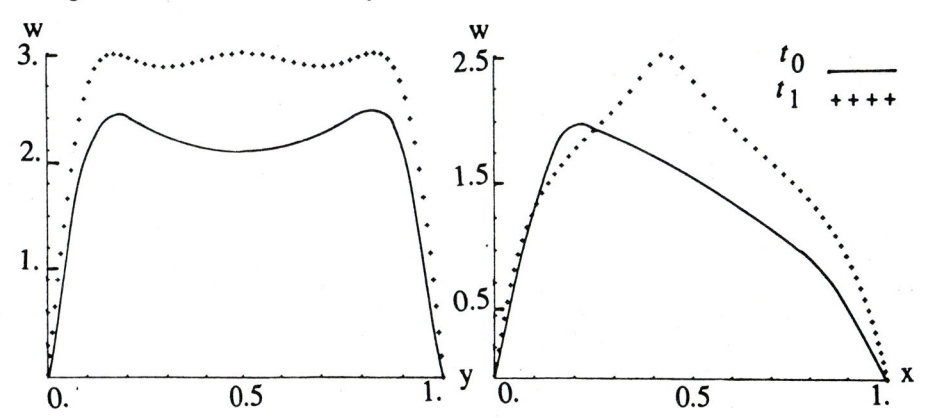

t_0 ———
t_1 ++++

Fig. 4 : Axial velocity profiles

The third case is a preliminary attempt to predict flow regimes at lower Ekman number and results shown in figure 5 for $Ro = 3$, $Ek = 0.001$ ($Re = 3000$) are, in our knowledge, an original contribution. The flow structure becomes more complex : the "primary" four-vortex structure remains but six secondary vortices appear near the corners. Furthermore, the hypothesis of axial invariance of the velocity field is no more verified : the axial flow structure being on the verge of destabilization, a complete three-dimensional computation is necessary.

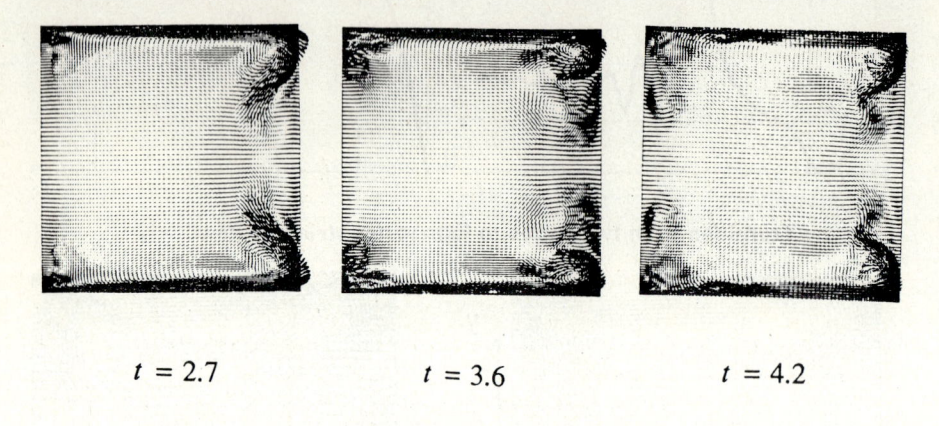

$t = 2.7$ $\qquad\qquad\qquad t = 3.6 \qquad\qquad\qquad t = 4.2$

Fig. 5 : Unsteady low Ekman number flow

CONCLUSIONS

A modified version of the *PEGASE* code has been developed and used for the calculations of unsteady viscous flows through a rotating square channel. Numerical results obtained assess the feasability of the method for accurate simulations of such flows and confirm the necessity of complete three-dimensional computations at high Reynolds numbers for engineering purposes.

In order to determine the stability of the third flow structure without any doubt and to clarify the complex mutual influence between the different forces in presence, further calculations are required in terms of spacial accuracy and long-term integration. In near future, these tasks will be done with a full conservative scheme [12] with the long term objective to reach Reynolds numbers of industrial interest about 30,000.

ACKNOWLEDGMENTS

This work was supported by the Direction des Recherches Etudes et Techniques. The authors express their thanks to Mr. D. Intès for his contribution to run the code.

REFERENCES

[1] H.S. Kheshgi, L.E. Scriven : "Viscous flow through a rotating square channel" Phys. Fluids 28(10), October 1985, pp. 2968-2979.

[2] K. Nandakumar, H. Raszillier, F. Durst : "Flow through rotating rectangular ducts" Phys. Fluids A 3(5), May 1991, pp. 770-781.

[3] B. Troff, T.H. Lê, T.P. Loc : "A numerical method for the three-dimensional unsteady incompressible Navier-Stokes equations" J. Compt. and Appl. Math., Vol. 35, 1991, pp. 311-318.

[4] K. Dang-Tran, G. Veber : " Direct Numerical Simulation of Laminar and Turbulent Flows in a Diverging Channel" AIAA 22nd Fluid Dynamics, Plasma Dynamics & Laser Conference, Honolulu, Hawai (USA), June 24-26, 1991. AIAA Paper 91-1776. T.P. ONERA n^0 1991-

[5] K. Dang-Tran, Y. Morchoisne : "Numerical methods for direct simulation of turbulent shear flows." Lecture Series 1989-3 on "Computational Fluid Dynamics" V.K.I., Rhodes-St-Genese (Belgium), February 6-10, 1989. T.P. ONERA n^0 1989-12.

[6] T. Tatsumi, T. Yoshimura : "Stability of the laminar flow in a rectangular duct" J. Fluid Mech. , vol 212, pp. 437-449, 1990.

[7] P.G. Esposito : "Numerical Simulation of a Three-Dimensional Lid-driven Cavity Flow" GAMM Workshop, Paris (France), June 13-14, 1991.

[8] T.H. Lê, Y. Morchoisne : "Traitement de la pression en incompressible visqueux." C. R. Acad. Sci. Paris, t. 312.Série II, p.1071-1076, 1991.

[9] J. Ryan, B. Troff, T.H. Lê, Y. Morchoisne : "Pressure calculation in incompressible flow" (to be submitted to Int. J. Numer. Methods Fluids).

[10] S. Rida, K. Dang-Tran : "Direct Simulation of Turbulent Pulsed Plane Channel Flows" 8th International Symposium on "Turbulent Shear Flow", Munich (Germany), September 9-11, 1991.

[11] J. Ryan, T.H. Lê, Y. Morchoisne : "Panels code solvers" 7th GAMM, Louvain (Belgique), 9-11 Septembre 1987.

[12] B. Cantaloube, T.H. Lê : "Numerical Simulation of Three-Dimensional Driven Cavity Flow" GAMM Workshop, Paris (France), June 13-14, 1991.

ANALYSIS OF SEMI-IMPLICIT TIME INTEGRATION SCHEMES FOR DIRECT NUMERICAL SIMULATION OF TURBULENT CONVECTION IN LIQUID METALS

M. Wörner, G. Grötzbach
Kernforschungszentrum Karlsruhe
Institut für Reaktorentwicklung
Postfach 3640, D-7500 Karlsruhe, Germany

SUMMARY

Fully explicit time integration schemes are very inefficient for numerical simulation of diffusion dominated problems. In case of natural convection flow in liquid metals an implicit treatment of the thermal diffusion terms allows for the use of substantially increased time steps without involving loss of physically relevant information. Two suitable semi-implicit time integration schemes are investigated analytically by a Von Neumann stability analysis and a spectral analysis of the numerical error. Numerical solutions by the semi-implicit schemes are compared to the exact solution of a 1D linear test problem. The results show the crucial influence of the discretization ratio $\lambda = \Delta t/\Delta x$ on the accuracy of the numerical solutions. First 3D time dependent numerical simulations of natural convection in liquid metals with the semi-implicit time integration schemes confirm the theoretically estimated gain in the time step width and result in CPU-time savings up to a factor of 50 compared to the fully explicit scheme.

1. INTRODUCTION

New developments for sodium cooled fast breeder reactors aim at more inherent safety features. The decay heat, e.g., shall be removed by natural convection only. In this context many experiments with model fluids are performed. Computer codes using turbulence models are developed to extrapolate the experimentally based knowledge to reactor conditions. To improve and to calibrate the problematic turbulence models for pure natural convection in liquid metals we intend to provide statistical turbulence data from direct numerical simulations.

In the code TURBIT [1] the time dependent, three-dimensional Navier Stokes equations and the thermal energy equation are solved in simple channel geometries using a finite volume method. The time integration is done by an explicit Euler-Leapfrog scheme. TURBIT has been successfully used for direct numerical simulation of different heat transfer problems, as for example Rayleigh-Bénard convection in air [2] and natural convection in an internally heated fluid layer [3]. However, application of the code to natural convection in liquid metals leads to enormous CPU-time requirements. The inefficiency of the code for this type of flow is due to the fully explicit time integration scheme. For numerical stability it requires the use of much smaller time steps than would be physically necessary to resolve even the highest frequencies of turbulence in time. Similar problems arise in the numerical simulation of isothermal channel flow with spectral methods [4]. Here very fine meshes have to be used near the walls to resolve the viscous boundary layers. In those cases the viscous diffusion terms are treated implicitly to avoid a physically irrelevant time step restriction.

In this paper we investigate two semi-implicit time integration schemes for the thermal energy equation. These are the Adams-Bashforth Crank-Nicolson scheme and the Leapfrog Crank-Nicolson scheme. In section 2 we will show that in case of direct numerical simulation of natural convection in liquid metals the strong time step restriction of explicit schemes is overcome by the implicit treatment of the thermal diffusion terms. In section 3 we will carry out a Von Neumann stability analysis and a spectral analysis of the numerical error. Comparison of numerical solutions of a one-dimensional linear test problem with the exact solution will be done in section 4. We will discuss the results in section 5 and make some remarks about the realization and practical experience with the semi-implicit schemes in TURBIT in section 6.

2. NECESSITY FOR SEMI-IMPLICIT TIME INTEGRATION

The stability criterion of the fully explicit time integration scheme used in TURBIT can be written, by using the summation convention, as

$$\Delta t \leq \left(\frac{|u_i|_{max}}{\Delta x_i} + 4 \frac{Max(v,a)}{\Delta x_i^2} \right)^{-1}. \qquad (1)$$

U_i denotes the components of the velocity vector and v and a the viscous and thermal diffusivity, respectively.

Liquid metals are characterized by very low Prandtl numbers $Pr = v/a$ (e.g. $Pr = 0.025$ for mercury and $Pr = 0.006$ for liquid sodium) and thus by a very high ratio of thermal to viscous diffusivity. So the temperature field in natural convection of liquid metals is governed by a large thermal conductivity allowing only for large scale structures and thick thermal boundary layers. In contrast the velocity field has very small spatial structures and very thin boundary layers near walls. For direct numerical simulation of turbulence it is essential to choose a grid which resolves all physically relevant length scales of the flow. Therefore we are enforced by the velocity field for the use of very fine mesh cells Δx_i. Thus in the stability criterion (1) the diffusion type terms become dominant. Because of the very low Prandtl-number of liquid metals it is not the viscous, but the highly efficient thermal diffusion process which requires the use of very small time steps. This is really a kind of paradoxon since the temperature field does not show any rapid variations which would justify the use of such small time steps. Substantially larger time steps (typically of one to two orders of magnitude) can be reached when the thermal diffusivity is removed from the stability criterion (1). This can be achieved by treating the diffusive terms in the thermal energy equation implicitly and the convective terms and the complete momentum equation still explicitly.

3. ANALYTICAL INVESTIGATION OF SEMI-IMPLICIT SCHEMES

Two semi-implicit time integration schemes which are suitable for diffusion dominated problems are recommended e.g. in [5]. Both handle the diffusive terms $L = a\nabla^2 T$ by the implicit Crank-Nicolson scheme, CN, whereas for the nonlinear convective terms $N = u\nabla T$ the explicit Adams-Bashforth, AB, or the Leapfrog scheme, LF, is used, respectively:

$$\text{ABCN–scheme:} \quad \frac{T^{n+1} - T^n}{\Delta t} = -\frac{1}{2}\left(3N^n - N^{n-1}\right) + \frac{1}{2}\left(L^{n+1} + L^n\right), \quad (2)$$

$$\text{LFCN–scheme:} \quad \frac{T^{n+1} - T^{n-1}}{2\Delta t} = -N^n + \frac{1}{2}\left(L^{n+1} + L^{n-1}\right). \quad (3)$$

Here T denotes the temperature and n the time level. Both schemes are of second order in time and exhibit only small numerical diffusion.

For simplicity we consider the following one-dimensional linearized model for the thermal energy equation

$$\frac{\partial T}{\partial t} + u_o \frac{\partial T}{\partial x} = a \frac{\partial^2 T}{\partial x^2} \qquad a, u_o = const. \quad (4)$$

An exact solution is

$$T_{ex}(x, t) = e^{-ik(x - u_o t)} \cdot e^{-ak^2 t} \quad (5)$$

where $i = \sqrt{-1}$ and k is representing a wavenumber.

3.1 Von Neumann Stability Analysis

For both schemes the time step criterion for the linearized problem can be derived by a Von Neumann stability analysis. It's main idea is to carry out a Fourier decomposition of the error, leading to an amplification matrix and thus to the eigenvalues of the numerical scheme. Then the stability criterion results from the demand that the spectral radius of the amplification matrix has to be less equal unity. This ensures that no single harmonic component of the error may grow in time. Further details of the method are given in [6].

$D = \frac{a \Delta t}{\Delta x^2}$

$C = \frac{u_o \Delta t}{\Delta x}$

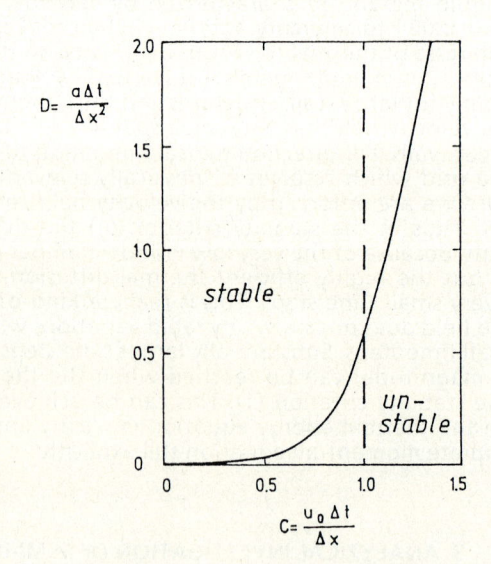

Fig. 1: Curves of stability for the ABCN- (solid line) and the LFCN-scheme (dashed line) according to [6].

The Von Neumann stability analysis of the ABCN- and LFCN-time integration schemes for the problem of equation (4) results in the curves of stability given in figure 1 where the Courant number $C = (u_0 \Delta t)/\Delta x$ and the diffusion number $D = (a\, \Delta t)/(\Delta x^2)$ are used. The stability criterion of the LFCN-scheme is just the well known Courant-Friedrichs-Lewy (CFL) condition $C \leq 1$. Thus the time step is, as desired, unaffected by the thermal diffusivity. In contrast the ABCN-scheme requires always a certain minimal diffusivity for numerical stability. On the other hand it is not limited to the CFL-condition for higher values of D. However, this is not an advantage compared with the LFCN-scheme since in TURBIT we have to meet the CFL-condition for reasons of numerical stability of the explicit integration of the momentum equations.

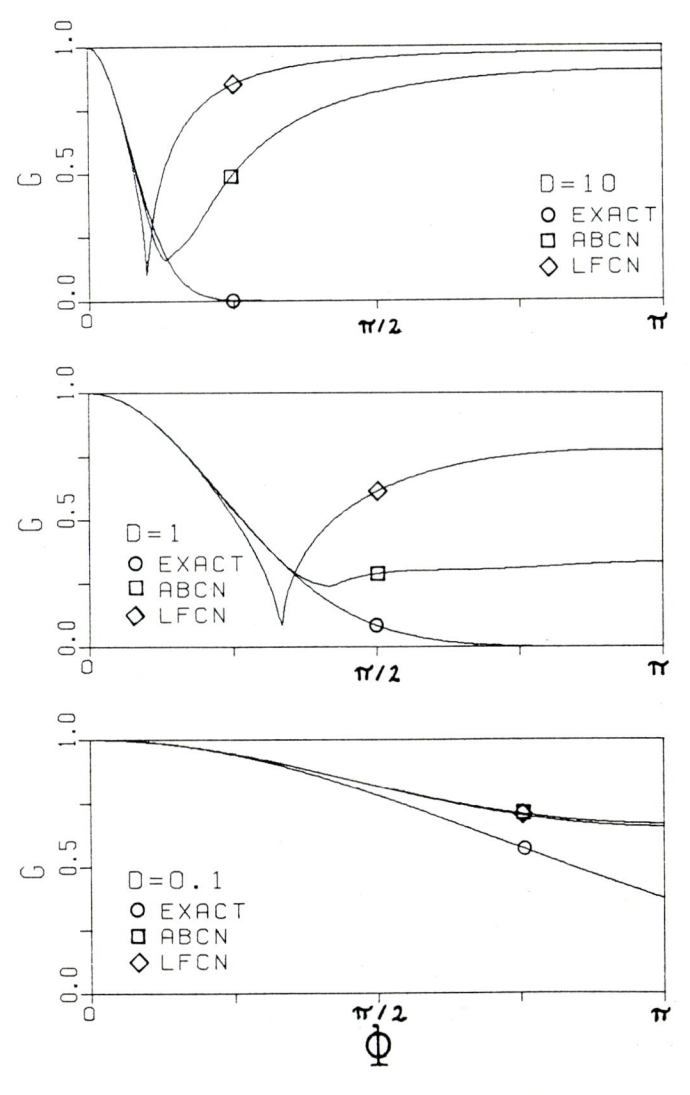

Fig. 2: Spectral analysis of the numerical error of ABCN- and LFCN-scheme for the Courant number $C = 0.1$ and three diffusion numbers.

3.2 Spectral Analysis of the Numerical Error

The damping in time of the amplitude of a spatial wave with wavenumber k follows from equation (5) to be

$$G_{ex} = \left| \frac{T(x, t + \Delta t)}{T(x, t)} \right| = e^{-ak^2 \Delta t} = e^{-D\phi^2}, \tag{6}$$

where $\phi = k \Delta x$ is representing a phase angle. By a spectral analysis of the numerical error [6] this physical damping due to diffusion is compared to that of a numerical scheme, which is falsified by numerical diffusion, for a wide range of wavenumbers. The damping in time by a numerical scheme is characterized by the maximum of the absolute value of its complex eigenvalues.

The results for a Courant number C = 0.1 and three different diffusion numbers D = 0.1, D = 1 and D = 10 are given in figure 2. The numerical schemes are working very well for small values of ϕ. This means that the time characteristics of structures associated with large wavelengths are well approximated by both numerical schemes. In contrast figure 2 shows that small scaled spatial structures ($\phi = \pi$ corresponds to the lowest resolvable wavelength 2 Δx) are damped too weakly by both schemes. Furthermore a clear tendency is outlined that the wavenumber range of which the time characteristic is well approximated becomes more and more limited as the diffusion number increases. The ABCN scheme works slightly better at high values of D than the LFCN scheme. Nevertheless it becomes evident that, with regard to the accuracy of the numerical results, a limitation with respect to the diffusion number will be necessary for both schemes.

However, one should keep in mind that we want to use these schemes for the solution of the thermal energy equation for turbulent natural convection in liquid metals. As stated above, for this type of flow the temperature field contains only large spatial structures. Thus both schemes may be expected to yield "physical" results.

4. NUMERICAL EXPERIMENTS

To get more information about the accuracy of the ABCN- and LFCN-scheme both are used for numerical solution of equation (4) on a domain $0 \leq x \leq L$. Corresponding to the initial condition

$$T(x, 0) = \sin(k \cdot x) \tag{7}$$

and the boundary conditions

$$T(0, t) = -\sin(ku_0 t) \cdot e^{-k^2 at}$$
$$T(L, t) = \sin[k(L - u_0 t)] \cdot e^{-k^2 at} \tag{8}$$

there exists the exact solution

$$T_{ex}(x, t) = \sin[k(x - u_0 t)] \cdot e^{-k^2 at} \tag{9}$$

which can be used for comparison to the numerical solutions.

The LFCN- and ABCN-scheme are both three level schemes and thus a new time plane is calculated using values of two past time planes. Therefore both schemes cannot be used for the first integration step where only one time plane - corresponding to the initial condition - is available. Therefore a two level scheme has to be used for the first integration step. Here a semi-implicit Euler scheme is used which is only of first order in time. The results we present in this paper are gained using the following parameters: $u_o = 0.1, a = 0.25, L = \pi$ and $k = 2$. The number of mesh cells is $M = 50$ resulting in a mesh width $\Delta x = L/(M-1) \approx 0.064$. To judge on the accuracy of the numerical solutions the absolute value of their deviation from the exact solution will be shown in a three-dimensional representation dependent on space and time:

$$z(x_i, t_i) = |T_{ex}(x, t) - T_{num}(x_i, t_i)|. \tag{10}$$

ABCN-scheme ($\lambda = 1$): $z_{max} = 4.1 \cdot 10^{-4}$

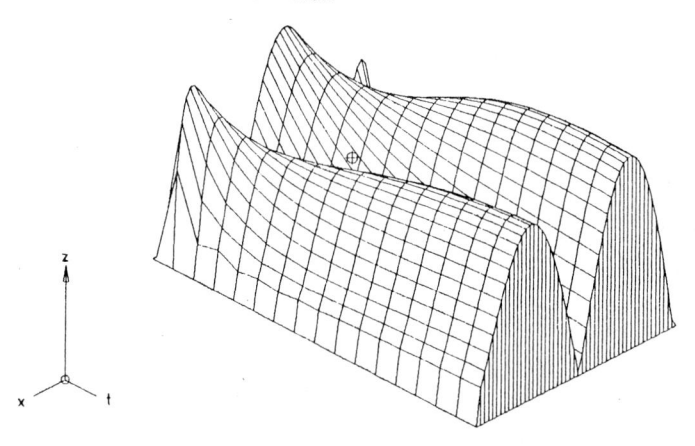

LFCN-scheme ($\lambda = 1$): $z_{max} = 3.9 \cdot 10^{-4}$

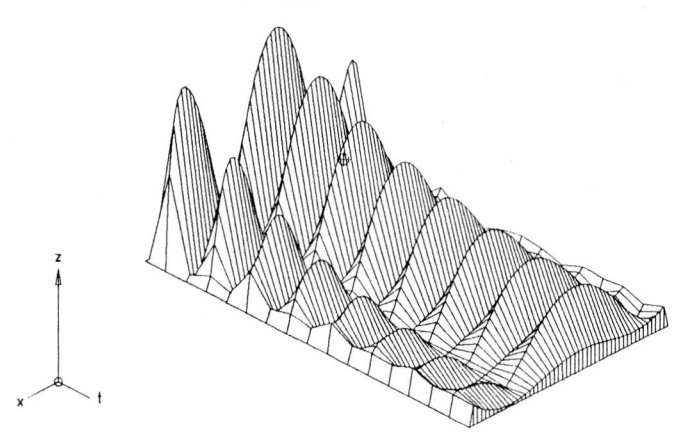

Fig. 3: Error $z(x_i, t_i) = |T_{ex}(x,t) - T_{num}(x_i, t_i)|$ for the test problem of equation (4) for $u_o = 0.1, a = 0.25, 0 \leq x \leq L = 3.14$ and $0 \leq t \leq 1.025$. The exact solution T_{ex} is given by equation (9) for $k = 2$. The discretization ratio used is $\lambda = \Delta t/\Delta x = 1$.

The main result of the numerical tests is the crucial influence of the discretization ratio $\lambda = \Delta t/\Delta x$ on the accuracy of the numerical solutions. Calculations done with $\lambda = 0.2$ show nearly identical error characteristics for the LFCN- and ABCN-scheme whereas they are clearly different if a larger time step corresponding to $\lambda = 1$ is used (Fig. 3). The results of the LFCN-scheme are more accurate not only as regards to the magnitude of the error but also with respect to statistical data of the solution. The reason is that the LFCN-scheme yields an alternating over- and underestimation of the exact solution whereas the ABCN-scheme does not show a change of sign in the deviation from the exact solution at a fixed spatial position for long times.

ABCN-scheme ($\lambda = 2$): $z_{max} = 1.5 \cdot 10^{-3}$

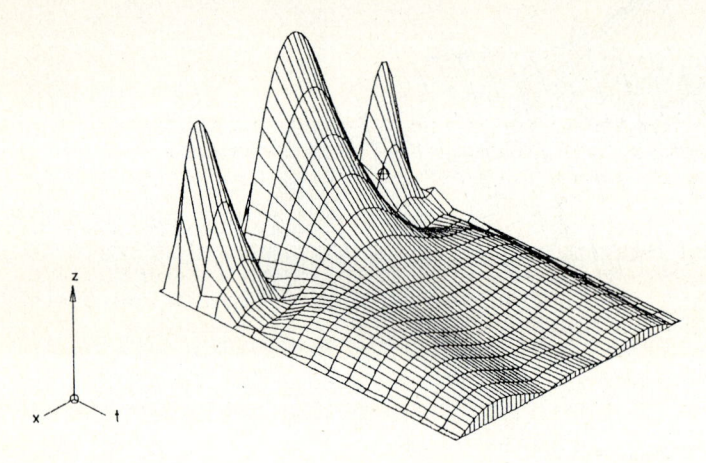

LFCN-scheme ($\lambda = 2$): $z_{max} = 1.7 \cdot 10^{-3}$

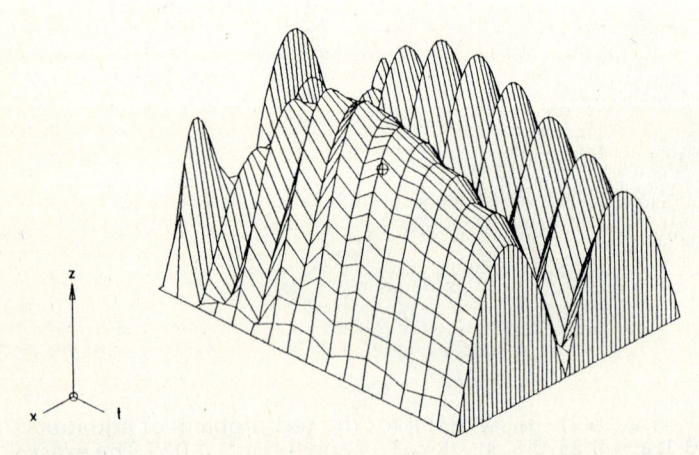

Fig.4: Error behaviour of test problem as in Fig. 3, but with discretization ratio $\lambda = \Delta t/\Delta x = 2$ and for $0 \leq t \leq 2.05$.

The error characteristics given in figure 4 result from a discretization ratio $\lambda = 2$. Now the solution of the ABCN-scheme is clearly superior to that of the LFCN-scheme. The latter exhibits strong oscillations in the error characteristic which are only weakly damped in time. To analyze this tendency towards $2\Delta t$ oscillations we consider the LFCN-scheme which can be written as:

$$\left| \left(2 + \frac{1}{D}\right) T_i - \left(T_{i+1} - T_{i-1}\right) \right|^{n+1} = -\frac{C}{D}\left(T_{i+1} - T_{i-1}\right)^n$$
$$+ \left| \left(T_{i+1} - T_{i-1}\right) - \left(2 - \frac{1}{D}\right) T_i \right|^{n-1}. \qquad (11)$$

It becomes evident that for small values of the mesh Peclet number

$$Pe_{\Delta x} = \frac{C}{D} = \frac{u_o \Delta x}{a} \qquad (12)$$

neighbouring time planes are only weakly coupled. This is the explanation why, once a time oscillation appeared, this will be damped only weakly.

The appearence of a first error oscillation can be explained by the Euler scheme which is used to calculate the first time plane out of the initial condition. Due to the decoupling of neighbouring time planes in case of a low mesh Peclet number, the first LFCN-step will also mainly use the data given by the initial condition for calculation of the second time plane. However, the error involved by the second order LFCN-scheme will be lower than that of the first order Euler-scheme. Especially the use of larger time steps Δt, or larger $\lambda = \Delta t/\Delta x$ respectively, will introduce quite different errors in the first two time planes and thus is leading to a first oscillation.

5. DISCUSSION

In the previous sections it has been shown that a decision between both schemes depends at least on the problem of interest. The ABCN-scheme works very well for diffusion dominated problems over the whole range of discretization ratios $\lambda = \Delta t/\Delta x$ investigated. Difficulties may arise due to numerical stability for more convection dominated problems. The main advantage of the LFCN-scheme is its superior numerical stability. However, in case of low mesh Peclet numbers it shows a tendency to time oscillations and should not be used with time steps corresponding to a discretization ratio $\lambda = \Delta t/\Delta x > 1$.

In TURBIT typical values of the discretization ratio λ may be expected to range between 0.1 and 1 when a semi-implicit solution scheme for the thermal energy equation is used. Thus application of both schemes is possible. However, since we use a Leapfrog scheme for the integration of the momentum equations it may be reasonable to use a Leapfrog-type scheme in the energy equation too, i.e. the LFCN- instead of the ABCN-scheme. By this the possibility of phase errors between the velocity- and the temperature field seems avoidable. These may be introduced by the use of such different schemes as explicit Euler-Leapfrog- and semi-implicit ABCN- scheme are representing. We took the decision to implement the ABCN-scheme in TURBIT too, since later on the necessity may arise to solve the momentum equation also semi-implicitly. Then time steps corresponding to $\lambda > 1$ become possible and therefore the ABCN-scheme has to be used for both, for the momentum and for the energy equation.

6. REALISATION IN TURBIT AND PRACTICAL EXPERIENCE

An important aspect according to semi-implicit time integration is the efficient solution of the arising linear equation system. The set of linear equations arising from the implicit treatment of the thermal diffusion terms is similar to that of a discretized Poisson equation [7]. Therefore in TURBIT we use a modified version of a direct FFT-based Poisson solver [8] for the solution of this set of equations. The additional CPU-time per time step of the semi-implicit scheme is about 10 to 20 percent compared to the fully explicit scheme. This is contrasted by a gain in the time step width of one to two orders of magnitude.

In case of direct numerical simulation of Rayleigh-Bénard convection in liquid sodium with the LFCN-scheme time steps can be used which are up to a factor of 50 larger than those allowed for the fully explicit scheme. However, in some of these applications numerical instability of both the ABCN- and the LFCN-scheme was observed in case of high diffusion number. Although the Von Neumann analysis is predicting numerical stability even for an infinite diffusion number this instability is not surprising due to the results of the spectral analysis of the numerical error discussed above. Further on one should mention that a stability criterion derived by the linearized version of a nonlinear problem can only give approximative results. To avoid these stability problems we restrict the diffusion number to a maximum value of $D_{max} = a\Delta t/(\Delta x_{min}^2) = 4$ and thus set an upper limit for the time step width.

In table 1 we present results for the 2D GAMM Benchmark [1, 9] on "Numerical Simulation of Oscillatory Convection in Low-Pr Fluids" calculated by the semi-implicit schemes on a Siemens VP 400 computer. Compared to calculations on a 30·4·64 grid with the explicit scheme the use of the ABCN-scheme results in a factor of 37 for the increase of the time step width. Furthermore with the LFCN-scheme calculations on a 50·4·102 grid could be realized, while these were impracticable with the fully explicit scheme. This mesh refinement results in a clear improvement of the requested results (see Table 1), namely the maximum horizontal velocity amplitude U^*_{max} and the frequency of oscillation f.

Table 1 Requested results for 2D GAMM Benchmark [1, 9] Case C, Gr = 40000, Pr = 0.015, R-R. (Remark: TURBIT is a 3D-code in which one has to use at least 4 mesh cells in the third direction even for a 2D problem)

Code (time integration scheme)	grid	Δt	t_{max}	CPU-time [min]	U^*_{max}	f
Reference-Code [9]	81·321	-	-	-	1.093	21.76
TURBIT (explizit [1])	30·4·64	$2.6 \cdot 10^{-4}$	72.1	1089 VP 50	0.987	22.35
TURBIT (semi-implicit ABCN)	30·4·64	$9.8 \cdot 10^{-3}$	228.1	90 VP 400	0.991	22.00
TURBIT (semi-implicit LFCN)	50·4·102	$4.2 \cdot 10^{-3}$	103.4	217 VP 400	1.026	21.86

7. CONCLUSIONS

Direct numerical simulation of natural convection in liquid metals using fully explicit time integration schemes results in strong time step restrictions. This is enforced by the numerical stability of the thermal diffusion process and can be overcome by semi-implicit time integration of the thermal energy equation. Two semi-implicit time integration schemes - the Adams-Bashforth Crank-Nicolson scheme and the Leapfrog Crank-Nicolson scheme, respectively - have been investigated analytically and numerically. For both schemes an increase of the time step width becomes possible without loosing physically relevant information. Dependent on the Prandtl number time steps can be used that are up to a factor of 50 larger than that allowed for fully explicit schemes. Since the arising set of linear equations is efficiently solved by an adapted direct FFT-based Poisson solver this time step increase results in a CPU-time saving of nearly the same magnitude. Results for a Benchmark problem [1, 9] gained by semi-implicit time integration show the validity of the new method. Thus direct numerical simulations of turbulent natural convection in liquid metals at moderate Rayleigh numbers become feasible with justifiable computational expense.

REFERENCES

[1] GRÖTZBACH, G: "Numerical simulation of oscillatory convection in low Prandtl number fluids with the TURBIT code", Notes on Numerical Fluid Mechanics, Vol. 27, pp. 57-64, Vieweg Verlag, Braunschweig, 1990.

[2] GRÖTZBACH, G: "Direct numerical simulation of laminar and turbulent Bénard convection", J. Fluid Mech., Vol. 119, pp. 27-53, 1982.

[3] GRÖTZBACH, G: "Direct numerical simulation of the turbulent momentum and heat transfer in an internally heated fluid layer", Proc. 7th Int. Heat Transfer Conf., München, Vol. 2, pp. 141-146, 1982.

[4] MOIN, P., REYNOLDS, W.C., FERZIGER, J.H.: "Large eddy simulation of incompressible turbulent channel flow", Dept. Mech. Engng., Stanford Univ., Rep. TF-12, 1978.

[5] PEYRET, R., TAYLOR, T.D.: "Computational methods for fluid flows", Springer Verlag, New York, 1983.

[6] HIRSCH, C.: "Numerical computation of internal and external flows", Vol. 1: "Fundamentals of numerical discretization", Wiley & Sons, New York, 1988.

[7] SCHUMANN, U.: Personal communication.

[8] SCHMIDT, H., SCHUMANN, U., VOLKERT, H., ULRICH, W.: "Three-dimensional, direct and vectorized elliptic solvers for various boundary conditions", DFVLR-Mitteilung 84-15, 1984.

[9] BEHNIA, M., DE VAHL DAVIS, G.: "Fine mesh solutions using stream function - vorticity formulation", Notes on Numerical Fluid Mechanics, Vol. 27, pp. 11-18, Vieweg Verlag, Braunschweig, 1990.

NUMERICAL TECHNIQUES FOR COMPRESSIBLE FLOWS

A Two-step Godunov-type Scheme for Multidimensional Compressible Flows

Andrea Di Mascio
INSEAN - Roma

Bernardo Favini
Dipartimento di Meccanica e Aeronautica
Università degli Studi di Roma *La Sapienza*

Abstract

A two-step second order extension of Godunov scheme is presented. The two-step algorithm, which is of the predictor-corrector type, exhibits a high stability limit, $CFL \leq 2$, coupled with an increase of computational efficiency when compared with classical second order Godunov-type schemes. Moreover, the predictor-corrector structure allows an accurate and efficient formulation of mixed initial boundary value problems. Some computations are reported to show the attainable accuracy of the scheme and its effectiveness in resolving complex flows both in cartesian and curvilinear grids.

1 Introduction

In a previous paper [1] we have developed a Godunov-type scheme for the solution of hyperbolic systems of conservation laws in integral form, which has been applied to one-dimensional initial value problems. The proposed scheme belongs to the class of second order ENO generalization [2] of the Godunov scheme [3], but we have adopted a two-step time integration of the predictor-corrector type.

The predictor-corrector algorithm was developed for hyperbolic systems in quasi-linear form and applied to the solution of the Euler equations coupled with shock-fitting techniques [4,5,6]. The main features of this scheme are: i) the space discretization is fully characteristic biased, ii) at each time step, a stencil of only two points is adopted. The peculiar structure of this scheme is responsible for both the high stability limit ($CFL \leq 2$) and the accuracy of the computation near boundaries. More precisely: i) the fully characteristic biasing avoids the introduction of artificial boundary conditions, and obtains the high stability limit; ii) the two point stencil ensures second order accuracy also at grid points close to the boundaries.

In Ref.[1] we have reformulated the original scheme [4,5,6] in conservation form and we have introduced a data reconstruction procedure that verifies a TVD condition in the scalar linear case. The two-step procedure makes the computation cheaper than in ENO schemes, requiring the same computer storage. In fact, the first stage coincides with the first order Godunov scheme and therefore it does not require any data reconstruction, while the second stage is computationally equivalent to a second order ENO scheme. A review of the scheme is presented in section 2.

As in the original scheme [4,5], the two-step Godunov-type algorithm prompts a simple and accurate procedure for the computation at cells close to the boundaries. In fact, in the corrector step, data for the slope reconstruction have to be only first order accurate, and these data are obtained at the boundaries from the solution (in the predictor step) of initial boundary value problems. When dealing instead with second order one-step schemes, the data for slope reconstruction have to be second order approximation of the exact solution, and problems arise at the boundaries.

In order to overcome this difficulty, it is customary to obtain the necessary additional informations by means of data extrapolation from the interior of the computational domain, a procedure that spoils the quality of the numerical approximation.

The formulation of the initial boundary value problem for the two-step algorithm is described in section 3. The extension of the formulation to multidimensional problems is obtained via operator splitting and it is presented in section 4.

2 Outline of the method : initial value problem

Consider the initial value problem

$$\frac{\partial U}{\partial t} + \frac{\partial F(U)}{\partial x} = 0 \qquad -\infty < x < \infty \quad t > 0 \tag{2.1}$$

$$U(x,0) = \tilde{U}(x)$$

where

$$U = \left\{ \begin{array}{c} \rho \\ \rho u \\ e \end{array} \right\} \quad F(U) = \left\{ \begin{array}{c} \rho u \\ \rho u^2 + p \\ (e+p)u \end{array} \right\}. \tag{2.2}$$

The meaning of the symbols is the usual one : "ρ" is the density, "p" the pressure, "e" the total energy per unit volume and "u" the velocity. We will consider only perfect gas satisfying the $\gamma = constant$ law.

The initial condition $\tilde{U}(x)$ is supposed to be piecewise smooth, i.e. with a finite number of discontinuities of the first kind in any bounded interval.

$U(x,t)$ is a weak solution of (2.1) if it satisfies

$$\int_{-\frac{h}{2}}^{\frac{h}{2}} U(x+\xi, t+\tau) d\xi - \int_{-\frac{h}{2}}^{\frac{h}{2}} U(x+\xi, t) d\xi + \int_{0}^{\tau} F(U(x+\frac{h}{2}, t+\eta)) d\eta - \int_{0}^{\tau} F(U(x-\frac{h}{2}, t+\eta)) d\eta = 0 \tag{2.3}$$

$\tau, h > 0$, or equivalently

$$\frac{\bar{U}(x, t+\tau) - \bar{U}(x,t)}{\tau} + \frac{\bar{F}(x+\frac{h}{2},t) - \bar{F}(x-\frac{h}{2},t)}{h} = 0 \tag{2.4}$$

where the following definitions are adopted:

$$\bar{U}(x,t) := \frac{1}{h} \int_{-\frac{h}{2}}^{\frac{h}{2}} U(x+\xi,t) d\xi \tag{2.5}$$

$$\bar{F}(x,t;U) := \frac{1}{\tau} \int_{0}^{\tau} F(U(x,t+\eta)) d\eta .$$

In the numerical approximation, the x-axis is divided into intervals $[x_{i-\frac{1}{2}}, x_{i+\frac{1}{2}}]$, $i = 0, \pm 1, \pm 2, \ldots$ and the solution is explicitly computed, at $x = x_i$ and $t^n = t_0 + n\tau$ from

$$V_j^{n+1} = V_j^n - \frac{\tau}{h} \left[\hat{F}_{j+\frac{1}{2}}^n - \hat{F}_{j-\frac{1}{2}}^n \right]. \tag{2.6}$$

In Ref.[2] it is shown that if $\tau = O(h)$ and

$$\hat{F}_{j+\frac{1}{2}}^n = \bar{F}(x_{j+\frac{1}{2}}, t^n; U) + d(x_{j+\frac{1}{2}})h^p \tag{2.7}$$

with $d(x)$ Lipschitz-continuous function of x, then the scheme is globally p-th order accurate.

In Ref.[1] we have shown how to recast a second order extension of Godunov scheme as a two-step algorithm :

predictor step

$$V_j^{n+\frac{1}{2}} = V_j^n - \frac{\tau}{2h}\left[\hat{F}_{j+\frac{1}{2}}^n - \hat{F}_{j-\frac{1}{2}}^n\right] \tag{2.8}$$

with

$$\hat{F}_{j+\frac{1}{2}}^n = F^R(V_j^n, V_{j+1}^n) \tag{2.9}$$

where $F^R(.,.)$ is the flux at $x = x_{j+\frac{1}{2}}$ of the Riemann problem with left and right states given by the arguments of the function;

corrector step

$$V_j^{n+1} = V_j^{n+\frac{1}{2}} - \frac{\tau}{2h}\left[\hat{F}_{j+\frac{1}{2}}^{n+\frac{1}{2}} - \hat{F}_{j-\frac{1}{2}}^{n+\frac{1}{2}}\right] \tag{2.10}$$

where

$$\hat{F}_{j+\frac{1}{2}}^{n+\frac{1}{2}} = F^R(V_j^L, V_{j+1}^R)$$
$$V_{j+\frac{1}{2}}^{R,L} = V(W_{j+\frac{1}{2}}^{R,L}) . \tag{2.11}$$

In the last formula, W may represent both primitive (u, p, ρ) or local characteristic variables at $t = t^{n+\frac{1}{2}}$, where "local" means that these variables are computed assuming the jacobian $A = \partial F/\partial U$ constant.

In both cases, the left and right states for the Riemann problem (2.11) are defined as

$$W_{j+\frac{1}{2}}^L = W_j^{n+\frac{1}{2}} + h\left.\frac{\delta W}{\delta x}\right|_j^{n+\frac{1}{2}} + \frac{\tau}{2}\left.\frac{\delta W}{\delta t}\right|_j^{n+\frac{1}{2}}$$

$$W_{j+\frac{1}{2}}^R = W_j^{n+\frac{1}{2}} - h\left.\frac{\delta W}{\delta x}\right|_{j+1}^{n+\frac{1}{2}} + \frac{\tau}{2}\left.\frac{\delta W}{\delta t}\right|_{j+1}^{n+\frac{1}{2}} \tag{2.12}$$

with

$$\left.\frac{\delta W}{\delta t}\right|_j^{n+\frac{1}{2}} = -C\left.\frac{\delta W}{\delta x}\right|_j^{n+\frac{1}{2}} \tag{2.13}$$

$$C = R^{-1}AR$$

where R is such that $\partial U = R\partial W$, and $\delta W/\delta x$ is a first order accurate approximation to $\partial W/\partial x$ given by

$$\left.\frac{\delta W}{\delta x}\right|_j^{n+\frac{1}{2}} = \frac{1}{h}\text{minmod}(W_j^{n+\frac{1}{2}} - W_{j-1}^{n+\frac{1}{2}}, W_{j+1}^{n+\frac{1}{2}} - W_j^{n+\frac{1}{2}}) \tag{2.14}$$

with

$$\mathrm{minmod}(x,y) = \begin{cases} \mathrm{sign}(x)\min(|x|,|y|) & \text{if } xy > 0 \\ 0 & \text{otherwise} \end{cases} \quad (2.15)$$

Adding the predictor and corrector steps, we get

$$V_j^{n+1} = V_j^n - \left[\tilde{F}_{j+\frac{1}{2}}^n - \tilde{F}_{j-\frac{1}{2}}^n\right] \quad (2.16)$$

where

$$\tilde{F}_{j+\frac{1}{2}}^n = \frac{1}{2}\left[\hat{F}_{j+\frac{1}{2}}^n + \hat{F}_{j+\frac{1}{2}}^{n+\frac{1}{2}}\right]. \quad (2.17)$$

It is easy to verify [1] that, far from points where U is no longer Lipschitz-continuous, the scheme (2.8-2.14) is second order accurate in both time and space, i.e.

$$\tilde{F}_{j+\frac{1}{2}}^n = \bar{F}(x_{j+\frac{1}{2}}, t^n; U) + O(h^2). \quad (2.18)$$

In the scalar linear case the scheme satisfies the TVD condition and it leads to oscillation free solutions in case of systems of non-linear conservation laws. The stability limit on the time step is

$$\tau_{max} = 2\min\frac{h_j}{a_j + |u_j|} \quad (2.19)$$

where a_j is the speed of sound at $x = x_j$.

Note that in the predictor step there is no data reconstruction, as the predictor step is identical to the first order Godunov scheme, whereas the computational effort in the corrector step is comparable to any second order Godunov-type scheme. Therefore the two-step algorithm reduces the computational work by about 40% with respect to second order ENO generalizations of Godunov scheme.

3 Initial boundary value problem

When the scheme (2.8-2.14) is applied to a cell near a solid boundary, the data reconstruction at the corrector level cannot be performed, since one of the first differences in (2.14) is missing. In fact, this difficulty is typical of any higher order scheme with at least a three-point stencil. In classical Godunov-type schemes this is overcome by continuing the solution outside the integration domain by means of internal value extrapolation. The missing first order difference can be calculated, and the right or left state for the Riemann problem at boundaries remains defined. Both first and second order extrapolation are commonly adopted. As a matter of fact, this procedure can work quite well for simple flow configurations as, for instance, one-dimensional flows, whereas for more general and realistic geometries the extrapolation of the flow variables from the interior can reduce the accuracy of the solution in the vicinity of the boundaries. Some numerical examples showing the effects induced by the extrapolation procedure are reported in the final section.

In the present method, the missing differences at the boundaries can be recovered knowing of the solution at the predictor step. Indeed, let us consider, for example, a one-dimensional problem with a rigid boundary at the left side ($x = 0$) of the domain. The associated boundary condition is

$$u(0,t) = 0 \qquad 0 \leq t < \infty. \quad (3.1)$$

From the predictor step, we know the numerical solution $V(x_1, t^{n+\frac{1}{2}})$ at $x = x_1 = h/2$ with first order accuracy, which can be used to compute the approximation of U at $t^{n+\frac{1}{2}^+}$ on the boundary $x = 0$ with the same accuracy. If we solve the following Riemann problem:

$$u_L = u_{-1}^{n+\frac{1}{2}} = -u_1^{n+\frac{1}{2}} = u_R$$

$$p_L = p_{-1}^{n+\frac{1}{2}} = p_1^{n+\frac{1}{2}} = p_R \tag{3.2}$$

$$\rho_L = \rho_{-1}^{n+\frac{1}{2}} = \rho_1^{n+\frac{1}{2}} = \rho_R$$

we obtain the state $V_0^{n+\frac{1}{2}^+}$, approximating $U(0, t^{n+\frac{1}{2}})$ with first order accuracy, if U is sufficiently smooth near $x = 0$. Using this value, the first difference $\delta W/\delta x$ in formulae (2.12) and (2.13) can be computed as

$$\left.\frac{\delta W}{\delta x}\right|_1^{n+\frac{1}{2}} = \text{minmod}\left(\frac{W_1^{n+\frac{1}{2}} - W_0^{n+\frac{1}{2}}}{h/2}, \frac{W_2^{n+\frac{1}{2}} - W_1^{n+\frac{1}{2}}}{h}\right). \tag{3.3}$$

Being $V_0^{n+\frac{1}{2}^+} = U(0, t^{n+\frac{1}{2}}) + O(h)$, the proof of second order accuracy at boundary follows the same lines of the proof at internal points illustrated in Ref.[1].

Although the above procedure is the correct one, numerical tests have shown that almost identical result are obtained if, at the boundary, we assume as $V_0^{n+\frac{1}{2}^+}$ the computed solution at $t = t^n$. Therefore, the solution of a new Riemann problem at $x = 0$, $t = t^{n+\frac{1}{2}^+}$ can be avoided, although the additional cost is a very small fraction of the total.

The above procedure can be considered the natural extension to conservation equations in integral form of the procedure developed in [5] for quasilinear forms. A similar formulation for one-step second-order flux-difference splitting schemes can be found in [7].

4 Two dimensional problems

The extension of the method presented in section 2 to two-dimensional problems is immediately obtained by means of the operator splitting procedure. Let us consider the problem

$$\frac{\partial U}{\partial t} + \frac{\partial F}{\partial x} + \frac{\partial G}{\partial y} = 0 \quad t > 0 \quad (x, y) \in \mathbf{R}^2$$

$$U(x, y, 0) = \tilde{U}(x, y) \tag{4.1}$$

where $\tilde{U}$ is supposed to be piecewise smooth and

$$U = \left\{\begin{array}{c} \rho \\ \rho u \\ \rho v \\ e \end{array}\right\} \quad F = \left\{\begin{array}{c} \rho u \\ \rho u^2 + p \\ \rho u v \\ (e+p)u \end{array}\right\} \quad G = \left\{\begin{array}{c} \rho v \\ \rho u v \\ \rho v^2 + p \\ (e+p)v \end{array}\right\}. \tag{4.2}$$

$U(x, y, t)$ is a weak solution of (4.1) if it satisfies

$$\frac{1}{S}\int_D U(x,y,t+\tau)dxdy - \frac{1}{S}\int_D U(x,y,t)dxdy + \frac{1}{\tau}\int_0^\tau \int_{\partial D} H(U(x,y,t+\eta))dcd\eta = 0. \tag{4.3}$$

D being any bounded domain in $\mathbf{R}^2$, S the measure of D, and dc the infinitesimal arc length. The flux $H(U)$ is defined as

$$H(U) = F(U)n_x + G(U)n_y \tag{4.4}$$

being $\{n_x, n_y\} = \vec{n}$ the outward normal to the boundary ∂D of D.

In the numerical approximation the physical domain is divided into finite volumes $D_{i,j}$, $i,j = 0, \pm 1, \pm 2, \ldots$. From here on, we limit the discussion to quadrilateral structured meshes, although no theoretical or practical difficulties arises with non-structured meshes (apart the fact that some extra information is required for connecting the nodes). The two-step algorithm for the numerical integration of (4.3) becomes:

Predictor

$$V_{i,j}^{n+\frac{1}{2}} = V_{i,j}^n - \frac{\tau}{2S_{i,j}} \left[\hat{F}_{i+\frac{1}{2},j}^n l_{i+\frac{1}{2},j} - \hat{F}_{i-\frac{1}{2},j}^n l_{i-\frac{1}{2},j} + \hat{F}_{i,j+\frac{1}{2}}^n l_{i,j+\frac{1}{2}} - \hat{F}_{i,j-\frac{1}{2}}^n l_{i,j-\frac{1}{2}} \right] \quad (4.5)$$

where $V_{i,j}^n$ is a second order approximation of the mean value of U in $D_{i,j}$ at t^n, i.e.

$$V_{i,j}^n = \frac{1}{S_{i,j}} \int_{D_{i,j}} U(x,y,t^n) dx dy + O(h^2) \quad (4.6)$$

and $\hat{F}_{i+\frac{1}{2},j}^n l_{i+\frac{1}{2},j}$ is a first order approximation to

$$\frac{1}{\tau} \int_{t^n}^{t^{n+\frac{1}{2}}} \int_{l_{i+\frac{1}{2},j}} (Fn_x + Gn_y) dc \,. \quad (4.7)$$

$l_{i+\frac{1}{2},j}$ being the interface between the finite volumes $D_{i,j}$ and $D_{i+1,j}$. In the above integral, $\rho, p, e, \vec{u}$ are computed as solutions of a Riemann problem along the normal to the cell interface, whose left and right states are $V_{i,j}$ and $V_{i+1,j}$ respectively. The other fluxes are computed in a similar way.

Corrector

$$V_{i,j}^{n+1} = V_{i,j}^{n+\frac{1}{2}} - \frac{\tau}{2S_{i,j}} \left[\hat{F}_{i+\frac{1}{2},j}^{n+\frac{1}{2}} l_{i+\frac{1}{2},j} - \hat{F}_{i-\frac{1}{2},j}^{n+\frac{1}{2}} l_{i-\frac{1}{2},j} + \hat{F}_{i,j+\frac{1}{2}}^{n+\frac{1}{2}} l_{i,j+\frac{1}{2}} - \hat{F}_{i,j-\frac{1}{2}}^{n+\frac{1}{2}} l_{i,j-\frac{1}{2}} \right] \quad (4.8)$$

where $\hat{F}_{i+\frac{1}{2},j}^{n+\frac{1}{2}}$ is again computed as the solution of a Riemann problem, but with right and left states given by

$$W_{j+\frac{1}{2}}^L = W_{i,j}^{n+\frac{1}{2}} + \delta_\xi W_{i,j}^{n+\frac{1}{2}} + \frac{\tau}{2} \left. \frac{\delta W}{\delta t} \right|_{i,j}^{n+\frac{1}{2}}$$

$$W_{j+\frac{1}{2}}^R = W_{i+1,j}^{n+\frac{1}{2}} - \delta_\xi W_{i+1,j}^{n+\frac{1}{2}} + \frac{\tau}{2} \left. \frac{\delta W}{\delta t} \right|_{i+1,j}^{n+\frac{1}{2}}$$

(4.9)

where if, for instance, primitive variables are considered,

$$W = \begin{Bmatrix} \rho \\ u \\ v \\ p \end{Bmatrix} = \begin{Bmatrix} w^1 \\ w^2 \\ w^3 \\ w^4 \end{Bmatrix} \quad (4.10)$$

$$\begin{cases} \delta_\xi w^k\big|_{i,j}^{n+\frac{1}{2}} = \text{minmod}(w^k_{i,j} - w^k_{i-1,j}, w^k_{i+1,j} - w^k_{i,j}) \\ \\ \delta_\eta w^k\big|_{i,j}^{n+\frac{1}{2}} = \text{minmod}(w^k_{i,j} - w^k_{i,j-1}, w^k_{i,j+1} - w^k_{i,j}) \end{cases} \quad k = 1,2,3,4 \qquad (4.11)$$

and

$$\frac{\delta w^1}{\delta t} = -\tilde{u}\rho_\xi - \tilde{v}\rho_\eta - \rho(\xi_x u_\xi + \eta_x u_\eta + \xi_y v_\xi + \eta_y v_\eta)$$

$$\frac{\delta w^2}{\delta t} = -\tilde{u} u_\xi - \tilde{v} u_\eta - \frac{1}{\rho}(\xi_x p_\xi + \eta_x p_\eta)$$

$$\frac{\delta w^3}{\delta t} = -\tilde{u} v_\xi - \tilde{v} v_\eta - \frac{1}{\rho}(\xi_y p_\xi + \eta_y p_\eta) \qquad (4.12)$$

$$\frac{\delta w^4}{\delta t} = -\tilde{u} p_\xi - \tilde{v} p_\eta - \gamma p(\xi_x u_\xi + \eta_x u_\eta + \xi_y v_\xi + \eta_y v_\eta)$$

$$\tilde{u} = \xi_x u + \xi_y v \quad \tilde{v} = \eta_x u + \eta_y v$$

where ρ_ξ means $\delta_\xi \rho / \delta \xi = \delta_\xi w^1 / \delta \xi$ and similarly for the other derivatives.

In the above relations, ξ and η are local curvilinear coordinates. The formulae (4.11) and (4.12) give first order approximation to $\partial W/\partial t$, $\partial W/\partial \xi$ and $\partial W/\partial \eta$.

As in the one-dimensional case [1], we can prove that the scheme is second order accurate in both time and space. Moreover, linear analysis and numerical tests show that the algorithm is numerically stable if

$$\tau \leq 2 \min\left(\frac{1}{\tilde{u} + a|\nabla \xi|}, \frac{1}{\tilde{v} + a|\nabla \eta|}\right). \qquad (4.13)$$

5 Numerical results

All the numerical results have been obtained by means of an exact Riemann solver.

5.1 Convex wall

The first test case considered consists of an external steady supersonic flow on a convex wall $y = \cos \pi x$, $0 < x < 0.8$; the total deviation of the flow is $\Delta \theta = 61.56^o$. The extension and the curvature of the wall (fig.1) are prescribed in such a way that the fluid undergoes an isentropic expansion without cavitation. Therefore the flow is smooth everywhere and the problems admits an analitycal solution. We have choosen this test case in order to compare the formulation of initial boundary value problems descrived in section 3 with first and second order extrapolations. The Mach number at inflow section has been chosen equal 2. The strength of the expansion wave is quite high, with a pressure ratio of about 7×10^{-2} between inflow and outflow conditions; therefore the test is rather severe for Godunov-type schemes. Indeed, Godunov-type schemes exhibits an incorrect thermodynamical behavior [8], with entropy variations depending on the velocity and its gradients. As a consequence, when a strong expansion wave occurs in a flow, the numerical approximation is degraded by the spurious production and/or destruction of entropy [9]. In the present computation, we are not trying to face directly this problem, but only to verify if a correct data reconstruction can reduce these undesirable effects. In table 1 a synthesis of the numerical results obtained with a orthogonal grid (60×30) is shown.

Table 1

	M	p_o/p_i	ρ_o/ρ_i	ΔS
First order	4.43	0.011	0.043	0.389
Second order	6.64	0.0074	0.061	−0.582
Pred.-Corr.	6.03	0.0077	0.050	−0.204
exact	5.67	0.0070	0.040	0

The first order extrapolation procedure completely spoils the numerical solution, as it can also be seen from the analysis of the iso-plots for Mach number and pressure (fig.2a and 2b). The second order extrapolation greatly improve the quality of the solution in terms of pressure values, but it overestimate the Mach number and exhibits a significant "entropy destruction" which is reflected also in the values of density. A globally better result has been obtained by means of the proposed formulation, with the smallest entropy error (fig.3a and 3b). If a non-orthogonal grid is adopted (fig.4), the differences between the present formulation and the first and second order extrapolation procedure increases as shown in table 2.

Table 2

	M	p_o/p_i	ρ_o/ρ_i	ΔS
First order	4.06	0.013	0.043	0.524
Second order	6.89	0.0074	0.068	−.675
Pred.-Corr.	6.01	0.0073	0.047	−0.177
exact	5.67	0.0070	0.040	0.

The better resolution obtained by the predictor-corrector with non-orthogonal grid depends on the higher resolution of the adopted grid, as it can be verified by inspection of fig.1 and 4.

5.2 NACA 0012 $\quad M_\infty = 0.85; \ \alpha = 1.^\circ$

This transonic flow is a classical test case for inviscid flowfield methods [10,11]. The outer boundary is located at 100 chords and the flow domain is discretized by means of an O-grid with 192 × 64 cells. The computed aerodynamic coefficients match well the results reported in [10,11]: $C_L = .360$, $C_D = 0.0588$ and $C_M = -0.124$. The histories of the lift coefficient and of the L_2 norm of the residual are shown in fig.5 and fig.6 . In fig.7 the C_p distribution is shown: we note the correct location of both upper and lower shocks and that the shocks are spread over three cells. The entropy distribution is plotted in fig.8: as we can expect from the analysis of the previous test case, the strong expansion occurring at the trailing edge induces a spurious entropy production. A global view of the iso-Mach about the profile is shown in fig.9 .

5.3 Mach 3 wind tunnel with a step

This test case is significant for the evaluation of inviscid flow models. The inflow Mach number is equal to 3 and a uniform flow is assumed as initial condition. Since the final steady state does not show relevant flow structures, the numerical solution is analyzed at an intermediate condition at t=4. Comparing the results shown in fig.10 with those obtained by P.Woodward and P. Colella [12] we notice that the present results compare perfectly with those obtained by means of MUSCL and PPM methods. More precisely, we note that the predictor-corrector scheme exhibits lower noise production at the strong shock ahead of the step than the MUSCL scheme, and is quite similar to the PPM method. This behavior depends on the tendency of the predictor-corrector scheme to spread a shock over two or three cells, as already remarked for the transonic flow past the NACA 0012 profile. In fact, as pointed out in [12], an excessively narrow shock structure is responsible for the noise which originates at the shift of the shock location from one cell to another.

References

[1] A. Di Mascio, B. Favini, to appear in *Meccanica*.

[2] A.Harten, B.Engquist, S.Osher, S.R.Chakravarthy, *J.Comput.Phys.*, **71**,231 (1987).

[3] S.K.Godunov, *Mat. Sb.*, **47**,271 (1959).

[4] Y-L. Zhu, B-M. Chen, *Computers and Fluids*, **9**,339 (1981).

[5] L.Zannetti, G.Moretti, *AIAA J.*, **20**,12 (1981).

[6] B.Gabutti, *Computers and Fluids*, **11**,3 (1983).

[7] M. Pandolfi, *AIAA Journal*, **22**,602 (1984).

[8] V. F. Kuropatenko, in *Numerical Methods in Fluid Dynamics*, ed. N. N. Yanenko and Yu. I. Shokin *MIR*, Moscow (1984).

[9] B. Favini, M. Di Giacinto, in *Proceedings XII ICNMFD, Oxford, England,1990*, ed. K. W. Morton, Springer-Verlag (1990).

[10] A. Dervieux et al., *Numerical Simulation of Compressible Euler Flows*, Vieweg (1989).

[11] AGARD Advisory Report no. 211, *Test Cases for Inviscid Flow Field Methods* (1985).

[12] P. Woodward, P. Colella *J. Comput. Phys.*, **54**, 115 (1984).

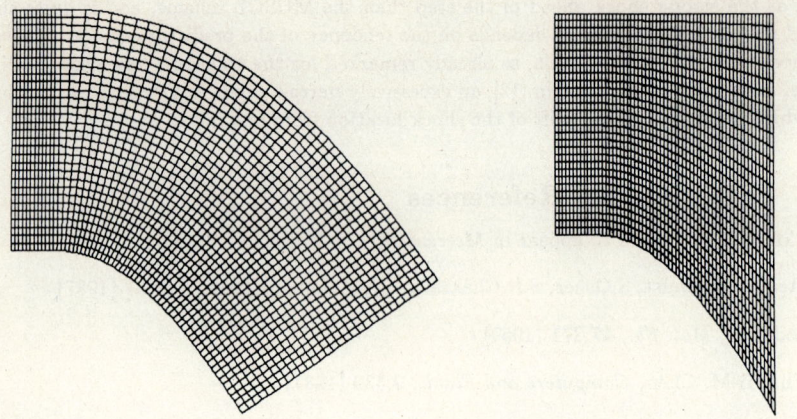

Fig. 1 Orthogonal grid.

Fig. 4 Non-orthogonal grid

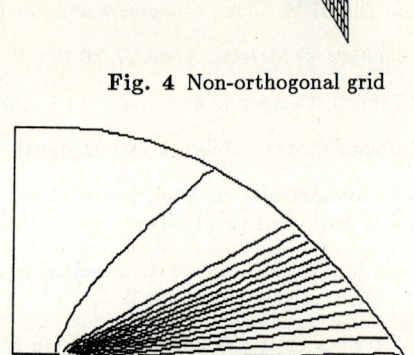

Fig. 2a First order: iso-Mach lines

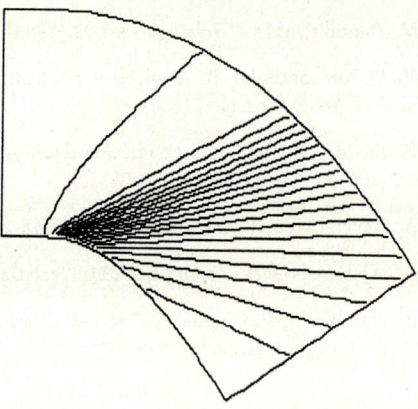

Fig. 2b First order: isobar lines

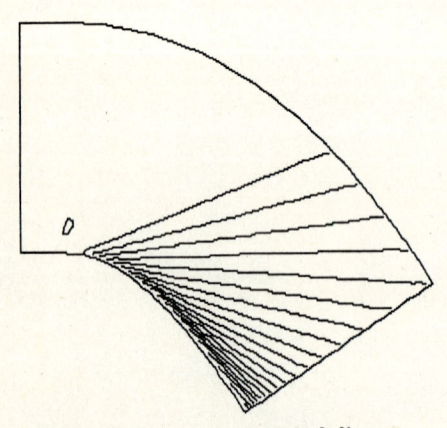

Fig. 3a Pred.-corr.: iso-Mach lines

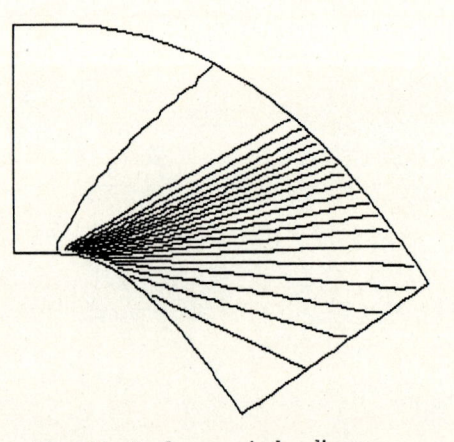

Fig. 3b Pred.-corr.: isobar lines

Fig. 5 Lift coefficient history

Fig. 6 Residual history

Fig. 7 C_p distribution

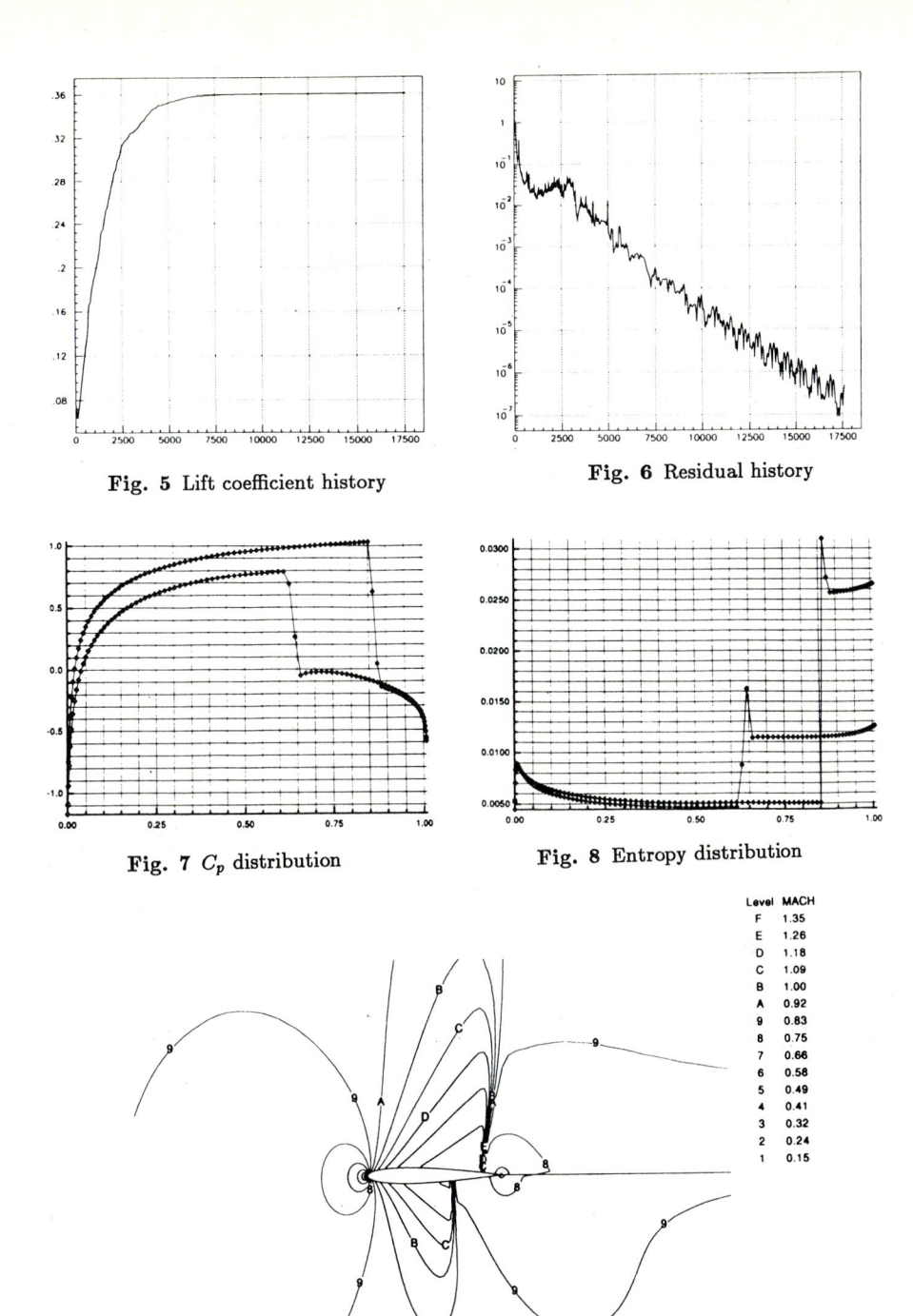

Fig. 8 Entropy distribution

Level	MACH
F	1.35
E	1.26
D	1.18
C	1.09
B	1.00
A	0.92
9	0.83
8	0.75
7	0.66
6	0.58
5	0.49
4	0.41
3	0.32
2	0.24
1	0.15

Fig. 9 Isomach lines for NACA 0012; $M_\infty = 0.85 \quad \alpha = 1.^o$

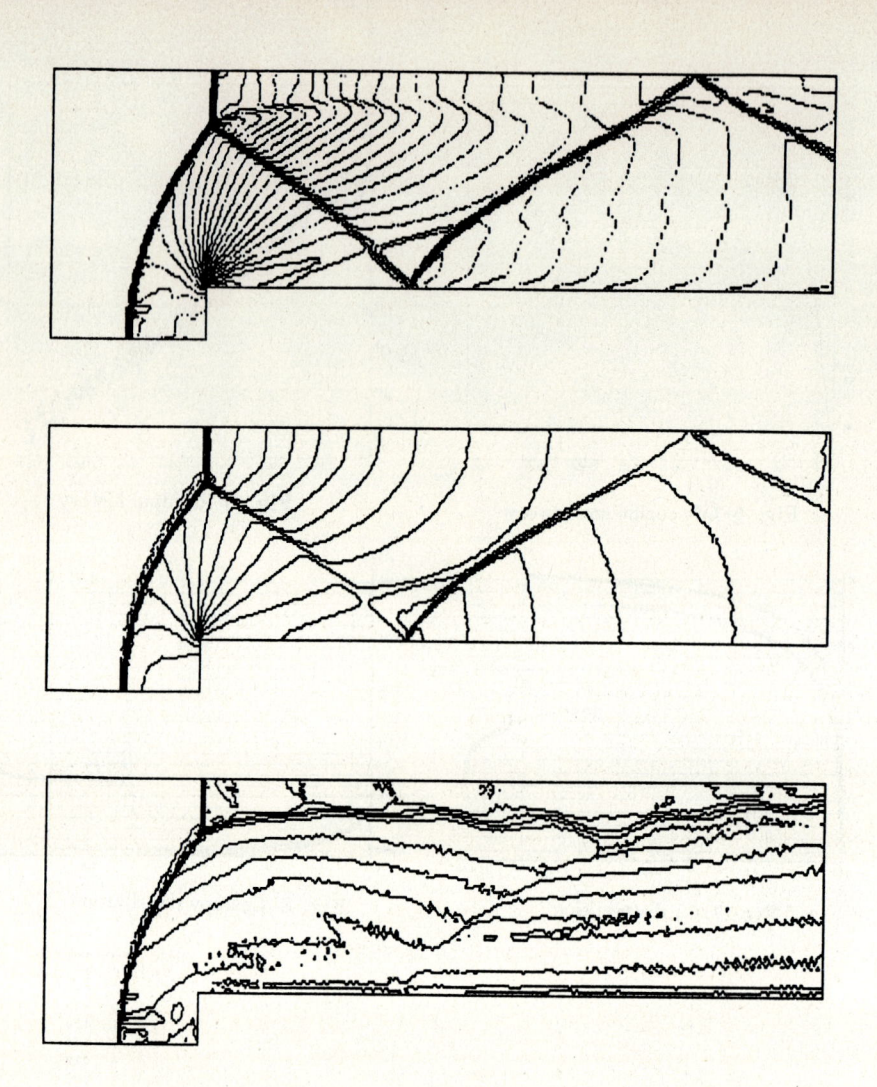

Fig. 10 Mach 3 wind tunnel with a step, grid 240×80, $CFL = 1.6$, time $= 4$:
  a) iso-density contours
  b) isobars coutours
  c) iso-entropy contours

FLUX BALANCE SPLITTING WITH ROTATED DIFFERENCES:
A SECOND ORDER ACCURATE CELL VERTEX UPWIND SCHEME

C.-C. Rossow
DLR Institut für Entwurfsaerodynamik
Flughafen, D-3300 Braunschweig, F. R. Germany

Abstract

A cell vertex upwind scheme using rotated differences for the solution of the two-dimensional Euler equations is presented. Upwinding is achieved by an unequal distribution of the transformed flux balances to the vertices of a computational cell. The equations are rotated with respect to the local velocity vector to determine the upwinding directions. A 1-D cell vertex upwind scheme is considered to give arguments for second order accurate steady state solutions of the one-sided, two-point discretization formulas used in cell vertex upwind schemes. Studies of successive mesh refinement confirm the conclusions from the 1-D example for the two-dimensional scheme. The 2-D scheme with rotated differences gives very good shock resolution for transonic flows, and the convergence rates are comparable to other schemes presently in use, despite the high nonlinearity introduced by the rotated differences.

1. INTRODUCTION

During the last two to three years considerable effort has been concentrated on the development of cell vertex upwind schemes. The basic idea of all schemes developed thus far is to employ an unequal distribution of the flux balances to the vertices of the computational cells, and therefore such methods will be called Flux Balance Splitting schemes in the following. In [1] a multidimensional approach according to [2] was used to achieve a diagonalization of the flux-jacobians independently of the direction of the grid coordinates. A splitting in streamwise and crossflow direction was employed in [3] to determine the weighting coefficients, and in [4] the directions of the fixed cartesian coordinate frame were used. In these methods the complete conservative flux balances of the Euler equations are used for the upwinding. Recently, a different concept has been used, where not the complete flux balances are employed, but simple wave solutions to the governing equations are considered, and the components of each wave are distributed to cell vertices with respect to the direction of wave propagation [5].

Since all schemes use one-sided, two-point finite difference formulas, a critical issue for cell vertex upwind methods is the question of their accuracy. The usual Taylor series analysis results in first order accurate schemes, which makes them inefficient for practical use. Another crucial point is the convergence to steady state, since all schemes become highly nonlinear for multi-dimensional flows. In order to establish a practical method, convergence rates comparable to other schemes presently in use have to be demonstrated.

The objective of the present contribution is therefore twofold: On the one hand, a 1-D cell vertex Flux Balance Splitting scheme will be outlined to give a general argument for second order accurate steady state solutions. On the other hand, the two-dimensional Flux Balance Splitting scheme of [4] is refined using rotated differences with respect to the local velocity vector rather than a fixed coordinate frame. Successive mesh refinement will be employed to assess the accuracy of the method, and the convergence properties of the scheme will be discussed for sub- and transonic flows.

2. CONSIDERATION OF 1-D SCHEME

2.1 Governing Equations

In the following, the basic idea of Flux Balance Splitting will be explained by an outline of the 1-D Flux Balance Splitting scheme according to [4]. For the one-dimensional case no conceptual difficulties arise, since the 1-D Euler equations may be completely diagonalized. The one-dimensional Euler equations for unsteady compressible inviscid flow may be written in a cartesian coordinate frame in integral form:

$$\frac{\partial}{\partial t} \int_V \vec{W} dV = - \int_{\partial V} \vec{F} \cdot \vec{n} dS \ , \qquad (1)$$

where

$$\vec{W} = \begin{bmatrix} \rho \\ \rho u \\ \rho E \end{bmatrix} \ , \ \vec{F} = \begin{bmatrix} \rho u \\ \rho u^2 + p \\ \rho(E+p) \end{bmatrix} \ , \ E = e + \frac{u^2}{2} \ , \ p = [\kappa - 1] \rho \left[E - \frac{u^2}{2} \right].$$

Application of Gauss' theorem to (1) leads to the system of differential equations in conservation form:

$$\frac{\partial \vec{W}}{\partial t} + \frac{\partial \vec{F}}{\partial x} = 0 \qquad (2)$$

with

$$\frac{\partial \vec{F}}{\partial x} = \overline{\overline{A}} \frac{\partial \vec{W}}{\partial x} \ , \ \overline{\overline{A}} = \frac{\partial \vec{F}}{\partial \vec{W}} \ .$$

Using the left and right modal matrices $\overline{\overline{M}}^{-1}$ and $\overline{\overline{M}}$, which consist of the left and right eigenvectors of the flux jacobian $\overline{\overline{A}}$, the system of equations (2) may be decoupled using the transformation

$$\overline{\overline{M}}^{-1} \frac{\partial \vec{W}}{\partial t} + \overline{\overline{\Lambda}} \overline{\overline{M}}^{-1} \frac{\partial \vec{W}}{\partial x} = 0 \ , \qquad (3)$$

where $\overline{\overline{\Lambda}}$ is the diagonal matrix containing the eigenvalues of $\overline{\overline{A}}$. Since

$$\overline{\overline{\Lambda}} = \overline{\overline{M}}^{-1} \overline{\overline{A}} \overline{\overline{M}}$$

it follows:

$$\overline{\overline{M}}^{-1} \frac{\partial \vec{W}}{\partial t} + \overline{\overline{M}}^{-1} \frac{\partial \vec{F}}{\partial x} = 0 \ . \qquad (4)$$

Note that due to this transformation the characteristic flux balances $\overline{\overline{M}}^{-1} \frac{\partial \vec{F}}{\partial x}$ are obtained directly as functions of the conservative flux balances.

2.2 Spatial Discretization

In a one-dimensional cell vertex scheme the discrete flow quantities are located at the nodes of the one-dimensional grid, as sketched in Figure 1. The discretization of equation (1) for each cell $i + 1/2$ with volume $V_{i+1/2}$ yields:

$$V_{i+1/2} \left(\frac{d}{dt} \vec{W} \right)_{i+1/2} + \vec{Q}_{i+1/2} = 0 \ . \qquad (5)$$

The conservative flux balance $\vec{Q}_{i+1/2}$ of cell $i + 1/2$ is defined by:

$$\vec{Q}_{i+1/2} = \vec{F}_{i+1} S_{i+1} - \vec{F}_i S_i + \vec{Q}^s_{i+1/2} \ , \qquad (6)$$

where S_i is the area of the cross section at node i, and the source term $\vec{Q}^s_{i+1/2}$, which arises for the quasi 1-D Euler equations, is defined as:

$$\vec{Q}^s_{i+1/2} = \begin{bmatrix} 0 \\ \frac{1}{2}(p_{i+1} + p_i)(S_{i+1} - S_i) \\ 0 \end{bmatrix} \ .$$

Note that the source term $\vec{Q}^s_{i+1/2}$ has been included into the flux balance $\vec{Q}_{i+1/2}$ of cell $i+1/2$, which is consistent with the integral form of the equations.

The conservative flux balances $\vec{Q}_{i+1/2}$ are transformed into characteristic flux balances using the local left modal matrix $\overline{\overline{M}}^{-1}$ according to equ. (4):

$$\vec{C}_{i+1/2} = \overline{\overline{M}}^{-1}_{i+1/2} \cdot \vec{Q}_{i+1/2} \quad , \tag{7}$$

where $\overline{\overline{M}}^{-1}_{i+1/2}$ is calculated by the averaging procedure of Roe [6]. Each cell sends its characteristic flux balances $\vec{C}_{i+1/2}$ to the DOWNWIND node of the corresponding cell. The downwind node is determined by the sign of the eigenvalue corresponding to the particular characteristic flux balance. Figure 2 gives a sketch of distributions of characteristic flux balances for subsonic and supersonic flows. The distribution step provides weighted characteristic flux balances $\vec{C}^L_{i+1/2}$ and $\vec{C}^R_{i+1/2}$ for the left and right node, respectively. From these, weighted conservative flux balances may be calculated by the inverse transformation with the right modal matrix $\overline{\overline{M}}$:

$$\vec{Q}^L_{i+1/2} = \overline{\overline{M}}_{i+1/2} \cdot \vec{C}^L_{i+1/2} \quad , \quad \vec{Q}^R_{i+1/2} = \overline{\overline{M}}_{i+1/2} \cdot \vec{C}^R_{i+1/2} \quad .$$

The contributions of left and right cells are summed up to yield the conservative flux balance $\vec{Q}_i$ at node i:

$$\vec{Q}_i = \vec{Q}^R_{i-1/2} + \vec{Q}^L_{i+1/2} \quad , \tag{9}$$

and the numerical analog of equation (1) for node i may be established by:

$$V_i \left(\frac{d}{dt} \vec{W} \right)_i + \vec{Q}^R_{i-1/2} + \vec{Q}^L_{i+1/2} = 0 \quad , \tag{10}$$

where $V_i = 0.5 \, (V_{i-1/2} + V_{i+1/2})$.

Note that for a central differencing scheme the discrete analog is given by:

$$V_i \left(\frac{d}{dt} \vec{W} \right)_i + 0.5 \left[\vec{Q}_{i-1/2} + \vec{Q}_{i+1/2} \right] = 0 \quad . \tag{11}$$

2.3 Solution Scheme

Division of equ. (10) by the volume V_i associated with node i leads to a set of ordinary differential equations of first order in time for the unknowns $\vec{W}$ at all nodes:

$$\frac{d}{dt} \vec{W}_i + \frac{1}{V_i} \left[\vec{Q}^R_{i-1/2} + \vec{Q}^L_{i+1/2} \right] = 0 \quad . \tag{12}$$

This equation is integrated in time by a simple forward Euler integration formula. Since the scheme is formally first order accurate due to the one-sided two-point difference formulas, forward Euler time integration is stable, and in the following a CFL number of 0.8 is used for all calculations presented. Convergence to steady state is accelerated by the use of local time steps. Note that no limiter functions or entropy switches are used.

2.4 Results

Results are shown for the transonic flow in a Laval nozzle with an inflow Mach number of $M_\infty = 0.45$. The geometry definition of the nozzle may be found in [7]. Figures 3a - 3c give results for this flow case obtained with the Flux Balance Splitting scheme, a TVD scheme [7], and a central differencing scheme according to [8]. Note that for the Mach number distributions in Figure 3a and the total pressure losses in Figure 3b, the results of the TVD scheme and the central differencing scheme are shifted to distinguish the different curves. The Flux Balance Splitting scheme captures the shock within one cell, whereas the TVD scheme and the central differencing scheme need 1.5 or 3 cells, respectively. Concerning total pressure losses, one observes that the Flux Balance Splitting scheme shows no oscillations but gives a perfect jump within one cell at the shock. Convergence histories look quite similar for all schemes, however for the TVD scheme and the central differencing scheme a 3-stage and 5-stage Runge-Kutta scheme were used.

2.5 Consideration of Accuracy

Performing a usual Taylor series analysis, one concludes that the present scheme is only first order accurate. This statement is confirmed by the fact that the integration with a forward Euler formula remains stable. However, it is only the accuracy at steady state which is of interest here, since the use of local time steps sacrifieces time accuracy.
In order to assess the accuracy of the steady state solution, at first the accuracy of the discretization of a flux balance inside a cell according to equ. (6) is considered. The surface integration is performed using the trapezoidal rule. Following [9], equ. (6) is divided by the volume of the particular cell, and a second order accurate discretization for the rate of change of $\vec{W}$ inside a cell is obtained. According to [9], it is the distribution of the flux balances towards the nodes, which causes a first order accurate discretization of the governing equations at the nodes. Now assume that at steady state all flux balances in a cell, i.e. the rates of change of $\vec{W}$ in all cells, are equal to zero. Then the distribution step, which provokes the one-sided differences (and therefore first order accuracy), has no influence on the steady state solution: it does not matter whether a flux balance equal to zero is distributed either to the left or to the right node as done in equ. (10), or whether it is distributed to both nodes using a weighting factor of 0.5 as performed in a central differencing scheme according to equ. (11). Therefore a steady state solution of the Flux Balance Splitting scheme is also a solution of a central differencing scheme without any artificial dissipation. Since central difference schemes provide second order accurate solutions, the particular steady state solution of the Flux Balance Splitting scheme is also second order accurate. Note that this argument only holds, if the source term is included into the discretization of the flux balance as given in equ. 6: It is only the complete flux balance with source term, which approaches zero at steady state.

Figure 4a shows the Mach number distribution of the subsonic flow in a Laval nozzle with an inflow Mach number of $M_\infty = 0.4$. In Figure 4b the convergence history of the maximum of the rate of change of $\vec{W}$ inside the computational cells is given. It can be seen from the figure that the rate of change is continously decreasing with the number of time steps. Thus, following the argument given above, the solution of this smooth flow field is second order accurate. However, for non-smooth flows with shocks, the rates of change of the state variables inside the cells directly up- and downstream of the shock remain at finite values. In smooth regions of the flow second order accuracy is retained, since the flux balances inside the cells vanish in these regions.

3. SOLUTION OF THE 2-DIMENSIONAL EULER EQUATIONS

3.1 Governing Equations

The two-dimensional Euler equations in integral form are given by:

$$\frac{\partial}{\partial t} \int_V \vec{W} dV = - \int_{\partial V} \bar{\bar{F}} \cdot \vec{n} dS , \qquad (13)$$

where

$$\vec{W} = \begin{bmatrix} \rho \\ \rho u \\ \rho v \\ \rho E \end{bmatrix} , \quad \bar{\bar{F}} = \begin{bmatrix} \rho \vec{q} \\ \rho u \vec{q} + p \vec{i}_x \\ \rho v \vec{q} + p \vec{i}_y \\ \rho H \vec{q} \end{bmatrix} , \quad \vec{q} = u \vec{i}_x + v \vec{i}_y , \quad H = E + \frac{P}{\rho} .$$

Application of Gauss' theorem to (13) leads to the conservative differential form:

$$\frac{\partial \vec{W}}{\partial t} + \frac{\partial \vec{F}}{\partial x} + \frac{\partial \vec{G}}{\partial y} = 0 ,\qquad(14)$$

with $\quad \dfrac{\partial \vec{F}}{\partial x} = \overline{\overline{A}}\, \dfrac{\partial \vec{W}}{\partial x} \;,\; \overline{\overline{A}} = \dfrac{\partial \vec{F}}{\partial \vec{W}} \quad$ and $\quad \dfrac{\partial \vec{G}}{\partial y} = \overline{\overline{B}}\, \dfrac{\partial \vec{W}}{\partial y} \;,\; \overline{\overline{B}} = \dfrac{\partial \vec{G}}{\partial \vec{W}}\;.$

In contrast to the one-dimensional Euler equations, a complete decoupling of the system of equations (14) is not possible, since the matrices $\overline{\overline{A}}$ and $\overline{\overline{B}}$ do not commute. Physically, this may be explained by observing that in multi-dimensional flows information is propagated in infinitely many directions. Nevertheless, equ. (14) may be transformed with respect to arbitrary directions. For example, a transformation with respect to the x-direction using the left modal matrix $\overline{\overline{M}}_A^{-1}$ of the flux-jacobian $\overline{\overline{A}}$ yields:

$$\overline{\overline{M}}_A^{-1}\, \frac{\partial \vec{W}}{\partial t} + \overline{\overline{M}}_A^{-1} \left(\frac{\partial \vec{F}}{\partial x} + \frac{\partial \vec{G}}{\partial y} \right) = 0 .\qquad(15)$$

In (15) the term $\dfrac{\partial \overline{\overline{G}}}{\partial y}$ prevents the complete decoupling of the system of equations.

3.2 Spatial discretization

For the two-dimensional scheme a structured mesh with quadilateral cells is considered. The flow quantities are located at the cell vertices, as sketched in Figure 5. Discretization of equ. (13) for a cell i + 1/2, j + 1/2 yields;

$$V_{i+1/2,\, j+1/2} \left(\frac{d}{dt}\, \vec{W} \right)_{i+1/2,\, j+1/2} + \vec{Q}_{i+1/2,\, j+1/2} = 0 ,\qquad(16)$$

where the numerical integration to obtain the flux balance $\vec{Q}_{i+1/2,\, j+1/2}$ is evaluated by the trapezoidal rule [9]. Since the 2-D Euler equations may not be completely decoupled, a problem arises in chosing the appropriate directions for the distribution of the flux balances towards the cell vertices. In [4] the flux balances $\vec{Q}_{i+1/2,\, j+1/2}$ have been transformed with respect to both, the x- and y-direction of the fixed cartesian coordinate frame, and the flux balances were distributed separately with respect to corresponding eigenvalues of the different directions.

The use of a fixed coordinate frame is viewed as insufficient for general problems, and the scheme is therefore refined by using the local velocity vector: The equations are rotated into a coordinate frame aligned with the local flow velocity. Then the flux balances $\vec{Q}_{i+1/2,\, j+1/2}$ are transformed into characteristic flux balances using the left modal matrix $\overline{\overline{M}}_s^{-1}$ of the streamwise direction. The characteristic flux balances are distributed to the vertices of the particular cell using the sign of the corresponding eigenvalues. After the distribution step, the inverse transformation is performed with the streamwise right modal matrix $\overline{\overline{M}}_s$, but only contributions to the continuity, streamwise momentum, and energy equation are retained; the contribution to the crossflow momentum equation is discarded. The whole process is repeated using the crossflow direction, but after the inverse transformation all contributions except those to the normal momentum equation are discarded. Note that for the crossflow direction the linear eigenvalues vanish, only +c and -c remain as eigenvalues, where c is the speed of sound.

To determine the weighting coefficients for the distribution of the characteristic flux balances, a concept similar to [4] is used. The coordinates of the cell vertices of a particular cell are projected onto the local velocity vector, yielding projected coordinates ξ as sketched in Figure 6. For information propagation into streamwise direction, i.e. positive eigenvalues, the weighting coefficient for a node n is determined by:

$$\Phi^n = \frac{\xi^n - \min(\xi^1, \xi^2, \xi^3, \xi_4)}{\sum_{N=1}^{4} (\xi^N - \min(\xi^1, \xi^2, \xi^3, \xi^4))} \quad , \quad \text{where} \quad \sum_{N=1}^{4} \Phi^N = 1 \text{ for conservation}. \tag{17}$$

For negative eigenvalues, all projected coordinates are multiplied by -1, and formula (17) is applied to the negative coordinates. The characteristic flux balances are then multiplied by the corresponding weighting factors. Formula (17) gives the most weight to the downwind nodes, and the weight of the most upwind node is zero. The crossflow direction is treated correspondingly by projecting the coordinates onto the direction normal to the local velocity.

The discrete analog of equ. (13) is obtained by collecting the contributions of all cells surrounding a node i,j :

$$V_{i,j} \left(\frac{d}{dt} \vec{W} \right)_{i,j} + \vec{Q}_{i,j} = 0 , \tag{18}$$

where

$$V_{i,j} = \frac{1}{4} [V_{i-1/2, j-1/2} + V_{i+1/2, j-1/2} + V_{i+1/2, j+1/2} + V_{i-1/2, j+1/2}] ,$$

and $\vec{Q}_{i,j}$ is given by the sum of the contributions of the four cells surrounding node i,j.

3.3 Solution Scheme

Equ. (18) is divided by the volume $V_{i,j}$ to obtain a set of ordinary differential equations in time:

$$\frac{d}{dt} \vec{W}_{i,j} + \frac{1}{V_{i,j}} \vec{Q}_{i,j} . \tag{19}$$

In the 1-D Flux Balance Splitting scheme the discrete equation has directly been integrated in time. However, in the two-dimensional case it was found that numerical oscillations occured which prevented convergence to machine zero, or even lead to divergence. Since the linear eigenvalues in crossflow direction vanish, the damping provided by the remaining two equations with eigenvalues +c and -c is insufficient. This is supported by the observation that the oscillations occured only in crossflow direction. The decoupling of streamlines was also reported in [3]. To remedy this problem, fourth differences of the flow variables scaled by the largest eigenvalues where added to provide the necessary damping. Following [8], a pressure switch was used to switch these terms off at shocks. Denoting the dissipative operator by D, equ. (19) now reads:

$$\frac{d}{dt} \vec{W}_{i,j} + \frac{1}{V_{i,j}} [\vec{Q}_{i,j} + \vec{D}_{i,j}] . \tag{20}$$

Equ. (20) is integrated by a 5-stage Runge-Kutta scheme. The dissipative operator is evaluated only in the first stage and then frozen for the remaining stages. Convergence to steady state is accelerated by the use of local time stepping and by implicit smoothing of residuals [10]. In all cases to be presented the CFL number is 7.5, and the user specified coefficient of the fourth differences dissipation is taken to 1/128.

3.4 Results for Subsonic Flow

In order to assess the accuracy of the two-dimensional Flux Balance Splitting scheme, the subsonic flow around the NACA 0012 airfoil has been calculated. The free stream conditions are chosen to $M_\infty = 0.63$ and $\alpha = 2.0°$, and the computational grids had an O-type topology. Since in steady subsonic flow no entropy is created, all deviations from free stream entropy are attributed to numerical errors of the scheme. Figure 7a-b show the distribution of entropy losses on the upper and lower surface of the airfoil on three successively refined meshes. It can clearly be seen from this figure that the loss of entropy is successively reduced by a factor of almost 4, thus demonstrating that the 2-D Flux Balance Splitting scheme presented here is second order accurate.

In the 2-D results, the same accuracy argument as for the 1-D example holds. Since the flux balances approach zero at steady state, the first order distribution step becomes insignificant. The accuracy of the steady state solution is only determined by the discretization of the flux balances, which are evaluated by the trapezoidal rule. According to [9] this yields second order accuracy on smooth meshes. The artificial dissipative operator $\vec{D}$ does not spoil the accuracy of the method. The addition of $\vec{D}$ prevents the flux balances to go exactly to zero at steady state, but since $\vec{D}$ is of third order for smooth flows, the influence of $\vec{D}$ diminishes rapidly with mesh refinement. Note that these arguments do only hold if the complete conservative flux balances $\vec{Q}_{i-1/2, j-1/2}$ are used, because only the complete flux balances approach zero at steady state.

Figure 8 shows the corresponding convergence histories on all three meshes to 10^{-6}. These rates are comparable to the rates obtained with a central differencing scheme according to [8], [9], provided no acceleration techniques such as multigrid are used.

3.5 Results for Transonic Flows

The next cases to be considered are transonic flows to demonstrate the shock capturing capabilities of the method. The computational grids are always O-meshes which consist of 128 cells around the airfoil and of 24 cells in normal direction.

Figures 9a-c show results of the calculation of the flow around the NACA 0012 at $M_\infty = 0.8$ and $\alpha = 1.25°$. It can be seen from the pressure distribution in Fig. 9a that the shock on the upper and even the weak shock on the lower surface are captured within one cell. In Fig. 9b lines of constant Mach number are shown together with the computational grid to demonstrate the very good shock resolution not only on the body contour but in the complete flow field. The convergence rate for this case is given in Fig. 9c. The rate is comparable to those obtained for the subsonic case.

Figure 10 shows the pressure distribution for the flow around the same airfoil, but for free stream conditions $M_\infty = 0.85$, $\alpha = 1.0°$. The strong shock on the upper surface is captured within two cells, while the weaker shock on the lower surface spaces just one cell.

Figures 11a-c give results obtained for the calculation of the transonic flow around the RAE 2822 airfoil at $M_\infty = 0.75$, $\alpha = 3.0°$. From the pressure distribution in Fig. 11a the strong acceleration of the flow around the leading edge is clearly visible. Again the shock on the upper side is captured within one cell. Figure 11b shows the corresponding entropy distribution on the surface of the airfoil. Note that despite the strong acceleration of the flow the entropy losses in the nose region are well below 0.5%. At the shock entropy shows an almost perfect jump with only few oscillations in front of the shock. Convergence for this case is also reasonable, as can be seen in Fig. 11c.

4. CONCLUSION

A cell vertex scheme based on Flux Balance Splitting has been presented. In contrast to classical cell centered upwind schemes, no Riemann problem is solved on the cell interfaces. Instead, characteristic flux balances are distributed according to the sign of corresponding eigenvalues, and contributions to the nodes of the computational grid are summed up to yield the discrete equations. The basic ideas of Flux Balance Splitting have been outlined for the solution of the quasi 1-D Euler equations, and results for steady flows demonstrated the remarkable shock capturing capabilities of the method. Due to the fact that the source term of the quasi 1-D Euler equations was consistently included into the discrete flux balances, arguments for second order steady state solutions could be given. Second order accuracy was obtained despite the fact that formally one-sided, two-point difference formulas were used.

The scheme has been extended to the solution of the two-dimensional Euler equations. Since the 2-D Euler equations can not be completely diagonalized, upwinding was performed along streamlines, only for the normal momentum equation the crossflow direction was used. Second order accuracy of the scheme has been demonstrated by successive mesh refinement, and the convergence rates of the scheme are comparable to other methods. The calculation of various transonic flow fields demonstrated the very good shock capturing capabilities of the scheme. Due to the fact that only transonic flows were considered, no conceptual difficulties arose at shocks, since in such flows only nearly normal shocks occur, and the quasi 1-D characteristic decomposition in the present approach remains valid.

5. REFERENCES

[1] Powell, K.G., Van Leer, B. *A Genuinely Multi-Dimensional Upwind Cell Vertex Scheme for the Euler Equations.* AIAA Paper No. 89-0095, 1989.

[2] Hirsch, Ch., Lacor, C., Deconinck, H. *Convection Algorithms Based on a Diagonalization Procedure for the Multidimensional Euler Equations.* AIAA Paper No. 87-1163, 1987.

[3] Giles, M., Anderson, W., Roberts, T. *Upwind Control Volumes: A New Approach.* AIAA Paper No. 90-0104, 1990.

[4] Rossow, C.-C. *Flux Balance Splitting - A New Approach for a Cell Vertex Upwind Scheme.* Proceedings of 12th Int. Conference on Numerical Methods in Fluid Dynamics, Oxford, Great Britain, 1990.

[5] Struijs, R., Deconinck, H., Roe, P.L. *Fluctuation splitting for the 2-D Euler equations.* VKI Lecture series 1991-01, 1991.

[6] Roe, P.L. *Approximate Riemann Solvers, Parameter Vectors, and Difference Schemes.* Journal of Computaional Physics 43, pp.357-372, 1981

[7] Li, H., Kroll, N. *Solution of One- and Two-Dimensional Euler Equations Using a TVD Scheme.* DLR IB 129 - 88/24, 1988.

[8] Jameson, A., Schmidt, W., Turkel, E. *Numerical Solution of the Euler Equations by Finite Volume Methods Using Runge-Kutta Time Stepping Schemes.* AIAA 81-1259, 1981.

[9] Rossow, C.-C. *Berechnung von Strömungsfeldern durch Lösung der Euler-Gleichungen mit einer erweiterten Finite-Volumen Diskretisierungsmethode* DLR FB 89/18, 1989.

[10] Jameson, A., Baker, T.J. *Solution of the Euler Equations for Complex Configurations.* AIAA Paper 83-1929, 1983.

6. FIGURES

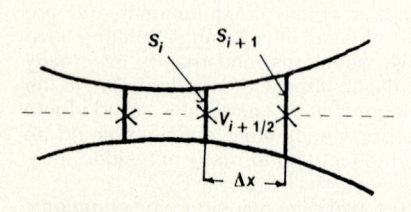

Figure 1: Sketch of 1-D control volume

× location of flow variables at nodes of 1-D mesh

Figure 2: Distribution of characteristic flux balances

Subsonic flow
$\lambda_1 = u - c < 0$
$\lambda_2 = u \quad > 0$
$\lambda_3 = u + c > 0$

$\vec{C}^L = \begin{bmatrix} C_1 \\ 0 \\ 0 \end{bmatrix}, \quad \vec{C}^R = \begin{bmatrix} 0 \\ C_2 \\ C_3 \end{bmatrix}$

Supersonic flow
$\lambda_1 = u - c > 0$
$\lambda_2 = u \quad > 0$
$\lambda_3 = u + c > 0$

$\vec{C}^L = \begin{bmatrix} 0 \\ 0 \\ 0 \end{bmatrix}, \quad \vec{C}^R = \begin{bmatrix} C_1 \\ C_2 \\ C_3 \end{bmatrix}$

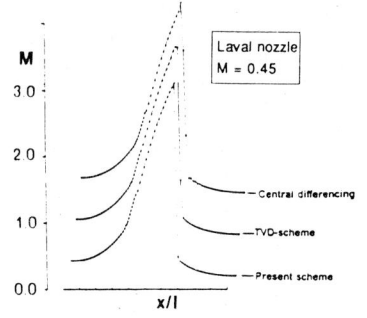

Figure 3a: Mach number distribution of Laval nozzle

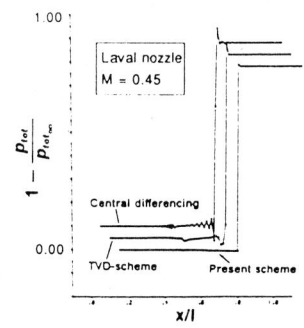

Figure 3b: Total pressure loss of Laval nozzle

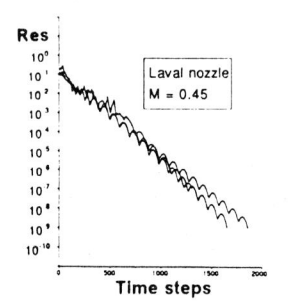

Figure 3c: Convergence histories

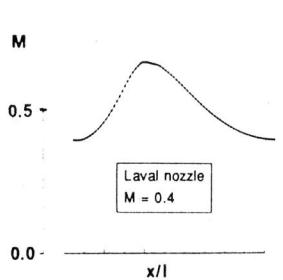

Figure 4a: Mach number distribution

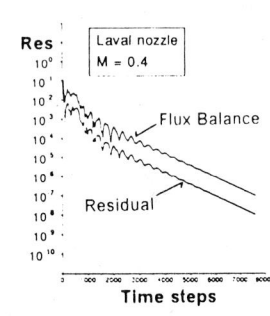

Figure 4b: Convergence histories

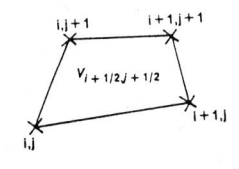

Figure 5: Sketch of 2-D control volume

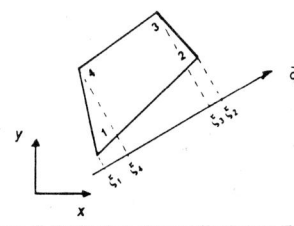

Figure 6: Projection of coordinates on flow direction

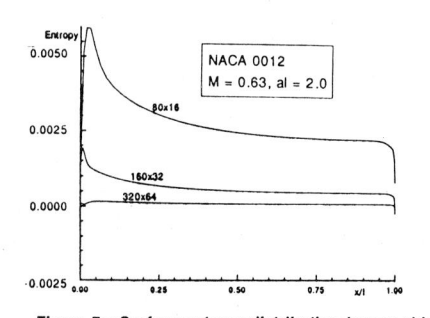

Figure 7a: Surface entropy distribution (upper side)

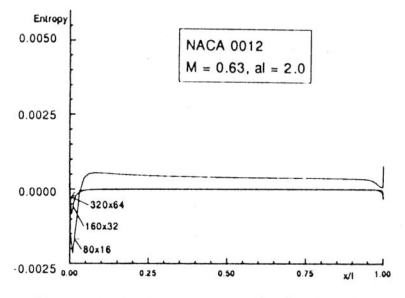

Figure 7b: Surface entropy distribution (lower side)

Figure 8: Convergence histories for mesh refinement

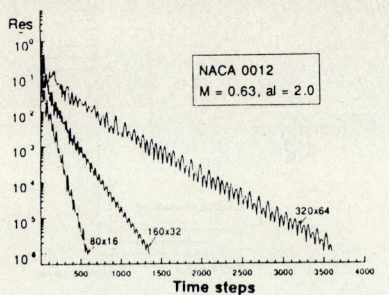

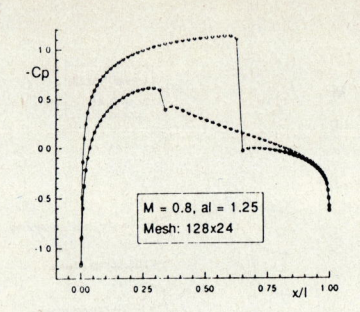

Figure 9: Surface pressure distribution for NACA 0012

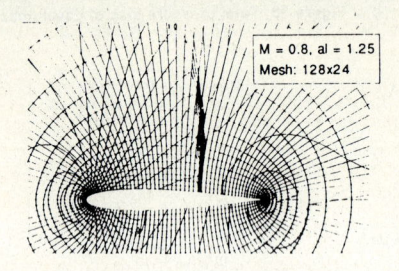

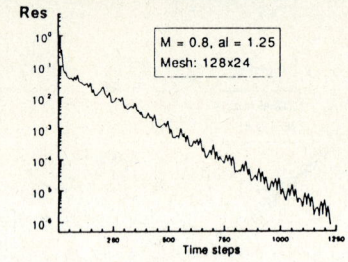

Figure 9: Mach number distribution for NACA 0012 ($\Delta M = 0.05$)

Figure 10: Convergence history for NACA 0012

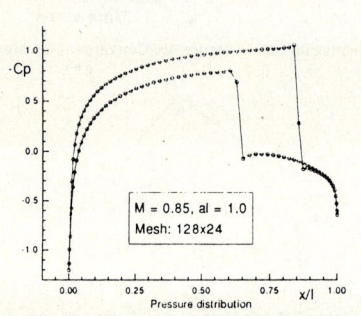

Figure 10: Surface pressure distribution for NACA 0012

Figure 11a: Surface pressure distribution for RAE 2822

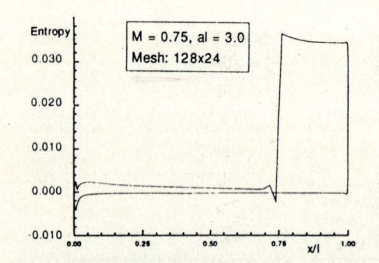

Figure 11b: Surface entropy distribution for RAE 2822

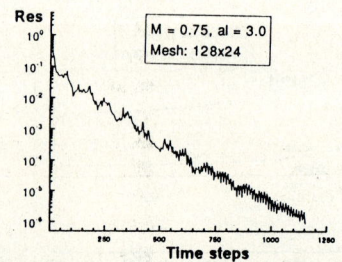

Figure 11c: Convergence history for RAE 2822

Hybrid Spectral Element Methods for Shock Wave Calculations

David Sidilkover, John Giannakouros, and George Em Karniadakis
Department of Mechanical and Aerospace Engineering
Program in Applied and Computational Mathematics
Princeton University
Princeton, NJ 08544
USA

1 Introduction

Spectral and spectral element methods have been used almost exlusively over the last decade in simulating turbulent incompressible flows. Their success however in simulating realistic problems of compressible flows has been somewhat limited despite some significant advances (see [3, 2, 4]). There are several issues to be addressed for a succesful implementation of spectral methods to compressible flow simulations. Treatment of shocks, recovery of spectral accuracy, robustness, boundary conditions, and application to complex geometries are perhaps the most important ones. While a specific implementation may be appropriate for resolving some of these issues (see, for instance [3]), usually this approach may result in a loss of generality of the method.

There has been an increasing trend over the last decade towards numerical schemes that are based on a hybrid construction of standard discretization algorithms, i.e. finite difference, finite elements, and spectral methods. A typical example of such a confluence of numerical algorithms is the spectral element method which is based on two weghted-residual techniques: finite element and spectral methods ([12]). This hybrid construction provides the desired flexibility of solving problems in complex geometries with spectral accuracy.

The work we propose in this paper stems from similar ideas: The objective is to obtain high-order (spectral-like) accuracy away from discontinuities while at the same time reprsent shocks by sharp transition layers. To this end we propose two hybrid spectral element methods: The first method is based on non-oscillatory approximations and is capable of resolving with high accuracy a very wide spectrum of scales in presence of shock waves in the solution. The second one is based on flux corrected transpost ideas (FCT, see [10, 11]) and it is more robust. Both one-dimensional and two-dimensional implementations are presented.

The paper is organized as follows: In section 2 we present a conservative formulation that employs staggered mesh discretizations. In section 3 we formulate the boundary and interfacial conditions. In sections 4 and 5 we present the spectral element-non oscillatory and spectral element-FCT algorithm, respectively. Finally, in section 6 we present numerical results obtained using both methods and we conclude in section 7 with a brief discussion.

2 Conservative Discretization

Let us consider for the sake of clarity the discretization of a scalar conservation law

$$u_t + f(u)_x = 0. \tag{1}$$

In the general case that we consider in this work the nodal points are distributed in a non-uniform manner and thus we need to define appropriate cell averaged quantities. In particular, adopting the terminology explained in Fig.1 the cell averaged velocity $\bar{u}_j$ is given by,

$$\bar{u}_j \equiv \bar{u}(x_j, t) = \frac{1}{x_{i^+} - x_{i^-}} \int_{x_{i^-}}^{x_{i^+}} u(x, t) dx. \tag{2}$$

Given this definition, equation (1) can be integrated along a cell extending from i^- to i^+ as follows,

$$\frac{d\bar{u}_j}{dt} + \frac{f(u_{i^+}) - f(u_{i^-})}{\Delta x_j} = 0 \tag{3}$$

where we have also defined

$$\Delta x_j \equiv x_{i^+} - x_{i^-}. \tag{4}$$

The above equation therefore suggests that the fluxes $f(u)$ should be evaluated at the ends of the cell using *de-averaged* (reconstructed) velocity values; this formulation leads to the conservative (or flux) form of the semi-discrete wave equation.

A spectral-Chebyshev expansion corresponds to a non-uniform distribution of points with cells of variable size Δx_j. Following the formulation of Cai et al [2] we select the set of points j to be the Gauss-Chebyshev points (see Fig.2) defined by,

$$x_j = \cos((j - 1/2)\Delta\theta) \quad \text{where} \quad \Delta\theta = \pi/N \quad \text{and} \quad 1 \leq j \leq N \tag{5}$$

while the end points i^+, i^- of each cell are the Gauss-Lobatto points defined as

$$x_i = \cos(i\Delta\theta) \quad 0 \leq i \leq N. \tag{6}$$

We refer to [2] for the details regarding averaging and reconstruction (of the point values from the cell averages) procedures. Note, that there are $N + 1$ point values, while only N cell averages. This means that that the degree of the "reconstructed" polynomial is one degree less than needed. Imposing boundary conditions is equivalent to increasing a degree of the recontstructed polynomial by 1.

The same averaging and reconstruction procedures can be repeated here for each element. To form a global interpolant however we need to impose a continuity condition at elemental interfaces as we explain in the next Section.

3 Boundary and Interfacial Conditions

The system of Euler equations for polytropic gas in one dimension is given by:

$$\mathbf{u}_t + \mathbf{f}(\mathbf{u})_x = 0, \tag{7}$$

whith

$$\mathbf{u} = \begin{pmatrix} \rho \\ m \\ E \end{pmatrix}, \quad \mathbf{f} = \begin{pmatrix} \rho q \\ qm + P \\ q(P + E) \end{pmatrix}, \tag{8}$$

$$P = (\gamma - 1)(E - \frac{1}{2}\rho q^2), \tag{9}$$

where ρ denotes density, q is velocity, P pressure, E total energy, $m = \rho q$ is the momentum and γ is the ratio of the specific heats of a polytropic gas.

While the discretization of the Euler system using the cell-averaging approach is straightforward, the imposition of the interfacial condition requires further discussion. A method for imposing characteristic boundary conditions on the conservative variables is presented in [5]. The interfacial condition can be imposed in a similar way. However, note that the reconstructed values are needed for the computation of the numerical fluxes. Therefore, the natural thing would be to impose boundary and interfacial conditions on the numerical fluxes directly. We describe the procedure below.

Consider the Jacobian matrix of the system given by $A(\mathbf{u}) = \partial \mathbf{f}/\partial \mathbf{u}$.

The right-eigenvectors of A are

$$\mathbf{r}_1(\mathbf{u}) = \begin{pmatrix} 1 \\ q - c \\ H - qc \end{pmatrix}, \quad \mathbf{r}_2(\mathbf{u}) = \begin{pmatrix} 1 \\ q \\ \frac{1}{2}q^2 \end{pmatrix}, \quad \mathbf{r}_3(\mathbf{u}) = \begin{pmatrix} 1 \\ q + c \\ H + qc \end{pmatrix}, \tag{10}$$

where $c = \sqrt{\gamma P/\rho}$ is the speed of sound and H represents enthalpy and is defined by

$$H = \frac{(E + P)}{\rho} = \frac{c^2}{\gamma - 1} + \frac{1}{2}q^2. \tag{11}$$

Let us denote the matrix of right-eigenvectors of the Jacobian $A = A(\tilde{\mathbf{u}})$ as,

$$R = (\mathbf{r}_1(\tilde{\mathbf{u}}), \mathbf{r}_2(\tilde{\mathbf{u}}), \mathbf{r}_3(\tilde{\mathbf{u}})), \tag{12}$$

where $\tilde{\mathbf{u}}$ is Roe-averaged state (see [14]) between the states $\mathbf{u}_N^k$ and $\mathbf{u}_0^{k+1}$ at the ends of two adjacent elements $k, k+1$. Here N and 0 denote the rightmost and the leftmost elemental nodes respectively.

Then

$$D \equiv R^{-1} \cdot A \cdot R = \begin{pmatrix} \lambda_1 & 0 & 0 \\ 0 & \lambda_2 & 0 \\ 0 & 0 & \lambda_3 \end{pmatrix}, \tag{13}$$

where the eigenvalues of A are given by

$$\lambda_1 = q - c, \quad \lambda_2 = q, \quad \lambda_3 = q + c. \tag{14}$$

Then the flux to be imposed at the interface γ between elements k and $k+1$ is determined by Roe's flux-splitting (see [14])

$$\mathbf{f}_\gamma = \frac{\mathbf{f}(\mathbf{u}_0^{k+1}) + \mathbf{f}(\mathbf{u}_N^k)}{2} - R \cdot |D| \cdot R^{-1} \frac{(\mathbf{u}_0^{k+1} - \mathbf{u}_N^k)}{2}. \tag{15}$$

Next, we briefly review the non-oscillatory spectral element method presented in [15]. Then we extend the FCT-spectral element method for the case of systems. This method was formulated previously for scalar conservation laws in [8].

4 Non-Oscillatory Method

The main difficulty in applying spectral methods to discontinuous problems is the Gibbs phenomenon. If a discontinuous function is approximated by a spectral expansion (Chebyshev, Fourier etc.), the approximation is only $O(1/N)$ accurate in smooth regions and contains $O(1)$ oscillations near the discontinuity. When spectral methods are applied to partial differential equations with discontinuous solutions, the Gibbs phenomenon may also lead to numerical instability.

An interesting approach to construct a non-oscillatory spectral approximation to a discontinuous function has been recently proposed in [3]. Let $u(x)$ be a piecewise C^∞ function with a jump discontinuity at point x_s and with a jump $[u]_{x_s}$. The key idea in [3] was to augment the Fourier spectral space with a saw-tooth function. It was shown that the approximation using the augmented spectral space will be non-oscillatory if the saw-tooth function approximates the strength and the location of the discontinuity with *second order* accuracy. In addition, a method for estimating the discontinuity parameters with specified accuracy based on the spectral expansion coefficients was suggested. More recently, it was pointed in [4] that a *first order* accurate approximation of discontinuity magnitude also leads to non-oscillatory behavior. Numerical experiments with discontinuous solutions using the Chebyshev spectral space augmented by a step function and cell averaging approach were reported in [2]. This approach was extended in [15] for the spectral element discretization. In addition, a subcell resolution method (see [9]) was used for estimation of the discontinuity parameters with an accuracy required for obtaining a non-oscillatory approximation. Some numerical experiments with this method are presented in Section 6.

5 FCT Method

The spectral element-FCT method was formulated for scalar conservation laws in [8]. Here we extend it for the systems of nonlinear conservation laws. This method consists of two stages: a transport - diffusive stage and an antidiffusive or corrective stage. In the first stage, a first order positive type scheme is implemented, while in the second a "limited" correction due to the spectral element discretization is made.

The main steps of the proposed algorithm are as follows:

- Step 1: Evaluate field of cell averages corresponding to the initial condition.

- Step 2: Compute the transportive fluxes corresponding to the low order scheme. The low order positive type scheme used here is Roe's scheme based on the cell-averaged values. The low order flux $\mathbf{F}_{I+}$ is defined as follows:

$$\mathbf{F}_{I+} = \frac{\mathbf{f}(\bar{\mathbf{u}}_{J+1}) + \mathbf{f}(\bar{\mathbf{u}}_J)}{2} - R \cdot |D| \cdot R^{-1} \frac{(\bar{\mathbf{u}}_{J+1} - \bar{\mathbf{u}}_J)}{2} \qquad (16)$$

where R is the matrix consisting of the right-eigenvectors of the Jacobian of the Euler system linearized around the Roe-averaged state between $\bar{\mathbf{u}}_{J+1}$ and $\bar{\mathbf{u}}_J$.

- Step 3: Advance (explicitly) cell averages in time using low order fluxes to obtain the low order transported and diffusive solution $\bar{\mathbf{u}}_J^{td}$. This is done using the third order Adams-Bashforth scheme ([6]).

- Step 4: Compute the transportive fluxes $\mathbf{f}_I$ corresponding to the spectral element discretization.

- Step 5: Compute the antidiffusive fluxes $\mathbf{A}_I = \mathbf{f}_I - \mathbf{F}_I$ and limit them to obtain $\mathbf{A}_I^c$. Here we use the same limiter as in [8] which is based on the original ideas by Boris and Book ([10]) and the extensions presented in [1]. We found it crucial that the limiter be applied to the characteristic antidiffusive fluxes and not componentwise.

- Step 6: Update (explicitly) the cell averages on the new time level using the limited antidiffusive fluxes (using the third order Adams-Bashforth scheme) $\bar{u}_j^{n+1}$.

- Step 7: Reconstruct point values from the cell averages at the new time level.

- Step 8: If the target time is not achieved go to Step 2.

In this section we describe in some detail the time stepping procedure, low order scheme and limiter.

6 Numerical Results

In this section we will present results of several numerical experiments including Euler system with Non-oscillatory method and Euler system with FCT method. Also we will report preliminary results concerning two-dimensional linear advection equation with spectral element-FCT method.

6.1 Non-Oscillatory Algorithm

Here we present our numerical experiments with the test problem considered in [2, 4]. We consider the following initial condition for (7),

$$\begin{cases} \rho_l = 3.857143, \quad q_l = 2.629367, \quad P_l = 10.33333 & -5 \leq x \leq -4, \\ \rho_r = 1 + \varepsilon \sin \pi x, \quad q_r = 0, \quad P_r = 1 & -4 < x \leq 5, \end{cases} \quad (17)$$

where $\varepsilon = 0.2$. The solution to (17) models the interaction between a moving shock and sinusoidal density disturbances (see [2, 4]). In Fig.4 we display the density profile for $N = 10$ and $K = 22$ (corresponding to 199 grid points) at time $t = 1.8$. The discontinuity cell was located using the momentum equation. We also plot for comparison the solution obtained by the second order MUSCL scheme with $N = 200$ in Fig.5. This example illustrates the superiority of high order method in resolving fine structures arising behind the moving shock wave.

6.2 FCT Algorithm

Here we consider the standard Riemann problems of (7) with the following initial data [16]:

1. Sod's problem

$$\begin{cases} \rho_l = 1, \quad q_l = 0, \quad P_l = 1 & -1 \leq x \leq 0, \\ \rho_r = 0.125, \quad q_r = 0, \quad P_r = 0.1 & 0 \leq x \leq 1, \end{cases} \quad (18)$$

2. Lax's problem

$$\begin{cases} \rho_l = 0.445, & q_l = 0.698, & P_l = 3.528 & -1 \leq x \leq 0, \\ \rho_r = 0.5, & q_r = 0, & P_r = 0.571 & 0 \leq x \leq 1. \end{cases} \quad (19)$$

The solution to the Sod's problem for $N = 75$, $K = 2$ elements at time $t = 4.0$ is plotted in Fig.6 The "reconstructed" polynomials in this case are of the order 74. The solution to the Lax's problem for for the same discretization, but at time $t = 4.0$ is plotted in Fig.7.

We also present in Fig.8 the solution to the Lax's problem obtained for $N = 3$, $K = 75$. This discretization corresponds to the "reconstructed" polynomials of the second order. The quantities plotted are cell-averaged values. Therefore, the solution is at most second order accurate in all the cases. The point we want to make is that the high order spectral element discretization is responsible for remarkably sharp representation of discontinuities in these cases. The comparison of Figures 7 and 8 (and also the results obtained by the second order accurate finite element methods presented in [17]) clearly illustrates this point.

We present in Fig.9 for comparison reasons density plots of the solutions to the Sod's and Lax's problems obtained by Cai and Shu using their hybrid ENO-spectral method developed in [4]. The representation of the discontinuities is dictated in this case by the third order finite difference ENO discretization. (The subcell resolution was used at the contact discontinuity in the Lax's problem.) We can conclude that the examples presented here demonstrate the superiority of the high order methods in representing discontinuities.

6.3 Two-Dimensional Linear Advection

We present here preliminary results on spectral element-FCT algorithm for two-dimensional linear advection. It can be shown that the averaging and reconstruction procedures in two dimensions can be reduced to the two subsequent application of one-dimensional procedures in each coordinate direction (for details see [7]). The problem considered here is

$$u_t + u_x + u_y = 0 \quad (20)$$

on the domain $[0, 2] \times [0, 2]$. Initial conditions are given by

$$u_0(x, y) = \begin{cases} 2, & \text{if} \quad x + y \leq 1 \\ .5, & \text{if} \quad x + y > 1 \end{cases} \quad (21)$$

The slices of the solution at time $t = 1$ are presented at Fig.10. The contact discontinuity in this case is represented by a sharp transition layer.

7 Discussion

In this work we have presented two new hybrid spectral element methods for solution of the systems of conservation laws.

One method is based on the non-oscillatory approximation concepts. The results show that this approach leads to a stable method. The new method is also capable of resolving very accurately fine structures arising from interactions of shocks with disturbances. The generalization of the method for the case of multiple discontinuities is straightforward.

However, the disadvantage of the method is that it requires to identify a discontinuity and to treat it in a special way (subcell resolution), and that it is not capable of treating rarefaction waves.

Another method we formulated here is a spectral element-FCT method. Unlike the previous one, this method is very robust. The numerical results demonstrate a clear superiority of high order shock capturing methods in resloving discontinuities (representing them by very sharp layers). However, this method is not as accurate in the smooth regions as the previous one.

A difficulty still to be resolved is the problem typical of any shock-capturing method on nonuniform meshes (see [13]) is $O(1)$ error behind a moving strong shock. Currently, we are working on a hybrid shock capturing method which allows to obtain sharp discontinuities, while at the same time to achieve good accuracy in the smooth regions.

Acknowledgements

We would like to thank Professor David Gottlieb for many helpful suggestions. This work wa supported by AFOSR Grant number 90-0261.

References

[1] McDonald B.E. Flux-corrected pseudospectral method for scalar hyperbolic conservation laws. *J. Comput. Phys.*, 82:413, 1989.

[2] W. Cai, D. Gottlieb, and A. Harten. Cell averaging Chebyshev methods for hyperbolic problems. Report No. 90-72, ICASE, 1990.

[3] W. Cai, D. Gottlieb, and C. Shu. Non-oscillatory spectral Fourier methods for shock wave calculations. *Math. Comp.*, 52:389–410, 1989.

[4] W. Cai and C. Shu. Uniform high order spectral methods for one and two dimensional Euler equations. Submitted for publication.

[5] W. S. Don and D. Gottlieb. Spectral simulations of an unsteady compressible flow past a circular cylinder. *Comp. Meth. Appl. Mech. and Eng.*, 80:39–58, 1990.

[6] C. W. Gear. *Numerical Initial Value Problems in Ordinary Differential Equations.* Prentice-Hall, 1973.

[7] J. Giannakouros. Spectral element-FCT method for compressible flow calculations. Master's thesis, Princeton University. in preparation.

[8] J. Giannakouros and G. E. Karniadakis. Spectral element FCT method for scalar hyperbolic conservation laws. to appear in *Int. J. Num. Meth. in Fluids*.

[9] A. Harten. ENO schemes with subcell resolution. Report No. 87-86, ICASE, 1987.

[10] Boris J.P. and Book D.L. Flux-corrected transport. I SHASTA, a transport algorithm that works. *J. Comput. Phys.*, 11:38, 1973.

[11] Boris J.P. and Book D.L. Flux-corrected transport. III minimal-error FCT algorithms. *J. Comput. Phys.*, 20:397, 1975.

[12] G. E. Karniadakis. Spectral element simulations of laminar and turbulent flows in complex geometries. *Appl. Num. Math.*, 6:85, 1989.

[13] P. L. Roe. Private communication.

[14] P. L. Roe. Approximate Riemann solvers, parameter vectors, and difference schemes. *J. Comp. Phys.*, 43:357–372, 1981.

[15] D. Sidilkover and G. E. Karniadakis. Non-oscillatory spectral element Chebyshev method for shock wave calculations. Submitted for publication.

[16] G. A. Sod. A survey of several finite difference methods for systems nonlinear hyperbolic conservation laws. *J. Comp. Phys.*, 27:1–31, 1978.

[17] J. Y. Yang, F. S. Lien, and C. A. Hsu. Non-oscillatory shock-capturing finite element methods for the one-dimensional compressible Euler equations. *Int. J. Num. Meth. in Fluids.*, 11:405–426, 1990.

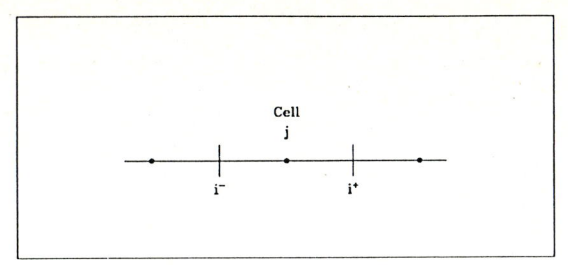

Figure 1: Cell average and point values.

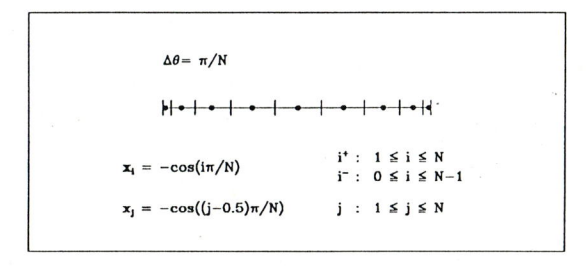

Figure 2: Spectral Chebyshev method. The set of points j defines the cell averaged quantities, while the point values i are used in evaluating the fluxes.

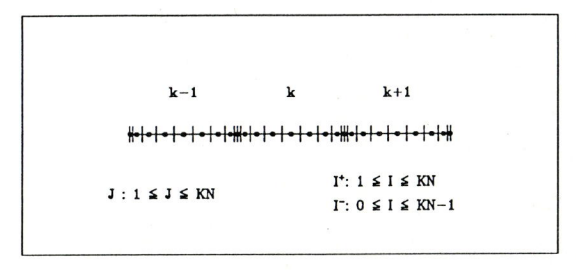

Figure 3: Spectral element Chebyshev discretization. J's are used for global indexing of cell averaged quantities and I's - for global indexing of point values.

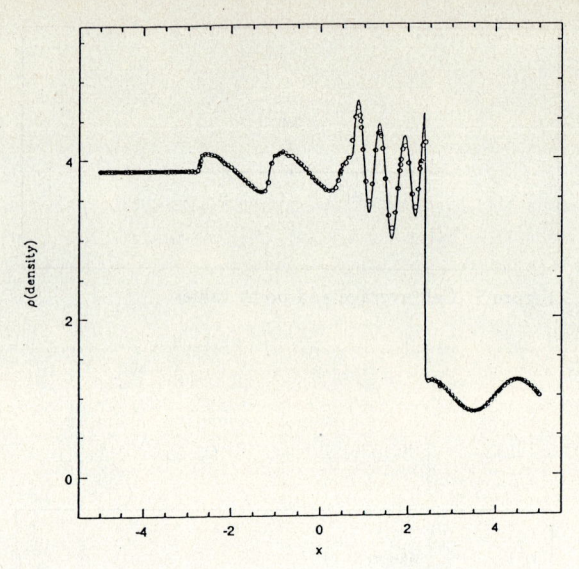

Figure 4: Spectral element non-oscillatory method. Moving shock interacting with sinusoidal density disturbances $K = 22$, $N = 10$ (199 grid points) at time $t = 1.8$ The solid represents the solution obtained by the third order ENO finite difference method with 1200 grid points (courtesy of Wai-Sun Don and David Gottlieb, Brown University).

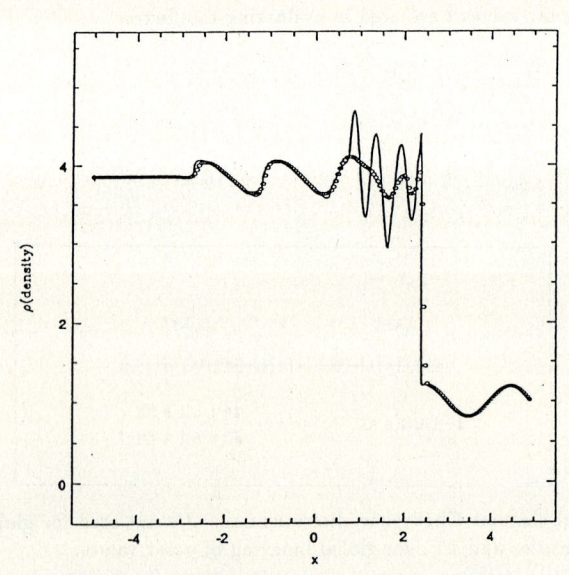

Figure 5: Solution to the same problem as in the previous figure obtained by MUSCL scheme using 200 grid points (courtesy of Wei Cai and Chi-Wang Shu).

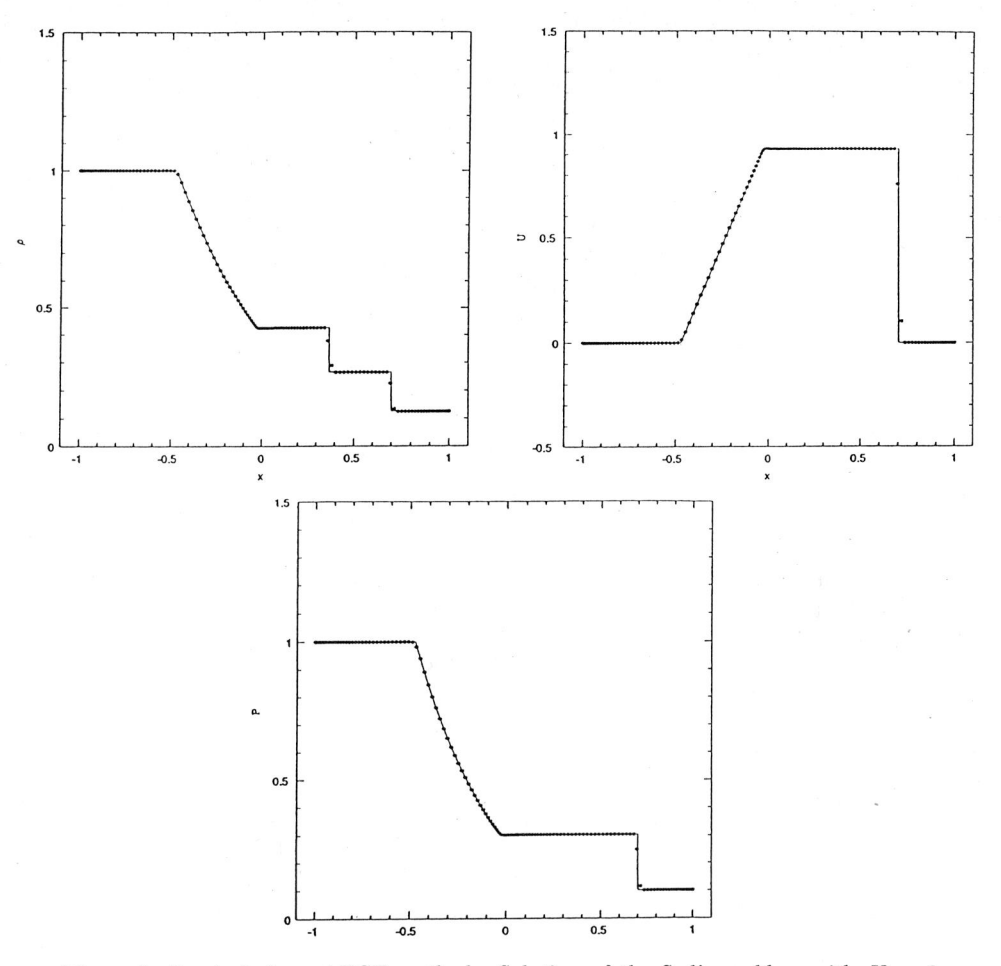

Figure 6: Spectral element-FCT method. Solution of the Sod's problem with $K = 2$, $N = 75$ at time $t = 0.4$. (a) density, (b) velocity, (c) pressure (the solid line represents the exact solution).

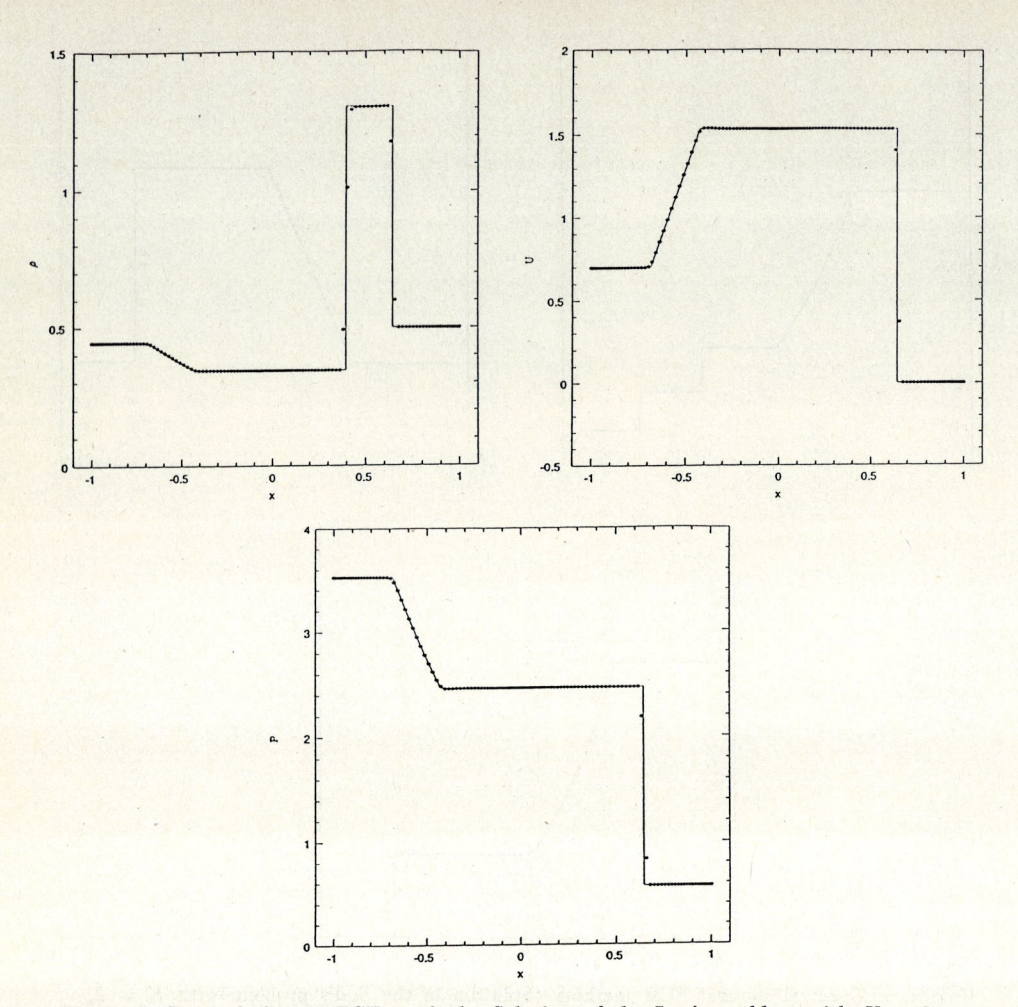

Figure 7: Spectral element-FCT method. Solution of the Lax's problem with $K = 2$, $N = 75$ at time $t = 0.26$. (a) density, (b) velocity, (c) pressure (the solid line represents the exact solution).

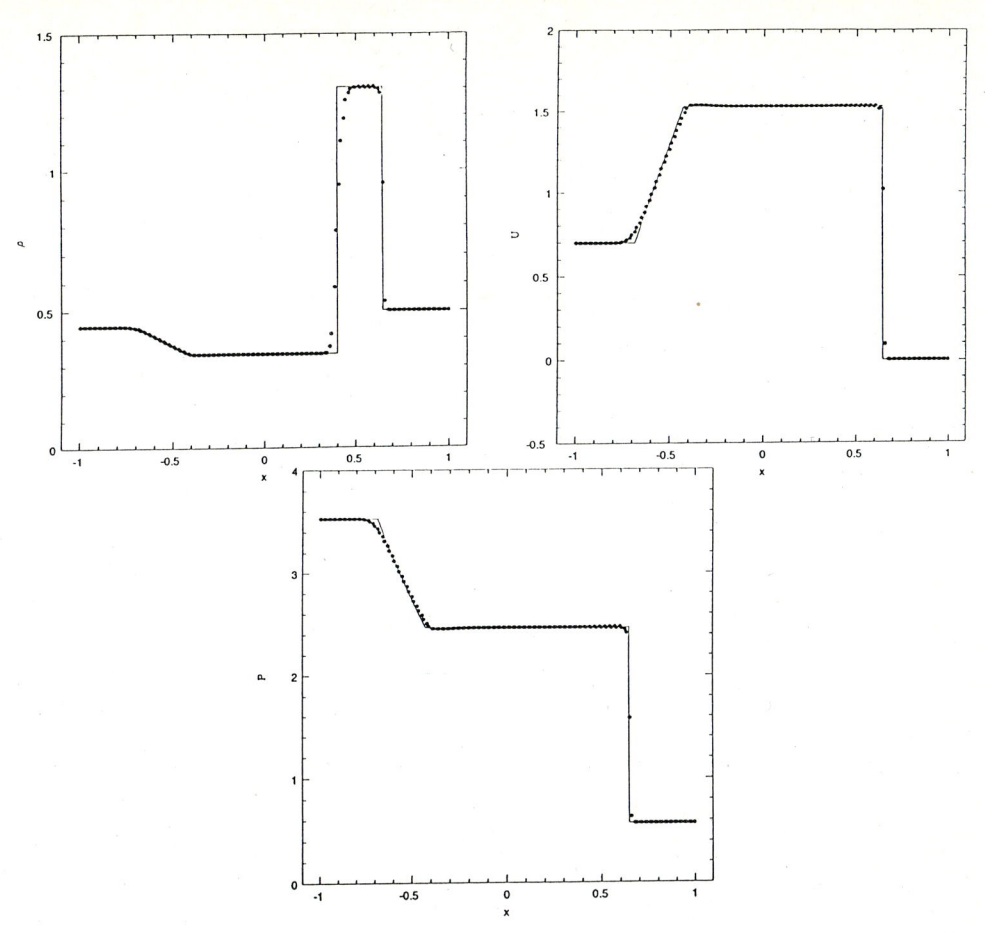

Figure 8: Spectral element-FCT method. Solution to the same problem as at the previous figure with $K = 75$, $N = 3$ (151 gridpoints). (a) density, (b) velocity, (c) pressure (the solid line represents the exact solution).

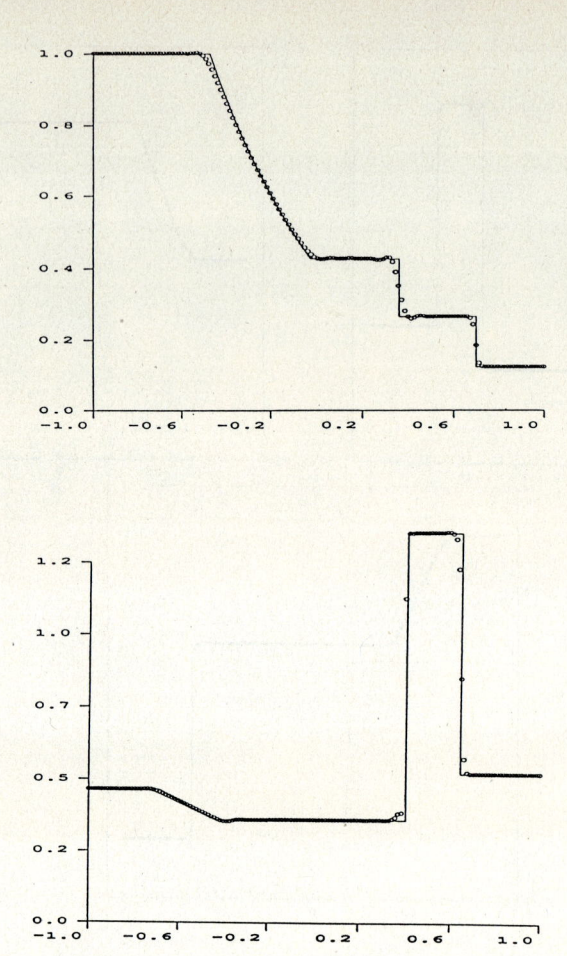

Figure 9: Density solutions obtained by the hybrid ENO-spectral method by Wei Cai and Chi-Wang Shu using 150 gridpoints (see [4]). (a) Sod's problem. (b) Lax's problem.

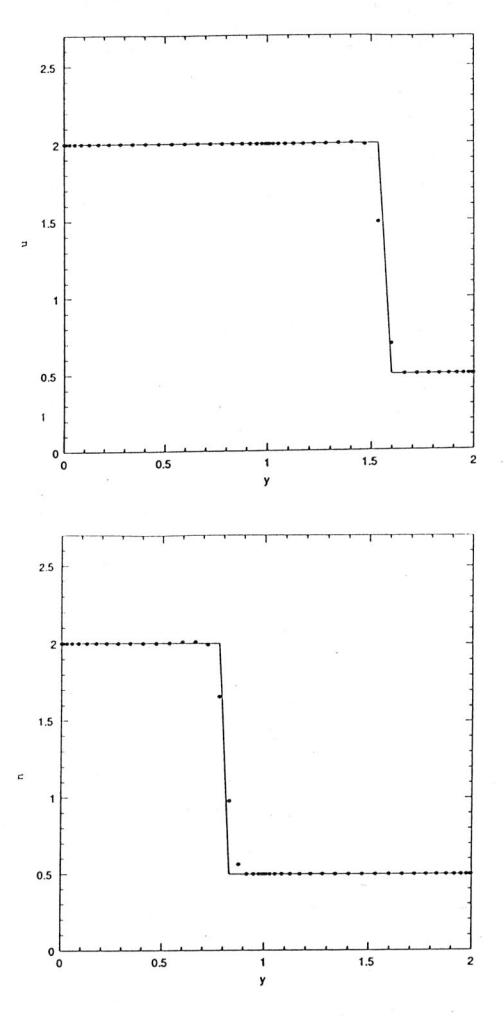

Figure 10: Two dimensional linear advection with $K_x = K_y = 2$, $N_x = N_y = 25$. Slices of the solution along: (a) the line $x = .885257$, (b) the line $x = 1.624345$ (the solid line represents the exact solution).

VISCOUS FLUX LIMITERS

E. F. Toro

Department of Aerospace Science
College of Aeronautics
Cranfield Institute of Technology
Cranfield, Beds MK43 0AL
England.

Abstract

We present **Numerical Viscosity Functions**, or NVFs, for use with Riemann-problem based shock-capturing methods as applied to viscous flows. In particular, **viscous flux limiters** are derived. The analysis pertain to a linear convection-diffusion model equation. Our NVFs combine the physical viscosity, the role of which is maximised, with numerical viscosity, whose role is minimised, to capture TVD solutions to viscous flows.

1 Introduction

The development of Riemann-problem based shock-capturing methods for hyperbolic problems has been a significant contribution to Computational Fluid Dynamics (CFD) in the last decade. These methods are high-order extensions of the first-order Godunov method [1]. Two main features are: they use solutions to local Riemann problems and introduce implicit numerical viscosity to capture oscillation-free inviscid shocks. Applications of these methods to a wide variety of inviscid flows have proved very successful.

Quite recently these methods have also been used to solve viscous flow problems such as the Navier-Stokes equations (eg. [2], [3]). The basic strategy is to deploy these methods for the convective terms (the Euler equations) and use central differences, for example, to discretise the viscous terms. The approach has lead to satisfactory results. This is not surprising since, in the absence of strong diffusion, it is the discretisation of the convective terms what is really crucial. Riemann-problem based methods are in many ways optimal for treating convective terms, particularly, for the way the implicit numerical dissipation to capture oscillation-free shocks is controlled. In despite of this property some workers (e.g. [2]) have reported the presence of excessive artificial viscosity in the computed solutions. The explanation for this is that numerical viscosity functions, in the form of flux limiters for example, introduce numerical viscosity as if the equations were actually inviscid. These limiters are blind to the fact that the equations do have physical viscosity which is (correctly) superimposed on the added (excessive) numerical viscosity. In this paper we present numerical viscosity functions that incorporate fully the physical viscosity so as to introduce the absolute minimum of artificial viscosity for capturing oscillation-free shocks. The analysis is based on a model convection-diffusion equation and we select a particular Riemann-problem based method, namely the Weigthed Average Flux (or WAF) Method [4]. The principles however apply to other methods of this kind.

The resulting NVFs can immediately be translated to more conventional functions such as flux limiters and slope limiters ([5], [6], [7]). Numerical experiments for the linear model and for Burgers' equation show the advantage of using the new viscous NVFs over traditional limiters.

The remaining part of this paper is divided as follows: section 2 contains a succinct description of the WAF method, its application to the model equation and a derivation of TVD regions. Section 3 deals with construction of numerical viscosity functions; section 4 contains numerical experiments and conclusions are drawn in section 5.

2 TVD regions for a model equation

We consider the model convection-diffusion equation

$$u_t + au_x = \alpha u_{xx} \qquad (1)$$

where a is a wave propagation speed and α is a viscosity coefficient. Both a and α are assumed constant. We interpret the left-hand side of Eq. 1 as a conservation law with flux $F(u) = au$. Applying an explicit conservative method to the convective terms in (1) and central differences to the viscous term gives

$$u_i^{n+1} = u_i^n + \frac{\Delta t}{\Delta x}(F_{i-1/2} - F_{i+1/2}) + d(u_{i+1}^n - 2u_i^n + u_{i-1}^n) \qquad (2)$$

where $d = \frac{\alpha \Delta t}{\Delta x^2}$ is the diffusion number and $F_{i+1/2}$ is the intercell numerical flux for the pair of cells $(i, i+1)$. In this paper we use the WAF [4] numerical flux

$$F_{i+1/2} = \frac{1}{2}(1 + A_{i+1/2})F_i + \frac{1}{2}(1 - A_{i+1/2})F_{i+1} \qquad (3)$$

where $A_{i+1/2}$ is a numerical viscosity function yet to be constructed. If $A_{i+1/2} = c$, where $c = \frac{a\Delta t}{\Delta x}$ is the Courant number, the WAF method reduces identically to the Lax-Wendroff method, which is second-order accurate in space and time. This identity between WAF and Lax-Wendroff is only true for the linear pure convection part of Eq. 1 (left-hand side). The quantities Δx and Δt define the mesh size in space and time respectively. In the first part of the analysis we assume $a > 0$.

As it is well known, the Lax-Wendroff method is upwind biased, for the coefficients (or weights) $W_1 = \frac{1+c}{2}$ and $W_2 = \frac{1-c}{2}$ in Eq. (3) control the upwind F_i and downwind F_{i+1} contributions to the intercell numerical flux $F_{i+1/2}$, with $W_1 > W_2$ for positive speed a. TVD solutions are ensured by increasing the upwind contribution and decreasing the downwind contribution. The objective here is to construct numerical viscosity functions $A = A(r, c, d)$ that depend on a flow parameter r, the Courant number c and the diffusion number d.

Insertion of (3) into (2) and after rearranging gives

$$H = \frac{u_i^{n+1} - u_i^n}{u_{i-1}^n - u_i^n} = \frac{1}{2}c[\frac{1}{r_i}(1 - A_{i+1/2}) + A_{i-1/2} + 1] + d(1 - \frac{1}{r_i}) \qquad (4)$$

where the flow parameter r_i is given by

$$\frac{u_i^n - u_{i-1}^n}{u_{i+1}^n - u_i^n} = \frac{\Delta_{upw}}{\Delta_{loc}} \qquad (5)$$

and is the ratio of the upwind jump Δ_{upw} to the local jump Δ_{loc} across the wave of speed a. We now impose the sufficient TVD condition $0 \leq H \leq 1$ in eq. (4). After rearranging we obtain

$$-1 - \frac{2}{R_{ce}} + \frac{2}{R_{ce}r_i} \leq \frac{1}{r_i}(1 - A_{i+1/2}) + A_{i-1/2} \leq \frac{2-c}{c} - \frac{2}{R_{ce}} + \frac{2}{R_{ce}r_i} \qquad (6)$$

where $R_{ce} = \frac{c}{d}$ is the cell Reynold's number.

So far we have assumed that the speed a is positive. For negative a the result is the same as that of inequialities (6) but c is changed to $|c|$ and the upwind jump in (5) is changed to $\Delta_{upw} = u_{i+2}^n - u_{i+1}^n$.

Hence in what follows we adopt the general case

$$-1 - \frac{2}{R_{ce}} + \frac{2}{R_{ce}r_i} \leq \frac{1}{r_i}(1 - A_{i+1/2}) + A_{i-1/2} \leq \frac{2-|c|}{|c|} - \frac{2}{R_{ce}} + \frac{2}{R_{ce}r_i} \qquad (7)$$

where now the cell Reynolds number is $R_{ce} = \frac{|c|}{d}$.

We now select two inequalities such that (7) holds automatically. These are

$$\frac{2}{R_{ce}r_i} - K \leq \frac{1}{r_i}(1 - A_{i+1/2}) \leq \frac{2-|c|}{|c|} - 1 + \frac{2}{R_{ce}r_i} \qquad (8)$$

$$-1 - \frac{2}{R_{ce}} + K \leq A_{i-1/2} \leq 1 - \frac{2}{R_{ce}} . \qquad (9)$$

For given c and d, or equivalently R_{ce}, inequalities (8)-(9) lead to TVD regions for A as a function of the flow parameter r (subscripts omitted) as shown in Fig. 1; dotted lines illustrate the boundaries of these regions for the inviscid case $d = 0$ which apply to the pure convection problem. The viscous TVD region is bounded by the thick full lines. There are two horizontal lines that define the upper and lower bounds A_M and L respectively. The lateral bounds for the left and right sides are given by the straight lines designated A_L and A_R. In the present study we keep the upper bound A_M fixed while all other bounds are allowed to vary.

The bounds are given by

$$A_M = 1 - \frac{2}{R_{ce}} \qquad (10)$$

$$L = -1 - \frac{2}{R_{ce}} \qquad (11)$$

$$A_L = 1 - \frac{2}{R_{ce}} + Kr \qquad (12)$$

$$A_R = 1 - \frac{2}{R_{ce}} - \frac{2(1-|c|)}{|c|}r . \qquad (13)$$

The parameter K is related to the minimum A_m by

$$K = A_m + 1 + \frac{2}{R_{ce}} . \qquad (14)$$

It is therefore the choice of K in the range

$$0 \leq K \leq |c| - 1 - \frac{2}{R_{ce}} \qquad (15)$$

that determines the value of the lower bound A_m as well as A_L. For $K = 0$ the boundary A_L coincides with A_M and the left TVD region coalesce to the single line A_M for negative r. Actually one could include the band bewteen A_M and 1 within the TVD region but this would result in excessive artificial viscosity. Note that the value $A = 1$ would give the Cole-Murmam first-order upwind method for the pure convection (inviscid) equation $u_t + au_x = 0$, which introduces too much artificial viscosity. The viscous version of this scheme is given by $A = A_M$, which reduces the amount of artificial viscosity of the Cole-Murmam scheme. The choice $A = |c|$ would give the Lax-Wendroff method for the inviscid equation which has no artificial viscosity. If $A_M \geq |c|$ then some artificial viscosity is needed to ensure TVD results. Otherwise the physical viscosity is sufficient to do this and one should select $A = |c|$ in that case. For $A \leq |c|$ one would be be adding negative artificial viscosity, that is to say the downwind contribution would be greater than that of the Lax-Wendroff scheme. The extreme case would be $A_M = A_m = -1$, which for the inviscid case leads to an unstable scheme.

The inclusion of the physical viscosity in the construction of the numerical viscosity functions A results in a reduction of the numerical viscosity introduced by the upwind contribution to the intercell flux. The physical viscosity compensates for that reduction.

3 Construction of numerical viscosity functions

There is virtually unlimited freedom in constructing NVFs A within the TVD regions shown in Fig. 1. The choice of the lower boundary A_m is significant. Here we select two cases, namely $A_m = |c|$ ($K = |c| - 1 - \frac{2}{R_{ce}}$) and $A_m = 2|c| - 1 - 2/Rce$. The first case generates NVFs that are associated with the well known flux limiter MINMOD or MINBEE while the second choice is associated with the flux limiter SUPERBEE. We thus call the respective families of NVFs the MINAD and SUPAD families.

Once a function A has been constructed we relate it to flux limiters via

$$B = \frac{A - 1}{|c| - 1} . \qquad (16)$$

3.1 The MINAD family

This family is shown in Fig. 2; it lies between the horizontal lines A_M and $A_m = |c|$ and the inclined lines

$$A_L = A_M + (|c| + 1 + \frac{2}{R_{ce}})r; A_R = A_M - (\frac{2}{|c|} + \frac{2}{R_{ce}})r . \qquad (17)$$

We select two members which we call MINAD1 and MINAD2. They are respectively shown by dotted and full lines. MINAD1 can be easily expressed as

$$MINAD1 = A_M + (|c| - A_M)rsign(r) for |r| \leq 1; MINAD1 = |c| \qquad (18)$$

otherwise MINAD2 passes through the points r_L and r_R which are given by

$$r_L = \frac{|c| + \frac{2}{R_{ce}} - 1}{|c| + \frac{2}{R_{ce}} + 1}; r_R = \frac{|c| + \frac{2}{R_{ce}} - 1}{\frac{2}{R_{ce}} - \frac{2}{|c|}} . \qquad (19)$$

3.2 The SUPAD family

This family is illustrated in Fig. 3. Only one member is selected and corresponds to the lowest boundary of the TVD region. For non-negative r and $d = 0$ this function can be identified with the inviscid flux limiter SUPERBEE. We therefore call it SUPAD.

4 Numerical experiments

Two test problems with exact solution are chosen for numerical experiments. Test 1 is the linear convection-diffusion equation (1) in the spacial domain $[0, 2]$, initial condition $u(0, x) = sin(\pi x)$ and periodic boundary conditions. The second test problem concerns Burgers' equation

$$u_t + uu_x = \alpha u_{xx} \tag{20}$$

in the spacial domain $[0, 1]$. The initial condition is $u(0, x) = 1.0$ for $x \leq 0.1$ and $u(0, x) = 0.0$ otherwise. Transmissive boundary conditions are applied. Of course the analysis of this paper is strictly valid only for the linear model equation, but empirical application to the (non-linear) Burgers' equation gives very satisfactory results. This is encouraging, since for realistic problems (eg. the Navier-Stokes equations) the extension of these ideas will necessarily be empirical. We use 100 computing cells for both tests and march the solution at a Courant number of 0.8; for α and a we take 0.001 and 1.0 respectively. We compare numerical results with the exact solution for MINMOD, MINAD2 SUPERBEE and SUPAD for positive r only; for negative r we set $A = A_M$.

Fig 4. shows results for Test 1; 4(a) shows the result of using MINMOD while 4(b) shows the result when using one of its viscous counterpart MINAD2; 4(c) and 4(d) are results obtained by SUPERBEE and its viscous counterpart SUPAD. For this case of a smooth solution one can clearly see the advantages of using the viscous functions of the MINAD type. Note that clipping is absent in 4(b). As expected SUPERBEE gives wrong results for smooth flows as seen in 4(c); its viscous counterpart does not improve matters as seen in 4(d). Note that MINMOD adds too much numerical viscosity while SUPERBEE underdiffuses the solution, which is also incorrect.

Fig. 5 shows the corresponding results for Test 2. For this case with a shock wave both families of viscous functions are superior to their inviscid counterparts.

5 Conclusions

Numerical viscosity functions for a model equation have been presented. These incorporate the physical viscosity so that the contribution from the numerical viscosity to ensure TVD results is minimised. Numerical experiments confirm the theoretical analysis. Extension of these ideas to realistic problems is a pending task.

ACKNOWLEDGEMENTS

The author is indebted to Nikolaos Nikiforakis and Stephen Billett for their invaluable contribution to the preparation of this paper.

References

[1] Godunov, S. K. A difference method for the numerical calculation of discontinuous solutions of hydrodynamic equations. Mat. Sb. 47, 1959; English translation US Department of Commerce, JPRS 7225, 1960.

[2] Hänel, D. On the numerical solution of the Navier-Stokes equations for compressible fluids. VKI Lecture Series on Computational Fluid Dynamics, March, 1989.

[3] Toro, E F and Brown, R E. The WAF method and splitting procedures for viscous shocked flows. Proc. 18th Inter. Symposium on Shock Waves. Tohoku University, Japan, July 1991.

[4] Toro, E F. A weighted average flux method for hyperbolic conservation laws. Proc. Royal Soc. London, A 423, pp 401-418, 1989.

[5] Sweeby, P. K. High resolution schemes using flux limiters for hyperbolic conservation laws. SIAM J. Numer. Analysis, 21, pp 995-1011, 1984.

[6] Roe, P. L. Some contributions to the modelling of discontinuous flows. Lectures in Applied Mathematics, Vol. 22, pp 163-193, 1985.

[7] Leveque, R. J. Numerical Methods for conservation laws. Birkhauser- Verlag, 1990.

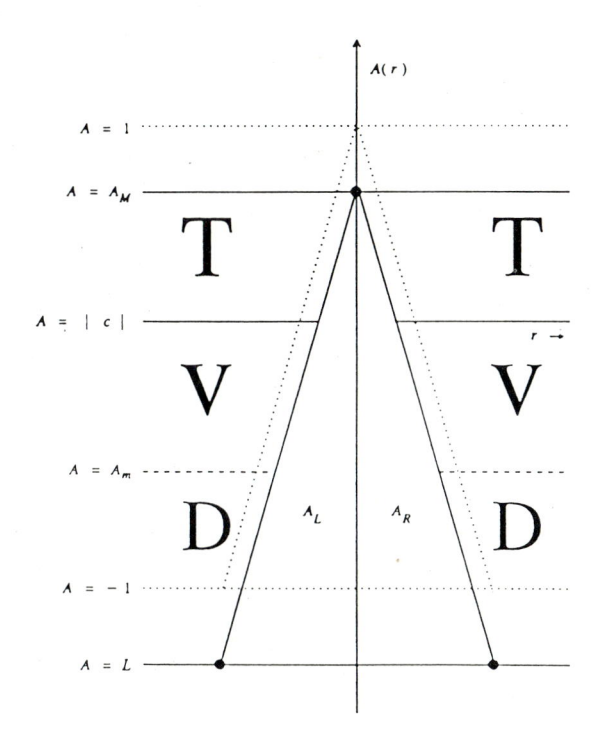

Figure 1: Viscous TVD regions for the model convection-diffusion equation.

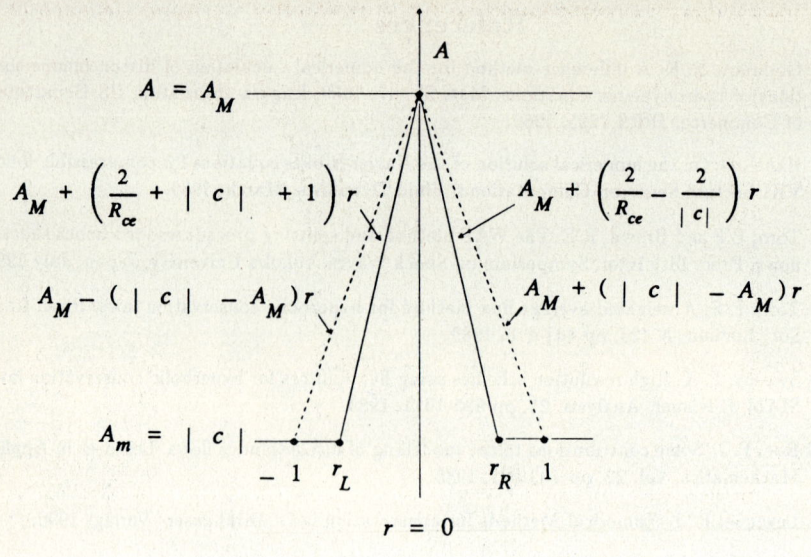

Figure 2: MINAD family of Numerical Viscosity Functions. MINAD1 is given by dotted lines and MINAD2 is given by full lines.

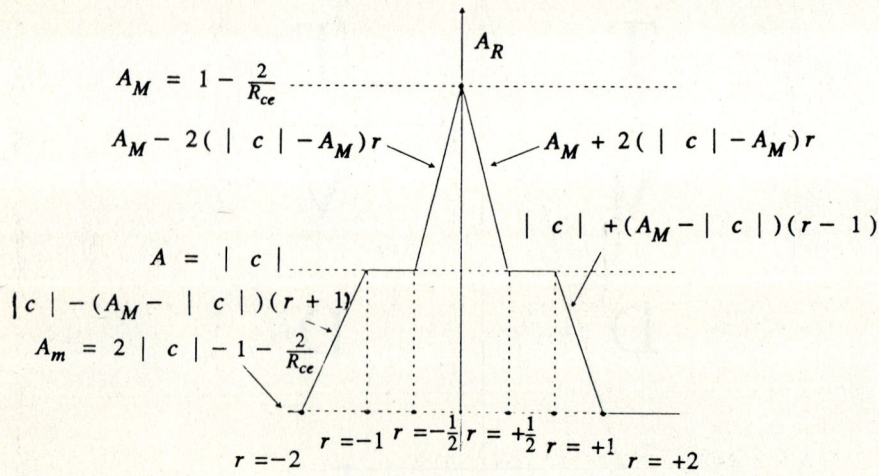

Figure 3: SUPAD family of Numerical Viscosity Functions. Only one member is selected and is given by full line.

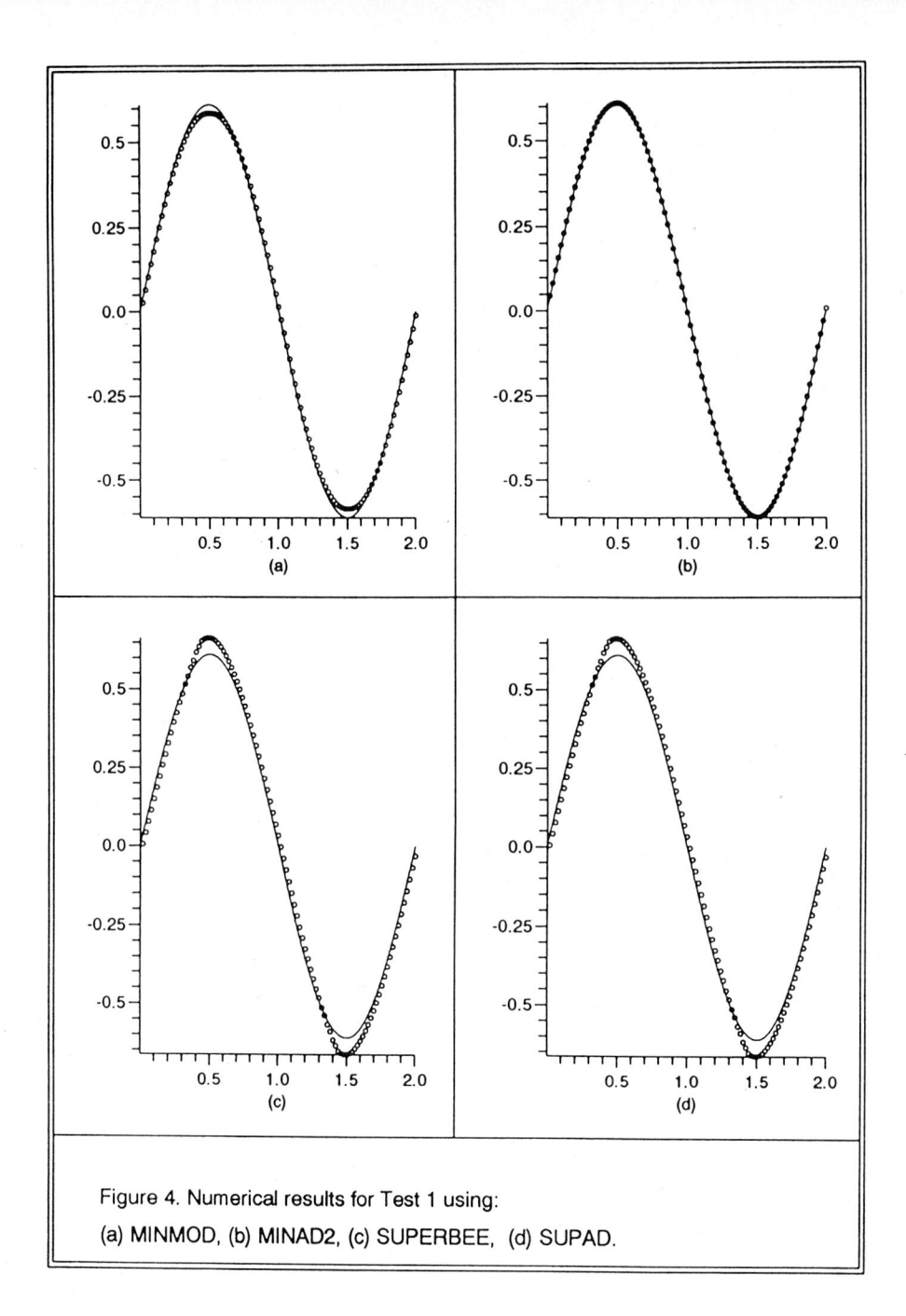

Figure 4. Numerical results for Test 1 using:
(a) MINMOD, (b) MINAD2, (c) SUPERBEE, (d) SUPAD.

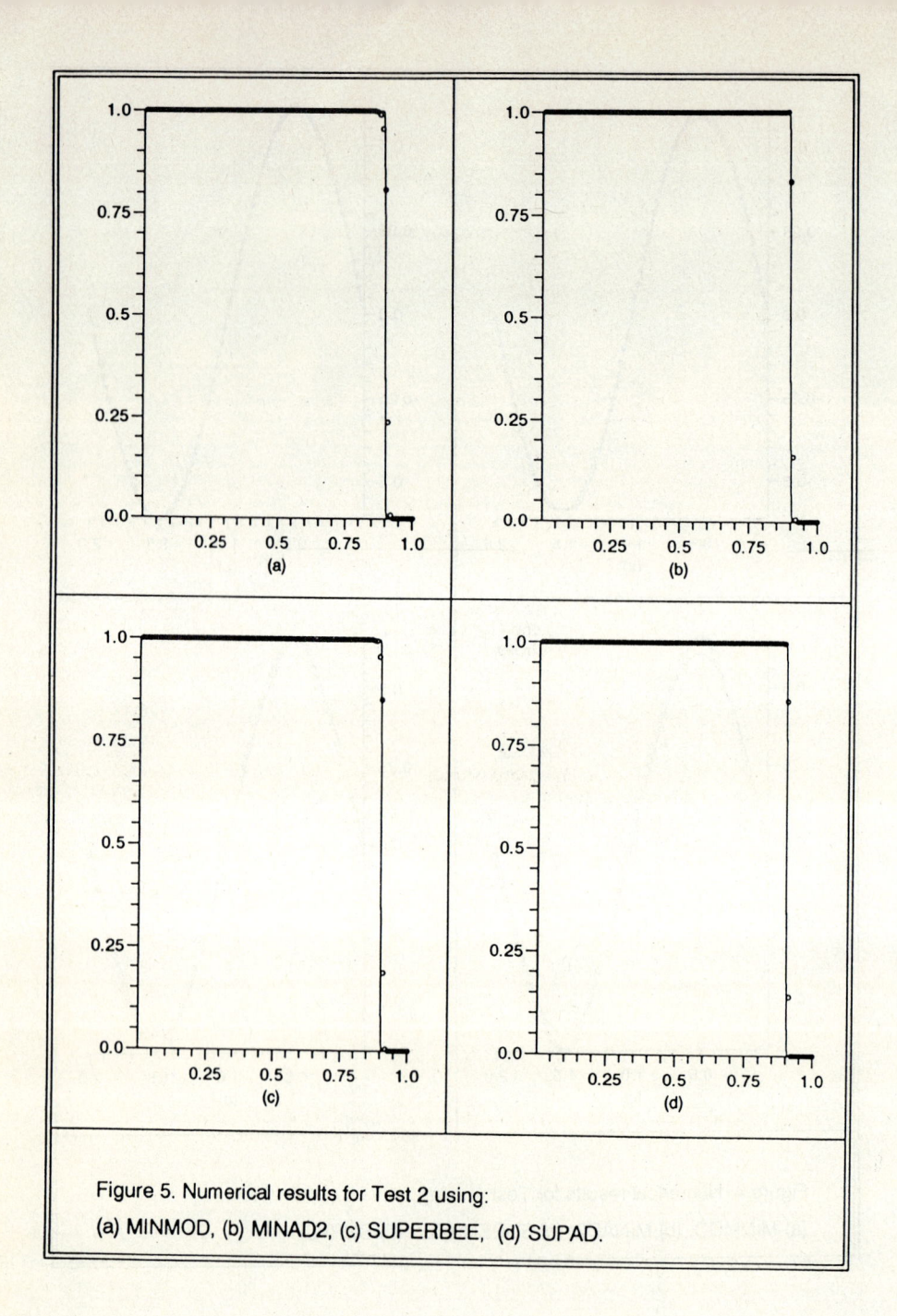

Figure 5. Numerical results for Test 2 using:
(a) MINMOD, (b) MINAD2, (c) SUPERBEE, (d) SUPAD.

Addresses of the Editors of the Series "Notes on Numerical Fluid Mechanics"

Prof. Dr. Ernst Heinrich Hirschel (General Editor)
Herzog-Heinrich-Weg 6
D-8011 Zorneding
Federal Republic of Germany

Prof. Dr. Kozo Fujii
High-Speed Aerodynamics Div.
The ISAS
Yoshinodai 3-1-1, Sagamihara
Kanagawa 229
Japan

Prof. Dr. Bram van Leer
Department of Aerospace Engineering
The University of Michigan
Ann Arbor, MI 48109-2140
USA

Prof. Dr. Keith William Morton
Oxford University Computing Laboratory
Numerical Analysis Group
8-11 Keble Road
Oxford OX1 3QD
Great Britain

Prof. Dr. Maurizio Pandolfi
Dipartimento di Ingegneria Aeronautica e Spaziale
Politecnico di Torino
Corso Duca Degli Abruzzi, 24
I-10129 Torino
Italy

Prof. Dr. Arthur Rizzi
FFA Stockholm
Box 11021
S-16111 Bromma 11
Sweden

Dr. Bernard Roux
Institut de Mécanique des Fluides
Laboratoire Associé au C. R. N. S. LA 03
1, Rue Honnorat
F-13003 Marseille
France

Brief Instruction for Authors

Manuscripts should have well over 100 pages. As they will be reproduced photomechanically they should be typed with utmost care on special stationary which will be supplied on request.
In print, the size will be reduced linearly to approximately 75 per cent. Figures and diagrams should be lettered accordingly so as to produce letters not smaller than 2 mm in print. The same is valid for handwritten formulae. Manuscripts (in English) or proposals should be sent to the general editor, Prof. Dr. E. H. Hirschel, Herzog-Heinrich-Weg 6, D-8011 Zorneding.